Advances in Biochemistry in Health and Disease

Series Editor: Naranjan S. Dhalla
Winnipeg, Manitoba, Canada

Volume 1: S. K. Cheema (ed), *Biochemistry of Atherosclerosis*

Volume 2: S. W. Schaffer and M-Saadeh Suleiman (eds), *Mitochondria: The Dynamic Organelle*

Volume 3: A. K. Srivastava and M. B. Anand-Srivastava (eds), *Signal Transduction in the Cardiovascular System in Health and Disease*

Volume 4: B. Ostadal and N. S. Dhalla (eds), *Cardiac Adaptations-Molecular Mechanisms*

Volume 5: B. I. Jugdutt and N. S. Dhalla (eds), *Cardiac Remodeling-Molecular Mechanisms*

For further volumes:
http://www.springer.com/series/7064

Bodh I. Jugdutt • Naranjan S. Dhalla
Editors

Cardiac Remodeling

Molecular Mechanisms

Springer

Editors
Bodh I. Jugdutt
Walter MacKenzie Health Sciences Centre
Division of Cardiology
University of Alberta
Edmonton, AB, Canada

Naranjan S. Dhalla
St. Boniface Hospital Research Centre
Institute of Cardiovascular Sciences
Winnipeg, MB, Canada

ISBN 978-1-4614-5929-3 ISBN 978-1-4614-5930-9 (eBook)
DOI 10.1007/978-1-4614-5930-9
Springer New York Heidelberg Dordrecht London

Library of Congress Control Number: 2013930229

Printed on acid-free paper

Springer is part of Springer Science+Business Media (www.springer.com)

This book is dedicated to my peers,
colleagues, teachers, students,
and friends who inspire and stimulate me
in lifelong search for knowledge that can be
applied to improve cardiovascular health
and benefit mankind;
And
to my family, who selflessly supported me in
my relentless pursuits of hidden truths:

Catherine Elizabeth (né Graham)
Bernadine Alexandra
Sunita Joanne
Asha Vivienne
Sunil Keith (1973–1992)

Bodh I. Jugdutt
Edmonton, Alberta, Canada

Preface

The concept of cardiac remodeling as a mechanism of heart disease leading to heart failure has evolved since the mid-1970s. The initial emphasis was on heart failure related to pressure and volume overload; this led to theories on adaptive and maladaptive structural and functional changes after life-threatening insults such as myocardial infarction and hypertensive heart disease. Later, the scope of cardiac remodeling expanded to pure and mixed pressure and volume overload states and a wide range of cardiomyopathies, inherited or acquired from infections or exposure to various therapeutic drugs with cardiotoxic pleiotropic effects and other cardiotoxic agents. Results of cardiovascular research at the bench and bedside levels and a host of population studies since the mid-1980s fueled the concept of adverse left ventricular remodeling during acute and subacute phases of myocardial infarction, with structural changes that have a negative impact on cardiac function. These studies have established that adverse cardiac remodeling is a major mechanism for progressive left ventricular enlargement, deterioration of ventricular function, increased suffering, and deaths from chronic heart failure. Concurrently over the last 4 decades, expanding knowledge of the basic molecular mechanisms and clinical implications of cardiac remodeling has identified several molecular pathways and potential targets, leading to drug discovery and development and improved therapies for major causes of adverse cardiac remodeling, such as myocardial infarction and hypertension. A major advance has been the appreciation that lifelong exposure to cardiovascular risk factors and cardiotoxic agents, beginning from the pediatric age through adulthood and old age, fuels the march to heart failure. This has opened up a new area of research into the biology of aging and its impact on cardiac remodeling.

Despite the advances, hearts continue to enlarge, and the heart failure burden continues to increase, especially after ST-segment-elevation myocardial infarction (STEMI). Many knowledge gaps exist. With the expanded spectrum of diseases that result in adverse cardiac remodeling, improved understanding of the underlying molecular mechanisms through research is crucial. During the last 20 years, attention focused on cellular and subcellular changes, including those at the molecular and biochemical levels. There has been an explosion in knowledge of molecular and cellular mechanisms, importance of oxidative stress, metabolic pathways, extracellular and intracellular matrix remodeling, and the far-reaching effects of infarct and non-infarct zone fibrosis in the progression to heart failure. This has led to a profusion of original scientific and review papers dealing with several aspects of molecular mechanisms of adverse cardiac remodeling. There is therefore a need to synthesize these ideas into one book on molecular mechanisms of cardiac remodeling.

The main objective of this book has been to summarize the major research advances in molecular, biochemical, and translational aspects of cardiac remodeling over the last 2 to 3 decades under one cover and touch on future directions. The invited leaders and established investigators in the field have generously contributed 30 chapters on key topics relating to molecular mechanisms, with emphasis on selected biochemical and translational aspects of cardiac remodeling. The authors have succinctly summarized large volumes of data on these key topics and highlighted novel pathways and key molecules that need to be further explored and possibly targeted. They provide integrative reviews of the basic mechanisms and clinical correlates as well as critical assessments of publications on the key topics by the leading investigators in the field. The reference lists are fairly comprehensive and include key papers that are currently not easily accessed from Pubmed or other search engines. The book is carefully organized into two sections: Section A contains 15 chapters that focus mainly on molecular mechanisms in pressure and volume overload hypertrophy, with some overlap into brief ischemia–reperfusion injury; Section B contains 15 chapters that focus on molecular mechanisms after myocardial injury and infarction. The list of topics is by no means comprehensive but addresses some major areas needing attention. To our knowledge, there is no other book on this topic to date.

In summary, this book provides a high-profile and valuable publication resource on molecular mechanisms of cardiac remodeling for both the present and future generations of researchers, teachers, students, and trainees. It should stimulate future translational research targeted towards discovery and development for preventing, limiting, and reversing bad remodeling over the next few decades, with the ultimate goal of preventing progression to systolic and/or diastolic heart failure. The chapters suggest potential novel strategies that should receive attention for translating basic research knowledge to application in patients at the bedside. We would like to thank all the authors for their excellent contributions. We would also like to express our deepest appreciation for the preparation and editorial help provided by Catherine E. Jugdutt, Eva Little, and Dr. Vijayan Elimban in assembling this book. Cordial thanks are also due to Ms. Portia Formento and Melanie Tucker, Springer, USA, for their continuous advice and understanding during the editorial process. We hope that the book will prove useful for scientists and clinicians, students and teachers, and the industry interested in drug and discovery research.

Edmonton, AB, Canada
Winnipeg, MB, Canada

Bodh I. Jugdutt
Naranjan S. Dhalla

Contents

Part I
Molecular Mechanisms of Remodeling in Pressure and Volume Overload Hypertrophy and Heart Failure

Chapter 1
β-Adrenergic Receptor Signaling in Heart Failure

Grace Jung Ah Lee, Lin Yan, Dorothy E. Vatner, and Stephen F. Vatner

Abstract Acute activation of the sympathetic system and resultant β-adrenergic receptor (β-AR) signaling are required to maintain homeostasis, providing inotropic support in times of need, as in "fight or flight" or response to any stress, such as cardiac dysfunction and heart failure. For most of the twentieth century, it was reasoned that sympathetic stimulation of β-ARs through administration of naturally occurring catecholamines or synthetic sympathomimetic amines could provide inotropic support and should be used in heart failure therapy. However, in heart failure, sympathetic drive to the heart is excessively increased, and chronic sympathetic stimulation is deleterious, since it increases $\dot{M}VO_2$, which cannot be met by appropriate increases in coronary blood flow, thereby creating subendocardial ischemia and intensifying the cardiac dysfunction. Furthermore, continued stimulation of the β-ARs also becomes problematic because it can activate multiple cellular processes including those involved in pathological remodeling seen in the development of cardiomyopathy. However, this reasoning took a diametrically opposite turn in the latter twentieth century when the adverse effects of chronic β-AR stimulation became apparent from experimental studies in transgenic mice with cardiac-specific overexpression of $G_{s\alpha}$ and β-ARs and also from clinical studies with poor outcomes for patients on chronic sympathomimetic amine therapy. At this time it was also found that internal compensatory physiological processes countering continued β-AR stimulation in the heart were cleverer than physicians. As a protective response, β-AR desensitize, which reduces the effectiveness of β-AR stimulation and the consequent increases in myocardial oxygen demands. Taken together, these factors were fundamental to the change in course from β-AR stimulation to β-AR blockade in the treatment of heart failure.

G.J.A. Lee • L. Yan • D.E. Vatner • S.F. Vatner, M.D. (✉)
Department of Cell Biology and Molecular Medicine, Cardiovascular Research Institute, New Jersey Medical School, University of Medicine and Dentistry of New Jersey, 185 South Orange Avenue, MSB G-609, Newark, NJ 07103, USA
e-mail: vatnersf@umdnj.edu

B.I. Jugdutt and N.S. Dhalla (eds.), *Cardiac Remodeling: Molecular Mechanisms*, Advances in Biochemistry in Health and Disease 5, DOI 10.1007/978-1-4614-5930-9_1,

Keywords β-Adrenergic receptor • Inotropic agonist • β-Adrenergic receptor blockers • Desensitization

1.1 Introduction

β-Adrenergic receptor (β-AR) signaling is central to all aspects of the pathophysiology of heart failure. The sympathetic nervous system including the neurohormones, epinephrine, and norepinephrine is rapidly called into action by any stress, such as cardiac dysfunction and heart failure. For most of the twentieth century, it was reasoned that sympathetic stimulation of β-ARs through administration of naturally occurring catecholamines or synthetic sympathomimetic amines could provide inotropic support and should be used in heart failure therapy. However, this reasoning took a diametrically opposite turn in the latter twentieth century when it was realized that patients with β-AR blocker therapy fared significantly better. The goal of this chapter is to document the scientific and clinical basis for the changing paradigm of the role of β-AR signaling in heart failure. To do this, the chapter has the following sections: Sect. 1.2 (The Discovery of β-ARs), Sect. 1.3 (Regulation of Cardiac Contractility by β-ARs), Sect. 1.4 (Targeting β-ARs in the Treatment of Heart Failure: Use of β-AR Inotropic Agonists), Sect. 1.5 (Adverse Effects of Chronic β-AR Stimulation in the Treatment of Heart Failure), Sect. 1.6 (Advent of β-AR Blockade Therapy), Sect. 1.7 (Mechanisms Mediating Salutary Effects of β-AR Blockade Therapy in Heart Failure), Sect. 1.8 (Future Directions), and Sect. 1.9 (Conclusions).

1.2 The Discovery of β-ARs

Although the concept of β-ARs mediating the signaling from the sympathetic nervous system to regulate cardiac function is axiomatic today, this was not always the case. Throughout much of the twentieth century, it was erroneously believed that adrenergic signaling was primarily mediated by two classes of neurotransmitters, sympathin E (excitatory) and sympathin I (inhibitory), classified according to their physiological response [1, 2]. This was due, in part, to the use of natural adrenalin, which contained variable mixtures of epinephrine and norepinephrine with quite different agonistic activities, resulting in obscured conclusions that masked their distinct effects. In retrospect, the fallacy of their results is clear. Not only do epinephrine and norepinephrine have different effects, e.g., norepinephrine has α-vasoconstrictor activity as well as β-vasodilator and inotropic activity, whereas epinephrine does not have much α-activity, but both elicit reflex effects in vivo with the most prominent mediated by the arterial baroreflex, which modulates the direct actions of the catecholamines on arterial pressure, heart rate, and peripheral vascular resistance.

In 1906, Dale first introduced the concept of receptors in connection with the sympathetic nervous system [3]. In his studies, he observed the actions of ergot alkaloid antagonists on the effects of epinephrine and proposed there are two distinct receptor types. One type, in which epinephrine mediated excitatory responses, was antagonized by ergot alkaloids, whereas in the second type, ergots had no effect on the inhibitory effects of epinephrine. Then in 1948, a major step was taken by Ahlquist, who challenged this idea of sympathins by characterizing two AR types, α and β, based on the rank order of catecholamine potencies rather than the nature of their physiological response (contraction vs. relaxation) [2].

However, the idea of ARs existing as physical entities received much skepticism [4, 5]. Even Ahlquist noted in his later paper that ARs are hypothetical structures that hold momentary value until the exact mechanism of adrenergic signaling is deciphered [6]. However, his seminal studies persevered and in 1967, Lands et al. extended his classification scheme by introducing two β-AR subtypes, β_1 and β_2, based on their affinities for epinephrine and norepinephrine [7]. Whereas β_1-ARs in cardiac and adipose tissue have approximately equal affinity for epinephrine and norepinephrine, β_2-ARs relax bronchial and vascular smooth muscle and have greater affinity for epinephrine than for norepinephrine. Then in 1972, Carlsson et al. provided pharmacological evidence that both β_1- and β_2-ARs are present and functional in the feline heart and that β_1-AR is the predominant subtype in both the atria and the ventricles [8].

From these findings, Lefkowitz developed highly specific radioligand-binding assays that allowed selective labeling of β-ARs, which was responsible for the most significant progress in the field in the latter half of the twentieth century [9]. Using this method, he and his colleagues physically identified cardiac β-ARs for the first time in the canine heart in 1975 [9]. Moreover, the radioligand-binding technique made possible the quantification of the relative proportions of β_1- and β_2-ARs and in 1983, it was reported that human left ventricle (LV) consists of 86% β_1-AR and 14% β_2-AR [10], thus confirming and extending the work of Carlsson. In addition, the interactions of β-ARs with various agonists and antagonists were explored based on the concept that the radioligand competes for the binding site with an agonist. In 1980, it was discovered that binding of an agonist and antagonist was affected by GTP [11], and taking into account that adenylyl cyclase systems require GTP for activation [12], the ternary complex model, consisting of the adrenergic receptor coupling to GTP-binding G protein to activate adenylyl cyclase (AC), was proposed [13].

The advances in molecular biology techniques that shortly followed led to the successful cloning of the β_2-AR, the very first G protein-coupled receptor to be cloned [14]. Then by the 1990s, six α-AR subtypes (α_{1A}, α_{1B}, α_{1C} and α_{2A}, α_{2B}, α_{2C}) [15, 16] and three β-AR subtypes (β_1, β_2, β_3) [17–19] were firmly established. Moreover, insights on the physiological actions of various AR subtypes were made possible through generation of transgenic mice models with targeted disruption of ARs [20–23]. Today, we now understand that α-ARs have positive inotropic activity (most prominent in rodents), but their primary role is regulating peripheral resistance [24], whereas β-ARs provide the strongest mechanism to regulate cardiac performance [25–28].

1.3 Regulation of Cardiac Contractility by β-ARs

In response to acute stress, the normal heart must be able to rapidly increase its output nearly fivefold to meet the higher metabolic demands [29]. This is primarily met by the sympathetic nervous system acting on the β-ARs to mediate positive chronotropic and inotropic responses in the heart. While all three β-ARs subtypes, namely β_1, β_2, and β_3, are expressed on the cardiomyocyte [9, 30], β_1-AR is the most abundant form (70–80% of total β-AR in the normal heart) and is one of the main regulators of cardiac performance, with β_2-AR being secondary, but also having powerful vasodilator properties in vessels [31]. The β_3-AR, however, is only minimally expressed, and its role in the heart remains controversial. Initially, the β_3-ARs were implicated to have a role in fat metabolism [32], but various studies suggest they have cardiac roles as well. For instance, in vivo stimulation of β_3-AR induces inotropic responses in rodents; however, no effect was observed in primates, suggesting its physiological response may be species specific [33]. In addition, these reports do not correspond to the findings of ex vivo studies, which showed negative inotropic effects of β_3-AR stimulation in the ventricular tissue of mice and humans potentially through the inhibitory G protein (G_i)–nitric oxide synthase (NOS) pathway [34, 35]. The apparent discrepancy between in vivo and ex vivo effects of β_3-AR signaling may depend on the type of agonist used as well as their differential effects under in vivo and ex vivo conditions. Interestingly, however, a recent study by Niu et al. demonstrated in vivo that β_3-AR has cardioprotective roles by inducing negative inotropic effects and maintaining NO and reactive oxygen species balance in the setting of catecholamine overstimulation in the failing heart [36]. Whether this effect is also observed in humans needs to be elucidated.

β_1- and β_2-ARs are the principal stimulatory G protein (G_s)-coupled receptors that drive heart rate and enhance myocardial contractility [25–28]. Upon amine binding, both β_1- and β_2-ARs activate adenylyl cyclase (AC) to increase the level of cyclic AMP (cAMP) (Fig. 1.1) [37]. The latter targets protein kinase A (PKA), which phosphorylates various proteins involved in excitation–contraction coupling. These include, but are not limited to L-type calcium channels (LTCC) [38], ryanodine receptors (RyR2) [39], phospholamban (PLB) [40, 41], and troponin I (TnI) [41], which together coordinate stronger contractions and hastened relaxation of the cardiac muscle. For instance, phosphorylation of LTCC increases the Ca^{2+} influx [42], which stimulates RyR2 to release Ca^{2+} from the sarcoplasmic reticulum (SR). The Ca^{2+}-induced Ca^{2+} release process is further enhanced by phosphorylation of RyR2 [43], thereby increasing its sensitivity to cytosolic [Ca^{2+}] and resulting in greater SR Ca^{2+} unloading necessary for stronger pumping of the myocardium [44]. The following relaxation phase is also accelerated through phosphorylation of PLB

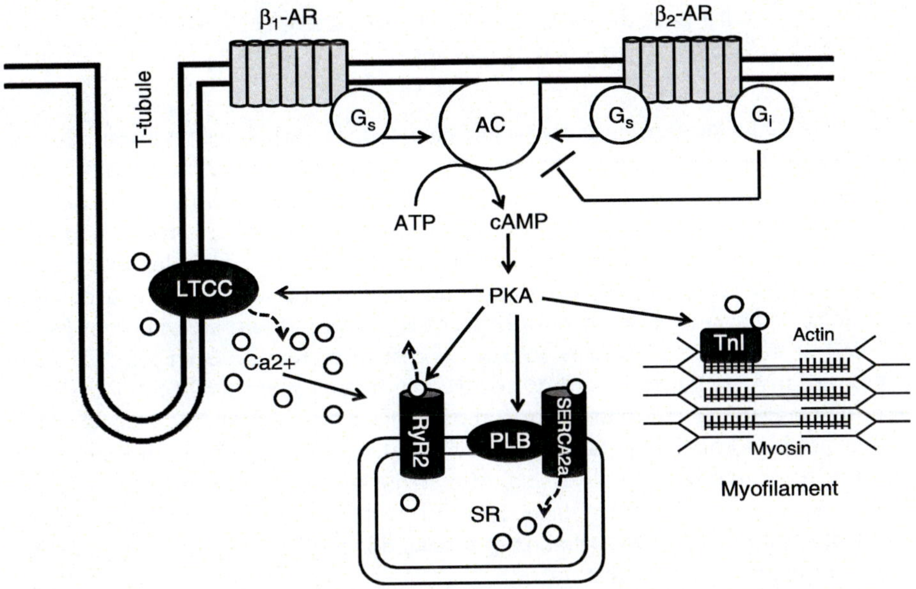

Fig. 1.1 Enhancement of cardiac contractility through β-AR signaling. Upon agonist binding, β_1- and β_2-ARs activate the stimulatory G protein (G_s)-adenylyl cyclase (AC)-protein kinase A (PKA) pathway to phosphorylate multiple calcium handling and myofilament proteins, including L-Type calcium channel (LTCC), ryanodine receptor (RyR2), phospholamban (PLB), and troponin I (TnI). Activation of these proteins by β-ARs enhances their function, resulting in stronger contractions as well as hastened relaxation. For instance, the Ca^{2+}-induced Ca^{2+} release process is further enhanced by phosphorylation of RyR2 by PKA, resulting in greater sarcoplasmic reticulum (SR) Ca^{2+} unloading during systole. The relaxation phase is also hastened by phosphorylation of PLB and TnI, resulting in faster reuptake of Ca^{2+} by the SR and faster relaxation of the myofilaments by preventing actin–myosin interaction. Interestingly, the β_2-AR also couples to inhibitory G protein (G_i), which could negate the effects of G_s–AC–PKA signaling, resulting in only modest inotropic response compared to β_1-AR

at Ser16 by PKA, which allows rapid sequestering of Ca^{2+} by relieving its inhibition of the SR Ca-ATPase (SERCA2a) pump [40, 45, 46], as well as phosphorylation of TnI, which decreases the sensitivity of myofilaments to Ca^{2+} [47–49], resulting in muscle relaxation.

Interestingly, studies from animal models reported that β_2-AR also couple to G_i, which could negate the effects of G_s–AC–PKA signaling, resulting in attenuation of enhanced contractility and hastened relaxation [50, 51]. For instance, (−)-noradrenaline hastened relaxation through β_1-ARs, but not with (−)-adrenaline through β_2-ARs in the feline ventricle [52], which was supported by the finding that β_2-ARs have modest inotropic effects compared to β_1-AR independent of receptor density in the rat heart [53, 54]. However, whether β_2-ARs couple to G_i in humans remains controversial. Kilts et al. demonstrated that β_2-ARs also couple to G_i in the human atria [55], whereas a later study by Molenaar et al. showed evidence that is inconsistent with the dual coupling feature of β_2-AR [56].

1.4 Targeting β-ARs in the Treatment of Heart Failure: Use of β-AR Inotropic Agonists

While β-AR blockers are currently widely employed in the treatment of heart failure, much of the therapeutic approaches in the twentieth century were based on the concept of increasing myocardial contractility through inotropic agents. The reasoning was simple: (1) The failing heart has poor contractility, (2) catecholamines are the most potent stimulators of myocardial contractility, and (3) catecholamines would be useful to increase contractility in patients with heart failure (quod erat demonstrandum). What was not known was either the concept of desensitization of β-AR [57–59] or the fact that there is an oxygen cost of increasing heart rate, LV wall stress, and myocardial contractility, i.e., the major determinants of $\dot{M}VO_2$, which, in turn, increase the requirement for coronary blood flow. The increased demand is easily met in normal hearts and coronary circulation, but not so in the setting of either hypertrophy or heart disease characterized by limited subendocardial coronary reserve [60, 61]. Under these conditions, the imbalance between coronary blood flow supply and myocardial oxygen demands results in myocardial ischemia, which exacerbates cardiac dysfunction.

Not recognizing these concepts, catecholamines and synthetic sympathomimetic amines were routinely administered to patients with heart failure; at first, norepinephrine or epinephrine was given [62], and then in the 1960s, isoproterenol (ISO), a nonselective β_1- and β_2-AR agonist with potent inotropic and chronotropic properties, was used in the treatment of heart failure [63]. Many patients who received this treatment developed fatal arrhythmias [63], and those with coronary artery disease had increased myocardial ischemia [64]. In an experimental correlate of the clinical situation, Hittinger et al. showed that isoproterenol infusion in conscious dogs with LV hypertrophy and failure further impaired both systolic and diastolic function [65]. The increased stress of inotropy as well as chronotropy under limited coronary flow reserve resulted in subendocardial hypoperfusion, which exacerbated the myocardial dysfunction.

In an effort to find a selective agent that could augment the inotropic state of the myocardium without affecting the chronotropic state, dopamine, a naturally occurring, nonselective α- and β-AR agonist, was used. While dopamine had less chronotropic effect compared to isoproterenol, the increase in cardiac index was still associated with elevated heart rate, resulting in ventricular tachyarrhythmias [66, 67]. Furthermore, the inotropic effect of dopamine was mediated in part by the release of endogenous noradrenaline [68–70]. In severe heart failure, where endogenous noradrenaline levels are low, the degree to which dopamine could increase cardiac output was often minimal and unpredictable [71, 72]. Moreover, infusion at high doses leads to stimulation of both α- and β-ARs with a predominant α-AR effect, resulting in excessive vasoconstriction [73, 74]. To minimize the effects on peripheral vasculature and heart rate, while retaining positive inotropic effects, dobutamine, a synthetic derivative of dopamine, was developed. Vatner et al. investigated the actions of dobutamine in healthy conscious dogs and found that it exerted potent inotropic effects with insignificant changes in heart rate [75]. However, the lack of increase in heart rate and effects on peripheral vascular resistance was due to mixed α- and β-adrenergic stimulating properties of the drug, coupled with counteracting influences induced by activation of autonomic reflexes. Accordingly, when autonomic reflexes were blocked or when one arm of the AR system (either α or β) was blocked, effects on heart rate and peripheral vascular resistance were identified. Furthermore, considering that many inotropic agents also have vasodilating effects, administration of nitroprusside, a vasodilating agent, was also explored. In the 1970s, there were various reports showing the promising effects of nitroprusside in improving heart function in the setting of heart failure [76–78]. A later study compared the effects of chronic infusion of dobutamine with those of nitroprusside in end-stage heart failure patients and showed that patients with nitroprusside treatment had significantly higher event-free survival rate than patients with dobutamine treatment [79]. From these findings, it was proposed that peripheral vasodilator therapy with nitroprusside was superior to dobutamine. However, the combined effects of vasodilators and inotropes resulted in a higher mortality rate than vasodilator therapy alone [80]. In the early 1980s, moderate success was observed with the β_1-AR selective inotropic agent, prenalterol. Although in an acute setting, it improved hemodynamics [81, 82], chronic administration showed detrimental effects [83]. Then in the 1990s, clinical trials with β_2-AR agonist, dopexamine, initially showed some clinical benefit over the placebo [84, 85]. However, the beneficial effects were due to the vasodilatory and cardiac unloading effects of this agonist.

In an attempt to improve cardiac function, β-AR downstream targets were also explored. Milrinone, a phosphodiesterase (PDE) inhibitor, was first approved for intravenous use in the late 1980s due to its effects on increasing cAMP half-life and subsequently increasing intracellular Ca^{2+} concentration and improving cardiac contractility. In addition, because increased cAMP levels also results in arterial and venous dilation, it may lead to hypotension. In clinical trials, milrinone had no clear benefit over placebo [86]. Then in 1986, enoximone, another PDE inhibitor, was studied. A year later, it was found to be superior to both dobutamine and nitroprusside in the management of heart failure [87]. However, in 1994, it was found that enoximone administration in patients with end-stage heart failure resulted in increased mortality [88]. Then in a larger phase III trial in 2009, ESSENTIAL-I and ESSENTIAL-II, enoximone treatment did not show any benefit over the placebo, which led to the termination of enoximone development [89].

1.5 Adverse Effects of Chronic β-AR Stimulation in the Treatment of Heart Failure

Acute activation of the sympathetic system is required to maintain homeostasis, providing inotropic support in times of need, such as in "fight or flight" situations. However, in heart failure, sympathetic drive to the heart is excessively increased, and as discussed in the last section, the chronic sympathetic stimulation is deleterious, since it increases $M\dot{V}O_2$, which cannot be met by appropriate increases in coronary blood flow. This results in subendocardial ischemia, which intensifies the cardiac dysfunction. Furthermore, continued stimulation of the β-ARs also becomes problematic because it can activate multiple cellular processes including those involved in pathological remodeling seen in the development of cardiomyopathy. Therefore, continued β-AR stimulation as occurs in heart failure induces β-AR desensitization [57–59], which is protective since it reduces the effectiveness of β-AR stimulation and the consequent increases in myocardial oxygen demands. The realization of these concepts was fundamental to the paradigm shift from β-AR stimulation to β-AR blockade in the treatment of heart failure. The background documenting this concept follows in both experimental and clinical settings.

1.5.1 Experimental Evidence

It was a considerable time before the adverse outcomes of chronic β-AR stimulation were demonstrated both in vitro and in vivo in numerous animal studies. Initially, it was thought that β-AR overexpression could be a novel therapy for heart failure. This concept was supported by studies that showed enhanced myocardial function in transgenic mice with up to 60-fold overexpression of β_2-ARs [90] without evidence of cardiac pathology [91]. Furthermore, transgene therapy with adenovirus coding the β_2-AR transgene significantly improved cardiac function in the failing rabbit hearts [92].

However, the error of this concept was realized when the chronic effects of β-AR stimulation were observed. The novel feature of the following studies was the continued monitoring of the mice as they aged. Iwase et al. reported that transgenic mice with cardiac-specific overexpression of $G_{s\alpha}$ ($G_{s\alpha}$ TG) have increased responsiveness to β-AR stimulation by ISO compared to WT mice as young adults, as evident by significantly higher left ventricular ejection fraction (LVEF) (Fig. 1.2a) [93]. However, upon aging, their function deteriorates, resulting in LV dilation, higher incidence of arrhythmias, and depression of LVEF (Fig. 1.2a) [93–95]. Furthermore, histological examination of these hearts revealed a picture of cardiomyopathy, including marked increase in hypertrophy, myocyte necrosis, and fibrosis (Fig. 1.2b) [93]. Confirming these findings, Engelhardt et al. showed that transgenic mice with cardiac-specific overexpression of β_1-AR (β_1-AR TG) have hyperfunction at a young age, but their function progressively worsens as the mice age [96], which was also further confirmed by the work of Peter et al. (Fig. 1.3a) [97]. Furthermore, premature mortality by 12 months of age was significantly higher in β_1-AR TG mice compared to WT mice (Fig. 1.3b) [98]. In support of these findings, in vitro studies in rat ventricular myocytes showed significant increases in apoptosis upon β_1-AR stimulation [100], and hypertrophy from ISO administration was abolished by a β_1-AR antagonist [101]. Furthermore, the concept of the beneficial effects of chronic β_2-AR overexpression was finally put to rest when Du et al. showed that

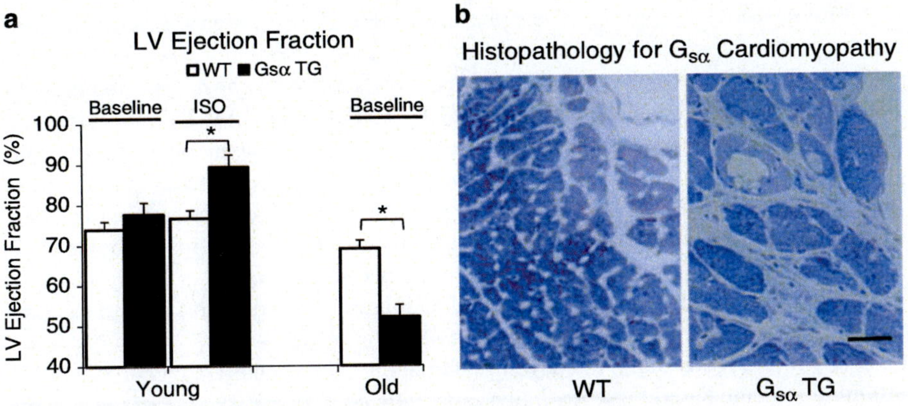

Fig. 1.2 Overexpression of $G_{s\alpha}$ in the heart induces cardiomyopathy. (**a**) $G_{s\alpha}$ TG mice have higher responsiveness to isoproterenol (ISO) compared to WT mice as young adults, but experience progressive deterioration of left ventricular ejection fraction (LVEF) with age (*$p<0.05$) [93, 94]. (**b**) LV subendocardium of 19-month-old WT mice and 15-month-old $G_{s\alpha}$ TG mice stained with toluidine blue. Old $G_{s\alpha}$ TG mice have marked cellular hypertrophy and fibrosis compared to WT mice. Bar=25 μm [93]. Figures adapted with permission from Lippincott Williams & Wilkins and Journal of Clinical Investigation

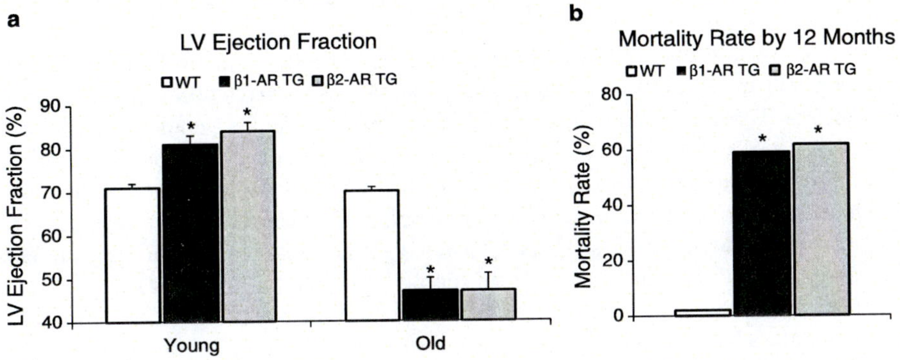

Fig. 1.3 Cardiac overexpression of β_1-AR or β_2-AR decreases heart function and increases mortality with age. (**a**) β_1- and β_2-AR TG animals have enhanced left ventricular ejection fraction (LVEF) as young adults, but their cardiac function significantly worsens as the animals age compared to the age-matched WT mice. (*$p<0.05$ vs. WT) [97]. (**b**) The mortality rate of β_1- and β_2-AR TG animals by 12 months of age is significantly higher compared to WT animals (*$p<0.05$ vs. WT) [98, 99]. Figures adapted with permission from Journal of Clinical Investigation, Lippincott Williams & Wilkins, and Oxford University Press

β_2-AR TG mice also develop progressive cardiomyopathy with increasing age, as reflected by deterioration of myocardial function [99], which was also confirmed by studies of Peter et al. (Fig. 1.3a) [97]. Similar to β_1-AR TG mice, β_2-AR TG mice also have significantly higher mortality by 12 months of age compared to the WT mice (Fig. 1.3b) [99].

Chronic stimulation of the β-AR pharmacologically with high doses of ISO also results in the rapid development of cardiomyopathy. Similar to transgenic mice models of cardiomyopathy, several studies in rodent models have reported marked increases in cellular necrosis and fibrosis [102–105] and impaired LV function [106, 107] upon administration of high doses of ISO, resulting in heart failure and increased mortality. In support of this, the effects of chronic catecholamine stress are exacerbated in mice with transgenic overexpression of components in the β-AR signaling pathway, such as AC5, one of the two major AC isoforms in the heart. It was shown that mice with cardiac-specific overexpression of AC5 (AC5 TG) have higher mortality, depressed LVEF, as well as increased fibrosis and apoptosis compared to the WT mice under chronic catecholamine stress [108].

It is now recognized that the development of cardiomyopathy from chronic β-AR stimulation is mediated by several signaling pathways (Fig. 1.4). One of these is the Ca^{2+} handling pathway. In animal models as well as heart failure patients, prolonged Ca^{2+} transients have been described [109, 110], and in humans, it was shown that this effect was due, in part, to the impairment of reuptake of Ca^{2+} into the SR during diastole [111, 112]. In fact, it was shown that in the failing heart, PLB is hypophosphorylated by PKA [113], resulting in delayed sequestering of Ca^{2+} by the SR and increased cytosolic Ca^{2+} during diastole, which is further exacerbated by enhanced activity of LTCC [114]. Engelhardt et al. showed that adverse effects of cardiac-specific overexpression of β_1-AR could be rescued by ablation of PLB [98]. The β_1-AR TG x $PLB^{-/-}$ bigenic mice had significantly lower diastolic calcium levels and upregulation of SERCA compared to β_1-AR TG mice. Moreover, the bigenic mice had significant improvement in mortality and cardiac function as well as reduction in hypertrophy and fibrosis. The cardiotoxic effects of increased intracellular Ca^{2+} could be attributed to activation of calcineurin [115] and Ca^{2+}/calmodulin-dependent protein kinase II (CaMKII) pathway [116], which could induce hypertrophy. However, it may also be due to the activation of the necrosis pathway. While Ca^{2+} is not the main regulator of the necrosis pathway, previous studies have suggested that increases in intracellular Ca^{2+} induced by β-AR stimulation by ISO could induce necrosis [117, 118].

Other distal signaling pathways that have been implicated involve several kinases, which mediate hypertrophy and cardiomyopathy. For example, inhibition of proto-oncogene serine/threonine-protein kinase (Raf-1), which activates MAPK/ERK kinase (MEK) and subsequently extracellular signal-regulated kinase (ERK), may contribute to the development of cardiomyopathy. Yan et al. showed that mice with AC5 disruption (AC5KO) were protected against aging cardiomyopathy, and the mechanism behind this protection was proposed to be enhanced pro-survival Raf-1/MEK/ERK signaling [119]. This is also supported by the finding that ERK activation could prevent myocardial necrosis and apoptosis [120–122] and ERK inhibition under ischemic insult results in increased size of myocardial infarction in vivo [120] as well as increased apoptosis in vitro [123]. Therefore, it may be possible that inhibition of this pro-survival pathway may contribute to the development of cardiomyopathy. Another pathway that contributes to the development of cardiomyopathy involves activation of proapoptotic protein p38α mitogen-activated protein kinase (p38α MAPK), which was shown to be elevated in $G_{s\alpha}$ TG cardiomyopathy

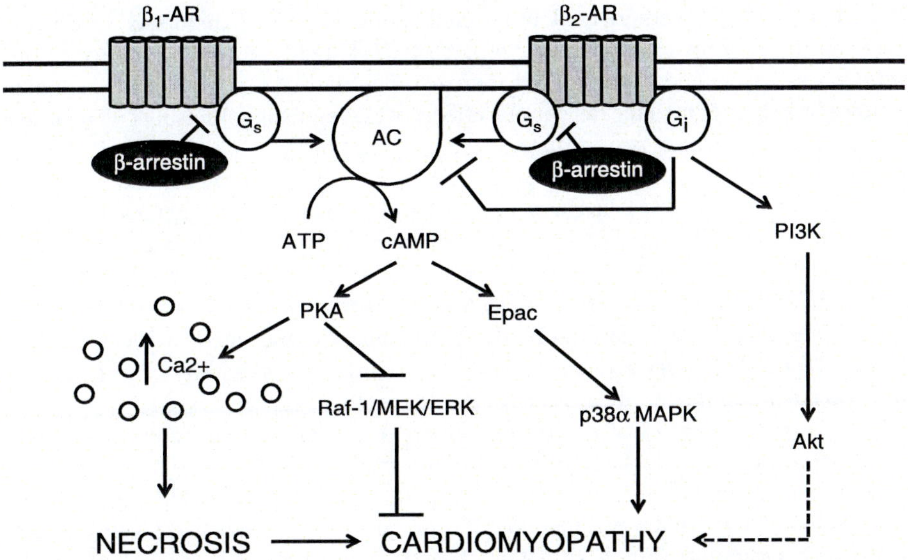

Fig. 1.4 β-AR signaling pathways involved in the development of cardiomyopathy. Chronic stimulation of β_1- and β_2-ARs drives cAMP levels via G_s-AC signaling to activate protein kinase A (PKA) and Epac. PKA induces enhanced activity of Ca^{2+} handling proteins, such as LTCC and RyR2, resulting in significant increases in cytosolic Ca^{2+} levels, which contributes to necrosis and subsequent development of cardiomyopathy. PKA could also inhibit the pro-survival proto-oncogene serine/threonine-protein kinase (Raf-1), mitogen-activated protein kinase/extracellular signal-regulated kinase (MEK), and extracellular signal-regulated kinase (ERK) pathway, thereby relieving this protective mechanism. Similarly, Epac activates p38α mitogen-activated protein kinase (p38α MAPK) to induce cardiomyopathy. The β_2-AR also couples to G_i to activate the phosphatidylinositol 3-kinase (PI3K)-protein kinase B (Akt) pathway. While acute stimulation of Akt may be beneficial, chronic stimulation of this protein may have adverse effects. As a protective measure, β-arrestins are recruited to decouple the receptor from downstream signaling, thereby desensitizing the β-AR

model [124]. Peter et al. reported that inhibition of p38α MAPK rescues the cardiomyopathy induced by β_2-AR [97]. While p38α MAPK disruption did not prevent β_1-AR-induced cardiomyopathy, there was upregulation of another proapoptotic protein, mammalian sterile 20-like kinase 1 (Mst-1) in the hearts of β_1-AR TG mice [97], suggesting that Mst-1 may play a role in the development of cardiomyopathy that is more associated with the β_1-AR. The last pathway, which involves phosphatidylinositol 3-kinase (PI3K)–protein kinase B (Akt) signaling, is a matter of debate. Various studies have shown that Akt mediates hypertrophy [125–127], but could also enhance cardiac contractility [128, 129], suggesting it has beneficial roles in the heart. Kim et al. showed that mice with cardiac-specific overexpression of active Akt had increased L-type Ca^{2+} density, as well as increased expression of SERCA2a protein, which resulted in enhanced myocardial contractility [128]. Furthermore, it has been well documented that Akt can protect the heart from apoptosis in vitro [130, 131] and in vivo [132]. Okumura et al. reported that mice with AC5 disruption (AC5KO) had significantly lower apoptosis compared to WT mice after ISO administration, which was associated with marked increase in active Akt [133]. However, other studies have highlighted the deleterious effects of Akt signaling. Matsui et al. reported that chronic Akt expression in mice results in a vast array of phenotypes, including moderate cardiac hypertrophy with preserved function to significant cardiac dilation as well as sudden death [134]. In a later study, Nagoshi et al. reported that transgenic mice with cardiac-specific Akt expression had significantly larger infarcts without any restoration of function under ischemia/reperfusion injury [135]. Furthermore, it was shown that activation of Akt results in a negative feedback mechanism to inhibit PI3K, concluding that in the heart, PI3K-dependent but Akt-independent pathways are required for full cardioprotection. Given these findings, the possibility of Akt contributing to the development of cardiomyopathy cannot be fully dismissed.

One of the key series of studies instrumental in changing the concept of β-AR stimulation to β-AR blockade in heart failure therapy involved the mouse models with cardiac-specific overexpression of either $G_{s\alpha}$ or β-AR. As noted earlier, these mice responded to increased sympathetic stimulation with enhanced levels of heart rate and contractility (Fig. 1.2a) [93]. This was not a transient process, as occurs with intravenous administration of catecholamines, but was a permanent fixture of their cardiac function through constant overexpression. The $G_{s\alpha}$ mice tolerated the chronically enhanced cardiac function when they were young, but developed cardiomyopathy as they aged, as reflected by cardiac dysfunction and heart failure, myocyte necrosis and apoptosis and cardiac fibrosis, as well as premature mortality from the cardiomyopathy. When these mice were put on continuous β-AR blockade therapy, they were protected from cardiomyopathy and showed improvement in LVEF as well as improvement in survival (Fig. 1.5a) [94]. These findings were not only novel at the time, but also played a significant role in the paradigm shift from β-AR agonist to β-AR blockade in the treatment of heart failure.

1.5.2 *Clinical Evidence*

A major deterrent to β-AR stimulation therapy was that patients who received inotropic agonists often experienced side effects, including elevated heart rate, arrhythmias, and peripheral vasoconstriction, which overshadowed any initial inotropic benefits gained. Moreover, patients with treatment experienced worsening of their condition and significantly higher mortality rates than the placebo group. For instance, ISO induced fatal arrhythmias and intensified myocardial ischemia in patients with coronary artery disease. Patients with dopamine treatment developed ventricular tachyarrhythmias. Moreover, prenalterol significantly increased mortality. In addition, while dobutamine had beneficial effects in the acute setting, chronic administration had adverse effects. In fact, the 1999 Flolan International Randomized Survival Trial proved dobutamine to increase mortality [137]. These findings were critical in the shift from beta agonists to β-AR blockers in the treatment of heart failure.

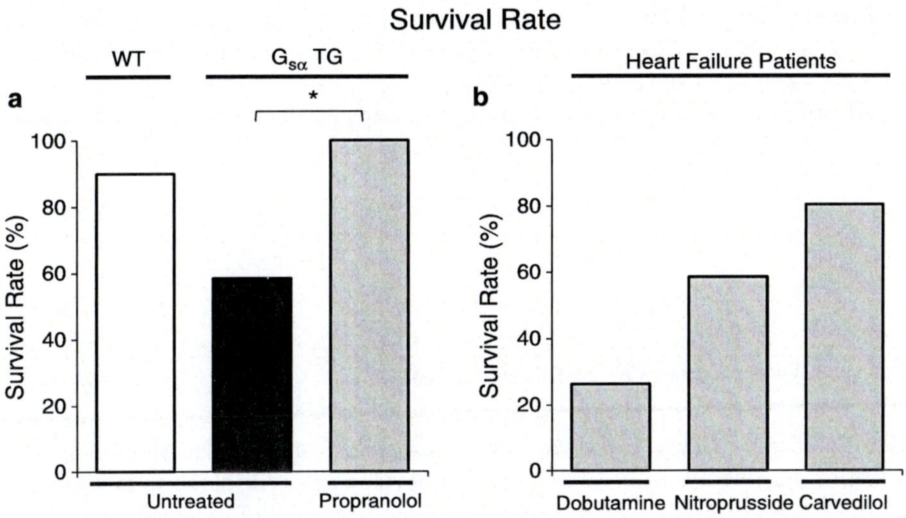

Fig. 1.5 β-AR blockade treatment increases survival rate. (**a**) Approximately 9-month-old WT and $G_{s\alpha}$ TG mice were followed for mean duration of 6.5 months with or without propranolol treatment. In the untreated group, $G_{s\alpha}$ TG mice have premature mortality, which was abolished by propranolol treatment. (*$p<0.05$ by log-rank test) [94]. (**b**) By 300 days of therapy, β-AR blocker, carvedilol, was shown to be superior in prolonging the lifespan of heart failure patients compared to both β-AR agonist (dobutamine) and vasodilator agent (nitroprusside) [79, 136] Figures moadapted with permission from Journal of Clinical Investigation and Oxford University Press

The other major clinical contribution was the concept of β-AR desensitization in patients with heart failure introduced by Bristow and his colleagues [57]. As an adaptive response to increased catecholamine levels, β-ARs desensitize through downregulation of the receptors and attenuation of downstream signaling. Bristow et al. demonstrated that isoproterenol (ISO) stimulation in failing human hearts resulted in 50–56% reduction in β-AR density, decreased AC activity, and decreased muscle contraction compared to normal hearts [57], and the half-equivalent dose of ISO was fivefold higher in the failing tissue compared to the normal tissue [138]. Dose–response study with dobutamine showed progressively lower inotropic response (dP/dt) in patients with severe heart failure compared to those with moderate heart failure [139]. The observed β-AR desensitization is mediated by two mechanisms. First, β-AR mRNA is downregulated [140, 141] potentially through degradation by A+U-rich element RNA-binding/degradation factor (AUF1) [142]. It was found that AUF1 expression is significantly elevated in individuals with heart failure, and its abundance was regulated by β-AR stimulation. In addition, AUF1 was able to interact with β-AR mRNA suggesting it may be responsible for its stability. Second, β-ARs are decoupled from downstream signaling [143–145].G protein-coupled receptor kinase 2 (GRK2), a member of the GRK family commonly known as β-AR kinase 1 (βARK1), phosphorylates agonist occupied β-AR [146] and subsequently recruits β-arrestin, which inhibits G protein-mediated signaling (Fig. 1.4) [147]. Studies in the left ventricles of patients with dilated or ischemic cardiomyopathy showed elevated expression and activity of GRK2 [141].

Therefore, in the setting where β-ARs are already desensitized, β-AR agonist treatment will not provide much inotropic response. In fact, it will exacerbate the condition, as seen in the clinical trials of β-AR agonists. For these reasons, the concept of β-AR desensitization was important in understanding the rationale for β-AR blockers in heart failure therapy.

1.6 Advent of β-AR Blockade Therapy

β-AR blockade therapy began with the seminal studies of Sir James Black, who was awarded the Nobel Prize in 1988 [148] for developing β-AR blockers in the treatment of cardiac disease. He began his studies with the goal of reducing myocardial oxygen demands [149], and by 1962, he and Stephenson introduced a nonselective β-AR blocker, pronethalol [150]. It was reported that pronethalol had strong antiarrhythmic effects in guinea pigs and dogs [151–154] and was effective in managing angina pectoris in patients [155]. Shortly after, Black developed another nonselective β-AR blocker, propranolol, in 1965 [156]. He compared the effects of pronethalol with propranolol and showed that propranolol was superior in reducing heart rate and blocking ISO-induced hypotension without any fall in blood pressure in dogs [156].

Furthermore, unlike inotropic agonists, propranolol treatment in mice did not induce myocardial damage [157]. In fact, it was shown to be effective in reducing mortality rates and ventricular fibrillation in experimental studies of coronary artery occlusion [158, 159]. Similar effects were seen in the clinical setting, and it was reported that propranolol was effective in improving myocardial oxygenation [160] and reducing mortality in patients with myocardial infarction [161]. Furthermore, it had strong antiarrhythmic effects as well as having beneficial effects in controlling angina pectoris in heart failure patients [162–164]. The success of propranolol also led to the development of β_1-AR selective β-AR blockers including practolol, atenolol, and metoprolol. While practolol had less β-AR blocking potency than propranolol, animals studies demonstrated that practolol was effective in preventing arrhythmias and other alterations in cardiac function [165]. Furthermore, administration of atenolol or metoprolol in mice was able to prevent myocardial injury induced by epinephrine, including inflammatory cell infiltration, fibrosis, and atrophy and necrosis [166]. Then in 1975, the beneficial effects of practolol and alprenolol were illustrated clinically [167]. Six patients were treated with oral β_1-AR selective blocker practolol, and one patient received the nonselective β-AR blocker, alprenolol, for 2–12 months. After treatment, all seven patients showed improvement in ventricular function and reduction in heart size. The long-term effect of β_1-AR selective blocker, metoprolol, was also evaluated in an international multicenter study, the metoprolol in dilated cardiomyopathy (MDC) trial [168]. Beginning from 1985, 383 patients were followed for at least 12 months. Although the beneficial effects of metoprolol were modest during the first year of treatment, the requirement for heart transplantation was dramatically decreased in the treatment group. In addition, metoprolol was well tolerated in the long term, indicated by the low withdrawal rate. In 1994, the trial on the β_1-AR selective blocker bisoprolol in the cardiac insufficiency bisoprolol study (CIBIS) was published [169] and was soon followed by the US Carvedilol (nonselective β/α_1-AR blocker) Study in 1996 [136]. Patients with carvedilol treatment had significant improvement in mortality, and the survival benefit compared to β-AR agonist, dobutamine, as well as to vasodilator agent, nitroprusside, is clear (Fig. 1.5b). The first large randomized clinical trial was the Cardiac Insufficiency Bisoprolol Study II (CIBIS-II), released in 1999, in which 2,647 heart failure patients were followed for a mean duration of 1.3 years [170–173]. The study was terminated early due to the clear benefit of bisoprolol in decreasing mortality rate (11.8% vs. 17.3% bisoprolol vs. placebo). In 2000, similar results were seen in the Metoprolol CR/XL Randomized Intervention Trial in congestive heart failure (MERIT-heart failure) [174].

1.7 Mechanisms Mediating Salutary Effects of β-AR Blockade Therapy in Heart Failure

While β-AR blockers improve cardiac function as well as hemodynamic response, the concept that β-AR blockade would be useful therapy in heart failure is counter-intuitive, with the inverse reasoning of why β-AR stimulation would be useful in heart failure, alluded to earlier. In addition, β-AR blockade has already been initiated by failing heart through desensitization and reduced β-AR density and regulation, also noted earlier. To put it simply the failing heart has an internal brain telling the physician how to design its therapy, i.e., if β-AR blockade is already initiated physiologically, then it follows that more of the same might be indicated. The counter argument is based on knowing that β-AR blockers have a profound cardiac depressant effect, which could not be tolerated in a heart already failing with marked cardiac depression. Indeed, early attempts at β-AR blockade often failed because of this. It was not until physicians recognized that by gradually and incrementally instituting β-AR blockade was it possible to achieve therapeutic levels without compounding the cardiac depression. This is combined with the fact that some of the current β-AR blockers used clinically, e.g., carvedilol, have actions other than simple β-AR blockade, such as vasodilating effects, and are as not as potent β-AR blockers as propranolol.

Empirically noting the above does not mean that all the mechanisms mediating the salutary action of β-AR blockade in heart failure are known. In fact, it remains unclear how β-AR blockers lower the risk of cardiac complications. However, it is apparent that β-blockers offer far more than simply blocking the receptor. First of all, again with inverse reasoning to why β-AR stimulation therapy failed clinically, β-AR blockers reduce heart rate, arguably the most important regulator of $M\dot{V}O_2$. As noted above, the imbalance between myocardial oxygen supply and demand is an important mechanism resulting in myocardial dysfunction through invocation of subendocardial ischemia. Since the most important regulator of myocardial oxygen demand is heart rate followed by LV wall stress, a drug which diminishes heart rate and prevents subendocardial ischemia will then reduce LV wall stress. These hemodynamic factors alone will eventually act to prevent a further decline in cardiac function followed by gradual recovery. This also results in a "resensitizing" effect. These include upregulating β-AR density, decreasing GRK2 activity, correcting the impairment of Ca^{2+} handling proteins, and reversing downstream signaling.

1.7.1 Upregulation of β-AR Density

Reduced contractility in chronic heart failure through desensitization mechanisms is protective, as it reduces myocardial oxygen demand, and more importantly, the sudden increases that occur with stress. As noted earlier, Bristow et al. initially reported that β-AR density is reduced by 50% in failing human ventricles, and later studies have confirmed this finding. Therefore, it is conceivable that increases in cardiac performance with β-blockade result, in part, from quantitative restoration of β-AR. In the normal rat heart, chronic infusion of nonselective β-AR blocker, propranolol, was associated with significant increases in both β_1- and β_2-AR density [175]. In addition, heart failure patients with β_1-AR blocker treatment, metoprolol, also showed upregulations of cardiac β-ARs [176, 177]. The mechanism behind this action is uncertain. However, the synthesis and degradation of β-ARs seem to be regulated on the RNA level. As mentioned earlier, AUF1 is significantly increased in heart failure and was shown to degrade β-ARs and thereby reducing its density [142]. In addition, the increase of β-ARs following of β-AR blockade was seen on the protein level as well as on the mRNA level [140]. Interestingly, however, carvedilol and bucindolol, nonselective β-AR blockers with vasodilating properties through α_1-AR inhibition, did not exert β-AR upregulating effects in heart failure patients, despite being as effective in improving cardiac function [176, 178]. It may be possible that these classes of β-AR blockers have other mechanisms mediating their beneficial effects.

1.7.2 Decreasing GRK2 Activity

The increase in inotropy seen in patients with β-AR blockade may indicate that there is an enhancement of β-AR signaling from its depressed state. GRK2 is a G protein-coupled receptor kinase responsible for phosphorylating the β-ARs and decoupling the receptor from downstream signaling via recruitment of β-arrestin. Patients with heart failure have elevated GRK2, consistent with the β-AR desensitization concept. In addition, studies in transgenic mice have shown that increase in GRK2 has negative inotropic response, which could be ameliorated by GRK2 inhibition [179]. Given this finding, it may be possible that improvement in contractile response seen in patients with β-AR blockade may be partly due to decreased GRK2 activity. In fact, studies in mice [180] and pigs [181] have shown that treatment with bisoprolol, atenolol, and carvedilol downregulate GRK. This effect was also seen in heart failure patients. Treatment with metoprolol or bisoprolol decreased GRK2 activity in the right atrium compared to patients who did not receive β-AR blockers [182, 183]. Therefore, downregulation of GRK2 may be an important mechanism by which β-AR blockers confer beneficial effects.

1.7.3 Correcting the Impairment of Ca^{2+} Handling Proteins

The decreased β-AR signaling also leads to subnormal phosphorylation of Ca^{2+} handling proteins. In fact, PLB and TnI have been shown to be hypophosphorylated in the failing hearts [113, 184, 185]. Decreased phosphorylation of PLB results in delayed SR Ca^{2+} uptake, which in part explains the relaxation deficit in heart failure patients, as well as decreased SR unloading, resulting in weaker contractions. Interestingly, however, RyR2 have been reported to be hyperphosphorylated in heart failure [39]. The result is a "leaky RyR2" that leads to higher diastolic Ca^{2+} concentrations and a decreased Ca^{2+} loading of the SR, which leads to higher propensity for arrhythmias. Given that β-AR signaling is attenuated in heart failure, it may be possible that β-AR blockade could correct these impairments through resensitizing the system, thereby restoring the phosphorylation of these proteins to normal levels. In fact, patients with carvedilol, metoprolol, or atenolol treatment had restoration of normal phosphorylation of RyR2 associated with improved cardiac muscle function [186].

1.7.4 Reversing Adverse Effects of Distal Signaling

It is also important to note that β-AR blockers could reverse the maladaptive signaling in distal mechanisms. For instance, the increase in levels of proteins involved in cellular growth and death, such as p38α MAPK, Jun N-terminal kinase (JNK), and Akt, seen in $G_{s\alpha}$ cardiomyopathy model was downregulated upon propranolol treatment [124]. In parallel to this evidence, patients who responded to either metoprolol or carvedilol had restoration of adult α-myosin heavy chain isoform and reduction of the fetal β-isoform levels [187, 188]. This effect could be due to decreased circulating catecholamine levels from β-AR blockade because re-induction of fetal genes and shift in myosin heavy chain isoform occur in parallel to the elevated levels of catecholamines seen in heart failure patients [189]. From these findings, it is clear that β-AR blockade does not simply antagonize the receptor, but also affects multiple levels of the signaling pathway to salvage the heart.

1.8 Future Directions

The treatment of heart failure and the reduction in mortality and morbidity have all improved markedly over the past several decades; however, this disease remains a leading cause of mortality and morbidity. Accordingly, there is considerable room for improvement in therapy. Advances will occur in finding new components of the β-AR signaling pathway to inhibit that are distal to the β-AR and have less adverse depressant effects on LV function. This could be at the level of inhibiting adenylyl cyclase [190] or even more distal signaling mechanisms, e.g., Raf/MEK/ERK, p38 MAPK, and other kinases yet to be defined. It is interesting that inotropic therapy is not dead, and new agents are being devised to stimulate the failing heart, without the adverse consequences of increasing $M\dot{V}O_2$. One example is the development of cardiac myosin activators, which improve the performance of the failing heart in chronically instrumented conscious dogs and patients with heart failure [191–193]. In contrast to sympathomimetic amines, such as dobutamine or dopamine, the cardiac myosin activators do not increase $M\dot{V}O_2$ (Fig. 1.6a), most likely because they reduce heart rate and do not increase LV wall stress. The most unique aspect of their action is to increase stroke volume and cardiac output by increasing the duration of cardiac contraction, and not by increasing the rate of LV pressure development (LV dP/dt) (Fig. 1.6b, c). These drugs along with those affecting Ca^{2+} and particularly at the level of the ryanodine receptor [39] could be responsible for the next advances in heart failure therapy.

1.9 Conclusions

This chapter summarized the development of β-AR stimulating agents for the treatment of heart failure from the discovery of β-ARs to how they regulate cardiac contractility and their use for most of the twentieth century in cardiac therapy. There are very few therapeutic approaches that demonstrated a diametrically opposite turn as did the transition from β-AR agonists to antagonists in the treatment of heart failure. The reasoning for treating heart failure with β-AR agonists was simple: (1) the failing heart has poor contractility, (2) catecholamines were the most potent stimulators of myocardial contractility, and (3) catecholamines would be useful to increase contractility in patients with heart failure (quod erat demonstrandum). What was not known was either the concept of desensitization of β-AR or the fact that there is an oxygen cost of increasing heart rate, LV wall stress, and myocardial contractility, i.e., the

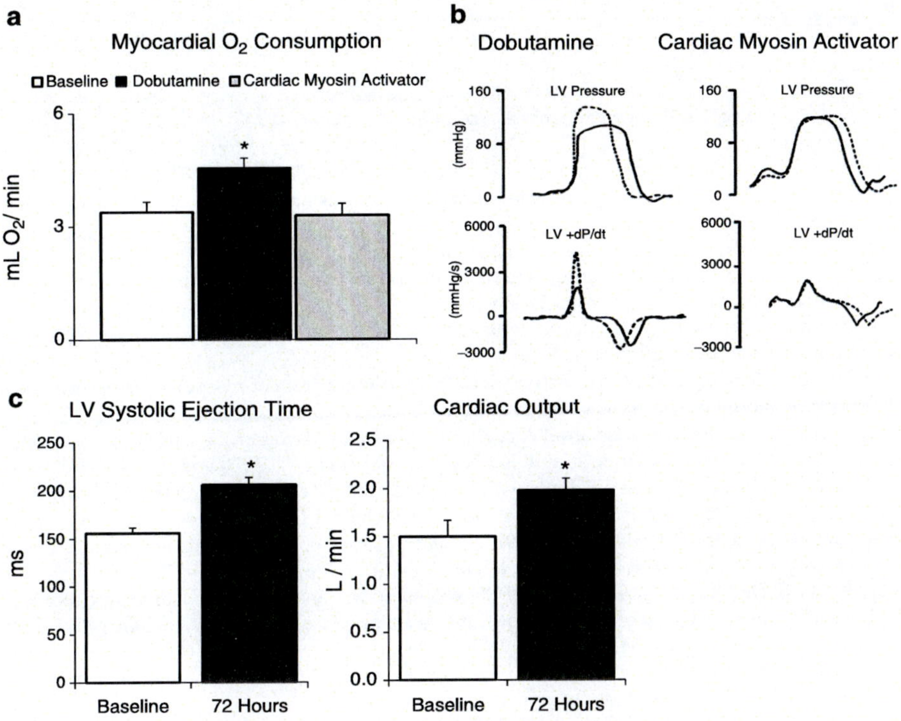

Fig. 1.6 Cardiac myosin activators improve cardiac function in hearts with systolic dysfunction. (**a**) Unlike dobutamine, cardiac myosin activators do not increase myocardial O_2 consumption (*$p<0.05$ vs. baseline) [191]. (**b**) Compared to dobutamine, cardiac myosin activators do not increase LV pressure development (solid line: before treatment, dashed line: after treatment) [191]. (**c**) 72-h infusion of cardiac myosin activator significantly increases LV systolic ejection time and cardiac output from the baseline (*$p<0.05$ vs. baseline) [191]. Figures adapted with permission from Lippincott Williams & Wilkins

major determinants of $M\dot{V}O_2$, which, in turn, increase the requirement for coronary blood flow. The increased myocardial metabolic demand is easily met in normal hearts and coronary circulations, but not so in the setting of either hypertrophy or heart disease characterized by limited subendocardial coronary reserve. Under these conditions, the imbalance between coronary blood flow supply and myocardial oxygen demands results in myocardial ischemia, which exacerbates cardiac dysfunction. All of these factors were fundamental to the change in course from β-AR stimulation to β-AR blockade in the treatment of heart failure, reinforced by poor clinical outcomes of patients on β-AR stimulation therapy. This chapter elucidates the experimental and clinical evidence delineating the mechanisms mediating increased cardiac function with β-AR agonists and conversely the deleterious effects of chronic stimulation, which can lead to heart failure, resulting in a transition for heart failure therapy from β-AR stimulation to blockade.

References

1. Cannon WB, Rosenbueth A (1933) Studies on conditions of activity in endocrine organs: XXIX. Sympathin E and sympathin I. Am J Physiol 104:557–574
2. Ahlquist RP (1948) A study of the adrenotropic receptors. Am J Physiol 153:586–600
3. Dale HH (1906) On some physiological actions of ergot. J Physiol 34:163–206
4. Dale H (1943) Modes of drug action. General introductory address. Trans Faraday Soc 39:319b–322b
5. De Jongh DK (1964) Some introductory remarks on the conception of receptors. In: Ariëns EJ (ed) Molecular pharmacology: the mode of action of biologically active compounds, vol 1. Academic, New York, NY, pp 13–15
6. Ahlquist RP (1967) Development of the concept of alpha and beta adrenotropic receptors. Ann N Y Acad Sci 139:549–552
7. Lands AM, Arnold A, McAuliff JP et al (1967) Differentiation of receptor systems activated by sympathomimetic amines. Nature 214:597–598
8. Carlsson E, Ablad B, Brandstrom A et al (1972) Differentiated blockage of the chronotropic effects of various adrenergic stimuli in the cat heart. Life Sci 11:953–958
9. Alexander RW, Williams LT, Lefkowitz RJ (1975) Identification of cardiac beta-adrenergic receptors by (-)[^{3}H]alprenolol binding. Proc Natl Acad Sci USA 72:1564–1568
10. Stiles GL, Taylor S, Lefkowitz RJ (1983) Human cardiac beta-adrenergic receptors: subtype heterogeneity delineated by direct radioligand binding. Life Sci 33:467–473
11. Kent RS, De Lean A, Lefkowitz RJ (1980) A quantitative analysis of beta adrenergic receptor interactions: resolution of high and low affinity states of the receptor by computer modeling of ligand binding data. Mol Pharmacol 17:14–23
12. Rodbell M (1980) The role of hormone receptors and GTP-regulating proteins in membrane transduction. Nature 284:17–22
13. Limbird LE, Gill DM, Lefkowitz RJ (1980) Agonist promoted coupling of the beta-adrenergic receptor with the guanine nucleotide regulatory protein of the adenylate cyclase system. Proc Natl Acad Sci USA 77:775–779
14. Dixon RA, Kobilka BK, Strader DJ et al (1986) Cloning of the gene and cDNA for mammalian beta-adrenergic receptor and homology with rhodopsin. Nature 321:75–79
15. Bylund DB (1992) Subtypes of alpha 1- and alpha 2-adrenergic receptors. FASEB J 6: 832–839
16. Graham RM, Perez DM, Hwa J et al (1996) Alpha 1-adrenergic receptor subtypes. Molecular structure, function and signaling. Circ Res 78:737–749
17. Frielle T, Collins S, Daniel KW et al (1987) Cloning of the cDNA for the human beta 1-adrenergic receptor. Proc Natl Acad Sci USA 84:7920–7924
18. Emorine LJ, Marullo S, Delavier-Klutchko C et al (1987) Structure of the gene for human beta 2-adrenergic receptor: expression and promoter characterization. Proc Natl Acad Sci USA 84:6995–6999
19. Emorine LJ, Marullo S, Briend-Sutren MM et al (1989) Molecular characterization of the human beta 3-adrenergic receptor. Science 245:1118–1121
20. Lowell BB (1998) Using gene knockout and transgenic techniques to study the physiology and pharmacology of beta3-adrenergic receptors. Endocr J 45(Suppl):S9–S13
21. Rohrer DK, Desai KH, Jasper JR et al (1996) Targeted disruption of the mouse beta1-adrenergic receptor gene: developmental and cardiovascular effects. Proc Natl Acad Sci USA 93:7375–7380
22. Rohrer DK, Bernstein D, Chruscinski A et al (1998) The developmental and physiological consequences of disrupting genes encoding beta 1 and beta 2 adrenoceptors. Adv Pharmacol 42:499–501
23. Link RE, Stevens MS, Kulatunga M et al (1995) Targeted inactivation of the gene encoding the mouse alpha 2c-adrenoceptor homolog. Mol Pharmacol 48:48–55

24. Triposkiadis F, Karayannis G, Giamouzis G et al (2009) The sympathetic nervous system in heart failure physiology, pathophysiology, and clinical implications. J Am Coll Cardiol 54:1747–1762
25. Rockman HA, Koch WJ, Lefkowitz RJ (2002) Seven-transmembrane-spanning receptors and heart function. Nature 415:206–212
26. Port JD, Bristow MR (2001) beta-Adrenergic receptors, transgenic mice, and pharmacological model systems. Mol Pharmacol 60:629–631
27. Lohse MJ, Engelhardt S, Eschenhagen T (2003) What is the role of beta-adrenergic signaling in heart failure? Circ Res 93:896–906
28. Brodde OE, Bruck H, Leineweber K (2006) Cardiac adrenoceptors: physiological and pathological relevance. J Pharmacol Sci 100:323–337
29. Vatner SF, Franklin D, Higgins CB et al (1972) Left ventricular response to severe exertion in untethered dogs. J Clin Invest 51:3052–3060
30. Gauthier C, Tavernier G, Charpentier F et al (1996) Functional beta3-adrenoceptor in the human heart. J Clin Invest 98:556–562
31. Molinari C, Battaglia A, Grossini E et al (1999) The role of beta 2-adrenergic vascular receptors in the peripheral vasodilation caused by 17 beta-estradiol in anesthetized pigs. Life Sci 65:1545–1552
32. Arch JS, Ainsworth AT, Cawthorne MA et al (1984) Atypical beta-adrenoceptor on brown adipocytes as target for anti-obesity drugs. Nature 309:163–165
33. Shen YT, Cervoni P, Claus T et al (1996) Differences in beta 3-adrenergic receptor cardiovascular regulation in conscious primates, rats, dogs. J Pharmacol Exp Ther 278:1435–1443
34. Tavernier G, Toumaniantz G, Erfanian M et al (2003) beta3-Adrenergic stimulation produces a decrease of cardiac contractility ex vivo in mice overexpressing the human beta3-adrenergic receptor. Cardiovasc Res 59:288–296
35. Gauthier C, Leblais V, Kobzik L et al (1998) The negative inotropic effect of beta3-adrenoceptor stimulation is mediated by activation of the nitric oxide synthase pathway in human ventricle. J Clin Invest 102:1377–1384
36. Niu X, Watts VL, Cingolani OH et al (2012) Cardioprotective effect of beta-3 adrenergic receptor agonism: role of neuronal nitric oxide synthase. J Am Coll Cardiol 59:1979–1987
37. Ishikawa Y, Homcy CJ (1997) The adenylyl cyclases as integrators of transmembrane signal transduction. Circ Res 80:297–304
38. Gerhardstein BL, Puri TS, Chien AJ et al (1999) Identification of the sites phosphorylated by cyclic AMP-dependent protein kinase on the beta-2 subunit of the L-type voltage-dependent calcium channels. Biochemistry 38:10361–10370
39. Marx SO, Reiken S, Hisamatsu Y et al (2000) PKA phosphorylation dissociates FKBP12.6 from the calcium release channel (ryanodine receptor): defective regulation in failing hearts. Cell 101:365–376
40. Simmerman HK, Jones LR (1998) Phospholamban: protein structure, mechanism of action, and role in cardiac function. Physiol Rev 78:921–947
41. Sulakhe PV, Vo XT (1995) Regulation of phospholamban and troponin-I phosphorylation in the intact rat cardiomyocytes by adrenergic and cholinergic stimuli: roles of cyclic nucleotides, calcium, protein kinases and phosphatases and depolarization. Mol Cell Biochem 149–150:103–126
42. De Jongh KS, Murphy BJ, Colvin AA et al (1996) Specific phosphorylation of a site in the full-length form of the alpha 1 subunit of the cardiac L-type calcium channel by adenosine 3′,5′-cyclic monophosphate-dependent protein kinase. Biochemistry 35:10392–10402
43. Ginsburg KS, Bers DM (2004) Modulation of excitation-contraction coupling by isoproterenol in cardiomyocytes with controlled SR Ca^{2+} load and Ca^{2+} current trigger. J Physiol 556:463–480
44. Shan J, Kushnir A, Betzenhauser MJ et al (2010) Phosphorylation of the ryanodine receptor mediates the cardiac fight or flight response in mice. J Clin Invest 120:4388–4398

45. Luo W, Grupp IL, Harrer J et al (1994) Targeted ablation of the phospholamban gene is associated with markedly enhanced myocardial contractility and loss of beta-agonist stimulation. Circ Res 75:401–409
46. MacLennan DH, Kranias EG (2003) Phospholamban: a crucial regulator of cardiac contractility. Nat Rev Mol Cell Biol 4:566–577
47. Solaro RJ, Moir AJ, Perry SV (1976) Phosphorylation of troponin I and the inotropic effect of adrenaline in the perfused rabbit heart. Nature 262:615–617
48. Zhang R, Zhao J, Mandveno A et al (1995) Cardiac troponin I phosphorylation increases the rate of cardiac muscle relaxation. Circ Res 76:1028–1035
49. Kaumann A, Bartel S, Molenaar P (1999) Activation of beta2-adrenergic receptors hastens relaxation and mediates phosphorylation of phospholamban, troponin I, and C-protein in ventricular myocardium from patients with terminal heart failure. Circulation 99:65–72
50. Xiao RP, Ji X, Lakatta EG (1995) Functional coupling of the β2-adrenoceptor to a pertussis toxin-sensitive G protein in cardiac myocytes. Mol Pharmacol 47:322–329
51. Xiao RP, Avdonin P, Zhou YY et al (1999) Coupling of beta2-adrenoreceptor to Gi proteins and its physiological relevance in murine cardiac myocytes. Circ Res 84:43–52
52. Lemoine H, Kaumann AJ (1991) Regional differences of beta1- and beta2-adrenoceptor-mediated functions in feline heart. A beta2-adrenoceptor mediated positive inotropic effects possibly unrelated to cyclic AMP. Naunyn Schmiedebergs Arch Pharmacol 334:56–69
53. McConville P, Fishbein KW, Lakatta EG et al (2003) Differences in the bioenergetic response of the isolated perfused rat heart to selective beta1- and beta2-adrenergic receptor stimulation. Circulation 107:2146–2152
54. McConville P, Spencer RG, Lakatta EG (2005) Temporal dynamics of inotropic, chronotropic and metabolic responses during beta1- and beta2-AR stimulation in the isolated, perfused rat heart. Am J Physiol Endocrinol Metab 289:E412–E418
55. Kilts JD, Gerhardt MA, Richardson MD et al (2000) Beta2-adrenergic and several other G protein-couple receptors in human atrial membranes activate both Gs and Gi. Circ Res 87:705–709
56. Molenaar P, Savarimuthu SM, Sarsero D et al (2007) (-)-Adrenaline elicits positive inotropic, lusitropic, and biochemical effects through beta2-adrenoceptors in human atrial myocardium from nonfailing and failing hearts, consistent with Gs coupling but not with Gi coupling. Naunyn Schmiedebergs Arch Pharmacol 375:11–28
57. Bristow MR, Ginsburg R, Minobe W et al (1982) Decreased catecholamine sensitivity and β-adrenergic-receptor density in failing human hearts. N Engl J Med 307:205–211
58. Bristow MR, Ginsburg R, Umans V et al (1986) Beta 1- and beta 2-adrenergic-receptor subpopulations in nonfailing and failing human ventricular myocardium: coupling of both receptor subtypes to muscle contraction and selective beta 1-receptor down-regulation in heart failure. Circ Res 59:297–309
59. Kiuchi K, Shannon RP, Komamura K et al (1993) Myocardial beta-adrenergic receptor function during the development of pacing-induced heart failure. J Clin Invest 91:907–914
60. Hittinger L, Mirsky I, Shen YT et al (1995) Hemodynamic mechanisms responsible for reduced subendocardial coronary reserve in dogs with severe left ventricular hypertrophy. Circulation 92:978–986
61. Nakano K, Corin WJ, Spann JF Jr et al (1989) Abnormal subendocardial blood flow in response to overload hypertrophy is associated with pacing-induced subendocardial dysfunction. Circ Res 65:1555–1564
62. Goldberg LI, Bloodwell RD, Braunwald E et al (1960) The direct effects of norepinephrine, epinephrine, and methoxamine on myocardial contractile force in man. Circulation 22: 1125–1132
63. Smith HJ, Oriol A, Morch J et al (1967) Hemodynamic studies in cardiogenic shock. Treatment with isoproterenol and metaraminol. Circulation 35:1084–1091
64. Maroko PR, Kjekshus JK, Sobel BE et al (1971) Factors influencing infarct size following experimental coronary artery occlusions. Circulation 43:67–82

65. Hittinger L, Shannon RP, Kohin S et al (1989) Isoproterenol-induced alterations in myocardial blood flow, systolic and diastolic function in conscious dogs with heart failure. Circulation 80:658–668
66. Lipp H, Fallicov RE, Resnekov L et al (1972) The effects of dopamine on depressed myocardial function following coronary embolization in the closed chest dog. Am Heart J 84:208–214
67. Gunnar RM, Loeb HS (1972) Use of drugs in cardiogenic shock due to acute myocardial infarction. Circulation 45:1111–1124
68. Tsai TH, Langer SZ, Trendelenburg U (1967) Effects of dopamine and alpha methyl-dopamine on smooth muscle and on the cardiac pacemaker. J Pharmacol Exp Ther 156:310–324
69. Nash CW, Wolff SA, Ferguson BA (1968) Release of tritiated noradrenaline from perfused rat hearts by sympathomimetic amines. Can J Physiol Pharmacol 46:35–42
70. Brodde OE, Inui J, Motomura S (1980) The mode of direct action of dopamine on the rabbit heart. J Cardiovasc Pharmacol 2:567–582
71. Chidsey CA, Braunwald E, Morrow AG et al (1963) Myocardial norepinephrine concentration in man: effects of reserpine and of congestive heart failure. N Engl J Med 269:653–658
72. Tuttle RR, Mills K (1975) Dobutamine. Development of a new catecholamine to selectively increase cardiac contractility. Circ Res 36:185–196
73. McNay JL, McDonald RH Jr, Goldberg LI (1965) Direct renal vasodilation produced by dopamine in the dog. Circ Res 16:510–517
74. Vatner SF, Millard RW, Higgins CB (1973) Coronary and myocardial effects of dopamine in the conscious dog: parasympatholytic augmentation of pressor and inotropic actions. J Pharmacol Exp Ther 187:280–295
75. Vatner SF, McRitchie RJ, Braunwald E (1974) Effects of dobutamine on left ventricular performance, coronary dynamics, and redistribution of cardiac output. J Clin Invest 53: 1265–1273
76. Franciosa JA, Limas CJ, Guiha NH et al (1972) Improved left ventricular function during nitroprusside infusion in acute myocardial infarction. Lancet 1:650–654
77. Guiha NJ, Cohn JN, Mikulic E et al (1974) Treatment of refractory heart failure with infusion of nitroprusside. N Engl J Med 291:587–592
78. Miller RR, Vismara LA, Zelis R et al (1975) Clinical use of sodium nitroprusside in chronic ischemic heart disease. Effects on peripheral vascular resistance and venous tone and on ventricular volume, pump and mechanical performance. Circulation 51:328–336
79. Capomolla S, Febo O, Opasich C et al (2001) Chronic infusion of dobutamine and nitroprusside in patients with end-stage heart failure awaiting heart transplantation: safety and clinical outcome. Eur J Heart Fail 3:601–610
80. Elkayam U, Tasissa G, Brinanay C et al (2007) Use and impact of inotropes and vasodilator therapy in hospitalized patients with severe heart failure. Am Heart J 153:98–104
81. Fitzpatrick D, Ikram H, Nicholls MG et al (1983) Hemodynamic, hormonal and electrolyte responses to prenalterol. Circulation 67:613–619
82. Petch MC, Wisbey C, Ormerod O et al (1984) Acute haemodynamic effects of oral prenalterol in severe heart failure. Br Heart J 52:49–52
83. Lambertz H, Meyer J, Erbel R (1984) Long-term hemodynamic effects of prenalterol in patients with severe congestive heart failure. Circulation 69:298–305
84. Gollub SB, Elkayam U, Young JB et al (1991) Efficacy and safety of a short-term (6-h) intravenous infusion of dopexamine in patients with severe congestive heart failure: a randomized, double-blind, parallel, placebo-controlled multicenter study. J Am Coll Cardiol 18: 383–390
85. Asanoi H, Sasayama S, Sakurai T et al (1995) Intravenous dopexamine in the treatment of acute congestive heart failure: results of a multicenter, double-blind, placebo-controlled withdrawal study. Cardiovasc Drugs Ther 9:791–797
86. Felker GM, Benza RL, Chandler AB et al (2003) Heart failure etiology and response to milrinone in decompensated heart failure: results from the OPTIME-CHF study. J Am Coll Cardiol 41:997–1003

87. Installé E, Gonzales M, Jacquemart JL et al (1987) Comparative effects on hemodynamics of enoximone (MDL 17, 043), dobutamine and nitroprusside in severe congestive heart failure. Am J Cardiol 60:46C–52C
88. Cowley AJ, Skene AM (1994) Treatment of severe heart failure: quantity or quality of life? A trial of enoximone. Enoximone investigators. Br Heart J 72:226–230
89. Metra M, Eichhorn E, Abraham WT et al (2009) Effects of low-dose oral enoximone administration on mortality, morbidity, and exercise capacity in patients with advanced heart failure: the randomized, double-blind, placebo-controlled, parallel group ESSENTIAL trials. Eur Heart J 30:3015–3026
90. Milano CA, Allen LF, Rockman HA et al (1994) Enhanced myocardial function in transgenic mice overexpressing the beta 2-adrenergic receptor. Science 264:582–586
91. Liggett SB, Tepe NM, Lorenz JN et al (2000) Early and delayed consequences of beta(2)-adrenergic receptor overexpression in mouse hearts: critical role for expression level. Circulation 101:1707–1714
92. Tevaearai HT, Eckhart AD, Walton GB et al (2002) Myocardial gene transfer and overexpression of beta2-adrenergic receptors potentiates the functional recovery of unloaded failing hearts. Circulation 106:124–129
93. Iwase M, Bishop SP, Uechi M et al (1996) Adverse effects of chronic endogenous sympathetic drive induced by cardiac Gs alpha overexpression. Circ Res 78:517–524
94. Asai K, Yang GP, Geng YJ et al (1999) Beta-adrenergic receptor blockade arrests myocyte damage and preserves cardiac function in the transgenic G(salpha) mouse. J Clin Invest 104:551–558
95. Iwase M, Uechi M, Vatner DE et al (1997) Cardiomyopathy induced by cardiac Gs alpha overexpression. Am J Physiol 272:H585–H589
96. Engelhardt S, Hein L, Wiesmann F et al (1999) Progressive hypertrophy and heart failure in beta1-adrenergic receptor transgenic mice. Proc Natl Acad Sci USA 96:7059–7064
97. Peter P, Brady J, Yan L et al (2007) Inhibition of p38 alpha MAPK rescues cardiomyopathy induced by overexpressed beta 2-adrenergic receptor, but not beta 1-adrenergic receptor. J Clin Invest 117:1335–1343
98. Engelhardt S, Hein L, Dyachenkow V et al (2004) Altered calcium handling is critically involved in the cardiotoxic effects of chronic beta-adrenergic stimulation. Circulation 109:1154–1160
99. Du XJ, Gao XM, Wang B et al (2000) Age-dependent cardiomyopathy and heart failure phenotype in mice overexpressing beta(2)-adrenergic receptors in the heart. Cardiovasc Res 48:448–454
100. Zaugg M, Xu W, Lucchinetti E et al (2000) Beta-adrenergic receptor subtypes differentially affect apoptosis in adult rat ventricular myocytes. Circulation 102:344–350
101. Morisco C, Zebrowski DC, Vatner DE et al (2001) Beta-adrenergic cardiac hypertrophy is mediated primarily by the beta(1) subtype in the rat heart. J Mol Cell Cardiol 33:561–573
102. Rona G, Chappel CI, Balazs T et al (1959) An infarct-like myocardial lesion and other toxic manifestations produced by isoproterenol in the rat. AMA Arch Pathol 67:443–445
103. Rona G (1985) Catecholamine cardiotoxicity. J Mol Cell Cardiol 17:291–306
104. Chappel CI, Rona G, Balaz T et al (1959) Severe myocardial necrosis produced by isoproterenol in the rat. Arch Int Pharmacodyn Ther 112:123–128
105. Jalil JE, Janicki JS, Pick R et al (1989) Fibrosis-induced reduction of endomyocardium in the rat after isoproterenol treatment. Circ Res 65:258–264
106. Tisné-Versailles J, Constantin M, Lamar JC et al (1985) Cardiotoxicity of high doses of isoproterenol on cardiac haemodynamics and metabolism in SHR and WKY rats. Arch Int Pharmacodyn Ther 273:142–154
107. Kudej RK, Iwase M, Uechi M et al (1997) Effects of chronic beta-adrenergic receptor stimulation in mice. J Mol Cell Cardiol 29:2735–2746
108. Lai L, Yan L, Hu CL et al (2009) Abstract 3290: increased Type 5 adenylyl cyclase expression mediates chronic catecholamine stress via increases in oxidative stress and down-regulation of MnSOD. Circulation 120:S781

109. Balke CW, Shorofsky SR (1998) Alterations in calcium handling in cardiac hypertrophy and heart failure. Cardiovasc Res 37:290–299
110. Gwathmey JK, Morgan JP (1985) Altered calcium handling in experimental pressure-overload hypertrophy in the ferret. Circ Res 57:836–843
111. Gwathmey JK, Copelas L, MacKinnon R et al (1987) Abnormal intracellular calcium handling in myocardium from patients with end-stage heart failure. Circ Res 61:70–76
112. Beuckelmann DJ, Näbauer M, Erdmann E (1992) Intracellular calcium handling in isolated ventricular myocytes from patients with terminal heart failure. Circulation 85:1046–1055
113. Bartel S, Stein B, Eschenhagen T et al (1996) Protein phosphorylation in isolated trabeculae from nonfailing and failing human hearts. Mol Cell Biochem 157:171–179
114. Schröder F, Handrock R, Beuckelmann DJ et al (1998) Increased availability and open probability of single L-type calcium channels from failing compared with nonfailing human ventricle. Circulation 98:969–976
115. Molkentin JD, Lu JR, Antos CL et al (1998) A calcineurin-dependent transcriptional pathway for cardiac hypertrophy. Cell 93:215–228
116. Zhang T, Johnson EN, Gu Y et al (2002) The cardiac-specific nuclear delta(B) isoform of Ca^{2+}/calmodulin-dependent protein kinase II induces hypertrophy and dilated cardiomyopathy associated with increased protein phosphatase 2A activity. J Biol Chem 227:1261–1267
117. Bloom S, Davis DL (1972) Calcium as mediator of isoproterenol-induced myocardial necrosis. Am J Pathol 69:459–470
118. Cheema Y, Sherrod JN, Zhao W et al (2011) Mitochondriocentric pathway to cardiomyocyte necrosis in aldosteronism: cardioprotective responses to carvedilol and nebivolol. J Cardiovasc Pharmacol 58:80–86
119. Yan L, Vatner DE, O'Connor JP et al (2007) Type 5 adenylyl cyclase disruption increases longevity and protects against stress. Cell 2:247–258
120. Strohm C, Barancik T, Bruehl ML et al (2000) Inhibition of the ER-Kinase by PD98059 and UO126 counteracts ischemic preconditioning in pig myocardium. J Cardiovasc Pharmacol 36:218–229
121. Ping P, Zhang J, Cao X et al (1999) PKC-dependent activation of p44/p42 MAPKs during myocardial ischemia-reperfusion in conscious rabbits. Am J Physiol Heart Circ Physiol 276:H1468–H1481
122. Fryer RM, Hsu AK, Gross GJ (1995) ERK and p38 MAP kinase activation are components of opioid-induced delayed cardioprotection. Basic Res Cardiol 96:136–142
123. Yue TL, Gu JL, Wang C et al (2000) Extracellular signal-regulated kinase plays an essential role in hypertrophic agonists, endothelin-1 and phenylephrine-induced cardiomyocyte hypertrophy. J Biol Chem 275:37895–37901
124. Karoor V, Vatner SF, Takagi G et al (2004) Propranolol prevents enhanced stress signaling in Gs alpha cardiomyopathy: potential mechanism for beta-blockade in heart failure. J Mol Cell Cardiol 36:305–312
125. Condorelli G, Drusco A, Stassi G et al (2002) Akt induces enhanced myocardial contractility and cell size in vivo in transgenic mice. Proc Natl Acad Sci USA 99:12333–12338
126. Walsh K (2006) Akt signaling and growth of the heart. Circulation 113:2032–2034
127. DeBosch B, Treskov I, Lupu TS et al (2006) Akt1 is required for physiological cardiac growth. Circulation 113:2097–2104
128. Kim YK, Kim SJ, Yatani A et al (2003) Mechanism of enhanced cardiac function in mice with hypertrophy induced by overexpressed Akt. J Biol Chem 278:47622–47628
129. Rota M, Boni A, Urbanek K (2005) Nuclear targeting of Akt enhances ventricular function and myocyte contractility. Circ Res 97:1332–1341
130. Matsui T, Li L, del Monte F et al (1999) Adenoviral gene transfer of activated phosphatidylinositol 3′-kinase and Akt inhibits apoptosis of hypoxic cardiomyocytes in vitro. Circulation 100:2373–2379
131. Fujio Y, Nguyen T, Wencker D et al (2000) Akt promotes survival of cardiomyocytes in vitro and protects against ischemia-reperfusion injury in mouse heart. Circulation 101:660–667

132. Matsui T, Tao J, del Monte F et al (2001) Akt activation preserves cardiac function and prevents injury after transient cardiac ischemia in vivo. Circulation 104:330–335
133. Okumura S, Vatner DE, Kurotani R et al (2007) Disruption of type 5 adenylyl cyclase enhances desensitization of cyclic adenosine monophosphate signal and increases Akt signal with chronic catecholamine stress. Circulation 116:1776–1783
134. Matsui T, Li L, Wu JC et al (2002) Phenotypic spectrum caused by transgenic overexpression of activated Akt in the heart. J Biol Chem 227:22896–22901
135. Nagoshi T, Matsui T, Aoyama T et al (2005) PI3K rescues the detrimental effects of chronic Akt activation in the heart during ischemia/reperfusion injury. J Clin Invest 115:2128–2138
136. Packer M, Bristow MR, Cohn JN et al (1996) The effect of carvedilol on morbidity and mortality in patients with chronic heart failure. U.S. carvedilol heart failure study group. N Engl J Med 334:1349–1355
137. O'Connor CM, Gattis WA, Uretsky BF et al (1999) Continuous intravenous dobutamine is associated with an increased risk of death in patients with advanced heart failure: insights from the Flolan International Randomized Survival Trial (FIRST). Am Heart J 138:78–86
138. Ginsburg R, Bristow MR, Billingham ME et al (1983) Study of the normal and failing isolated human heart: decreased response of failing heart to isoproterenol. Am Heart J 106:535–540
139. Fowler MB, Laser JA, Hopkins GL et al (1986) Assessment of the beta-adrenergic receptor pathway in the intact failing human heart: progressive down-regulation and subsensitivity to agonist response. Circulation 74:1290–1302
140. Bristow MR, Minobe WA, Raynolds MV et al (1993) Reduced beta 1 receptor messenger RNA abundance in the failing human heart. J Clin Invest 92:2737–2745
141. Ungerer M, Böhm M, Elce JS et al (1993) Altered expression of beta-adrenergic receptor kinase and beta 1-adrenergic receptors in the failing human heart. Circulation 87:454–463
142. Pende A, Tremmel KD, DeMaria CT et al (1996) Regulation of the mRNA-binding protein AUF1 by activation of the beta-adrenergic receptor signal transduction pathway. J Biol Chem 271:8493–8501
143. Hausdorff WP, Caron MG, Lefkowitz RJ (1990) Turning off the signal: desensitization of beta-adrenergic receptor function. FASEB J 4:2881–2889
144. Lohse MJ (1992) Mechanisms of β-adrenergic receptor desensitization. In: Hargrave PA, Hofmann KP, Kaupp UB (eds) Signal Transmission in Photoreceptor Systems. Berlin, Springer-Verlag, pp 160–171
145. Lohse MJ, Benovic JL, Caron MG et al (1990) Multiple pathways of rapid beta 2-adrenergic receptor desensitization. J Biol Chem 265:3202–3211
146. Benovic JL, Mayor F Jr, Staniszewski C (1987) Purification and characterization of the beta-adrenergic receptor kinase. J Biol Chem 262:9026–9032
147. Lohse MJ, Benovic JL, Codina J et al (1990) beta-Arrestin: a protein that regulates beta-adrenergic receptor function. Science 248:1547–1550
148. Gerskowitch VP, Hull RAD, Shankley NP (1988) The pharmacological toolmakers rational approach to drug design—an appreciation of Sir James Black. Trends Pharmacol Sci 9:435–437
149. Black JW (1976) Ahlquist and the development of beta-adrenoceptor antagonists. Postgrad Med J 52(4):11–13
150. Black JW, Stephenson JS (1962) Pharmacology of a new adrenergic beta-receptor-blocking compound (Nethalide). Lancet 2:311–314
151. Elij D, Medez R (1964) The modification of digitalis intoxication by excluding adrenergic influences on the heart. J Pharmacol Exp Ther 144:97–103
152. Williams EM, Sekiya A (1963) Prevention of arrhythmias due to cardiac glycosides by block of sympathetic beta-receptors. Lancet 1:420–421
153. Somani P, Hardman HF, Lum BKB (1963) Antagonism of cardiac arrhythmias by nethalide. Phamacologist 5:260
154. Murray WJ, McKnight RL, Davis DA (1963) Antagonism of hydrocarbon anaesthetic-epinephrine arrhythmias in dogs by nethalide, a dichloroisoproterenol analogue. Proc Soc Exp Biol med 113:439–441

155. Prichard BN, Dickson CJ, Alleyne GA et al (1963) Effect of pronethalol in angina pectoris. Br Med J 2:1226–1229
156. Black JW, Duncan WA, Shanks RG (1965) Comparison of some properties of pronethalol and propranolol. Br J Phamacol Chemother 25:577–591
157. Allin EF, Miller JM, Rowe GG et al (1974) Effects of intraperitoneal administration of propranolol on the mouse heart. Histochemical and electron microscopic observations. Am J Cardiol 33:639–642
158. Fearson RE (1967) Propranolol in the prevention of ventricular fibrillation due to experimental coronary artery occlusion. Observation on its mode of action. Am J Cardiol 20:222–228
159. Khan MI, Hamilton JT, Manning GW (1972) Protective effect of beta adrenoceptor blockade in experimental coronary occlusion in conscious dogs. Am J Cardiol 30:832–837
160. Mueller HS, Ayres SM, Religa A et al (1974) Propranolol in the treatment of acute myocardial infarction. Circulation 49:1078–1087
161. Hawkins CM, Richardson DW, Vokonas PS (1983) Effect of propranolol in reducing mortality in older myocardial infarction patients. The Beta-Blocker Heart Attack Trial experience. Circulation 67:194–197
162. Hamer J, Grandjean T, Melendez L et al (1964) Effect of propranolol (Inderal) in angina pectoris: preliminary report. Br Med J 2:720–723
163. Keelan P (1965) Double-blind trial of propranolol (Inderal) in angina pectoris. Br Med J 1:897–898
164. Besterman EM, Friedlander DH (1965) Clinical experiences with propranolol. Postgrad Med J 41:526–535
165. Dunlop D, Shanks RG (1968) Selective blockade of adrenoceptive beta receptors in the heart. Brit J Pharmacol Chemother 32:201
166. Kubota T, Yamazaki N, Sudo J et al (1990) Protective effects of adrenoceptor-blocking agents on myocardial injury induced by epinephrine in mice. J Toxicol Sci 15:1–13
167. Waagstein F, Hjalmarson A, Varnauskas E et al (1975) Effect of chronic beta-adrenergic receptor blockade in congestive cardiomyopathy. Br Heart J 37:1022–1036
168. Waagstein F, Bristow MR, Swedberg K et al (1993) Beneficial effects of metoprolol in idiopathic dilated cardiomyopathy. Metoprolol in dilated cardiomyopathy (MDC) trial study group. Lancet 342:1441–1446
169. CIBIS investigators and committees (1994) A randomized trial of beta-blockade in heart failure. The Cardiac Insufficiency Bisoprolol Study (CIBIS). Circulation 90:1765–1773
170. CIBIS-II investigators and committees (1999) The cardiac insufficiency bisoprolol study II (CIBIS-II): a randomised trial. Lancet 353:9–13
171. Segev A, Mekori YA (1999) The cardiac insufficiency bisoprolol study II. Lancet 353:1361
172. Drummond GA, Squire IB (1999) The cardiac insufficiency bisoprolol study II. Lancet 353:1361
173. Poole-Wilson PA (1999) The cardiac insufficiency bisoprolol study II. Lancet 353:1360–1361
174. Hjalmarson A, Goldstein S, Fagerberg B et al (2000) Effects of controlled-release metoprolol on total mortality, hospitalizations, and well-being in patients with heart failure: the Metoprolol CR/XL Randomized Intervention Trial in congestive heart failure (MERIT-HF). MERIT-HF Study Group. JAMA 283:1295–1302
175. Aarons RD, Molinoff PB (1982) Changes in the density of beta adrenergic receptors in rat lymphocytes, heart and lung after chronic treatment with propranolol. J Pharmacol Exp Ther 221:439–443
176. Gilbert EM, Abraham WT, Olsen S et al (1996) Comparative hemodynamic, left ventricular functional and antiadrenergic effects of chronic treatment with metoprolol versus carvedilol in the failing heart. Circulation 94:2817–2825
177. Sigmund M, Jakob H, Becker H et al (1996) Effects of metoprolol on myocardial beta-adrenoceptors and Gi alpha-proteins in patients with congestive heart failure. Eur J Clin Pharmacol 51:127–132
178. Bristow MR (2000) beta-Adrenergic receptor blockade in chronic heart failure. Circulation 101:558–569

179. Hata JA, Williams ML, Koch WJ (2004) Genetic manipulation of myocardial beta-adrenergic receptor activation and desensitization. J Mol Cell Cardiol 37:11–21
180. Iaccarino G, Tomhave ED, Lefkowitz RJ et al (1998) Reciprocal in vivo regulation of myocardial G protein-coupled receptor kinase expression by beta-adrenergic receptor stimulation and blockade. Circulation 98:1783–1789
181. Ping P, Gelzer-Bell R, Roth DA et al (1995) Reduced beta-adrenergic receptor activation decreases G-protein expression and beta-adrenergic receptor kinase activity in porcine heart. J Clin Invest 95:1271–1280
182. Leineweber K, Roche P, Beilfuss A et al (2005) G-Protein-coupled receptor kinase activity in human heart failure: effects of beta-adrenoceptor blockade. Cardiovasc Res 66:512–519
183. Brodde OE (2007) Beta-adrenoceptor blocker treatment and the cardiac beta-adrenoceptor-G-protein(s)-adenylyl cyclase system in chronic heart failure. Naunyn Schmiedebergs Arch Pharmacol 374:361–372
184. Bodor GS, Oakeley AE, Allen PD et al (1997) Troponin I phosphorylation in the normal and failing adult human heart. Circulation 96:1495–1500
185. Messer AE, Jacques AM, Marston SB (2007) Troponin phosphorylation and regulatory function in human heart muscle: dephosphorylation of Ser23/24 on troponin I could account for the contractile defect in end-stage heart failure. J Mol Cell Cardiol 42:247–259
186. Reiken S, Wehrens XH, Vest JA et al (2003) Beta-blockers restore calcium release channel function and improve cardiac muscle performance in human heart failure. Circulation 107:2459–2466
187. Lowes BD, Gilbert EM, Abraham WT et al (2002) Myocardial gene expression in dilated cardiomyopathy treated with beta-blocking agents. N Engl J Med 346:1357–1365
188. Abraham WT, Gilbert EM, Lowes BD et al (2002) Coordinate changes in myosin heavy chain isoform gene expression are selectively associated with alterations in dilated cardiomyopathy phenotype. Mol Med 8:750–760
189. Frankenstein L, Nelles M, Slavutsky M et al (2007) Beta-blockers influence the short-term and long-term prognostic information of natriuretic peptides and catecholamines in chronic heart failure independent from specific agents. J Heart Lung Transplant 26:1033–1039
190. Iwatsubo K, Bravo C, Uechi M et al (2012) Prevention of heart failure in mice by an antiviral agent that inhibits type 5 cardiac adenylyl cyclase. Am J Physiol Heart Circ Physiol 302:H2622–H2628
191. Shen TY, Malik FI, Zhao X et al (2010) Improvement of cardiac function by a cardiac Myosin activator in conscious dogs with systolic heart failure. Circ Heart Fail 3:522–527
192. Malik FI, Hartman JJ, Elias KA et al (2011) Cardiac myosin activation: a potential therapeutic approach for systolic heart failure. Science 331:1439–1443
193. Cleland JG, Teerlink JR, Senior R et al (2011) The effects of the cardiac myosin activator, omecamtiv mecarbil, on cardiac function in systolic heart failure: a double-blind, placebo-controlled, crossover, dose-ranging phase 2 trial. Lancet 378:676–683

Chapter 2
Remodeling of Potassium Channels in Cardiac Hypertrophy

Tetsuo Sasano and Junko Kurokawa

Abstract The potassium channel is a major target of remodeling in cardiac hypertrophy. To maintain physiological cardiac function in the face of increased workloads, hypertrophied cardiac myocytes undergo downregulation of K^+ channels that results in a prolongation of action potential duration (APD) and upregulation of Ca^{2+} entry channels. Increased intracellular calcium in cardiac hypertrophy activates calcineurin/nuclear factor of activated T cell pathway to permit remodeling of the K^+ channels, resulting in a positive feedback between the K^+ channel remodeling and alteration of Ca^{2+} handling. Although the I_{to} channel is the major target of the K^+ channel remodeling in hypertrophied cardiomyocytes, alteration of other K^+ channels and/or K^+ channel regulators plays an important role in the remodeling and arrhythmogenicity. In this chapter, we list types of K^+ channels and their mRNA that undergo remodeling in cardiac hypertrophy and discuss molecular mechanisms of the remodeling.

Keywords Cardiac remodeling • Hypertrophy • Potassium channels • I_{to} channels • Gene expression • Ca^{2+} handling • Action potential prolongation • Arrhythmias

2.1 Introduction

Hypertrophic growth of the heart is an adaptive process in response to increased workloads. Hypertrophy has been shown to be associated with arrhythmias which can be caused by an alteration in cellular electrophysiology and cardiac remodeling,

T. Sasano • J. Kurokawa, Ph.D. (✉)
Department of Bio-informational Pharmacology, Medical Research Institute, Tokyo Medical and Dental University, 1-5-45 Yushima Bunkyo-ku, Tokyo 113-8510, Japan
e-mail: junkokuro.bip@mri.tmd.ac.jp

B.I. Jugdutt and N.S. Dhalla (eds.), *Cardiac Remodeling: Molecular Mechanisms*, Advances in Biochemistry in Health and Disease 5, DOI 10.1007/978-1-4614-5930-9_2,

including a prolongation of the action potential duration (APD). Although changes in several ionic currents have been reported in hypertrophied cardiomyocytes, accumulated experimental data have highlighted an importance of K^+ channel remodeling in the APD prolongation in models of hypertrophy [1]. Although a reduction in the transient outward potassium current (I_{to}) is the most consistent ionic current change in cardiac hypertrophy, the target type of K^+ channels varies between species and experimental conditions for hypertrophy. Furthermore, it is believed that heterogeneous reduction, either transmurally or regionally, in I_{to} density associates with QT dispersion that accounts for a mechanism to develop arrhythmias.

Subsequent studies employing molecular biological approaches revealed that modulations of Ca^{2+}-responsive signaling pathways underlie the K^+ channel remodeling [1–3]. Recently, involvement of other pathways has been reported. These molecular mechanisms may provide possible new therapeutic targets against lethal arrhythmias and sudden death in cardiac hypertrophy.

2.2 Electrical Remodeling in Cardiac Hypertrophy

Electrical remodeling in cardiac hypertrophy is considered as a compensatory response to increased workloads. Although the activated signal mechanisms differ in experimental model, it has been reported that renin–angiotensin–aldosterone system, sympathetic nervous system, and other neurohumoral mechanism are activated in cardiac hypertrophy. Mechanical stretch on myocyte also activated signal cascades in heart tissue and cardiomyocyte. These activated signal transductions induce changes in gene expression and function of ion channels, initially to maintain cardiac function. Once these compensatory mechanisms are disrupted, the maladapted responses cause several pathological features including cardiac arrhythmias.

2.2.1 *Changes in Action Potential Duration*

The most striking feature of excitable cell is its action potential profile. The cardiac myocyte is characterized as having long APD with plateau of maintained depolarization before repolarization. Changes in APD result from alterations in the expression and function of depolarizing and repolarizing currents. Electrical remodeling in cardiac hypertrophy is primarily characterized by prolongation of APD. The prolongation of APD is a consistent finding in hypertrophy by several experimental models including pressure and/or volume overload by aortic constriction [4–13] and hypertension [14–18]. Since APD is primarily responsible for the time course of repolarization, this APD prolongation represents delays in cardiac repolarization. The prolongation of APD is not a unique phenotype in cardiac hypertrophy. Several experimental studies utilizing failing heart induced by myocardial infarction [19–21], pacing tachycardia [22–24], and genetically modified animal model [25–28] exhibit the APD prolongation as observed in hypertrophy. In addition, myocardial infarction model has higher arrhythmogenicity [29, 30]. Thus, some of findings in failing heart are also useful to comprehend the mechanisms of remodeling in hypertrophy. Since the heart tissue consists of myocyte layers with different action potential profiles, APD varies across the myocardial wall and the region of the heart [31–34]. The regional heterogeneity in APD is enhanced by hypertrophy [35]. In addition, cardiac hypertrophy also enhanced temporal dispersion of APD. The enhanced spatial and temporal dispersions contribute to the arrhythmogenicity in hypertrophied animal models. This finding is consistent with increased dispersion of monophasic action potential duration and electrocardiographic QT intervals and increased prevalence of ventricular arrhythmia in human [36–38].

2.2.2 *Experimental Models of Cardiac Hypertrophy*

2.2.2.1 Experimental Models to Induce Hypertrophy

Cardiac hypertrophy has been induced by various experimental techniques. Although the prolongation of APD is a common electrophysiological change, signals and ion currents contributing to this APD prolongation differ in each model (Table 2.1).

The pressure overload model by constriction of aorta is widely used to induce left ventricular hypertrophy (LVH). The extent of hypertrophy depends on the site and degree of aortic constriction, resulting in a variety of changes in electrical remodeling. Hypertension model is also considered as pressure and/or volume overload and is another tool to investigate electrical remodeling in hypertrophy. The hypertension model has slower progression of LVH than aortic constriction, which contributes different response of remodeling. Monocrotaline-induced pulmonary hypertension is used as a right ventricular hypertrophy (RVH) model, mostly in rats [8, 39]. Constriction of the pulmonary artery is also utilized for inducing RVH in large mammals [10].

In most of models, hypertrophy is accompanied with a variable amount of heart failure. The balance of heart failure and hypertrophy may explain the difference among several models including complete atrioventricular block [40–43], catecholamine treatment [44], and pacing-induced tachycardia model [22–24] in addition to pressure overload.

2.2.2.2 Animal Species and Expression of Channels

The species of animal is an important factor to understand electrical remodeling in cardiac hypertrophy. APD prolongation is observed regardless of animal species, and it is caused by alterations in K^+ channels in most studies. However, the mechanism of APD prolongation is different between small and large mammals due to the different expression of subtype of K^+ channel (Table 2.1).

Table 2.1 K⁺ channel remodeling in cardiac hypertrophy

Model		Currents					
Methods	Species	APD	I_{to} ($I_{to\text{-}f}$, $I_{to\text{-}s}$)	I_{K1}	I_K (I_{Kr}, I_{Ks}, I_{Kur})	Comment	References
Pressure/volume overload							
Ao (abdominal)	Feline	↑		↔	↓	I_K slowed activation, faster deactivation	[4]
PAC (RVH)	Feline			↑	↓	I_K slowed activation, faster deactivation	[7]
Ao	Dog	↑	↑				[9]
Ao	Rabbit	↑	↑	↓	↔		[5]
Ao	Guinea pig	↑		↔	↔	$I_{Ca,L}$↑	[11]
PAC (RVH)	Ferret		↓			Slowed TTP, decay and recovery	[10]
Ao (ascending)	Rat	↑ epi	$I_{to\text{-}f}$ ↓	↔	↔		[13]
Ao (abdominal)	Rat	↑	↓apex, free wall				[6]
Ao (abdominal)	Rat	↑	↓				[12]
DOCA/salt	Rat	↑	↓			Small negative shift gating	[15]
DOCA/salt	Rat	↑	$I_{to\text{-}f}$ ↓		I_{Kur} ↔		[17]
SHR	Rat	↑	↓	↔			[14]
RVHTN	Rat	↑	↑			Slowed I_{to} decay	[16]
RVHTN	Rat					↓Kv4.2/4.3 mRNA, ↔Kv1.2, 1.4, 1.5, 2.1, KCNQ1	[18]
Monocrotaline	Rat	↑ RV	↓ RV				[8]
Monocrotaline	Rat	↑ RV > LV	↓			↓KChIP2 mRNA	[39]
AV block							
	Canine	↑			I_{Ks} ↓, I_{Kr} in RV↓	↓ KCNQ1/KCNE1 mRNA	[41, 43]
	Rabbit	↑QT	$I_{to\text{-}f}$ ↔	↑	I_{Ks} ↓, I_{Kr} ↓		[42]
	Mouse	↑QT	$I_{to\text{-}f}$ ↓			↓Kv4.2/KChIP2 mRNA	[40]
Pacing tachycardia							
	Dog	↑	↓	↓			[23]
	Rabbit	↑	↓	↔			[24]
Catecholamine treatment							
Iso	Rat	↑	↓				[44]
Iso	Rat	↑	↓ Subepi > Subendo			Indirect assessment by 4AP	[35]

(continued)

Table 2.1 (continued)

Model			Currents				
Methods	Species	APD	I_{to} ($I_{to\text{-}f}$, $I_{to\text{-}s}$)	I_{K1}	I_K (I_{Kr}, I_{Ks}, I_{Kur})	Comment	References
Genetic							
Growth hormone-secreting tumor	Rat	↑	↓				[28]
Calsequestrin	Mouse	↑	↓	↓		I_{NCX}↑, I_{Na}↓, $I_{Ca,L}$↓	[25]
H-Ras-v12	Mouse	↑	↓			I_{NCX}↑	[27]
Clinical							
Severe HF	Human	↑	↓	↓	↔		[100]
Failing	Human	↑	↓ Subendo				[69]
Failing LV	Human		$I_{to\text{-}f}$↓ Subepi			$I_{to\text{-}f}$ slowed recovery in subendo	[70]
HF	Human					↓Kv4.3 mRNA, ↔Kv1.4, Kvβ1, Kir2.1, herg	[46]
HF	Human	↑	$I_{to\text{-}f}$↓	↓	I_{Ks} ↓		[47]

↑ increase, ↓ decrease,↔ unchanged, *RVH* right ventricular hypertrophy, *AoC* aortic constriction, *PAC* pulmonary artery constriction, *TTP* time to peak, *SHR* spontaneously hypertensive rat, *DOCA* deoxycorticosterone acetate, *RVHTN* renovascular hypertension, *GH* growth hormone, *RAS* renin–angiotensin system, *Iso* isoproterenol, *HF* heart failure

Rodents (mice and rats) have shorter cardiac APD than large mammals to adapt to the extremely high heart rate (~600 bpm in mice). I_{to} is a major repolarizing current throughout this short action potential, but the contribution of delayed rectifier potassium current (I_K) and inward rectifier potassium current (I_{K1}) remains subtle. The hypertrophy model in rodents revealed reduction of I_{to} in most of the studies. I_{to} consists of two components: fast ($I_{to\text{-}f}$) and slow ($I_{to\text{-}s}$) components according to the fast and slow rates of inactivation, respectively. In rodents, $I_{to\text{-}f}$ is considered as the major target channel rather than $I_{to\text{-}s}$ [45]. Reduction of I_{to} is achieved by change in gating kinetics and/or the change in expression level of channels.

In large mammals, I_{to} is responsible for the initial rapid phase of action potential repolarization, representing a notch preceding plateau phase. The major contributor for repolarization is I_K (I_{Kr} and I_{Ks}). The hypertrophy model in large mammals showed reduction of I_K, but the change in I_{to} varies among experimental models.

Although data on K^+ channel remodeling in human cardiac hypertrophy are lacking, reports using human failing heart showed downregulation of I_{to} in left and right ventricle [46] and downregulated I_{Ks} in right ventricle [47].

2.3 Remodeling of Potassium Channels

In models of hypertrophied myocardium, previous investigations have highlighted the importance of remodeling of potassium channels in the APD prolongation. The APD of cardiomyocytes is determined by a plateau phase that is controlled by a fine balance between small inward (calcium) and outward (potassium) ionic currents (Fig. 2.1). Functional downregulation of K^+ currents has consistently been observed especially in severe hypertrophy and failure regardless of species, and has been implicated as a major contributor to the APD prolongation. Distinct molecular complexes of potassium channels involve in hypertrophy depending on experimental conditions and species as described in Sect. 2.2. In this section, previous and recent publications regarding remodeling of potassium channels in channel transcript levels are summarized.

2.3.1 *Transient Outward Potassium Current* (I_{to})

Downregulation of I_{to} is arguably the most consistent alteration of potassium currents in hypertrophied myocardium. Because I_{to} plays an important role in the early repolarization phase of action potential in most species, it is believed that suppression of I_{to} provides an explanation of QT prolongation and QT dispersion leading to fatal ventricular arrhythmias. Exceptions are studies of some compensatory pressure overload hypertrophy models, in which there was no change or increase in I_{to} densities [16, 48, 49]. I_{to} has been classified into $I_{to\text{-}f}$ and $I_{to\text{-}s}$ as described in Sect. 2.2. These two components are encoded with distinct channel genes and are modulated differently and regionally in response to hypertrophic stresses.

2.3.1.1 The Fast Transient Outward Potassium Current ($I_{to\text{-}f}$)

In mammalian hearts, the fast transient outward potassium current ($I_{to\text{-}f}$) is conducted by channels comprised of voltage-gated α-pore-forming subunits (Kv4.3 in human/canine and Kv4.2 and Kv4.3 in rodents/ferrets) and an auxiliary β-subunit (Kv channel interacting protein 2; KChIP2). Expression of Kv4 channels correlates with regional heterogeneities in $I_{to\text{-}f}$, which is reflected by a predominant expression in the mid-myocardium and epicardium. In many hypertrophy models, reduction of $I_{to\text{-}f}$ density generally correlates with decreased Kv4 channel transcript levels. It is notable that hypertrophic stress induced by non-myocyte cell-conditioned growth medium or phenylephrine ablates regulated expression of I_{to} and Kv4 mRNA, which is seen during normal heart development [50].

KChIP2 has at least two functions, which are to assist transport of Kv4 channels to the plasma membrane and to regulate gating kinetics of Kv4 channels [51]. The former function is in line with no $I_{to\text{-}f}$ functional expression in KChIP2 knockout

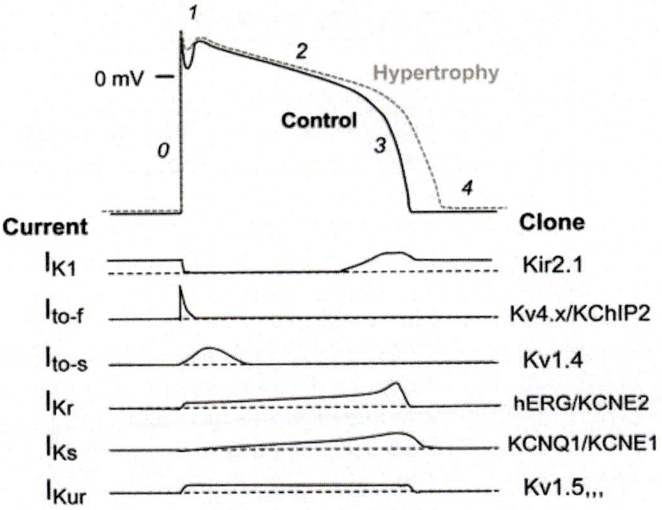

Fig. 2.1 Schematic diagram illustrating the major potassium currents that underlie the action potential in the mammalian ventricle. Top panel shows a control (*solid line*) and hypertrophied (*gray dotted line*) action potential profile. The phases of the action potential are labeled. *Lower panels* show schematic time courses of major K^+ currents as well as clones that encode the currents

mice [52], and KChIP2 accelerates recovery from inactivation in the latter case [51]. Hypertrophy stimulations decrease gene expression of KChIP2 as well as Kv4 α-subunit, which results in reductions in I_{to-f} [10, 53]. Actually, recent data in hypertrophied rat neonatal cardiomyocytes suggest that decrease in KChIP2 expression controls α1AR-induced reduction in I_{to-f} through regulation of NF-κB [54].

2.3.1.2 The Slow Transient Outward Potassium Current (I_{to-s})

Kv1.4 is the predominant gene which encodes the voltage-gated α-pore-forming subunit of the endocardial I_{to-s} channel. In contrast to I_{to-f}, responses of I_{to-s} to hypertrophic stimulations are controversial. In LVH/RVH rats, there was no change in Kv1.4 mRNA, although Kv4.2 and Kv4.3 mRNA are the target of remodeling [8, 18, 55]. In a transgenic mouse model of hypertrophy with overexpressed L-type Ca^{2+} channels, Kv1.4 mRNA is increased, whereas Kv4.2 and Kv4.3 mRNA are reduced [56]. Interestingly, Kv1.4 mRNA levels are high at birth, increase during postnatal days (P0 to P10), and subsequently decrease to very low levels in adult rat ventricles, while Kv4.2 increase between birth and adult, becoming the predominant K^+ channel protein [57]. Such K^+ channel isoform switch during postnatal development is diminished by incubation with non-myocyte cell-conditioned growth medium or phenylephrine that induces cardiomyocyte hypertrophy, resulting in reversion of fetal phenotype of transient K^+ channel currents [50].

2.3.2 Delayed Rectifier Potassium Current (I_K)

I_K plays an important role in cardiac repolarization process. Reduction of I_K currents in hypertrophied cardiomyocytes has been reported in different models, which may contribute to delayed repolarization in heart failure. In feline ventricular myocytes, hypertrophic stimulation has reduced I_K current density with slowed activation and accelerated deactivation, which may result in a greater predisposition to developing early after depolarization [4, 7]. In hypertrophy models in rabbits, both two components of the I_K currents, I_{Kr} and I_{Ks}, are downregulated [42]. Actually, most of the studies consistently show reduction of I_{Ks} [42, 47, 58], while I_{Kr} (or HERG) is reduced or unchanged [46, 57]. The different results of I_{Kr} changes are model specific. Reduction of I_{Ks} may be associated with alteration of transmural APD heterogeneity. In ventricles of canine failing hearts, the percentage of I_{Ks} reduction was greater in epicardial and endocardial cells than that in midmyocardial cells, which may result in elimination of the transmural I_{Ks} gradient [58].

2.3.3 Other Potassium Current

Kv1.5 reduction in hypertrophy is also observed [59–61]. Kv1.5 encodes the ultra-rapid delayed rectifier K^+ currents (I_{Kur}) which is dominantly found in atrium not in ventricles. Kv1.5 remodeling in post-MI LVH appears to be related to thyroid hormone (T3) level [61]. Because the Kv1.5 gene has a T3 responsive element in the promoter region, T3 enhances Kv1.5 expression in rat heart [62]. Downregulation of KChIP2 also contributes to the Kv1.5 remodeling, because KChIP2 interacts with Kv1.5 to assist cell surface expression [63]. Antisense oligodeoxynucleotides directed against Kv1.5 mRNA revealed atrial-specific Kv1.5 remodeling [64]. Thus, role of Kv1.5 remodeling may be more significant in atrial fibrillation rather than in ventricular hypertrophy [65]. There are several studies to show downregulation of I_{K1} or alteration of kinetics in heart failure [23], but several studies reported no change of Kir2.1 mRNA in failing heart compared with normal heart [46, 66].

2.4 Regional Heterogeneity of Electrical Remodeling

Regional heterogeneity of cardiac repolarization times results in the T wave in the electrocardiogram. Transmural and/or interregional heterogeneities of potassium channel expression are major contributors to the regional heterogeneity of repolarization times. The densities of $I_{to\text{-}f}$ vary transmurally in the heart, that is, more prominent in epicardium than in endocardium [32, 67, 68]. Hypertrophic stimulation alters the transmural heterogeneity [69, 70]. In rats with LVH induced by isoprenaline [71] and aortic constriction [6], APD is predominantly prolonged in epicardial cells resulting from greater reduction in $I_{to\text{-}f}$ amplitude compared to endocardial cells. In rat ventricular myocytes that are remote from infarct zone, hypertrophy in epicardium reduced current densities of $I_{to\text{-}f}$ and protein expression of Kv4.2 and Kv4.3 greater than in endocardium [72].

There is interregional heterogeneity of left ventricular repolarization [6, 73]. APD and $I_{to\text{-}f}$ density vary gradually from apex to septum. APD is shortest in left ventricular apex, longest in septum, and intermediate in left ventricular free wall, reflecting from differential $I_{to\text{-}f}$ amplitude [6]. Hypertrophied rat cardiomyocytes induced by abdominal aorta constriction ablate the interregional heterogeneity by greater $I_{to\text{-}f}$ reduction in apex and free wall [6]. Myocardial infarction also eliminates the interregional heterogeneity of APD, I_{to}, and Kv4.2 expression between RV free wall and intraventricular septum [74].

2.5 Possible Mechanisms for Remodeling of Potassium Channels

The involvement of the calcineurin/NFAT pathway in reducing $I_{to\text{-}f}$ has been reported [75–79] (Fig. 2.2). Calcineurin is a Ca^{2+}/calmodulin-dependent protein phosphatase, and its activity is increased in hypertrophied hearts [80–82]. A calcineurin inhibitor cyclosporine A abolishes reduction of $I_{to\text{-}f}$ and Kv4.2/Kv4.3 mRNA expression in hypertrophied rat ventricle [83]. In fetal rodents, calcineurin/NFAT increased $I_{to\text{-}f}$ by transcriptional upregulation of Kv4.2 [84].

On the other hand, APD prolongation by I_{to} reduction in mice and rats leads to calcineurin activation via increase in $[Ca^{2+}]_{in}$ [76, 85–87]. In case of human or canine, reduction in I_{to}, I_{K1}, and I_{Ks} contributes to APD prolongation [47, 58], leading to increase in $[Ca^{2+}]_{in}$. But contribution of the regulation of channels to activation of calcineurin in human or canine is unclear.

The ubiquitous transcriptional factor NFAT has four reported isoforms, which are designated as NFAT1, NFAT2, NFAT3, and NFAT4. Unlike NFAT1, –2, or –4 in immune tissues, NFAT3 is the prominent isoform in the heart. Ca^{2+}-dependent activation of calcineurin induces NFAT dephosphorylation and NFAT translocation to the nucleus where it can interact with GATA4 to activate transcription of hypertrophic responsive genes [88]. Cardiac K^+ channel genes (Kv1.5, Kv2.1, Kv4.2, Kv4.3, and KChIP2) which have putative NFAT binding sites in the promoter regions are downregulated through NFAT3 pathways after myocardial infarction [89]. Among these K^+ channel genes, downregulation of Kv4.2 is the most sensitive to increase in NFAT3 activity compared with other genes (Kv1.5, Kv2.1, Kv4.3, and KChIP2) which require robust increase in NFAT3 activity [78]. Magnitudes of NFAT3 activation vary among hypertrophic models, which can partly explain why some studies do not show downregulation of Kv1.5 and Kv2.1. In a murine hypertrophic model, activation of NFAT3 by calcineurin transduces variations in $[Ca^{2+}]_i$ into differences in I_{to} density in endocardial and epicardial myocytes [89]. $[Ca^{2+}]_i$ and calcineurin/NFAT3

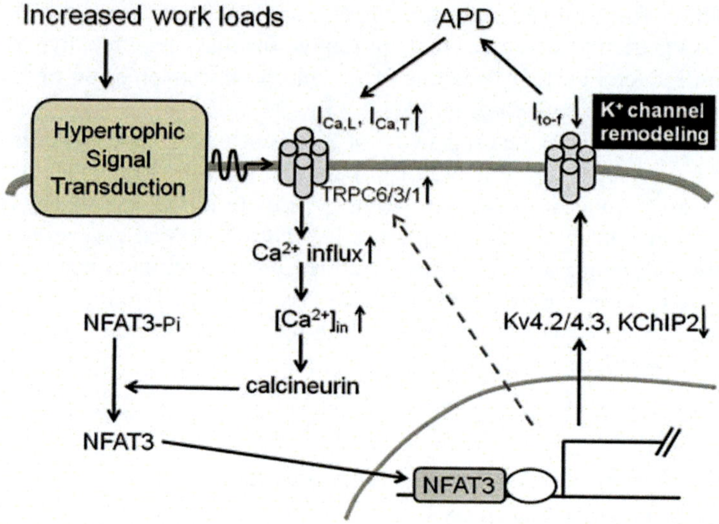

Fig. 2.2 Potential positive feedback between increased Ca^{2+} influx and K^+ channel remodeling

activity is higher in endocardium than in epicardial myocytes, resulting in lower Kv4.2 expression and $I_{to\text{-}f}$ density in endocardial cells, at least in mice [90, 91]. This hypothesis is supported by the finding that suppression of Kv4 expression and I_{to} density in post-myocardial infarction hypertrophy is abolished by calcineurin inhibitor cyclosporine A in a rat model [83] and NFAT knockout mice [89].

Involvement of signaling pathways other than the calcineurin/NFAT pathway in K^+ channel remodeling has been also reported. Oxidative stress through an Rac-dependent NADPH oxidase activation and superoxide production is involved in decreases in Kv4.3 mRNA [92, 93]. Kv4.2 and Kv1.4 mRNA expression is also regulated by endogenous oxidoreductase systems [72, 94, 95]. Involvement of NF-κB signaling in downregulation of $I_{to\text{-}f}$ through control of KChIP2 and Kv4.2 expression has been reported [54]. Downregulation of $I_{to\text{-}f}$ by either α1 adrenergic receptor or TNF-α stimulation may require NF-κB signaling-dependent decreases KChIP2 expression [54]. Mitogen-activated protein kinase (MAPK)/MEK and CaMK II are also involved in reduction of KChIP2 mRNA expression [53, 96]. One can speculate that TRPC6 may affect K^+ channel remodeling by increase in pathological calcineurin/NFAT signaling [97].

2.6 Conclusions

Interruption of the Ca^{2+}-dependent positive feedback loop is a potential new therapeutic target against arrhythmias in cardiac hypertrophy [1, 98]. Considerable progress has been made in understanding the molecular mechanisms of cardiac remodeling of K^+ channels. We now realize that these mechanisms are highly complex reflecting the crosstalk between a variety of regulatory proteins, which can be altered in diseased hearts [99]. Clearly, future research is still needed to understand the clinical relevance of these regulatory proteins.

References

1. Furukawa T, Kurokawa J (2006) Potassium channel remodeling in cardiac hypertrophy. J Mol Cell Cardiol 41:753–761
2. Tomaselli GF, Marban E (1999) Electrophysiological remodeling in hypertrophy and heart failure. Cardiovasc Res 42:270–283
3. Wickenden AD, Kaprielian R, Parker TG et al (1997) Effects of development and thyroid hormone on K^+ currents and K^+ channel gene expression in rat ventricle. J Physiol 504(Pt 2):271–286
4. Furukawa T, Bassett AL, Furukawa N et al (1993) The ionic mechanism of reperfusion-induced early after depolarizations in feline left ventricular hypertrophy. J Clin Invest 91:1521–1531
5. Gillis AM, Geonzon RA, Mathison HJ et al (1998) The effects of barium, dofetilide and 4-aminopyridine (4-AP) on ventricular repolarization in normal and hypertrophied rabbit heart. J Pharmacol Exp Ther 285:262–270
6. Gomez AM, Benitah JP, Henzel D et al (1997) Modulation of electrical heterogeneity by compensated hypertrophy in rat left ventricle. Am J Physiol 272:H1078–H1086
7. Kleiman RB, Houser SR (1989) Outward currents in normal and hypertrophied feline ventricular myocytes. Am J Physiol 256:H1450–H1461
8. Lee JK, Kodama I, Honjo H et al (1997) Stage-dependent changes in membrane currents in rats with monocrotaline-induced right ventricular hypertrophy. Am J Physiol 272: H2833–H2842
9. Lue WM, Boyden PA (1992) Abnormal electrical properties of myocytes from chronically infarcted canine heart. Alterations in Vmax and the transient outward current. Circulation 85:1175–1188
10. Potreau D, Gomez JP, Fares N (1995) Depressed transient outward current in single hypertrophied cardiomyocytes isolated from the right ventricle of ferret heart. Cardiovasc Res 30:440–448
11. Ryder KO, Bryant SM, Hart G (1993) Membrane current changes in left ventricular myocytes isolated from guinea pigs after abdominal aortic coarctation. Cardiovasc Res 27: 1278–1287
12. Tomita F, Bassett AL, Myerburg RJ, Kimura S (1994) Diminished transient outward currents in rat hypertrophied ventricular myocytes. Circ Res 75:296–303
13. Volk T, Nguyen TH, Schultz JH et al (2001) Regional alterations of repolarizing K^+ currents among the left ventricular free wall of rats with ascending aortic stenosis. J Physiol 530:443–455
14. Cerbai E, Barbieri M, Li Q, Mugelli A (1994) Ionic basis of action potential prolongation of hypertrophied cardiac myocytes isolated from hypertensive rats of different ages. Cardiovasc Res 28:1180–1187
15. Coulombe A, Momtaz A, Richer P et al (1994) Reduction of calcium-independent transient outward potassium current density in DOCA salt hypertrophied rat ventricular myocytes. Pflugers Arch 427:47–55
16. Li Q, Keung EC (1994) Effects of myocardial hypertrophy on transient outward current. Am J Physiol 266:H1738–H1745

17. Momtaz A, Coulombe A, Richer P et al (1996) Action potential and plateau ionic currents in moderately and severely DOCA-salt hypertrophied rat hearts. J Mol Cell Cardiol 28:2511–2522
18. Takimoto K, Li D, Hershman KM et al (1997) Decreased expression of Kv4.2 and novel Kv4.3 K^+ channel subunit mRNAs in ventricles of renovascular hypertensive rats. Circ Res 81:533–539
19. Bril A, Forest MC, Gout B (1991) Ischemia and reperfusion-induced arrhythmias in rabbits with chronic heart failure. Am J Physiol 261:H301–H307
20. Gidh-Jain M, Huang B, Jain P, el Sherif N (1996) Differential expression of voltage-gated K^+ channel genes in left ventricular remodeled myocardium after experimental myocardial infarction. Circ Res 79:669–675
21. Qin D, Zhang ZH, Caref EB et al (1996) Cellular and ionic basis of arrhythmias in postinfarction remodeled ventricular myocardium. Circ Res 79:461–473
22. Gao Z, Barth AS, DiSilvestre D et al (2008) Key pathways associated with heart failure development revealed by gene networks correlated with cardiac remodeling. Physiol Genomics 35:222–230
23. Kaab S, Nuss HB, Chiamvimonvat N et al (1996) Ionic mechanism of action potential prolongation in ventricular myocytes from dogs with pacing-induced heart failure. Circ Res 78:262–273
24. Rozanski GJ, Xu Z, Whitney RT et al (1997) Electrophysiology of rabbit ventricular myocytes following sustained rapid ventricular pacing. J Mol Cell Cardiol 29:721–732
25. Knollmann BC, Knollmann Ritschel BE, Weissman NJ et al (2000) Remodelling of ionic currents in hypertrophied and failing hearts of transgenic mice overexpressing calsequestrin. J Physiol 525(Pt 2):483–498
26. Ripplinger CM, Li W, Hadley J et al (2007) Enhanced transmural fiber rotation and connexin 43 heterogeneity are associated with an increased upper limit of vulnerability in a transgenic rabbit model of human hypertrophic cardiomyopathy. Circ Res 101:1049–1057
27. Ruan H, Mitchell S, Vainoriene M et al (2007) Gi alpha 1-mediated cardiac electrophysiological remodeling and arrhythmia in hypertrophic cardiomyopathy. Circulation 116:596–605
28. Xu XP, Best PM (1991) Decreased transient outward K^+ current in ventricular myocytes from acromegalic rats. Am J Physiol 260:H935–H942
29. Sasano T, Kelemen K, Greener ID, Donahue JK (2009) Ventricular tachycardia from the healed myocardial infarction scar: validation of an animal model and utility of gene therapy. Heart Rhythm 6:S91–S97
30. Sasano T, McDonald AD, Kikuchi K, Donahue JK (2006) Molecular ablation of ventricular tachycardia after myocardial infarction. Nat Med 12:1256–1258
31. Furukawa T, Kimura S, Furukawa N et al (1992) Potassium rectifier currents differ in myocytes of endocardial and epicardial origin. Circ Res 70:91–103
32. Litovsky SH, Antzelevitch C (1988) Transient outward current prominent in canine ventricular epicardium but not endocardium. Circ Res 62:116–126
33. Liu DW, Antzelevitch C (1995) Characteristics of the delayed rectifier current (I_{Kr} and I_{Ks}) in canine ventricular epicardial, midmyocardial, and endocardial myocytes. A weaker IKs contributes to the longer action potential of the M cell. Circ Res 76:351–365
34. Sicouri S, Antzelevitch C (1995) Electrophysiologic characteristics of M cells in the canine left ventricular free wall. J Cardiovasc Electrophysiol 6:591–603
35. Shipsey SJ, Bryant SM, Hart G (1997) Effects of hypertrophy on regional action potential characteristics in the rat left ventricle: a cellular basis for T-wave inversion? Circulation 96:2061–2068
36. Atiga WL, Fananapazir L, McAreavey D et al (2000) Temporal repolarization lability in hypertrophic cardiomyopathy caused by beta-myosin heavy-chain gene mutations. Circulation 101:1237–1242
37. Barr CS, Naas A, Freeman M et al (1994) QT dispersion and sudden unexpected death in chronic heart failure. Lancet 343:327–329

38. Piccirillo G, Germano G, Quaglione R et al (2002) QT-interval variability and autonomic control in hypertensive subjects with left ventricular hypertrophy. Clin Sci (Lond) 102:363–371
39. Hardziyenka M, Campian ME, Verkerk AO et al (2012) Electrophysiologic remodeling of the left ventricle in pressure overload-induced right ventricular failure. J Am Coll Cardiol 59:2193–2202
40. Bignolais O, Quang KL, Naud P et al (2011) Early ion-channel remodeling and arrhythmias precede hypertrophy in a mouse model of complete atrioventricular block. J Mol Cell Cardiol 51:713–721
41. Ramakers C, Vos MA, Doevendans PA et al (2003) Coordinated down-regulation of KCNQ1 and KCNE1 expression contributes to reduction of I_{Ks} in canine hypertrophied hearts. Cardiovasc Res 57:486–496
42. Tsuji Y, Opthof T, Yasui K et al (2002) Ionic mechanisms of acquired QT prolongation and torsades de pointes in rabbits with chronic complete atrioventricular block. Circulation 106:2012–2018
43. Volders PG, Sipido KR, Vos MA et al (1998) Cellular basis of biventricular hypertrophy and arrhythmogenesis in dogs with chronic complete atrioventricular block and acquired torsade de pointes. Circulation 98:1136–1147
44. Meszaros J, Ryder KO, Hart G (1996) Transient outward current in catecholamine-induced cardiac hypertrophy in the rat. Am J Physiol 271:H2360–H2367
45. Brahmajothi MV, Campbell DL, Rasmusson RL et al (1999) Distinct transient outward potassium current (I_{to}) phenotypes and distribution of fast-inactivating potassium channel alpha subunits in ferret left ventricular myocytes. J Gen Physiol 113:581–600
46. Kaab S, Dixon J, Duc J et al (1998) Molecular basis of transient outward potassium current downregulation in human heart failure: a decrease in Kv4.3 mRNA correlates with a reduction in current density. Circulation 98:1383–1393
47. Li GR, Lau CP, Leung TK, Nattel S (2004) Ionic current abnormalities associated with prolonged action potentials in cardiomyocytes from diseased human right ventricles. Heart Rhythm 1:460–468
48. Brooksby P, Levi AJ, Jones JV (1993) The electrophysiological characteristics of hypertrophied ventricular myocytes from the spontaneously hypertensive rat. J Hypertens 11: 611–622
49. Ten Eick RE, Zhang K, Harvey RD, Bassett AL (1993) Enhanced functional expression of transient outward current in hypertrophied feline myocytes. Cardiovasc Drugs Ther 7(Suppl 3):611–619
50. Guo W, Kamiya K, Hojo M et al (1998) Regulation of Kv4.2 and Kv1.4 K^+ channel expression by myocardial hypertrophic factors in cultured newborn rat ventricular cells. J Mol Cell Cardiol 30:1449–1455
51. Nerbonne JM, Kass RS (2005) Molecular physiology of cardiac repolarization. Physiol Rev 85:1205–1253
52. Kuo HC, Cheng CF, Clark RB et al (2001) A defect in the Kv channel-interacting protein 2 (KChIP2) gene leads to a complete loss of I_{to} and confers susceptibility to ventricular tachycardia. Cell 107:801–813
53. Jia Y, Takimoto K (2006) Mitogen-activated protein kinases control cardiac KChIP2 gene expression. Circ Res 98:386–393
54. Panama BK, Latour-Villamil D, Farman GP et al (2011) Nuclear factor kappaB downregulates the transient outward potassium current Ito, f through control of KChIP2 expression. Circ Res 108:537–543
55. Capuano V, Ruchon Y, Antoine S et al (2002) Ventricular hypertrophy induced by mineralocorticoid treatment or aortic stenosis differentially regulates the expression of cardiac K^+ channels in the rat. Mol Cell Biochem 237:1–10
56. Bodi I, Muth JN, Hahn HS et al (2003) Electrical remodeling in hearts from a calcium-dependent mouse model of hypertrophy and failure: complex nature of K^+ current changes and action potential duration. J Am Coll Cardiol 41:1611–1622

57. Xu H, Dixon JE, Barry DM et al (1996) Developmental analysis reveals mismatches in the expression of K^+ channel alpha subunits and voltage-gated K^+ channel currents in rat ventricular myocytes. J Gen Physiol 108:405–419
58. Li GR, Lau CP, Ducharme A et al (2002) Transmural action potential and ionic current remodeling in ventricles of failing canine hearts. Am J Physiol Heart Circ Physiol 283:H1031–H1041
59. Heath BM, Xia J, Dong E et al (1998) Overexpression of nerve growth factor in the heart alters ion channel activity and beta-adrenergic signalling in an adult transgenic mouse. J Physiol 512(Pt 3):779–791
60. Matsubara H, Suzuki J, Inada M (1993) Shaker-related potassium channel, Kv1.4, mRNA regulation in cultured rat heart myocytes and differential expression of Kv1.4 and Kv1.5 genes in myocardial development and hypertrophy. J Clin Invest 92:1659–1666
61. Ojamaa K, Kenessey A, Shenoy R, Klein I (2000) Thyroid hormone metabolism and cardiac gene expression after acute myocardial infarction in the rat. Am J Physiol Endocrinol Metab 279:E1319–E1324
62. Abe A, Yamamoto T, Isome M et al (1998) Thyroid hormone regulates expression of shaker-related potassium channel mRNA in rat heart. Biochem Biophys Res Commun 245: 226–230
63. Li H, Guo W, Mellor RL, Nerbonne JM (2005) KChIP2 modulates the cell surface expression of Kv 1.5-encoded K^+ channels. J Mol Cell Cardiol 39:121–132
64. Feng J, Wible B, Li GR et al (1997) Antisense oligodeoxynucleotides directed against Kv1.5 mRNA specifically inhibit ultrarapid delayed rectifier K^+ current in cultured adult human atrial myocytes. Circ Res 80:572–579
65. Van Wagoner DR, Pond AL, McCarthy PM et al (1997) Outward K^+ current densities and Kv1.5 expression are reduced in chronic human atrial fibrillation. Circ Res 80:772–781
66. Fernandez-Velasco M, Ruiz-Hurtado G, Delgado C (2006) I_{K1} and I_f in ventricular myocytes isolated from control and hypertrophied rat hearts. Pflugers Arch 452:146–154
67. Fedida D, Giles WR (1991) Regional variations in action potentials and transient outward current in myocytes isolated from rabbit left ventricle. J Physiol 442:191–209
68. Furukawa T, Myerburg RJ, Furukawa N et al (1990) Differences in transient outward currents of feline endocardial and epicardial myocytes. Circ Res 67:1287–1291
69. Wettwer E, Amos GJ, Posival H, Ravens U (1994) Transient outward current in human ventricular myocytes of subepicardial and subendocardial origin. Circ Res 75:473–482
70. Nabauer M, Beuckelmann DJ, Uberfuhr P, Steinbeck G (1996) Regional differences in current density and rate-dependent properties of the transient outward current in subepicardial and subendocardial myocytes of human left ventricle. Circulation 93:168–177
71. Bryant SM, Shipsey SJ, Hart G (1999) Normal regional distribution of membrane current density in rat left ventricle is altered in catecholamine-induced hypertrophy. Cardiovasc Res 42:391–401
72. Rozanski GJ, Xu Z, Zhang K, Patel KP (1998) Altered K^+ current of ventricular myocytes in rats with chronic myocardial infarction. Am J Physiol 274:H259–H265
73. Cheng J, Kamiya K, Liu W et al (1999) Heterogeneous distribution of the two components of delayed rectifier K^+ current: a potential mechanism of the proarrhythmic effects of methanesulfonanilideclass III agents. Cardiovasc Res 43:135–147
74. Kaprielian R, Sah R, Nguyen T et al (2002) Myocardial infarction in rat eliminates regional heterogeneity of AP profiles, I_{to} K^+ currents, and $[Ca^{2+}]_i$ transients. Am J Physiol Heart Circ Physiol 283:H1157–H1168
75. Perrier E, Perrier R, Richard S, Benitah JP (2004) Ca^{2+} controls functional expression of the cardiac K^+ transient outward current via the calcineurin pathway. J Biol Chem 279: 40634–40639
76. Lebeche D, Kaprielian R, Hajjar R (2006) Modulation of action potential duration on myocyte hypertrophic pathways. J Mol Cell Cardiol 40:725–735
77. Dong D, Duan Y, Guo J et al (2003) Overexpression of calcineurin in mouse causes sudden cardiac death associated with decreased density of K^+ channels. Cardiovasc Res 57:320–332

78. Rossow CF, Dilly KW, Yuan C et al (2009) NFATc3-dependent loss of I_{to} gradient across the left ventricular wall during chronic beta adrenergic stimulation. J Mol Cell Cardiol 46:249–256
79. Xiao L, Coutu P, Villeneuve LR et al (2008) Mechanisms underlying rate-dependent remodeling of transient outward potassium current in canine ventricular myocytes. Circ Res 103:733–742
80. Molkentin JD, Lu JR, Antos CL et al (1998) A calcineurin-dependent transcriptional pathway for cardiac hypertrophy. Cell 93:215–228
81. Liu HB, Yang BF, Dong DL (2010) Calcineurin and electrical remodeling in pathologic cardiac hypertrophy. Trends Cardiovasc Med 20:148–153
82. Hill JA (2003) Electrical remodeling in cardiac hypertrophy. Trends Cardiovasc Med 13:316–322
83. Deng L, Huang B, Qin D et al (2001) Calcineurin inhibition ameliorates structural, contractile, and electrophysiologic consequences of postinfarction remodeling. J Cardiovasc Electrophysiol 12:1055–1061
84. Gong N, Bodi I, Zobel C et al (2006) Calcineurin increases cardiac transient outward K^+ currents via transcriptional up-regulation of Kv4.2 channel subunits. J Biol Chem 281: 38498–38506
85. Sah R, Oudit GY, Nguyen TT et al (2002) Inhibition of calcineurin and sarcolemmal Ca^{2+} influx protects cardiac morphology and ventricular function in K_v4.2N transgenic mice. Circulation 105:1850–1856
86. Lebeche D, Kaprielian R, del Monte F et al (2004) In vivo cardiac gene transfer of Kv4.3 abrogates the hypertrophic response in rats after aortic stenosis. Circulation 110:3435–3443
87. Kassiri Z, Zobel C, Nguyen TT et al (2002) Reduction of I_{to} causes hypertrophy in neonatal rat ventricular myocytes. Circ Res 90:578–585
88. Morin S, Charron F, Robitaille L, Nemer M (2000) GATA-dependent recruitment of MEF2 proteins to target promoters. EMBO J 19:2046–2055
89. Rossow CF, Minami E, Chase EG et al (2004) NFATc3-induced reductions in voltage-gated K^+ currents after myocardial infarction. Circ Res 94:1340–1350
90. Dilly KW, Rossow CF, Votaw VS et al (2006) Mechanisms underlying variations in excitation-contraction coupling across the mouse left ventricular free wall. J Physiol 572:227–241
91. Rossow CF, Dilly KW, Santana LF (2006) Differential calcineurin/NFATc3 activity contributes to the Ito transmural gradient in the mouse heart. Circ Res 98:1306–1313
92. Zhang TT, Takimoto K, Stewart AF et al (2001) Independent regulation of cardiac Kv4.3 potassium channel expression by angiotensin II and phenylephrine. Circ Res 88:476–482
93. Zhou C, Ziegler C, Birder LA et al (2006) Angiotensin II and stretch activate NADPH oxidase to destabilize cardiac Kv4.3 channel mRNA. Circ Res 98:1040–1047
94. Li X, Li S, Xu Z et al (2006) Redox control of K^+ channel remodeling in rat ventricle. J Mol Cell Cardiol 40:339–349
95. Flescher E, Tripoli H, Salnikow K, Burns FJ (1998) Oxidative stress suppresses transcription factor activities in stimulated lymphocytes. Clin Exp Immunol 112:242–247
96. El-Haou S, Balse E, Neyroud N et al (2009) Kv4 potassium channels form a tripartite complex with the anchoring protein SAP97 and CaMKII in cardiac myocytes. Circ Res 104:758–769
97. Kuwahara K, Nakao K (2010) New molecular mechanisms for cardiovascular disease:transcriptional pathways and novel therapeutic targets in heart failure. J Pharmacol Sci 116:337–342
98. Wasson S, Reddy HK, Dohrmann ML (2004) Current perspectives of electrical remodeling and its therapeutic implications. J Cardiovasc Pharmacol Ther 9:129–144
99. Niwa N, Nerbonne JM (2010) Molecular determinants of cardiac transient outward potassium current (I_{to}) expression and regulation. J Mol Cell Cardiol 48:12–25
100. Beuckelmann DJ, Nabauer M, Erdmann E (1993) Alterations of K^+ currents in isolated human ventricular myocytes from patients with terminal heart failure. Circ Res 73:379–385

Chapter 3
Role of Gender in Ca^{2+} Cycling and Cardiac Remodeling Due to Heart Failure

Naranjan S. Dhalla, Amrit Malik, Shelly Zieroth, and Paramjit S. Tappia

Abstract Heart failure (HF) is invariably preceded by cardiac hypertrophy. Contractile dysfunction has been linked to ventricular remodeling and progressive defects in cardiomyocyte Ca^{2+} handling as well as changes in the sensitivity of myofibrils to Ca^{2+}. Since the renin–angiotensin system and the sympathetic nervous system are activated in HF, it is likely that subcellular alterations, at the level of the sarcoplasmic reticulum and sarcolemma as well as molecular changes that affect Ca^{2+} homeostasis in the heart, may be due to prolonged exposure to angiotensin II and excess production of catecholamines. In addition, activation of different neurohormonal systems is also linked to cardiac remodeling and subsequent insufficiency in contractile performance. Female sex hormones such as estrogen and progesterone are considered to play a cardioprotective role; however, women lose this protection against various risk factors of heart disease at postmenopausal stage. Accordingly, this chapter highlights some of the mechanisms that cause Ca^{2+}-handling abnormalities in the pathogenesis of HF and the role of gender and menopausal status in influencing the nature and extent of these alterations. Furthermore, the impact of different pharmacological agents including angiotensin-converting enzyme inhibitors, angiotensin receptor blockers, β-adrenoceptor blockers, and Ca^{2+} antagonists on the changes in Ca^{2+}-cycling

N.S. Dhalla (✉)
Department of Physiology, Faculty of Medicine, Institute of Cardiovascular Sciences, St. Boniface Hospital Research Centre, University of Manitoba, 351 Tache Avenue, Winnipeg, MB, Canada R2H 2A6
e-mail: nsdhalla@sbrc.ca

A. Malik • S. Zieroth
Department of Internal Medicine, Faculty of Medicine, St. Boniface Hospital, University of Manitoba, Winnipeg, MB, Canada

P.S. Tappia
Asper Clinical Research Institute, St. Boniface Hospital Research, Winnipeg, MB, Canada

B.I. Jugdutt and N.S. Dhalla (eds.), *Cardiac Remodeling: Molecular Mechanisms*, Advances in Biochemistry in Health and Disease 5, DOI 10.1007/978-1-4614-5930-9_3,

proteins in HF is discussed. Finally, areas where there is a deficiency in knowledge with respect to the mechanisms of subcellular remodeling in HF that need to be addressed are also identified.

Keywords Cardiac hypertrophy • Heart failure • Subcellular remodeling • Ca^{2+}-cycling proteins • Sarcolemma • Sarcoplasmic reticulum • Myofibrils • Sex differences • Estrogen

3.1 Introduction

Heart failure (HF) is a complex problem which is invariably preceded by cardiac hypertrophy as an adaptive mechanism mainly due to the activation of renin–angiotensin system (RAS), sympathetic nervous system (SNS), and several other neurohormonal systems [1–11] (Fig. 3.1). Although the mechanisms for the transition of cardiac hypertrophy to HF are poorly understood, ventricular remodeling due to marked changes in extracellular matrix has been suggested to be closely associated with the development of contractile dysfunction [12, 13]. On the other hand, in view of the ability of sarcoplasmic reticulum (SR) and sarcolemma (SL) to transport Ca^{2+} and regulate the contractile apparatus [14–16], it has been suggested that cardiac dysfunction in HF is also due to progressive defects in the abilities of SR and SL to handle Ca^{2+} as well as changes in the sensitivity of myofibrils (MF) to Ca^{2+} in cardiomyocytes [14–19]. In fact, Ca^{2+}-handling abnormalities have been demonstrated to occur in cardiomyocytes isolated from failing hearts [19–25]. Furthermore, depression in the sensitivity of MF to Ca^{2+} and MF Ca^{2+}-stimulated ATPase activity in the failing heart has been shown to be due to changes in myosin isozyme composition, myosin light chain kinase (MLCK), troponin phosphorylation, and troponin Ca^{2+}-binding activities [14–16, 18, 26]. Such subcellular alterations are considered to be due to occurrence of intracellular Ca^{2+} overload, elevated levels of tumor necrosis factor-alpha (TNF-α), development of oxidative stress, and activation of proteases [2, 14–16, 27, 28] (Fig. 3.2). Furthermore, varying degrees of changes in SR Ca^{2+}-pump ATPase, Ca^{2+}-release channels, and phospholamban (PLB) as well as mRNA levels for these SR proteins are found to occur in failing hearts [14–16, 20, 21]. Alterations in SL enzyme and their gene expression in different types of failing hearts are related to Na^+–K^+ ATPase, Ca^{2+}-channels, and Na^+–Ca^{2+} exchange [22–26, 29, 30]. Likewise, changes in MF proteins and their gene expression for α-myosin heavy chain, β-myosin heavy chain, and regulatory proteins (MLCK, troponin I, and troponin T) are observed in HF [14, 31–34]. While the activation of protein kinase C (PKC) has been shown to depress the SR Ca^{2+}-pump ATPase expression [35–39], PKC and other kinases such as protein kinase A, Ca^{2+}-calmodulin-dependent kinase (CaMK), and mitogen-activated protein kinase [40–43] are involved in the pathway of signal transduction for subcellular remodeling in failing hearts (Fig. 3.3). Since RAS and SNS are activated in HF [44–48] and

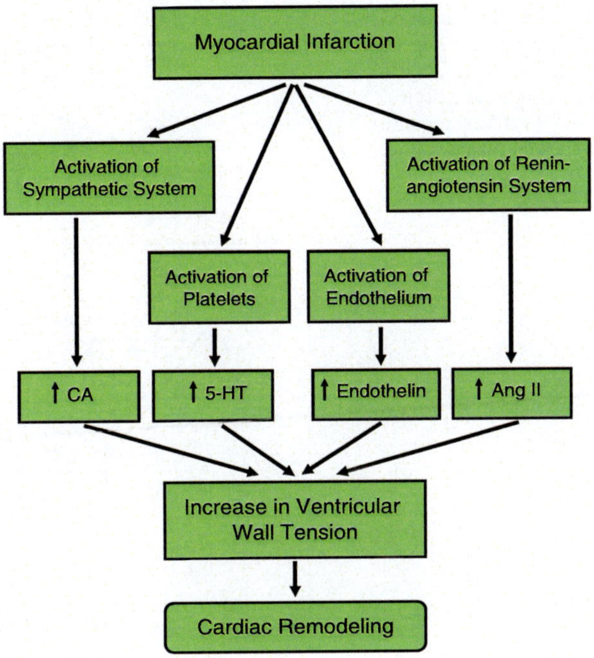

Fig. 3.1 Different neurohormones involved in cardiac remodeling in response to myocardial infarction. *CA* catecholamines, *5-HT* 5-hydroxytryptamine, *Ang II* angiotensin II

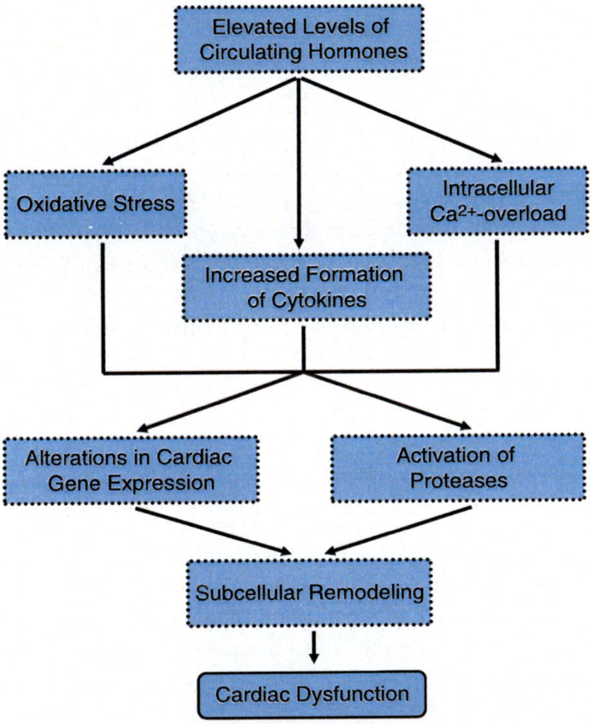

Fig. 3.2 Role of elevated circulating hormones in subcellular remodeling and subsequent cardiac dysfunction

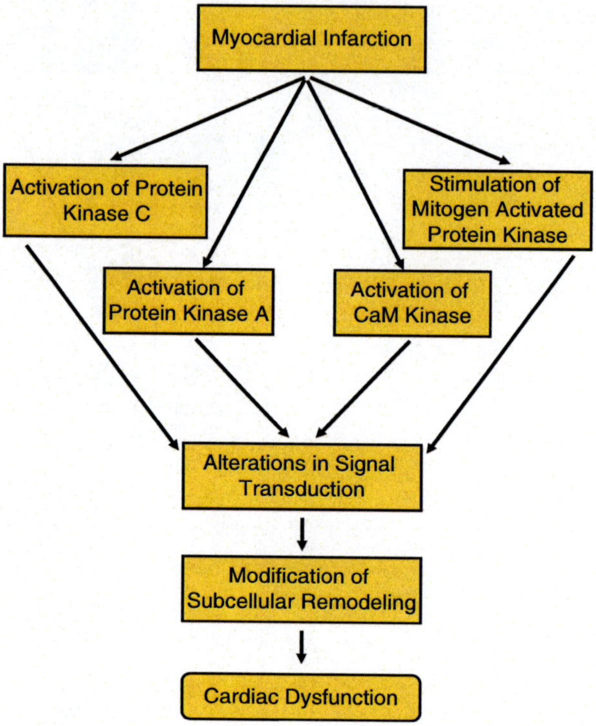

Fig. 3.3 Activation of different kinase signal transduction systems due to myocardial infarction. *CaM kinase* Ca^{2+}/calmodulin-dependent kinase

prolonged exposure of the heart to angiotensin II (Ang II) as well as catecholamines causes changes in cardiac gene expression [49–52], it is likely that subcellular alterations in the hearts may be due to the activation of both RAS and SNS. It is noteworthy that most of the information on the mechanisms of HF is obtained from males whereas little is known about the cellular and molecular mechanisms of cardiac dysfunction during the development of HF in females. Figure 3.4 summarizes the subcellular compartments that undergo alterations during cardiac remodeling in the development of HF.

3.2 Gender Difference and Pathophysiology of HF

Several epidemiological studies have established that gender is a major determinant of cardiovascular risk factors and heart disease [53–57]. Premenopausal females have a lower risk of developing coronary artery disease compared to men; however, this female advantage is lost after menopause [58–60]. Furthermore, women show more diastolic HF with normal ejection fraction, whereas men have more systolic

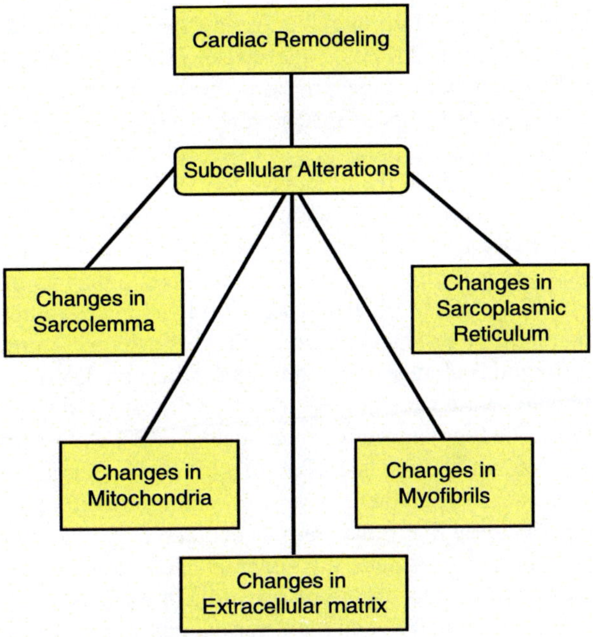

Fig. 3.4 Cardiac remodeling due to alterations in the different subcellular compartments

HF with depressed ejection fraction [61–63]. Although HF in men, unlike women, is normally preceded by eccentric cardiac hypertrophy [62], gender difference in several types of HF is associated with differences in cardiac remodeling and cardiac dysfunction. HF in male rats due to volume overload was accompanied by a greater degree of cardiac dysfunction, cardiac remodeling, and apoptosis in comparison to females [64–68]. Gender differences in aortic regurgitation-induced cardiac function and cardiac remodeling have also been observed in patients [69, 70]. Volume overload in boys with HF was found to produce a greater decrease in mRNA levels for SR Ca^{2+}-pump protein than that seen in girls [71]. Gender and sex-related differences were observed for cardiac function in patients with HF due to aging [72] as well as for apoptosis in HF patients with ischemic and dilated cardiomyopathy [73, 74]. HF due to pressure overload in males was also observed to produce greater changes with respect to cardiac remodeling and cardiac dysfunction [75, 76] as well as a greater reduction in gene expression in SR Ca^{2+}-pump protein and Na^+–Ca^{2+} exchanger [77] in comparison to those seen in females. Gender differences in the expression of Na^+–Ca^{2+} exchanger current and β-AR responsiveness were also seen in HF due to tachycardia in pigs [78]. Induction of myocardial infarction (MI) for 4–6 weeks in female rats produced a lesser increase in ventricular wall thickness and fewer abnormalities in diastolic filling compared to male rats without any difference in infarct size, left ventricle (LV) cavity enlargement, or systolic dysfunction [79–81]. Sex-related differences in changes in cardiac function as well as cardiac remodeling, without any difference in infarct size, were also seen in mice upon inducing MI for 2–12 weeks [82–84]. It is pointed out that while gender differences with respect to both cardiac dysfunction and cardiac remodeling have been shown to occur in different types of HF, very little information regarding molecular and cellular mechanisms for the gender-related difference in HF due to MI is available in the literature.

It is generally held that after menopause, females lose advantage for protection against various risk factors of heart disease and the female sex hormones such as estrogen and progesterone are considered to play an important role in cardioprotection. Conversely, greater susceptibility of males to heart disease may be due to the markedly high levels of circulating testosterone compared to females. Sex-related differential changes in gene expression and DNA synthesis in normal male and female hearts were considered to be due to the regulatory effects of sex hormones [85]. The regulatory influence of sex hormones in normal male and female animals was shown in studies on the effects of ovariectomy with or without estrogen and progesterone treatment as well as castration with or without testosterone treatment. Ovariectomy in normal animals depressed cardiac function and induced changes in SR Ca^{2+}-pump protein, Ca^{2+}-release channels, and SL Na^{+}–Ca^{2+} exchange activities; these effects were abolished by estrogen replacement [86, 87]. Both ovariectomy in female and castration in male normal rats were also found to decrease cardiac function and produce a shift in myosin enzymes; these effects were attenuated by treatment with estrogen and testosterone, respectively [88]. Furthermore, castration in normal male rats also decreased mRNA levels for SL Na^{+}–Ca^{2+} exchanger, L-type Ca^{2+} channels, and β-ARs; varying degrees of alterations in these effects in castrated animals were observed upon treatment with testosterone [89–91]. On the other hand, ovariectomy attenuated the volume overload-induced cardiac dysfunction and cardiac remodeling, and these effects were prevented by estrogen treatment [65–68, 92, 93]. Alterations in β-AR signal transduction and apoptosis due to volume overload in ovariectomized animals were also attenuated by estrogen [66–68]. The cardioprotective effect of estrogen was seen against volume overload-induced structural and functional changes in males [94]. Ovariectomy augmented pressure-induced hypertrophy and changes in Akt–eNOS signaling pathway in females [95] as well as chronic β-AR stimulation-induced calpain activation and impairment in Akt–eNOS pathway [96]. Chronic administration of estrogen normalized MI-induced cardiac remodeling and the increase in endothelin-1 receptor expression [97]. These observations support the view that estrogen is cardioprotective in attenuating cardiac dysfunction and cardiac remodeling in HF; however, the role of other sex hormones in this regard is poorly understood. Furthermore, it should be noted that MI is the major cause of HF when compared with other risk factors such as pressure overload, volume overload, valvular defects, and different types of cardiomyopathies [28, 98, 99], and thus, it is of critical importance that molecular and cellular mechanisms of MI-induced HF in both males and females be carefully examined to gain further information on gender difference with respect to HF.

3.3 Pathogenesis of Cardiac Dysfunction in HF Due to MI

Several animals such as rabbits, cats, and dogs exhibit varying degrees of MI upon occlusion of the coronary artery [100–107]; however, due to high mortality and relatively small infarct size, these animals do not form good experimental models of HF. Some investigators used mice as a model for inducing MI [82–84], but the viable LV from these animals is not sufficient for biochemical analysis of subcellular organelles during the development of HF. On the other hand, rat was found to be an excellent model of HF for studying changes in the viable myocardium following occlusion of the left coronary artery; a wide spectrum (15–60% of the LV) of scar occurred and healed completely [108–115]. The degree of depression in the LV function was observed to depend upon myocardial infarct size as well as duration of MI [116, 117]. Ventricular remodeling due to an increase in collagenous material in the extracellular matrix was found to occur following MI [118–123], and this was suggested to explain cardiac dysfunction in HF. In addition, both plasma and cardiac RAS were activated in the infarcted animal and promoted the formation of Ang II [124]. We have examined time-course changes in heart function and observed that mild, moderate, and severe stages of HF occurred at 4, 8, and 16 weeks after inducing MI in rats, respectively [125]. Furthermore, the number of Ca^{2+} channels, which permit Ca^{2+} entry into the cardiac cell, as well as K^+ channels, was depressed at the moderate (8 weeks) and severe (16 weeks) stages of HF following MI in rats [112, 125]. SR Ca^{2+} uptake activity was depressed at 4, 8, and 16 weeks after inducing MI [126, 127]; this change was associated with a depression in SR Ca^{2+}-pump ATPase as well as gene expression [128–130]. Likewise, the density of beta-adrenergic receptors (β-ARs) decreased without any changes in the affinity ($1/K_d$) at 4 and 8 weeks whereas the number of alpha-adrenergic receptors (α-ARs) increased at 8 and 16 weeks after inducing MI [131]. Further experiments revealed depressions in β_1-AR and adenylyl cyclase activities in failing LV of the infarcted animals [132]. The activities of SL Na^+–K^+ ATPase [133, 134] and SL Na^+–Ca^{2+} exchange [135] were depressed without any change in the SL Ca^{2+}-pump activity at 8 and 16 weeks of MI. Alterations in membrane potentials as well as Ca^{2+} currents, Na^+–Ca^{2+} exchange currents, and expression of K^+ channels were also observed in the infarcted myocardium [136–140]. Furthermore, decrease in cardiac contractility of the infarcted heart was shown to be partly due to changes in myosin isozymes as well as actomyosin ATPase activity [114, 141]. Alterations in myosin light chain have been found [142, 143], and a decrease in the sensitivity of the viable myocardium to Ca^{2+} has been reported [144, 145] in HF due to MI. We also reported marked changes in MF ATPase as well as myosin protein and gene expression in hearts failing due to MI [146, 147]. These studies suggest that heart dysfunction in rats with HF due to MI is not only associated with changes in extracellular matrix, but alterations in SL and SR as well as MF in cardiomyocytes are also involved intimately.

Treatment of infarcted animals with HF by various angiotensin-converting enzyme (ACE) inhibitors and angiotensin receptor (AT_1R) antagonists prevented ventricular remodeling, improved heart function, changed hormone profile, and reduced mortality [148–160]. Likewise, the treatment of infarcted rats with captopril attenuated the loss of cardiac β-AR responses and enhanced intracellular Ca^{2+} handling in the viable LV [161, 162]. Blockade of RAS by agents such as imidapril, enalapril, and losartan also ameliorated changes in the β-AR signal transduction system in HF due to MI [163, 164]. Different β-AR blockers, propranolol and atenolol, also partially prevented MI-induced changes in cardiac performance as well as MF and SR activities [165, 166]. Favorable effects in reducing LV remodeling and improving heart function in rats with MI were obtained upon exercise training as well as upon treatment with growth hormone, thyroid hormones, and antioxidants [167–170]. Treatment of animals with antiplatelet agents, sarpogrelate (a 5-HT_{2A} receptor antagonist) and cilostazol (a phosphodiesterase inhibitor), revealed attenuation of LV remodeling, cardiac dysfunction, and remodeling of both SR and MF in HF due to MI [171–173]. Since Ca^{2+} antagonists, metabolic therapy, and treatment with Na^+–H^+ exchange inhibitors have also been reported to produce beneficial effects in the infarcted animals [174–179], it is suggested that various molecular targets in cardiomyocytes are involved for improving cardiac function in HF. Extensive studies in our laboratory on treatment of MI-induced HF with different agents [180–192] have shown that improvement of cardiac performance in the failing heart is associated with the prevention of alterations in cardiac gene expression specific for SL, SR, and MF proteins. Furthermore, our work on MI-induced HF has indicated that changes in SR Ca^{2+}-cycling proteins are more dramatic in comparison to alterations in the function of both SL and MF.

3.4 Conclusions

From the foregoing review of literature, it is clear that most of the data regarding the molecular and cellular mechanisms of cardiac dysfunction in HF due to MI, which represents the major form of HF, are obtained by employing male hearts. It is also evident that gender differences with respect to development of cardiac remodeling and cardiac dysfunction in different types of HF have been identified; however, the mechanisms for such differences are poorly understood. It is therefore emphasized that a comprehensive study on the mechanisms of gender difference in HF should be undertaken. Since abnormalities in the function of SR, by virtue of its ability to release and accumulate Ca^{2+} for the occurrence of cardiac contraction and relaxation, are considered to be the major determinants of cardiac dysfunction in HF [2, 14, 16, 26], a careful examination of SR Ca^{2+}-release channels and Ca^{2+}-pump activities, protein content, and gene expression in the MI-induced HF in both males and females should be carried out. In view of the early, moderate, and severe stages of HF in male rats at 4, 8, and 16 weeks of inducing MI, respectively [125, 126, 131–133], both male and female hearts will be used for the assessment of changes in Ca^{2+}-cycling proteins at different times of inducing MI to establish delayed transition of cardiac hypertrophy to HF in females compared to males. Since the activation of CaMK is known to increase SR Ca^{2+}-release channel and Ca^{2+}-pump activities [40, 182], CaMK activity and protein content should be determined to establish if gender difference in MI-induced HF is due to difference in the regulation of Ca^{2+}-cycling proteins by CaMK. In view of the role of prolonged activation of RAS in the development of HF, various components of peripheral and cardiac RAS should be monitored in MI-induced HF to investigate if gender difference in HF is determined by difference in the activation of RAS. Since sex hormones are known to impart gender difference [85–88], the beneficial effects of estrogen and progesterone in modifying MI-induced HF and changes in Ca^{2+}-cycling proteins should be examined in females with or without ovariectomy whereas the adverse effects of testosterone need to be studied in males with or without castration. The mechanisms of the beneficial effects of estrogen in females with or without ovariectomy should be investigated by monitoring alterations in signal transduction in HF due to MI whereas the role of estrogen receptors should be examined by measuring gene expression for Ca^{2+}-cycling proteins in cardiomyocytes.

Acknowledgment The research reported in this chapter was supported by a grant from the Canadian Institute of Health Research. Infrastructural support was provided by the St. Boniface Hospital Research Foundation.

References

1. Cohn JN, Ferrari R, Sharpe N (2000) Cardiac remodeling–concepts and clinical implications: a consensus paper from an international forum on cardiac remodeling. Behalf of an International Forum on Cardiac Remodeling. J Am Coll Cardiol 35:569–582
2. Dhalla NS, Temsah RM, Netticadan T (2000) Role of oxidative stress in cardiovascular diseases. J Hypertens 18:655–673
3. Deschamps AM, Spinale FG (2006) Pathways of matrix metalloproteinase induction in heart failure: bioactive molecules and transcriptional regulation. Cardiovasc Res 69:666–676
4. Mudd JO, Kass DA (2008) Tackling heart failure in the twenty-first century. Nature 451:919–928
5. Rich MW (2011) The year in quality of care in heart failure. J Card Fail 17:443–450
6. Goldstein S (2001) Heart failure therapy at the turn of the century. Heart Fail Rev 6:7–14
7. Jessup M, Brozena S (2003) Heart failure. N Engl J Med 348:2007–2018
8. Lee CS, Tkacs NC (2008) Current concepts of neurohormonal activation in heart failure: mediators and mechanisms. AACN Adv Crit Care 19:364–385
9. Zhang W, Elimban V, Nijjar MS et al (2005) Role of renin-angiotensin system in the development of cardiac hypertrophy and heart failure. Adapt Biol Med 4:239–257
10. Rehsia NS, Dhalla NS (2010) Mechanisms of the beneficial effects of β-adrenoceptor antagonists in congestive heart failure. Exp Clin Cardiol 15:e86
11. Rehsia NS, Dhalla NS (2010) Potential of endothelin-1 and vasopressin antagonists for the treatment of congestive heart failure. Heart Fail Rev 15:85–101
12. Briest W, Holzl A, Rassler B et al (2003) Significance of matrix metalloproteinases in norepinephrine-induced remodelling of rat hearts. Cardiovasc Res 57:379–387
13. Spinale FG (2007) Myocardial matrix remodeling and the matrix metalloproteinases: influence on cardiac form and function. Physiol Rev 87:1285–1342
14. Dhalla NS, Dent MR, Tappia PS et al (2006) Subcellular remodeling as a viable target for the treatment of congestive heart failure. J Cardiovasc Pharmacol Ther 11:31–45

15. Babick AP, Dhalla NS (2007) Role of subcellular remodeling in cardiac dysfunction due to congestive heart failure. Med Princ Pract 16:81–89
16. Dhalla NS, Saini-Chohan HK, Rodriguez-Leyva D et al (2009) Subcellular remodelling may induce cardiac dysfunction in congestive heart failure. Cardiovasc Res 81:429–438
17. Prestle J, Quinn FR, Smith GL (2003) Ca^{2+}-handling proteins and heart failure: novel molecular targets? Curr Med Chem 10:967–981
18. Machackova J, Barta J, Dhalla NS (2006) Myofibrillar remodeling in cardiac hypertrophy, heart failure and cardiomyopathies. Can J Cardiol 22:953–968
19. Gajarsa JJ, Kloner RA (2011) Left ventricular remodeling in the post-infarction heart: a review of cellular, molecular mechanisms, and therapeutic modalities. Heart Fail Rev 16:13–21
20. Shannon TR, Pogwizd SM, Bers DM (2003) Elevated sarcoplasmic reticulum Ca^{2+} leak in intact ventricular myocytes from rabbits in heart failure. Circ Res 93:592–594
21. Babick AP, Cantor EJF, Babick JT et al (2004) Cardiac contractile dysfunction in J2N-k cardiomyopathic hamsters is associated with impaired SR function and regulation. Am J Physiol 287:C1202–C1208
22. Schwinger RH, Bundgaard H, Müller-Ehmsen J et al (2003) The Na, K-ATPase in the failing human heart. Cardiovasc Res 57:913–920
23. Verdonck F, Volders PG, Vos MA et al (2003) Intracellular Na^{+} and altered Na^{+} transport mechanisms in cardiac hypertrophy and failure. J Mol Cell Cardiol 35:5–25
24. Despa S, Islam MA, Weber CR et al (2002) Intracellular Na^{+} concentration is elevated in heart failure but Na/K pump function is unchanged. Circulation 105:2543–2548
25. Gomez AM, Schwaller B, Porzig H et al (2002) Increased exchange current but normal Ca^{2+} transport via Na^{+}-Ca^{2+} exchange during cardiac hypertrophy after myocardial infarction. Circ Res 91:323–330
26. Dhalla NS, Rangi S, Babick AP et al (2012) Cardiac remodeling and subcellular defects in heart failure due to myocardial infarction and aging. Heart Fail Rev 17:671–681
27. Kaur K, Dhingra S, Slezak J et al (2009) Biology of TNFα and IL-10, and their imbalance in heart failure. Heart Fail Rev 14:113–123
28. Müller AL, Dhalla NS (2012) Role of various proteases in cardiac remodeling and progression of heart failure. Heart Fail Rev 17:395–409
29. Sweadner KJ, Herrera VL, Amato S et al (1994) Immunologic identification of Na^{+}, K^{+}-ATPase isoforms in myocardium Isoform change in deoxycorticosterone acetate-salt hypertension. Circ Res 74:669–678
30. Studer R, Reinecke H, Bilger J et al (1994) Gene expression of the cardiac Na^{+}-Ca^{2+} exchanger in end-stage human heart failure. Circ Res 75:443–453
31. Morano I, Hadicke K, Haase H et al (1997) Changes in essential myosin light chain isoform expression provide a molecular basis for isometric force regulation in the failing human heart. J Mol Cell Cardiol 29:1177–1187
32. Chen Z, Higashiyama A, Yaku H et al (1997) Altered expression of troponin T isoforms in mild l eft ventricular hypertrophy in the rabbit. J Mol Cell Cardiol 29:2345–2354
33. Solaro RJ, Burkart EM (2002) Functional defects in troponin and the systems biology of heart failure. J Mol Cell Cardiol 34:689–693
34. Torrealba JR, Lozano E, Griffin M et al (2002) Maximal ATPase activity and calcium sensitivity of reconstituted myofilaments are unaltered by the fetal troponin T re-expressed during human heart failure. J Mol Cell Cardiol 34:797–805
35. Rouet-Benzineb P, Gontero B, Dreyfus P et al (2000) Angiotensin II induces nuclear factor-κB activation in cultured neonatal rat cardiomyocytes through protein kinase C signaling pathway. J Mol Cell Cardiol 32:1767–1778
36. Baines CP, Song CX, Zheng YT et al (2003) Protein kinase C epsilon interacts with and inhibits the permeability transition pore in cardiac mitochondria. Circ Res 92:873–880
37. Sabri A, Steinberg SF (2003) Protein kinase C isoform-selective signals that lead to cardiac hypertrophy and the progression of heart failure. Mol Cell Biochem 251:97–101

38. Wang J, Liu X, Arneja AS et al (1999) Alterations in protein kinase A and protein kinase C levels in heart failure due to genetic cardiomyopathy. Can J Cardiol 15:683–690
39. Sentex E, Wang X, Liu X et al (2006) Expression of protein kinase C isoforms in cardiac hypertrophy and heart failure due to volume overload. Can J Physiol Pharmacol 84:227–238
40. English JM, Cobb MH (2002) Pharmacological inhibitors of MAPK pathways. Trends Pharmacol Sci 23:40–45
41. Hagemann C, Blank JL (2003) The ups and downs of MEK kinase interactions. Cell Signal 13:863–875
42. Zhang W, Elimban V, Nijjar MS et al (2003) Role of mitogen-activated protein kinase in cardiac hypertrophy and heart failure. Exp Clin Cardiol 8:173–183
43. Kirchhefer U, Schmitz W, Scholz H et al (1999) Activity of cAMP-dependent protein kinase and Ca^{2+}/calmodulin-dependent protein kinase in failing and nonfailing human hearts. Cardiovasc Res 42:254–261
44. Basu S, Sinha SK, Shao Q et al (1996) Neuropeptide Y modulation of sympathetic activity in myocardial infarction. J Am Coll Cardiol 27:1796–1803
45. Dostal DE, Baker KM (1999) The cardiac renin-angiotensin system: conceptual, or a regulator of cardiac function? Circ Res 85:643–650
46. Kim S, Iwao H (2001) Molecular and cellular mechanisms of angiotensin II-mediated cardiovascular and renal diseases. Pharmacol Rev 52:11–34
47. Wang X, Dhalla NS (2000) Modification of beta-adrenoceptor signal transduction pathway by genetic manipulation and heart failure. Mol Cell Biochem 214:131–155
48. Adameova A, Abdellatif Y, Dhalla NS (2009) Role of the excessive amounts of circulating catecholamines and glucocorticoids in stress-induced heart disease. Can J Physiol Pharmacol 87:493–514
49. Ju H, Scammel-LaFleur T, Dixon IMC (1996) Altered mRNA abundance of calcium transport genes in cardiac myocytes induced by angiotensin II. J Mol Cell Cardiol 28:1119–1128
50. Shizukuda Y, Buttrick PM, Greenen DL et al (1998) β-adrenergic stimulation causes cardiocyte apoptosis: influence of tachycardia and hypertrophy. Am J Physiol Heart Circ Physiol 275:H961–H968
51. Linck B, Boknik P, Baba HA et al (1998) Long-term β adrenoceptor-mediated alteration in contractility and expression of phospholamban and sarcoplasmic reticulum Ca^{++}-ATPase in mammalian ventricle. J Pharmacol Exp Ther 286:531–538
52. Lai LP, Raju VS, Delehanty JM et al (1998) Altered sarcoplasmic reticulum Ca^{2+} ATPase gene expression in congestive heart failure: effect of chronic norepinephrine infusion. J Mol Cell Cardiol 30:175–185
53. Bello N, Mosca L (2004) Epidemiology of coronary heart disease in women. Prog Cardiovasc Dis 46:287–295
54. Mercuro G, Deidda M, Piras A et al (2010) Gender determinants of cardiovascular risk factors and diseases. J Cardiovasc Med 10:207–220
55. Solimene MC (2010) Coronary heart disease in women: a challenge for the 21st century. Clinics 65:99–106
56. Juutilainen A, Kortelainen S, Lehto S et al (2004) Gender difference in the impact of type 2 diabetes on coronary heart disease risk. Diabetes Care 27:2898–2904
57. Takenouchi Y, Kobayashi T, Taguchi K et al (2009) Gender differences in endothelial function in aortas from type 2 diabetic model mice. J Pharmacol Sci 11:91–99
58. Vitale C, Miceli M, Rosano GM (2007) Gender-specific characteristics of atherosclerosis in menopausal women: risk factors, clinical course and strategies for prevention. Climacteric 10:16–20
59. Thomas CM, Smart EJ (2007) Gender as a regulator of atherosclerosis in murine models. Curr Drug Targets 8:1172–1180
60. Campbell DJ, Somaratne JB, Jenkins AJ et al (2011) Differences in myocardial structure and coronary microvasculature between men and women with coronary artery disease. Hypertension 57:186–192

61. Regitz-Zagrosek V, Overteit-Prigione S, Seeland U et al (2010) Sex and gender differences in myocardial hypertrophy and heart failure. Circ J 74:1265–1273
62. Regitz-Zagrosek V, Brokat S, Tschope C (2007) Role of gender in heart failure with normal left ventricular ejection fraction. Prog Cardiovasc Dis 49:241–251
63. Hoppe BL, Hermann DD (2003) Sex differences in the causes and natural history of heart failure. Curr Cardiol Rep 5:193–199
64. Gardner JD, Brower GL, Janicki JS (2002) Gender differences in cardiac remodeling secondary to chronic volume overload. J Card Fail 8:101–107
65. Dent MR, Tappia PS, Dhalla NS (2010) Gender differences in cardiac dysfunction and remodeling due to volume overload. J Card Fail 16:439–449, erratum: J Card Fail 17:179, 2011
66. Dent MR, Tappia PS, Dhalla NS (2010) Gender differences in apoptotic signaling in heart failure due to volume overload. Apoptosis 15:499–510, erratum: Apoptosis 16:757, 2011
67. Dent MR, Tappia PS, Dhalla NS (2011) Gender differences in β-adrenoceptor system in cardiac hypertrophy due to arteriovenous fistula. J Cell Physiol 226:181–186
68. Dent MR, Tappia PS, Dhalla NS (2012) Gender related alterations of β-adrenoceptor mechanisms in heart failure due to arteriovenous fistula. J Cell Physiol 227:3080–3087
69. Klodas E, Enriquez M, Tajik AJ et al (1996) Surgery for aortic regurgitation in women Contrasting indications and outcomes compared with men. Circulation 94:2472–2478
70. Drolet MC, Lachance D, Plante E et al (2006) Gender-related differences in left ventricular remodeling in chronic severe aortic valve regurgitation in rats. J Heart Valve Dis 15: 345–351
71. Pavlovic M, Schaller A, Steiner B et al (2005) Gender modulates the expression of calcium-regulating proteins in pediatric atrial myocardium. Exp Biol Med 230:853–859
72. Tasevska-Dinevska G, Kennedy LM, Cline-Iwarson A et al (2011) Gender differences in variables related to B-natriuretic peptide, left ventricular ejection fraction and mass, and peak oxygen consumption, in patients with heart failure. Int J Cardiol 149:364–371
73. Guerra S, Leri A, Wang X et al (1999) Myocyte death in the failing human heart is gender dependent. Circ Res 85:856–866
74. Pierdominici M, Ortona E, Franconi F et al (2011) Gender specific aspects of cell death in the cardiovascular system. Curr Pharm Des 17:1046–1055
75. Villari B, Campbell SE, Schneider J et al (1995) Sex-dependent differences in left ventricular function and structure in chronic pressure overload. Eur Heart J 16:1410–1419
76. Tamura T, Said S, Gerdes M (1999) Gender-related differences in myocyte remodeling in progression to heart failure. Hypertension 33:676–680
77. Weinberg EO, Thienelt CD, Katz SE et al (1999) Gender differences in molecular remodeling in pressure overload hypertrophy. J Am Coll Cardiol 34:264–273
78. Wei S-K, Mecurley JM, Hanlon SU et al (2007) Gender differences in Na/Ca exchanger current and β-adrenergic responsiveness in heart failure in pig myocytes. Ann N Y Acad Sci 1099:183–189
79. Litwin SE, Katz SE, Litwin CM et al (1999) Gender differences in postinfarction left ventricular remodeling. Cardiology 91:173–183
80. Jain M, Liao R, Podesser BK et al (2002) Influence of gender on the response to hemodynamic overload after myocardial infarction. Am J Physiol Heart Circ Physiol 283: H2544–H2550
81. Kanashiro-Takeuchi RM, Heidecker B, Lamirault G et al (2009) Sex-specific impact of aldosterone receptor antagonism on ventricular remodeling and gene expression after myocardial infarction. Clin Transl Sci 2:134–142
82. Cavasin MA, Tao Z, Menon S et al (2004) Gender differences in cardiac function during early remodeling after acute myocardial infarction in mice. Life Sci 75:2181–2192
83. Bridgman P, Aronovitz MA, Kakkar R et al (2005) Gender-specific patterns of left ventricular and myocyte remodeling following myocardial infarction in mice deficient in the angiotensin II type 1a receptor. Am J Physiol Heart Circ Physiol 289:H586–H592

84. Shioura KM, Geenan DL, Goldspink PH (2008) Sex-related changes in cardiac function following myocardial infarction in mice. Am J Physiol Regul Integr Comp Physiol 295:R528–R534
85. Rosenkranz-Weiss P, Tomek RJ, Mathew J et al (1994) Gender-specific differences in expression of mRNAs for functional and structural proteins in rat ventricular myocardium. J Mol Cell Cardiol 26:261–270
86. Bupha-Intr T, Wattanapermpool J (2006) Regulatory role of ovarian sex hormones in calcium uptake activity of cardiac sarcoplasmic reticulum. Am J Physiol Heart Circ Physiol 291: H1101–H1108
87. Kravtsov GM, Kam KW, Liu J et al (2007) Altered Ca^{2+} handling by ryanodine receptor and Na^{+}-Ca^{2+} exchange in the heart from ovariectomized rats: role of protein kinase A. Am J Physiol Cell Physiol 292:C1625–C1635
88. Scheuer J, Malhotra A, Schaible TF et al (1987) Effects of gonadectomy and hormonal replacement on rat hearts. Circ Res 61:12–19
89. Golden KL, Marsh JD, Jaing Y (2002) Castration reduces mRNA levels for calcium regulatory proteins in rat heart. Endocrine 19:339–344
90. Golden KL, Marsh JD, Jiang Y et al (2003) Gonadectomy of adult male rats reduces contractility of isolated cardiac myocytes. Am J Physiol Endocrinol Metab 285:E449–E453
91. Golden KL, Marsh JD, Jiang Y (2004) Testosterone regulates mRNA levels of calcium regulatory proteins in cardiac myocytes. Horm Metab Res 36:197–202
92. Brower GL, Gardner JD, Janicki JS (2003) Gender mediated cardiac protection from adverse ventricular remodeling is abolished by ovariectomy. Mol Cell Biochem 251:89–95
93. Gardner JD, Brower GL, Voloshenyuk TG et al (2008) Cardioprotection in female rats subjected to chronic volume overload: synergistic interaction of estrogen and phytoestrogens. Am J Physiol Heart Circ Physiol 294:H198–H204
94. Gardner JD, Murray DB, Voloshenyuk TG et al (2010) Estrogen attenuates chronic volume overload induced structural and functional remodeling in male rat hearts. Am J Physiol Heart Circ Physiol 298:H497–H504
95. Bhuiyan MS, Shioda N, Fukunaga K (2007) Ovariectomy augments pressure overload-induced hypertrophy associated with changes in Akt and nitric oxide synthase signaling pathways in female rats. Am J Physiol Endocrinol Metab 293:E1606–E1614
96. Bhuiyan MS, Shioda N, Fukunaga K (2009) Chronic β-AR activation-induced calpain activation and impaired eNOS-Akt signaling mediates cardiac injury in ovariectomized female rats. Expert Opin Ther Targets 13:275–286
97. Smith PJW, Ornatsky D, Stewart DJ et al (2000) Effects of estrogen replacement on infarct size, cardiac remodeling, and the endothelin system after myocardial infarction in ovariectomized rats. Circulation 102:2983–2989
98. Yousef ZR, Redwood SR, Marber MS (2000) Postinfarction left ventricular remodelling: where are the theories and trials leading us? Heart 83:76–80
99. Duhamel TA, Dhalla NS (2007) New insights into the causes of heart failure. Drug Discov Today Dis Mech 4:175–184
100. McDonald KM, Rector T, Carlyle TF et al (1994) Angiotensin-converting enzyme inhibition and β-adrenoceptor blockade regress established ventricular remodeling in a canine model of discrete myocardial damage. J Am Coll Cardiol 24:1762–1768
101. Cox MM, Berman I, Myerburg RJ et al (1991) Morphometric mapping of regional myocyte diameters after healing of myocardial infarction in cats. J Mol Cell Cardiol 23:127–135
102. Vanoli E, DeFerrari GM, Stramba-Badiale M et al (1991) Vagal stimulation and prevention of sudden death in conscious dogs with a healed myocardial infarction. Circ Res 68: 1471–1481
103. Sabbah HN, Kono T, Stein PD et al (1992) Left ventricular shape changes during the course of evolving heart failure. Am J Physiol Heart Circ Physiol 263:H266–H270
104. Jugdutt BI, Schwarz-Michorowski BL, Khan MI (1992) Effect of long-term captopril therapy on left ventricular remodeling and function during healing of canine myocardial infarction. J Am Coll Cardiol 19:713–721

105. Connelly CM, Ngoy S, Schoen FJ et al (1992) Biomechanical properties of reperfused transmural myocardial infarcts in rabbits during the first week after infarction. Implications for left ventricular rupture. Circ Res 71:401–413
106. Knowlton AA, Connelly CM, Romo GM et al (1992) Rapid expression of fibronectin in the rabbit heart after myocardial infarction with and without reperfusion. J Clin Invest 89:1060–1068
107. Kozlovskis PL, Gerdes AM, Smets M et al (1991) Regional increase in isolated myocyte volume in chronic myocardial infarction in cats. J Mol Cell Cardiol 23:1459–1466
108. Hugel S, Horn M, de Groot M et al (1999) Effects of ACE inhibition and beta-receptor blockade on energy metabolism in rats postmyocardial infarction. Am J Physiol Heart Circ Physiol 277:H2167–H2175
109. Wagner K, Kamkin A, Kiseleva I et al (2000) Effects of metoprolol and ramipril on action potentials after myocardial infarction in rats. Eur J Pharmacol 388:263–266
110. Theres HP, Wagner KD, Romberg D et al (2000) Combined treatment with ramipril and metoprolol prevents changes in the creatine kinase isoenzyme system and improves hemodynamic function in rat hearts after myocardial infarction. Cardiovasc Drugs Ther 14:597–606
111. Geenen DL, Malhotra A, Liang D et al (1991) Ventricular function and contractile proteins in the infarcted overloaded rat heart. Cardiovasc Res 25:330–336
112. Gopalakrishnan M, Triggle DJ, Rutledge A et al (1991) Regulation of K^+ and Ca^{2+} channels in experimental cardiac failure. Am J Physiol 261:H1979–H1987
113. Chasteney EA, Liang CS, Jr Hood WB (1992) β-adrenoceptor and adenylate cyclase function in the infarct model of rat heart failure. Proc Soc Exp Biol Med 200:90–94
114. Ontkean M, Gay R, Greenberg B (1991) Diminished endothelium-derived relaxing factor activity in an experimental model of chronic heart failure. Circ Res 69:1088–1096
115. Rowley KG, Tung LH, Hodsman GP et al (1991) Altered α1-adrenoceptor-mediated responses in atria of rats with chronic left ventricular infarction. J Cardiovasc Pharmacol 17:474–479
116. Fletcher PJ, Pfeffer JM, Pfeffer MA et al (1981) Left ventricular diastolic pressure-volume relations in rats with healed myocardial infarction. Effects on systolic function. Circ Res 49:618–626
117. DeFelice A, Frering R, Horan P (1989) Time course of hemodynamic changes in rats with healed severe myocardial infarction. Am J Physiol 257:H289–H296
118. DeTombe PP, Wannenburg T, Fan D et al (1996) Right ventricular contractile protein function in rats with left ventricular myocardial infarction. Am J Physiol 271:H73–H79
119. Pelouch V, Dixon IMC, Sethi R et al (1994) Alteration of collagenous protein profile in congestive heart failure secondary to myocardial infarction. Mol Cell Biochem 129:121–131
120. Liu K, Jackson M, Sowa MG et al (1996) Modification of the extracellular matrix following myocardial infarction monitored by FTIR spectroscopy. Biochim Biophys Acta 1315:73–77
121. Omens JH, Miller TR, Covell JW (1997) Relationship between passive tissue strain and collagen uncoiling during healing of infarcted myocardium. Cardiovasc Res 33:351–358
122. Eberli FR, Sam F, Ngoy S et al (1998) Left-ventricular structural and functional remodeling in the mouse after myocardial infarction: assessment with the isovolumetrically-contracting Langendorff heart. J Mol Cell Cardiol 30:1443–1447
123. Gosselin H, Qi X, Rouleau JL (1998) Correlation between cardiac remodelling, function, and myocardial contractility in rat hearts 5 weeks after myocardial infarction. Can J Physiol Pharmacol 76:53–62
124. Shao Q, Takeda N, Temsah R et al (1996) Prevention of hemodynamic changes due to myocardial infarction by early treatment of rats with imidapril. Cardiovasc Pathobiol 1:180–186
125. Dixon IMC, Lee SL, Dhalla NS (1996) Nitrendipine binding in congestive heart failure due to myocardial infarction. Circ Res 66:782–788
126. Afzal N, Dhalla NS (1992) Differential changes in left and right ventricular SR calcium transport in congestive heart failure. Am J Physiol 262:H868–H874
127. Yamaguchi F, Sanbe Y, Takeo S (1997) Cardiac sarcoplasmic reticular function in rats with chronic heart failure following myocardial infarction. J Mol Cell Cardiol 29:753–763

128. Afzal N, Dhalla NS (1996) Sarcoplasmic reticular Ca^{2+} pump ATPase activity in congestive heart failure due to myocardial infarction. Can J Cardiol 12:1065–1073
129. Zarain-Herzberg A, Afzal N, Elimban V et al (1996) Decreased expression of cardiac sarcoplasmic reticulum Ca^{2+}-pump ATPase in congestive heart failure due to myocardial infarction. Mol Cell Biochem 163(164):285–290
130. Yoshiyama M, Takeuchi K, Hanatani A et al (1997) Differences in expression of sarcoplasmic reticulum Ca^{2+}-ATPase and Na^+-Ca^{2+} exchanger genes between adjacent and remote noninfarcted myocardium after myocardial infarction. J Mol Cell Cardiol 29:255–264
131. Dixon IMC, Dhalla NS (1991) Alterations in cardiac adrenoceptors in congestive heart failure secondary to myocardial infarction. Coron Artery Dis 2:805–814
132. Sethi R, Dhalla KS, Beamish RE et al (1997) Differential changes in left and right ventricular adenylyl cyclase activities in congestive heart failure. Am J Physiol 272:H884–H893
133. Dixon IMC, Hata T, Dhalla NS (1992) Sarcolemmal Na^+-K^+-ATPase activity in congestive heart failure due to myocardial infarction. Am J Physiol 262:C664–C671
134. Semb SO, Lunde PK, Holt E et al (1998) Reduced myocardial Na^+, K^+-pump capacity in congestive heart failure following myocardial infarction in rats. J Mol Cell Cardiol 30:1311–1328
135. Dhalla NS, Wang X, Sethi R et al (1997) β-adrenergic linked signal transduction mechanisms in failing hearts. Heart Fail Rev 2:55–65
136. Kiseleva I, Kamkin A, Pylaev A et al (1998) Electrophysiological properties of mechanosensitive atrial fibroblasts from chronic infarcted rat heart. J Mol Cell Cardiol 30:1083–1093
137. Zhang XQ, Moore RL, Tillotson DL et al (1995) Calcium currents in postinfarction rat cardiac myocytes. Am J Physiol 269:C1464–C1473
138. Aggarwal R, Pu J, Boyden PA (1997) Ca^{2+}-dependent outward currents in myocytes from epicardial border zone of 5-day infarcted canine heart. Am J Physiol 273:H1386–H1394
139. Zhang XQ, Tillotson DL, Moore RL et al (1996) Na^+/Ca^{2+} exchange currents and SR Ca^{2+} contents in postinfarction myocytes. Am J Physiol 271:C1800–C1807
140. Gidh-Jain M, Huang B, Jain P et al (1996) Differential expression of voltage-gated K^+ channel genes in left ventricular remodeled myocardium after experimental myocardial infarction. Circ Res 79:669–675
141. Geenen DL, Malhotra A, Scheuer J (1989) Regional variation in rat cardiac myosin isoenzymes and ATPase activity after infarction. Am J Physiol 256:H745–H750
142. Liu X, Shao Q, Dhalla NS (1995) Myosin light chain phosphorylation in cardiac hypertrophy and failure due to myocardial infarction. J Mol Cell Cardiol 27:2613–2621
143. Akiyama K, Akopian G, Jinadasa P et al (1997) Myocardial infarction and regulatory myosin light chain. J Mol Cell Cardiol 29:2641–2652
144. Li P, Hoffmann PA, Li B et al (1997) Myocardial infarction alters myofilament calcium sensitivity and mechanical behavior of myocytes. Am J Physiol 272:H360–H370
145. Fellenius E, Hansen CA, Miøs O et al (1985) Chronic infarction decreases maximum cardiac work and sensitivity of heart to extracellular calcium. Am J Physiol 249:H80–H87
146. Wang J, Liu X, Ren B et al (2002) Modification of myosin gene expression by imidapril in failing heart due to myocardial infarction. J Mol Cell Cardiol 34:847–857
147. Wang J, Guo X, Dhalla NS (2004) Modification of myosin protein and gene expression in failing hearts due to myocardial infarction by enalapril or losartan. Biochim Biophys Acta 1690:177–184
148. Ren B, Lukas A, Shao Q et al (1998) Electrocardiographic changes and mortality due to myocardial infarction in rats with or without imidapril treatment. J Cardiovasc Pharmacol Ther 3:11–22
149. Schoemaker RG, Debets JJ, Struyker-Boudier HA et al (1991) Delayed but not immediate captopril therapy improves cardiac function in conscious rats, following myocardial infarction. J Mol Cell Cardiol 23:187–197
150. Ren B, Shao Q, Ganguly PK et al (2004) Influence of long-term treatment of imidapril on mortality, cardiac function, and gene expression in congestive heart failure due to myocardial infarction. Can J Physiol Pharmacol 82:1118–1127

151. Dixon IMC, Ju H, Jassal DS et al (1996) Effect of ramipril and losartan on collagen expression in right and left heart after myocardial infarction. Mol Cell Biochem 165:31–45
152. van Krimpen C, Smits JF, Cleutjens JP et al (1991) DNA synthesis in the non-infarcted cardiac interstitium after left coronary artery ligation in the rat: effects of captopril. J Mol Cell Cardiol 23:1245–1253
153. McDonald KM, Chu C, Francis GS et al (1997) Effect of delayed intervention with ACE-inhibitor therapy on myocyte hypertrophy and growth of the cardiac interstitium in the rat model of myocardial infarction. J Mol Cell Cardiol 29:3203–3210
154. Gonzalez W, Beslot F, Laboulandine I et al (1996) Inhibition of both angiotensin-converting enzyme and neutral endopeptidase by S21402 (RB105) in rats with experimental myocardial infarction. J Pharmacol Exp Ther 278:573–581
155. Kuizinga MC, Smits JF, Arends JW et al (1998) AT2 receptor blockade reduces cardiac interstitial cell DNA synthesis and cardiac function after rat myocardial infarction. J Mol Cell Cardiol 30:425–434
156. Ford WR, Khan MI, Jugdutt BI (1998) Effect of the novel angiotensin II type 1 receptor antagonist L-158,809 on acute infarct expansion and acute anterior myocardial infarction in the dog. Can J Cardiol 14:73–80
157. Raya TE, Gay RG, Aguirre M et al (1989) Importance of venodilatation in prevention of left ventricular dilatation after chronic large myocardial infarction in rats: a comparison of captopril and hydralazine. Circ Res 64:330–337
158. Michel JB, Lattion AL, Salzmann JL et al (1988) Hormonal and cardiac effects of converting enzyme inhibition in rat myocardial infarction. Circ Res 62:641–650
159. Wollert KC, Studer R, von Bülow B et al (1994) Survival after myocardial infarction in the rat. Role of tissue angiotensin-converting enzyme inhibition. Circulation 90:2457–2467
160. Liu YH, Yang XP, Sharov VG et al (1997) Effects of angiotensin-converting enzyme inhibitors and angiotensin II type 1 receptor antagonists in rats with heart failure. Role of kinins and angiotensin II type 2 receptors. J Clin Invest 99:1926–1935
161. Sanbe A, Takeo S (1995) Long-term treatment with angiotensin I-converting enzyme inhibitors attenuates the loss of cardiac β-adrenoceptor responses in rats with chronic heart failure. Circulation 92:2666–2675
162. Litwin SE, Morgan JP (1992) Captopril enhances intracellular calcium handling and β-adrenergic responsiveness of myocardium from rats with postinfarction failure. Circ Res 71: 797–807
163. Sethi R, Shao Q, Takeda N et al (2003) Attenuation of changes in G_i-proteins and adenylyl cyclase in heart failure by an ACE inhibitor, imidapril. J Cell Mol Med 7:277–286
164. Sethi R, Shao Q, Ren B et al (2004) Changes in β-adrenoceptors in heart failure due to myocardial infarction are attenuated by blockade of renin-angiotensin system. Mol Cell Biochem 263:11–20
165. Machackova J, Sanganalmath S, Barta J et al (2010) Amelioration of cardiac remodeling in congestive heart failure by β-adrenoceptor blockade is associated with depression in sympathetic activity. Cardiovasc Toxicol 10:9–16
166. Machackova J, Sanganalmath S, Elimban V et al (2011) β-adrenergic blockade attenuates cardiac dysfunction and myofibrillar remodelling in congestive heart failure. J Cell Mol Med 15:545–554
167. Orenstein TL, Parker TG, Butany JW et al (1995) Favorable left ventricular remodeling following large myocardial infarction by exercise training. Effect on ventricular morphology and gene expression. J Clin Invest 96:858–866
168. Cittadini A, Grossman JD, Napoli R et al (1997) Growth hormone attenuates early left ventricular remodeling and improves cardiac function in rats with large myocardial infarction. J Am Coll Cardiol 29:1109–1116
169. Tomanek RJ, Zimmerman MB, Suvarna PR et al (1998) A thyroid hormone analog stimulates angiogenesis in the post-infarcted rat heart. J Mol Cell Cardiol 30:923–932

170. Hill MF, Singal PK (1996) Antioxidant and oxidative stress changes during heart failure subsequent to myocardial infarction in rats. Am J Pathol 148:291–300
171. Sanganalmath SK, Barta J, Takeda N et al (2008) Antiplatelet therapy mitigates cardiac remodeling and dysfunction in congestive heart failure due to myocardial infarction. Can J Physiol Pharmacol 86:180–189
172. Barta J, Sanganalmath SK, Kumamoto H et al (2008) Antiplatelet agents sarpogrelate and cilostazol affect experimentally-induced ventricular arrhythmias and mortality. Cardiovasc Toxicol 8:127–135
173. Sanganalmath SK, Babick AP, Barta J et al (2008) Antiplatelet therapy attenuates subcellular remodelling in congestive heart failure. J Cell Mol Med 12:1728–1738
174. Shimoyama H, Sabbah HN, Tanimura M et al (1998) Short-term hemodynamic effects of mibefradil in dogs with chronic heart failure: comparison with diltiazem. J Pharmacol Exp Ther 285:746–752
175. Mengi SA, Dhalla NS (2004) Carnitine palmitoyltransferase-I, a new target for the treatment of heart failure: perspectives on a shift in myocardial metabolism as a therapeutic intervention. Am J Cardiovasc Drugs 4:201–209
176. Gumina RJ, Mizumura T, Beier N et al (1998) A new sodium/hydrogen exchange inhibitor, EMD 85131, limits infarct size in dogs when administered before or after coronary artery occlusion. J Pharmacol Exp Ther 286:175–183
177. Karmazyn M, Gan XT, Humphreys RA et al (1999) The myocardial Na^+–H^+ exchange: structure, regulation, and its role in heart disease. Circ Res 85:777–786
178. Yoshida H, Karmazyn M (2000) Na^+/H^+ exchange inhibition attenuates hypertrophy and heart failure in 1-wk postinfarction rat myocardium. Am J Physiol 278:H300–H304
179. Chen L, Gan XT, Haist JV et al (2001) Attenuation of compensatory right ventricular hypertrophy and heart failure following monocrotaline-induced pulmonary vascular injury by the Na^+–H^+ exchange inhibitor cariporide. J Pharmacol Exp Ther 298:469–476
180. Shao Q, Ren B, Zarain-Herzberg A et al (1999) Captopril treatment improves the sarcoplasmic reticular Ca^{2+} transport in heart failure due to myocardial infarction. J Mol Cell Cardiol 31:1663–1672
181. Wang J, Liu X, Sentex E et al (2003) Increased expression of protein kinase C isoforms in heart failure due to myocardial infarction. Am J Physiol Heart Circ Physiol 284: H2277–H2287
182. Netticadan T, Temsah RM, Kawabata K et al (2000) Sarcoplasmic reticulum Ca^{2+}/Calmodulin-dependent protein kinase is altered in heart failure. Circ Res 86:596–605
183. Guo X, Chapman D, Dhalla NS (2003) Partial prevention of changes in SR gene expression in congestive heart failure due to myocardial infarction by enalapril or losartan. Mol Cell Biochem 254:163–172
184. Sethi R, Dhalla KS, Ganguly PK et al (1999) Beneficial effects of propionyl L-carnitine on sarcolemmal changes in congestive heart failure due to myocardial infarction. Cardiovasc Res 42:607–615
185. Sethi R, Wang X, Ferrari R et al (2004) Improvement of cardiac function and β-adrenergic signal transduction by propionyl L-carnitine in congestive heart failure due to myocardial infarction. Coronary Artery Dis 15:65–71
186. Yu C-H, Panagia V, Tappia PS et al (2002) Alterations of sarcolemmal phospholipase D and phosphatidate phosphohydrolase in congestive heart failure. Biochim Biophys Acta 1584:65–72
187. Tappia PS, Liu S-Y, Shatadal S et al (1999) Changes in sarcolemmal PLC isoenzymes in postinfarct congestive heart failure: partial correction by imidapril. Am J Physiol 277: H40–H49
188. Tappia PS, Yu C-H, Di Nardo P et al (2001) Depressed responsiveness of phospholipase C isoenzymes to phosphatidic acid in congestive heart failure. J Mol Cell Cardiol 33:431–440
189. Guo X, Wang J, Elimban V et al (2008) Both enalapril and losartan attenuate sarcolemmal Na^+-K^+-ATPase remodeling in failing rat heart due to myocardial infarction. Can J Physiol Pharmacol 86:139–147

190. Shao Q, Ren B, Saini HK et al (2005) Sarcoplasmic reticulum Ca^{2+} transport and gene expression in congestive heart failure are modified by imidapril treatment. Am J Physiol 288:H1674–H1682
191. Shao Q, Ren B, Elimban V et al (2005) Modification of sarcolemmal Na^{+}-K^{+}-ATPase and Na^{+}/Ca^{2+} exchanger expression in heart failure by blockade of renin-angiotensin system. Am J Physiol 288:H2637–H2646
192. Saini HK, Shao Q, Musat S et al (2005) Imidapril treatment improves the attenuated inotropic and intracellular calcium responses to ATP in heart failure due to myocardial infarction. Br J Pharmacol 144:202–211

Chapter 4
The Failing Heart: Is It an Inefficient Engine or an Engine Out of Fuel?

Waleed G.T. Masoud, Alexander S. Clanachan, and Gary D. Lopaschuk

Abstract To meet its high-energy demand, the heart is very flexible in its choice of energy substrates. It can use a variety of energy substrates which include fatty acids, glucose, lactate, pyruvate, ketones, and amino acids. In the failing heart, significant changes in cardiac energy substrate metabolism occur, although there is no consensus as to exactly what these changes are. Energy starvation in heart failure has been extensively discussed, where reduced oxygen and energy substrate delivery to the heart, reduced cardiac energy substrate uptake, reduced mitochondrial oxidative phosphorylation, and decreased metabolic flexibility have been implicated as contributing factors to the declining mechanical function in heart failure. In addition to energy starvation, there is also the possibility of inefficient energy utilization in the failing heart. This inefficiency can occur at the level of ATP production where the preferential dependence on fatty acids consumes more oxygen per unit ATP and/or the overexpression of uncoupling proteins can increase energy loss as heat rather than ATP production. Increased ATP utilization for non-contractile purposes, such as ionic homeostasis and futile cycling of fatty acids, can also contribute to

W.G.T. Masoud • A.S. Clanachan
Department of Pharmacology, Faculty of Medicine and Dentistry, University of Alberta, Edmonton, AB, Canada

G.D. Lopaschuk (✉)
Department of Pharmacology, Faculty of Medicine and Dentistry, University of Alberta, Edmonton, AB, Canada

Department of Pediatrics, Faculty of Medicine and Dentistry, University of Alberta, Edmonton, AB, Canada

Mazankowski Alberta Heart Institute, Edmonton, AB, Canada

423 Heritage Medical Research Centre, Faculty of Medicine and Dentistry, University of Alberta, Edmonton, AB, Canada
e-mail: Gary.Lopaschuk@ualberta.ca

B.I. Jugdutt and N.S. Dhalla (eds.), *Cardiac Remodeling: Molecular Mechanisms*, Advances in Biochemistry in Health and Disease 5, DOI 10.1007/978-1-4614-5930-9_4,

inefficiency in the failing heart. Impaired phosphocreatine/creatine kinase shuttle activity may also contribute to inefficient transport of ATP from the mitochondria to the contractile myofibrils. The degree and type of energy inefficiency in the failing heart are likely dependent on the pathogenesis and severity of heart failure. In this chapter, we review the various contributors to energy inefficiency in heart failure and discuss the potential to optimize cardiac energy metabolism as a potential treatment for heart failure.

Keywords Heart failure • Energy starvation • Cardiac inefficiency • Fatty acid oxidation • Glycolysis • Glucose oxidation

4.1 Introduction

The heart is a continuously contracting organ with a great need for energy. This is illustrated by the high turnover rates of adenosine triphosphate (ATP) in the heart. Although cardiac ATP content is very low (5 μmol/g wet weight), the heart has high ATP hydrolysis rates (~30 μmol/g wet weight/min). Thus, the heart has virtually no energy reserves, and the ATP pool of the heart is essentially turned over every ten beats [1–4]. To secure the energy needed to perform its tasks (such as mechanical work, ionic homeostasis, cardiac signaling, and synthetic function), the heart is flexible in the choice of energy substrates it utilizes. Many different energy substrates can be used, which include fatty acids, glucose, lactate, pyruvate, ketone bodies, and amino acids. The relative contribution of each of these energy substrates to overall cardiac ATP production depends on substrate availability, the workload of the heart, circulating hormones (such as insulin and catecholamines), age, oxygen availability, and pathophysiologic changes that accompany various diseases including heart failure [5, 6], hypertrophy [7], cardiomyopathy [8], and ischemia–reperfusion injuries [9]. However, under physiologic conditions, the adult heart normally derives most its ATP from fatty acid oxidation, with the remainder primarily being derived from the oxidation of carbohydrates (glucose, lactate, and pyruvate) [10–14].

Heart failure is defined as failure of the heart to pump an adequate amount of blood to meet the peripheral needs of the body. It develops following various insults to the heart including, but not limited to, ischemic insults, hypertrophy following long-standing untreated hypertension, genetic or acquired cardiomyopathies, storage diseases, or hyperdynamic circulation such as severe anemia or hyperthyroidism [15]. During the progression to heart failure, adaptive changes occur in the heart, including structural and metabolic remodeling. Systemic neurohumoral compensatory mechanisms develop to maintain adequate cardiac output. This includes activation of renin–angiotensin–aldosterone system (RAAS) and sympathetic system [16]. However, as heart failure progresses, these initially adaptive mechanisms become maladaptive leading to increased workloads and energy demand of the heart as well as increased energy cost of work ending in decompensation [17].

The study of mechanisms contributing to heart failure is complicated by the many causes of heart failure. However, numerous animal models of heart failure have been studied, with each mimicking a subtype such as pressure overload (e.g., transverse aortic constriction and abdominal aortic constriction), volume overload (e.g., aortocaval shunt), stress overload (e.g., rapid ventricular pacing), genetic alterations (transgenic animals), or post-infarction (coronary artery ligation) (see [18] for excellent review). As will be discussed, each of these experimental models, as well as complementary human studies, has provided valuable information as to how alterations in energy metabolism contribute to the severity of heart failure.

4.2 Energy Generation and Utilization in the Heart

The heart is a highly dynamic organ. The human heart pumps about 10 t of blood on a daily basis and beats about 100,000 times a day [19]. To accomplish this task, the heart uses multiple energy substrates to generate the needed energy. In this regard, the heart is the highest energy substrate consumer in the body on a gram weight basis [19]. The process of energy generation and utilization in the heart involves several processes, namely, energy substrate delivery, glycolysis, carbohydrate oxidation, β-oxidation, tricarboxylic acid cycle (TCA cycle) activity, mitochondrial oxidative phosphorylation, and the transfer of the high-energy phosphates to the contracting myofilaments.

4.2.1 Energy Substrate Delivery

This process entails the uptake of various energy substrates by the cardiomyocytes. These substrates are then metabolized, either via glycolysis or via mitochondrial oxidative metabolism to form acetyl CoA that feeds into the TCA cycle. Normally, TCA cycle activity is coupled to oxidative phosphorylation for ATP production [19, 20].

Fatty acids are primarily delivered to the heart either as free fatty acids bound to albumin or as triacylglycerols (TGs) present in chylomicrons and very low density lipoproteins [1, 20]. Fatty acid supply to the heart is a key determinant of fatty acid use by the heart [1]. In early heart failure, levels of fatty acids can be elevated compared with non-heart failure patients [16]. In contrast, circulating levels of glucose, lactate, ketones, pyruvate, and amino acids are not that different in heart failure patients (see [20] for review). As a result, while alterations in fatty acid supply may alter energy metabolism in heart failure, potential differences in energy substrate use in the failing heart are more likely due to alterations in energy substrate uptake and utilization, as opposed to differences in energy substrate supply to the heart.

4.2.2 Glycolysis

Glucose is transported into the cardiomyocyte primarily by either GLUT1 or insulin-dependent GLUT4 transporters. Once inside the cardiomyocytes, glucose is phosphorylated to glucose-6-P, with the majority of this glucose-6-P then entering glycolysis to eventually produce pyruvate [20]. The advantage of glycolysis is that it produces ATP without the expense of oxygen. However, the amount of ATP generated is small in comparison to the amount produced by mitochondrial oxidative phosphorylation, and as a result, glycolysis normally produces less than 5% of the total ATP requirements of the heart [1]. However, as will be discussed, this proportion can change in the failing heart.

4.2.3 Fatty Acid β-Oxidation, Glucose Oxidation, and the TCA Cycle Activity

Fatty acids, delivered to the heart as free fatty acids bound to albumin or liberated from TGs by the activity of lipoprotein lipase, are taken up by the heart, esterified to CoA, and the acyl group transported into the mitochondria via carnitine palmitoyl transferase-1 (CPT1) to undergo fatty acid β-oxidation. Each cycle of fatty acid β-oxidation shortens the fatty acid moiety by two carbons and produces one acetyl CoA molecule. Acetyl CoA then feeds into the TCA cycle that is, in turn, coupled with electron transfer chain for ATP generation. Oxidation of fatty acids produces more ATP per molecule than any other energy substrate. However, this process consumes more oxygen; hence, fatty acids are an inefficient source of energy. The contribution of fatty acid β-oxidation to cardiac energy generation varies, depending on fatty acid availability, energy demand, cardiac pathologic status, and competition with the other energy substrates (see [20] for review).

Glucose oxidation is involved in the decarboxylation of glycolytically derived pyruvate by the pyruvate dehydrogenase complex (PDC) to form acetyl CoA. Two acetyl CoA molecules are produced from each glucose molecule. Pyruvate can also be produced from lactate via the activity of lactate dehydrogenase. Acetyl CoA from the PDC then feeds into the TCA cycle. A major product of the TCA cycle is nicotinamide adenine dinucleotide (NADH), which then feeds into the electron transport chain. Glucose oxidation produces less ATP per molecule than fatty acid oxidation. However, glucose oxidation consumes less oxygen; hence, it is considered a more efficient energy substrate (see [2–4, 11, 21] for review).

4.2.4 Oxidative Phosphorylation for ATP Production

Under aerobic conditions about 95% of the hearts, ATP requirements are met through the mitochondrial oxidative phosphorylation of ADP. The reducing equivalents are transferred to the mitochondria via flavine adenine dinucleotide (FADH2) and NADH, which are generated via dehydrogenase reactions in fatty acid β-oxidation, the TCA cycle, the glycolytic pathway, and from the oxidation of pyruvate derived from glucose or lactate. The relative contribution of the various energy substrates to ATP production is affected by many factors including the developmental stage of the heart, cardiac workload (pre- and afterload), heart rate and force of contraction, and the presence of various cardiac pathologies (e.g., hypertrophy, failure, or ischemia–reperfusion injury) [19, 20, 22, 23].

4.2.5 ATP Transfer and Phosphocreatine/Creatine Kinase Shuttle

Transfer of the high-energy phosphate bond in ATP that is generated inside the mitochondria (via oxidative phosphorylation) to the site of ATPases (i.e., the myofibrils, sarcoplasmic reticulum, and sarcolemma) is facilitated by the creatine kinase shuttle system. Creatine kinase (CK) is a reversible kinase that initially phosphorylates creatine to phosphocreatine (PCr) using the high-energy phosphate bond in ATP [24]. The CK shuttle is facilitated by mitochondrial and cytoplasmic CKs, which eventually results in the rephosphorylation of ADP to ATP in the cytoplasm.

In cardiomyocytes, about two-thirds of creatine are phosphorylated and act as a reservoir for ATP. Thus, when energy production declines, PCr levels decline while ATP levels are initially maintained. This is accompanied by an increase in ADP levels [25, 26]. Because of this, total creatine levels and PCr/ATP ratio can be used as indices of cardiac energetic status. In general, both indices decline in early phases of heart failure before an actual decline in total ATP levels occurs [24–26].

Since creatine is not synthesized de novo in cardiac muscle [27], the activity of creatine transporters plays a regulatory role in the intracellular availability of creatine in cardiomyocytes. Reduction of creatine transporter expression and/or activity can be one of the contributing factors to deterioration of PCr/CK shuttle observed in many cardiac diseases, including heart failure [24–26].

The association of various isoforms of CK with subcellular structures ensures adequate transfer of ATP-derived energy for adequate coupling of energy generation and utilization. For example, the tight association of myofibrillar CK (MM-CK) with smooth endoplasmic reticulum ATPase (SERCA) is thought to enable efficient energy transfer for Ca^{2+} uptake [28]. Similarly, the mitochondrial CK (miCK) being located in the inner mitochondrial membrane near the adenine nucleotide translocase is thought to help export the high-energy phosphate of ATP through the formation of ADP and PCr, thus maintaining mitochondrial respiration by prevention of intramitochondrial ATP accumulation [29] (for further review, see [30]).

4.3 Nature of the Metabolic Changes in Heart Failure

Heart failure is associated with a number of changes in energy metabolism [19, 20, 23, 30, 31]. Whether these changes contribute to the contractile failure of these hearts, hence considered as maladaptive and should be discouraged [32, 33], or whether they represent an adaptive response that should be encouraged is not yet firmly established [34]. Moreover, there is no agreement regarding the actual nature of the energy metabolic changes that occur in heart failure, and two main concepts have emerged (Fig. 4.1). The first concept is that the failing heart is energetically starved, similar to an engine out of fuel [19, 35]. This implies lower rates of energy metabolism and lower rates of production of the energy currency, ATP. The second concept is that the failing heart may not necessarily be energy starved, but rather is inefficient in its use of energy [36].

4.3.1 Heart Failure and Cardiac Inefficiency

Cardiac energy is expended for mechanical work generation, ionic homeostasis, and other vital cellular functions, such as synthesis and degradation of various intracellular molecules and their trafficking to various cellular compartments or to the extracellular compartment [37–39]. In 1949 Bing et al. [40] defined cardiac mechanical efficiency as the "cardiac work generated per energy consumed." Since most of cardiac energy generated (about 95% under aerobic conditions) is derived from fatty acid and carbohydrate metabolism [21], and the fact that metabolic rates in the cardiac muscle are tightly coupled with energy demand [41, 42], oxygen consumption (MVO_2) by the cardiac muscle can be used as a measure of energy production. If energy production increases without a parallel increase in cardiac work, then a decrease in cardiac efficiency would occur. A decrease in the efficiency in utilizing energy for mechanical function has the potential to be a major contributor to cardiac dysfunction in heart failure [21, 43–49]. This cardiac inefficiency can develop primarily at three different levels: (1) increased oxygen cost of acetyl CoA and ATP production, (2) increased ATP consumption for homeostatic activities, and (3) impaired energy transfer to myofibrils.

4.3.1.1 Increased Oxygen Cost of ATP Production Due to an Increased Dependence on Fatty Acids

As mentioned, under aerobic conditions, the heart derives most of its energy requirements from fatty acids [21, 43–49]. Moreover, the activated neurohumoral mechanisms during early stages of heart failure favor lipolysis, leading to increased circulating fatty acids and more exposure of the cardiomyocytes to fatty acids [16]. When one molecule of palmitate is fully metabolized, it yields 105 ATP molecules, compared to the metabolism of glucose where only 31 ATP molecules are produced.

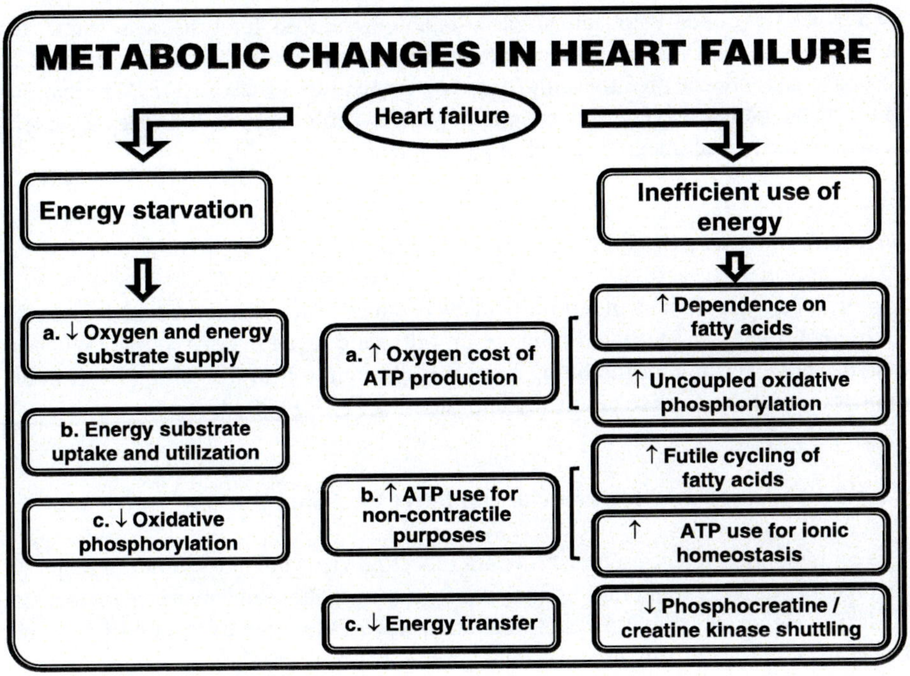

Fig. 4.1 Energy metabolic changes in heart failure. The energy starvation theory describes the failing heart as an engine out of fuel with (**a**) reduction of oxygen and substrate delivery to the heart, (**b**) reduction in energy substrate uptake and utilization, and (**c**) reduced mitochondrial oxidative phosphorylation. The cardiac inefficiency theory considers the inefficient energy utilization for mechanical work due to (**a**) increased oxygen cost of ATP production which in turn can be due to the preferential dependence on fatty acids or uncoupling of oxidation phosphorylation, (**b**) increased use of ATP for homeostatic activities that include futile cycling of fatty acids and ionic homeostasis secondary to increased Na^+ and Ca^{2+} accumulation induced by mismatched glucose metabolism, and (**c**) reduced energy transfer to myofibrils due to impaired phosphocreatine/creatine kinase shuttle

However, despite this higher ATP yield of palmitate, a higher MVO_2 is also needed [20]. This is because while the initial cytoplasmic metabolism of glucose (i.e., glycolysis) produces two ATP, the initial cytoplasmic metabolism of fatty acids (i.e., the formation of long chain acyl CoA from fatty acids and CoA) actually consumes two high-energy phosphates (i.e., ATP to AMP). Also, the fact that palmitate oxidation yields both $FADH_2$ and NADH, as compared to only NADH produced during glucose metabolism, contributes to palmitate being an inefficient source of energy. Because $FADH_2$ bypasses complex I of the electron transfer chain, it pumps less protons and generates less ATP than NADH (reviewed in [20]). This means that a preferential dependence on fatty acids for energy generation in early stages of heart failure could lead to more MVO_2 for the same energy yield, hence causing inefficiency in ATP generation. This effect was shown many years ago when increasing fatty acid supply to the heart secondary to adrenergic stimulation-induced lipolysis was found to increase cardiac MVO_2 without changing the external work [46, 49]. In theory, a complete dependence on fatty acids for energy generation leads to 10–13% reduction in the calculated efficiency. However, the observed differences in efficiency are even higher, indicative of involvement of other mechanisms [20, 23].

In heart failure, there is not a uniform consensus as to what happens to cardiac fatty acid oxidation rates. Both human and experimental studies have shown an increase in fatty acid β-oxidation rates, a decrease in fatty acid β-oxidation rates, or no change in fatty acid β-oxidation rates (see references [11, 20], and [23] for reviews). As a result, it cannot be conclusively stated as to whether the heart is more inefficient due to an increased reliance on fatty acid as an energy source. What is evident, however, is that inhibiting fatty acid β-oxidation in the failing heart does appear to improve cardiac efficiency and function [11, 19, 20, 23].

4.3.1.2 Uncoupled Mitochondrial Oxidative Phosphorylation

The mitochondrial electron transfer chain (also known as the mitochondrial respiratory chain) utilizes reducing equivalents in the mitochondrial matrix to generate ATP. The electrochemical gradient of protons across the inner mitochondrial membrane is essential for this process. This gradient is generated by the movement of protons from the mitochondrial matrix to the intermembrane space by complexes I, III, and IV. The protons then move down the electrochemical gradient to the mitochondrial matrix to provide the energy needed for ATP synthase activity, with the subsequent phosphorylation of ADP to ATP [20, 50, 51].

The uncoupling proteins (UCP) provide a potential alternative route for transport of protons across the inner mitochondrial membrane without the involvement in ATP synthesis [51]. These UCPs were originally discovered in brown adipose tissue where they help oxidatively generated energy to dissipate in the form of heat secondary to failure of coupling with phosphorylation of ADP for ATP generation [52, 53]. In the heart, UCP2 and UCP3 are preferentially expressed [54]. UCP3 has been suggested to mediate fatty acid-induced uncoupling of oxidation phosphorylation [55]. Moreover, high circulating fatty acid levels can increase UCP3 expression in the heart [56]. The aforementioned observation provides a potential explanation why the reported difference in efficiency due to dependence on fatty acid oxidation is higher than the theoretically calculated difference. Several reports from heart failure animal models describe overexpression of UCPs (see [57] for review). In an abdominal aorta constriction model of heart failure in the rat, decreases in ATP, ADP, AMP, and PCr contents in the failing heart were associated with a significant increase in UCP2 expression [58]. Adenovirus-mediated overexpression of UCP2 in neonatal cardiomyocytes was also found to increase basal oxygen consumption without affecting total ATP content [59], indicative of an increase in cardiac inefficiency. Moreover, UCP2 overexpression disrupted mitochondrial membrane potential secondary to failure to control calcium-induced calcium release from sarco/endoplasmic reticulum calcium ATPase (SERCA) [59]. In cardiac mitochondria isolated 10 weeks following coronary artery ligation in rats, UPC3 was also shown to be upregulated, mitochondria were less coupled (lower ADP/oxygen ratio), and a significant reduction of efficiency was observed [60]. Furthermore, a positive correlation between UPC3 levels and circulating non-fasting fatty acid levels was observed, an observation which supports the regulatory role of circulating fatty acids on cardiac UPC3 expression and thus increased inefficiency [60].

4.3.1.3 Futile Cycling in the Heart

Normally, the heart consumes generated ATP for mechanical work, ionic homeostasis, and synthesis of various cellular molecules [22, 37, 39]. However, utilization of ATP can be preferentially directed to additional homeostatic activities leading to further deterioration of inefficiency in heart failure.

As fatty acid supply to the myocardium increases early in the disease progression secondary to increased sympathetic activity, excess fatty acyl CoA has the potential to accumulate in the mitochondrial matrix. A mitochondrial thioesterase has been identified, which may cleave this unneeded fatty acyl CoA into fatty acid anion and free CoA [61, 62]. This process maintains the CoA pool in the mitochondria for further activation of fatty acids for β-oxidation. However, not only is the high-energy CoA bond lost due to the thioesterase activity, the resulting fatty acid anions have been proposed to be exported to the cytosolic compartment via the activity of UCP3 [63, 64]. While this process prevents intramitochondrial accumulation of potentially harmful fatty acid anions, it increases the futile utilization of ATP for the subsequent esterification of these fatty anions prior to subsequent fatty acid β-oxidation [20, 23].

Another source of futile ATP utilization is the glycerolipid/TG cycling which is another fate for fatty acids taken up by the heart. In this case, fatty acids cycle between the intracellular TG pool and the free fatty acid state. Excess ATP is thus utilized at the esterification step prior to incorporation of fatty acids into the TG pool or prior to fatty acid β-oxidation [65].

Whether these two pathways of futile fatty acid cycling are involved in the cardiac inefficiency in heart failure is not yet clear. However, as indicated above, the increased sympathetic activity in heart failure is expected to mobilize fatty acids from adipose tissue, increasing the circulating levels of fatty acids and cardiac exposure to fatty acids [17].

4.3.1.4 Reversion to Fetal Glycolytic Metabolism Favors Mismatched Glucose Metabolism and the Subsequent Higher Energy Cost of Ionic Homeostasis

It is known that the fetal heart is largely glycolytic [66]. Studies with newborn rabbit hearts have shown that rapidly after birth, the expression of fatty acid oxidation regulating enzymes such as peroxisome proliferator-activated receptor α (PPARα) increases, while those regulating glycolysis decline. These changes mark the conversion to fatty acid oxidation as the main source of energy [20, 67–69].

In experimental models of heart failure in which an initial pathologic hypertrophy develops (such as salt-sensitive hypertensive rats [7], pressure overload models such as transverse aortic constriction [70], and volume overload models such as aortocaval fistula [66, 71, 72]), reversion to a fetal glycolytic metabolism occurs. Proteomic analysis of mitochondrial proteins also confirmed this shift to glycolytic metabolism in rats 8 weeks after coronary artery ligation [73].

The reversion to glycolytic metabolism that is associated with a decline in mitochondrial oxidative capacity is thought to help produce energy from a substrate that utilizes less oxygen per ATP; hence, efficiency is improved. On the other hand, this increase in glycolysis without a parallel increase in glucose oxidation can favor mismatched metabolism, whereby hydrolysis of glycolytically derived ATP can accelerate proton production with the development of intracellular acidosis [74]. This can subsequently lead to an increase in Na^+ and Ca^{2+} accumulation [74], resulting in an impaired contractility (Fig. 4.2). Correction of intracellular accumulation of Na^+ and Ca^{2+} increases ATP utilization for ionic homeostasis, rather than for contractile activity.

Reversion to fetal glycolytic metabolism is associated with changes in the expression of gene and protein levels and activity of enzymes known to regulate both glycolysis and fatty acid β-oxidation. Members of the peroxisome proliferator-activated receptor (PPAR) family are specifically involved. PPARα, which is predominantly expressed in the heart and skeletal muscles, is known to regulate fatty acid oxidation, and thus, its inhibition can be responsible for the reversion to fetal glycolytic metabolism [75]. In this regard, genetic deletion of PPARα has been shown to result in hypertrophy and inhibition of fatty acid β-oxidation in mice exposed to transverse aortic constriction [76]. This is associated with decreased expression of fatty acid oxidation genes [77] and a decline of fatty acid oxidation rates accompanied by a stimulation of glucose oxidation [78]. Previous findings led to the notion that stimulating PPARα in failing hearts could improve function through prevention of the development of fetal glycolytic metabolic profile, via prevention of the downregulation of fatty acid β-oxidation [79]. As a result, pharmacologic approaches to activate PPARα (i.e., with the use of fenofibrate) have been used in an attempt to decrease heart failure severity. This was found to produce a uniform upregulation of PPARα-regulated genes, although results have been variable regarding the improvement of heart function [80, 81]. A worsening of ex vivo LV function was observed in hypertrophied hearts derived from rats exposed to transverse aortic constriction [80]. In post-infarct rat hearts [81] and in dogs with heart failure due to excessive tachycardia [82], there was no improvement of LV function or reduction of chamber volume following chronic fenofibrate treatment. On the other hand, the use of fenofibrate in a porcine tachycardia model of heart failure resulted in lessened deterioration of LV dysfunction [83]. The variability in the reported functional consequences of PPARα activation hints at the lower likelihood of PPARα involvement in progression to heart failure. However, these findings do not appear to be a concern in heart failure patients who are already using

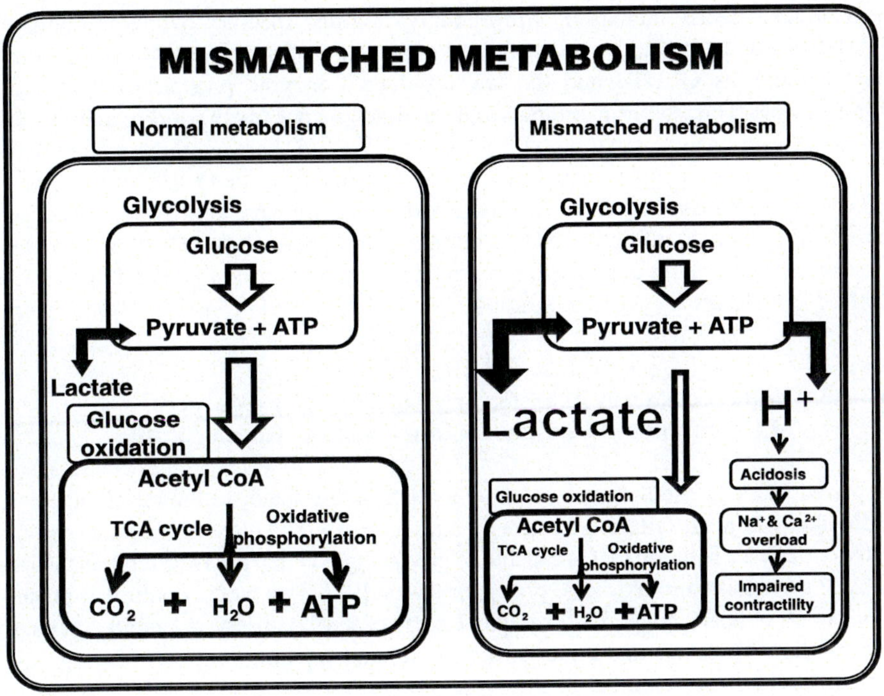

Fig. 4.2 Normal versus mismatched glucose metabolism. Intracellular proton production arises when the rate of glycolysis exceeds the rate of glucose oxidation. If glycolysis increases and/or glucose oxidation decreases (mismatched rates of glucose metabolism), proton production accelerates and causes intracellular acidosis. Efflux of proton leads to Na^+ overload and ultimately Ca^{2+} overload that causes cardiac mechanical dysfunction and inefficiency

fenofibrate to reduce cholesterol. In fact, fenofibrate may possess cardioprotective effects mediated by non-metabolic pathways. For example, it was reported that fenofibrate lessens cardiac fibrosis and diastolic dysfunction in salt-sensitive hypertensive rats, probably via non-metabolic effects involving the suppression of inflammation via inhibiting NF-*k*B [84].

PPARβ/δ is another member of the PPAR family which may be involved in the reversion of the heart to a fetal glycolytic metabolism. Although not specific for cardiac muscle, it is expressed in high amounts in the heart [85]. The role of PPARβ/δ in cardiac metabolism has been studied by cardiac-specific deletion of PPARβ/δ in mice. These mice develop severe cardiomyopathy ending in heart failure and premature death associated with decreased cardiac expression of genes responsible for fatty acid oxidation, decreased cardiac fatty acid oxidation rates, and increased cardiac lipid deposition [86]. In contrast, cardiac-specific PPARβ/δ overexpression increases expression of genes involved in fatty acid β-oxidation. However, fatty acid oxidation rates in these mice were normal, indicative of possible post-translational modifications. Neither genes involved in TG synthesis nor those involved in fatty acid uptake were increased. Surprisingly, cardiac-specific overexpression of PPARβ/δ was associated with increased expression of genes involved in glucose uptake (such as GLUT4 and the key regulatory enzyme phosphofructokinase), which was associated with an increase in cardiac glucose uptake and oxidation [75, 87–89]. Despite the fact that both PPARα and PPARβ/δ share overlapping pathways for control of fatty acid oxidation [75], cardiomyocyte PPARβ/δ restriction in PPARα null mice resulted in no further inhibition of fatty acid β-oxidation, but rather a pronounced inhibition of mitochondrial biogenesis [90]. Whether the reversion to fetal glycolytic phenotype is associated with increased PPARβ/δ expression in the failing hearts is yet to be studied.

4.3.1.5 Impaired Energy Transfer to Myofibrils: Consequences of Compromised Phosphocreatine/Creatine Kinase Shuttle

Impaired PCr/CK shuttling may also be a contributing factor to the development of inefficiency in heart failure. PCr acts as a reservoir to replenish ATP content via the activity of CK, a reaction that is capable of producing ATP ten times faster than the rate of ATP synthesis from oxidative phosphorylation [91]. This fact is confirmed in various models of heart failure. For example, Hearse et al. [92] observed a rapid decline in myocardial PCr levels and contractility in isolated working rat hearts during post-ischemic reperfusion. Similar results were reported by Whitman et al. [93] using isolated perfused rabbit hearts. The same concept was further confirmed in a porcine model of transverse aortic constriction-induced cardiac hypertrophy and failure [94].

PCr/ATP ratio is usually used as an index of the energy status of the heart [95, 96], which can also predict possible protection against supervening ischemic insults and the subsequent development of heart failure. Changes in PCr levels have been reported in many heart failure patient studies. Winter et al. [97] reported an earlier decline in PCr levels in non-ischemic heart failure patients using magnetic resonance spectroscopy, where a significant reduction in cardiac creatine levels was shown by a significant reduction of water/creatine index. This decrease in PCr was also reported early in patients with dilated cardiomyopathy [98]. These findings conform to the studies of Neubauer et al. [99], who used ^{31}P-NMR spectroscopy and found a 70% reduction in PCr levels in heart failure patients with dilated cardiomyopathy while ATP levels were unaltered. The PCr/ATP ratio was thus markedly reduced in these patients. The authors found that the decline of PCr/ATP ratio correlated with the severity of heart failure and that the ratio improved by recompensation. The authors also found that PCr/ATP ratio is a valid predictor of mortality in these patients. They found that patients with normal PCr/ATP (more than 1.6) had an all-cause mortality of 10% following a 2.5-year surveillance. On the other hand, patients with low PCr/ATP ratio had 40% all-cause mortality [99].

Not only do PCr levels decline in heart failure but the ability of the failing heart to maintain adequate ATP levels in response to exercise challenge is also compromised to various degrees depending on heart failure etiology. Using the handgrip exercise, Weiss et al. [100] observed that PCr/ATP ratio in the left ventricular wall is maintained in non-ischemic heart failure patients, but is greatly reduced in coronary artery disease heart failure patients. Similar changes were also observed by Yabe et al. [101] at the level of subendocardium.

4.3.2 Heart Failure and Energy Starvation

The energy starvation theory was initially proposed by Hermann in 1939 [102], who reviewed the chemical nature of heart failure and addressed the reduction of myocardial creatine in the heart. From this perspective, the failing heart can be described as an engine out of fuel [19]. This implies an ability to improve or at least prevent further deterioration of cardiac function by stimulating the various stages of high-energy generation [35]. Because this theory gained much interest and is described in detail elsewhere, we will only briefly highlight this concept (refer to [19] for review).

As heart failure progresses to decompensation, the reduced cardiac output reduces peripheral tissue perfusion including the heart tissue itself. The associated pulmonary congestion reduces oxygen saturation leading to further deterioration of oxygen supply. In general, reduced intracellular oxygen availability does not limit peak oxygen utilization. However, as heart failure progresses, the workload increases, and cardiac MVO_2 can reach a maximum and becomes limited leading to subsequent deficiency of energy generation and utilization [103]. This is accompanied by reduced mitochondrial oxidative phosphorylation [104]. In addition, flexibility in substrate utilization is lost [105]. Moreover, membrane expression of the insulin-sensitive glucose transporter GLUT4 is reduced in advanced heart failure [106]. On the other hand, expression of fatty acid uptake protein (CD36) and intracellular fatty acid transporter (fatty acid binding protein) are increased in both ischemic and dilated cardiomyopathy patients [107].

4.4 Proof of Principle for Cardiac Inefficiency and Energy Starvation in Heart Failure

Energy starvation can be confirmed by the generalized decline in metabolic rates seen in advanced cases of heart failure, whether in patients or experimental models [5, 108]. Lending support to this theory is the observation that mechanical function in advanced heart failure can be improved by using metabolic modulators. In this regard, stimulation of glucose oxidation (e.g., by dichloroacetate) in patients with congestive heart failure was found to improve mechanical efficiency and hemodynamic parameters [109]. Also, intracoronary infusion of pyruvate in heart failure patients can induce a short-term functional improvement in heart function [110]. Furthermore, cardiac-specific overexpression of GLUT1 can prevent the development of left ventricular dysfunction after transverse aortic banding [111], and induction of mitochondrial biogenesis can prevent cardiomyopathy in mice [35]. Treatments that limit the energy demand of the failing heart by reducing mechanical work such as β-adrenoceptor antagonists are also known to be useful adjuncts to the traditional heart failure therapy [112–115].

Support for cardiac inefficiency in heart failure arises from the beneficial effect of fatty acid β-oxidation inhibitors to treat heart failure. Inhibition of fatty acid oxidation improves mechanical function in heart failure, evidence which clearly stands against the energy starvation theory (see [11, 19, 20], and [22] for reviews). A number of approaches to inhibit fatty acid β-oxidation have been used, which include (1) reduction of circulating fatty acid levels with β-adrenoceptor antagonists such as nebivolol that inhibit lipolysis [113, 114], (2) the use of PPARγ agonists such as thiazolidinediones that favor sequestration of fatty acids in adipose tissue [116–118], (3) inhibition of mitochondrial fatty acid uptake by etomoxir [119] or perhexiline [120–122], and (4) inhibition of mitochondrial β-oxidation with agents such as trimetazidine [123–128]. The resulting indirect stimulation of glucose oxidation as a consequence of lower rates of fatty acid β-oxidation improves the coupling of glycolysis to glucose oxidation, leading to reduced protons production by hydrolysis of glycolytically derived ATP (Fig. 4.2). Also, shifting substrate preference from fatty acid β-oxidation is expected to increase efficiency of energy production without changes in MVO_2. Indeed, the observed improvement of mechanical function in patient with congestive heart failure with the use of glucose oxidation stimulators such as dichloroacetate [109] can be explained on basis of improved matching between glycolytic flux and glucose oxidation. In this regard, the increased glycolytic flux observed in severe heart failure was associated with downregulation of glucose oxidation [5], indicative of an even greater mismatch in the pathways of glucose metabolism.

4.5 Conclusion

Both cardiac energy starvation and cardiac inefficiency may contribute to the severity of heart failure. These two contributing mechanisms are also not mutually exclusive, and may coexist. However, it is likely that energy starvation occurs later in the course of heart failure, while inefficiency may be present throughout the course of heart failure, including the very early stages where metabolic interventions are expected to yield maximum benefit. It would thus be desirable to consider cardiac inefficiency when designing metabolic modulators to treat heart failure, since stimulation of substrate utilization may not always be beneficial. This is supported by reduced efficiency observed with acceleration of fatty acid β-oxidation or of glycolysis.

References

1. Neely JR, Morgan HE (1974) Relationship between carbohydrate and lipid metabolism and the energy balance of heart muscle. Annu Rev Physiol 36:413–459
2. Opie LH (1968) Metabolism of the heart in health and disease. I. Am Heart J 76:685–698
3. Opie LH (1969) Metabolism of the heart in health and disease. II. Am Heart J 77:100–122 contd
4. Opie LH (1969) Metabolism of the heart in health and disease. III. Am Heart J 77:383–410 contd
5. Lei B, Lionetti V, Young ME et al (2004) Paradoxical downregulation of the glucose oxidation pathway despite enhanced flux in severe heart failure. J Mol Cell Cardiol 36:567–576
6. Osorio JC, Stanley WC, Linke A et al (2002) Impaired myocardial fatty acid oxidation and reduced protein expression of retinoid X receptor-alpha in pacing-induced heart failure. Circulation 106:606–612
7. Dodd MS, Ball DR, Schroeder MA et al (2012) In vivo alterations in cardiac metabolism and function in the spontaneously hypertensive rat heart. Cardiovasc Res 95(1):69–76
8. Xu J, Nie HG, Zhang XD et al (2011) Down-regulated energy metabolism genes associated with mitochondria oxidative phosphorylation and fatty acid metabolism in viral cardiomyopathy mouse heart. Mol Biol Rep 38:4007–4013
9. Li X, Arslan F, Ren Y et al (2012) Metabolic adaptation to a disruption in oxygen supply during myocardial ischemia and reperfusion is underpinned by temporal and quantitative changes in the cardiac proteome. J Proteome Res 11:2331–2346
10. Gertz EW, Wisneski JA, Stanley WC, Neese RA (1988) Myocardial substrate utilization during exercise in humans. Dual carbon-labeled carbohydrate isotope experiments. J Clin Invest 82:2017–2025
11. Stanley WC, Lopaschuk GD, Hall JL, McCormack JG (1997) Regulation of myocardial carbohydrate metabolism under normal and ischaemic conditions otential for pharmacological interventions. Cardiovasc Res 33:243–257
12. Wisneski JA, Gertz EW, Neese RA et al (1985) Dual carbon-labeled isotope experiments using D-[6-14C] glucose and L-[1,2,3-13C3] lactate: a new approach for investigating human myocardial metabolism during ischemia. J Am Coll Cardiol 5:1138–1146
13. Wisneski JA, Gertz EW, Neese RA et al (1985) Metabolic fate of extracted glucose in normal human myocardium. J Clin Invest 76:1819–1827
14. Wisneski JA, Stanley WC, Neese RA, Gertz EW (1990) Effects of acute hyperglycemia on myocardial glycolytic activity in humans. J Clin Invest 85:1648–1656
15. Maciver DH, Dayer MJ, Harrison AJ (2012) A general theory of acute and chronic heart failure. Int J Cardiol. E pub ahead of print. DOI 10.1016/j.ijcard.2012.03.093
16. Francis GS, McDonald KM, Cohn JN (1993) Neurohumoral activation in preclinical heart failure Remodeling and the potential for intervention. Circulation 87(suppl 5):IV90–IV96
17. Kemp CD, Conte JV (2012) The pathophysiology of heart failure. Cardiovasc Pathol 21(5):365–371
18. Houser SR, Margulies KB, Murphy AM et al (2012) Animal models of heart failure: a scientific statement from the American Heart Association. Circ Res 111(1):131–150
19. Neubauer S (2007) The failing heart–an engine out of fuel. N Engl J Med 356:1140–1151
20. Lopaschuk GD, Ussher JR, Folmes CD et al (2010) Myocardial fatty acid metabolism in health and disease. Physiol Rev 90:207–258
21. Stanley WC, Recchia FA, Lopaschuk GD (2005) Myocardial substrate metabolism in the normal and failing heart. Physiol Rev 85:1093–1129
22. Ingwall JS (2009) Energy metabolism in heart failure and remodelling. Cardiovasc Res 81:412–419
23. Jaswal JS, Keung W, Wang W et al (2011) Targeting fatty acid and carbohydrate oxidation - a novel therapeutic intervention in the ischemic and failing heart. Biochim Biophys Acta 1813(7):1333–1350

24. Strumia E, Pelliccia F, D'Ambrosio G (2012) Creatine phosphate: pharmacological and clinical perspectives. Adv Ther 29:99–123
25. Guzun R, Timohhina N, Tepp K et al (2011) Systems bioenergetics of creatine kinase networks: physiological roles of creatine and phosphocreatine in regulation of cardiac cell function. Amino Acids 40:1333–1348
26. Sahlin K, Harris RC (2011) The creatine kinase reaction: a simple reaction with functional complexity. Amino Acids 40:1363–1367
27. Boehm E, Chan S, Monfared M et al (2003) Creatine transporter activity and content in the rat heart supplemented by and depleted of creatine. Am J Physiol Endocrinol Metab 284:E399–E406
28. Rossi AM, Eppenberger HM, Volpe P et al (1990) Muscle-type MM creatine kinase is specifically bound to sarcoplasmic reticulum and can support Ca2+ uptake and regulate local ATP/ADP ratios. J Biol Chem 265:5258–5266
29. Wallimann T, Wyss M, Brdiczka D et al (1992) Intracellular compartmentation, structure and function of creatine kinase isoenzymes in tissues with high and fluctuating energy demands: the 'phosphocreatine circuit' for cellular energy homeostasis. Biochem J 281(Pt 1):21–40
30. Ventura-Clapier R, Garnier A, Veksler V, Joubert F (2011) Bioenergetics of the failing heart. Biochim Biophys Acta 1813:1360–1372
31. Tuunanen H, Knuuti J (2011) Metabolic remodelling in human heart failure. Cardiovasc Res 90:251–257
32. Ohte N, Narita H, Iida A et al (2009) Impaired myocardial oxidative metabolism in the remote normal region in patients in the chronic phase of myocardial infarction and left ventricular remodeling. J Nucl Cardiol 16:73–81
33. Hasegawa S, Yamamoto K, Sakata Y et al (2008) Effects of cardiac energy efficiency in diastolic heart failure: assessment with positron emission tomography with 11C-acetate. Hypertens Res 31:1157–1162
34. van Bilsen M, van Nieuwenhoven FA, van der Vusse GJ (2009) Metabolic remodelling of the failing heart: beneficial or detrimental? Cardiovasc Res 81:420–428
35. Schulz TJ, Westermann D, Isken F et al (2010) Activation of mitochondrial energy metabolism protects against cardiac failure. Aging (Albany NY) 2:843–853
36. Jullig M, Hickey AJ, Chai CC et al (2008) Is the failing heart out of fuel or a worn engine running rich? A study of mitochondria in old spontaneously hypertensive rats. Proteomics 8:2556–2572
37. Ingwall JS, Weiss RG (2004) Is the failing heart energy starved? On using chemical energy to support cardiac function. Circ Res 95:135–145
38. Ingwall JS (2006) Energetics of the failing heart: new insights using genetic modification in the mouse. Arch Mal Coeur Vaiss 99:839–847
39. Taegtmeyer H, Wilson CR, Razeghi P, Sharma S (2005) Metabolic energetics and genetics in the heart. Ann N Y Acad Sci 1047:208–218
40. Bing RJ, Hammond MM (1949) The measurement of coronary blood flow, oxygen consumption, and efficiency of the left ventricle in man. Am Heart J 38:1–24
41. Suga H (2003) Cardiac energetics: from E(max) to pressure-volume area. Clin Exp Pharmacol Physiol 30:580–585
42. Suga H (1990) Ventricular energetics. Physiol Rev 70:247–277
43. Burkhoff D, Weiss RG, Schulman SP et al (1991) Influence of metabolic substrate on rat heart function and metabolism at different coronary flows. Am J Physiol 261:H741–H750
44. Korvald C, Elvenes OP, Myrmel T (2000) Myocardial substrate metabolism influences left ventricular energetics in vivo. Am J Physiol Heart Circ Physiol 278:H1345–H1351
45. Lammerant J, Huynh-Thu T, Kolanowski J (1985) Inhibitory effects of the D(–)isomer of 3-hydroxybutyrate on cardiac non-esterified fatty acid uptake and oxygen demand induced by norepinephrine in the intact dog. J Mol Cell Cardiol 17:421–433
46. Mjos OD (1971) Effect of free fatty acids on myocardial function and oxygen consumption in intact dogs. J Clin Invest 50:1386–1389

47. Mjos OD (1971) Effect of inhibition of lipolysis on myocardial oxygen consumption in the presence of isoproterenol. J Clin Invest 50:1869–1873
48. Mjos OD, Kjekshus J (1971) Increased local metabolic rate by free fatty acids in the intact dog heart. Scand J Clin Lab Invest 28:389–393
49. Simonsen S, Kjekshus JK (1978) The effect of free fatty acids on myocardial oxygen consumption during atrial pacing and catecholamine infusion in man. Circulation 58:484–491
50. Hinkle PC (2005) P/O ratios of mitochondrial oxidative phosphorylation. Biochim Biophys Acta 1706:1–11
51. Kadenbach B (2003) Intrinsic and extrinsic uncoupling of oxidative phosphorylation. Biochim Biophys Acta 1604:77–94
52. Bouillaud F, Combes-George M, Ricquier D (1983) Mitochondria of adult human brown adipose tissue contain a 32 000-Mr uncoupling protein. Biosci Rep 3:775–780
53. Enerback S (2010) Brown adipose tissue in humans. Int J Obes (Lond) 34(Suppl 1): S43–S46
54. McLeod CJ, Aziz A, Hoyt RF et al (2005) Uncoupling proteins 2 and 3 function in concert to augment tolerance to cardiac ischemia. J Biol Chem 280:33470–33476
55. Boehm EA, Jones BE, Radda GK et al (2001) Increased uncoupling proteins and decreased efficiency in palmitate-perfused hyperthyroid rat heart. Am J Physiol Heart Circ Physiol 280:H977–H983
56. Cole MA, Murray AJ, Cochlin LE et al (2011) A high fat diet increases mitochondrial fatty acid oxidation and uncoupling to decrease efficiency in rat heart. Basic Res Cardiol 106:447–457
57. Laskowski KR, Russell RR (2008) Uncoupling proteins in heart failure. Curr Heart Fail Rep 5:75–79
58. Li N, Wang J, Gao F et al (2009) The relationship between uncoupling protein 2 expression and myocardial high energy phosphates content in abdominal aorta constriction induced heart failure rats. Zhonghua Xin Xue Guan Bing Za Zhi 37:1108–1112
59. Turner JD, Gaspers LD, Wang G, Thomas AP (2010) Uncoupling protein-2 modulates myocardial excitation-contraction coupling. Circ Res 106:730–738
60. Murray AJ, Cole MA, Lygate CA et al (2008) Increased mitochondrial uncoupling proteins, respiratory uncoupling and decreased efficiency in the chronically infarcted rat heart. J Mol Cell Cardiol 44:694–700
61. Saddik M, Lopaschuk GD (1991) Myocardial triglyceride turnover and contribution to energy substrate utilization in isolated working rat hearts. J Biol Chem 266:8162–8170
62. Saddik M, Lopaschuk GD (1992) Myocardial triglyceride turnover during reperfusion of isolated rat hearts subjected to a transient period of global ischemia. J Biol Chem 267:3825–3831
63. Himms-Hagen J, Harper ME (2001) Physiological role of UCP3 may be export of fatty acids from mitochondria when fatty acid oxidation predominates: an hypothesis. Exp Biol Med (Maywood) 226:78–84
64. Schrauwen P, Hoeks J, Hesselink MK (2006) Putative function and physiological relevance of the mitochondrial uncoupling protein-3: involvement in fatty acid metabolism? Prog Lipid Res 45:17–41
65. Rame JE, Barouch LA, Sack MN et al (2011) Caloric restriction in leptin deficiency does not correct myocardial steatosis: failure to normalize PPAR{alpha}/PGC1{alpha} and thermogenic glycerolipid/fatty acid cycling. Physiol Genomics 43:726–738
66. Oka T, Lam VH, Zhang L et al (2012) Cardiac hypertrophy in the newborn delays the maturation of fatty acid beta-oxidation and compromises postischemic functional recovery. Am J Physiol Heart Circ Physiol 302:H1784–H1794
67. Razeghi P, Young ME, Alcorn JL et al (2001) Metabolic gene expression in fetal and failing human heart. Circulation 104:2923–2931
68. Rajabi M, Kassiotis C, Razeghi P, Taegtmeyer H (2007) Return to the fetal gene program protects the stressed heart: a strong hypothesis. Heart Fail Rev 12:331–343

69. Taegtmeyer H, Sen S, Vela D (2010) Return to the fetal gene program: a suggested metabolic link to gene expression in the heart. Ann N Y Acad Sci 1188:191–198
70. Dai DF, Hsieh EJ, Liu Y et al (2012) Mitochondrial proteome remodelling in pressure overload-induced heart failure: the role of mitochondrial oxidative stress. Cardiovasc Res 93:79–88
71. Moravec J, El Alaoui-Talibi Z, Moravec M, Guendouz A (1996) Control of oxidative metabolism in volume-overloaded rat hearts: effect of pretreatment with propionyl-L-carnitine. Adv Exp Med Biol 388:205–212
72. El Alaoui-Talibi Z, Guendouz A, Moravec M, Moravec J (1997) Control of oxidative metabolism in volume-overloaded rat hearts: effect of propionyl-L-carnitine. Am J Physiol 272:H1615–H1624
73. Wang J, Bai L, Li J et al (2009) Proteomic analysis of mitochondria reveals a metabolic switch from fatty acid oxidation to glycolysis in the failing heart. Sci China C Life Sci 52:1003–1010
74. Baartscheer A, Schumacher CA, Coronel R, Fiolet JW (2011) The driving force of the Na/Ca-exchanger during metabolic inhibition. Front Physiol 2:10
75. Madrazo JA, Kelly DP (2008) The PPAR trio: regulators of myocardial energy metabolism in health and disease. J Mol Cell Cardiol 44:968–975
76. Smeets PJ, Teunissen BE, Willemsen PH et al (2008) Cardiac hypertrophy is enhanced in PPAR alpha–/– mice in response to chronic pressure overload. Cardiovasc Res 78:79–89
77. Watanabe K, Fujii H, Takahashi T et al (2000) Constitutive regulation of cardiac fatty acid metabolism through peroxisome proliferator-activated receptor alpha associated with age-dependent cardiac toxicity. J Biol Chem 275:22293–22299
78. Campbell FM, Kozak R, Wagner A et al (2002) A role for peroxisome proliferator-activated receptor alpha (PPARalpha) in the control of cardiac malonyl-CoA levels: reduced fatty acid oxidation rates and increased glucose oxidation rates in the hearts of mice lacking PPARalpha are associated with higher concentrations of malonyl-CoA and reduced expression of malonyl-CoA decarboxylase. J Biol Chem 277:4098–4103
79. Sarma S, Ardehali H, Gheorghiade M (2010) Enhancing the metabolic substrate: PPAR-alpha agonists in heart failure. Heart Fail Rev 17(1):35–43
80. Young ME, Laws FA, Goodwin GW, Taegtmeyer H (2001) Reactivation of peroxisome proliferator-activated receptor alpha is associated with contractile dysfunction in hypertrophied rat heart. J Biol Chem 276:44390–44395
81. Morgan EE, Rennison JH, Young ME et al (2006) Effects of chronic activation of peroxisome proliferator-activated receptor-alpha or high-fat feeding in a rat infarct model of heart failure. Am J Physiol Heart Circ Physiol 290:H1899–H1904
82. Labinskyy V, Bellomo M, Chandler MP et al (2007) Chronic activation of peroxisome proliferator-activated receptor-alpha with fenofibrate prevents alterations in cardiac metabolic phenotype without changing the onset of decompensation in pacing-induced heart failure. J Pharmacol Exp Ther 321:165–171
83. Brigadeau F, Gele P, Wibaux M et al (2007) The PPARalpha activator fenofibrate slows down the progression of the left ventricular dysfunction in porcine tachycardia-induced cardiomyopathy. J Cardiovasc Pharmacol 49:408–415
84. Ogata T, Miyauchi T, Sakai S et al (2004) Myocardial fibrosis and diastolic dysfunction in deoxycorticosterone acetate-salt hypertensive rats is ameliorated by the peroxisome proliferator-activated receptor-alpha activator fenofibrate, partly by suppressing inflammatory responses associated with the nuclear factor-kappa-B pathway. J Am Coll Cardiol 43:1481–1488
85. Alvarez-Guardia D, Palomer X, Coll T et al (2011) PPARbeta/delta activation blocks lipid-induced inflammatory pathways in mouse heart and human cardiac cells. Biochim Biophys Acta 1811:59–67
86. Cheng L, Ding G, Qin Q et al (2004) Cardiomyocyte-restricted peroxisome proliferator-activated receptor-delta deletion perturbs myocardial fatty acid oxidation and leads to cardiomyopathy. Nat Med 10:1245–1250

87. Opie LH, Sack MN (2002) Metabolic plasticity and the promotion of cardiac protection in ischemia and ischemic preconditioning. J Mol Cell Cardiol 34:1077–1089
88. Mitra B, Panja M (2005) Myocardial metabolism: pharmacological manipulation in myocardial ischaemia. J Assoc Physicians India 53:552–560
89. Lopaschuk GD (1998) Treating ischemic heart disease by pharmacologically improving cardiac energy metabolism. Presse Med 27:2100–2104
90. Liu J, Wang P, He L et al (2011) Cardiomyocyte-restricted deletion of PPARbeta/delta in PPARalpha-null mice causes impaired mitochondrial biogenesis and defense, but no further depression of myocardial fatty acid oxidation. PPAR Res 2011:372854
91. Ingwall JS (2006) On the hypothesis that the failing heart is energy starved: lessons learned from the metabolism of ATP and creatine. Curr Hypertens Rep 8:457–464
92. Hearse DJ (1979) Oxygen deprivation and early myocardial contractile failure: a reassessment of the possible role of adenosine triphosphate. Am J Cardiol 44:1115–1121
93. Whitman GJ, Kieval RS, Seeholzer S et al (1985) Recovery of left ventricular function after graded cardiac ischemia as predicted by myocardial P-31 nuclear magnetic resonance. Surgery 97:428–435
94. Ye Y, Gong G, Ochiai K et al (2001) High-energy phosphate metabolism and creatine kinase in failing hearts: a new porcine model. Circulation 103:1570–1576
95. Edwards LM, Ashrafian H, Korzeniewski B (2011) In silico studies on the sensitivity of myocardial PCr/ATP to changes in mitochondrial enzyme activity and oxygen concentration. Mol Biosyst 7:3335–3342
96. Fragasso G, Perseghin G, De Cobelli F et al (2006) Effects of metabolic modulation by trimetazidine on left ventricular function and phosphocreatine/adenosine triphosphate ratio in patients with heart failure. Eur Heart J 27:942–948
97. Winter JL, Castro P, Meneses L et al (2010) Myocardial lipids and creatine measured by magnetic resonance spectroscopy among patients with heart failure. Rev Med Chil 138:1475–1479
98. Hardy CJ, Weiss RG, Bottomley PA, Gerstenblith G (1991) Altered myocardial high-energy phosphate metabolites in patients with dilated cardiomyopathy. Am Heart J 122:795–801
99. Neubauer S, Krahe T, Schindler R et al (1992) 31P magnetic resonance spectroscopy in dilated cardiomyopathy and coronary artery disease. Altered cardiac high-energy phosphate metabolism in heart failure. Circulation 86:1810–1818
100. Weiss RG, Bottomley PA, Hardy CJ, Gerstenblith G (1990) Regional myocardial metabolism of high-energy phosphates during isometric exercise in patients with coronary artery disease. N Engl J Med 323:1593–1600
101. Yabe T, Mitsunami K, Inubushi T, Kinoshita M (1995) Quantitative measurements of cardiac phosphorus metabolites in coronary artery disease by 31P magnetic resonance spectroscopy. Circulation 92:15–23
102. Herrmann G, Decherd M (1939) The chemical nature of heart failure. Ann Intern Med 12:1233–1244
103. Gong G, Liu J, Liang P et al (2003) Oxidative capacity in failing hearts. Am J Physiol Heart Circ Physiol 285:H541–H548
104. Leong HS, Brownsey RW, Kulpa JE, Allard MF (2003) Glycolysis and pyruvate oxidation in cardiac hypertrophy–why so unbalanced? Comp Biochem Physiol A Mol Integr Physiol 135:499–513
105. Taegtmeyer H (2000) Genetics of energetics: transcriptional responses in cardiac metabolism. Ann Biomed Eng 28:871–876
106. Dutka DP, Pitt M, Pagano D et al (2006) Myocardial glucose transport and utilization in patients with type 2 diabetes mellitus, left ventricular dysfunction, and coronary artery disease. J Am Coll Cardiol 48:2225–2231
107. Garcia-Rua V, Otero MF, Lear PV et al (2012) Increased expression of Fatty-Acid and calcium metabolism genes in failing human heart. PLoS One 7:e37505
108. Tuunanen H, Engblom E, Naum A et al (2006) Decreased myocardial free fatty acid uptake in patients with idiopathic dilated cardiomyopathy: evidence of relationship with insulin resistance and left ventricular dysfunction. J Card Fail 12:644–652

109. Bersin RM, Wolfe C, Kwasman M et al (1994) Improved hemodynamic function and mechanical efficiency in congestive heart failure with sodium dichloroacetate. J Am Coll Cardiol 23:1617–1624
110. Hermann HP, Pieske B, Schwarzmuller E et al (1999) Haemodynamic effects of intracoronary pyruvate in patients with congestive heart failure: an open study. Lancet 353: 1321–1323
111. Liao R, Jain M, Cui L et al (2002) Cardiac-specific overexpression of GLUT1 prevents the development of heart failure attributable to pressure overload in mice. Circulation 106:2125–2131
112. Ong HT, Ong LM, Kow FP (2012) Beta-blockers for heart failure: an evidence based review answering practical therapeutic questions. Med J Malaysia 67:7–11
113. Dery AS, Hamilton LA, Starr JA (2011) Nebivolol for the treatment of heart failure. Am J Health Syst Pharm 68:879–886
114. Riva N, Lip GY (2011) Nebivolol for the treatment of heart failure. Expert Opin Investig Drugs 20:1733–1746
115. Klapholz M (2009) Beta-blocker use for the stages of heart failure. Mayo Clin Proc 84: 718–729
116. Zhu P, Lu L, Xu Y, Schwartz GG (2000) Troglitazone improves recovery of left ventricular function after regional ischemia in pigs. Circulation 101:1165–1171
117. Yue TL, Bao W, Gu JL et al (2005) Rosiglitazone treatment in Zucker diabetic Fatty rats is associated with ameliorated cardiac insulin resistance and protection from ischemia/reperfusion-induced myocardial injury. Diabetes 54:554–562
118. Sidell RJ, Cole MA, Draper NJ et al (2002) Thiazolidinedione treatment normalizes insulin resistance and ischemic injury in the zucker Fatty rat heart. Diabetes 51:1110–1117
119. Schmitz FJ, Rosen P, Reinauer H (1995) Improvement of myocardial function and metabolism in diabetic rats by the carnitine palmitoyl transferase inhibitor Etomoxir. Horm Metab Res 27:515–522
120. Abozguia K, Elliott P, McKenna W et al (2010) Metabolic modulator perhexiline corrects energy deficiency and improves exercise capacity in symptomatic hypertrophic cardiomyopathy. Circulation 122:1562–1569
121. Horowitz JD, Chirkov YY (2010) Perhexiline and hypertrophic cardiomyopathy: a new horizon for metabolic modulation. Circulation 122:1547–1549
122. Lee L, Campbell R, Scheuermann-Freestone M et al (2005) Metabolic modulation with perhexiline in chronic heart failure: a randomized, controlled trial of short-term use of a novel treatment. Circulation 112:3280–3288
123. Zhang L, Lu Y, Jiang H et al (2012) Additional use of trimetazidine in patients with chronic heart failure: a meta-analysis. J Am Coll Cardiol 59:913–922
124. Fragasso G, Salerno A, Lattuada G et al (2011) Effect of partial inhibition of fatty acid oxidation by trimetazidine on whole body energy metabolism in patients with chronic heart failure. Heart 97:1495–1500
125. Gao D, Ning N, Niu X et al (2011) Trimetazidine: a meta-analysis of randomised controlled trials in heart failure. Heart 97:278–286
126. Wenmeng W, Qizhu T (2011) Early administration of trimetazidine may prevent or ameliorate diabetic cardiomyopathy. Med Hypotheses 76:181–183
127. Gunes Y, Guntekin U, Tuncer M, Sahin M (2009) Improved left and right ventricular functions with trimetazidine in patients with heart failure: a tissue Doppler study. Hear Vessel 24:277–282
128. Sisakian AS, Torgomian AL, Barkhudarian AL (2006) The effects of trimetazidine on left ventricular function and physical exercise tolerance in patients with ischemic cardiomyopathy. Klin Med (Mosk) 84:55–58

Chapter 5
Regulation of Cardiac Hypertrophic Remodeling by the USP15/SLIM1 Pathway

Hiroto Nakajima

Abstract Many lines of evidence support the important role of the ubiquitin (Ub)–proteasome system (UPS) in the development of heart diseases. Cardiomyopathy or cardiac responses to pressure or volume overload are also accompanied by alterations in the UPS. At the beginning of this review, we provide an overview of recent studies regarding the features and roles of the UPS and its components in cardiac hypertrophic remodeling. Then, we focus on our animal model of cardiac remodeling, which was induced by cardiac-targeted overexpression of Ub-specific protease 15 (USP15) in mice. Skeletal muscle LIM protein 1 (SLIM1), a direct interactor and substrate of USP15, was increased in the remodeling heart of this model. We discuss possible mechanisms of the hypertrophic responses in the model on the basis of several novel findings on USP15 and SLIM1.

Keywords Deubiquitination • DUB • FHL1 • NFATC1 • TGFβ • BMP • SMAD • Cardiac remodeling • Cardiac hypertrophy • Cardiomyopathy

5.1 Introduction

The ubiquitin (Ub)–proteasome system (UPS) is an ATP-dependent proteolytic system that requires the polyubiquitination of a protein intended for degradation by the 26S proteasome (reviewed in [1]).

Protein ubiquitination consists of the covalent linkage of Ub to one or more lysine residues of a target protein or the preceding Ub. In general, polyubiquitination of a target protein through lysine 48 results in degradation, while polyubiquitination

---

H. Nakajima (✉)
Japan Research Foundation for Clinical Pharmacology (JRFCP),
3-14-10, Nihonbashi, Chuo-ku, Tokyo 103-8234, Japan
e-mail: nakajima.hiroto.py@daiichisankyo.co.jp

B.I. Jugdutt and N.S. Dhalla (eds.), *Cardiac Remodeling: Molecular Mechanisms*,
Advances in Biochemistry in Health and Disease 5, DOI 10.1007/978-1-4614-5930-9_5,

through lysine 63 or monoubiquitination of lysine residues can be used for other purposes such as DNA repair, re-localization, endocytosis, cellular signaling, or regulation of protein function [1]. The process of ubiquitination involves the concerted action of Ub enzymes composed of E1 (Ub-activating enzyme), E2 (Ub-conjugating enzymes), and E3 (Ub ligases). The E3 Ub ligases confer substrate specificity to the whole system, and around a thousand E3 enzymes have been identified in human genome based on specific, commonly shared structural motifs [1].

On the other hand, removal of covalently attached Ub from proteins is called deubiquitination and is catalyzed by deubiquitinating enzymes (DUBs). Functions of DUBs are generally categorized as follows in the Ub pathway (reviewed in [2]):

1. Processing of Ub precursors
2. Editing or rescue of Ub conjugates, which are generally adducts to other proteins in the cell but can also be ligated to abundant small nucleophiles such as glutathione
3. Recycling of Ub or Ub oligomers from Ub–protein conjugates targeted for degradation
4. Disassembly of unanchored Ub oligomers

The human genome encodes approximately 90 DUBs, and they fall into at least five distinct subfamilies based on their sequence similarities and likely mechanisms of action (reviewed in [3, 4]). The largest one of these subfamilies consists of the Ub-specific proteases (USPs), cysteine proteases containing a conserved catalytic triad within a defined catalytic domain [4].

Recently, much attention has been made on the UPS in the development of cardiac diseases. In the following sections, we provide an overview of the recent studies regarding the features and roles of the UPS and its components in cardiac hypertrophic remodeling, and then focus on our animal model of cardiac remodeling, which was induced by cardiac-targeted overexpression of USP15.

5.2 Profile of the Ub Pathways in Cardiac Hypertrophic Remodeling

A number of studies have shown that the accumulation of ubiquitinated proteins and lower levels of transcription of the proteasome components in human heart failure progressed from compensated hypertrophy, compared with non-failing hearts [5, 6], suggesting impaired removal of abnormal proteins by UPS. Additionally, in an experimental mouse model of cardiac hypertrophy and heart failure induced by transverse aortic constriction (TAC), both high levels of ubiquitinated proteins and depression of proteasome activities have been shown [7].

However, Depre and colleagues have reported the upregulation of the levels of UPS components and proteasome activities during the development of cardiac hypertrophy caused by either pressure overload or cardiac-specific expression of Hsp22 in mice [8, 9]. The authors have demonstrated that pharmacological inhibition of proteasome reverses cardiac remodeling in overloaded hearts and stabilizes the contractile function by repressing collagen accumulation via reduced NF-kB activity [10]. Proteasome inhibition also results in beneficial effects on cardiac remodeling induced by hypertension in rat hypertension models [11, 12].

Several E3 Ub ligases have been described as restrictedly expressed in cardiac and skeletal muscles (reviewed in [13, 14]), and transcripts of some of these genes are upregulated, although to small extents, in pressure-overloaded rat hearts [15]. These E3 ligases protect against the development of physiological or pathological cardiac hypertrophy. Overexpression of muscle atrophy F-box protein (MaFBx, also called atrogin-1), muscle ring finger protein-1 (MuRF1), or minute double minute 2 (Mdm2) inhibits phenylephrine- or endothelin-1-induced hypertrophy in cardiomyocytes [16–18]. Overexpression of atrogin-1 in hearts of transgenic (TG) mice reduces the levels of substrate protein calcineurin and blunts cardiac hypertrophy induced by TAC [16]. Conversely, whereas atrogin-1-deficient mice exhibit no cardiac phenotype under basal conditions, they develop marked physiological cardiac hypertrophy after voluntary running exercises [19]. Similar to atrogin-1-deficient mice, MuRF1-deficient mice show no signs of cardiac hypertrophy under basal conditions, but develop exaggerated hypertrophy after TAC [20].

Regarding DUB, upregulation of the expression of USP18 and UCHL1 has been reported in a mouse in vivo model of cardiac hypertrophy induced by TAC ([21], cited in [22]) as well as an in vitro model of cardiac myocyte hypertrophy induced by the stimulation with fibronectin [22]. In the latter model, transcripts of several USPs, including USP15, are upregulated besides USP18 and UCHL1 [22].

The expression of USP7 [also called herpes virus-associated ubiquitin-specific protease (HAUSP)] and USP20 (also called VDU2) was investigated in human dilated cardiomyopathy, and both were shown to be upregulated [23, 24]. Birks and colleagues investigated the activity of multiple DUBs in myocardial samples from patients with dilated cardiomyopathy and non-failing donors for transplantation by using a thiol-reactive ubiquitin-derived probe. They detected the activity of USP15 and several other DUBs in both samples [23].

5.3 USP15-Mediated Cardiac Remodeling

5.3.1 Interaction of USP15 and SLIM1

Expression of the transcripts of USP15 has been shown in skeletal muscle, heart, and a few other tissues [25, 26], and its enzymatic activity has been confirmed in human hearts as described above [23].

The biological roles of USP15 have been unknown until recently. By analysis of the COP9 signalosome complex, USP15 has been revealed to associate with the complex and stabilize the E3 Ub ligase, ring-box 1 (RBX1, also called ROC1) [27–29]. Several substrates of USP15 have been successively reported such as NF-κB inhibitor

IκBα [30], the tumor suppressor protein adenomatous polyposis coli (APC) [31], and procaspase-3 [32]. The entire scope of the biological roles of USP15, however, remains unclear, as a large number of previously unknown protein interactions with USP15 have been proteomically discovered [33], and their functions are not yet fully investigated.

We have identified a novel interaction of USP15 with skeletal muscle LIM protein 1 (SLIM1, also called FHL1, FHL1A, KyoT, or KyoT1) in an effort to find novel protein–protein interactions by computational prediction and its biochemical confirmation [34]. We have shown that the interaction is direct because USP15 and SLIM1 directly bound under cell-free conditions and co-immunoprecipitated from the lysates of the cells, where they were co-expressed. SLIM1 was detected in both polyubiquitinated and non-ubiquitinated forms when SLIM1 and Ub were co-expressed in mammalian cells in the presence of a proteasome inhibitor. Polyubiquitinated forms of SLIM1 were markedly reduced with the co-expression of USP15, but not the catalytically inactive mutant USP15. Furthermore, co-expression of USP15 increased the protein levels of SLIM1 depending on its deubiquitinating activity, suggesting an escape from the proteasomal degradation.

Thus, the protein levels of SLIM1 are regulated by the UPS, and USP15 interacts physically and functionally with SLIM1. SLIM1 is a member of the four-and-a-half LIM protein (FHL) family [35, 36], which is comprised of an N-terminal half LIM or single zinc finger followed by four LIM domains, and is highly expressed in skeletal muscle with an intermediate expression in the heart [36–38]. Accumulating evidence indicates a key role of SLIM1 in the pathogenesis of myopathies and cardiomyopathies. Gene expression analyses have shown the downregulation of SLIM1 in dilated cardiomyopathy [39] and upregulation in hypertrophic cardiomyopathy [40–42]. Additionally, in mouse models of hypertrophic and dilated cardiomyopathy, upregulation of cardiac ventricular expression of SLIM1, but not the related proteins in FHL family, SLIM2 or SLIM3, has been reported [38]. More recently, mutations in the SLIM1 gene have been identified in various types of human X-linked myopathies associated with or without cardiomyopathy (reviewed in [43, 44]). Furthermore, a critical role of SLIM1 as a mediator of hypertrophic biomechanical stress responses has been demonstrated by using SLIM1-deficient mice [45], and the involvement of SLIM1 in skeletal muscle mass control has been shown by using TG mice which overexpress SLIM1 in skeletal muscle [46].

5.3.2 *Hypertrophic Responses of the Heart in USP15-TG Mice*

The above results and information suggest that USP15 may have the potential to induce cardiac or skeletal muscle remodeling. To test this hypothesis, we generated TG mice with cardiac-specific overexpression of human USP15, which is more than 97% identical and almost 100% similar to murine USP15, under the control of α-myosin heavy chain (MHC) promoter (USP15-TG mice) [34].

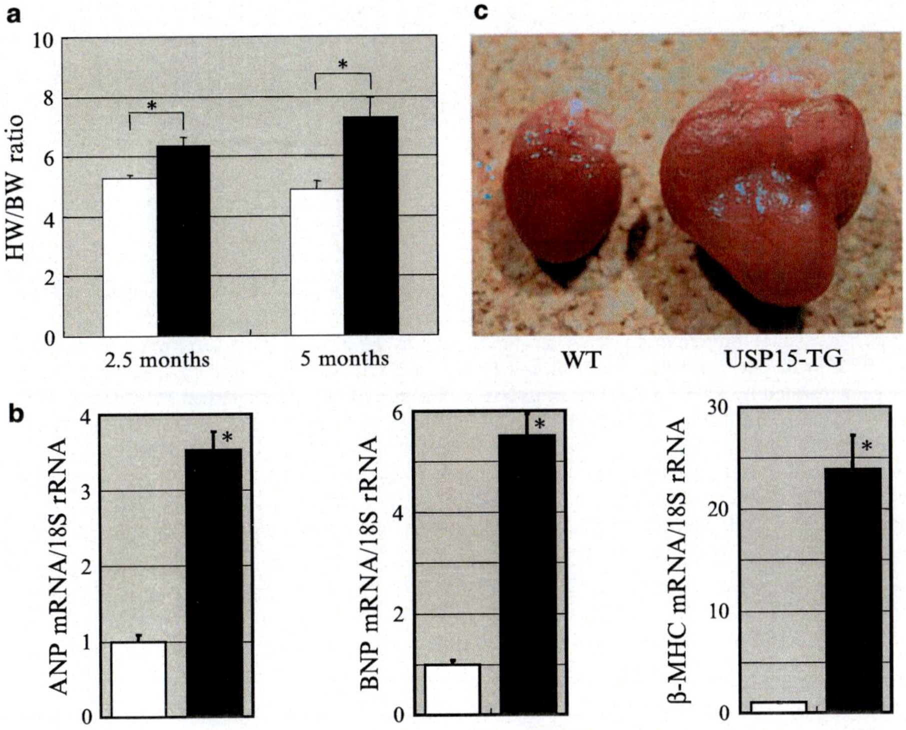

Fig. 5.1 Hypertrophic responses in the heart of USP15-TG mice (reproduced from [34] with permission). (**a**) HW/BW ratios (mg/g) in WT (*open bars*) and USP15-TG (*filled bars*) mice were determined at the indicated months after birth. Means ± SE are shown (n=6–11). * Significantly different between WT and USP15-TG mice ($p<0.05$). (**b**) Total RNA was isolated from the hearts of WT (*open bars*) and USP15-TG (*filled bars*) mice, and mRNA levels for ANP (*left*), BNP (*center*), and β-MHC (*right*) normalized to 18S rRNA were determined by real-time quantitative RT-PCR. mRNA levels in hearts of WT mice were arbitrarily assigned to 1, and fold differences were calculated. Means ± SE are shown (n=5). * Significantly different between WT and USP15-TG mice ($p<0.05$). (**c**) Hearts were removed from WT and USP15-TG mice at the age of 8 months and photographed

As a result, cardiac-specific overexpression of USP15 in TG mice induced cardiac remodeling with massive enlargement in heart size. As shown in Fig. 5.1a, b, heart weight to body weight (HW/BW) ratios and mRNA levels of fetal gene markers, such as atrial natriuretic peptide (ANP), B-type natriuretic peptide (BNP), and β-MHC, in the heart turned out to be significantly higher in USP15-TG mice compared with wild-type (WT) mice. These characteristics are features of cardiac hypertrophy, although we need histological examination to say clearly. The hypertrophic responses in the heart of USP15-TG mice depend on the deubiquitinating activity of USP15 because another line of TG mice, in which transfected human USP15 was spontaneously multi-mutated to lose deubiquitinating activity, showed neither an increase of HW/BW ratio nor an increase of these fetal gene markers.

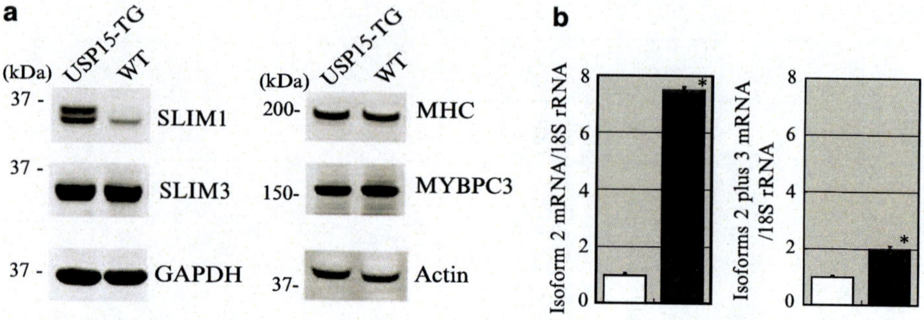

Fig. 5.2 SLIM1 protein and mRNA levels are high in the heat of USP15-TG mice (reproduced from [34] with permission). (**a**) Total proteins were extracted from the hearts of WT and USP15-TG mice, separated by SDS-PAGE, and analyzed by Western blotting with anti-SLIM1, anti-SLIM3, anti-GAPDH, anti-MHC, anti-MYBPC3, and anti-actin primary antibodies. (**b**) Total RNA was isolated from the hearts of WT (*open bars*) and USP15-TG (*filled bars*) mice and mRNA levels for SLIM1 isoform 2 (*left*) and SLIM1 isoforms 2 plus 3 (*right*) normalized to 18S rRNA were determined by real-time quantitative RT-PCR. mRNA levels in hearts of WT mice were arbitrarily assigned to 1, and fold differences were calculated. Means ± SE are shown ($n = 5$). (*Astritch*) denotes significantly different between WT and USP15-TG mice ($p < 0.05$)

USP15-TG mice showed no morphologic abnormalities in appearance; however, they started to die after the age of approximately 8 months. Figure 5.1c shows the morphology of the heart from a moribund USP15-TG mouse at the age of 8 months, exhibiting gross enlargement, which is more in atrium than in ventricle. Although the expression of human USP15 transgene, driven by α-MHC promoter, is supposed to be uniformly distributed over the heart according to the literature [47], the effect may not be homogenous throughout the heart (see below).

5.3.3 *Increase of SLIM1 in the Heart of USP15-TG Mice*

The cardiac remodeling in USP15-TG mice is accompanied with a marked increase of endogenous SLIM1, as shown in Fig. 5.2a [34]. However, the levels of SLIM3 (another member of the FHL family), MHC (mostly composed of α-MHC), myosin-binding protein C3 (MYBPC3, SLIM1-interacting sarcomeric protein [48]) are almost the same as those of normal hearts in WT mice. Immunoblot with anti-SLIM1 antibody in Fig. 5.2a also shows that one isoform of SLIM1 (resident type) is present in the heart extract of WT mice, while two isoforms of SLIM1 are in that of USP15-TG mice: the resident type and an extra one. On the basis of the molecular weight of these isoforms and the epitope of the anti-SLIM1 antibody, we determined the resident SLIM1 as isoform 3 in Entrez Gene and the extra one as SLIM1 isoform 2, which has extra 16 amino acids fused with the N terminus of isoform 3.

As we have shown that USP15 interacts physically and functionally with SLIM1 and stabilizes SLIM1 by deubiquitination in the transient expression system in mammalian cells, we suppose this mechanism works to increase endogenous SLIM1 in the heart in USP15-TG mice, although this is yet to be proven. From the results of real-time quantitative RT-PCR, however, the increase of isoform 2 is attributable, at least partly, to the transcriptional activation [34]. Figure 5.2b shows that the levels of mRNA encoding specifically the isoform 2 in the heart of USP15-TG mice are seven times those in the heart of WT mice, but only twice as much difference for the levels of mRNA encoding either of the two isoforms (isoforms 2 plus 3). It is of note that USP15 associates with some of the components of U5/U6-snRNP such as PRPF4 and SART3 and suggests the involvement of USP15 in mRNA splicing [33]. Overexpression of USP15 may influence the stoichiometric balance of these components or accompanying regulatory factors in the splicing machinery, resulting in the increased expression of the splicing variant isoform 2.

SLIM1 has been shown to play a crucial role in cardiac hypertrophy by sensing biomechanical stress responses as a component within the cardiomyocyte sarcomere complex and by initiating changes in the titin- and mitogen-activated protein kinase (MAPK)-mediated responses [45]. Moreover, skeletal muscle-specific overexpression of SLIM1 induced skeletal muscle hypertrophy in TG mice showed an increase in fiber size and oxidative slow fiber-type expression, which led to increased muscle strength and endurance [46]. The authors also showed that SLIM1 directly complexes with nuclear factor of activated T cells, cytoplasmic, calcineurin-dependent 1 (NFATC1) and enhances its transcriptional activity and that SLIM1 overexpression in a myoblastic cell line results in hypertrophic myotube formation, which is sensitive to cyclosporine A, the inhibitor of the upstream regulator calcineurin [46]. NFATC1 and four other members of this family are all present in the heart [49]. NFATC4, another member of this family, induces cardiac hypertrophy in TG mice, whereby constitutively nuclear mutant of NFATC4 is overexpressed [50]. Meanwhile, NFATC3, but not NFATC4, has been claimed to mediate calcineurin-mediated cardiac hypertrophic responses [49]. Any or multiple factors of this family are most likely involved in the hypertrophic responses in the remodeling heart of USP15-TG mice; however, further work is needed for confirmation.

Expression of SLIM1 is spatially controlled in adult normal hearts: higher in the atrium than in the ventricle in mice [51] and only detectable in the atrium but not the ventricle in rabbits [37]. Nevertheless, SLIM1 expression is upregulated in the TAC-induced hypertrophied ventricles in mice [38]. Viewed together, it may be speculated that the increased levels of SLIM1 are still skewed for the atrium over the ventricle in the heart of USP15-TG mice, which may influence the morphology change of the remodeling heart (see above).

5.3.4 Possible Signals to Cardiac Remodeling in USP15-TG Mice

Several substrates of USP15 have been reported: the NF-κB inhibitor IκBα [30] and the tumor suppressor protein APC [31] and procaspase-3 [32]. Deubiquitination and stabilization of IκBα result in the downregulation of NF-κB, which has ambivalent effects on cellular survival [13], and those of the latter two gene products are generally considered to lead to a cytostatic or apoptotic effect.

Quite recently, studies have shown that USP15 is involved in the signaling pathways of transforming growth factor (TGF) β superfamily at two different phases [52, 53]. Eichhorn et al. have reported using glioblastoma cells that USP15 forms a complex with small mother against decapentaplegic (SMAD) 7 and SMAD-specific E3 ubiquitin protein ligase 2 (SMURF2) and is recruited to the TGFβ receptor complex, where it deubiquitinates and stabilizes type I TGFβ receptor (TβRI), leading to an enhanced TGFβ signal [52]. Both TGFβ and its receptors are present in the heart and expressed in cardiac myocytes and nonmyocytes (reviewed in [54]). TGFβ has clearly been implicated in cardiomyocyte hypertrophy and may participate in the pathogenesis of human hypertrophic cardiomyopathy (reviewed in [55]). Noncanonical pathways via the activation of TGFβ-activated kinase TAK1 are implicated in TGFβ-mediated hypertrophic response, although more proof is required [55].

On the other hand, Inui and colleagues have shown that USP15 deubiquitinates a mono- or di-ubiquitinated form of SMAD1, 2, and 3 [53]. These SMADs, classified as receptor-regulated SMAD (R-SMAD), act as signal transducers and transcriptional modulators in the downstream of the TGFβ superfamily members such as TGFβ and bone morphogenetic proteins (BMP) (reviewed in [56]). Upon stimulation with TGFβ or BMP, R-SMAD is phosphorylated by the receptors and oligomerized with SMAD4, common partner SMAD (co-SMAD), and the complex translocates into the nucleus, acting as a transcriptional modulator (canonical pathways) [56]. Inui et al. have shown that monoubiquitination of R-SMADs targets the DNA-binding domains of R-SMADs and prevents promoter recognition and that R-SMAD deubiquitination by USP15 is required for full responsiveness to TGFβ or BMP [53]. These isoforms of SMADs are all present in the cardiovascular system (reviewed in [57]). Activation of SMAD2 and 3, downstream of TGFβ stimulation, is considered to be related to deleterious development of cardiac fibrosis and failure [57], and activated SMAD3 complexed with SMAD4 has been shown to induce apoptosis in cardiomyocytes [58]. Meanwhile, BMP-2 and BMP receptors are expressed in freshly cultured neonatal rat cardiac myocytes [59]. In the downstream of BMP stimuli, activated SMAD1 complexes with SMAD4 and promotes survival of cardiac myocytes [59].

Although unproven, it is conceivable that these SMAD-mediated signals as well as noncanonical signals are intensified by overexpressed USP15 and strongly affect the cardiac remodeling in USP15-TG mice. In the heart, expression of BMP-10 is restricted only in the right atrium in adult normal mice [60] although upregulated in the hypertrophied ventricle of hypertensive rats [61], and BMP-2, BMP-4, and BMP-5 are shown to be present in human aortic valves [62]. These features of spatially differentiated expression of BMP genes may also influence the morphology of the remodeling heart in USP15-TG mice (see above).

In addition, SLIM1 has been shown recently to interact with SMAD2, 3, and 4 [63]. Ding et al. have shown that upregulation of SLIM1 induces the phosphorylation of SMAD2 and 3, depending on casein kinase (CK) 1δ but not on TGFβ receptors, and increases the formation and nuclear accumulation of the oligomeric complex with SMAD4, resulting in transcriptional modulation of their target genes, e.g., activation of the cyclin-dependent kinase inhibitor *p21* and suppression of the oncogene c-myc [63]. Thus, USP15 may activate SMAD signaling not only directly but also indirectly through the increase of SLIM1 by closely linked interactions.

5.4 Conclusion

Cardiac-specific overexpression of USP15 increases SLIM1 in the heart and induces cardiac hypertrophic remodeling. SLIM1 plays an essential role in hypertrophic responses of cardiac myocytes probably through activation of NFATs. Moreover, USP15 interacts with TβRI and SMADs and reinforces their activities, which may possibly contribute to the cardiac hypertrophic responses in USP15-TG mice. In addition, the interaction of SLIM1 with SMADs may also possibly modify the cardiac remodeling in USP15-TG mice. Importantly, catalytic activity of USP15 has been detected in human hearts, and its transcription is upregulated in an in vitro model of cardiac myocyte hypertrophy. Although more research is needed, these lines of evidence implicate USP15 in the processes of cardiac remodeling and may provide a novel target for therapeutics in heart failure.

References

1. Ciechanover A, Iwai K (2004) The ubiquitin system: from basic mechanisms to the patient bed. IUBMB Life 56:193–201
2. Amerik AY, Hochstrasser M (2004) Mechanism and function of deubiquitinating enzymes. Biochim Biophys Acta 1695:189–207
3. Nijman SM, Luna-Vargas MP, Velds A et al (2005) A genomic and functional inventory of deubiquitinating enzymes. Cell 123:773–786
4. Komander D, Clague MJ, Urbe S (2009) Breaking the chains: structure and function of the deubiquitinases. Nat Rev Mol Cell Biol 10:551–563
5. Hein S, Arnon E, Kostin S et al (2003) Progression from compensated hypertrophy to failure in the pressure-overloaded human heart: structural deterioration and compensatory mechanisms. Circulation 107:984–991
6. Weekes J, Morrison K, Mullen A et al (2003) Hyperubiquitination of proteins in dilated cardiomyopathy. Proteomics 3:208–216
7. Tsukamoto O, Minamino T, Okada K et al (2006) Depression of proteasome activities during the progression of cardiac dysfunction in pressure-overloaded heart of mice. Biochem Biophys Res Commun 340:1125–1133
8. Depre C, Wang Q, Yan L et al (2006) Activation of the cardiac proteasome during pressure overload promotes ventricular hypertrophy. Circulation 114:1821–1828
9. Hedhli N, Wang L, Wang Q et al (2008) Proteasome activation during cardiac hypertrophy by the chaperone H11 Kinase/Hsp22. Cardiovasc Res 77:497–505
10. Hedhli N, Lizano P, Hong C et al (2008) Proteasome inhibition decreases cardiac remodeling after initiation of pressure overload. Am J Physiol Heart Circ Physiol 295:H1385–H1393

11. Meiners S, Hocher B, Weller A et al (2004) Downregulation of matrix metalloproteinases and collagens and suppression of cardiac fibrosis by inhibition of the proteasome. Hypertension 44:471–477
12. Meiners S, Dreger H, Fechner M et al (2008) Suppression of cardiomyocyte hypertrophy by inhibition of the ubiquitin-proteasome system. Hypertension 51:302–308
13. Zolk O, Schenke C, Sarikas A (2006) The ubiquitin-proteasome system: focus on the heart. Cardiovasc Res 70:410–421
14. Mearini G, Schlossarek S, Willis MS, Carrier L (2008) The ubiquitin–proteasome system in cardiac dysfunction. Biochim Biophys Acta 1782:749–763
15. Razeghi P, Baskin KK, Sharma S et al (2006) Atrophy, hypertrophy, and hypoxemia induce transcriptional regulators of the ubiquitin proteasome system in the rat heart. Biochem Biophys Res Commun 342:361–364
16. Li HH, Kedar V, Zhang C et al (2004) Atrogin-1/muscle atrophy F-box inhibits calcineurin-dependent cardiac hypertrophy by participating in an SCF ubiquitin ligase complex. J Clin Invest 114:1058–1071
17. Arya R, Kedar V, Hwang JR et al (2004) Muscle ring finger protein-1 inhibits PKC{epsilon} activation and prevents cardiomyocyte hypertrophy. J Cell Biol 167:1147–1159
18. Toth A, Nickson P, Qin LL, Erhardt P (2006) Differential regulation of cardiomyocyte survival and hypertrophy by MDM2, an E3 ubiquitin ligase. J Biol Chem 281:3679–3689
19. Li HH, Willis MS, Lockyer P et al (2007) Atrogin-1 inhibits Akt-dependent cardiac hypertrophy in mice via ubiquitin-dependent coactivation of Forkhead proteins. J Clin Invest 117:3211–3223
20. Willis MS, Ike C, Li L et al (2007) Muscle ring finger 1, but not muscle ring finger 2, regulates cardiac hypertrophy in vivo. Circ Res 100:456–459
21. Chen D, Izumo S, Kohane I et al (2002) Microarray-based analysis of gene expression during pressure-overload induced cardiac hypertrophy. In: 67th Cold Spring Harbor symposia on quantitative biology: the cardiovascular system, Cold Spring Harbor Laboratory Press, Cold Spring Harbor, 2002
22. Chen H, Huang XN, Stewart AF, Sepulveda JL (2004) Gene expression changes associated with fibronectin-induced cardiac myocyte hypertrophy. Physiol Genomics 18:273–283
23. Birks EJ, Latif N, Enesa K et al (2008) Elevated p53 expression is associated with dysregulation of the ubiquitin-proteasome system in dilated cardiomyopathy. Cardiovasc Res 79:472–480
24. Kääb S, Barth AS, Margerie D et al (2004) Global gene expression in human myocardium–oligonucleotide microarray analysis of regional diversity and transcriptional regulation in heart failure. J Mol Med 82:308–316
25. Nagase T, Ishikawa K, Miyajima N et al (1998) Prediction of the coding sequences of unidentified human genes. IX. The complete sequences of 100 new cDNA clones from brain which can code for large proteins in vitro. DNA Res 5:31–39
26. Angelats C, Wang XW, Jermiin LS et al (2003) Isolation and characterization of the mouse ubiquitin-specific protease Usp15. Mamm Genome 14:31–46
27. Zhou C, Wee S, Rhee E et al (2003) Fission yeast COP9/signalosome suppresses cullin activity through recruitment of the deubiquitylating enzyme Ubp12p. Mol Cell 11:927–938
28. Wee S, Geyer RK, Toda T, Wolf DA (2005) CSN facilitates Cullin-RING ubiquitin ligase function by counteracting autocatalytic adapter instability. Nat Cell Biol 7:387–391
29. Hetfeld BK, Helfrich A, Kapelari B et al (2005) The zinc finger of the CSN-associated deubiquitinating enzyme USP15 is essential to rescue the E3 ligase Rbx1. Curr Biol 15:1217–1221
30. Schweitzer K, Bozko PM, Dubiel W, Naumann M (2007) CSN controls NF-kappaB by deubiquitinylation of IkappaBalpha. EMBO J 26:1532–1541
31. Huang X, Langelotz C, Hetfeld-Pechoc BK et al (2009) The COP9 signalosome mediates beta-catenin degradation by deneddylation and blocks adenomatous polyposis coli destruction via USP15. J Mol Biol 391:691–702
32. Xu M, Takanashi M, Oikawa K et al (2009) USP15 plays an essential role for caspase-3 activation during Paclitaxel-induced apoptosis. Biochem Biophys Res Commun 388:366–371

33. Sowa ME, Bennett EJ, Gygi SP, Harper JW (2009) Defining the human deubiquitinating enzyme interaction landscape. Cell 138:389–403
34. Isumi Y, Hirata T, Saitoh H et al (2011) Transgenic overexpression of USP15 in the heart induces cardiac remodeling in mice. Biochem Biophys Res Commun 405:216–221
35. Morgan MJ, Madgwick AJ (1996) Slim defines a novel family of LIM-proteins expressed in skeletal muscle. Biochem Biophys Res Commun 225:632–638
36. Lee SM, Tsui SK, Chan KK et al (1998) Chromosomal mapping, tissue distribution and cDNA sequence of four-and-a-half LIM domain protein 1 (FHL1). Gene 216:163–170
37. Brown S, Biben C, Ooms LM et al (1999) The cardiac expression of striated muscle LIM protein 1 (SLIM1) is restricted to the outflow tract of the developing heart. J Mol Cell Cardiol 31:837–843
38. Chu PH, Ruiz-Lozano P, Zhou Q et al (2000) Expression patterns of FHL/SLIM family members suggest important functional roles in skeletal muscle and cardiovascular system. Mech Dev 95:259–265
39. Yang J, Moravec CS, Sussman MA et al (2000) Decreased SLIM1 expression and increased gelsolin expression in failing human hearts measured by high-density oligonucleotide arrays. Circulation 102:3046–3052
40. Hwang DM, Dempsey AA, Lee CY, Liew CC (2000) Identification of differentially expressed genes in cardiac hypertrophy by analysis of expressed sequence tags. Genomics 66:1–14
41. Hwang DM, Dempsey AA, Wang RX et al (1997) A genome-based resource for molecular cardiovascular medicine: toward a compendium of cardiovascular genes. Circulation 96:4146–4203
42. Lim DS, Roberts R, Marian AJ (2001) Expression profiling of cardiac genes in human hypertrophic cardiomyopathy: insight into the pathogenesis of phenotypes. J Am Coll Cardiol 38: 1175–1180
43. Shathasivam T, Kislinger T, Gramolini AO (2010) Genes, proteins and complexes: The multifaceted nature of FHL family proteins in diverse tissues. J Cell Mol Med 14:2702–2720
44. Cowling BS, Cottle DL, Wilding BR et al (2011) Four and a half LIM protein 1 gene mutations cause four distinct human myopathies: a comprehensive review of the clinical, histological and pathological features. Neuromuscul Disord 21:237–251
45. Sheikh F, Raskin A, Chu PH et al (2008) An FHL1-containing complex within the cardiomyocyte sarcomere mediates hypertrophic biomechanical stress responses in mice. J Clin Invest 118:3870–3880
46. Cowling BS, McGrath MJ, Nguyen MA et al (2008) Identification of FHL1 as a regulator of skeletal muscle mass: implications for human myopathy. J Cell Biol 183:1033–1048
47. Subramaniam A, Jones WK, Gulick J et al (1991) Tissue-specific regulation of the alpha-myosin heavy chain gene promoter in transgenic mice. J Biol Chem 266:24613–24620
48. McGrath MJ, Cottle DL, Nguyen MA et al (2006) Four and a half LIM protein 1 binds myosin-binding protein c and regulates myosin filament formation and sarcomere assembly. J Biol Chem 281:7666–7683
49. Wilkins BJ, De Windt LJ, Bueno OF et al (2002) Targeted disruption of NFATc3, but not NFATc4, reveals an intrinsic defect in calcineurin-mediated cardiac hypertrophic growth. Mol Cell Biol 22:7603–7613
50. Molkentin JD, Lu JR, Antos CL et al (1998) A calcineurin-dependent transcriptional pathway for cardiac hypertrophy. Cell 93:215–228
51. Tabibiazar R, Wagner RA, Liao A, Quertermous T (2003) Transcriptional profiling of the heart reveals chamber-specific gene expression patterns. Circ Res 93:1193–1201
52. Eichhorn PJ, Rodón L, Gonzàlez-Juncà A et al (2012) USP15 stabilizes TGF-β receptor I and promotes oncogenesis through the activation of TGF-β signaling in glioblastoma. Nat Med 18:429–435
53. Inui M, Manfrin A, Mamidi A et al (2011) USP15 is a deubiquitylating enzyme for receptor-activated SMADs. Nat Cell Biol 13:1368–1375
54. Brand T, Schneider MD (1995) The TGF beta superfamily in myocardium: ligands, receptors, transduction, and function. J Mol Cell Cardiol 27:5–18

55. Dobaczewski M, Chen W, Frangogiannis NG (2011) Transforming growth factor (TGF)-β signaling in cardiac remodeling. J Mol Cell Cardiol 51:600–606
56. Miyazono K, ten Dijke P, Heldin CH (2000) TGF-beta signaling by smad proteins. Adv Immunol 75:115–157
57. Euler Taimor G, Heger J (2006) The complex pattern of SMAD signaling in the cardiovascular system. Cardiovasc Res 69:15–25
58. Schneiders D, Heger J, Best P et al (2005) SMAD proteins are involved in apoptosis induction in ventricular cardiomyocytes. Cardiovasc Res 67:87–96
59. Izumi M, Fujio Y, Kunisada K et al (2001) Bone morphogenetic protein-2 inhibits serum deprivation-induced apoptosis of neonatal cardiac myocytes through activation of the Smad1 pathway. J Biol Chem 276:31133–31141
60. Chen H, Shi S, Acosta L et al (2004) BMP10 is essential for maintaining cardiac growth during murine cardiogenesis. Development 131:2219–2231
61. Nakano N, Hori H, Abe M et al (2007) Interaction of BMP10 with Tcap may modulate the course of hypertensive cardiac hypertrophy. Am J Physiol Heart Circ Physiol 293: H3396–H3403
62. Ankeny RF, Thourani VH, Weiss D et al (2011) Preferential activation of SMAD1/5/8 on the fibrosa endothelium in calcified human aortic valves - association with low BMP antagonists and SMAD6. PLoS One 6:e20969
63. Ding L, Wang Z, Yan J et al (2009) Human four-and-a-half LIM family members suppress tumor cell growth through a TGF-beta-like signaling pathway. J Clin Invest 119:349–361

Chapter 6
Role of Galectin-3 Pathways in the Pathogenesis of Cardiac Remodeling and Heart Failure

Lili Yu and Rudolf A. de Boer

Abstract Myocardial injuries stemming from pressure overload or myocardial infarction lead to cardiac remodeling and represent major health problems worldwide. An ever accumulating body of experimental and clinical research appoints galectin-3, a β-galactoside-binding lectin, as a key player in this maladaptive response to myocardial injury. Herein, a specific role for galectin-3 in inflammation and fibrogenesis has been elucidated in experimental and clinical studies. Galectin-3 was first associated with pathological conditions leading to cardiac remodeling, such as inflammation and fibrosis. Then, as the carbohydrate recognition domain of galectin-3 reacts with glycosylated proteins such as laminin, fibronectin, and tenascin, a multifunctional role of galectin-3 in the extracellular matrix was postulated. Notably, experimental animal studies clearly showed that galectin-3 is a mediator of crucial steps in fibrogenesis and further induces cardiac inflammation, hypertrophy, and dysfunction. Possible mechanisms pertaining to galectin-3 inflammatory and fibrotic properties have been suggested to involve macrophage activation, galectin-3-induced chemotaxis, and activation of the TGF-β–Smad3 signaling pathways. Additionally, the link between plasma galectin-3 and fibrosis was also established in clinical biomarker studies. Galectin-3 and its pathways may be explored further in order to develop more efficient strategies to target cardiac remodeling in heart failure leading to fibrosis.

Keywords Galectin-3 • Heart failure • Remodeling • Biomarker

L. Yu • R.A. de Boer, M.D., Ph.D., F.E.S.C. (✉)
Department of Cardiology, University Medical Center Groningen,
University of Groningen, PO Box 30.001, Groningen 9700, RB, The Netherlands
e-mail: r.a.de.boer@umcg.nl

B.I. Jugdutt and N.S. Dhalla (eds.), *Cardiac Remodeling: Molecular Mechanisms*,
Advances in Biochemistry in Health and Disease 5, DOI 10.1007/978-1-4614-5930-9_6,

6.1 Structure, Expression, and Function of Galectin-3

Galectin-3 is a 29–35 kDa chimera-type galectin. Galectins form a large family of galactosidase-binding lectins, but galectin-3 is the only member of the galectin family with an extended N-terminal domain (110–130 amino acids). This N-terminal domain is composed of tandem repeat sequences comprising nine amino acid residues and is connected to a C-terminal carbohydrate recognition domain (CRD) of about 130 amino acids [1, 2]. The CRD interacts with various glycosylated proteins and modulates cell–cell adhesion and cell signaling in the extracellular compartment [1]. Further, the galectin-3 CRD and collagen-like domains communicate with a variety of extracellular matrix proteins (ECM), carbohydrates (e.g., *N*-acetyllactosamine), and also unglycosylated proteins such as cell surface receptors (macrophage antigens CD11b/CD18) and extracellular receptors (collagen IV) [3–6], and they furthermore modulate cell–cell adhesion and cell signaling in the extracellular compartment.

Galectins are expressed in various cells and tissues and are important for diverse physiological and pathological processes, such as immune and inflammatory responses, tumor development and progression, neural degeneration, atherosclerosis, diabetes, as well as wound repair. Galectin-3 has been detected in many proliferative cells including tumor cells, eosinophils, neutrophils, and activated macrophages [7, 8]. Notably, many of these cell types are operative in the inflammatory response and the formation of fibrosis (fibrogenesis).

Galectin-3 is predominantly located in the cytoplasm. While intracellular galectin-3 plays a pivotal role in diverse cell growth, anti-apoptosis signaling, and mRNA splicing, extracellular (or cell surface bound) galectin-3, on the other hand, participates in cell–cell and cell–matrix adhesion, by binding to glycosylated ECM components, including laminin, fibronectin, tenascin, and Mac-2-binding protein [3, 4, 9–12]. Thus, extracellular galectin-3 appears in tight communication with factors involved in fibrogenesis.

Differential expression of galectin-3 has been reported for different murine organs. Herein, lower expression was found in cerebrum, heart, and pancreas, while moderate expression was found in liver, ileum, kidney, and adrenal gland and high expression in lung, spleen, stomach, colon, uterus, and ovary [13]. Figure 6.1 displays an overview of galectin-3 distribution in various healthy tissues of different species—our data are in concert with results published by Kim and colleagues [13]. Notably, a growing body of evidence reveals that high expression levels of galectin-3 are closely associated with pathological conditions, specifically conditions pertaining to inflammation and fibrosis which are also key conditions in cardiac remodeling.

6.1.1 Localization of Galectin-3 in Damaged Tissue

Inflammation and fibrosis are intertwined pathological states, and galectin-3 has consistently been observed to be involved in these damaged states.

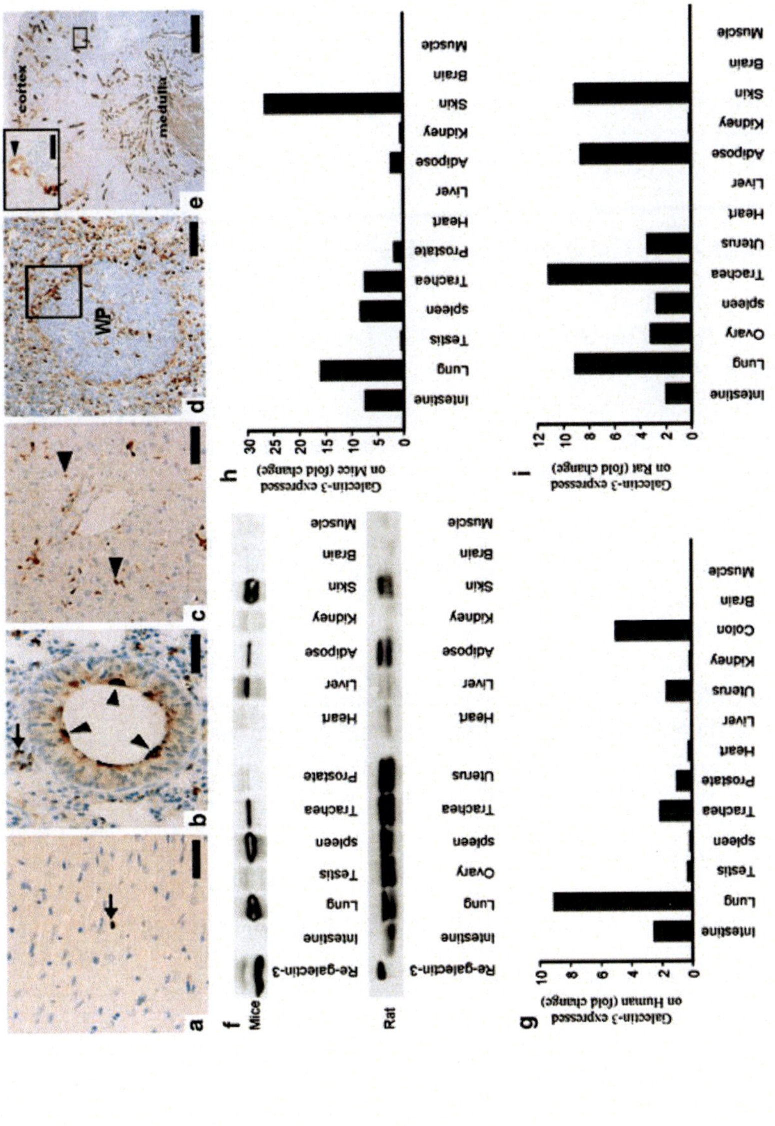

Fig. 6.1 Immunohistochemical localization and protein and mRNA expression of galectin-3 in various tissues of different species. Galectin-3 immunoreactivity has been observed in heart (**a**), lung (**b**), liver (**c**), spleen (**d**), and kidney (**e**). (**f–i**): Galectin-3 protein expression (assessed by Western blot) and mRNA expression (assessed by qPCR) have been predominantly observed in (mouse, rat, human) lung, testis/ovary, spleen, prostate, adipose tissue, and skin. Picture (**a–e**) reprinted from [13] with permission from the publisher; figure (**f–i**): unpublished data (Yu et al.). Scale bars: 50 μm (**a** and **b**); 60 μm (**c**); 90 mm (**d**); 400 μm (**e**)

Galectin-3 is not only secreted by activated macrophages, but in situ hybridization and immunohistochemistry analysis showed that galectin-3 is highly distributed in the fibrotic area of myocardium and co-localizes to macrophages [14]. Moreover, an ever growing body of experimental evidence found that galectin-3 manifests in various damaged tissue, in particular in tissues with increased collagen deposition and localization in the fibrotic area [14–23]. Further, galectin-3 was expressed in proliferating fibroblast and was also found in nucleus when cells were exposed to apoptotic stimuli in vitro [24]. Moreover, galectin-3 expression has been detected in isolated cardiac fibroblast to localize galectin-3-binding sites. Interestingly, contrary to cardiac fibroblast, galectin-3-binding sites were absent from cardiomyocytes [7, 8, 14, 25–27].

6.2 Galectin-3 in Experimental Fibrosis

6.2.1 Early Experimental Support in Liver, Kidney, and Lung Fibrosis

A prominent role for galectin-3 in fibrogenesis has been elucidated in recent years. Herein, work has specifically described fibrotic states in the liver, kidneys, pancreas, and lungs. For example, galectin-3 showed to be temporarily and spatially associated with fibrosis with a minimal expression in healthy rat liver, while highest expression was found at peak fibrosis, and virtually no galectin-3 expression was present after recovery from fibrosis [19]. Then, galectin-3 deficiency in bile duct-ligated (BDL) rats was recently shown to significantly diminish BDL-induced hepatic TGF-β1 and procollagen expression and associated hepatic fibrosis [28]. Additionally, in a murine model of cirrhosis, bone marrow cell transplantation significantly decreased liver fibrosis which was associated with decreased hepatic galectin-3 expression [29]. A recent report by Mackinnon et al. [23] extended the involvement of galectin-3 to fibrosis in the lung. The authors reported a significantly reduced TGF-β1- and bleomycin-induced lung fibrosis in galectin-3-deficient mice.

Likewise, galectin-3 expression and secretion by macrophages has been identified as a major contributor to renal fibrosis. Mice with galectin-3 deficiency did not show macrophage recruitment upon interferon gamma/LPS stimulation [19]. Henderson et al. also observed that galectin-3 mediated TGF-β-induced myofibroblast activation, a crucial step in the fibrogenesis cascade [17]. Lastly, the therapeutic potential of targeting galectin-3 to relieve fibrotic burden has recently been investigated by Kolatsi-Joannou and colleagues [30] who demonstrated that folic acid induced kidney fibrosis and associated galectin-3 expression was significantly reduced by 1% treatment with modified citrus pectin (MCP). MCP is a pectin derivative binding to the galectin-3 CRD, thereby inhibiting galectin-3 aforementioned effects.

Altogether, a growing body of literature consistently supports a role of galectin-3 in fibrogenesis. This appears a generalized effect, not confined to one organ.

Fibrosis is also central to the maladaptive response to myocardial injury, such as pressure overload and myocardial infarction leading to cardiac remodeling [14, 21, 22]. Supporting the notion that galectin-3 constitutes a global player in fibrogenesis, it is not surprising that galectin-3 has also been shown to be involved in myocardial fibrogenesis.

6.2.2 Galectin-3 in Cardiac Remodeling Pathway: Recent Experimental Support

The first evidence showing an involvement of galectin-3 in heart failure stems from a landmark study of Sharma and colleagues [14]. In a comprehensive microarray study, galectin-3 was identified as the most robustly overexpressed gene in heart failure-prone hypertrophied hearts compared to functionally compensated hearts in homozygous transgenic TGFmRen2-27(Ren-2) rats. Ren-2 rats overexpress the mouse Ren-2d renin gene and spontaneously develop heart failure after 13–15 weeks. Further, continuous low-dose infusion of recombinant galectin-3 into the pericardial sac caused left ventricular dysfunction in healthy Sprague–Dawley rats, associated with collagen deposition and other signs of cardiac remodeling [14]. These initial observations warranted galectin-3 to be considered as a new target for intervention in heart failure.

Then, more recently, Sharma et al. [21] found a significant and high expression of galectin-3 in cardiac tissue of Ang-II-induced hypertension in mice. In this model, galectin-3 was released by infiltrating macrophages in the myocardium and led to interstitial collagen deposition. Treatment with *N*-Acetyl-Ser-Asp-Lys-Pro (Ac-SDKP), an endogenous tetrapeptide specifically degraded by angiotensin-converting enzyme (ACE), reduced macrophage activation and galectin-3 expression herein and prevented galectin-3-induced cardiac inflammation, fibrosis, and remodeling.

Additionally, Liu et al. showed that in galectin-3-induced cardiac remodeling, galectin-3 increased the number of macrophages and mast cell infiltration and activated the TGF-β/Smad3 pathway. Ac-SDKP partially or near completely prevented galectin-3-induced cardiac inflammation, fibrosis, hypertrophy, and dysfunction, possibly by inhibition of the TGF-β/Smad3 signaling pathway [22].

Recent studies by Thandavarayan et al. [31], by Kamal et al. [32], and by Psarras et al. [33] further reveal a specific association of galectin-3 with left ventricular dysfunction and fibrosis. Specifically, Thandavarayan et al. [31] found a significant increase in myocardial hypertrophy and fibrosis, as well as apoptosis, all associated with upregulated galectin-3 in 14-3-3η protein mutant (DN-14-3-3η) mice after induction of diabetes. 14-3-3η protein has a regulatory role in apoptosis, adhesion, cellular proliferation, differentiation, survival, and signal transduction pathways [34]. As the authors stated, upregulated galectin-3 appears to be a general phenomenon in

LV dysfunction [31]. Moreover, Kamal et al. [32] demonstrated that cardiac hypertrophy, progressive fibrotic cardiac remodeling, and increased collagen deposition were accompanied with significantly increased galectin-3 expression in cardiomyopathic hearts in a rat model of myosin-induced experimental autoimmune myocarditis (EAM). Notably, herein galectin-3 overexpression was dramatically reduced by treatment with T-3999, a novel phenylpyridazinone, indicating an inhibitory function of T-3999 on galectin-3.

Then, Psarras et al. [33] found that desmin-deficient (des–/–) mice exhibit marked myocardial degeneration and fibrosis, which were associated with osteopontin (OPN) and galectin-3 overexpression (226× for OPN and 26× for galectin-3). OPN, like galectin-3, has chemotactic properties and is thus recruited to inflammatory sites [35, 36]. Psarras et al. [33] further compared des–/– OPN –/– mice with des–/– OPN +/+ mice and found that des–/– OPN –/– mice not only displayed remarkable improvements in ventricular function (53%) but also in myocardial fibrosis (44%) while also significantly reducing galectin-3 gene expression (by ≈80%) compared to des–/– OPN +/+ mice, indicating that the observed diminished inflammatory and fibrotic response in OPN-deficient des–/– mice could be partly explained by the significant reduced myocardial galectin-3 level [33].

So, accumulating experimental evidence implicates a role of galectin-3 in the development of organ fibrosis. Whether galectin-3 is a potential therapeutic target in left ventricular (LV) remodeling and heart failure is unknown. We have conducted experimental studies, and our results suggest that galectin-3 may be a target for therapy. Genetic disruption and pharmacological inhibition of galectin-3 attenuated adverse LV remodeling, fibrosis, and subsequent HF development. We perturbed mice with angiotensin II (Ang-II) and transverse aortic constriction (TAC), causing LV hypertrophy, decreased LV contractility, and increased LV end-diastolic pressure, associated with increased fibrosis in wild-type (WT) mice. However, galectin-3 knockout (Gal3-KO) mice did not develop LV dysfunction and fibrosis. Additionally, in homozygous TGR(mREN)27 rats, pharmacological inhibition of galectin-3 with an oligosaccharide almost completely prevented LV dysfunction and fibrosis [54, 55].

This indeed suggests that drugs binding to galectin-3 may be potential therapeutic candidates for the prevention of heart failure with extensive fibrosis. It remains unclear what mechanisms underpin these effects.

6.3 Role of Galectin-3 in Modulation of Fibrosis

6.3.1 Potential Mechanism of Galectin-3 in Extracellular Matrix

Fibroblasts, myofibroblasts, and macrophages have been identified as important cells in the initiation and progression of fibrogenesis, scar formation, and tissue remodeling [37–39]. Extracellular matrix remodeling (ECM) is a crucial aspect in fibrogenesis, and galectin-3 seems to play a multifunctional role in the ECM environment, as its CRD reacts with glycosylated proteins in the ECM, such as laminin, fibronectin, tenascin [3, 4], as well as membrane proteins, such as αM/β2 (CD11b/18) [40].

6.3.2 *Potential Mechanisms of Galectin-3 in Myofibroblast Differentiation*

Fibroblast to myofibroblast differentiation and activation by inflammatory cytokines, like TGF-β, preceded by influx of cells such as macrophages are some of the initial steps in the process of fibrogenesis. A large body of research supports a role for galectin-3 in this process. Herein, it appears that macrophages and TGF-β induce myofibroblast activation via galectin-3, but that macrophage recruitment and TGF-β expression are independent of galectin-3.

First, galectin-3 was visualized in the fibrotic area co-localizing with fibroblasts and macrophages [14]. Second, it was shown that infusion of recombinant galectin-3 into pericardial sac leads to inflammatory cell infiltration, cardiac fibroblast proliferation, collagen synthesis, and deposition, essentially contributing to interstitial and perivascular fibrosis [14, 21, 22]. Then, Dvorankova et al. demonstrated myofibroblast activation in vitro upon treatment with a moderately high-dose recombinant galectin-3 [41]. Further, galectin-3 deficiency significantly reduced myofibroblast activation in carbon tetrachloride (CCL4)-induced hepatic fibrosis and renal fibrosis (in a model of unilateral ureter obstruction, UUO) [17, 19].

Then, macrophages regulate fibroblast and myofibroblast activation in ECM, and macrophage-derived galectin-3 presents in various fibrotic pathologies. Injured tissue displays a marked increase in galectin-3 expression by activated macrophages and also an increased TGF-β expression; all these factors promote fibroblast proliferation and myofibroblast activation [17]. Macrophage depletion, then, significantly inhibits myofibroblast activation and decreases fibrosis [19]. For example, Henderson and colleagues showed that galectin-3-deficient macrophage recruitment could not drive myofibroblast accumulation and activation. Utilizing a crossover experiment with wild-type or galectin-3-deficient macrophage supernatant and galectin-3-deficient renal fibroblast, these authors further observed that proliferation of galectin-3-deficient renal fibroblasts was activated by wild-type macrophages and attenuated by a galectin-3 inhibitor bis-(3-deoxy-3-{3-methoxybenzamido}-β-D-galactopyranosyl-sulfane), while galectin-3-deficient macrophages did not induce proliferation in galectin-3-deficient renal fibroblasts.

On the other hand, galectin-3 deficiency markedly reduces activated myofibroblast but does not affect macrophage recruitment or pro-inflammatory cytokine profiles in injured tissue, such as IL-6 and TNF-α. Additionally, while galectin-3 deficiency led to reduced collagen deposition and reduced myofibroblast activation, TGF-β expression, or Smad2/3 phosphorylation was not influenced [19].

Furthermore, our unpublished galectin-3 chemotaxis assay results show that recombinant galectin-3 significantly induces monocyte migration, which could be markedly attenuated after treatment with galectin-3 inhibitors including modified citrus pectin (MCP) and lactose. These inhibitors act as a ligand, binding to galectin-3's CRD.

Altogether, these events clearly indicate galectin-3 to be a key player in the signal axis of fibrosis generation, specifically inducing macrophage and TGF-β-induced myofibroblast activation [17, 19, 23, 42].

6.4 Clinical Utility of Galectin-3

Clinical trials have consistently indicated a potential clinical utility of galectin-3 as a biomarker for prognosticating heart failure. Herein, van Kimmenade et al. were the first to evaluate the prognostic and predictive value of galectin-3 as a biomarker in acute heart failure [43], in the pro-BNP investigation of dyspnea in the emergency department (PRIDE). While N-terminal pro-brain natriuretic peptide (NT-pro-BNP) was a superior predictor for diagnosis of acute heart failure compared to galectin-3 and apelin (herein galectin-3 was a better predictor than apelin) (Fig. 6.2a), galectin-3 was the superior predictor compared to NT-pro-BNP and apelin for prognosis in acute heart failure. Multivariate logistic regression analysis revealed that elevated plasma levels of galectin-3 were indeed the most powerful predictor for death or the combination of death and recurrent heart failure within 60 days.

Plasma galectin-3 levels were related to detailed echocardiographic examinations in a substudy ($N=115$) of the PRIDE [44]. Galectin-3 levels were not strongly related to markers of LV structure or systolic function, but related to measures of RV function and diastolic dysfunction, and highest galectin-3 concentrations were strongly associated with a higher risk of 4-year mortality, which is independent from LV dimensions, function, and RV pressure, supporting the role galectin-3 may play in fibrosis and progressive cardiac failure.

Further and specifically relating galectin-3 to fibrosis, plasma galectin-3 levels were significantly correlated with several serum markers of cardiac ECM turnover, such as PIIINP, MMP-2, and TIMP-1, in 106 patients with chronic heart failure (New York Heart Association class II–III; mean LV ejection fraction [LVEF], 35+9%) [45]. Subsequently, Milting et al. described the kinetics of galectin-3 in 55 end-stage heart failure patients with the need for mechanical circulatory support (MCS). Notably, this study found that fibrosis-related biomarkers, such as tissue inhibitor of metalloproteinase (TIMP), tenascin C (TNC), OPN, BNP, and galectin-3, were all increased in patients with terminal heart failure. Interestingly, MCS only reduced the loading-related biomarker BNP, but none of the other fibrosis-related biomarkers. Additionally, patients who did not survive on MCS had higher baseline galectin-3 levels when compared with patients who lived until transplantation [45].

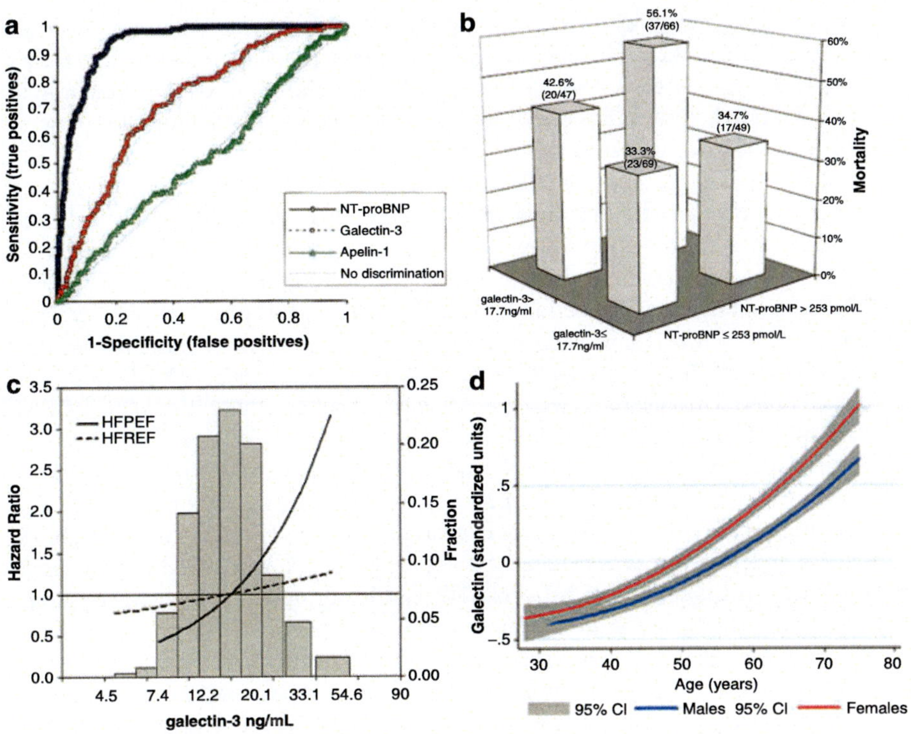

Fig. 6.2 Galectin-3 as a biomarker; data from several clinical trials. (**a**) Combined receiver-operating characteristic (ROC) curves for amino-terminal pro-brain natriuretic peptide (NT-pro-BNP), galectin-3, and apelin for the diagnosis of heart failure in dyspneic patients. The ROC analysis for NT-pro-BNP showed an area *under* the *curve* (AUC) for NT-pro-BNP of 0.94 ($p=0.0001$). The ROC analysis for galectin-3 showed an AUC of 0.72 ($p=0.0001$). The AUC for apelin for diagnosis of acute heart failure was 0.52 ($p=0.23$). This figure was reprinted from van Kimmenade et al. [43], with permission. (**b**) Mortality as a function of baseline galectin-3 and NT-pro-BNP categories. The median value of NT-pro-BNP (253 pmol/L) was used to define two levels of NT-pro-BNP concentration. Of the 232 subjects, 231 had both galectin-3 and NT-pro-BNP measurements. The number of patients in each category is as follows: high galectin-3 and high NT-pro-BNP ($n=66$), low galectin-3 and low NT-pro-BNP ($n=69$), low galectin-3 and high NT-pro-BNP ($n=49$), and high galectin-3 and low NT-pro-BNP ($n=47$). Reprinted from [46] by Lok et al., with permission. (**c**) Graphical depiction of the risk estimates for experiencing the primary outcome in patients with HFPEF and HFREF with increasing levels of plasma galectin-3. The distribution of (log-transformed) galectin-3 is depicted in the background in *brown bars*. A similar increase in galectin-3 causes a much more pronounced increase in risk in patients with HFPEF compared to patients with HFREF. Figure reprinted from [47] by de Boer et al. with permission. (**d**) Graph showing galectin-3 levels in male (*blue line*) and female subjects (*red line*) from the general population. *Gray-shaded areas* indicate 95% confidence intervals. Galectin-3 levels increase with increasing age, particularly in female subjects. This figure is reprinted from [49] by de Boer et al. with permission

Then, a larger study by Lok et al. comprising 232 patients with *chronic* heart failure (New York Heart Association function class III or IV) demonstrated that patients with high baseline levels of both galectin-3 and NT-pro-BNP had around 1.5- to 2-fold higher mortality rate (Fig. 6.2b) [46]. This study demonstrated incremental value of galectin-3 over NT-pro-BNP alone.

Additionally, a large study of 592 patients with heart failure (Coordinating Study Evaluating Outcomes of Advising and Counseling in Heart Failure, COACH trial, [47]), with mean follow-up of 18 months supported prognostic value of galectin-3 to predict re-hospitalization and death after correction for age, gender, BNP, eGFR, and diabetes, but not after correction for LVEF. A subanalysis revealed that increased plasma galectin-3 levels represents a stronger incremental risk in patients with preserved LVEF (HFPEF) compared to the patients with reduced LVEF (HFREF) ($P<0.001$) even when absolute galectin-3 levels did not differ between patients with HFPEF and HFREF [47] (Fig. 6.2c).

Furthermore, in the HF-ACTION study where plasma galectin-3 levels were assessed in 895 subjects with heart failure from a randomized, controlled trial of exercise training in patients with chronic heart failure with NYHA class II, III, or IV symptoms, galectin-3 was associated with NYHA class, lower systolic blood pressure, higher creatinine, higher NT-pro-BNP, and lower maximal oxygen consumption. However, this association diminished after adjustment for NT-pro-BNP [48].

Finally, recent data show that small increases in galectin-3 may confer increased CV risk in the general population, in subjects at risk for heart failure development. Briefly, 7,968 subjects were included in this study from the Prevention of Renal and Vascular End-Stage Disease (PREVEND) cohort (mean age of 50 ± 13 years, median follow-up of ~10 years). Plasma galectin-3 levels correlated very strong with age and sex (Fig. 6.2d) and weakly with a wide range of risk factors of CV disease, including blood pressure, serum lipids, body mass index, renal function, and NT-pro-BNP. After correction for classical CV risk factors (smoking, blood pressure, cholesterol, and diabetes), increased plasma galectin-3 levels independently predicted all-cause mortality in a large community-based cohort [49].

Altogether, these available clinical studies have so far confirmed that plasma galectin-3 levels were significantly upregulated in acutely decompensated heart failure [43, 44, 50, 51], chronic heart failure [46–49], and end-stage heart failure with the need for mechanical circulatory support (MCS) [52]. Furthermore, clinical results from our group further demonstrated the predictive and prognostic value of galectin-3 [46, 47, 49]. A relationship between galectin-3 and cardiovascular (CV) risk factors was investigated in the general population of PREVEND study, and a strong gender-specific interaction was revealed in the correlation between galectin-3 and cardiovascular risk factors [49]. Notably, the established link between plasma galectin-3 and fibrosis was also established in clinical biomarker studies [52] and needs to be explored further in order to develop more efficient strategies to target cardiac remodeling in heart failure leading to fibrosis.

6.5 Conclusion and Mapping of Galectin-3 Pathways

In conclusion, galectin-3 is highly expressed in the fibrotic area of the failing or stressed heart [14, 17, 19, 21, 22], and cardiac fibroblasts and macrophages are the main sources of galectin-3. Further, galectin-3 was shown to activate the TGF-β/Smad3 pathway [22], while macrophages and the inflammatory factor TGF-β demonstrated to elevate galectin-3 expression. Then, galectin-3 inhibition or deficiency was found to not affect macrophage activation and TGF-β expression levels. Therefore, galectin-3 may be considered as an independent participant in macrophage and TGF-β/Smad modulation pathway.

Other experimental animal studies reported that galectin-3 was not only significantly associated with myofibroblast-induced collagen synthesis and deposition but was also markedly correlated with ECM fibrosis markers, such as α-SMA, COL1A1, COL3A1, TIMP, and MMP [53] (Fig. 6.3).

Lastly, relevant clinical studies identified that plasma galectin-3 is significantly correlated with serum extracellular fibrosis turnover biomarkers, like PINP, PIIINP, TIMP, and MMP. Collectively, galectin-3 may be suggested as "culprit

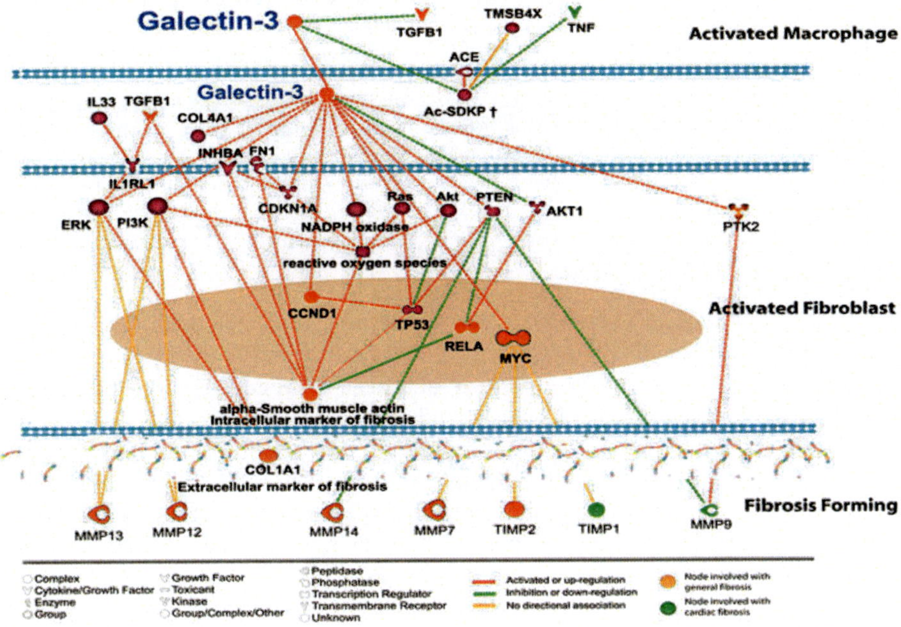

Fig. 6.3 Galectin-3 pathways. The network represents molecular relationships between different gene products. *Node shapes* indicate the functional class of the gene product, whereas *node colors* indicate a role in general fibrosis (*orange*) or cardiac fibrosis (*green*). Edge colors indicate upregulation or activation (*red*), downregulation or inhibition (*green*), or involvement without clear directionality (*yellow*). This figure is reprinted from [53] by de Boer et al. with permission

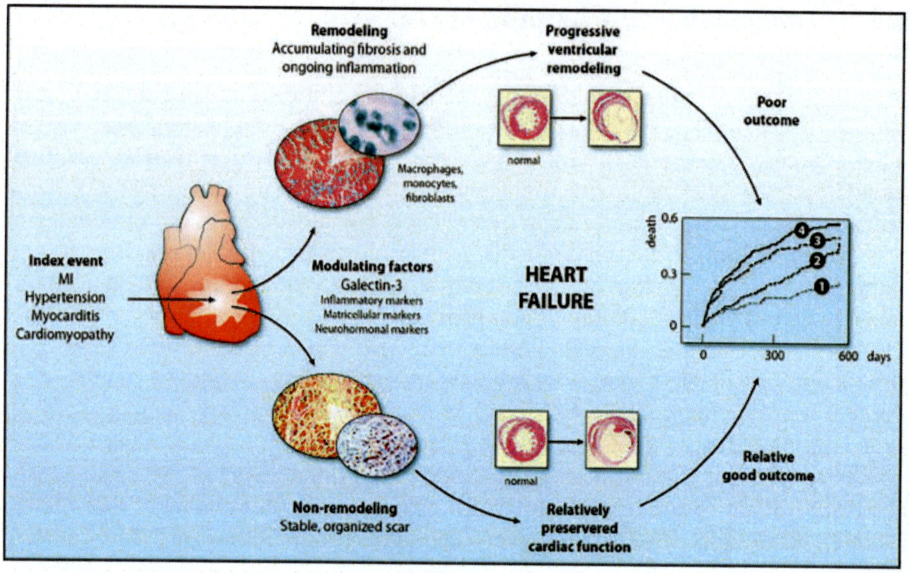

Fig. 6.4 Working scheme representing the sequence of events following an index event (such as myocardial infarction (MI), hypertension, myocarditis, and cardiomyopathy) leading to remodeling and non-remodeling heart failure. The graph within this graphic is taken from de Boer et al. [56] and represents the adjusted Cox regression curves for quartiles of plasma galectin-3 showing the cumulative risk for the combined endpoint, death. The *back circles* with the *white numbers* represent quartile 1 through 4, respectively. Galectin-3 is displayed as a central modulating factor involved in the remodeling process which leads to ongoing damage and eventually poor heart failure outcome. Therapeutically, galectin-3 inhibition could favor non-remodeling heart failure, thereby potentially improving heart failure outcome. Figure reprinted from [56] by de Boer et al. with permission

biomarker" involved in pathophysiology circle of cardiac remodeling and heart failure. A suggested pathway of galectin-3 is displayed in Fig. 6.4 [56].

Data from experimental renal damage and cancer suggest that galectin-3 is a feasible target for therapy. Our pilot data lend support to the notion that also in heart failure galectin-3 may be a target for therapy. More research is warranted herein, specifically, at what stage galectin-3 comes into play and what the ideal window would be for intervention. Most data show that preventative regimen might work, but if fibrosis, once ensued, could be attenuated or reversed is also unknown. From a clinical point of view, this is of utmost importance.

Disclosures BG Medicine Inc. (BGM, Waltham, MA, USA) holds certain rights with respect to the use of galectin-3 in heart failure. The UMCG, Department of Cardiology, which employs the authors, received research grants from BGM. Dr. de Boer received consultancy fees from BGM.

Acknowledgment We thank Maxi Meissner, Ph.D., for her assistance in preparing this manuscript.

References

1. Seetharaman J, Kanigsberg A, Slaaby R et al (1998) X-ray crystal structure of the human galectin-3 carbohydrate recognition domain at 2.1-A resolution. J Biol Chem 273:13047–13052
2. Henrick K, Bawumia S, Barboni EA et al (1998) Evidence for subsites in the galectins involved in sugar binding at the nonreducing end of the central galactose of oligosaccharide ligands: sequence analysis, homology modeling and mutagenesis studies of hamster galectin-3. Glycobiology 8:45–57
3. Rosenberg I, Cherayil BJ, Isselbacher KJ et al (1991) Mac-2-binding glycoproteins. Putative ligands for a cytosolic beta-galactoside lectin. J Biol Chem 266:18731–18736
4. Sato S, Hughes RC (1992) Binding specificity of a baby hamster kidney lectin for H type I and II chains, polylactosamine glycans, and appropriately glycosylated forms of laminin and fibronectin. J Biol Chem 267:6983–6990
5. Ochieng J, Furtak V, Lukyanov P (2004) Extracellular functions of galectin-3. Glycoconj J 19:527–535
6. de Boer RA, Yu L, van Veldhuisen DJ (2010) Galectin-3 in cardiac remodeling and heart failure. Curr Heart Fail Rep 7:1–8
7. Hughes RC (1997) The galectin family of mammalian carbohydrate-binding molecules. Biochem Soc Trans 25:1194–1198
8. Hughes RC (1999) Secretion of the galectin family of mammalian carbohydrate-binding proteins. Biochim Biophys Acta 1473:172–185
9. Woo HJ, Shaw LM, Messier JM et al (1990) The major non-integrin laminin binding protein of macrophages is identical to carbohydrate binding protein 35 (Mac-2). J Biol Chem 265:7097–7099
10. Koths K, Taylor E, Halenbeck R et al (1993) Cloning and characterization of a human Mac-2-binding protein, a new member of the superfamily defined by the macrophage scavenger receptor cysteine-rich domain. J Biol Chem 268:14245–14249
11. Probstmeier R, Montag D, Schachner M (1995) Galectin-3, a beta-galactoside-binding animal lectin, binds to neural recognition molecules. J Neurochem 64:2465–2472
12. Ochieng J, Warfield P (1995) Galectin-3 binding potentials of mouse tumor EHS and human placental laminins. Biochem Biophys Res Commun 217:402–406
13. Kim H, Lee J, Hyun JW et al (2007) Expression and immunohistochemical localization of galectin-3 in various mouse tissues. Cell Biol Int 31:655–662
14. Sharma UC, Pokharel S, van Brakel TJ et al (2004) Galectin-3 marks activated macrophages in failure-prone hypertrophied hearts and contributes to cardiac dysfunction. Circulation 110:3121–3128
15. Kasper M, Hughes RC (1996) Immunocytochemical evidence for a modulation of galectin 3 (Mac-2), a carbohydrate binding protein, in pulmonary fibrosis. J Pathol 179:309–316
16. Wang L, Friess H, Zhu Z et al (2000) Galectin-1 and galectin-3 in chronic pancreatitis. Lab Invest 80:1233–1241
17. Henderson NC, Mackinnon AC, Farnworth SL et al (2006) Galectin-3 regulates myofibroblast activation and hepatic fibrosis. Proc Natl Acad Sci USA 103:5060–5065
18. Sasaki S, Bao Q, Hughes RC (1999) Galectin-3 modulates rat mesangial cell proliferation and matrix synthesis during experimental glomerulonephritis induced by anti-Thy1.1 antibodies. J Pathol 187:481–489
19. Henderson NC, Mackinnon AC, Farnworth SL et al (2008) Galectin-3 expression and secretion links macrophages to the promotion of renal fibrosis. Am J Pathol 172:288–298
20. Eis V, Luckow B, Vielhauer V et al (2004) Chemokine receptor CCR1 but not CCR5 mediates leukocyte recruitment and subsequent renal fibrosis after unilateral ureteral obstruction. J Am Soc Nephrol 15:337–347
21. Sharma U, Rhaleb NE, Pokharel S et al (2008) Novel anti-inflammatory mechanisms of N-Acetyl-Ser-Asp-Lys-Pro in hypertension-induced target organ damage. Am J Physiol Heart Circ Physiol 294:H1226–H1232

22. Liu YH, D'Ambrosio M, Liao TD et al (2009) N-acetyl-seryl-aspartyl-lysyl-proline prevents cardiac remodeling and dysfunction induced by galectin-3, a mammalian adhesion/growth-regulatory lectin. Am J Physiol Heart Circ Physiol 296:H404–H412
23. Mackinnon AC, Gibbons MA, Farnworth SL et al (2012) Regulation of Transforming Growth Factor-beta1-driven Lung Fibrosis by Galectin-3. Am J Respir Crit Care Med 185:537–546
24. Takenaka Y, Fukumori T, Yoshii T et al (2004) Nuclear export of phosphorylated galectin-3 regulates its antiapoptotic activity in response to chemotherapeutic drugs. Mol Cell Biol 24:4395–4406
25. Sato S, Burdett I, Hughes RC (1993) Secretion of the baby hamster kidney 30-kDa galactose-binding lectin from polarized and nonpolarized cells: a pathway independent of the endoplasmic reticulum-Golgi complex. Exp Cell Res 207:8–18
26. Sato S, Hughes RC (1994) Regulation of secretion and surface expression of Mac-2, a galactoside-binding protein of macrophages. J Biol Chem 269:4424–4430
27. Nickel W (2003) The mystery of nonclassical protein secretion. A current view on cargo proteins and potential export routes. Eur J Biochem 270:2109–2119
28. Jiang JX, Chen X, Hsu DK et al (2012) Galectin-3 modulates phagocytosis-induced stellate cell activation and liver fibrosis in vivo. Am J Physiol Gastrointest Liver Physiol 302:G439–G446
29. de Oliveira SA (2012) de Freitas Souza BS, Sa Barreto EP, et al. Reduction of galectin-3 expression and liver fibrosis after cell therapy in a mouse model of cirrhosis. Cytotherapy 14:339–349
30. Kolatsi-Joannou M, Price KL, Winyard PJ et al (2011) Modified citrus pectin reduces galectin-3 expression and disease severity in experimental acute kidney injury. PLoS One 6:e18683
31. Thandavarayan RA, Watanabe K, Ma M et al (2008) 14-3-3 protein regulates Ask1 signaling and protects against diabetic cardiomyopathy. Biochem Pharmacol 75:1797–1806
32. Kamal FA, Watanabe K, Ma M et al (2011) A novel phenylpyridazinone, T-3999, reduces the progression of autoimmune myocarditis to dilated cardiomyopathy. Heart Vessels 26:81–90
33. Psarras S, Mavroidis M, Sanoudou D et al (2011) Regulation of adverse remodelling by osteopontin in a genetic heart failure model. Eur Heart J 33:1954–1963
34. Wilker E, Yaffe MB (2004) 14–3–3 Proteins–a focus on cancer and human disease. J Mol Cell Cardiol 37:633–642
35. Banerjee A, Apte UM, Smith R et al (2006) Higher neutrophil infiltration mediated by osteopontin is a likely contributing factor to the increased susceptibility of females to alcoholic liver disease. J Pathol 208:473–485
36. Fisher AD (1992) Maxillofacial surgery should become a specialty of medicine. Br Dent J 172:46–47
37. Friedman SL (2000) Molecular regulation of hepatic fibrosis, an integrated cellular response to tissue injury. J Biol Chem 275:2247–2250
38. Brown RD, Ambler SK, Mitchell MD et al (2005) The cardiac fibroblast: therapeutic target in myocardial remodeling and failure. Annu Rev Pharmacol Toxicol 45:657–687
39. de Cavanagh EM, Ferder M, Inserra F et al (2009) Angiotensin II, mitochondria, cytoskeletal, and extracellular matrix connections: an integrating viewpoint. Am J Physiol Heart Circ Physiol 296:H550–H558
40. Krzeslak A, Lipinska A (2004) Galectin-3 as a multifunctional protein. Cell Mol Biol Lett 9:305–328
41. Dvorankova B, Szabo P, Lacina L et al (2011) Human galectins induce conversion of dermal fibroblasts into myofibroblasts and production of extracellular matrix: potential application in tissue engineering and wound repair. Cells Tissues Organs 194:469–480
42. Henderson NC, Sethi T (2009) The regulation of inflammation by galectin-3. Immunol Rev 230:160–171
43. van Kimmenade RR, Januzzi JL Jr, Ellinor PT et al (2006) Utility of amino-terminal pro-brain natriuretic peptide, galectin-3, and apelin for the evaluation of patients with acute heart failure. J Am Coll Cardiol 48:1217–1224

44. Shah RV, Chen-Tournoux AA, Picard MH et al (2010) Galectin-3, cardiac structure and function, and long-term mortality in patients with acutely decompensated heart failure. Eur J Heart Fail 12:826–832
45. Lin YH, Lin LY, Wu YW et al (2009) The relationship between serum galectin-3 and serum markers of cardiac extracellular matrix turnover in heart failure patients. Clin Chim Acta 409:96–99
46. Lok DJ, Van Der Meer P, de la Porte PW et al (2010) Prognostic value of galectin-3, a novel marker of fibrosis, in patients with chronic heart failure: data from the DEAL-HF study. Clin Res Cardiol 99:323–328
47. de Boer RA, Lok DJ, Jaarsma T et al (2011) Predictive value of plasma galectin-3 levels in heart failure with reduced and preserved ejection fraction. Ann Med 43:60–68
48. Felker GM, Fiuzat M, Shaw LK et al (2012) Galectin-3 in ambulatory patients with heart failure: results from the HF-ACTION study. Circ Heart Fail 5:72–78
49. de Boer RA, van Veldhuisen DJ, Gansevoort RT et al (2012) The fibrosis marker galectin-3 and outcome in the general population. J Intern Med 272:55–64
50. Christenson RH, Duh SH, Wu AH et al (2010) Multi-center determination of galectin-3 assay performance characteristics: Anatomy of a novel assay for use in heart failure. Clin Biochem 43:683–690
51. Grandin EW, Jarolim P, Murphy SA et al (2012) Galectin-3 and the development of heart failure after acute coronary syndrome: pilot experience from PROVE IT-TIMI 22. Clin Chem 58:267–273
52. Kramer F, Milting H (2011) Novel biomarkers in human terminal heart failure and under mechanical circulatory support. Biomarkers 16(Suppl 1):S31–S41
53. de Boer RA, Voors AA, Muntendam P et al (2009) Galectin-3: a novel mediator of heart failure development and progression. Eur J Heart Fail 11:811–817
54. Yu L, Ruifrok WP, Sillje HHW et al (2011) Genetic disruption of galectin-3 prevents adverse cardiac remodeling. Eur Heart J 32(Abstract Supplement):1094
55. Yu L, Ruifrok WP, Sillje HHW et al (2011) Pharmacological inhibition of galectin-3 attenuates cardiac remodeling and heart failure. Eur Heart J 32(Abstract Supplement):1097
56. de Boer RA, Meissner M, van Veldhuisen DJ (2012) Galectin-3. In: Alan S Maisel (ed) Cardiac biomarkers: expert advice for clinicians, 1st edn. New Delhi, pp 193–207. ISBN 978-93-5025-564-3

Chapter 7
A Mitochondriocentric Pathway to Cardiomyocyte Necrosis: An Upstream Molecular Mechanism in Myocardial Fibrosis

Adedayo A. Adeboye, Kevin P. Newman, Dwight A. Dishmon, Shadwan Alsafwah, Syamal K. Bhattacharya, and Karl T. Weber

Abstract The pathophysiologic origins of heart failure can be attributed to a pathologic remodeling of myocardium, including necrotic loss of cardiomyocytes and consequent reparative fibrosis. Hypertensive heart disease with concentric left ventricular hypertrophy and fibrosis represents a major etiologic factor accounting for diastolic heart failure. Herein, we focus on molecular mechanisms to the precursor of fibrosis, namely, cardiomyocyte necrosis, whose pathogenic origin resides in a mitochondriocentric signal-transducer–effector pathway. Its major components include intracellular Ca^{2+} overloading of cytosolic and mitochondrial domains, the induction of oxidative stress by these organelles which overwhelms endogenous antioxidant defenses, and the increased opening potential of the mitochondrial permeability transition pore. Novel cardioprotective strategies aimed at preventing the progressive remodeling of the failing heart should target upstream molecular mechanisms that prevent cardiomyocyte necrosis rather than downstream events involving collagen turnover related to fibrosis.

Keywords Mitochondria • Calcium overload • Oxidative stress • Necrosis • Fibrosis • Aldosteronism

Abbreviations

ALDO	Aldosterone
ALDOST	Aldosterone/salt treatment
CHF	Congestive heart failure

A.A. Adeboye • K.P. Newman • D.A. Dishmon • S. Alsafwah
S.K. Bhattacharya • K.T. Weber, M.D. (✉)
Division of Cardiovascular Diseases, University of Tennessee Health Science Center, 956 Court Avenue, Suite A312, Memphis, TN, USA
e-mail: KTWeber@uthsc.edu

B.I. Jugdutt and N.S. Dhalla (eds.), *Cardiac Remodeling: Molecular Mechanisms*, Advances in Biochemistry in Health and Disease 5, DOI 10.1007/978-1-4614-5930-9_7,

CVF	Collagen volume fraction
DOCST	Deoxycorticosterone/salt treatment
HHD	Hypertensive heart disease
mPTP	Mitochondrial permeability transition pore
MSTE	Mitochondriocentric signal-transducer–effector
PAC	Plasma aldosterone concentration
PTH	Parathyroid hormone
RAAS	Renin–angiotensin–aldosterone system
ROS	Reactive oxygen species
SHPT	Secondary hyperparathyroidism
Spiro	Spironolactone

7.1 Introduction

The congestive heart failure (CHF) syndrome has reached epidemic proportions; its debilitating symptoms and signs account for the number one admitting diagnosis to US hospitals. CHF strains our workforce and drains our health-care economy. Of necessity, improved management strategies have become national priorities.

A pathologic remodeling of myocardium contributes to the failure of this muscular pump during systolic and/or diastolic phases of the cardiac cycle and is related to (a) the loss of contractile parenchyma or cardiomyocytes and (b) their consequent replacement by extracellular matrix or reparative fibrosis. Foci of fibrosis, or microscopic scarring, are found scattered throughout the right and left heart of the explanted failing human heart and are considered the major cause of myocardial remodeling in the cardiomyopathic heart of either ischemic or nonischemic origin [1–4]. Fibrous tissue consists predominantly of type I (75 %) and type III (20 %) fibrillar collagens aligned either in parallel or in series with muscle fibers that adversely influence ventricular function [4–6]. Type I fibrillar collagen has the tensile strength of steel, and therefore, its accumulation adversely impacts myocardial stiffness. Moreover, fibrillar collagen encircling muscle may lead to muscle fiber atrophy [7].

Hypertensive heart disease (HHD) represents a major etiologic factor that accounts for heart failure in which the failing heart may manifest as either diastolic or systolic predominance. The clinical syndrome CHF, with its characteristic signs and symptoms, frequently accompanies either diastolic or systolic heart failure. Its origins are rooted in a salt-avid state mediated by effector hormones of the renin–angiotensin–aldosterone system (RAAS). Concentric left ventricular hypertrophy is a distinguishing feature of HHD, so too is fibrosis, which is found throughout the right and left heart [8–10]. It is not the quantity, rather the quality (structural remodeling), of myocardium that contributes to diastolic heart failure in HHD.

The failing heart's inability to sustain adequate systemic blood flow and organ perfusion elicits adaptive homeostatic responses. In the case of underperfused kidneys, this evokes neurohormonal activation of the RAAS. The attendant effector hormone-mediated retention of salt and water now accounts for the appearance of the clinical CHF syndrome [11]. Prolonged elevations in circulating angiotensin II and aldosterone each have pathologic consequences on diverse tissues, including the heart as well as skeletal muscle and peripheral blood mononuclear cells (monocytes and lymphocytes) [12–15].

Herein, we focus on the pathologic remodeling of myocardium by reparative fibrosis and molecular mechanisms leading to its precursor, cardiomyocyte necrosis. Ensuing tissue repair at sites of necrosis and resultant scarring preserves the structural integrity of the myocardium, but at the cost of increased stiffness. Cardioprotective strategies aimed at preventing the progressive nature of heart failure with ongoing cardiomyocyte loss and consequent incremental fibrosis should target the upstream molecular mechanisms to prevent necrosis, rather than downstream events related to collagen turnover. In this context, we propose that the specific cascade and the pathogenic origins to necrosis is a mitochondriocentric signal-transducer–effector (MSTE) pathway.

7.2 Animal Model of Aldosteronism

We identified the MSTE pathway to cardiomyocyte necrosis using a rat model of aldosteronism, where plasma aldosterone (ALDO) concentrations (PAC) are raised to those found in human CHF together with 1 % NaCl in drinking water. Aldosterone/salt treatment (ALDOST) allows us to address the pathologic remodeling of the previously normal myocardium that accompanies the secondary aldosteronism of CHF or the primary aldosteronism attendant with autonomous ALDO production associated with adrenal pathology. We have found this remodeling to include cardiomyocyte necrosis and myocardial scarring, a footprint of necrosis. Subcellular and molecular mechanisms responsible for necrosis were then sequentially and systematically investigated.

7.2.1 Cardiac Remodeling in Aldosteronism

Much like the cardiomyopathic heart of ischemic or nonischemic origins, the myocardium in primary aldosteronism includes foci of scarring found scattered throughout the nonpressure-overloaded, nonhypertrophied right and the pressure-overloaded, hypertrophied left ventricle [16]. Cardiac pathology is found at week 4 of ALDOST. It includes multiple foci of microscopic scarring found scattered throughout the right and left atria and ventricles [10, 15, 17]. The accumulation of collagen in the myocardium is a *reparative* fibrosis. It replaces cardiomyocytes lost to necrotic cell death and in so doing preserves the structural integrity of myocardium (reviewed in [18]). In contrast, apoptotic cell death is devoid of inflammatory cell or fibroblast responses and, therefore, is not accompanied by fibrosis.

Myocardial fibrosis can be defined as an increase in its hydroxyproline concentration, an amino acid specific to collagen, and morphologically the volume of myocardium occupied by fibrillar collagen. Furthermore, collagen fiber cross-linking and the relative abundance of type I and III collagens are features of the collagen matrix that determine abnormalities in myocardial stiffness [19]. Fibrosis increases myocardial stiffness leading to diastolic heart failure, also termed heart failure with preserved ejection fraction (reviewed in [20–22]). Diastolic failure is commonly seen with the concentric hypertrophy that accompanies arterial hypertension [23].

7.2.2 *Cardiac Remodeling: Role of Hemodynamic Factors*

A conspicuous heterogeneity in myofiber size is found in the heart during chronic ALDOST. In the pressure-overloaded left ventricle, myofibers are hypertrophied while they remain normal in diameter in the nonpressure-overloaded right ventricle. Surrounded by fibrillar collagen at sites of scarring, myofibers in either ventricle become atrophic.

In the case of myocardial fibrosis, hemodynamic factors are not directly involved (reviewed in [11]). This conceptual framework and role of a circulating factor is primarily based on the following finding: (a) the presence of fibrosis in the nonpressure-overloaded right atria and ventricle, (b) the absence of fibrosis when a pressure overload on the LV is caused by infrarenal aortic banding without subsequent RAAS activation or when ALDO treatment is combined with a low-Na^+ diet or when 1 % NaCl alone is given, and (c) the prevention of fibrosis with either a small (nondepressor) or large (depressor) dose of Spiro, an ALDO receptor antagonist, which respectively fails to or does prevent hypertension. Furthermore, an intracerebroventricular infusion of a mineralocorticoid receptor antagonist prevents hypertension, but not fibrosis [24], whereas upregulation of ALDO synthase in the heart accounts for increased tissue levels of ALDO, but is not accompanied by fibrosis [25]. Thus, the cumulative evidence gathered to date indicates that the adverse myocardial remodeling during ALDOST is both independent of hypertension and unrelated to plasma or tissue-derived ALDO per se. Therefore, another circulating factor which accompanies aldosteronism must be operative (vide infra).

7.2.3 *Hypercalciuria and Hypermagnesuria in Aldosteronism*

Metabolic studies demonstrated an early and persistent elevation in urinary and fecal excretion of Ca^{2+} and Mg^{2+} in rats during ALDOST [26, 27]. The marked loss of Ca^{2+} and Mg^{2+} was evident throughout weeks 1–4 of ALDOST and involved both renal and gastrointestinal sites of excretion, each of which are sites of high-density ALDO receptor binding [28]. Hypercalciuria accompanies the short-term treatment of man or animals with a mineralocorticoid plus dietary salt [29–34]. It is present in patients with primary aldosteronism, where it is accentuated by dietary Na^+ loading [33, 35]. The increment in urinary Ca^{2+} excretion, which occurs in the distal segment of the nephron, is dietary Na^+-dependent. Urinary Mg^{2+} excretion also rises with dietary Na^+ loading in rats treated with a mineralocorticoid, while the hypermagnesuria found in patients with primary aldosteronism is abrogated by either Spiro or surgical removal of diseased adrenal tissue [36–38]. In rats receiving 4 weeks ALDOST, we found Spiro cotreatment to attenuate enhanced urinary and fecal excretion of each cation beginning at week 1 onward [26, 27].

7.2.4 *Secondary Hyperparathyroidism*

The sustained urinary and fecal loss of Ca^{2+} and Mg^{2+}, which accompanies ALDOST, leads to a progressive fall in plasma ionized concentrations of these divalent cations. Plasma ionized $[Ca^{2+}]_o$ and $[Mg^{2+}]_o$ were markedly reduced at week 4, while each of these cations had already begun to fall at weeks 1 and 2 ALDOST. Ionized hypocalcemia and hypomagnesemia each regulate parathyroid hormone (PTH) secretion [39, 40]. Bone and its mineral stores are the primary reserve for these cations with bone resorption facilitated by PTH. Elevations in PTH, elaborated in response to reduced plasma ionized $[Ca^{2+}]_o$ and $[Mg^{2+}]_o$, are expected in the setting of sustained hypercalciuria and hypermagnesuria. We indeed found elevated plasma PTH levels, consistent with secondary hyperparathyroidism (SHPT), during weeks 1–4 ALDOST. Both serum ionized $[Ca^{2+}]_o$ and total plasma Ca^{2+} are reduced in response to deoxycorticosterone/salt treatment (DOCST) [34], together with increased serum PTH and urinary excretion of cAMP, a biomarker of parathyroid activity. SHPT has been reported in patients with primary aldosteronism [38, 41–43] where accompanying aberrations in serum ionized and total Ca^{2+}, together with elevations in PTH, were normalized by either adrenal surgery or Spiro [38, 43].

Our second major finding, and further evidence in keeping with persistent SHPT, was the marked and progressive reduction in bone mineral density and content of tibia and femur that appeared by week 4 ALDOST and became more evident at weeks 5 and 6. This fall in bone mineral density was rapid and accompanied by a corresponding reduction in bone strength. Urinary hydroxyproline, a marker of bone resorption, was increased during ALDOST or DOCST [44], while the hypercalciuria seen with 8 % NaCl loading alone was likewise accompanied by SHPT with a loss of bone Ca^{2+} and Mg^{2+} and increased urinary excretion of various markers of bone resorption [45–47].

7.3 A Molecular Pathway Leading to Cardiomyocyte Necrosis

A molecular mechanism accounting for cardiomyocyte necrosis and subsequent scarring of myocardium found at 4 weeks ALDOST was identified and the pathogenic role of circulating hormone elucidated.

7.3.1 A Proinflammatory Phenotype

Evidence of a proinflammatory phenotype and oxidative stress in the myocardium is found at 4 weeks ALDOST [15, 48–51]. This includes the presence of 3-nitrotyrosine, a by-product of the reaction involving superoxide and nitric oxide; an activation of the gp91^{phox} subunit of NADPH oxidase found in inflammatory cells invading the injured myocardium that contributes to superoxide generation; upregulated redox-sensitive nuclear transcription factor-κB and a proinflammatory gene cascade it regulates that includes intercellular adhesion molecule-1, monocyte chemoattractant protein-1, and tumor necrosis factor-alpha; and increased tissue levels of 8-isoprostane and malondialdehyde, biomarkers of lipid peroxidation [15, 51]. There is also considerable evidence of oxidative stress in blood and urine consistent with the systemic nature of an altered redox state during chronic aldosteronism. This proinflammatory phenotype during ALDOST was prevented by cotreatment with either Spiro or *N*-acetylcysteine, an antioxidant.

7.3.2 Intracellular Ca^{2+} Overloading and SHPT

Our hypothesis for the induction of oxidative stress during ALDOST draws upon Albrecht Fleckenstein's original concept that intracellular Ca^{2+} overloading is an integral pathophysiologic feature leading to myocardial necrosis [52]. In rats receiving 1 and 4 weeks ALDOST, we monitored intracellular Ca^{2+} in cardiomyocytes and mitochondria harvested from the heart. We found increased Ca^{2+} levels in the myocardium during preclinical, clinical, and pathologic stages of ALDOST, accompanied by biomarker evidence of oxidative stress that included increased levels of malondialdehyde and 8-isoprostane and increased H_2O_2 production [14, 27, 51, 53].

The calcium-sensing receptor of the parathyroid glands responds to hypocalcemia with increased secretion of PTH. Accordingly, plasma PTH levels were elevated at weeks 1–4 ALDOST [27] with SHPT as evidenced by bone resorption [26]. We therefore hypothesized that the intracellular Ca^{2+} overloading and induction of oxidative stress that accompanies ALDOST leading to cardiomyocyte necrosis and fibrosis are mediated by the calcitropic hormone, PTH, and not ALDO. This represents an example of the SHPT-associated Ca^{2+} paradox as characterized by Fujita and Palmieri [54]. The elegant studies of Massry and coworkers have demonstrated PTH-mediated intracellular Ca^{2+} overloading of cardiomyocytes that included cardiac myocytes incubated with PTH [55], cells harvested from normal rats receiving a 2-week infusion of PTH, and rats with chronic renal failure having SHPT [56]. In each case, cotreatment with verapamil, a Ca^{2+} channel blocker, prevented the rise in intracellular Ca^{2+}. PTH also regulates cardiomyocyte Ca^{2+} channel opening [57]. Using radiolabeled PTH, Nordquist and Palmieri found that PTH not only penetrates plasmalemma but also localizes within the cytoplasm of renal tubular cells and predominantly in mitochondria, where its mechanism of transport and function remain to be fully elucidated [58]. Rasmussen and others found PTH to alter mitochondrial $[Ca^{2+}]_m$ and respiration [59–63]. Excessive intracellular Ca^{2+} accumulation in cardiomyocytes and consequent generation of reactive oxygen species (ROS) alter intracellular signaling events, including their perpetuation of intracellular Ca^{2+} overloading via L-type Ca^{2+} channel entry and inhibition of Ca^{2+} efflux by Ca^{2+}-ATPase [64, 65]. A Na^+/Ca^{2+} exchanger involved in regulating Na^+-dependent Ca^{2+} efflux from mitochondria may also be contributory [66, 67].

7.3.3 Mitochondria-Based Oxidative Stress

PTH-mediated intracellular Ca^{2+} overloading is coupled to induction of oxidative stress in diverse tissues that includes cardiomyocytes and their mitochondria. The generation of ROS and reactive nitrogen species appears to overwhelm their rate of detoxification by the cumulative capacity of antioxidant defenses. Together with lost ATP synthesis, Ca^{2+} overloading and oxidative stress of mitochondria lead to a nonphysiologic opening of their mitochondrial permeability transition pore (mPTP). The ensuing osmotic-based structural and functional degeneration of these organelles triggers the downhill final common cell death pathway leading to cardiomyocyte necrosis and subsequent replacement fibrosis [68].

7.3.4 The Cellular–Molecular Cascade to Cardiomyocyte Necrosis

A series of site-directed, sequential pharmacologic interventions targeted along the cellular–molecular cascade was used to block downstream events leading to cardiomyocyte necrosis. These observations collectively validated our hypothesis regarding the pathologic sequelae of events leading to adverse structural remodeling of myocardium in rats with chronic aldosteronism. Various cotreatments were used. They included (1) Spiro, which attenuated the enhanced urinary and fecal losses of these cations to prevent hypocalcemia and hypomagnesemia and thereby ensuing SHPT [27]; (2) a Ca^{2+}- and Mg^{2+}-supplemented diet, together with vitamin D_3 to enhance Ca^{2+} absorption, which prevented hypocalcemia and SHPT [69]; (3) parathyroidectomy prevented SHPT [70] and vascular lesions and the rise in aortic tissue Ca^{2+} content during DOCST [71]; (4) cinacalcet, a calcimimetic that resets the threshold of the parathyroid glands' Ca^{2+}-sensing receptor to prevent SHPT despite hypocalcemia [72]; (5) amlodipine, a Ca^{2+} channel blocker, which prevents intracellular Ca^{2+} overloading [53]; and (6) *N*-acetylcysteine, an antioxidant that abrogated oxidative stress [15].

Taken together, the multitude of evidence gathered to date congruently supports that PTH-mediated intracellular Ca^{2+} overloading is the most tenable mechanism that leads to the induction of oxidative stress during aldosteronism where ROS, primarily derived from mitochondria in cardiomyocytes and NADPH oxidase in vascular tissue, overwhelm cellular antioxidant defenses. This scenario begs the question whether the overall consequence of an excessive generation of prooxidants or cumulative endogenous antioxidant defenses in combating ROS was compromised.

7.3.5 Subsarcolemmal Mitochondria-Targeted Cardioprotection

The signal-transducer–effector pathway leading to necrotic cell death during ALDOST revolves around intramitochondrial Ca^{2+} overloading, together with an induction of oxidative stress and opening of the mPTP. To further validate this concept, we hypothesized mitochondria-targeted interventions would prove cardioprotective. Accordingly, 8-week-old male Sprague–Dawley rats receiving 4 weeks ALDOST were cotreated with either quercetin, a flavonoid with mitochondrial antioxidant properties, or cyclosporine A, an mPTP inhibitor, and compared to ALDOST alone and untreated, age-/sex-matched controls (see Fig. 7.1). We monitored mitochondrial free Ca^{2+} and biomarkers of oxidative stress, including 8-isoprostane and H_2O_2 production; mPTP opening; total Ca^{2+} in cardiac tissue; collagen volume fraction (CVF) to quantify replacement fibrosis, a biomarker of cardiomyocyte necrosis; and employed TUNEL assay to address apoptosis in coronal sections of ventricular myocardium. Compared to controls, at 4 weeks ALDOST, we found a marked increase in mitochondrial H_2O_2 production and 8-isoprostane levels, a greater propensity for mPTP opening, and higher concentrations of mitochondrial free $[Ca^{2+}]_m$ and total tissue Ca^{2+}, coupled with a fivefold rise in CVF without any TUNEL-based evidence of cardiomyocyte apoptosis. Each of these pathophysiologic responses to ALDOST was prevented by quercetin or cyclosporine A cotreatment. Thus, mitochondria play a central role in initiating the cellular–molecular pathway that leads to necrotic cell death and consequent myocardial scarring. This destructive cycle can be interrupted and myocardium salvaged with its structure preserved by mitochondria-targeted cardioprotective strategies.

7.4 Summary and Conclusions

Intracellular Ca^{2+} overloading accompanies inappropriate (relative to dietary Na^+) elevations in PAC, termed aldosteronism, and leads to a prooxidant/proinflammatory phenotype involving the heart. The dual deleterious consequences of elevated PAC and dietary Na^+ are linked to the induction of oxidative stress via PTH-mediated intracellular Ca^{2+} overloading. The genesis of this SHPT occurs in response to plasma ionized hypocalcemia and hypomagnesemia whose appearance is a consequence of marked urinary and fecal excretory losses of these divalent cations that accompany chronic aldosteronism.

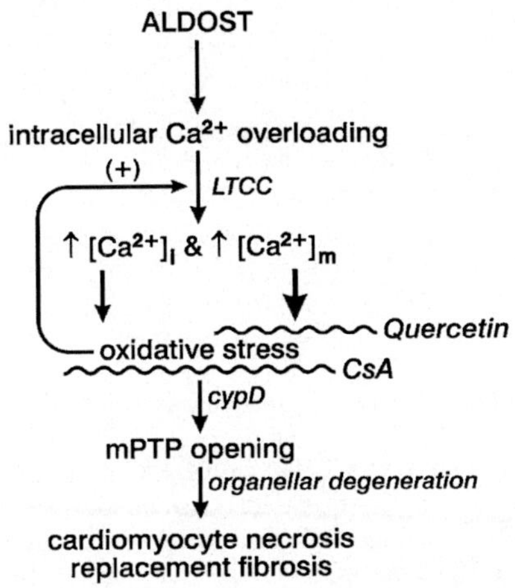

Fig. 7.1 A central hypothesis representing the pathophysiologic sequence of events in chronic aldosterone/salt treatment (ALDOST) that perpetuate cardiomyocyte necrosis with replacement fibrosis (or scarring). Mediated via L-type Ca^{2+} channels (LTCC), intracellular Ca^{2+} overloading of cardiomyocytes, including cytosolic $[Ca^{2+}]_i$ and mitochondrial $[Ca^{2+}]_m$ components, leads to the induction of oxidative stress and generation of reactive oxygen species predominantly by mitochondria (*heavy arrow*) and which feeds back to perpetuate LTCC activity. This Ca^{2+} overloading-mediated redox injury pathway involves the pathologic opening of the mitochondrial permeability transition pore (mPTP)-cyclophilin D (cypD) complex. The ensuing osmotic swelling of these organelles and their degeneration, coupled with impaired mitochondrial generation of high energy phosphates, leads to necrosis and subsequent wound healing with fibrous tissue formation. Quercetin serves as a potent mitochondria-targeted antioxidant, whereas cyclosporine A (CsA) inhibits the opening of the mPTP (Reprinted with permission from Shahbaz AU, Kamalov G, Zhao W, Zhao T, Johnson PL, Sun Y, Bhattacharya SK, Ahokas RA, Gerling IC, Weber KT. Mitochondria-targeted cardioprotection in aldosteronism. *J Cardiovasc Pharmacol.* 2011;57:37–43)

Sustained Ca^{2+} overloading has its pathologic consequences. In the case of cardiac myocytes and their mitochondria, intracellular Ca^{2+} overloading leads to an induction of oxidative stress and opening of the mPTP with ensuing organellar destruction and cellular necrosis with subsequent wound healing response leading to a replacement fibrosis or scarring.

Pharmaceutical and/or nutriceuticals used as targeted interventions of the cellular and molecular pathway leading to excessive intracellular Ca^{2+} accumulation protect against adverse myocardial remodeling and hence are cardioprotective. These cumulative salutary responses raise the prospect that therapeutic interventions, capable of favorably influencing extra- and intracellular Ca^{2+} homeostasis and blocking the MSTE pathway, could potentially optimize the management of the secondary aldosteronism of CHF and patients with primary aldosteronism.

Acknowledgments This work was supported, in part, by NIH grants R01-HL73043 and R01-HL90867 (KTW). Its contents are solely the responsibility of the authors and do not necessarily represent the official views of the NIH.

References

1. Beltrami CA, Finato N, Rocco M et al (1994) Structural basis of end-stage failure in ischemic cardiomyopathy in humans. Circulation 89:151–163
2. de Leeuw N, Ruiter DJ, Balk AH et al (2001) Histopathologic findings in explanted heart tissue from patients with end-stage idiopathic dilated cardiomyopathy. Transpl Int 14:299–306
3. Waller TA, Hiser WL, Capehart JE, Roberts WC (1998) Comparison of clinical and morphologic cardiac findings in patients having cardiac transplantation for ischemic cardiomyopathy, idiopathic dilated cardiomyopathy, and dilated hypertrophic cardiomyopathy. Am J Cardiol 81:884–894
4. Schaper J, Speiser B (1992) The extracellular matrix in the failing human heart. Basic Res Cardiol 87(Suppl 1):303–309
5. Yoshikane H, Honda M, Goto Y et al (1992) Collagen in dilated cardiomyopathy–scanning electron microscopic and immunohistochemical observations. Jpn Circ J 56:899–910
6. Marijianowski MM, Teeling P, Mann J, Becker AE (1995) Dilated cardiomyopathy is associated with an increase in the type I/type III collagen ratio: a quantitative assessment. J Am Coll Cardiol 25:1263–1272
7. Jalil JE, Janicki JS, Pick R, Abrahams C, Weber KT (1989) Fibrosis-induced reduction of endomyocardium in the rat after isoproterenol treatment. Circ Res 65:258–264
8. Pearlman ES, Weber KT, Janicki JS, Pietra GG, Fishman AP (1982) Muscle fiber orientation and connective tissue content in the hypertrophied human heart. Lab Invest 46:158–164
9. Huysman JAN, Vliegen HW, Van der Laarse A, Eulderink F (1989) Changes in nonmyocyte tissue composition associated with pressure overload of hypertrophic human hearts. Pathol Res Pract 184:577–581
10. Sun Y, Ramires FJA, Weber KT (1997) Fibrosis of atria and great vessels in response to angiotensin II or aldosterone infusion. Cardiovasc Res 35:138–147
11. Weber KT (2001) Aldosterone in congestive heart failure. N Engl J Med 345:1689–1697
12. Delafontaine P, Akao M (2006) Angiotensin II as candidate of cardiac cachexia. Curr Opin Clin Nutr Metab Care 9:220–224
13. Gerling IC, Sun Y, Ahokas RA et al (2003) Aldosteronism: an immunostimulatory state precedes the proinflammatory/fibrogenic cardiac phenotype. Am J Physiol Heart Circ Physiol 285:H813–H821
14. Ahokas RA, Warrington KJ, Gerling IC et al (2003) Aldosteronism and peripheral blood mononuclear cell activation. A neuroendocrine-immune interface. Circ Res 93:e124–e135
15. Sun Y, Zhang J, Lu L et al (2002) Aldosterone-induced inflammation in the rat heart. Role of oxidative stress. Am J Pathol 161:1773–1781
16. Campbell SE, Farb A, Weber KT (1993) Pathologic remodeling of the myocardium in a weightlifter taking anabolic steroids. Blood Press 2:213–216
17. Brilla CG, Pick R, Tan LB, Janicki JS, Weber KT (1990) Remodeling of the rat right and left ventricle in experimental hypertension. Circ Res 67:1355–1364
18. Laurent GJ (1987) Dynamic state of collagen: pathways of collagen degradation *in vivo* and their possible role in regulation of collagen mass. Am J Physiol 252:C1–C9
19. López B, González A, Díez J (2010) Circulating biomarkers of collagen metabolism in cardiac diseases. Circulation 121:1645–1654
20. Zile MR, Brutsaert DL (2002) New concepts in diastolic dysfunction and diastolic heart failure: Part I: diagnosis, prognosis, and measurements of diastolic function. Circulation 105:1387–1393
21. Zile MR, Brutsaert DL (2002) New concepts in diastolic dysfunction and diastolic heart failure: Part II: causal mechanisms and treatment. Circulation 105:1503–1508
22. Díez J (2009) Towards a new paradigm about hypertensive heart disease. Med Clin North Am 93:637–645
23. Shapiro LM, McKenna WJ (1984) Left ventricular hypertrophy: relation of structure to diastolic function in hypertension. Br Heart J 51:637–642

24. Young M, Fullerton M, Dilley R, Funder J (1994) Mineralocorticoids, hypertension, and cardiac fibrosis. J Clin Invest 93:2578–2583
25. Garnier A, Bendall JK, Fuchs S et al (2004) Cardiac specific increase in aldosterone production induces coronary dysfunction in aldosterone synthase-transgenic mice. Circulation 110:1819–1825
26. Chhokar VS, Sun Y, Bhattacharya SK et al (2004) Loss of bone minerals and strength in rats with aldosteronism. Am J Physiol Heart Circ Physiol 287:H2023–H2026
27. Chhokar VS, Sun Y, Bhattacharya SK et al (2005) Hyperparathyroidism and the calcium paradox of aldosteronism. Circulation 111:871–878
28. Sheppard K, Funder JW (1987) Mineralocorticoid specificity of renal type I receptors: in vivo binding studies. Am J Physiol 252(2 Pt 1):E224–E229
29. Suki WN, Schwettmann RS, Rector FC Jr, Seldin DW (1968) Effect of chronic mineralocorticoid administration on calcium excretion in the rat. Am J Physiol 215:71–74
30. Massry SG, Coburn JW, Chapman LW, Kleeman CR (1968) The effect of long-term desoxycorticosterone acetate administration on thc renal excretion of calcium and magnesium. J Lab Clin Med 71:212–219
31. Gehr MK, Goldberg M (1986) Hypercalciuria of mineralocorticoid escape: clearance and micropuncture studies in the rat. Am J Physiol 251(5 Pt 2):F879–F888
32. Cappuccio FP, Markandu ND, MacGregor GA (1988) Renal handling of calcium and phosphate during mineralocorticoid administration in normal subjects. Nephron 48:280–283
33. Rastegar A, Agus Z, Connor TB, Goldberg M (1972) Renal handling of calcium and phosphate during mineralocorticoid "escape" in man. Kidney Int 2:279–286
34. Zikos D, Langman C, Gafter U, Delahaye B, Lau K (1986) Chronic DOCA treatment increases Ca absorption: role of hypercalciuria and vitamin D. Am J Physiol 251(3 Pt 1):E279–E284
35. Rossi E, Perazzoli F, Negro A et al (1998) Acute effects of intravenous sodium chloride load on calcium metabolism and on parathyroid function in patients with primary aldosteronism compared with subjects with essential hypertension. Am J Hypertens 11:8–13
36. Berthelot A, Pernot F, Gairard A (1983) Influence of the thyroid and parathyroid glands on magnesium metabolism during mineralocorticoid treatment (DOCA+NaCl) in the rat. Ann Nutr Metab 27:349–354
37. Horton R, Biglieri EG (1962) Effect of aldosterone on the metabolism of magnesium. J Clin Endocrinol Metab 22:1187–1192
38. Rossi E, Sani C, Perazzoli F et al (1995) Alterations of calcium metabolism and of parathyroid function in primary aldosteronism, and their reversal by spironolactone or by surgical removal of aldosterone-producing adenomas. Am J Hypertens 8:884–893
39. Morrissey JJ, Cohn DV (1978) The effects of calcium and magnesium on the secretion of parathormone and parathyroid secretory protein by isolated porcine parathyroid cells. Endocrinology 103:2081–2090
40. Mayer GP, Hurst JG (1978) Comparison of the effects of calcium and magnesium on parathyroid hormone secretion rate in calves. Endocrinology 102:1803–1814
41. Fertig A, Webley M, Lynn JA (1980) Primary hyperparathyroidism in a patient with Conn's syndrome. Postgrad Med J 56:45–47
42. Hellman DE, Kartchner M, Komar N, Mayes D, Pitt M (1980) Hyperaldosteronism, hyperparathyroidism, medullary sponge kidneys, and hypertension. JAMA 244:1351–1353
43. Resnick LM, Laragh JH (1985) Calcium metabolism and parathyroid function in primary aldosteronism. Am J Med 78:385–390
44. Palmieri GM, Hawrylko J (1977) Effects of aldosterone on the urinary excretion of total and non-dialyzable hydroxyproline. Horm Metab Res 9:507–509
45. Sellmeyer DE, Schloetter M, Sebastian A (2002) Potassium citrate prevents increased urine calcium excretion and bone resorption induced by a high sodium chloride diet. J Clin Endocrinol Metab 87:2008–2012
46. Goulding A, Campbell D (1983) Dietary NaCl loads promote calciuria and bone loss in adult oophorectomized rats consuming a low calcium diet. J Nutr 113:1409–1414

47. Creedon A, Cashman KD (2000) The effect of high salt and high protein intake on calcium metabolism, bone composition and bone resorption in the rat. Br J Nutr 84:49–56
48. Somers MJ, Mavromatis K, Galis ZS, Harrison DG (2000) Vascular superoxide production and vasomotor function in hypertension induced by deoxycorticosterone acetate-salt. Circulation 101:1722–1728
49. Pu Q, Neves MF, Virdis A, Touyz RM, Schiffrin EL (2003) Endothelin antagonism on aldosterone-induced oxidative stress and vascular remodeling. Hypertension 42:49–55
50. Virdis A, Neves MF, Amiri F et al (2002) Spironolactone improves angiotensin-induced vascular changes and oxidative stress. Hypertension 40:504–510
51. Gandhi MS, Deshmukh PA, Kamalov G et al (2008) Causes and consequences of zinc dyshomeostasis in rats with chronic aldosteronism. J Cardiovasc Pharmacol 52:245–252
52. Fleckenstein A, Frey M, Fleckenstein-Grun G (1983) Consequences of uncontrolled calcium entry and its prevention with calcium antagonists. Eur Heart J 4(Suppl H):43–50
53. Ahokas RA, Sun Y, Bhattacharya SK, Gerling IC, Weber KT (2005) Aldosteronism and a proinflammatory vascular phenotype. Role of Mg^{2+}, Ca^{2+} and H_2O_2 in peripheral blood mononuclear cells. Circulation 111:51–57
54. Fujita T, Palmieri GM (2000) Calcium paradox disease: calcium deficiency prompting secondary hyperparathyroidism and cellular calcium overload. J Bone Miner Metab 18:109–125
55. Smogorzewski M, Zayed M, Zhang YB, Roe J, Massry SG (1993) Parathyroid hormone increases cytosolic calcium concentration in adult rat cardiac myocytes. Am J Physiol 264:H1998–H2006
56. Perna AF, Smogorzewski M, Massry SG (1989) Effects of verapamil on the abnormalities in fatty acid oxidation of myocardium. Kidney Int 36:453–457
57. Rampe D, Lacerda AE, Dage RC, Brown AM (1991) Parathyroid hormone: an endogenous modulator of cardiac calcium channels. Am J Physiol 261(6 Pt 2):H1945–H1950
58. Nordquist RE, Palmieri MA (1974) Intracellular localization of parathyroid hormone in the kidney. Endocrinology 95:229–237
59. Deluca HF, Engstrom GW, Rasmussen H (1962) The action of vitamin D and parathyroid hormone in vitro on calcium uptake and release by kidney mitochondria. Proc Natl Acad Sci USA 48:1604–1609
60. Sallis JD, Deluca HF, Rasmussen H (1963) Parathyroid hormone-dependent uptake of inorganic phosphate by mitochondria. J Biol Chem 238:4098–4102
61. Sallis JD, DeLuca HF (1966) Action of parathyroid hormone on mitochondria. Magnesium- and phosphate-independent respiration. J Biol Chem 241:1122–1127
62. Kimmich GA, Rasmussen H (1966) The effect of parathyroid hormone on mitochondrial ion transport in the terminal portion of the cytochrome chain. Biochim Biophys Acta 113: 457–466
63. Rasmussen H, Ogata E (1966) Parathyroid hormone and the reactions of mitochondria to cations. Biochemistry 5:733–745
64. Massry SG, Smogorzewski M (1994) Mechanisms through which parathyroid hormone mediates its deleterious effects on organ function in uremia. Semin Nephrol 14:219–231
65. Touyz RM (2005) Reactive oxygen species as mediators of calcium signaling by angiotensin II: implications in vascular physiology and pathophysiology. Antioxid Redox Signal 7: 1302–1314
66. Palty R, Silverman WF, Hershfinkel M et al (2010) NCLX is an essential component of mitochondrial Na^+/Ca^{2+} exchange. Proc Natl Acad Sci USA 107:436–441
67. Kuo TH, Zhu L, Golden K et al (2002) Altered Ca^{2+} homeostasis and impaired mitochondrial function in cardiomyopathy. Mol Cell Biochem 238:119–127
68. Kamalov G, Ahokas RA, Zhao W et al (2010) Temporal responses to intrinsically coupled calcium and zinc dyshomeostasis in cardiac myocytes and mitochondria during aldosteronism. Am J Physiol Heart Circ Physiol 298:H385–H394

69. Goodwin KD, Ahokas RA, Bhattacharya SK et al (2006) Preventing oxidative stress in rats with aldosteronism by calcitriol and dietary calcium and magnesium supplements. Am J Med Sci 332:73–78
70. Vidal A, Sun Y, Bhattacharya SK et al (2006) Calcium paradox of aldosteronism and the role of the parathyroid glands. Am J Physiol Heart Circ Physiol 290:H286–H294
71. Yang F, Nickerson PA (1988) Effect of parathyroidectomy on arterial hypertrophy, vascular lesions, and aortic calcium content in deoxycorticosterone-induced hypertension. Res Exp Med (Berl) 188:289–297
72. Selektor Y, Ahokas RA, Bhattacharya SK et al (2008) Cinacalcet and the prevention of secondary hyperparathyroidism in rats with aldosteronism. Am J Med Sci 335:105–110

Chapter 8
The ACE2/Ang (1–7) Pathway in Cardiac Remodeling Due to Pressure Overload

Seyyed M.R. Kazemi-Bajestani, Vaibhav B. Patel, Wang Wang, and Gavin Y. Oudit

Abstract The renin–angiotensin system (RAS) is a critical regulator of cardiovascular and renal functions, primarily through the angiotensin-converting enzyme (ACE)-mediated generation of angiotensin (Ang) II. Ang II mediates majority of its actions through Ang II type I receptor (AT1R) activation. Discovery of homolog of ACE, the ACE2, at the start of the twenty-first century, provided insight into the endogenous negative regulation of RAS. ACE2, a type I transmembrane protein, is a carboxypeptidase which converts Ang II into Ang 1–7. ACE2 has shown blood pressure-lowering effects, largely via reduction of Ang II-mediated pressor effects, along with the suppression of pathological cardiac remodeling. Ang 1–7 acts through its receptor, Mas receptor, and shows antiproliferative, anti-fibrotic, and thus anticardiac remodeling effects. ACE2 also converts Ang I into Ang 1–9, which is recently known to possess antiproliferative effects via AT2R activation. This chapter largely focuses on pathological remodeling as a result of pressure overload, role of Ang II/AT1R axis in the remodeling process, and negative regulation of RAS by ACE2/Ang 1–7/Mas receptor and Ang 1–9/AT2R axis.

Keywords Renin–angiotensin system • ACE2 • Ang 1–7 • Pressure overload • Cardiac remodeling

---

S.M.R. Kazemi-Bajestani • W. Wang
Department of Physiology, Mazankowski Alberta Heart institute, University of Alberta, Edmonton, AB, Canada

V.B. Patel
Department of Medicine, Mazankowski Alberta
Heart institute, University of Alberta, Edmonton, AB, Canada

G.Y. Oudit, M.D., Ph.D., F.R.C.P.(C) (✉)
Department of Physiology, Mazankowski Alberta Heart institute, University of Alberta, Edmonton, AB, Canada

Division of Cardiology, Department of Medicine, Mazankowski Alberta Heart institute, University of Alberta, Edmonton, AB, Canada T6G 2S2
e-mail: gavin.oudit@ualberta.ca

B.I. Jugdutt and N.S. Dhalla (eds.), *Cardiac Remodeling: Molecular Mechanisms*, Advances in Biochemistry in Health and Disease 5, DOI 10.1007/978-1-4614-5930-9_8,

8.1 Renin–Angiotensin System: An Essential Regulator of Cardiovascular Function

The renin–angiotensin system (RAS) is an essential regulator of cardiovascular and renal functions. Enhanced angiotensin II (Ang II) formation, as a product of dysregulated RAS, in plasma and several organs including heart and kidney precedes various pathologic cardiovascular processes [1]. Elevated Ang II induces hypertension, myocardial hypertrophy, fibrosis, and diastolic dysfunction, which might be exacerbated by angiotensin-converting enzyme 2 (ACE2) deficiency [2]. The vasoconstrictor effect of Ang II is associated with development of hypertension/pressure overload and its cardiac hypertrophy/remodeling consequences [1]. Furthermore, local expression of Ang II in the heart is a critical mediator of pressure-overload-induced cardiac hypertrophy [3]. Ang II overexpression can also mediate kidney inflammation/fibrosis [4], increase thrombotic potential, augment release of aldosterone from the adrenal cortex, increase central neural secretion of antidiuretic hormone/adrenocorticotropic hormone, and elevate sympathetic tone which may aggravate the induced hypertension [5]. Ang II activates several myocardial hypertrophic/fibrosis signaling pathways, including mitogen-activated protein kinase (MAPK), protein kinase C (PKC)α, PKCβ1, Janus Kinase 2 and signal transducer and activator of transcription 3 (JAK2-STAT3), and extracellular signal-regulated kinases (ERK) 1/2 [2, 6]. Increased inflammatory and fibrotic responses of myocardium to Ang II by increasing type I and III collagen, transforming growth factor (TGF)-β1, and tumor necrosis factor (TNF) α confirm the key role of Ang II in development of inflammation and subsequent fibrosis [7]. Ang II by induction of monocyte chemoattractant protein (MCP)-1 leads to maladaptive cardiac fibrosis [7]. Ang II induces cardiomyocyte apoptosis which is mediated via nicotinamide adenine dinucleotide phosphate (NADPH) oxidase activation through p38 MAPK activation, decrease in B-cell lymphoma 2(Bcl-2) protein, and caspase activation [8]. In the presence of Ang II, p38 MAPK activation is mediated by reactive oxygen species (ROS)-dependent Ca(2+)/calmodulin-dependent protein kinase (CaMK)II activation [9].

Ang II by enhancement of cardiac inflammation, apoptosis, fibrosis, hypertrophy, and ROS production leads to remodeling and dysfunction. ACE2 is an important enzymatic component of RAS which causes degradation of Ang II into Ang 1–7. Degradation of Ang II by ACE2 can apparently attenuate vascular homeostasis through the continuous control of vascular tone and constitutive inhibition of Ang II-mediated cardiomyopathic effects [1]. ACE2 deficiency augments Ang II-mediated oxidative stress/inflammation and subsequent age-dependent cardiomyopathy [10]. Ang 1–7 can be generated both from Ang I and Ang II and can offset the effects of Ang II [11]. All consequent ligands of RAS may affect through three different receptors: MasR, AT1R, and AT2R. Ang 1–7 function through its receptor, Mas, composes vasodilatory, antiproliferative, and antihypertrophic peptide. Genetic and pharmacological augmentation of ACE2/Ang 1–7/Mas axis seems an endogenous counter-regulatory pathway that results in improvement of heart failure.

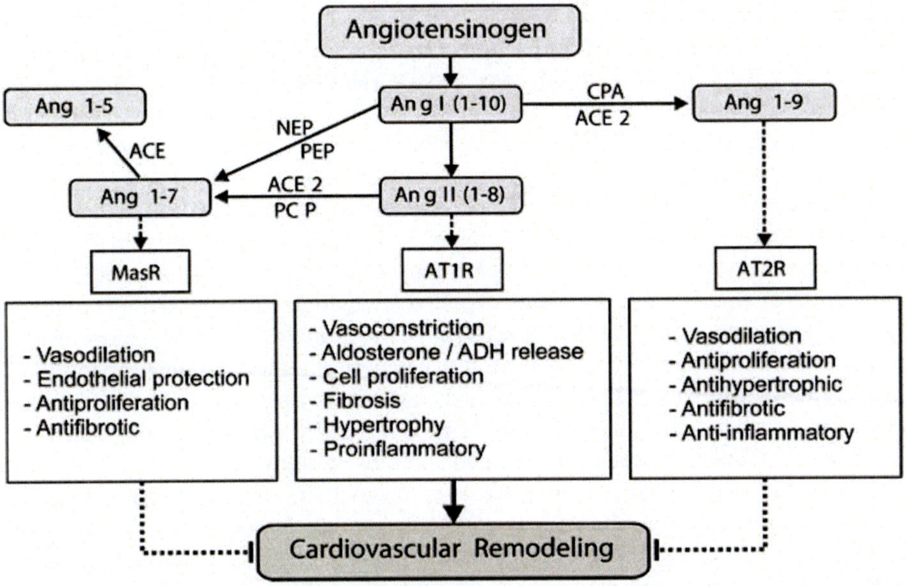

Fig. 8.1 The enzymatic cascade involved in the renin–angiotensin system, key receptor systems, and the biological effects mediated by Ang II, Ang 1–7, and Ang 1–9. *Solid lines*, enzymatic pathways; *broken lines*, peptide agonist interacting with its key receptor. *Bold line*, stimulatory effect; *dotted lines*, inhibitory effects; *ACE2* angiotensin-converting enzyme 2, *ADH* antidiuretic hormone, *Ang* angiotensin, *APA* aminopeptidase A, *AT1R and AT2R* Ang II type 1 and type 2 receptor, respectively; *CPA* carboxypeptidase A, *Mas R* Ang 1–7 receptor, *NEP* neutral endopeptidase, *PCP* prolyl carboxypeptidase (also known as angiotensinase C), *PEP* prolyl endopeptidase

Vasoconstrictive, proliferative, hypertrophic, fibrotic, and proinflammatory effects of Ang II are mediated by AT1R activation [1].

In addition of the effects of ACE2 on conversion of Ang II to Ang 1–7, Ang 1–9 also can be produced in the RAS system through the affects of ACE2 on Ang I [1]. Recently, it has been shown that Ang 1–9 can antagonize pro-hypertrophic signaling in cardiomyocytes via the AT2R activation [12] (Fig. 8.1).

8.2 Pressure Overload Induces Cardiac Hypertrophy/ Remodeling: Role of RAS

Pressure overload results in mechanical stress, which is transduced into intracellular signals that regulate gene expression and protein synthesis in cardiac myocytes [13, 14]. Pressure overload induces cardiac hypertrophy by increasing the expression of proto-oncogenes (c-fos and c-myc) [13]; growth-stimulating substances [15] enhanced phosphatidylinositol turnover and activation of MAPK subsequent to PKC [16]. Mice with reduced cardiac transcription factor CCAAT/enhancer-binding protein (C/EBPβ) levels showed considerable resistance to cardiac failure upon pressure overload [17]. Downregulation of ERK1/2, protein kinase B (Akt), and glycogen synthase kinase (GSK)-3β is involved in the transition from pressure-overload compensated hypertrophy to heart failure [18].

RAS activation plays a key role in pressure-overload-induced cardiac hypertrophy [19]. Ang II, especially when it is expressed locally, is an essential mediator of pressure-overload-induced cardiac hypertrophy. In a pressure-overload model, local expression of Ang II leads to cardiac hypertrophy by upregulating NADPH oxidase expression and promoting ROS synthesis [20]. Antagonizing Ang II receptors results in regression of hypertensive left ventricular hypertrophy and inhibits the intracellular signaling pathway of stretch-mediated cardiomyocyte hypertrophy [21]. A considerable analogy exists between the signaling pathways that mediate cardiac hypertrophy induced by pressure overload and those induced by Ang II. Either Ang II or pressure-overload mechanical stress activates PKC. PKC activation in these parallel process increases the activity of MAPK, which subsequently induces c-fos mRNA expression. Accumulation of these signaling pathways eventually lead to augmented protein synthesis in cardiac myocytes/hypertrophy [21, 22].

Pressure overload induces mechanical stress, which leads to cardiac hypertrophy through AT1R independently of Ang II [23]. AT1R is considered as a "mechanical sensor" and converts mechanical stress into biochemical signals inside the cells [23]. AT1R is involved in the mechanical stress-induced activation of ERKs and phosphorylation of Jak2 [23]. Mechanical stress evoked by activation of AT1R also induces cardiac hypertrophy through calcineurin, a Ca^{2+}-dependent phosphatase, pathway independent of Ang II effects [24]. Furthermore, activating transcription factor 3 (ATF3) plays a crucial role in development of either Ang II or pressure-overload-induced cardiac hypertrophy [25]. The effects of ATF3 on cardiac remodeling are mediated by negative feedback to the ERK and C-jun-N-terminal kinase (JNK) pathways and regulation of pro-fibrotic cytokines [26] (Fig. 8.2).

Oxidative stress and inflammation play important roles in development of activated RAS-induced cardiac remodeling. Pressure overload combined with ACE2 deficiency results in activation of Ang II-stimulated signaling pathways in myocardium which is associated with increased expression and phosphorylation of $p47^{phox}$, NADPH oxidase activity, and superoxide generation, leading to enhanced matrix metalloproteinase (MMP)-mediated degradation of the extracellular matrix [27]. Pressure-overload-induced Ang II elevation augments cardiovascular inflammation through the activation of nuclear factor kappa-light-chain-enhancer of activated B cells (NFκB) and the subsequent production of inflammatory mediators like interleukin (IL)-6 and IL-1β, which leads to exacerbated hypertension [28].

Impaired RAS is associated with hypertension development. Plasma levels of Ang II in hypertensive patients are considerably higher than control individuals [29]. Infusion of Ang II for 14 days leads to significant increase in peak systolic blood pressure over vehicle-infused wild-type mice [2]. According to animal models, it is clear that a pressure overload trigger like transverse aortic constriction is a

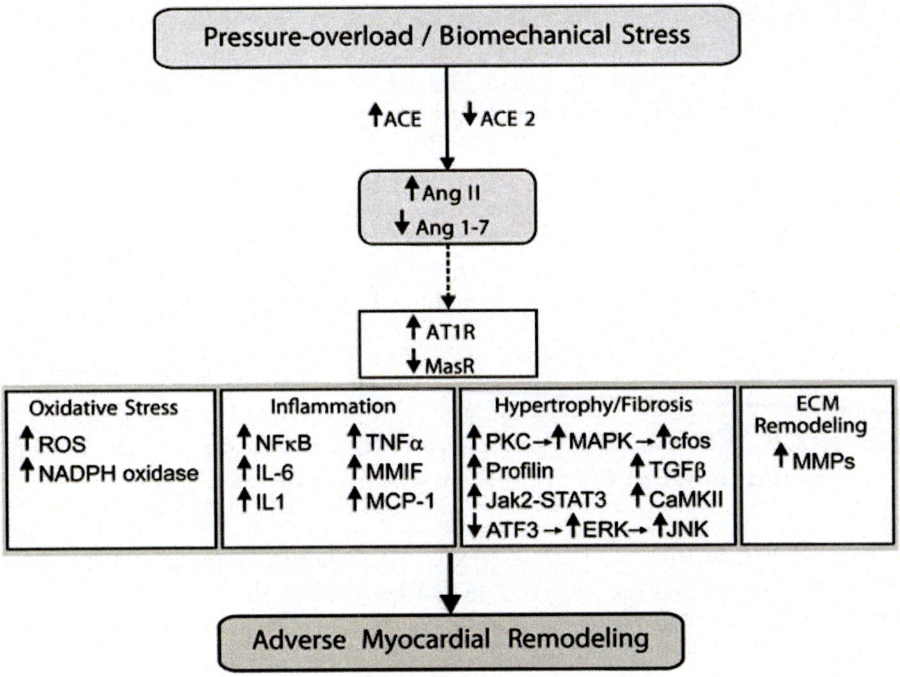

Fig. 8.2 Pressure-overload-induced Ang II signaling pathways which result in adverse myocardial remodeling. *Solid line*, enzymatic pathway; *broken lines*, peptide agonist interacting with its key receptor. *Bold line*, stimulatory effect; *ACE* angiotensin-converting enzyme, *AT1R* Ang II type 1 receptor, *ATF3* Activating transcription factor 3, *CaMKII* Ca^{2+}/calmodulin-dependent protein kinases II, *ERK* extracellular signal-regulated kinase, *IL-6* interleukin-6, *JAK2-STAT* Janus Kinase 2-signal transducer and activator of transcription system, *JNK* C-jun-N-terminal kinase, *MAPK* mitogen-activated protein kinase, *MasR* Ang 1–7 receptor, *MCP-1* monocyte chemoattractant protein 1, *MMIF* macrophage migration inhibitory factor, *MMP* matrix metalloproteinase, *NADPH* nicotinamide adenine dinucleotide phosphate, *NFκB* nuclear factor kappa-light-chain-enhancer of activated B cells, *PKC* protein kinase C, *ROS* reactive oxygen species, *TGFβ1* transforming growth factorβ1, *TNF α* tumor necrosis factor α

primary factor of Ang II increase. However, in humans, it is not apparent that Ang II elevation is a cause or consequence of hypertension. Long-term treatment of the hypertensive patients with Ang II antagonists is associated with lower risk of clinical events [30]. Ang II, as a cause or consequence of hypertension, is a peptide which worsens the cardiovascular status of the patients and may aggravate the cardiac consequences of hypertension.

Various regulator components of RAS have been targeted in recent decades for hypertension treatment. Direct renin inhibition has been proposed to be a comprehensive RAS inhibitor by suppression of the plasma renin activity at its rate-limiting step [31]. For several years, ACE inhibition has been recognized to reduce blood pressure and offset hypertension consequences. ACE inhibitor can inhibit Ang I to

Ang II conversion and therefore prevents the cardiovascular side effects of Ang II. Chronic ACE inhibition decreases the extent of left ventricular hypertrophy and improves cardiac function and survival [32]. ACE inhibition can limit both myocyte and interstitial remodeling despite ongoing cardiac pressure via blunting the induction of collagen I and III, laminin B, and fibronectin at both the mRNA and protein levels [33]. Angiotensin receptor blockers (ARBs) are other group of pharmaceuticals which can modulate RAS in hypertensive patients. Left ventricular hypertrophy can be prevented by ARBs in a pressure-overload model [34]. Gene polymorphism studies about critical components of RAS showed the possible association to the development of hypertension [35, 36] although some studies have not been able to show a strong dependency [37].

8.3 ACE2 Effects on Hypertension: Blood Pressure Reduction and Cardiac Remodeling Suppression

Treatment with ACE2 was found to significantly reduce blood pressure and attenuate cardiac remodeling in various hypertensive models [1, 38]. Reduced ACE2 mRNA and protein expression in hypertensive rats and association of ACE2 impairment with cardiac dysfunction was first reported in 2002 [39]. Generally, the possible mechanism for direct involvement of ACE2 in hypertension attenuation includes Ang II degradation to Ang 1–7. ACE2 antagonizes the pressor, hypertrophic, and pro-oxidative effects of Ang II by decreasing the levels of Ang II [2]. Increased Ang 1–7/Mas results in vasodilation, preserved endothelial function, and cardiac remodeling/fibrosis suppression [2]. Other evidence showed that decrease in the circulatory plasma Ang II was the driving cause of decreased hypertension coupled with a partial contribution from the increased levels of Ang 1–7 [40].

ACE2 not only use RAS substrate but has several other biological substrates. ACE2 as a monocarboxypeptidase can mediate other parallel reactions in apelin and bradykinin family which results in further antihypertensive/remodeling outcomes [1]. Recombinant human ACE2 (rhACE2) reduces Ang II-induced hypertension in mice, by reducing Ang II-mediated activation of enhanced ERK1/2, PKC pathways, and renal fibrosis [2]. ACE2 can decrease Ang II-induced expression of profilin-1 and its subsequent hypertrophic signaling, MAPK, ERK1/2, and JNK, in an Ang 1–7-dependent manner. Overexpression of ACE2 and decreased profilin-1 levels results in suppression of ERK1/2 and JNK phosphorylation in aortas of spontaneously hypertensive rats [41]. In response to pressure overload, activation of MAPK is found to be more in mice with ACE2 deficiency compared to those with intact ACE2. Cardiomyocytes MAPK response to Ang II is considerably enhanced in lack of ACE2 [42]. In a pressure-overload model, ACE2 provides therapeutic benefits against pathological myocardial remodeling as evident by the levels of pathological hypertrophy markers: α-skeletal actin, brain natriuretic peptide, β-myosin heavy chain, and procollagen type Iα1 and procollagen type IIIα1 [2].

In the pressure-overload context, ACE2 is a factor which can attenuate biomechanical stress-induced NADPH oxidase activation, superoxide generation, and enhanced MMP-mediated degradation of cardiac extracellular matrix [27]. It can be concluded that ACE2 is an important enzyme which can decrease the susceptibility of myocardium to mechanical stress. It is important to mention that most of the animal models of pressure overload only investigated the pure effect of pressure overload on RAS system; however, in humans, the process of hypertension is more complex. Hypertension is a multifactorial disease and is due to genetic and environmental determinants [43]. Combination of several affective factors, including atherosclerosis, on hypertension makes the interpretation complicated. RAS predominantly regulates blood pressure, endothelial function, and atherosclerosis, three essential risk factors of cardiovascular diseases and heart failure [1]. The effects of ACE2 on attenuation of these exacerbating factors may attenuate the consequences of hypertension and prevent cardiac remodeling. ACE2 can suppress Ang II- induced inflammation by reducing macrophage migration inhibitory factor [44] and T-lymphocyte-mediated inflammation [4]. ACE2 deficiency contributes to increased proinflammatory cytokines, including IL-6, MCP-1, and vascular cell adhesion molecule-1 [44].

8.4 Ang 1–7/Mas Axis in Pressure-Overload-Induced Heart Failure

Ang 1–7 is produced primarily by the enzymatic breakdown of either Ang I or Ang II. Ang 1–7 can be formed directly from Ang I by endopeptidases including neutral endopeptidase and prolyl endopeptidase. ACE2, as a carboxypeptidase, also can metabolize Ang II from its C terminus and produce Ang 1–7. Ang 1–7 by antagonizing Ang II results in beneficial vasodilation and anticardiac remodeling outcomes [1]. Vasodilator effect of Ang 1–7 in the heart is suggested to be dependent on vasodilator prostaglandins and nitric oxide (NO) release [45]. Evidence from in vivo and ex vivo studies in humans and several animal models shows a key role of Ang 1–7 in blood pressure modulation, probably through a combination of bradykinin and NO signaling [46].

Treatment of cardiomyocytes with Ang 1–7 attenuates Ang II-induced hypertrophy by modulating calcineurin/nuclear factor of activated T-cell signaling cascade via the NO/cyclic guanosine monophosphate pathway [47]. Therapeutic application of Ang 1–7 is limited due to its short half-life and rapid turnover; therefore, ACE2 is a good alternative with acceptable half-life which ultimately leads to Ang 1–7 over production in pathologic conditions [1]. Another limitation for Ang 1–7 supplementation is Mas receptor desensitization and internalization following Ang 1–7 stimulation [48].

Ang 1–7/Mas axis may offset Ang II pathologic effects through other diverse mechanism. Mas is considered to be a physiological antagonist of the Ang II AT1R [49]. Ang 1–7 showed beneficial effects in several experimental models of cardiovascular disease, including attenuating the pressure-overload-induced heart failure [50]. Ang 1–7 can reduce hypertension-induced cardiac remodeling through a direct effect on the heart. Interestingly, Ang 1–7 showed selective effects on intracellular signaling pathways that mediate cardiac remodeling. Ang 1–7-mediated effects in the heart result in significant decrease in phosphorylation of c-Src and p38 kinase seen after Ang II infusion [50]. Ang 1–7 upregulates MAPK phosphatase dual-specificity phosphatase in association with a decrease in Ang II-mediated phosphorylation of ERK1/2 [51]. Treatment of the spontaneously hypertensive rats with Ang 1–7 showed attenuation of elevated mean arterial pressure and significant increase in MMP-2 mRNA expression. Furthermore, increased tissue inhibitor of metalloproteinases (TIMP-1) mRNA expression in left ventricles of hypertensive rats was significantly attenuated by Ang 1–7 [52]. This novel effect of Ang 1–7 on normalization of expression of pro-fibrotic factors and enhancement of anti-fibrotic factors shows the beneficial effects on recovery of pressure-overload-induced remodeling.

The effect of Ang 1–7 treatment on upregulation of cardiac nitric oxide synthase (NOS) in hypertensive state can be considered as another cardioprotective mechanism of Ang 1–7 against pressure overload [53]. In an ACE2-deficient state, supplementation with Ang 1–7 suppresses the increased NADPH oxidase activity, resulting in rescue of the early dilated cardiomyopathy and normalization of the atrial natriuretic factor, B type natriuretic peptide (BNP), and α-skeletal actin in a pressure-overload model [27]. Interestingly, blocking the Ang II/AT1R axis with irbesartan or enhancing Ang 1–7/Mas receptor axis via Ang 1–7 treatment in pressure-overloaded ACE2-deficient mice showed similar degree of cardioprotection against pressure-overload-induced cardiac remodeling. This illustrated that, there is a marked degree of redundancy in Ang II/AT1R axis and Ang 1–7/Mas receptor axis, counter-regulatory pathways, in ACE2-deficient environment and Ang 1–7 can be a potential therapeutic regimen to counteract cardiac remodeling [54] (Fig. 8.3).

8.5 Ang 1–9/AT2R: Another Valuable Product of ACE2 Activity in RAS

Recently, an interesting role of Ang 1–9 in reduction of cardiac fibrosis and altered aortic contraction in hypertensive rats via the AT2R has been shown [55]. Ang 1–9 supplementation did not affect the blood pressure, cardiac function, or left ventricular mass index; however, Ang 1–9 reduced cardiac fibrosis by 50% through modulating collagen I expression via AT2R mediation and improved NO bioavailability in this model [55]. Application of a selective AT2R inhibitor could significantly increase the MMP-14 gene expression in hypertensive hearts [55]. Activation of intracellular signaling pathway Rho A and its Rho kinase [Rho-associated, coiled-coil containing protein kinase (ROCK)] pathway is a novel mechanism of vasoconstriction and cardiovascular remodeling in hypertensive patients [56]. ROCK inhibition results in reduced systolic blood pressure and increased plasma and aortic ACE2 enzymatic activity, aortic NOS mRNA levels, and Ang 1–9 plasma levels.

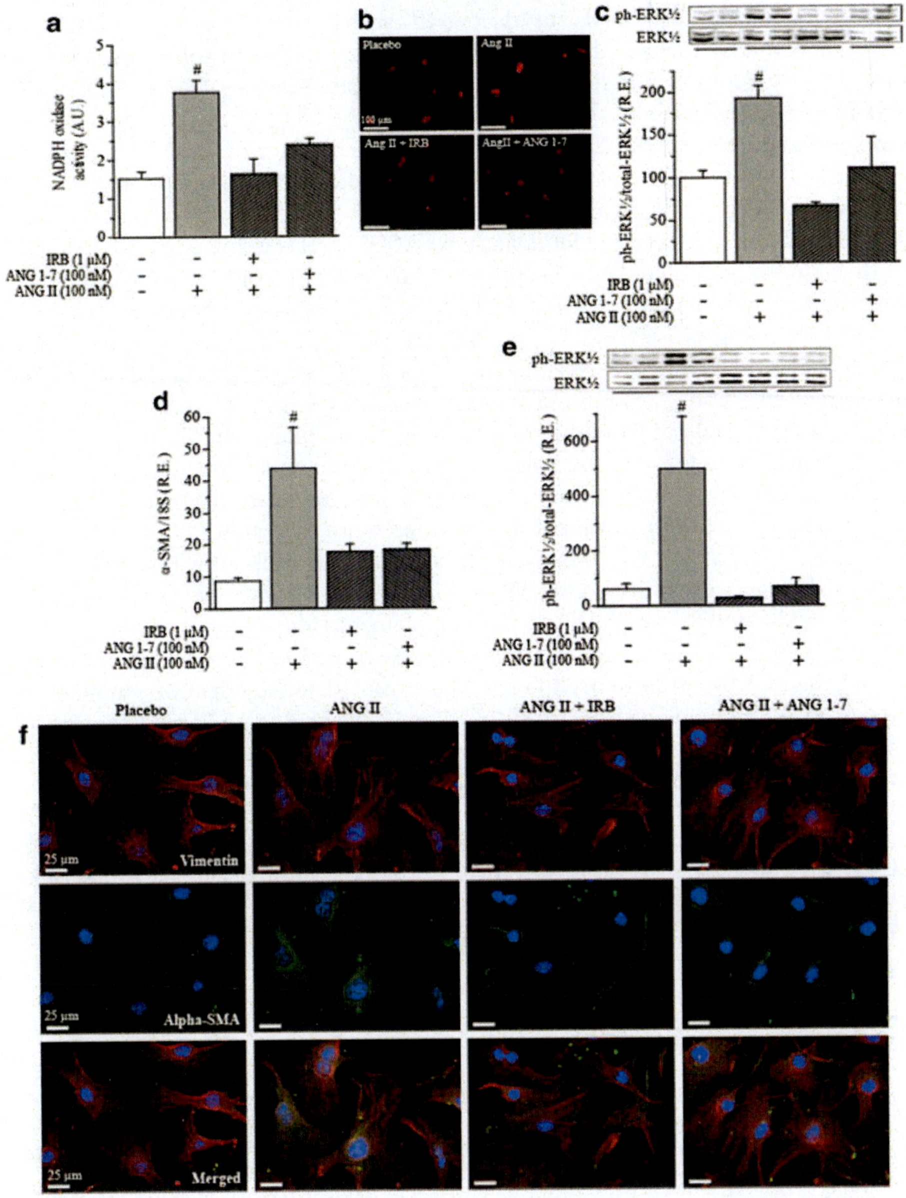

Fig. 8.3 Irbesartan (IRB), AT1R blocker, and Ang 1–7 suppresses oxidant stress, signaling, and remodeling induced by Ang II in ACE2-null cardiomyocytes and cardiofibroblasts isolated from pressure-overloaded hearts. Ang II increases oxidant stress, as evident by increased NADPH oxidase activity (**a**) and DHE fluorescence (**b**) along with phosphorylation of ERK 1/2 (**c**) in isolated cardiomyocytes which were lowered by IRB and Ang 1–7 pretreatment. In cardiofibroblasts, Ang II exposure stimulated mRNA expression of α-SMA (**d**) with increased phosphorylation of ERK1/2 (**e**) resulting in increased immunostaining for α-SMA (**f**) which was prevented by pretreatment with IRB and Ang 1–7 (**d–f**). *A.U.* arbitary unit, *R.E.* relative expression, *ERK ½* extracellular signal-regulated kinase 1/2, *α-SMA* alpha-smooth muscle actin. #$p<0.05$ compared to all other groups (reproduced with permission [54])

ROCK inhibition also leads to decreased aortic mRNA levels of TGF-β1, plasminogen activator inhibitor, and MCP-1 as well as decrease aortic NADPH oxidase activity and ROS production [57]. This increased levels of ACE2 activity in accordance with elevated Ang 1–9 as well as decreased levels of hypertrophic mediators reveal the beneficial effects of Ang 1–9 as an indicator of ROCK inhibition.

The beneficial signaling pathways against cardiovascular remodeling that are affected by Ang 1–7 or Ang 1–9 are similar. It is not clear that which of these peptides will be produced predominantly by ACE2 effects in hypertensive patients. Further studies may elucidate this point that can be very helpful in clinical settings, because then we can augment the ACE2 effects by adding another factor which boosts either Ang 1–9 or Ang 1–7 production. However, in the myocardial infarction model, this comparison has been performed. Administration of Ang 1–9 after myocardial infarction prevents the increase in plasma Ang II levels, left ventricular ACE activities, and cardiac myocyte hypertrophy development. These effects are not mediated by Ang 1–7 or the MasR. In experimental myocardial infarction models, increased Ang 1–9 levels were associated with attenuation of left ventricular hypertrophy even after adjustment for blood pressure reduction [58]. In summary, Ang 1–9, as a product of ACE2 activity, is produced from Ang I. Furthermore, increased Ang 1–9 may have a feedback effect on ACE activity which can prevent Ang II production in parallel. Ang 1–9 supplementation not only neutralizes the pressure-overload-induced fibrosis but also shows anti-ischemia-induced cardiac remodeling. Further studies need to be done to elucidate the effect of combination therapy of ACE2 with either Ang 1–9/AT2R agonists or Ang 1–7/MasR agonists in the pressure-overload models. Their probable feedback effects on the RAS component make the judgment complex at this time.

Acknowledgments GYO is a Clinician-Investigator of the Alberta Innovates – Health Solutions (AI-HS) and the Distinguish Clinician Scientist of the Heart and Stroke Foundation of Canada and Canadian Institutes of Health Research. WW is supported by the Mazankowski Alberta Heart Institute Graduate Scholarship, and SMKB is supported by the STEADHI HF Training Program.

References

1. Oudit GY, Penninger JM (2011) Recombinant human angiotensin-converting enzyme 2 as a new renin-angiotensin system peptidase for heart failure therapy. Curr Heart Fail Rep 8:176–183
2. Zhong J, Basu R, Guo D et al (2010) Angiotensin-converting enzyme 2 suppresses pathological hypertrophy, myocardial fibrosis, and cardiac dysfunction. Circulation 122:717–728
3. Uechi M, Tanaka Y, Aramaki Y et al (2008) Evaluation of the renin-angiotensin system in cardiac tissues of cats with pressure-overload cardiac hypertrophy. Am J Vet Res 69:343–348
4. Zhong J, Guo D, Chen CB et al (2011) Prevention of angiotensin II-mediated renal oxidative stress, inflammation, and fibrosis by angiotensin-converting enzyme 2. Hypertension 57:314–322
5. Weir MR, Dzau VJ (1999) The renin-angiotensin-aldosterone system: a specific target for hypertension management. Am J Hypertens 12:205S–213S

6. Mehta PK, Griendling KK (2007) Angiotensin II cell signaling: physiological and pathological effects in the cardiovascular system. Am J Physiol Cell Physiol 292:C82–C97
7. Haudek SB, Cheng J, Du J et al (2010) Monocytic fibroblast precursors mediate fibrosis in angiotensin-II-induced cardiac hypertrophy. J Mol Cell Cardiol 49:499–507
8. Qin F, Patel R, Yan C et al (2006) NADPH oxidase is involved in angiotensin II-induced apoptosis in H9C2 cardiac muscle cells: effects of apocynin. Free Radic Biol Med 40:236–246
9. Palomeque J, Rueda OV, Sapia L et al (2009) Angiotensin II-induced oxidative stress resets the Ca^{2+} dependence of Ca^{2+}–calmodulin protein kinase II and promotes a death pathway conserved across different species. Circ Res 105:1204–1212
10. Oudit GY, Kassiri Z, Patel MP et al (2007) Angiotensin II-mediated oxidative stress and inflammation mediate the age-dependent cardiomyopathy in ACE2 null mice. Cardiovasc Res 75:29–39
11. Tom B, Dendorfer A, Danser AH (2003) Bradykinin, angiotensin-(1–7), and ACE inhibitors: how do they interact? Int J Biochem Cell Biol 35:792–801
12. Flores-Munoz M, Smith NJ, Haggerty C et al (2011) Angiotensin1–9 antagonises pro-hypertrophic signalling in cardiomyocytes via the angiotensin type 2 receptor. J Physiol 589:939–951
13. Komuro I, Kurabayashi M, Takaku F et al (1988) Expression of cellular oncogenes in the myocardium during the developmental stage and pressure-overloaded hypertrophy of the rat heart. Circ Res 62:1075–1079
14. Guo D, Thiyam G, Bodiga S et al (2011) Uncoupling between enhanced excitation-contraction coupling and the response to heart disease: lessons from the PI3Kgamma knockout murine model. J Mol Cell Cardiol 50:606–612
15. Zhang R, Crump J, Reisin E (2003) Regression of left ventricular hypertrophy is a key goal of hypertension management. Curr Hypertens Rep 5:301–308
16. Komuro I, Katoh Y, Kaida T et al (1991) Mechanical loading stimulates cell hypertrophy and specific gene expression in cultured rat cardiac myocytes. Possible role of protein kinase C activation. J Biol Chem 266:1265–1268
17. Bostrom P, Mann N, Wu J et al (2010) C/EBPbeta controls exercise-induced cardiac growth and protects against pathological cardiac remodeling. Cell 143:1072–1083
18. Li XM, Ma YT, Yang YN et al (2009) Downregulation of survival signalling pathways and increased apoptosis in the transition of pressure overload-induced cardiac hypertrophy to heart failure. Clin Exp Pharmacol Physiol 36:1054–1061
19. Baker KM, Chernin MI, Wixson SK et al (1990) Renin-angiotensin system involvement inpressure-overload cardiac hypertrophy in rats. Am J Physiol 259:H324–H332
20. Liu J, Zhou J, An W et al (2010) Apocynin attenuates pressure overload-induced cardiac hypertrophy in rats by reducing levels of reactive oxygen species. Can J Physiol Pharmacol 88:745–752
21. Kojima M, Shiojima I, Yamazaki T et al (1994) Angiotensin II receptor antagonist TCV-116 induces regression of hypertensive left ventricular hypertrophy in vivo and inhibits the intracellular signaling pathway of stretch-mediated cardiomyocyte hypertrophy in vitro. Circulation 89:2204–2211
22. Tsuda T, Kawahara Y, Ishida Y et al (1992) Angiotensin II stimulates two myelin basic protein/microtubule-associated protein 2 kinases in cultured vascular smooth muscle cells. Circ Res 71:620–630
23. Zou Y, Akazawa H, Qin Y et al (2004) Mechanical stress activates angiotensin II type 1 receptor without the involvement of angiotensin II. Nat Cell Biol 6:499–506
24. Zhou N, Li L, Wu J et al (2010) Mechanical stress-evoked but angiotensin II-independent activation of angiotensin II type 1 receptor induces cardiac hypertrophy through calcineurin pathway. Biochem Biophys Res Commun 397:263–269
25. Zou Y, Lin L, Ye Y et al (2012) Qiliqiangxin inhibits the development of cardiac hypertrophy, remodeling and dysfunction during four weeks of pressure overload in mice. J Cardiovasc Pharmacol 59:268–280

26. Zhou H, Shen DF, Bian ZY et al (2011) Activating transcription factor 3 deficiency promotes cardiac hypertrophy, dysfunction, and fibrosis induced by pressure overload. PLoS One 6:e26744. doi:10.1371/journal.pone.0026744
27. Bodiga S, Zhong JC, Wang W et al (2011) Enhanced susceptibility to biomechanical stress in ACE2 null mice is prevented by loss of the p47(phox) NADPH oxidase subunit. Cardiovasc Res 91:151–161
28. Lahera V, Cachofeiro V, de Las Heras N (2011) Interplay of hypertension, inflammation, andangiotensin II. Am J Hypertens 24:1059. doi:10.1038/ajh.2011.142
29. Swartz SL, Williams GH, Hollenberg NK et al (1979) Converting enzyme inhibition in essential hypertension: the hypotensive response does not reflect only reduced angiotensin II formation. Hypertension 1:106–111
30. Swindle JP, Buzinec P, Iorga SR et al (2011) Long-term clinical and economic outcomes associated with angiotensin II receptor blocker use in hypertensive patients. Curr Med Res Opin 27:1719–1731
31. Abdellatif AA (2012) The role of renin inhibition in treating the hypertensive patient with diabetes: a summary of preclinical and clinical evidence. Expert Rev Cardiovasc Ther 10:251–263
32. Weinberg EO, Schoen FJ, George D et al (1994) Angiotensin-converting enzyme inhibition prolongs survival and modifies the transition to heart failure in rats with pressure overload hypertrophy due to ascending aortic stenosis. Circulation 90:1410–1422
33. Grimm D, Kromer EP, Bocker W et al (1998) Regulation of extracellular matrix proteins inpressure-overload cardiac hypertrophy: effects of angiotensin converting enzyme inhibition. J Hypertens 16:1345–1355
34. Liu L, Wang W, Meng X et al (2010) Left ventricular hypertrophy induced by abdominal aortic banding and its prevention by angiotensin receptor blocker telmisartan–a proteomic analysis. J Physiol Biochem 66:329–338
35. Lu N, Yang Y, Wang Y et al (2012) ACE2 gene polymorphism and essential hypertension: an updated meta-analysis involving 11,051 subjects. Mol Biol Rep 39:6581–6589
36. Yanyan L (2012) The angiotensin converting enzyme gene insertion/deletion (I/D) polymorphism and essential hypertension in the Chinese population: a meta-analysis including 21058 participants. Intern Med J 42:439–444
37. Ned RM, Yesupriya A, Imperatore G et al (2012) The ACE I/D polymorphism in US adults: limited evidence of association with hypertension-related traits and sex-specific effects by race/ethnicity. Am J Hypertens 25:209–215
38. Wang W, Bodiga S, Das SK et al (2012) Role of ACE2 in diastolic and systolic heart failure. Heart Fail Rev 17:683–691
39. Crackower MA, Sarao R, Oudit GY et al (2002) Angiotensin-converting enzyme 2 is an essential regulator of heart function. Nature 417:822–828
40. Wysocki J, Ye M, Rodriguez E et al (2010) Targeting the degradation of angiotensin II with recombinant angiotensin-converting enzyme 2: prevention of angiotensin II-dependent hypertension. Hypertension 55:90–98
41. Zhong JC, Ye JY, Jin HY et al (2011) Telmisartan attenuates aortic hypertrophy in hypertensive rats by the modulation of ACE2 and profilin-1 expression. Regul Pept 166:90–97
42. Yamamoto K, Ohishi M, Katsuya T et al (2006) Deletion of angiotensin-converting enzyme 2 accelerates pressure overload-induced cardiac dysfunction by increasing local angiotensin II. Hypertension 47:718–726
43. Carretero OA, Oparil S (2000) Essential hypertension. Part I: definition and etiology. Circulation 101:329–335
44. Zhong JC, Yu XY, Lin QX et al (2008) Enhanced angiotensin converting enzyme 2 regulates the insulin/Akt signalling pathway by blockade of macrophage migration inhibitory factor expression. Br J Pharmacol 153:66–74
45. Castro CH, Santos RA, Ferreira AJ et al (2005) Evidence for a functional interaction of the angiotensin-(1–7) receptor Mas with AT1 and AT2 receptors in the mouse heart. Hypertension 46:937–942

46. Katovich MJ, Grobe JL, Raizada MK (2008) Angiotensin-(1–7) as an antihypertensive, antifibrotic target. Curr Hypertens Rep 10:227–232
47. Gomes ER, Lara AA, Almeida PW et al (2010) Angiotensin-(1–7) prevents cardiomyocyte pathological remodeling through a nitric oxide/guanosine 3′,5′-cyclic monophosphate-dependent pathway. Hypertension 55:153–160
48. Gironacci MM, Adamo HP, Corradi G et al (2011) Angiotensin (1–7) induces MAS receptor internalization. Hypertension 58:176–181
49. Kostenis E, Milligan G, Christopoulos A et al (2005) G-protein-coupled receptor Mas is a physiological antagonist of the angiotensin II type 1 receptor. Circulation 111:1806–1813
50. Mercure C, Yogi A, Callera GE et al (2008) Angiotensin(1–7) blunts hypertensive cardiac remodeling by a direct effect on the heart. Circ Res 103:1319–1326
51. McCollum LT, Gallagher PE, Ann Tallant E (2012) Angiotensin-(1–7) attenuates angiotensin II- induced cardiac remodeling associated with upregulation of dual-specificity phosphatase 1. Am J Physiol Heart Circ Physiol 302:H801–H810
52. Pei Z, Meng R, Li G et al (2010) Angiotensin-(1–7) ameliorates myocardial remodeling and interstitial fibrosis in spontaneous hypertension: role of MMPs/TIMPs. Toxicol Lett 199:173–181
53. Costa MA, Lopez Verrilli MA, Gomez KA et al (2010) Angiotensin-(1–7) upregulates cardiac nitric oxide synthase in spontaneously hypertensive rats. Am J Physiol Heart Circ Physiol 299:H1205–H1211
54. Patel VB, Bodiga S, Fan D et al (2012) Cardioprotective effects mediated by angiotensin II type 1 receptor blockade and enhancing angiotensin 1–7 in experimental heart failure in angiotensin- converting enzyme 2-null mice. Hypertension 59:1195–1203
55. Flores-Munoz M, Work LM, Douglas K et al (2012) Angiotensin-(1–9) attenuates cardiac fibrosis in the stroke-prone spontaneously hypertensive rat via the angiotensin type 2 receptor. Hypertension 59:300–307
56. Johns DG, Dorrance AM, Leite R et al (2000) Novel signaling pathways contributing to vascular changes in hypertension. J Biomed Sci 7:431–443
57. Ocaranza MP, Rivera P, Novoa U et al (2011) Rho kinase inhibition activates the homologous angiotensin-converting enzyme-angiotensin-(1–9) axis in experimental hypertension. J Hypertens 29:706–715
58. Ocaranza MP, Lavandero S, Jalil JE et al (2010) Angiotensin-(1–9) regulates cardiac hypertrophy in vivo and in vitro. J Hypertens 28:1054–1064

Chapter 9
Local Actions of Natriuretic Peptides and Nitric Oxide in Cardiac Remodeling: Implications for Therapy

Michaela Kuhn and Hitoshi Nakagawa

Abstract Cardiac remodeling in response to sustained pressure overload involves cardiomyocyte hypertrophy and contractile dysfunction together with interstitial changes such as fibrosis and diminished capillary density. These changes are modulated by mechanical forces and factors secreted between and even within cells, exerting paracrine and auto-/intracrine actions. Among these factors are the natriuretic peptides atrial and B-type NPs (ANP and BNP, mainly released from myocytes), C-type NP (CNP, possibly secreted from fibroblasts), as well as nitric oxide (NO) (produced by different myocardial cell types). These factors, which signal through specific guanylyl cyclase receptors and intracellular cyclic GMP as second messenger, may act as key local antihypertrophic, antifibrotic, and proangiogenic factors during cardiac remodeling. This chapter will therefore summarize the insights obtained from experimental and clinical studies to highlight (1) natriuretic peptide- and NO-dependent modulation of myocyte growth and secretion of profibrotic factors, (2) the role of direct fibroblast actions of natriuretic peptides and NO during cardiac hypertrophy, (3) potential local proangiogenic actions of ANP and BNP, (4) alterations of these ligand/receptor systems in cardiac remodeling, and (5) potential therapeutical implications.

Keywords Cardiac remodeling • Natriuretic peptides • Nitric oxide • Cyclic GMP

Abbreviations

ANP	Atrial natriuretic peptide
BNP	B-type natriuretic peptide

M. Kuhn (✉) • H. Nakagawa
Institute of Physiology, University of Würzburg, Röntgenring 9, 97070 Würzburg, Germany
e-mail: michaela.kuhn@mail.uni-wuerzburg.de

B.I. Jugdutt and N.S. Dhalla (eds.), *Cardiac Remodeling: Molecular Mechanisms*, Advances in Biochemistry in Health and Disease 5, DOI 10.1007/978-1-4614-5930-9_9,

cGKI	cGMP-dependent protein kinase type I
CNP	C-type natriuretic peptide
GC	Guanylyl cyclase receptor
GC	Guanylyl cyclase
NO	Nitric oxide
PDE	Phosphodiesterase

9.1 Introduction

Cyclic guanosine 3′-5′-monophosphate (cGMP) is a ubiquitous intracellular signaling molecule that controls many physiological processes, from cell contractility, secretion, and permeability to cell differentiation, growth, and survival [1]. In mammals, cGMP can be generated from GTP by two families of guanylyl cyclases (GCs), intracellular nitric oxide (NO)-responsive GCs ("soluble" GCs), and transmembrane GCs ("particulate" GCs or pGCs; GC-A, -B, -C, -E, -F) [2]. GC-A, -B, and -C are receptors for extracellular peptide hormones; GC-A binds atrial and B-type natriuretic peptides (ANP with approximately tenfold higher affinity as BNP), GC-B binds C-type natriuretic peptide (CNP), and GC-C is stimulated by the intestinal peptides guanylin and uroguanylin. The activity of the retinal GC-E and GC-F instead is modulated by intracellular calcium-regulated proteins termed guanylyl cyclase activating proteins (GCAPs). These cytosolic proteins and their guanylyl cyclase receptors are specifically expressed in photoreceptor cells and they are essential to phototransduction and vision. In general, cyclic GMP exerts its actions through nucleotide-gated (CNG) ion channels, cGMP-dependent protein kinases (cGKs type I and II), and cGMP-regulated phosphodiesterases (PDEs) [3–5]. These different third messengers of cGMP are differentially expressed in different types of cells and operate via different effector proteins to modulate cell functions. Intracellular cGMP concentrations are controlled by the activities of these cGMP-forming receptors, as well as by the activity of cGMP-hydrolyzing phosphodiesterases (mainly PDE type 5; less PDEs 1, 2, and 9) [6].

Dysfunctions at various steps of the cGMP signaling cascade have been implicated in cardiovascular, neuronal, and skeletal diseases such as arterial hypertension, cardiac hypertrophy and insufficiency, metabolic syndrome, retinal degeneration, Alzheimer's disease, schizophrenia, and skeletal deformities [2, 7, 8]. Drugs that increase the intracellular cGMP concentration are used in the clinics, for instance, the NO-releasing organic nitrates for the treatment of angina pectoris or the PDE5 inhibitor sildenafil (Viagra®) for erectile dysfunction and pulmonary hypertension. (Pre) clinical studies showed that PDE5 inhibitors inhibit the immunosuppressive effects of inflammation; improve antitumor immunity, for example, in melanoma; and possibly improve cardiac function in patients with cardiac insufficiency. Hence, the cGMP signaling system comprises several potential targets for novel pharmacological therapies.

The critical role of the NO/sGC/cGMP and ANP/BNP/GC-A/cGMP systems in the regulation of arterial blood pressure is known for a long time. Importantly, a recent genetic study showed that even subtle expression changes in components of these pathways (e.g., sGC, ANP, BNP) significantly influence blood pressure and cardiovascular disease risk in humans [9]. The main hypotensive paracrine action of endothelial NO is vasodilatation. In contrast, the cardiac peptides ANP and BNP exert much more pleiotropic hypotensive and hypovolemic endocrine actions which include the following: stimulation of renal function, vasodilatation, mildly enhanced endothelial albumin permeability in the microcirculation of the skin and skeletal muscle, increased fluid efflux from the intravascular to the lymphatic system within the spleen, inhibition of the renin–angiotensin–aldosterone (RAA) system by direct actions on juxtaglomerular cells and the adrenal glomerulosa, and central nervous effects that decrease salt appetite and water drinking [2, 8, 10].

While ANP and BNP expression is found primarily in the heart, CNP expression is more diffuse. Expression has been described in vascular endothelial cells, growth plate cartilage, uterus, ovaries, brain, and other tissues [11]. CNP does not circulate in the blood in appreciable amounts and therefore may act locally. The GC-B receptor is also expressed in many different tissues, in particular in vascular endothelial and smooth muscle cells, regions of brain and bone, and at high density in fibroblasts [11]. The most critical functions of CNP/GC-B have been unmasked by the severe consequences of genetic dysfunctions of this ligand/receptor system. Ablation of CNP, GC-B, or cGMP-dependent protein kinase II (PKG II) in mice all resulted in severe dwarfism as a result of impaired endochondral ossification [12–14]. In humans, homozygous GC-B mutations are the cause of acromesomelic dysplasia, Maroteaux type (AMDM), a severe form of autosomal recessive, short-limbed dwarfism with characteristic skeletal deformities [15]. These and other studies demonstrated that C-type "natriuretic peptide" in fact is unlikely to regulate renal sodium excretion under physiological conditions. Instead, it exerts important autocrine/paracrine GC-B-mediated modulatory effects on cellular proliferation and differentiation within the heart, during angiogenesis and fundamentally during bone formation.

Regarding the cardiovascular system, it is widely accepted that an *acute* increase in the intracellular cGMP concentration has beneficial short-term effects such as vasodilatation and inhibition of platelet aggregation, and the molecular mechanisms of these cGMP effects have been worked out in great detail. However, the long-term actions of cGMP on cell growth, differentiation, viability, and associated (patho)physiological processes are less well understood and controversially discussed. As described below, in experimental studies, cGMP *inhibits* pathological cardiomyocyte hypertrophy and proliferation/differentiation of cardiac fibroblasts. On the other hand, cGMP *stimulates* the regenerative proliferation of vascular endothelia or insulin-secreting pancreatic β-cells, and it protects cardiomyocytes from ischemic apoptosis. Hence, obviously natriuretic peptides and NO, via their cGMP-forming receptors, can exert opposite effects on cell growth and survival depending on the cell type, and even in the same cell, they might inhibit growth/proliferation or prevent cell death depending on the cGMP generators, effectors, and cGMP concentrations involved. The exact intracellular cascades that mediate these effects or that provoke alterations of cGMP signaling during diseases are only partly known.

The chapter focuses on the complex actions of natriuretic peptides and NO and their shared second messenger cGMP in the control of cell growth, differentiation, and survival in the heart, especially during the development of cardiac diseases provoked by pressure overload or after tissue ischemia. A great challenge in understanding cGMP's pleiotropic activities within and outside the heart is to improve our knowledge of the intracellular compartmentalization of cGMP signaling complexes and their functional relevance. Some answers to these questions have been provided by studies in various transgenic mouse models with conventional and conditional, cell-restricted deletions of specific cGMP generators (sGC and membrane GCs such as GC-A and GC-B), cGMP effectors (CNG channels, cGMP-dependent protein kinases), and their downstream targets. These mouse models were instrumental to getting new insights into cytoprotective versus deleterious cGMP signaling. The results of this experimental research could also identify new therapeutical indications (e.g., prevention of cardiac remodeling and diastolic dysfunction, improvement of postischemic angiogenesis) for known drugs that modulate the cGMP pathway, such as the NO-independent stimulators and activators of sGC or PDE5 inhibitors. Therefore, this chapter reviews the cardiac functions of sGC and the membrane GC receptors for natriuretic peptides (ANP, BNP, CNP), with special focus on the insights gained to date from genetically modified mice, the role of alterations of these ligand/receptor systems in cardiac remodeling, and potential therapeutical implications.

9.2 Monogenetic Mouse Models Indicate That Local Cardiac Functions of Natriuretic Peptides and NO Can Counteract Hypertensive Cardiac Remodeling

The cardiac pathological response to sustained pressure overload involves cardiomyocyte hypertrophy and contractile dysfunction along with interstitial changes such as fibrosis and altered capillary density. These changes are orchestrated by mechanical forces and factors secreted between and even within cells, exerting paracrine and auto/intracrine actions. Among these factors are the natriuretic peptides ANP and BNP (mainly released from myocytes), CNP (possibly secreted from fibroblasts), as well as nitric oxide (NO) (produced by different myocardial cell types), which may act as key local antihypertrophic, antifibrotic, and proangiogenic factors during cardiac remodeling. This section will therefore summarize the insights obtained from studies in genetic mouse models with global or conditional, cell-specific deletion of GC-A, GC-B, NO synthases, or cGMP-dependent protein kinase type I (cGKI) to highlight (1) natriuretic peptide- and NO-dependent modulation of myocyte growth and secretion of profibrotic factors (such as transforming growth factor beta, TGF-β, and connective tissue growth factor, CTGF), (2) the role of direct fibroblast actions of natriuretic peptides and NO during cardiac hypertrophy, and (3) potential local proangiogenic actions of ANP and BNP.

9.2.1 Atrial and B-Type Natriuretic Peptides and Their Common Guanylyl Cyclase-A Receptor

During chronic hemodynamic overload, the expression levels of ANP and especially BNP in the cardiac ventricles significantly increase [16]. Studies with cultured neonatal rat cardiomyocytes and cardiac fibroblasts suggested that NPs in this situation may act not only as circulating endocrine factors but also as local antihypertrophic (ANP) and antifibrotic (BNP) cardiac factors [17]. Indeed, mice with global deletion of ANP (ANP$^{-/-}$) or its GC-A receptor (GC-A$^{-/-}$) not only have chronic arterial hypertension and hypervolemia but also exhibit very pronounced cardiac remodeling [18, 19]. To test the hypothesis that NPs locally modulate cardiomyocyte growth in vivo, we generated mice with selective deletion of GC-A in cardiomyocytes through Cre-Lox recombination [20]. Already under resting conditions, these mice exhibit mild cardiomyocyte hypertrophy and elevation of known hypertrophy-marker genes. In addition, they respond to pressure overload (provoked by surgical transverse aortic constriction) or systemic Angiotensin II administration with exacerbated cardiac hypertrophy and fibrosis [20–23]. These observations were in line with studies of Patel and colleagues [23], showing that cardiac overexpression of a dominant-negative form of GC-A in transgenic animals also resulted in a modest accentuation of the hypertrophic response to pressure overload, together with a dramatically increased cardiac fibrotic response. Intriguingly, at difference to the ANP$^{-/-}$ mice, the cardiac phenotype of mice lacking the BNP gene was fibrosis, without cardiomyocyte hypertrophy [24]. Extending these observations, two later reports demonstrated that, in cultured cardiac fibroblasts, BNP stimulates selective matrix metalloproteinases that degrade collagen and suppress profibrotic extracellular matrix genes activated by TGF-ß [25, 26]. Hence, endogenous BNP might participate in an intramyocardial myocyte–fibroblast communication, modulating the secretory activity of fibroblast and extracellular matrix production.

Within the heart, the GC-A receptor is not only expressed in cardiomyocytes and fibroblasts but also and even at much higher density in coronary endothelial cells [27]. Various in vitro studies showed that ANP and BNP can stimulate proliferation and migration of cultured endothelial cells (such as human umbilical vein endothelial cells, HUVEC) and promote the migration and differentiation of endothelial progenitor cells (EPC), suggesting that these peptides exert proangiogenic actions [28, 29]. Postnatally, (patho)physiological angiogenesis is critical for tissue regeneration after ischemia but also during adaptive tissue growth. In particular, myocyte hypertrophy and angiogenesis are coordinately regulated during adaptive cardiac growth, which is at least in part due to the secretion of proangiogenic factors by cardiac myocytes [30]. Disruption of the balanced growth of myocytes and endothelia leads to contractile dysfunction and heart failure [30]. Based on these reports, we hypothesized that ANP and/or BNP, released from hypertrophic ventricular cardiomyocytes, stimulates proangiogenic responses of neighboring capillary endothelial cells. To follow this hypothesis, we generated mice with conditional, endothelial-restricted deletion of the GC-A receptor. Notably, these mice responded to pressure overload (transverse aortic constriction) with myocardial capillary rarefaction, fibrosis, and diastolic dysfunction [31], suggesting that endogenous ANP and BNP indeed function as stress-responsive local regulators of angiogenesis in the murine hypertrophic heart.

These experimental studies taken together support an important autocrine/paracrine role for the cardiac ANP–BNP–GC-A signaling pathway, which moderates cardiomyocyte growth and fibroblast proliferation/differentiation, mildly stimulates coronary angiogenesis, and reduces extracellular matrix production. Thereby, ANP and BNP are "best friends of the heart," attenuating cardiac remodeling in chronic pressure overload or tissue ischemia [32].

9.2.2 C-Type Natriuretic Peptide and its cGMP-Forming Guanylyl Cyclase B Receptor

Whereas the (patho)physiological cardiac roles of ANP/BNP and their GC-A receptor have been widely studied, the cardiovascular functions of the CNP/GC-B system are less clear. Severe bone abnormalities and premature death of mice with global ablation of CNP or GC-B impeded detailed cardiovascular phenotyping [14]. To study the cardiac effects of endogenous CNP, Langenickel et al. [33] generated a transgenic rat model overexpressing a dominant-negative mutated GC-B receptor in cardiomyocytes. Already under baseline conditions, these animals displayed age-dependent but blood pressure-independent cardiac hypertrophy and tachycardia. These cardiac alterations were exacerbated by chronic cardiac volume overload. In line with these results, cardiomyocyte-restricted overexpression of CNP attenuated cardiac hypertrophy and inflammation induced by myocardial infarction in mice [34]. In addition to these modulating effects on cardiomyocyte growth, CNP might modulate myocyte contractile–relaxation functions. In isolated myocytes, the CNP/GC-B/cGMP pathway, via cGMP-dependent protein kinase I (cGKI), stimulated phosphorylation of the sarcoplasmic reticulum (SR) regulatory protein phospholamban, thereby enhancing $[Ca^{2+}]_i$ handling and contractility. In contrast, ANP had no effect [35].

Notably, the expression levels of CNP and of its cGMP-forming GC-B receptor are upregulated in failing hearts [36, 37]. Cardiac fibroblasts might be a local source for CNP in this situation [36]. Hence, endogenous CNP might participate in intramyocardial fibroblast–myocyte communication during chamber remodeling, modulating myocyte growth and contractility as well as inflammatory infiltration.

9.2.3 The Nitric Oxide/GC System

Cardiomyocytes express both the endothelial (NOS3) and neuronal (NOS1) isoforms of NO synthases [38]. NOS3 localizes to caveolae, where compartmentalization with ß-adrenergic receptors and L-type Ca^{2+} channels allows NO to attenuate ß-adrenergic inotropy. NOS1, however, is targeted to SR, and NOS1-derived NO can stimulate SR Ca^{2+} release via activating the ryanodine receptor. Mice with global deletion of either NOS1 ($NOS1^{-/-}$) or NOS3 ($NOS3^{-/-}$) both develop age-related cardiac hypertrophy by 5 months, although only NOS3-deficient mice are hypertensive. These data suggest that NOS1-derived and NOS3-derived intracrine NO mediates opposite effects on cardiac contractile functions, but similar inhibitory effects on pathological cardiomyocyte growth [39]. These intracrine effects of NO in cardiomyocytes might involve cGMP production by sGC but also cGMP-independent signaling pathways, such as S-nitrosylation of target proteins [39].

9.3 Intracellular Pathways Mediating the cGMP-Dependent Cardiac Effects of Natriuretic Peptides or Nitric Oxide

9.3.1 *Cardiomyocytes*

The downstream signaling pathways which transduce the raises in myocyte cGMP levels in changes of myocyte growth and contraction/relaxation functions are much less clear than that leading to increased cGMP levels. At least two cGMP-stimulated proteins (as third messengers) are expressed in myocytes: cGMP-dependent protein kinase type I (cGKI) and phosphodiesterase (PDE) 2, a dual substrate esterase, which appears to hydrolyze cGMP under resting conditions but targets cAMP in the presence of ß-adrenergic stimulation [6]. Studies in cultured neonatal cardiomyocytes indicated that *cGKI* is the main third messenger activated by cGMP in cardiomyocytes, mediating antihypertrophic effects of locally produced natriuretic peptides or nitric oxide [40]. However, conclusive in vivo studies about the cardiac role of cGKI were missing for a long time because the cardiac expression level of this kinase is rather low. Even more, studies of cardiac hypertrophy in mice with global deletion of cGKI were hampered by their severe systemic phenotype, with gastrointestinal dysfunction and early lethality [41]. To circumvent these limitations, recently we generated mice with cardiomyocyte (CM)—restricted deletion (KO) of cGKI [35]. Absence of cGKI in myocytes did not alter baseline cardiac growth and function or cardiac hypertrophic and contractile responses of mice to chronic ß-adrenergic stimulation with isoproterenol. However, CM cGKI KO mice reacted to chronic Angiotensin II administration or chronic pressure overload (induced by surgical aortic constriction) with dilated cardiomyopathy, which was never observed in respective control littermates with unaltered cGKI expression levels. Overall, our data indicated that at least in mice, cGKI activation is involved in myocyte processes which prevent dilated cardiomyopathy and cardiac dysfunction in response to sustained pressure overload and/or neurohumoral stressors such as Angiotensin II.

Several proteins centrally involved in cardiomyocyte calcium transients and electromechanical coupling have been shown to be regulated by cGKI, such as the L-type Ca^{2+} channel (inhibition), phospholemman (inhibition), phospholamban (inhibition, resulting in enhanced SR calcium recycling by SERCA), troponin I (diminished Ca^{2+} sensitivity of sarcomeric myofilaments), and titin (improved ventricular diastolic distensibility) [42]. However, this section will focus on three specific targets of cGKI which are critically and primarily involved in the regulation of pathological growth responses of cardiomyocyte to hormones signaling via $G_{q/11}$-coupled receptors, such as Angiotensin II, endothelin-1, or catecholamines: the regulator of G protein signaling (RGS)-2 and the transient receptor potential canonical channels TRPC3 and TRPC6.

RGS-2 is a GTPase-activating protein for $G_{q/11\alpha}$ subunits. RGS-2 deficiency is linked to arterial hypertension in humans and mice. Causative mechanisms are not fully understood but possibly involve increased peripheral vascular resistance [43]. RGS-2 is specifically phosphorylated and thereby activated by cGKI [44, 45]. For instance, in vascular smooth muscle cells, the NO/sGC/cGMP/cGKI pathway enhances the association of RGS-2 with the plasma membrane and thereby attenuates the calcium and vasoconstrictory effects of vasopressin or endothelin [44, 45]. Similarly, in cardiac myocytes, cGMP/cGKI-mediated modulation of RGS-2 and subsequent inhibition of $AT_1/G_{q/11\alpha}$-signaling contribute to the counterregulation of the calcium raising—and hypertrophic responses to Angiotensin II by natriuretic peptides [22, 46].

Studies in neonatal rat cardiomyocytes and in transgenic mice have indicated that growth hormones acting via $G_{q/11\alpha}$-coupled receptors, such as Angiotensin II, activate the Ca^{2+}-permeable cation channels *TRPC3/C6* through phospholipase C (PLC)/DAG signaling pathways [22, 47]. In neonatal rat and adult mouse, cardiomyocyte activation of TRPC3/C6 ultimately leads to activation of L-type Ca^{2+} channels and pathologically increased $[Ca^{2+}]_i$, which ultimately translates into cardiomyocyte hypertrophy [22]. cGMP-stimulated, cGKI-mediated inhibitory phosphorylation of TRPC3 and TRPC6 provides another mechanism mediating the specific counterregulation of the calcium and hypertrophic responses of cardiomyocytes by natriuretic peptides and NO [48, 49].

9.3.2 *Fibroblasts*

Experimental observations in vitro/in vivo indicated that natriuretic peptides, via GC-A/cGMP/cGKI signaling, inhibit the stimulatory effects of Angiotensin II and TGF-β both on fibroblasts [50] and also on cardiomyocyte secretion of profibrotic factors such as CTGF, thereby moderating interstitial fibrosis accompanying hypertensive heart disease [35]. In particular, molecular studies by Li et al. [50] showed that the cGMP pathway interferes with downstream signaling from TGF-β in cardiac fibroblasts by cGKI-mediated phosphorylation of the transcription factor Smad3. The specific phosphorylation of two amino acids (S309 and T368) prevents the heterodimerization of Smad3 with Smad4 and thereby disrupts their nuclear translocation, resulting in repression of transcriptional activation. Ultimately this signaling cascade attenuates the stimulatory effects of TGF-β on fibroblast proliferation, myofibroblast transformation, and synthesis of extracellular matrix proteins.

9.3.3 Endothelial Cells

The intracellular pathways which link the activation of ANP/BNP/GC-A/cGMP signaling to stimulation of coronary endothelial proliferation and migration can be assumed only based on in vitro/in vivo studies with peripheral endothelia, for example, HUVECs. Intriguingly, based on these studies, cGKI, which mediates growth-inhibiting effects of cGMP in cardiac myocytes and fibroblasts, elicits growth-promoting effects in endothelial cells and their progenitors [51–53]. One effect of endothelial NP/cGMP/cGKI signaling is the phosphorylation of vasodilator-stimulated phosphoprotein (VASP), which is associated with focal adhesion sites and adherens junctions [31, 54]. This pathway was shown to be important for reorganization of the actin cytoskeleton and endothelial tube formation [29]. Members of the mitogen-activated protein kinase (MAPK) family, including ERK1/2 and p38 MAPK and the serine/threonine kinase Akt/PKB, are also important mediators of the endothelial growth and migration responses to natriuretic peptides [28]. Taken together, these in vitro data suggest that activation of VASP, MAPK/ERK, and Akt/PKB cooperates in the NP/GC-A/cGKI-dependent stimulation of endothelial cell proliferation and migration, processes which are required for coronary angiogenesis accompanying adaptive cardiac hypertrophy.

9.4 Chronic Cardiac Hypertrophy Is Accompanied by Impaired Signaling of Natriuretic Peptides and Nitric Oxide

9.4.1 Altered Cardiac NP/GC-A Signaling

Recent genetic studies examined the association of common variants at the human *ANP* and *BNP gene* loci with circulating concentrations of ANP/BNP and arterial blood pressure [9, 55]. The results demonstrate that genetically determined small variations in natriuretic peptide concentrations correlate with significant inverse changes in blood pressure, emphasizing the critical role of these hormones in the maintenance of chronic arterial blood pressure homeostasis. Indeed, gene polymorphisms of ANP or GC-A have been associated with essential hypertension and also with hypertensive cardiac hypertrophy [56–59].

Apart from these genetic variations, functional alterations of ANP, BNP, or GC-A might also be involved in cardiovascular diseases, especially in the aggravation of hypertensive cardiac hypertrophy. Patients with heart hypertrophy/insufficiency have markedly elevated plasma levels of ANP and BNP, with these levels being highly related to the severity of the disease [60]. However, several observations implicate that despite these high hormone levels, patients with heart failure of different etiologies may be in fact in a state of NP deficiency. For instance, mass spectrometry analyses revealed that most of the immunoreactive BNP in heart failure is proBNP, which possesses much lower activity than the processed 32-amino acid hormone [61]. In addition, experimental together with clinical studies have shown that chronic high NP concentrations lead to homologous desensitization of their receptor, which is due to a complex pattern of GC-A (de)phosphorylation [2, 62, 63]. Even more, in this situation of impaired cGMP formation, ANP or BNP can stimulate a cGMP-independent signaling pathway of the GC-A receptor in cardiac myocytes, ultimately resulting in pathologically elevated intracellular Ca^{2+} levels. This pathway involves the allosteric activation of TRPC3/C6 channels by GC-A which forms a stable complex with TRPC3/C6 channels. The concomitant rise in myocyte $[Ca^{2+}]_i$ can aggravate cardiac hypertrophy and possibly increase the propensity to arrhythmias [64].

9.4.2 *Diminished Intracrine Production and Activity of Myocyte NO*

In cardiomyocytes, NOS3 (endothelial NO synthase) is colocalized with the cGMP-producing sGC receptor in caveolin-3 containing lipid raft microdomains. Studies in rats and mice have shown that pressure overload-induced cardiac hypertrophy is accompanied both with diminished NOS activity through functional uncoupling of the enzyme [65] and with depressed NO/heme-dependent activation of the NO receptor, sGC, which is due to enhanced oxidation [66]. The latter apparently is linked to translocation of sGC out of the Cav3 microdomain of hypertrophied cardiomyocytes, which seems to favor sGC oxidation and reduces cGMP formation in response to NO [66].

9.4.3 *Enhanced Myocardial Expression and Activity of PDE5*

Lastly, several experimental and clinical studies have shown that the cGMP-degrading phosphodiesterase (PDE) 5 is induced in left and right hypertensive heart disease [67–69]. This could hasten inactivation of cGMP formed by NP/GC-A, CNP/GC-B, or NO/sGC signaling.

9.5 Implications for the Treatment of Hypertensive Heart Disease

The here presented studies in genetic mouse and rat models with global or conditional GC-A, GC-B, NO synthase, or cGKI deletion suggest that an impairment of the myocardial effects of natriuretic peptides or NO might aggravate and accelerate the progression of hypertensive cardiac remodeling and heart failure. As mentioned above, several experimental and clinical observations implicate that despite high endogenous levels of NPs or NO, patients with heart failure of different etiologies may be in fact in a state of hormone and cGMP deficiency, due to the production of inactive hormones, dysfunction of the respective cGMP-forming guanylyl cyclase receptors, or enhanced cGMP degradation. These findings suggest that synthetic natriuretic peptides, NO donors or mimetics, their analogs, or drugs preventing cGMP degradation could be effective in preventing the progression of cardiac remodeling and dysfunction.

9.5.1 Synthetic Natriuretic Peptides and Their Analogs

Mainly for "patent and historical reasons," Anaritide® (synthetic ANP) is clinically used in Japan, Neseritide® (synthetic BNP) in the USA, and Uralitide® (synthetic urodilatin, the renal isoform of ANP) in Europe [70, 71]. Anaritide® has been used for the treatment of different cardiovascular, renal, and metabolic diseases [72], but large placebo-controlled clinical trials demonstrating its clear efficacy are missing. Presently, the clinical use of synthetic BNP (Neseritide®) and synthetic urodilatin (Uralitide®) is restricted to the treatment of acute heart failure [73]. This indication is based on the renal natriuretic and venodilatory actions of the peptides and their counterregulatory effects on the renin–angiotensin–aldosterone system. Due to these acute effects, NPs diminish systemic fluid retention and cardiac pre- and afterload. However, based on the results of the recent Acute Study of Clinical Effectiveness of Neseritide® in Decompensated Heart Failure Trial (ASCEND-HF), the benefit of short-term NP administration in these patients remains questionable [73]. On the other hand, experimental studies have shown that long-term ANP or BNP administration by local adenoviral-driven expression can prevent the development of hypertensive heart disease in rats [74]. Of course, long-term administration of synthetic NPs or their modified analogs (chimeric peptides) to patients remains challenging due to their short in vivo half-life and peptidergic structure. However, novel technologies are being developed to allow oral or subcutaneous delivery of proteins, making the long-term use of these hormones feasible. In particular, a recent case report showed the efficacy of subcutaneous BNP administration in the treatment of a patient with essential hypertension [75]. Based on the here presented experimental observations, the development of strategies for long-term therapeutical administration of synthetic ANP or BNP or their synthetic modified analogs (such as the novel chimeric peptides CD-NP and CU-NP [76]) could represent a novel, elegant therapeutical approach to prevent the progression of hypertensive heart disease and cardiac dysfunction if used early in the course of hypertension.

9.5.2 *NO-Independent Activators and Stimulators of SGC*

Activation of sGC has traditionally been achieved with nitrovasodilators. However, long-term application of these drugs is associated with the development of tolerance and potentially deleterious cGMP-independent actions. Even more, as already mentioned, in hypertensive heart disease, sGC–cGMP signaling is impaired due to reduced NO bioavailability and an alteration in the redox state of SGC, making it unresponsive to NO. Accordingly, increasing directly the activity of sGC is an attractive pharmacologic strategy. With the development of two novel classes of drugs, sGC stimulators (BAY 41-2272 and riociguat, which stimulate the enzyme synergistically with NO) and sGC activators (cinaciguat), the hypothesis that restoration of NO–sGC–cGMP signaling is beneficial in patients with cardiac hypertrophy/heart failure can now be tested [77]. Characterization of these agents in preclinical and clinical studies has begun with investigations suggesting both hemodynamic effects and organ-protective properties independent of hemodynamic changes. Both sGC stimulators and activators had antihypertensive actions and attenuated cardiac hypertrophy and fibrosis in rodent models of systemic arterial hypertension induced by Angiotensin II infusion or surgical aortic constriction [78–80]. However, clinical studies with the intravenous sGC activator cinaciguat for treatment of acute heart failure (COMPOSE program) were terminated early due to an excess of nonfatal adverse hypotension without improvements in dyspnea or cardiac index (please see a recent comprehensive review by Gheorghiade et al. [81]). However, given the preclinical anti-remodeling properties, the possibility of sGC stimulators and activators improving outcomes in chronic heart failure remains. An oral formulation at a dose with low risk of hypotension would be desirable and would make these agents applicable to the outpatient setting where they could be tested for effects on the progression of heart failure and subsequent rehospitalization.

9.5.3 *Inhibitors of Phosphodiesterase 5*

Cyclic GMP formed by natriuretic peptides or NO is hydrolyzed by members of the phosphodiesterase (PDE) family, mainly PDE5 and possibly PDE9, as well as by the dual substrate, cGMP/cAMP hydrolyzing PDEs 1 and 2. PDE5 remains the best characterized, and its inhibition by drugs such as sildenafil (Viagra®) is clinically used to treat erectile dysfunction and pulmonary hypertension [82]. In particular, many small clinical studies and a large randomized controlled (Super-1) trial have demonstrated the safety and efficacy of sildenafil in improving mean pulmonary artery pressure, pulmonary vascular resistance, cardiac index, and exercise tolerance in patients with pulmonary arterial hypertension [83]. Despite its expression in cardiomyocytes, the role of PDE5 within the heart has long been considered minimal. PDE5 expression and activity in the heart are two orders of magnitude lower than in the lungs. Nevertheless, experimental studies have indicated potent direct cardiac anti-remodeling effects from selective PDE5 inhibitors, ameliorating cardiac ischemia/reperfusion [84] and effectively preventing and even reversing *left* and *right* ventricular hypertrophy, fibrosis, and dilatative cardiomyopathy in mice subjected to constant pressure overload induced by surgical aortic or pulmonary arterial constriction [85]. Based on these and other data, the National Institutes of Health initiated a multicenter, placebo-controlled phase 3 trial (RELAX; http://clinicaltrials.gov/ct2/show/NCT00763867) to test the utility of sildenafil for treating heart failure patients with a preserved ejection fraction (diastolic heart failure). These patients commonly display systemic arterial hypertension and substantial ventricular hypertrophy. The first results of this study will be available in 2013.

9.5.4 Rescue/Resensitization of Natriuretic Peptide Receptors

As another important aspect, this book chapter emphasizes that cardiac hypertrophy/insufficiency is accompanied by dephosphorylation/desensitization of the GC-A and (less) GC-B receptors for natriuretic peptides [62, 86] and oxidation/dysfunction of the sGC receptor for NO [66]. Of note, the enzymes regulating the activity and responsiveness of these guanylyl cyclase receptors are largely unknown. Therefore, clarification of the responsible receptor or postreceptor defects and identification of proteins that regulate the activity of GC-A, GC-B, and sGC in cardiomyocytes, fibroblasts, and endothelial cells may have important pathophysiological and therapeutical implications. This could lead to the development of drugs that could revolutionize the treatment of some forms of cardiac disease.

9.6 Conclusions and Future Directions

Although the observations in rodent systems cannot be directly extrapolated to human beings, the application of gene-targeting technology in mice has provided valuable information regarding the cardiac (patho)physiology and diverse local functions of natriuretic peptides, NO, their guanylyl cyclase receptors, and cGMP signaling in different types of myocardial cells. They also pointed out the potential implications of specific dysfunctions in these systems for human diseases. In fact, as reviewed in this chapter, specific links between polymorphisms or mutations in genes coding for NPs and NO and their receptors and human polygenetic diseases (essential hypertension and hypertensive heart disease) were very recently found and reemphasized the value of genetic mouse models—and in general the important physiological role of the second messenger cGMP in all types of cells of the heart.

Although our general knowledge about cGMP formation and its cellular effects has substantially increased in the last years, critical gaps remain. The gaps mostly concern the posttranslational modifications and regulation of the guanylyl cyclase receptors and the distal signaling pathways that mediate the positive versus negative effects of cGMP on cell growth and survival, for instance, in endothelial cells and cardiomyocytes. These questions are particularly difficult to address, but the answers are needed to understand and potentially improve repair mechanisms in the diseased state, such as cardiac and vascular remodeling and ischemic heart disease.

One of the major open questions concerns the subcellular compartmentalization of cGMP formation and signaling. sGCs are thought to generate global cytoplasmic cGMP signals, whereas transmembrane GCs (GC-A, GC-B) possibly establish local membrane-associated cGMP microdomains, especially in cardiomyocytes. Moreover, our recent observation of distinct effects of ANP (via GC-A) and CNP (via GC-B) on myocyte calcium homeostasis or contractile functions suggests that even different membrane-bound guanylyl cyclase receptors (GC-A and GC-B) generate different cGMP microdomains [35]. Hence, a major focus of future experimental and clinical studies in cardiovascular medicine will be to dissect the role of global versus local cGMP and associated signaling complexes in the control of cardiac myocyte, fibroblast, and endothelial cell growth, survival, and differentiation. This knowledge will be central to the development of new therapeutical strategies.

Acknowledgments MK's work reviewed in this chapter was supported by grants from the Deutsche Forschungsgemeinschaft (SFB 688) and from the comprehensive heart failure center (CHFC) in Würzburg. The authors would like to apologize to all colleagues whose studies were not cited for lack of space.

References

1. Beavo JA, Brunton LL (2002) Cyclic nucleotide research—still expanding after half a century. Nat Rev Mol Cell Biol 3:710–718
2. Kuhn M (2003) Structure, regulation, and function of mammalian membrane guanylyl cyclase receptors, with a focus on guanylyl cyclase-A. Circ Res 93:700–709
3. Biel M, Michalakis S (2009) Cyclic nucleotide-gated channels. Handb Exp Pharmacol 191:111–136
4. Hofmann F, Bernhard D, Lukowski R, Weinmeister P (2009) cGMP regulated protein kinases (cGK). Handb Exp Pharmacol 191:137–162
5. Francis SH, Blount MA, Corbin JD (2011) Mammalian cyclic nucleotide phosphodiesterases: molecular mechanisms and physiological functions. Physiol Rev 91:651–690
6. Conti M, Beavo J (2007) Biochemistry and physiology of cyclic nucleotide phosphodiesterases: essential components in cyclic nucleotide signaling. Annu Rev Biochem 76: 481–511
7. Bender AT, Beavo JA (2006) Cyclic nucleotide phosphodiesterases: molecular regulation to clinical use. Pharmacol Rev 58:488–520
8. Kuhn M (2009) Function and dysfunction of mammalian membrane guanylyl cyclase receptors: lessons from genetic mouse models and implications for human diseases. Handb Exp Pharmacol 191:47–69

9. Ehret GB, International Consortium for Blood Pressure Genome-Wide Association Studies et al (2011) Genetic variants in novel pathways influence blood pressure and cardiovascular disease risk. Nature 478:103–109
10. Brenner BM, Ballermann BJ, Gunning ME, Zeidel ML (1990) Diverse biological actions of atrial natriuretic peptide. Physiol Rev 70:665–699
11. Olney RC (2006) C-type natriuretic peptide in growth: a new paradigm. Growth Horm IGF Res 16:S6–S14
12. Pfeifer A, Aszodi A, Seidler U et al (1996) Intestinal secretory defects and dwarfism in mice lacking cGMP-dependent protein kinase II. Science 274:2082–2086
13. Chusho H, Tamura N, Ogawa Y et al (2004) Dwarfism and early death in mice lacking C-type natriuretic peptide. Proc Natl Acad Sci USA 98:4016–4021
14. Tamura N, Doolittle LK, Hammer RE et al (2004) Critical roles of the guanylyl cyclase B receptor in endochondral ossification and development of female reproductive organs. Proc Natl Acad Sci USA 101:17300–17305
15. Bartels CF, Bükülmez H, Padayatti P et al (2004) Mutations in the transmembrane natriuretic peptide receptor NPR-B impair skeletal growth and cause acromesomelic dysplasia, type Maroteaux. Am J Hum Genet 75:27–34
16. de Bold AJ, Ma KK, Zhang Y et al (2001) The physiological and pathophysiological modulation of the endocrine function of the heart. Can J Physiol Pharmacol 79:705–714
17. Calderone A, Thaik CM et al (1998) Nitric oxide, atrial natriuretic peptide, and cGMP inhibit the growth-promoting effects of norepinephrine in cardiac myocytes and fibroblasts. J Clin Invest 101:812–818
18. Lopez MJ, Wong SK, Kishimoto I et al (1995) Salt-resistant hypertension in mice lacking the guanylyl cyclase-A receptor for atrial natriuretic peptide. Nature 378:65–68
19. John SW, Veress AT, Honrath U et al (1996) Blood pressure and fluid-electrolyte balance in mice with reduced or absent ANP. Am J Physiol 271:R109–R114
20. Holtwick R, Van Eickels M, Skryabin BV et al (2003) Pressure-independent cardiac hypertrophy in mice with cardiomyocyte-restricted inactivation of the atrial natriuretic peptide receptor guanylyl cyclase-A. J Clin Invest 111:1399–1407
21. Kilic A, Bubikat A, Gaßner B, Baba HA, Kuhn M (2007) Local actions of atrial natriuretic peptide counteract angiotensin II stimulated cardiac remodeling. Endocrinology 148: 4162–4416
22. Klaiber M, Kruse M, Völker K et al (2010) Novel insights into the mechanisms mediating the local antihypertrophic effects of cardiac atrial natriuretic peptide: role of cGMP-dependent protein kinase and RGS2. Basic Res Cardiol 5:583–595
23. Patel JB, Valencik ML, Pritchett AM et al (2005) Cardiac-specific attenuation of natriuretic peptide A receptor activity accentuates adverse cardiac remodeling and mortality in response to pressure overload. Am J Physiol Heart Circ Physiol 289:H777–H784
24. Tamura N, Ogawa Y, Chusho H et al (2000) Cardiac fibrosis in mice lacking brain natriuretic peptide. Proc Natl Acad Sci USA 97:4239–4244
25. Tsuruda T, Boerrigter G, Huntley BK et al (2002) Brain natriuretic peptide is produced in cardiac fibroblasts and induces matrix metalloproteinases. Circ Res 91:1127–1134
26. Kapoun AM, Liang F, O'Young G et al (2004) B-type natriuretic peptide exerts broad functional opposition to transforming growth factor-beta in primary human cardiac fibroblasts: fibrosis, myofibroblast conversion, proliferation, and inflammation. Circ Res 94:453–461
27. Kuhn M (2012) Endothelial actions of atrial and B-type natriuretic peptides. Br J Pharmacol 166:522–531
28. Kook H, Itoh H, Choi BS et al (2003) Physiological concentration of atrial natriuretic peptide induces endothelial regeneration in vitro. Am J Physiol 284:H1388–H1397
29. Chen H, Levine YC, Golan DE, Michel T, Lin AJ (2008) Atrial natriuretic peptide-initiated cGMP pathways regulate vasodilator-stimulated phosphoprotein phosphorylation and angiogenesis in vascular endothelium. J Biol Chem 283:4439–4447

30. Walsh K, Shiojima I (2007) Cardiac growth and angiogenesis coordinated by intertissue interactions. J Clin Invest 117:3176–3179
31. Kuhn M, Völker K, Schwarz K et al (2009) The natriuretic peptide/guanylyl cyclase-A system functions as a stress-responsive regulator of angiogenesis in mice. J Clin Invest 119: 2019–2030
32. Molkentin JD (2003) A friend within the heart: natriuretic peptide receptor signaling. J Clin Invest 111:1275–1277
33. Langenickel TH, Buttgereit J, Pagel-Langenickel I et al (2006) Cardiac hypertrophy in transgenic rats expressing a dominant-negative mutant of the natriuretic peptide receptor B. Proc Natl Acad Sci USA 12:4735–4740
34. Wang Y, de Waard MC, Sterner-Kock A et al (2007) Cardiomyocyte-restricted over-expression of C-type natriuretic peptide prevents cardiac hypertrophy induced by myocardial infarction in mice. Eur J Heart Fail 9:548–557
35. Frantz S, Klaiber M, Baba HA et al (2011) Stress-dependent dilated cardiomyopathy in mice with cardiomyocyte-restricted inactivation of cyclic GMP-dependent protein kinase I. Eur Heart J [Epub ahead of print]
36. Kalra PR, Clague JR, Bolger AO et al (2003) Myocardial production of C-type natriuretic peptide in chronic heart failure. Circulation 4:571–573
37. Dickey DM, Flora DR et al (2007) Differential regulation of membrane guanylyl cyclases in congestive heart failure: natriuretic peptide receptor (NPR)-B, not NPR-A, is the predominant natriuretic peptide receptor in the failing heart. Endocrinology 7:3518–3522
38. Hammond J, Balligand JL (2012) Nitric oxide synthase and cyclic GMP signaling in cardiac myocytes: from contractility to remodeling. J Mol Cell Cardiol 52:330–340
39. Barouch LA, Harrison RW, Skaf MW et al (2002) Nitric oxide regulates the heart by spatial confinement of nitric oxide synthase isoforms. Nature 416:337–339
40. Fiedler B, Lohmann SM, Smolenski A et al (2002) Inhibition of calcineurin-NFAT hypertrophy signaling by cGMP-dependent protein kinase type I in cardiac myocytes. Proc Natl Acad Sci USA 99:11363–11368
41. Feil R, Gappa N, Rutz M et al (2002) Functional reconstitution of vascular smooth muscle cells with cGMP-dependent protein kinase I isoforms. Circ Res 90:1080–1086
42. Francis SH, Busch JL, Corbin JD, Sibley D (2010) cGMP-dependent protein kinases and cGMP phosphodiesterases in nitric oxide and cGMP action. Pharmacol Rev 62:525–563
43. Osei-Owusu P, Sabharwal R, Kaltenbronn KM et al (2012) Regulator of G protein signaling 2 deficiency causes endothelial dysfunction and impaired endothelium-derived hyperpolarizing factor-mediated relaxation by dysregulating Gi/o signaling. J Biol Chem 287:12541–12549
44. Heximer SP, Knutsen RH, Sun X et al (2003) Hypertension and prolonged vasoconstrictor signaling in RGS2-deficient mice. J Clin Invest 111:445–452
45. Tang KM, Wang GR, Lu P et al (2003) Regulator of G-protein signaling-2 mediates vascular smooth muscle relaxation and blood pressure. Nat Med 9:1506–1512
46. Takimoto E, Koitabashi N, Hsu S et al (2009) Regulator of G protein signaling 2 mediates cardiac compensation to pressure overload and antihypertrophic effects of PDE5 inhibition in mice. J Clin Invest 119:408–420
47. Wu X, Eder P, Chang B, Molkentin JD (2010) TRPC channels are necessary mediators of pathologic cardiac hypertrophy. Proc Natl Acad Sci USA 107:7000–7005
48. Kinoshita H, Kuwahara K, Nishida M et al (2010) Inhibition of TRPC6 channel activity contributes to the antihypertrophic effects of natriuretic peptides-guanylyl cyclase-A signaling in the heart. Circ Res 106:1849–1860
49. Nishida M, Watanabe K, Sato Y et al (2010) Phosphorylation of TRPC6 channels at Thr69 is required for anti-hypertrophic effects of phosphodiesterase 5 inhibition. J Biol Chem 285:13244–13253
50. Li P, Wang D, Lucas J et al (2008) Atrial natriuretic peptide inhibits transforming growth factor ß-induced Smad signaling and myofibroblast transformation in mouse cardiac fibroblasts. Circ Res 102:185–192

51. Yamahara K, Itoh H, Chun TH et al (2003) Significance and therapeutical potential of the natriuretic peptides/cGMP/cGMP-dependent protein kinase pathway in vascular regeneration. Proc Natl Acad Sci USA 100:3404–3409
52. Aicher A, Heeschen C, Feil S et al (2009) cGMP-dependent protein kinase I is crucial for angiogenesis and postnatal vasculogenesis. PLoS One 4:e4879
53. Koika V, Zhou Z, Vasileiadis I et al (2010) PKG-I inhibition attenuates vascular endothelial growth factor-stimulated angiogenesis. Vascul Pharmacol 53:215–222
54. Smolenski A, Poller W, Walter U, Lohmann SM (2000) Regulation of human endothelial cell focal adhesion sites and migration by cGMP-dependent protein kinase I. J Biol Chem 275:25723–25732
55. Newton-Cheh C, Johnson T, Gateva V et al (2009) Genome-wide association study identifies eight loci associated with blood pressure. Nat Genet 41:666–676
56. Rutledge DR, Sun Y, Ross EA (1995) Polymorphisms within the atrial natriuretic peptide gene in essential hypertension. J Hypertens 13:953–955
57. Beige J, Ringel J, Hohenbleicher H et al (1997) HpaII-polymorphism of the atrial-natriuretic-peptide gene and essential hypertension in whites. Am J Hypertens 10:1316–1318
58. Rubattu S, Bigatti G, Evangelista A et al (2006) Association of atrial natriuretic peptide and type a natriuretic peptide receptor gene polymorphisms with left ventricular mass in human essential hypertension. J Am Coll Cardiol 48:499–505
59. Ellis KL, Newton-Cheh C, Wang TJ et al (2011) Association of genetic variation in the natriuretic peptide system with cardiovascular outcomes. J Mol Cell Cardiol 50:695–701
60. McKie PM, Cataliotti A, Sangaralingham SJ et al (2011) Predictive utility of atrial, N-terminal pro-atrial, and N-terminal pro-B-type natriuretic peptides for mortality and cardiovascular events in the general community: a 9-year follow-up study. Mayo Clin Proc 86:1154–1160
61. Menon SG, Mills RM, Schellenberger U, Saqhir S, Protter AA (2009) Clinical implications of defective B-type natriuretic peptide. Clin Cardiol 32:E36–E41
62. Potter LR (2011) Regulation and therapeutic targeting of peptide-activated receptor guanylyl cyclases. Pharmacol Ther 130:71–82
63. Schröter J, Zahedi RP, Hartmann M et al (2010) Homologous desensitization of guanylyl cyclase A, the receptor for atrial natriuretic peptide, is associated with a complex phosphorylation pattern. FEBS J 277:2440–2453
64. Klaiber M, Dankworth B, Kruse M et al (2011) A cardiac pathway of cyclic GMP-independent signaling of guanylyl cyclase A, the receptor for atrial natriuretic peptide. Proc Natl Acad Sci USA 108:18500–18505
65. Takimoto E, Champion HC, Li M et al (2005) Oxidant stress from nitric oxide synthase-3 uncoupling stimulates cardiac pathologic remodeling from chronic pressure load. J Clin Invest 115:1221–1231
66. Tsai EJ, Liu Y, Koitabashi N et al (2012) Pressure-overload-induced subcellular relocalization/oxidation of soluble guanylyl cyclase in the heart modulates enzyme stimulation. Circ Res 110:295–303
67. Nagendran J, Archer SL, Soliman D et al (2007) Phosphodiesterase type 5 is highly expressed in the hypertrophied human right ventricle, and acute inhibition of phosphodiesterase type 5 improves contractility. Circulation 116:238–248
68. Pokreisz P, Vandenwijngaert S, Bito V et al (2009) Ventricular phosphodiesterase-5 expression is increased in patients with advanced heart failure and contributes to adverse ventricular remodeling after myocardial infarction in mice. Circulation 119:408–516
69. Shan X, Quaile MP, Monk JK et al (2012) Differential expression of PDE5 in failing and non-failing human myocardium. Circ Heart Fail 5:79–86
70. Hayek S, Nemer M (2011) Cardiac natriuretic peptides: from basic discovery to clinical practice. Cardiovasc Ther 29:362–376
71. Gassanov N, Biesenbach E, Caglayan E et al (2012) Natriuretic peptides in therapy for decompensated heart failure. Eur J Clin Pharmacol 68:223–230

72. Saito Y (2010) Roles of atrial natriuretic peptide and its therapeutic use. J Cardiol 56:262–270
73. Ezekowitz JA, Hernandez AF, O'Connor CM et al (2012) Assessment of dyspnea in acute decompensated heart failure: insights from ASCEND-HF (Acute Study of Clinical Effectiveness of Nesiritide in Decompensated Heart Failure) on the contributions of peak expiratory flow. J Am Coll Cardiol 59:1441–1448
74. Cataliotti A, Tonne JM, Bellavia D et al (2011) Long-term cardiac pro-B-type natriuretic peptide gene delivery prevents the development of hypertensive heart disease in spontaneously hypertensive rats. Circulation 123:1297–1305
75. Cataliotti A, Costello-Boerrigter LC, Chen HH, Textor SC, Burnett JC Jr (2012) Sustained blood pressure-lowering actions of subcutaneous B-type natriuretic peptide (nesiritide) in a patient with uncontrolled hypertension. Mayo Clin Proc 87:413–415
76. Zakeri R, Burnett JC (2011) Designer natriuretic peptides: a vision for the future of heart failure therapeutics. Can J Physiol Pharmacol 89:593–601
77. Evgenov OV, Pacher P, Schmidt PM et al (2006) NO-independent stimulators and activators of soluble guanylate cyclase: discovery and therapeutic potential. Nat Rev Drug Discov 5: 755–768
78. Boerrigter G, Costello-Boerrigter LC, Cataliotti A et al (2003) Cardiorenal and humoral properties of a novel direct soluble guanylate cyclase stimulator BAY 41–2272 in experimental congestive heart failure. Circulation 107:686–689
79. Masuyama H, Tsuruda T, Kato J et al (2006) Soluble guanylate cyclase stimulation on cardiovascular remodeling in angiotensin II-induced hypertensive rats. Hypertension 48:972–978
80. Masuyama H, Tsuruda T, Sekita Y et al (2009) Pressure-independent effects of pharmacological stimulation of soluble guanylate cyclase on fibrosis in pressure-overloaded rat heart. Hypertens Res 32:597–603
81. Gheorghiade M, Marti CN, Sabbah HN et al (2012) Soluble guanylate cyclase: a potential therapeutic target for heart failure. Heart Fail Rev [Epub ahead of print]
82. Kass DA, Champion HC, Beavo JA (2007) Phosphodiesterase type 5: expanding roles in cardiovascular regulation. Circ Res 101:1084–1095
83. Galiè N, Ghofrani HA, Torbicki A, Sildenafil Use in Pulmonary Arterial Hypertension (SUPER) Study Group et al (2005) Sildenafil citrate therapy for pulmonary arterial hypertension. N Engl J Med 353:2148–2157
84. Salloum FN, Abbate A, Das A et al (2008) Sildenafil (Viagra) attenuates ischemic cardiomyopathy and improves left ventricular function in mice. Am J Physiol Heart Circ Physiol 294:H1398–H1406
85. Takimoto E, Champion HC, Li M et al (2005) Chronic inhibition of cyclic GMP phosphodiesterase 5A prevents and reverses cardiac hypertrophy. Nat Med 11:214–222
86. Dickey DM, Dries DL, Margulies KB, Potter LR (2012) Guanylyl cyclase (GC)-A and GC-B activities in ventricles and cardiomyocytes from failed and non-failed human hearts: GC-A is inactive in the failed cardiomyocyte. J Mol Cell Cardiol 52:727–732

Chapter 10
Modulating G Protein-Coupled Receptors to Effect Reverse Cardiac Remodeling

Cinzia Perrino and Howard A. Rockman

Abstract G protein-coupled receptors (GPCRs) represent the largest known family of transmembrane receptors and therapeutic targets in cardiovascular medicine, accounting for a large number of marketed cardiovascular pharmaceuticals. Traditionally, GPCR stimulation promotes G protein signaling and, to limit unrestrained stimulation, activation of G protein-coupled receptor kinases (GRKs), leading to agonist-dependent receptor phosphorylation. In turn, GPCR phosphorylation promotes β-arrestin binding to the receptors, which sterically prevents further G protein signaling and scaffold receptors to the internalization machinery. However, novel aspects of GPCR signaling have been recently appreciated, including G protein modulators, G protein-independent pathways, and GRK adrenal modulation of adrenergic drive. Since all currently used drugs have been developed using assays only testing G protein-dependent effects, the discovery of such novel signal transduction pathways might represent an important opportunity to identify additional therapeutic approaches to reverse or prevent cardiac remodeling and failure.

Keywords G proteins • Receptors • Cardiac remodeling • Arrestins • Hypertrophy • Heart failure • Signaling

C. Perrino, M.D., Ph.D. (✉)
Department of Clinical Medicine, Cardiovascular and Immunological Sciences, Federico II University, Via Pansini 5, 89131 Naples, Italy
e-mail: perrino@unina.it

H.A. Rockman
Department of Clinical Medicine, Cardiovascular and Immunological Sciences, Department of Medicine, Department of Cell Biology, Department of Molecular Genetics, Duke University Medical Center, Durham, NC, USA

B.I. Jugdutt and N.S. Dhalla (eds.), *Cardiac Remodeling: Molecular Mechanisms*, Advances in Biochemistry in Health and Disease 5, DOI 10.1007/978-1-4614-5930-9_10,

10.1 Introduction

Heart failure is a common final phenotype of many forms of heart disease. Initiated by structural and molecular remodeling of the heart, it frequently leads to contractile dysfunction. Despite optimal treatment, heart failure is a relentless, progressive disease, and novel therapeutic approaches are being continuously explored. Molecular, cellular, and interstitial changes are critical in the development of cardiac remodeling and failure. Among these signals, those activated by plasma membrane G protein-coupled receptors (GPCRs, also known as seven-transmembrane receptors) have been undoubtedly the most well investigated, currently accounting for the majority of prescriptions for cardiovascular diseases.

To date, there are nearly 200 GPCRs with known cognate ligands and a much larger number of "orphan" receptors, i.e., receptors without a known "parent" ligand [1]. A substantial number of orphan receptors are predicted to be expressed in the heart [2]. Among cardiovascular GPCRs, the most commonly investigated and clinically targeted include the nine different adrenergic receptors ($\alpha_{1A, 1B, 1D}$; $\alpha_{2A,2B, 2C}$; and $\beta_{1, 2, 3}$ARs), followed by the angiotensin II (AT_1R, AT_2R), endothelin (ET-A, ET-B), muscarinic (M2, M3), dopamine, adenosine, glucagon, prostanoid, bradykinin, histamine (H_1, H_2, H_3), NPY (Y_1, Y_2), opioid (μ, κ, δ), and serotonin receptors [3]. Undoubtedly, the most extensively examined cardiac GPCR is the β-adrenergic receptor (βAR), which preferentially couples to the G_s–adenylyl cyclase–cAMP signaling cascade [4].

In response to cardiac stress or hemodynamic overload, several neurohumoral adaptations occur, including the activation of the renin–angiotensin–aldosterone and the sympathetic nervous systems, leading to heightened and continuous GPCR stimulation in the heart. Although neurohumoral activation contributes to the maintenance of perfusion of vital organs, it causes impaired GPCR signaling and several maladaptive intracellular consequences, leading to cardiac remodeling and eventually failure [5]. Indeed, βAR blockers, angiotensin receptor blockers, and angiotensin-converting enzyme inhibitors have been shown to reverse cardiac remodeling, reduce cardiac hypertrophy, slow the progression of heart failure, and improve survival in heart failure patients [6–11]. Importantly, all currently used drugs have been developed using assays only testing "conventional" GPCR signaling. Thus, the recent discovery of novel GPCR signal transduction pathways might represent an important opportunity to identify additional therapeutic approaches to reverse or prevent cardiac remodeling and failure.

10.2 Old and New Paradigms in GPCR Signaling

10.2.1 Classical G Protein-Dependent Signaling

Activation of all types of GPCRs results in G protein-mediated generation of second messengers and/or activation of ion channels (Fig. 10.1) [4]. By inducing

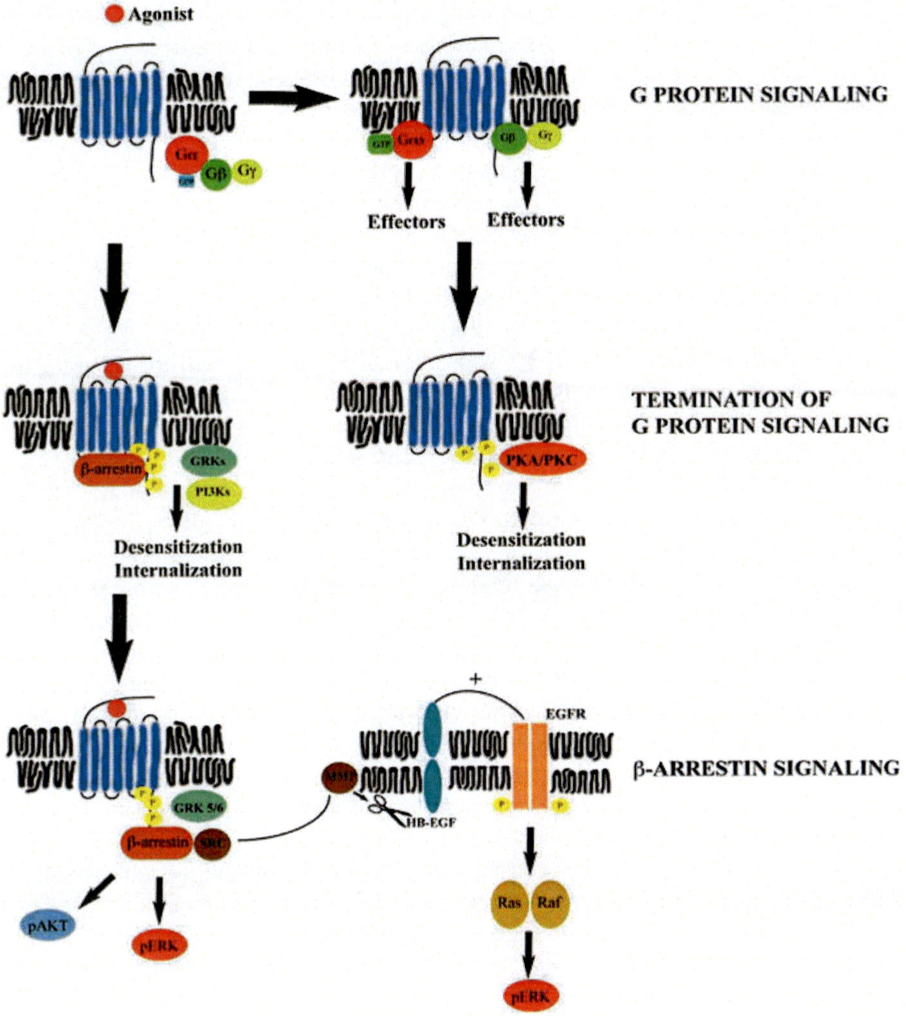

Fig. 10.1 Schematic representation of signaling pathways activated by GPCRs. In response to agonist stimulation, GPCRs undergo a conformational change, leading to the activation and dissociation of G protein subunits and subsequent activation of downstream effectors ("classical" G protein signaling). In the presence of continuous agonist stimulation, termination of G protein signaling is obtained through receptor phosphorylation by second messenger-activated kinases (e.g., protein kinase A or protein kinase C) or by G protein-coupled receptor kinases (GRKs). GPCR phosphorylation by GRK 5 or 6 promotes β-arrestin-dependent signaling that further stimulates a plethora of downstream signaling pathways (MAPKs, PI3K, AKT, EGFR, etc.). Abbreviations: *MMP* matrix metalloproteinase, *HB-EGF* heparin-binding epidermal growth factor, *EGFR* epidermal growth factor receptor

conformational changes in the receptor, agonist stimulation allows interaction of the activated receptor with the cognate heterotrimeric G proteins and promotes the dissociation of G proteins into G_α and $G_{\beta\gamma}$ subunits. Activated G_α subunits couple with an effector, an enzyme, or an ion channel [12]. Dissociated $G_{\beta\gamma}$ subunits also target a wide range of signaling pathways involved in desensitization, downregulation, apoptosis, and ion channel activation [12].

In the presence of continuous agonist stimulation, a rapid dampening of receptor responsiveness (i.e., receptor "desensitization") is obtained through receptor phosphorylation by second messenger-activated kinases (e.g., protein kinase A or protein kinase C) in a process known as heterologous desensitization or by a different family of serine/threonine GPCR kinases named GRKs (Fig. 10.1) [13]. Phosphorylation mediated by GRKs is promoted by occupancy of the receptor by agonist and triggers an "agonist-specific" or homologous desensitization. Of the seven known GRKs, only GRKs 2, 3, and 5 are appreciably expressed in the heart, and GRK2/GRK5 are the most abundant [13]. Membrane targeting of all the GRKs is critical to their function and is conferred by the C-terminal tail domain. While GRKs 2 and 3 are primarily cytoplasmic and are targeted to the plasma membrane by binding $G_{\beta\gamma}$ subunits through their C-terminal pleckstrin homology domain, GRK5 constitutively binds phospholipids and displays preferential membrane localization [13]. Although individual GRKs might promote specific patterns of GPCR phosphorylation that might affect downstream signaling (see below) [14], the phosphorylation signal invariably determines the uncoupling of the receptor from its signal-transducing G protein and enhances its affinity for a family of cytosolic proteins known as the β-arrestins [15]. GRK-mediated GPCR phosphorylation and binding of β-arrestin to agonist-occupied receptor facilitate the interaction of the receptor with proteins of the clathrin-coated pit machinery to promote internalization into intracellular compartments.

After internalization, GPCR trafficking is regulated by the variable binding affinity to the different β-arrestin isoforms and their relative intracellular localization [16]. According to the differential affinities for β-arrestins, GPCRs can be divided into two major classes: class A receptors (such as β_1ARs and D_1 receptors), binding β-arrestin 2 with greater affinity than β-arrestin 1 and losing their interaction during internalization, and class B receptors (such as AT_{1A} receptors), binding β-arrestins 1 and 2 with equal affinity and the interaction remaining intact during internalization [17]. Importantly, while β-arrestin 1 can be found in both the cytoplasm and the nucleus, β-arrestin 2 is localized only in the cytoplasm [17].

Internalized receptors can be sorted for degradation to lysosomes or for recycling to acidified vesicles where they are dephosphorylated and recycled back to the plasma membrane [18]. It has been previously suggested that, for some GPCRs such as βARs, a continuous equilibrium between internalization and recycling to the sarcolemma exists in vivo under physiological conditions [19]. Consistent with these assumptions, failing human hearts are characterized by large intracellular pools of βARs within endosomal compartments [20], and these findings are consistent with similar observations in animal models of heart failure [21]. Interestingly, changes in environmental conditions associated with reverse cardiac remodeling (e.g., changes in cardiac load by left ventricular assist devices or neurohumoral stimulation) seem to promote reverse trafficking of "sequestered" βARs back to the plasma membrane [20]. Although it is not yet known how long βARs reside within the endosomal compartments with continuous agonist stimulation, the mechanisms for net loss of cell surface receptors (i.e., receptor downregulation) are due to a decrease in receptor synthesis, a destabilization of receptor messenger RNA, and an increase in receptor degradation.

After removal of the agonist, the attenuated GPCR responsiveness is reversed in a process known as "resensitization," which primarily involves dephosphorylation of the receptors by the serine–threonine protein phosphatase 2A (PP2A) [22]. Compared to desensitization and internalization, much less is currently known about the resensitization and dephosphorylation processes. Although sequestration into vesicles seems not to be required for desensitization, it appears to be important for resensitization, since agents blocking sequestration do not resensitize [22]. These data led to a model proposing that sequestration of the receptors into internalized vesicles might be required for their dephosphorylation within an acidic environment [22]. However, this view has been recently challenged (see below) [23].

10.2.2 Role of Phosphoinositide 3-Kinases in GPCR Function

Several recent studies have highlighted the important role of the family of protein and lipid kinases named phosphoinositide 3-kinases (PI3Ks) in the modulation of GPCR internalization and signaling at multiple levels and with different molecular mechanisms. Firstly, it has been demonstrated that Gβγ and phosphatidylinositol 3-kinases (PI3Ks) physically interact in the cytosol through the helical phosphoinositide kinase domain (PIK) of PI3Ks [24]. Upon agonist binding and release of $G_{\beta\gamma}$ subunits, Gβγ mediates PI3K membrane translocation, wherein it generates phosphoinositides regulating receptor internalization [25]. PI3Kγ inhibition by overexpressing a kinase-dead mutant markedly attenuates βAR endocytosis [26, 27], demonstrating that production of phosphatidylinositol 3,4,5-trisphosphate (PIP3) at the receptor complex is necessary for efficient receptor endocytosis. In particular, PIP_3 is needed for AP-2 adaptor recruitment at the plasma membrane and for the consequent organization of clathrin-coated pits [24]. Subsequent studies have also shown that overexpression of the minimal PIK domain of PI3Ks disrupts the complex between GRK2 and endogenous PI3Ks, resulting in the targeted reduction of PI3K activity at the site of agonist-occupied receptors and inhibition in the formation of cytosolic aggregates following agonist stimulation [24]. Since the PIK domain is conserved in all PI3K isoforms, GRK2 will complex with either PI3Kα or PI3Kγ depending on the availability of the isoform in a given tissue, in heart tissues the predominant isoforms being PI3Kα, β, and γ [28].

Interestingly, it has been recently shown that the cytoskeletal protein non-muscle tropomyosin is a substrate for PI3K protein kinase activity [29]. A constitutively dephosphorylated tropomyosin mutant blocked agonist-dependent βAR internalization, while a tropomyosin mutant mimicking constitutive phosphorylation reversed the defective βAR internalization induced by PI3K inhibition [29], suggesting that non-muscle tropomyosin might exert an important role in GPCR internalization and that its phosphorylation by PI3K might represent a crucial event in its regulation.

In addition to its effects on βAR internalization and trafficking, PI3Kγ might also negatively regulate βAR resensitization by phosphorylating I2PP2A, an endogenous inhibitor of PP2A, and thus modulating PP2A activity [23]. Indeed, I2PP2A phosphorylation by PI3Ks might result in enhanced binding to PP2A catalytic subunit, inhibiting its phosphatase activity and promoting βAR desensitization [23]. These studies suggest that GPCRs might be resensitized at the plasma membrane without necessarily undergoing internalization as previously postulated, and this process might be regulated by PI3Kγ.

Finally, recent evidence suggests that PI3Kγ might also integrate PIP_3 and cAMP signaling in the heart [30]. In contrast to mice with genetic deletion of PI3Kγ ($PI3K\gamma^{-/-}$), mice carrying a targeted mutation in the PI3Kγ gene causing loss of the kinase activity ($PI3K\gamma^{KD/KD}$) displayed normal basal cardiac contractility or cAMP levels and, after pressure overload, were protected from pathological cardiac remodeling [31]. Indeed, $PI3K\gamma^{KD/KD}$ mutants can still regulate intracellular cAMP levels by interacting with the phosphodiesterase PDE3B, and this interaction, likely based on the formation of a large multiprotein complex, does not require the kinase activity of PI3Kγ [31]. In addition, PI3Kγ directly binds protein kinase A (PKA), the main effector molecule of βAR signaling, and a recognized activator of PDE3B [30]. PI3Kγ-anchored PKA activates PDE3B to enhance cAMP degradation and phosphorylates p110γ to inhibit PIP_3 production, providing local feedback control of PIP_3 and cAMP signaling events in the heart.

10.2.3 G Protein-Independent, Beta-Arrestin-Mediated Signaling

In addition to the well-known role in GPCR desensitization, internalization, and trafficking, β-arrestins and GRKs participate in further signal propagation, assembling macromolecular complexes in the receptor environment, activating different signal transduction pathways, and even regulating other receptor families, such as tyrosine kinase receptors (Fig. 10.1).

While it has been known for several years that GPCRs activate mitogen-activated protein kinases (MAPKs), it has now been well documented that GPCRs can also utilize β-arrestins to activate a signaling [32–35]. The kinetics of ERK activation, as well as its cellular distribution, are quite distinct from those of ERK phosphorylation stimulated by conventional G protein-mediated processes [36]. Indeed, while G protein activation leads to rapid (2–5 min) and both nuclear and cytoplasmic ERK activation, β-arrestin-mediated ERK activation has different kinetics, with a slower and more prolonged pattern of ERK activation, only in the cytoplasm [33, 34].

It has been recently disclosed that β-arrestin-mediated ERK signaling following catecholamine stimulation is also dependent, at least in part, on transactivation of the

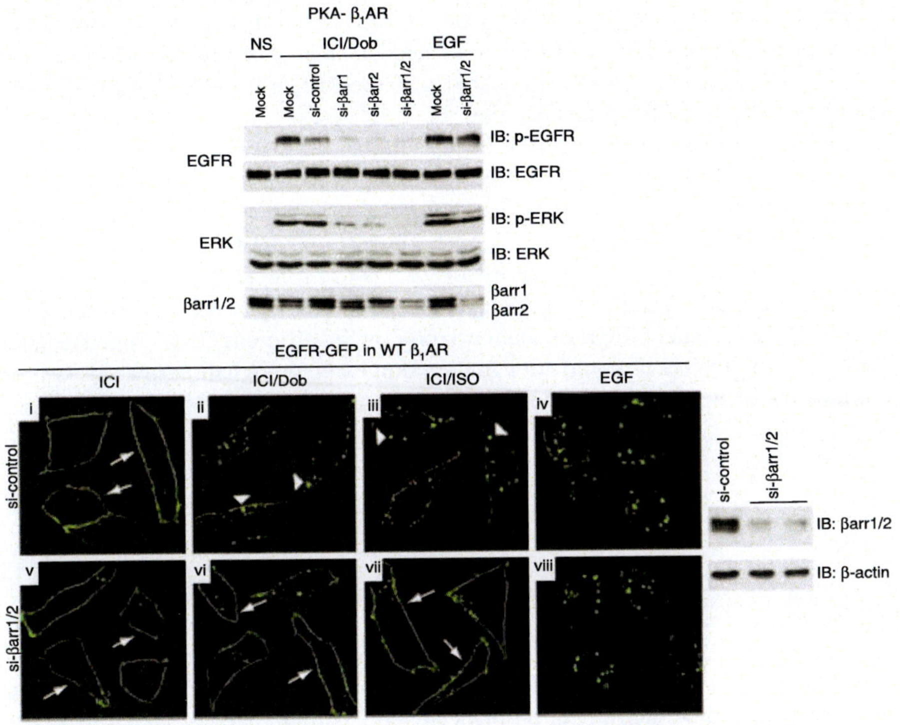

Fig. 10.2 β-arrestins are required for β_1AR-mediated transactivation of the epidermal growth factor receptor. *Top*. HEK293 cells stably expressing β_1AR mutants lacking PKA phosphorylation sites (PKA$^-$ β_1AR) were transfected with FLAG–EGFR alone (Mock) or with siRNAs targeting β-arrestin1 (si-βarr1), β-arrestin2, β-arrestin1/2, or scrambled siRNA (si-control). Cells were then treated with the β_2AR-specific antagonist ICI-118,551 (ICI), ICI+dobutamine (ICI/Dob), or epidermal growth factor (EGF), and phospho-ERK/phospho-EGFR levels were assessed. *Bottom*. HEK293 cells stably expressing wild-type (WT) β_1AR were transfected with EGFR–GFP and si-control or si-βarr1/2 to knock down expression (*right panel*). In the absence of agonist, EGFR–GFP was located at the membrane (i and v, *arrows*), while EGF stimulation induced EGFR–GFP redistribution into aggregates (iv and viii). Treatment of si-control-transfected cells with dobutamine (Dob) or isoproterenol (ISO) also resulted in redistribution of EGFR into aggregates (ii and iii, *arrowheads*), an effect that was diminished in si-βarr1/2-transfected cells, where EGFR–GFP remained at the membrane (vi and vii, *arrows*) (Reproduced with permission from [39])

epidermal growth factor receptor (EGFR), which then mediates ERK activation via the canonical Ras pathway [37]. Multiple mechanisms seem to connect GPCRs to EGFR, including the formation of GPCR–EGFR signaling complexes [38] and the matrix-metalloproteinase-mediated release of heparin-binding epidermal growth factor (HB-EGF) to stimulate EGFR phosphorylation, internalization, and initiation of ERK signaling (Figs. 10.1 and 10.2) [40, 41]. In addition to ERK activation, several GPCRs have been also shown to activate the PI3K–AKT pathways in a β-arrestin-dependent manner [42]. Interestingly, β-arrestin-dependent activation of both ERK and PI3K–AKT seems to cooperatively promote phosphorylation and inactivation of the proapoptotic protein BAD (BCL2-associated agonist of cell death) [43].

It has been only recently appreciated that the different GRK isoforms, while phosphorylating GPCRs at specific sites, create a sort of "bar code" responsible for the initiation of specific signaling via the alternative β-arrestin-biased pathways [44]. Indeed, while inhibition of GRK 5 or 6 attenuates β-arrestin-mediated ERK activation via the β1AR and $AT1_A$ receptor, this latter remains unaffected (or may even increase) upon GRK 2 or 3 knockdown (Figs. 10.1 and 10.3) [14, 45]. Furthermore, GRK 5 or 6 overexpression seems to promote β-arrestin-dependent ERK signaling after $AT_{1A}R$ or β_2AR in vitro stimulation [41, 46]. Taken together, these data suggest that GPCR phosphorylation by the different GRKs together with β-arrestin binding is a critical factor in directing receptor signaling towards G protein-dependent or G protein-independent pathways.

10.2.4 GPCR Oligomerization

According to the "traditional" paradigm, monomeric GPCRs interact with single heterotrimeric G protein units, and, in response to agonist-mediated conformational changes in the receptor, G protein activation occurs, activating/inhibiting effector molecules. However, a growing body of evidence shows that many, if not all, GPCRs might exist as homo- and/or heterodimers [47–49]. Although the exact nature of the dimer interface is still unknown, several distinctive dimerization interfaces have been described within the transmembrane helices, the NH_2-terminal or the COOH-terminal [50–52]. Two basic models of interaction have been proposed (but not yet demonstrated): one in which dimerization occurs simply by superficial contact between monomers, and another in which several domains from different monomers are tangled between each other [48]. Importantly, heterodimerization between GPCRs might induce intracellular responses that differ from those activated by homogeneous populations of GPCRs, thereby generating previously unrealized diversity of function (Fig. 10.4) [48]. Furthermore, the development of cell- or tissue-specific GPCR populations might be extremely clinically relevant, since it might allow the development of agents specific for a certain heterodimer and regulate its signaling only in certain tissues.

In the heart, accumulating data regarding the existence of direct interactions among adrenergic receptors, and between adrenergic receptors and other GPCR families indicate a much more complex role for each receptor in regulating cardiac function. β_1AR and β_2AR homo- and heterodimers have been shown to be constitutive [53, 54] or agonist dependent [55]. When expressed in cells, β_1AR–β_2AR heterodimers activate specific intracellular responses compared with the respective homodimers [56], and this is particularly important, since cardiomyocytes express both β_1 and β_2ARs, and their expression levels can be differently regulated in response to cardiac overload and remodeling, leading to different proportions of homo- or heterodimers. Furthermore, βARs have also been shown to heterodimerize

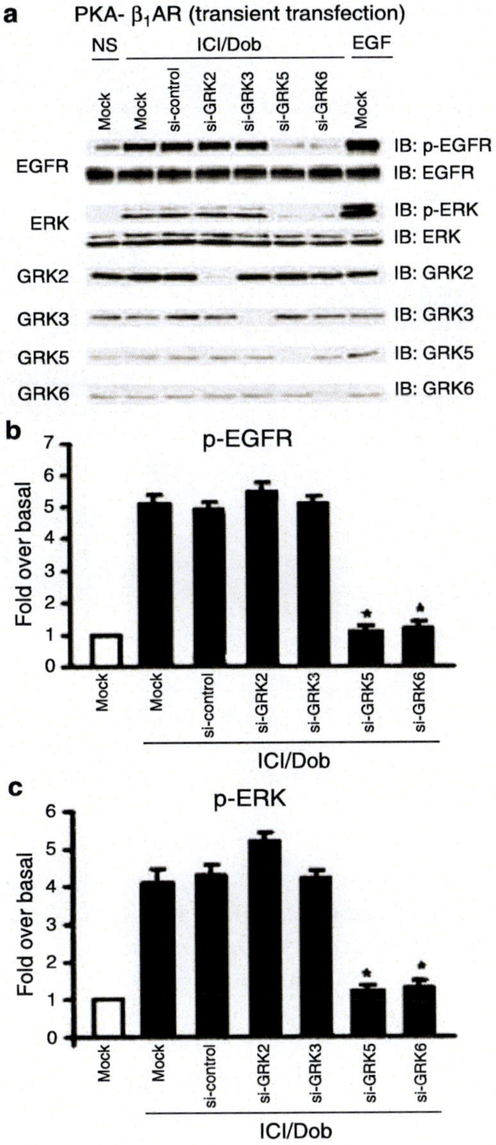

Fig. 10.3 GRK5 and 6 are required for β-arrestin-mediated EGFR transactivation. (**a**) HEK293 cells transiently expressing β_1AR mutants lacking PKA phosphorylation sites (PKA$^-$ β_1AR) were transfected with FLAG–EGFR alone (Mock) or with siRNAs targeting ubiquitous GRKs (si-GRK2, si-GRK3, si-GRK5, and si-GRK6) or a scrambled siRNA sequence (si-control). Summary of multiple independent experiments showing significant inhibition of EGFR transactivation (**b**) and ERK1/2 activation (**c**) upon dobutamine (Dob) stimulation in the cells transfected with siRNA targeting GRK5 or 6; $^*P<0.001$ versus Dob-stimulated Mock, si-control, si-GRK2, and si-GRK3 (reproduced with permission from [39])

with α_{2A}ARs [57], suggesting that GPCRs might display variable activation of signaling pathways depending on the dimer partner and/or subcellular segregation.

In addition to the interaction between same family members, interfamily interactions between endogenous βARs and AT_1Rs were recently demonstrated in the heart [58]. Interestingly, the blockade of one of the two receptors in complex was sufficient to inhibit signaling and trafficking of both receptors. Indeed, treatment of murine myocytes with a βAR blocker inhibited angiotensin/Gq coupling and contractility,

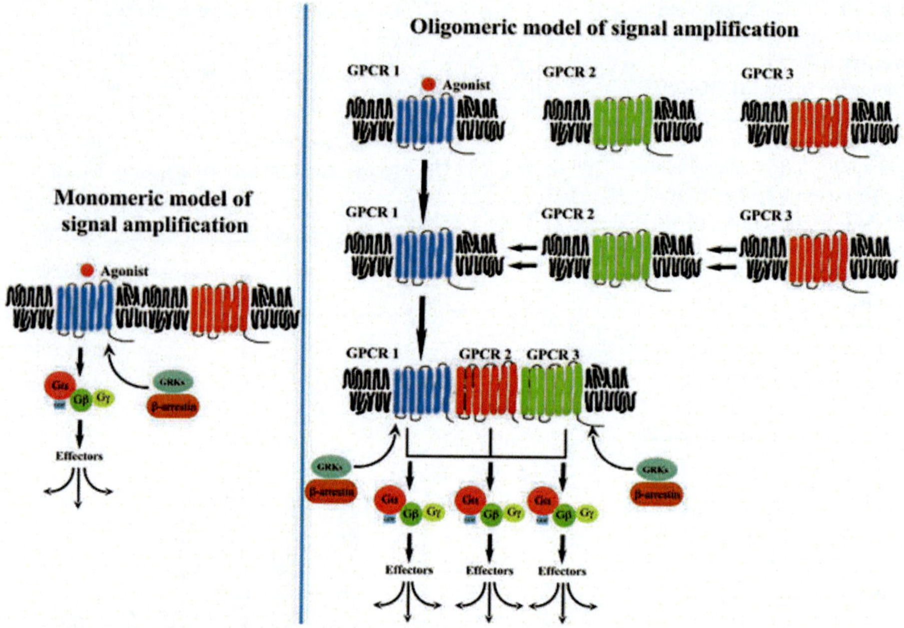

Fig. 10.4 Schematic representation of signaling pathways activated by GPCR oligomerization. According to the traditional paradigm, a single agonist activates a single receptor, its cognate G protein, and effectors (monomeric model of signal amplification, *left*). Ligand binding to one receptor might promote the formation of an oligomeric complex (*right*), leading to novel, unexpected intracellular responses that differ from those activated by homogeneous populations of GPCRs

while treatment of mice with a selective angiotensin receptor blocker attenuated heart rate response to a βAR agonist [58]. Whether, and how, receptor ligands might affect the dimerization process, and whether pharmacology of monomeric receptors is distinct from that of homodimers, is still not clear.

10.3 Novel Molecular Approaches to Modulate GPCR Signaling and Cardiac Remodeling

10.3.1 Deorphanization of Cardiac GPCRs

It has been postulated that among cardiac GPCRs, approximately one third may be "orphan" receptors, with no currently known endogenous agonists [2]. Very little information is currently available regarding the expression, regulation, and physiological or pathophysiological functions of such orphan receptors, and this might be a fertile area for future research, leading to the discovery of new cardiovascular drugs to ameliorate cardiac remodeling [3].

Orphan GPCR research has been previously used to identify novel hormone, peptide, and neurotransmitter ligands [59, 60]. Recently, large compound libraries have been used to identify chemicals that interact with orphan GPCRs that may have therapeutic potential [61]. Recently, a "reverse" approach has been undertaken to shed light on the pathophysiological role of GPR22 in cardiac remodeling [62]. GPR22 is expressed at high levels in the heart, signaling through G_i/G_o activation. In response to TAC, cardiac pathological remodeling and dysfunction were enhanced in GPR22 knockout mice (GRP22$^{-/-}$) compared to their wild-type littermates, suggesting that this orphan receptor might exert an important role in cardiac remodeling. Although GRP22 downstream kinases are still not known and will be subject for future research, one of the most intriguing aspects of these studies is the search for novel, unrecognized endogenous ligands. These exciting studies might identify novel therapeutic targets and a novel strategy in orphan GPCR research, by defining a phenotype of an orphan receptor without immediately deorphanizing it.

10.3.2 Modulators of G Protein Signaling

In response to excessive GPCR stimulation, perturbations in G protein signaling also occur and are major contributors to pathological cardiac remodeling [63]. A long-standing strategy to target GPCR signaling at the level of the G proteins has been to mitigate Gβ-γ signaling, initially using large peptide inhibitors and more recently small molecule inhibitors [64]. However, the magnitude and duration of G protein activation are mainly regulated by the rate of GTP hydrolysis by Gα subunits, since the resulting GDP-bound inactive Gα reassociates with the Gβ-γ dimer and can enter a new activation cycle. Thus, the recent discovery of a new large family of proteins regulating G protein signaling (RGS proteins) and capable of decreasing amplitude and duration of both Gα- and Gβ-γ-mediated signals might represent novel additional therapeutic targets [65, 66]. According to sequence homology within the RGS domain and other non-RGS domains, the superfamily of RGS proteins can be divided into subfamilies, and several of these canonical RGS protein isoforms are expressed in the myocardium [67]. So far, several different strategies have been undertaken to modulate RGS expression or function in vitro: inhibitory peptides or competitive antibodies have been used to disrupt the Gα–RGS interface, and RNAi, ribozymes, or antisense oligonucleotides have been used to reduce RGS expression [67]. In addition, gain and loss of function RGS in vivo models [67] and an RGS-insensitive Gα model have been produced [68]. Undoubtedly, the R4 is one of the best characterized subfamilies in the heart, with RGS4 being the prototypical member [67]. RGS4 transgenic mice exhibited no basal phenotype, but were characterized by rapid heart failure and increased mortality in response to pressure overload [69] while transiently ameliorated cardiac dysfunction in $G\alpha_q$ transgenic mice [70].

More recently, RGS2 has been shown to modulate cardiac remodeling in response to pressure overload, since mice with global genetic deletion of RGS2 were characterized by enhanced Gα_q signaling, cardiac hypertrophy, and accelerated dysfunction after transverse aortic constriction [71]. Similarly, RGS5 has been demonstrated to reduce cardiac hypertrophy and fibrosis in response to pressure overload [72]. Therapeutic benefits could be derived from inhibition or enhancement of RGS protein function, depending on the nature of the targeted isoform, its regulatory function, and the cellular and pathophysiological context [67]. Importantly, closely related RGS isoforms with similar sequence and structure seem to have different responsiveness to the same inhibitors, suggesting that RGS inhibitors might have the advantage of increasing the specificity of exogenous GPCR agonists by targeting specific RGS proteins. In contrast, enhancing RGS protein function could be beneficial in settings where reduction in RGS protein expression or activity is associated with pathophysiological consequences. Although RGS proteins are clearly promising targets for therapeutic development, much work still must be done to develop strategies that can eventually be successfully used in vivo.

10.3.3 Cardiac or Adrenal GRK2 Inhibition

The chronic elevation in circulating catecholamine (CA) levels under conditions of heart failure is considered a crucial event in the development of pathological cardiac remodeling and contractile dysfunction, causing several molecular abnormalities in the heart. Upregulation of GRK2 contributes significantly to βAR dysfunction and ventricular pathological remodeling [4, 13], and a large number of previous studies have shown that cardiac GRK2 inhibition effectively prevents cardiac dysfunction and ameliorates survival in HF models (recently reviewed in [13]). However, recent studies have shown that, in addition to the myocardium, GRK2 is upregulated in the adrenal gland in animal models of HF, leading to enhanced CA release via desensitization/downregulation of α_2ARs in chromaffin cells [73]. Adrenal GRK2 seems to be not only an important trophic factor for the adrenal gland in health but also a major driving force behind adrenal hypertrophy and hyperfunction in HF. Interestingly, adrenal-restricted GRK2 deletion indeed leads to a significant reduction of circulating CAs in vivo in post-myocardial infarction HF, and the resultant reduction of both circulating norepinephrine and epinephrine prevents the deterioration of cardiac function and βAR signaling. The beneficial effects of adrenal-specific GRK2 deletion also include downregulation of cardiac GRK2, an important indicator of cardiac dysfunction in HF, even without a dramatic reduction in CA levels, which could be detrimental for cardiac hemodynamic support of the circulation under critical conditions. In addition, adrenal GRK2 inhibition via adenoviral-mediated in vivo gene therapy using a GRK2 inhibitory peptide decreased circulating CAs and improved cardiac inotropic reserve and function in rats with already established HF [74]. These results suggest that lowered GRK2 expression and activity in the adrenal gland might have a significant beneficial effect on cardiac remodeling by exerting an indirect sympatholytic effect on CA secretion from the adrenal medulla [75] and that reduction of sympathetic outflow via adrenal GRK2 inhibition might represent a novel attractive therapeutic strategy in the prevention of cardiac pathological remodeling [76, 77].

10.3.4 Targeted PI3K Displacement

Cardiac remodeling induced by various pathological causes is consistently associated to profound βAR abnormalities and activation of both GRK2 and PI3Ks [4, 20, 26, 78]. Several lines of evidence now suggest that selective inhibition of receptor-localized PI3K activation ameliorates βAR dysfunction and cardiac remodeling in different experimental settings [20, 26, 78]. Transgenic cardiac overexpression of the kinase-dead PI3K mutant in mice to reduce GRK2-associated PI3K activity preserved βAR levels and function after chronic pressure overload [28]. Similarly, receptor-localized inhibition of GRK2-associated PI3K activity preserved βAR signaling, improved cardiac function, and prolonged survival in a murine model of severe heart failure induced by cardiac-specific calsequestrin overexpression [27]. Interestingly, in a mouse model of intermittent pressure overload characterized by a mild hypertrophic response but profound molecular, structural, and functional abnormalities, selective inhibition of receptor-localized PI3K prevented βAR abnormalities, vascular rarefaction, and diastolic dysfunction while not remarkably affecting cardiac remodeling [79].

Consistent with in vitro studies showing that PIP_3 generation by PI3K is required for efficient internalization of βARs [24] in pigs with pacing-induced heart failure or left ventricular samples from failing human hearts, βARs underwent redistribution to intracellular compartments with downregulation from the plasma membrane, and this was accompanied by a significant increase in membrane-targeted PI3K activity [21]. Similar to these findings in failing hearts, chronic isoproterenol (ISO) infusion in mice induced a rapid redistribution of βARs into endosomal compartments [21]. Importantly, in mice with transgenic overexpression of the PIK domain of PI3K, the redistribution of βARs into early and late endosomes was inhibited despite prolonged ISO stimulation, resulting in normal βAR plasma membrane levels and ability to couple [21]. These data indicate that disruption of the GRK2/PI3K complex by overexpressing the PIK domain peptide alters the intracellular trafficking of βARs following prolonged agonist stimulation by preventing membrane translocation of PI3Ks and βAR-targeted generation of PIP3.

Very recently, pharmacological inhibitors of PI3Kγ have been shown to reduce arterial blood pressure levels by a vasorelaxant Akt/L-type calcium channel mechanism [80] and ameliorate cardiac remodeling in response to pressure overload [30]. Importantly, the molecular approach of displacing the kinase activity of PI3K from GRK2 might selectively restore agonist-dependent βAR signaling without inducing chronic βAR stimulation or affecting total PI3K activity [21].

10.3.5 GPCR Biased Ligands

For both $AT_{1A}Rs$ and β_1ARs, it is well known that chronic G protein signaling is detrimental in cardiac remodeling, whereas accumulating experimental evidence now suggests that β-arrestin-mediated signaling, in particular EGFR transactivation, exerts beneficial effects in cardiac remodeling [39, 81]. The physiological relevance of this pathway in cardioprotection was first demonstrated using transgenic mice overexpressing wild-type β_1ARs or mutant β_1ARs lacking GRK phosphorylation sites (β_1AR GRK^-). Under conditions of chronic catecholamine stress, β_1AR GRK^- mice were unable to induce β-arrestin-mediated EGFR transactivation, and, compared to mice expressing wild-type β_1ARs or mutant β_1ARs lacking PKA phosphorylation sites, these mice showed increased cardiomyocyte apoptosis and left ventricular dysfunction [39].

Recently, it has been also shown that mechanical stretch triggered a conformational change in $AT_{1A}Rs$, leading to β-arrestin recruitment to the receptor in the absence of ligand and antiapoptotic ERK signaling in the absence of detectable G protein activation [81]. Interestingly, hearts from mice lacking β-arrestin or $AT_{1A}Rs$ displayed blunted ERK and Akt activation, impaired transactivation of the EGFR, and enhanced myocyte apoptosis [81], suggesting that the heart responds to acute increases in mechanical stress by activating β-arrestin-mediated cell survival signals.

Based on these studies, it is now becoming increasingly clear that distinct conformations of activated GPCRs might preferentially allow binding of β-arrestin to promote downstream protective signaling in the absence of deleterious excessive G protein activation [82]. It is conceivable that specific ligands might differentially or even exclusively activate one or another signaling mechanism by inducing specific conformations of a wild-type endogenous receptor, suggesting that it should be possible to develop drugs that can activate one or the other pathway preferentially or alternatively (Fig. 10.5). This assumption led to the hypothesis of "biased agonism," which is a property of the ligand–receptor complex, whereby a ligand or a receptor may be biased towards a particular signaling pathway. Thus, "biased ligands" or "biased receptors" are those that preferentially exploit either β-arrestin or G protein signaling pathways [36]. Since many of the currently available therapies to treat heart failure, such as angiotensin receptor blockers or βAR blockers, block both G protein- and β-arrestin-mediated signaling, this concept, also described as "functional selectivity" or "collateral/pluridimensional efficacy," might have major clinical implications for pharmacological therapeutics targeting GPCRs to effect reverse cardiac remodeling in the near future [15].

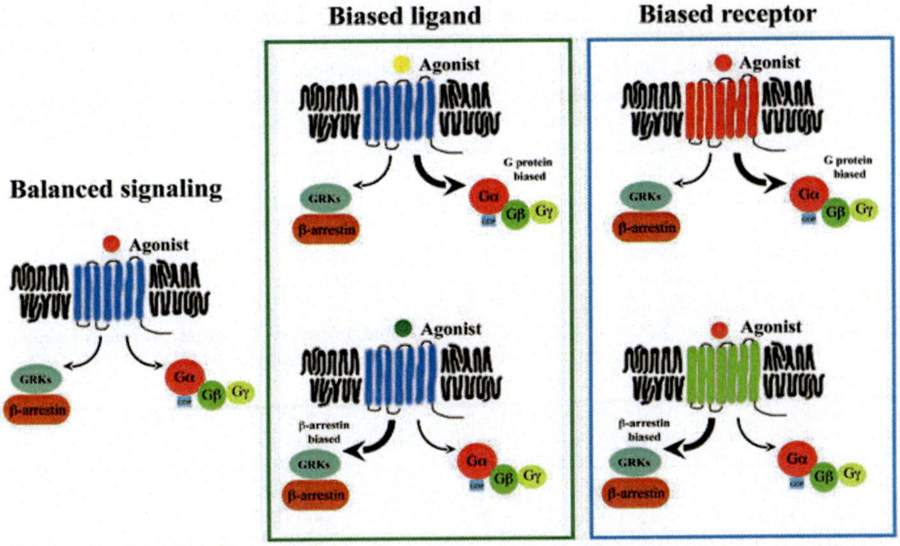

Fig. 10.5 Balanced or biased signaling of G protein-coupled receptors. Binding of a ligand results in activation of G protein signaling and β-arrestins, as well as desensitization and internalization (balanced signaling, *left panel*). Two possible mechanisms underlying biased signaling are also schematically represented. In the case of a biased ligand, binding of a ligand (G protein biased or β-arrestin biased) to an unbiased receptor results in a biased response. In the case of a biased receptor, binding of an unbiased ligand to the biased receptor (G protein biased or β-arrestin biased) might also result in a biased response

References

1. Vassilatis DK, Hohmann JG, Zeng H et al (2003) The g protein-coupled receptor repertoires of human and mouse. Proc Natl Acad Sci USA 100:4903–4908
2. Hakak Y, Shrestha D, Goegel MC, Behan DP, Chalmers DT (2003) Global analysis of g-protein-coupled receptor signaling in human tissues. FEBS Lett 550:11–17
3. Tang CM, Insel PA (2004) Gpcr expression in the heart; "new" receptors in myocytes and fibroblasts. Trends Cardiovasc Med 14:94–99
4. Rockman HA, Koch WJ, Lefkowitz RJ (2002) Seven-transmembrane-spanning receptors and heart function. Nature 415:206–212
5. Port JD, Bristow MR (2001) Altered beta-adrenergic receptor gene regulation and signaling in chronic heart failure. J Mol Cell Cardiol 33:887–905
6. Cohn JN, Tognoni G (2001) Valsartan Heart Failure Trial I. A randomized trial of the angiotensin-receptor blocker valsartan in chronic heart failure. N Engl J Med 345:1667–1675
7. Kober L, Torp Pedersen C, Carlsen JE et al (1995) A clinical trial of the angiotensin-converting-enzyme inhibitor trandolapril in patients with left ventricular dysfunction after myocardial infarction. Trandolapril cardiac evaluation (trace) study group. N Engl J Med 333:1670–1676
8. Lindholm LH, Ibsen H, Dahlof B et al (2002) Cardiovascular morbidity and mortality in patients with diabetes in the losartan intervention for endpoint reduction in hypertension study (life): A randomised trial against atenolol. Lancet 359:1004–1010

9. Pfeffer MA, Braunwald E, Moye LA et al (1992) Effect of captopril on mortality and morbidity in patients with left ventricular dysfunction after myocardial infarction. Results of the survival and ventricular enlargement trial. The save investigators. N Engl J Med 327:669–677
10. Pitt B, Poole-Wilson PA, Segal R et al (2000) Effect of losartan compared with captopril on mortality in patients with symptomatic heart failure: Randomised trial–the losartan heart failure survival study elite ii. Lancet 355:1582–1587
11. Packer M, Bristow MR, Cohn JN et al (1996) The effect of carvedilol on morbidity and mortality in patients with chronic heart failure. U.S. Carvedilol heart failure study group. N Engl J Med 334:1349–1355
12. Neves SR, Ram PT, Iyengar R (2002) G protein pathways. Science 296:1636–1639
13. Belmonte SL, Blaxall BC (2011) G protein coupled receptor kinases as therapeutic targets in cardiovascular disease. Circ Res 109:309–319
14. Kim J, Ahn S, Ren XR et al (2005) Functional antagonism of different g protein-coupled receptor kinases for beta-arrestin-mediated angiotensin ii receptor signaling. Proc Natl Acad Sci USA 102:1442–1447
15. Noor N, Patel CB, Rockman HA (2011) Beta-arrestin: A signaling molecule and potential therapeutic target for heart failure. J Mol Cell Cardiol 51:534–541
16. Luttrell LM, Lefkowitz RJ (2002) The role of beta-arrestins in the termination and transduction of g-protein-coupled receptor signals. J Cell Sci 115:455–465
17. Oakley RH, Laporte SA, Holt JA, Caron MG, Barak LS (2000) Differential affinities of visual arrestin, beta arrestin1, and beta arrestin2 for g protein-coupled receptors delineate two major classes of receptors. J Biol Chem 275:17201–17210
18. Tan CM, Brady AE, Nickols HH, Wang Q, Limbird LE (2004) Membrane trafficking of g protein-coupled receptors. Annu Rev Pharmacol Toxicol 44:559–609
19. Gregory KN, Hahn H, Haghighi K et al (2004) Increased particulate partitioning of pkc epsilon reverses susceptibility of phospholamban knockout hearts to ischemic injury. J Mol Cell Cardiol 36:313–318
20. Perrino C, Rockman HA (2007) Reversal of cardiac remodeling by modulation of adrenergic receptors: A new frontier in heart failure. Curr Opin Cardiol 22:443–449
21. Perrino C, Naga Prasad SV, Schroder JN et al (2005) Restoration of beta-adrenergic receptor signaling and contractile function in heart failure by disruption of the betaark1/phosphoinositide 3-kinase complex. Circulation 111:2579–2587
22. Krueger KM, Daaka Y, Pitcher JA, Lefkowitz RJ (1997) The role of sequestration in g protein-coupled receptor resensitization. Regulation of beta2-adrenergic receptor dephosphorylation by vesicular acidification. J Biol Chem 272:5–8
23. Vasudevan NT, Mohan ML, Gupta MK, Hussain AK, Naga Prasad SV (2011) Inhibition of protein phosphatase 2a activity by pi3kgamma regulates beta-adrenergic receptor function. Mol Cell 41:636–648
24. Naga Prasad SV, Laporte SA, Chamberlain D et al (2002) Phosphoinositide 3-kinase regulates beta2-adrenergic receptor endocytosis by ap-2 recruitment to the receptor/beta-arrestin complex. J Cell Biol 158:563–575
25. Naga Prasad SV, Barak LS, Rapacciuolo A, Caron MG, Rockman HA (2001) Agonist-dependent recruitment of phosphoinositide 3-kinase to the membrane by beta-adrenergic receptor kinase 1. A role in receptor sequestration. J Biol Chem 276:18953–18959
26. Naga Prasad SV, Perrino C, Rockman HA (2003) Role of phosphoinositide 3-kinase in cardiac function and heart failure. Trends Cardiovasc Med 13:206–212
27. Perrino C, Naga Prasad SV, Patel M, Wolf MJ, Rockman HA (2005) Targeted inhibition of beta-adrenergic receptor kinase-1-associated phosphoinositide-3 kinase activity preserves beta-adrenergic receptor signaling and prolongs survival in heart failure induced by calsequestrin overexpression. J Am Coll Cardiol 45:1862–1870
28. Nienaber JJ, Tachibana H, Naga Prasad SV et al (2003) Inhibition of receptor-localized pi3k preserves cardiac beta-adrenergic receptor function and ameliorates pressure overload heart failure. J Clin Invest 112:1067–1079

29. Naga Prasad SV, Jayatilleke A, Madamanchi A, Rockman HA (2005) Protein kinase activity of phosphoinositide 3-kinase regulates beta-adrenergic receptor endocytosis. Nat Cell Biol 7:785–796
30. Perino A, Ghigo A, Ferrero E et al (2011) Integrating cardiac pip3 and camp signaling through a pka anchoring function of p110gamma. Mol Cell 42:84–95
31. Patrucco E, Notte A, Barberis L et al (2004) Pi3kgamma modulates the cardiac response to chronic pressure overload by distinct kinase-dependent and -independent effects. Cell 118:375–387
32. Luttrell LM, Ferguson SS, Daaka Y et al (1999) Beta-arrestin-dependent formation of beta2 adrenergic receptor-src protein kinase complexes. Science 283:655–661
33. Luttrell LM, Roudabush FL, Choy EW et al (2001) Activation and targeting of extracellular signal-regulated kinases by beta-arrestin scaffolds. Proc Natl Acad Sci USA 98:2449–2454
34. Wei H, Ahn S, Shenoy SK et al (2003) Independent beta-arrestin 2 and g protein-mediated pathways for angiotensin ii activation of extracellular signal-regulated kinases 1 and 2. Proc Natl Acad Sci USA 100:10782–10787
35. Ahn S, Shenoy SK, Wei H, Lefkowitz RJ (2004) Differential kinetic and spatial patterns of beta-arrestin and g protein-mediated erk activation by the angiotensin ii receptor. J Biol Chem 279:35518–35525
36. Rajagopal S, Rajagopal K, Lefkowitz RJ (2010) Teaching old receptors new tricks: biasing seven-transmembrane receptors. Nat Rev Drug Discov 9:373–386
37. Daub H, Weiss FU, Wallasch C, Ullrich A (1996) Role of transactivation of the egf receptor in signalling by g-protein-coupled receptors. Nature 379:557–560
38. Maudsley S, Pierce KL, Zamah AM et al (2000) The beta(2)-adrenergic receptor mediates extracellular signal-regulated kinase activation via assembly of a multi-receptor complex with the epidermal growth factor receptor. J Biol Chem 275:9572–9580
39. Noma T, Lemaire A, Naga Prasad SV et al (2007) Beta-arrestin-mediated beta1-adrenergic receptor transactivation of the egfr confers cardioprotection. J Clin Invest 117:2445–2458
40. Tilley DG, Kim IM, Patel PA, Violin JD, Rockman HA (2009) Beta-arrestin mediates beta1-adrenergic receptor-epidermal growth factor receptor interaction and downstream signaling. J Biol Chem 284:20375–20386
41. Prenzel N, Zwick E, Daub H et al (1999) Egf receptor transactivation by g-protein-coupled receptors requires metalloproteinase cleavage of prohb-egf. Nature 402:884–888
42. DeWire SM, Ahn S, Lefkowitz RJ, Shenoy SK (2007) Beta-arrestins and cell signaling. Annu Rev Physiol 69:483–510
43. Ahn S, Kim J, Hara MR, Ren XR, Lefkowitz RJ (2009) {beta}-arrestin-2 mediates anti-apoptotic signaling through regulation of bad phosphorylation. J Biol Chem 284:8855–8865
44. Violin JD, Ren XR, Lefkowitz RJ (2006) G-protein-coupled receptor kinase specificity for beta-arrestin recruitment to the beta2-adrenergic receptor revealed by fluorescence resonance energy transfer. J Biol Chem 281:20577–20588
45. Kim IM, Tilley DG, Chen J et al (2008) Beta-blockers alprenolol and carvedilol stimulate beta-arrestin-mediated egfr transactivation. Proc Natl Acad Sci USA 105:14555–14560
46. Shenoy SK, Drake MT, Nelson CD et al (2006) Beta-arrestin-dependent, g protein-independent erk1/2 activation by the beta2 adrenergic receptor. J Biol Chem 281:1261–1273
47. Bouvier M (2001) Oligomerization of g-protein-coupled transmitter receptors. Nat Rev Neurosci 2:274–286
48. Breitwieser GE (2004) G protein-coupled receptor oligomerization: Implications for g protein activation and cell signaling. Circ Res 94:17–27
49. George SR, O'Dowd BF, Lee SP (2002) G-protein-coupled receptor oligomerization and its potential for drug discovery. Nat Rev Drug Discov 1:808–820
50. Overton MC, Blumer KJ (2002) The extracellular n-terminal domain and transmembrane domains 1 and 2 mediate oligomerization of a yeast g protein-coupled receptor. J Biol Chem 277:41463–41472

51. Tarasova NI, Rice WG, Michejda CJ (1999) Inhibition of g-protein-coupled receptor function by disruption of transmembrane domain interactions. J Biol Chem 274:34911–34915
52. Calver AR, Robbins MJ, Cosio C et al (2001) The c-terminal domains of the gaba(b) receptor subunits mediate intracellular trafficking but are not required for receptor signaling. J Neurosci 21:1203–1210
53. Angers S, Salahpour A, Joly E et al (2000) Detection of beta 2-adrenergic receptor dimerization in living cells using bioluminescence resonance energy transfer (bret). Proc Natl Acad Sci USA 97:3684–3689
54. Mercier JF, Salahpour A, Angers S, Breit A, Bouvier M (2002) Quantitative assessment of beta 1- and beta 2-adrenergic receptor homo- and heterodimerization by bioluminescence resonance energy transfer. J Biol Chem 277:44925–44931
55. Hebert TE, Moffett S, Morello JP et al (1996) A peptide derived from a beta2-adrenergic receptor transmembrane domain inhibits both receptor dimerization and activation. J Biol Chem 271:16384–16392
56. Lavoie C, Mercier JF, Salahpour A et al (2002) Beta 1/beta 2-adrenergic receptor heterodimerization regulates beta 2-adrenergic receptor internalization and erk signaling efficacy. J Biol Chem 277:35402–35410
57. Xu J, He J, Castleberry AM et al (2003) Heterodimerization of alpha 2a- and beta 1-adrenergic receptors. J Biol Chem 278:10770–10777
58. Barki-Harrington L, Luttrell LM, Rockman HA (2003) Dual inhibition of beta-adrenergic and angiotensin ii receptors by a single antagonist: A functional role for receptor-receptor interaction in vivo. Circulation 108:1611–1618
59. Chung S, Funakoshi T, Civelli O (2008) Orphan gpcr research. Br J Pharmacol 153(Suppl 1):S339–S346
60. Sakurai T (2005) Reverse pharmacology of orexin: From an orphan gpcr to integrative physiology. Regul Pept 126:3–10
61. Civelli O, Saito Y, Wang Z, Nothacker HP, Reinscheid RK (2006) Orphan gpcrs and their ligands. Pharmacol Ther 110:525–532
62. Adams JW, Wang J, Davis JR et al (2008) Myocardial expression, signaling, and function of gpr22: A protective role for an orphan g protein-coupled receptor. Am J Physiol Heart Circ Physiol 295:H509–H521
63. Salazar NC, Chen J, Rockman HA (2007) Cardiac gpcrs: Gpcr signaling in healthy and failing hearts. Biochim Biophys Acta 1768:1006–1018
64. Kamal FA, Smrcka AV, Blaxall BC (2011) Taking the heart failure battle inside the cell: Small molecule targeting of gbetagamma subunits. J Mol Cell Cardiol 51:462–467
65. Ross EM, Wilkie TM (2000) Gtpase-activating proteins for heterotrimeric g proteins: Regulators of g protein signaling (rgs) and rgs-like proteins. Annu Rev Biochem 69:795–827
66. Berman DM, Wilkie TM, Gilman AG (1996) Gaip and rgs4 are gtpase-activating proteins for the gi subfamily of g protein alpha subunits. Cell 86:445–452
67. Zhang P, Mende U (2011) Regulators of g-protein signaling in the heart and their potential as therapeutic targets. Circ Res 109:320–333
68. Huang X, Fu Y, Charbeneau RA et al (2006) Pleiotropic phenotype of a genomic knock-in of an rgs-insensitive g184s gnai2 allele. Mol Cell Biol 26:6870–6879
69. Rogers JH, Tamirisa P, Kovacs A et al (1999) Rgs4 causes increased mortality and reduced cardiac hypertrophy in response to pressure overload. J Clin Invest 104:567–576
70. Rogers JH, Tsirka A, Kovacs A et al (2001) Rgs4 reduces contractile dysfunction and hypertrophic gene induction in galpha q overexpressing mice. J Mol Cell Cardiol 33:209–218
71. Takimoto E, Koitabashi N, Hsu S et al (2009) Regulator of g protein signaling 2 mediates cardiac compensation to pressure overload and antihypertrophic effects of pde5 inhibition in mice. J Clin Invest 119:408–420
72. Li H, He C, Feng J et al (2010) Regulator of g protein signaling 5 protects against cardiac hypertrophy and fibrosis during biomechanical stress of pressure overload. Proc Natl Acad Sci USA 107:13818–13823

73. Mudd JO, Kass DA (2008) Tackling heart failure in the twenty-first century. Nature 451:919–928
74. Lymperopoulos A, Rengo G, Funakoshi H, Eckhart AD, Koch WJ (2007) Adrenal grk2 upregulation mediates sympathetic overdrive in heart failure. Nat Med 13:315–323
75. Lymperopoulos A, Rengo G, Gao E et al (2010) Reduction of sympathetic activity via adrenal-targeted grk2 gene deletion attenuates heart failure progression and improves cardiac function after myocardial infarction. J Biol Chem 285:16378–16386
76. Liggett SB, Mialet-Perez J, Thaneemit-Chen S et al (2006) A polymorphism within a conserved beta(1)-adrenergic receptor motif alters cardiac function and beta-blocker response in human heart failure. Proc Natl Acad Sci USA 103:11288–11293
77. Bristow MR, Krause-Steinrauf H, Nuzzo R et al (2004) Effect of baseline or changes in adrenergic activity on clinical outcomes in the beta-blocker evaluation of survival trial. Circulation 110:1437–1442
78. Perrino C, Esposito G, Rockman HA (2005) Defects in cardiomyocyte function: Role of beta-adrenergic receptor dysfunction. Panminerva Med 47:143–155
79. Perrino C, Naga Prasad SV, Mao L (2006) Intermittent pressure overload triggers hypertrophy-independent cardiac dysfunction and vascular rarefaction. J Clin Invest 116:1547–1560
80. Carnevale D, Vecchione C, Mascio G et al (2012) Pi3kgamma inhibition reduces blood pressure by a vasorelaxant akt/l-type calcium channel mechanism. Cardiovasc Res 93:200–209
81. Rakesh K, Yoo B, Kim IM et al (2010) Beta-arrestin-biased agonism of the angiotensin receptor induced by mechanical stress. Sci Signal 3:ra46
82. Violin JD, Lefkowitz RJ (2007) Beta-arrestin-biased ligands at seven-transmembrane receptors. Trends Pharmacol Sci 28:416–422

Chapter 11
Role of Inflammation and Matrix Proteinases in Cardiac Remodeling Following Stress and Injury

Davy Vanhoutte and Stephane Heymans

Abstract Whether the myocardial damage is initially caused by ischemia or viral infection of the heart, a tightly regulated inflammatory response is provoked in an attempt to restore cardiac homeostasis and to preserve its proper function. However, a growing body of evidence suggests that uncontrolled cardiac inflammation—due to mechanisms we do not fully understand—can lead to adverse cardiac remodeling and facilitates the disease progression to its sequelae, dilated cardiomyopathy, congestive heart failure, or even sudden death. Discoveries over the last years have made it clear that an important regulator of the inflammatory pathways may be found in the cardiac extracellular matrix (ECM), known as the matrix metalloproteinase (MMP) system. It is now becoming clear that degradation and turnover of the cardiac ECM is neither the sole nor the main function of these proteinases. In fact, a growing body of evidence suggests that a complex bidirectional cross talk exists between the MMP system and a wide variety of cytokines, chemokines, growth factors, and other bioactive molecules that regulates diverse and sometimes opposing aspects of the inflammatory response in the injured heart. Furthermore, the use of single MMP-mutant mice seems to indicate MMP inhibition as potentially cardioprotective after myocardial infarction while detrimental during the pathogenesis of coxsackievirus B3-induced viral myocarditis. These findings

D. Vanhoutte, Ph.D. (✉)
Department of Pediatrics, Molecular Cardiovascular Biology, Cincinnati Children's Hospital Medical Center, 240 Albert Sabin Way, MLC 7020, Cincinnati, OH 45229-3039, USA

Department of Cardiovascular Sciences, K.U. Leuven, Leuven, Belgium
e-mail: Davyvanhoutte@gmail.com

S. Heymans
Department of Cardiovascular Sciences, K.U. Leuven, Leuven, Belgium

Center for Heart Failure Research, Cardiovascular Research Institute Maastricht (CARIM), Maastricht University, Maastricht, The Netherlands

B.I. Jugdutt and N.S. Dhalla (eds.), *Cardiac Remodeling: Molecular Mechanisms*, Advances in Biochemistry in Health and Disease 5, DOI 10.1007/978-1-4614-5930-9_11,

enhance our knowledge in the MMP field and in that of inflammatory cardiomyopathy and will without a doubt allow for novel and highly specific therapeutic interventions in the future.

Keywords Matrix metalloproteinases • Extracellular matrix • Inflammation • Cardiac remodeling • Myocardial infarction • Viral myocarditis

11.1 Cardiac Injury, Inflammation, and Repair

Whether myocardial injury is induced by myocardial infarction (MI) or viral infection of the heart (e.g., coxsackievirus B3 (CVB3)-induced viral myocarditis), a tightly regulated orchestra of inflammatory mechanisms is provoked [1]. Cardiac inflammation functions to limit tissue damage and clears the wound from dead or infected cells and matrix debris while activating reparative pathways essential for proper scar formation, and this in an attempt to restore cardiac homeostasis and to preserve its proper function [1–3]. Interestingly, although both pathologies have their own unique features when it comes to etiology, pathogenesis, and disease progression, they both reflect inflammatory mechanisms that are partially shared.

11.1.1 Initiation of the Inflammatory Response

In contrast to cells undergoing apoptosis, necrotic cell death resulting from cardiac ischemia or viral infection leads to disruption of cell membranes and the release of cytoplasmatic and nuclear components that contain damage-associated molecular patterns (DAMPs) [2, 4]. Together with ECM fragments and pathogen-associated molecular patterns (PAMPs), DAMPs have the ability to activate the innate immune system through (1) toll-like receptor-mediated pathways, (2) the complement cascade, and (3) reactive oxygen species (ROS) [2, 4]. These three molecular "alarm" pathways converge on activation of the nuclear factor (NF)-Kβ system that drives the production of pro-inflammatory cytokines, chemokines, and growth factors by cardiac resident cells.

11.1.2 Pro-inflammatory Phase: "A Partnership Between the Innate and Adaptive Immune System"

At this point, an intense inflammatory chain reaction is initiated (Fig. 11.1). Pro-inflammatory cytokines and chemokines, including interleukin (IL)-6, IL-1β, tumor necrosis factor (TNF)-α, and interferon (INF)-γ, fuel the inflammatory process by

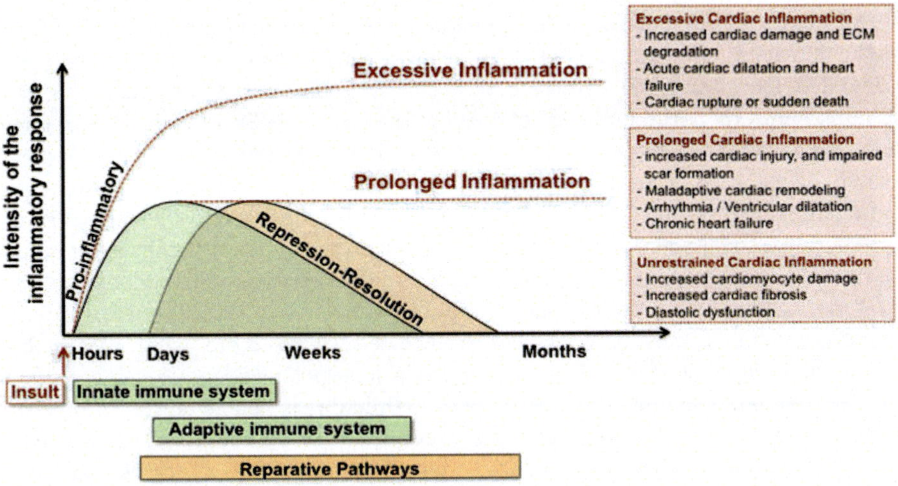

Fig. 11.1 *Cardiac inflammation: "when friend turns foe."* A complex inflammatory reaction is provoked in response to cardiac injury. In normal conditions, the pro-inflammatory phase is followed by an active phenomenon of inflammatory repression and resolution that involves both innate and adaptive immune system. In the meantime, reparative pathways are activated in order to replace the dead tissue by a fibrotic scar to restore cardiac homeostasis and preserve proper function of the injured heart. Unfortunately, in a subset of patients, cardiac inflammation becomes excessive, prolonged, or unrestrained and facilitates the disease progression to its sequelae

activating resident monocytes and macrophages [2, 5, 6]. In addition, they also upregulate cell adhesion molecules (e.g., vascular cell adhesion molecule (VCAM)-1 and intercellular adhesion molecule (ICAM)-1), monocyte- and neutrophil-attracting chemokines (CC, e.g., monocyte chemotactic protein-1 (MCP-1)/CCL2 and CXC, e.g., IL-8/CXCL8, respectively), and chemokine receptors (CXCRs), thereby generating directional cues for leukocyte subpopulations to egress from the vasculature into the injured or infected myocardium [2, 3].

Neutrophils and monocytes are recruited very early into the injured myocardium [2]. Once recruited and exposed to macrophage colony-stimulating factor, TNF-α, INF-γ, and others, monocytes differentiate into pro-inflammatory M1 phagocytotic macrophages [7, 8]. Together with neutrophils, they release matrix-degrading MMPs, which allow them to migrate to the site of injury where they jointly remove necrotic cardiomyocytes and secrete large amounts of pro-inflammatory mediators to further amplify the pro-inflammatory signaling cascade [2, 3, 9]. At the same time, MMPs released by neutrophils, macrophages, fibroblasts, and/or endothelial cells cleave a multitude of inflammatory mediators, consequently amplifying or dampening the inflammatory response. Besides neutrophils and monocytes, natural killer (NK) cells also infiltrate the injured heart. Although the significance of NK cells after MI is only beginning to be recognized [10], they play an important role in the pathogenesis of viral myocarditis by directly recognizing and killing virus-infected cells [6].

Subsequently, immune cells from the adaptive immune system also accumulate in the damaged myocardium (Fig. 11.1). During viral myocarditis, B-lymphocytes, $CD4^+$ T-helper, and $CD8^+$ cytotoxic T-lymphocytes are recruited to maximize the elimination of the virus-infected cells and to limit viral replication [3, 6]. The contribution of adaptive immunity in the context of MI, on the other hand, has been widely disregarded because it contradicts the classic notion that adaptive immunity is not stimulated by self-antigens. However, recent studies revealed the presence of conventional $CD4^+$ and $CD8^+$ T-lymphocytes as well as regulatory T cells in infarcted hearts [11], indicating that antigens that are normally not accessible (e.g., troponin I) or neoantigens created during ischemic injury are released from the myocardium and presented to T cells. Nevertheless, further investigations are mandatory to elucidate the full extent of the adaptive immune system post-MI.

The immune response is now at its peak. While this is beneficial for viral clearance and removal of dead cells and matrix debris, it can also contribute to irreversible myocardial damage due to the nonproliferating nature of cardiomyocytes (Fig. 11.1).

11.1.3 Suppression and Resolution of the Inflammatory Response

Once the initial inflammatory stimulus has been eliminated, a set of "brakes" is initiated to suppress and resolve the inflammatory infiltrate and promote the reparative pathways (Fig. 11.1). First, after their short life-span, apoptotic neutrophils themselves release mediators to abrogate further neutrophil recruitment and promote their resolution. A well-studied example, annexin A1, dampens neutrophil accumulation by several mechanisms, including downregulation of neutrophil transendothelial migration, increasing neutrophil apoptosis, and promoting macrophage-mediated phagocytosis of apoptotic neutrophils [2, 5]. Secondly, the clearance of apoptotic neutrophils by phagocytes launches an anti-inflammatory, pro-resolving transcriptional program with release of IL-10, IL-13, and transforming growth factor-β (TGF-β) [2, 5]. IL-10, for example, inhibits IL-1, IL-6, IL-8, and TNF-α production and has been reported to modulate the phenotypic switch from pro-inflammatory M1 to anti-inflammatory M2 macrophages [7, 8]. Furthermore, IL-10 may play a significant role in ECM remodeling by promoting TIMP-1 synthesis that contributes to the stabilization of the cardiac ECM [2]. Together with immunosuppressive $CD4^+CD25^+Foxp3^+$ regulatory T-lymphocytes, M2 macrophages further amplify the array of suppressive and pro-resolving cytokines, chemokines, and growth factors such as IL-10, decoy type II IL-1 receptor, IL-1 receptor antagonist (IL-1Ra), and TGF-β that dampens the immune response and promotes the replacement of dead tissue by a fibrotic scar [2, 5, 11]. TGF-β also modulates (myo)fibroblast phenotype and function, inhibits MMP production, induces TIMPs, and stimulates collagen synthesis [2]. Next, (myo)fibroblasts migrate into the empty spaces and produce large amounts of ECM proteins, and an extensive microvascular network is established. Finally, (myo)fibroblasts and vascular cells undergo apoptosis, and a mature highly organized collagen-based scar is formed that provides the necessary mechanical support to the injured heart [3, 9].

11.1.4 Cardiac Inflammation: "When Friend Turns Foe!"

Unfortunately, in a subset of patients, the inflammatory response becomes excessive and chronic or expands into the noninjured myocardium and can facilitate the disease progression to its sequelae (Fig. 11.1) [2, 3, 6, 9]. An array of studies using various approaches to manipulate either the innate or adoptive immune response after MI and during viral myocarditis has revealed that (1) excessive inflammation due to enhanced expression of pro-inflammatory mediators or the absence of proper repression may activate proapoptotic pathways and augment matrix degradation causing cardiac rupture or sudden death; (2) prolongation of the inflammatory reaction—due to defective repression or resolution—may exaggerate cardiac injury, promote a disorganized collagen scar, and thereby alter the mechanical properties of the heart, resulting in exaggerated maladaptive cardiac remodeling which is associated with higher incidence of arrhythmia, ventricular dilatation, chronic heart failure, and mortality; and, finally, (3) defective containment of the inflammatory reaction may lead to extension of the inflammatory infiltrate into the healthy myocardium, thereby enhancing (myo)fibroblast-mediated collagen deposition and fibrosis, myocardial stiffness, and diastolic dysfunction. Moreover, in the setting of viral myocarditis, it is believed that not the direct infection of myocytes but rather the subsequent immune response plays a major role in the myocardial damage and ensuing morbidity [6].

Taken together, there is little doubt that uncontrolled cardiac inflammation is deleterious and codetermines the patients' clinical outcome. Unfortunately, the complex nature of myocardial inflammation has hindered the development of innovative inflammatory-based therapeutic strategies and underlines our need to better understand the regulatory mechanisms of the inflammatory response in cardiac diseases.

11.2 Matrix Metalloproteinases and Their Tissue Inhibitors: "The Basics"

Although originally discovered in 1962 as collagen-degrading enzymes responsible for tadpole morphogenesis, the MMP system is becoming widely accepted as a tightly regulated proteolytic system that modulates a wide variety of molecular and cellular processes during development, homeostasis, and disease [12–14].

11.2.1 The MMP Family

The MMPs constitute a superfamily of structurally related and highly conserved zinc-dependent endopeptidases (Table 11.1 and Fig. 11.2a) [13, 14]. To date, at least 25 MMPs have been identified. MMPs are either secreted or membrane bound, thereby confining their catalytic activity to membrane proteins in the secretory pathway or pericellular space. The MMP family is traditionally divided according to their sequence homology, ECM substrate specificity, and cellular localization in several subclasses: (1) the collagenases, (2) the gelatinases, (3) the matrilysins, (4) the stromelysins, and (5) the membrane-type MMPs [13, 14]. The first four groups of MMPs are secreted as latent zymogens, termed pro-MMPs. In marked contrast, the fifth group represents a unique subcategory that consists of all membrane-bound MMPs, the MT-MMPs, which are non-substrate related. A final subcategory comprises of all MMPs that do not fall in any of the categories above.

11.2.2 Regulation of MMP Activity

Similar to all secreted proteinases, the catalytic activity of MMPs is tightly regulated at four points: gene expression, compartmentalization, pro-MMP activation, and inhibition [18]. Typically, MMP expression is low in healthy tissues yet increases in all repair and remodeling processes, including those of the heart [3, 9].

MMP transcription is regulated by specific signals that are temporally limited and spatially confined. In the heart, their transcription can be regulated by pro-inflammatory cytokines (e.g., IL and TNF-α), growth factors (e.g., TGF-β), and other bioactive molecules (angiotensin II and endothelin), as well as ROS and mechanical stimuli [13, 19]. Moreover, all cell types found in the myocardium, either in basal conditions (myocytes, fibroblasts, endothelial cells) or in response to an inflammatory stimulus (neutrophils and macrophages), express one or more types of MMP species (Table 11.1) [5].

Upon secretion, MMPs are initially kept in a catalytically inactive state. In fact, a large reservoir of recruitable pro-MMPs exists within the myocardium—compartmentalization. MMPs can either be embedded in the cardiac ECM or bound to accessory molecules (e.g., MMP-9 to CD44) [18]. Subsequent activation can therefore result in a localized yet rapid surge of proteolytic activity. MMP activation requires proteolytic cleavage of the amino terminal pro-peptide domain by serine proteinases such as plasmin and urokinase plasminogen activator (uPA), thereby exposing their Zn^{2+}-binding site in the catalytic domain (a process known as the "cysteine-switch mechanism"; Fig. 11.2b) [13]. The formation of plasmin itself is under control of the plasminogen system (plasminogen activators (PA), tissue-type PA (t-PA), urokinase-type PA (uPA), and the PA inhibitor (PAI)) and thereby forms an important point of control of MMP activity (Fig. 11.3).

Table 11.1 Matrix metalloproteinases and their tissue inhibitors

	Cellular source	Substrates[a] Collagen	Pro-MMPs	ECM related	Relevant to inflammation[a]
Collagenases					
MMP-1	CMC, Fb, ECs, and macrophages	I, II, III, VII, VIII, X	2, 9	Agg, Gel, PG	IL-1β; TNF-α; MCP-1, MCP-2, MCP-3, and MCP-4; SDF-1
MMP-8	Neutrophils, macrophages, and VSMC	I, II, III, V, VII, VIII, X		Agg, EL, FN, Gel, LN	CXCL5; CXCL8/IL-8; CXCL9
MMP-13	Fb	I, II, III, IV, VII, IX, X, XIV	9	Agg, FN, Gel	MCP-3; SDF-1; CXCL8/IL-8
Gelatinases					
MMP-2	CMC, Fb, mFb, ECs, VSMCs, macrophages, and lymphocytes	I, III, IV, V, VII, X, XI, XIV	1, 9, 13	Agg, EL, FN, Gel, LN, PG, VN	IL-1β; TNF-α; MCP-3; SDF-1; CX_3CL1; latent TGF-β1, TGF-β2, and TGF-β3
MMP-9	CMC, Fb, neutrophils, macrophages, lymphocytes, EC, and VSMC	IV, V, VII, X, XIV		Agg, EL, FN, Gel, PG, VN	IL-1β; TNF-α; IFN-β; MCP-1; SDF-1; CXCL1; CXCL4, CXCL8/IL-8; CXCL9; latent VEGF; latent TGF-β1, TGF-β2, and TGF-β3; TSP-1; IL-1R; IL-2R
Stromelysins					
MMP-3	CMC, Fb, macrophages, and VSMCs	III, IV, IX, X, XI	1, 7, 8, 9, 13	Agg, EL, FN, Gel, LN, PG, VN	IL-1β; TNF-α; MCP-1, MCP-2, MCP-3, and MCP-4; SDF-1; latent TGF-β1; OPN; E-cadherin
MMP-7	CMC and macrophages	IV, X	1, 2, 9	Agg, Casein, EL, FN, Gel, LN, PG, VN	Latent TNF-α; synd-1; OPN

(continued)

Table 11.1 (continued)

		Substrates[a]			
	Cellular source	Collagen	Pro-MMPs	ECM related	Relevant to inflammation[a]
Membrane-type MMPs					
MT1-MMP	CMC, Fb, mFb, ECs, and VSMCs	I, II, III	2, 13	Agg, EL, FN, Gel, LN	MCP-3; SDF-1; CXCL8/IL-8; synd-1, latent TGF-β
Other MMPs					
MMP-12	Macrophages	IV		Casein, EL, FN, Gel, LN, PG, VN	Latent TNF-α; MCP-1, MCP-2, MCP-3, and MCP-4

TIMPs	Cellular source	Substrates
TIMP-1	CMC, Fb, macrophages, ECs, and VSMCs	Inhibits all MMPs, except MMP-2 and MT1MMP
TIMP-2	CMC, Fb, macrophages, EC, and VSMCs	Inhibits all MMPs, except MMP-2; activates pro-MMP-2
TIMP-3	CMC, Fb, and VSMCs	Inhibits MMP-1, MMP-2, MMP-3, and MMP-9 and inhibits TNF-α-converting enzyme
TIMP-4	CMC, Fb, and VSMCs	Inhibits MMP-1, MMP-3, MMP-7, and MMP-9

The data presented in this table are restricted to the MMPs and TIMPs relevant to this chapter and are collated from [12, 15–17]
MMP matrix metalloproteinase, *TIMP* tissue inhibitor of MMP, *ECM* extracellular matrix, *CMC* cardiomyocyte, *(m)Fb* (myo)fibroblast, *ECs* endothelial cells, *VSMCs* vascular smooth muscle cells, *Agg* aggrecan, *Gel* gelatin, *PG* proteoglycan link protein, *EL* elastin, *FN* fibronectin, *LN* laminin, *VN* vitronectin, *IL* interleukin, *IL-1R* IL-1 receptor, *IFN* interferon, *TNF* tumor necrosis factor, *MCP* monocyte chemotactic protein, *SDF-1* stromal cell-derived factor-1 (= CXCL12, CXC-chemokine ligand-12), *E-cadherin* epithelial cadherin, *VEGF* vascular endothelial growth factor, *TGF* transforming growth factor, *Synd* syndecan, *OPN* osteopontin, *TSP* thrombospondin
[a]The list of substrates has been limited to proteins that are cleaved by MMPs and relevant to the regulation of the inflammatory response

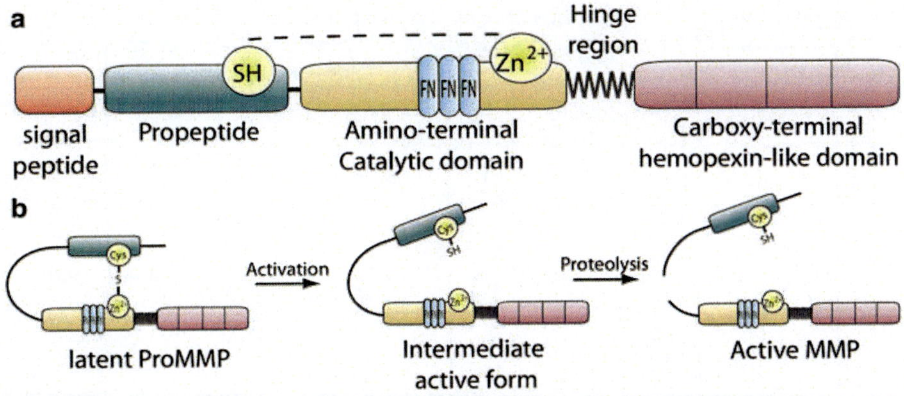

Fig. 11.2 *MMPs: "the basics."* (**a**) Domain structure of the mammalian MMP family. The general structure consists of (1) a signal peptide (SP), (2) a pro-domain, (3) a catalytic domain, and (4) a hemopexin-like domain which functions in substrate recognition. To be classified as an MMP, a protein needs to have at least the conserved pro-domain and catalytic domain. The catalytic domain typically contains a zinc ion (Zn^{2+}) in the active site (therefore the prefix "metallo") that interacts with a conserved cysteine residue (thiol, SH) in the pro-domain. Several MMPs, for example, MMP-2 and MMP-9, contain multiple fibronectin type II repeats, whereas the membrane-type MMPs (MT-MMPs) contain a transmembrane domain, a glycosylphosphatidylinositol anchor, or a signal anchor attached to the hemopexin-like domain. MT-MMPs also contain a furin-cleavage site at the C terminus of the pro-domain and are therefore activated intracellularly. (**b**) The cysteine-switch mechanism. The pro-peptide maintains an MMP inactive. When the interaction between the conserved cysteine residue in the pro-domain and the active site Zn^{2+} ion is disrupted, the active site becomes accessible, resulting in an intermediate active enzyme. Next, the pro-domain is removed by autocatalytic or other proteases, resulting in a fully active enzyme. Nevertheless, only the disruption of the Zn^{2+}–thiol interaction is an absolute requirement

Furthermore, several pro-MMPs (e.g., pro-MMP-9 and pro-MMP-2) are substrate for other MMPs (e.g., MMP-3 and MT1-MMP, respectively), leading to autocatalytic activation of MMPs and providing a mechanism for quantum bursts of proteolytic activity (Fig. 11.3). In contrast, MT-MMPs (e.g., MMP-14, also known as MT-MMP-1) contain a furin-cleavage site to allow intracellular activation before insertion in the cell membrane by members of the proprotein convertase family of proteases [18].

A final point of control is the delicate balance between MMPs and their endogenous inhibitors. The latter has been shown to be critical during cardiac homeostasis and in the development of heart failure [15]. Four endogenous tissue inhibitors of metalloproteinases (TIMP-1 to TIMP-4) exist that firmly regulate MMP activity (Fig. 11.3). TIMPs bind to the catalytic site of specific MMPs in a stoichiometric 1:1 molar ratio, thereby blocking the access to their substrates. In addition, TIMPs can also bind pro-MMPs at the aminoterminus and regulate MMP function prior to activation. Finally, TIMP binding does not exclusively lead to MMP inhibition. Indeed, the binding of TIMP-2 to MMP-2 is required for activation of the enzyme in a complex with MT1-MMP [15].

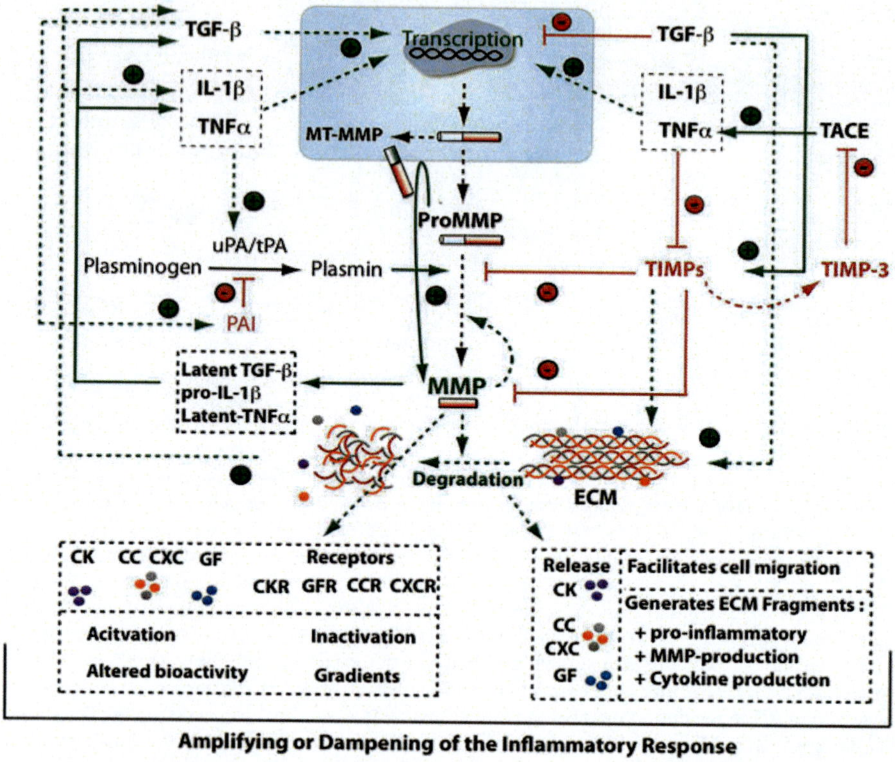

Fig. 11.3 *MMPs and inflammation: "a bidirectional cross talk."* A simplified scheme summarizing (1) MMP production, activation, and inhibition as well as (2) the complex interactions between the MMP system and inflammation as discussed in Sects. 11.2 and 11.3 of this chapter. *MMP* matrix metalloproteinases, *MT-MMP* membrane-type MMP, *TIMP* tissue inhibitor of MMP, *IL* interleukin, *TGF-β* transforming growth factor-β, *TNF-α* tumor necrosis factor-α, *uPA* urokinase plasminogen activator, *tPA* tissue PA, *PAI* PA inhibitor, *TACE* TNF-α-converting enzyme, *CK* cytokine, *CC and CXC* chemokine, *GF* growth factor, *R* receptor, + stimulatory, – inhibitory

11.2.3 MMPs in the Heart: "Much More than ECM-Degrading Enzymes"

As illustrated in Table 11.1, MMPs are collectively capable of degrading most structural components of the basement membrane and the cardiac ECM. In this aspect, increased MMP activity—in response to cardiac injury and stress—was rapidly associated with cardiac ECM remodeling and the development of progressive ventricular dilatation and dysfunction or even fatal cardiac rupture, for example, after MI [9, 15]. However, it is important to note that MMPs and TIMPs have far-reaching properties well beyond ECM turnover. Recent studies are continuously identifying novel MMP substrates, implicating the MMP family in a wide variety of biological functions (Table 11.1). MMPs mobilize growth factors, cytokines, and chemokines, not only by liberating them from the ECM but also by cleavage and dissociation of cytokine- and growth factor-binding proteins [12]. In this capacity, MMPs can contribute to a multitude of localized signaling pathways and cellular responses that are varied (e.g., modulating cell proliferation, migration, apoptosis, angiogenesis, and inflammation) and complex in nature, evoking seemingly paradoxical effects (e.g., pro- or anti-inflammatory and pro- and anti-angiogenic responses) [14, 18]. In addition, intracellular roles are being discovered. For example, MMP-2 is involved in the proteolysis of sarcomeric and cytoskeletal proteins in cardiomyocytes under oxidative stress and may have additional biological actions in other subcellular locations including caveolae, nuclei, and mitochondria [20]. Finally, TIMPs also exhibit MMP-independent, cytokine-like properties that are beginning to be characterized (reviewed by Vanhoutte et al. [15]). Taken together, it is critical to recognize both ECM- and non-ECM-related aspects of MMPs and TIMPs when considering their role in the heart.

11.3 Inflammation and Matrix Metalloproteinases: "A Bidirectional Cross Talk"

Evidently, by physical degradation of the basement membrane and the cardiac interstitium, MMPs facilitate inflammatory cell migration to the site of injury [9]. However, a far more complex cross talk, including both positive and negative feedback and forward loops, has been observed between MMPs, inflammation, and ECM turnover and is summarized in Fig. 11.3.

First, although inflammatory mediators, including cytokines, chemokines, and growth factors (e.g., NF-Kβ, IL-1β, IL-6, TNF-α, MCP-1, and TGF-β), have distinct functions during the inflammatory process, they also regulate the expression of multiple MMPs and TIMPs [2, 19, 21] (Fig. 11.3). For example, IL-1β and TNF-α cause increased expression of MMP-2, MMP-9, and MMP-13 by cardiac fibroblasts [22]. Furthermore, they both have the capacity of inducing MMP activator uPA [21]. TGF-β, on the other hand, has a dual effect. TGF-β can suppress the transcription of MMP-1 and MMP-3, whereas it increases the expression of MMP-2 and MMP-9 [23, 24]. Conversely, inflammatory mediators can also control MMP activity by influencing the expression of endogenous protease inhibitors. IL-1 and TNF-α, for example, inhibit the expression of TIMPs, thereby tipping the net proteolytic balance towards increased MMP activity, whereas TGF-β can induce the synthesis of protease inhibitors, including PAI-1 and TIMPs (Fig. 11.3) [19].

Secondly, active MMPs can cleave a multitude of inflammatory mediators—including cytokines, chemokines, growth factors, and their receptors—and/or affect their bioavailability by releasing them from the ECM (extended overview in

Table 11.1). Examples of MMP substrates relevant to the inflammatory response include MCP-1, MCP-2, MCP-3, and MCP-4, CXCL12 (stromal cell-derived factor-1), fractalkine (CX_3CL1), pro-IL-1β, latent TNF-α, interferon-β, and latent TGF-β [12, 18]. Interestingly, cleavage of these molecules can either result in their activation, inactivation, or conversion to other molecules with altered bioactivity, consequently amplifying or dampening the inflammatory response. As such, MMP-2, MMP-3, and MMP-9 can proteolyze pro-IL-1β to activate pro-inflammatory cytokine IL-1β, while MMP-1, MMP-2, MMP-3, MMP-7, MMP-9, MMP-12, MMP-14, and MMP-17 have TNF-α shedding activities [18, 25]. Similarly, MMPs can also establish chemokine gradients that enhance leukocyte trafficking to the site of injury [18, 25]. Conversely, MMPs are also able to suppress of the inflammatory response. MMP-1, MMP-3, and MMP-12, for example, have the ability to cleave MCP-1 to MCP-4 and CXCL1, thereby abrogating chemotaxis and putting a brake on leukocyte infiltration [26, 27], whereas MMP-2, MMP-3, MMP-9, and MT1-MMP can liberate latent TGF-β from the ECM, consequently increasing the bioavailability of the mature anti-inflammatory and pro-fibrotic TGF-β [18, 25]. Other MMP-mediated anti-inflammatory actions include proteolysis of IL-1 receptor (IL-1R), IL-2R, and IFN-β by MMP-9 rendering them inactive [25, 28]. Hence, MMPs form a negative feedback loop for controlling inflammatory cell recruitment and activation.

Next, through means of MMP-mediated proteolytic degradation, ECM fragments are generated that exert important immunomodulatory properties (Fig. 11.3), referred to as matrikines [29]. For example, various fragments of collagen type IV, elastins, and laminins have been shown to elicit neutrophil and/or macrophage chemotaxis. On the other hand, several fibronectin fragments influence the expression and secretion of MMPs (MMP-9 and MMP-12) and cytokines (IL-1 and IL-6) by monocytes/macrophages. Thus, besides their structural roles, ECM proteins can function as important actors in the cardiac inflammatory response [30].

Finally, studies indicate that pro-MMPs can associate with integrins or other cell surface receptors on leukocytes and promote their motility independent from their activation [31]. Furthermore, there are lines of evidence indicating TIMP-1, TIMP-2, and TIMP-3 might influence the inflammatory response through MMP-independent pathways [15]. As such, TIMP-3 was identified as one of the main inhibitors of TNF-α-converting enzyme (TACE), an enzyme primarily responsible for the activation of TNF-α (Fig. 11.3), thus adding yet another level of complexity to the MMP/TIMP system and their delicate regulatory role in the setting of inflammation.

Taken together, these studies expose the increasingly complex network of interactions between the MMP/TIMP system and inflammation. Moreover, they support the notion of MMPs and TIMPs as an integral part of the immune response. Nevertheless, it is important to note that many of these immunomodulatory functions are the result of in vitro studies and are yet to be validated in an in vivo setting.

11.4 MMPs, Inflammation, and Cardiac Remodeling: "Friend and Foe"

The discovery of the MMP family has been an enormous breakthrough in the field of ECM metabolism and cardiac remodeling [13]. Furthermore, keeping in mind the complex molecular cross talk between the MMP and inflammatory system as well as the importance of cardiac inflammation in the disease progression and clinical outcome of MI and viral myocarditis, it is not surprising that the MMP system has evolved to being a major focus of research in the pathophysiology of both diseases. As summarized in Table 11.2, genetic approaches have clearly demonstrated that any deviation from the delicate balance between MMPs and TIMPs can result in diverse and sometimes opposing consequences on the pathogenesis of MI and viral myocarditis. In fact, whereas genetic deletion of individual MMPs seems to be cardioprotective after MI, the deficiency of exactly the same family members produces catastrophic consequences during the pathogenesis of viral myocarditis (Table 11.2).

11.4.1 Spatiotemporal Expression Profile of MMPs/TIMPs in Response to Cardiac Injury

During MI or viral myocarditis, the homeostatic balance between MMPs and TIMPs present in the myocardium is largely disrupted. Both human and animal studies have revealed increased levels of MMP-1, MMP-2, MMP-3, MMP-7, MMP-8, MMP-9, MMP-12, MMP-13, MMP-14, and TIMP-1 and TIMP-2, while TIMP-3 and TIMP-4 levels decrease after an ischemic insult to the heart [9, 16]. Similarly, MMP-2, MMP-3, MMP-8, MMP-9, MMP-13, TIMP-1, and TIMP-2 are upregulated in a mouse model of CVB3-induced viral myocarditis, while others, such as MMP-11 and MMP-7, remain unchanged or such as TIMP-4 decreased [3, 17]. Importantly, in both diseases, MMPs and TIMPs follow a unique spatiotemporal localization, thus pointing towards highly specific effects on molecular and cellular responses during the acute and chronic phases in response to cardiac injury. Nevertheless, to date, the potential roles of many MMPs have not yet been evaluated, underscoring the continuing need to unravel the full complement of MMPs and TIMPs in response to MI or viral infection of the heart.

11.4.2 MMPs/TIMPs and Myocardial Infarction

Heymans and colleagues were the first to indicate that acute—yet transient—protease inhibition may represent a viable therapeutic target to prevent cardiac rupture after MI, whereas prolonged inhibition leads to cardiac failure [35]. Using

Table 11.2 The role of individual MMPs and TIMPs in myocardial infarction and viral myocarditis

MMPs	Mouse model	Cardiac phenotype	Proposed mechanism	References
Collagenases				
MMP-8	MMP-8$^{-/-}$, CVB3	Decreased T-lymphocyte migration	Lack of collagenase activity necessary for lymphocyte migration	[32]
Gelatinases				
MMP-2	MMP-2$^{-/-}$, MI	Decreased leukocyte infiltration and collagen degradation, slower wound healing, decreased LV dilatation and cardiac rupture	Lack of laminin- and fibronectin-peptide fragments mediated macrophage migration	[33, 34]
	MMP-2$^{-/-}$, CVB3	Increased susceptibility to viral myocarditis	Lack of CCL7 (MCP-3) degradation	[17]
MMP-9	MMP-9$^{-/-}$, MI	Decreased leukocyte infiltration and collagen deposition, increased angiogenesis, decreased LV dilatation and cardiac rupture	Impaired ECM degradation	[35, 36]
	MMP-9$^{-/-}$, CVB3	Increased viral load, CD3^{+} T-lymphocytes, myocardial damage, and fibrosis, yet preserved cardiac output	Lack of IFN-β/IFN-γ inactivation by MMP-9	[32]
Stromelysins				
MMP-3	MMP-3$^{-/-}$, MI	No differences in cardiac rupture, leukocyte infiltration, and collagen deposition	NA	[35]
MMP-7	MMP-7$^{-/-}$, MI	Increased macrophage infiltration, decreased arrhythmia, and improved survival	Impaired connexin-43 processing	[37]
Membrane-type MMPs				
MT1-MMP	MT1-MMP$^{+/-}$, MI	Decreased collagenolytic potential of cardiac fibroblasts, decreased dilatation and dysfunction, and increased long-term survival	Attenuated collagen type I catabolism	[38]
Other MMPs				
MMP-12	MMP-12$^{-/-}$, MI	No differences in cardiac rupture, leukocyte infiltration, and collagen deposition	NA	[35]
	MMP-12$^{-/-}$, CVB3	High mortality early post infection	Lack of innate antiviral immune response	[39]

TIMPs	Mouse model	Cardiac phenotype	Proposed mechanism	References
TIMP-1	TIMP-1$^{-/-}$, MI	Increased hypertrophic response, reduced fibrillar collagen content and adverse LV remodeling	Increased MMP-13 activity and decreased collagen deposition	[40, 41]
	AdTIMP-1, MI	Decreased leukocyte infiltration, wound healing, and angiogenesis; decreased cardiac rupture	Increased MMP inhibition	[35]
	TIMP-1$^{-/-}$, CVB3	Increased survival and decreased myocardial apoptosis	Impaired retention of immune cells at sites of infection and increased MMP-mediated cell death	[42]
	AdTIMP-1, CVB3	Reduced cardiac inflammation, necrosis, fibrosis, LV dilatation, and dysfunction	No cleavage of MMP zymogen	[43]
TIMP-2	TIMP-2$^{-/-}$, MI	Increased inflammation and IL-6 and MCP-1 expression; augmented MMP-8, MMP-9, MMP-13, and MT1-MMP activity; decreased collagen density; exaggerated infarct expansion and LV remodeling	Loss of TIMP-2-mediated MT1-MMP inhibition	[44]
	AdTIMP-2, MI	Decreased MMP-2 and MMP-9 activity, leukocyte infiltration, and LV dilatation and improved long-term survival	Decreased MMP-mediated leukocyte infiltration	[45]
TIMP-3	TIMP-3$^{-/-}$, M	Increased neutrophil infiltration and MMP-2 and MMP-9 activity; suppressed TGF-β1 expression and collagen synthesis; increased LV dilatation and dysfunction; and higher incidence of cardiac rupture	Increased EGF/EGFR signaling downregulates TGF-β1 expression and collagen synthesis	[46, 47]
TIMP-4	TIMP-4$^{-/-}$, MI	ND	ND	NA
	TIMP-4$^{-/-}$, CVB3	ND	ND	NA

MMP matrix metalloproteinase, *TIMP* tissue inhibitor of MMP, *AdTIMP* adenoviral-mediated TIMP overexpression, *MI* mouse myocardial infarction, *CVB3* human coxsackievirus B3 mouse model of viral myocarditis, *LV* left ventricle, *CCL* CC-chemokine ligand, *MCP* monocyte chemotactic protein, *IFN* interferon, *IL* interleukin, *TGF* transforming growth factor, *EGF/EGFR* epidermal growth factor/EGF receptor, *ND* not done, *NA* not applicable various genetic approaches to reduce protease activity, including uPA-deficient mice and adenoviral-mediated overexpression of protease inhibitors TIMP-1 or PAI-1, the authors revealed significantly reduced leukocyte infiltration and protection against acute cardiac rupture in a mouse model of myocardial infarction. Furthermore, lack of uPA suppressed the expression of IL-6 and MPO, indicating that PA/MMPs influence the cytokine response and oxidative stress, thereby potentially promoting cardiomyocyte injury in the infarcted heart [35]. However, these acute cardioprotective effects were followed by delayed wound healing, decreased angiogenesis, and contractile dysfunction in uPA-deficient hearts, indicating that a timely tuned balance between inflammation and proteases is essential for proper wound healing and functional recovery of the infarcted heart [35]. Similarly, Creemers et al. [48] revealed that after MI, the absence of MMP-activating plasminogen depressed MMP-2 and MMP-9 activity and macrophage infiltration and completely abolished wound healing, thereby promoting altered ventricular remodeling and contractile dysfunction. PAI-1-null mice, on the other hand, had increased leukocyte infiltration and died of ventricular wall rupture within 7 days post-MI [49].

Subsequently, several genetic studies have tried to expand our knowledge on the individual roles of MMPs and TIMPs post-MI (Table 11.2). Lack of MMP-2 or MMP-9 in mice almost completely protected against cardiac rupture post-MI [33, 35]. Both phenotypes were marked by reduced inflammatory influx—macrophages and total leukocytes, respectively—and a slower removal of necrotic debris from the infarcted left ventricle, whereas others demonstrated attenuated left ventricular dilatation and dysfunction in both MMP-2- and MMP-9-deficient mice, without affecting collagen scar formation 14 days post-MI [34, 36]. In contrast, similar cardiac rupture rates were observed in MMP-7-null and WT mice after MI [37]. The absence of MMP-7 did, however, improve long-term post-MI survival by taking MMP-7-mediated cleavage of gap junction protein connexin-43 out of the equation, consequently leading to improved electrical conduction patterns in the infarcted heart. Interestingly, this phenotype was associated with an increased macrophage density in MMP-7-null hearts post-MI. In contrast, membrane-anchored MT1-MMP-null mice display multiple organ defects and die within weeks of birth. Given this morbid status, Koenig et al. [38] were limited to the use of mice heterozygous for an MT1-MMP-null allele (MT1MMP$^{+/-}$ mice) to investigate the role of MT1MMP post-MI. MT1MMP$^{+/-}$ mice display reduced numbers of macrophages, a significant post-MI survival benefit, and retain myocardial structure and function, without affecting MMP-2 and MMP-9 activity post-MI. Instead, the beneficial effects observed in MT1MMP$^{+/-}$ hearts were attributed to a decrease in MT1-MMP-mediated interstitial collagen type I catabolism. Finally, no differences were noted in the incidence of cardiac rupture, leukocyte infiltration, and collagen deposition within the first 14 days post-MI in MMP-3 and MMP-12 knockout mice, indicating either functional redundancy, compensation by other MMPs, or that these MMPs might only be important in the later stages after MI [35].

On the other side of the spectrum, gene deletion of several TIMPs convincingly showed that further activation of MMPs results in a higher risk of acute cardiac rupture or heart failure post-MI (Table 11.2). Following MI, TIMP-1-null mice had elevated activity levels of MMP-13, lower fibrillar collagen content, and more severe LV dilatation and dysfunction, however, without significantly changing rupture rates as compared to WT controls [40, 41]. Kandalam et al. [44, 50] revealed that TIMP-2 deficiency leads to exacerbated LV dilatation and dysfunction post-MI, associated with increased neutrophil and macrophage infiltration, expression of pro-inflammatory mediators IL-6 and MCP-1, and decreased fibrillar collagen content. Although MMP-8, MMP-9, and MMP-13 activity levels were increased in absence of TIMP-2, the authors ascribed the protective role of TIMP-2 to the inhibition of MT1-MMP-mediated degradation of fibrillar collagen in the infarcted heart. Concordantly, adenoviral-mediated TIMP-2 overexpression reduced MMP-2 and MMP-9 gelatinase activity, reduced leukocyte infiltration, preserved cardiac structure and function, and improved long-term survival post-MI [45]. Finally, two independent studies reported a high incidence of post-MI cardiac rupture and poor long-term survival after MI in TIMP-3-null mice. The latter was associated with higher density of neutrophil infiltration as well as overt activity of MMP-2 and MMP-9 in the infarcted myocardium [46, 47]. Furthermore, Hammoud et al. [46] indicated that lack of TIMP-3 resulted in increased EGF/EGFR signaling which downregulates TGF-β and collagen synthesis.

Taken together, these studies collectively indicate a cardioprotective effect of MMP inhibition in the setting of MI. With exception of MMP-7, many of the authors tried to pinpoint the mechanism solely to the effects of the MMP system on ECM metabolism or to the generation of ECM-peptide fragments that elicit chemotactic properties (e.g., MMP-2; Table 11.2) [33]. However, given our current knowledge, one must wonder whether ECM degradation is really the primary target of MMP inhibition post-MI and to what extent MMPs and TIMPs directly influence the pro-inflammatory and/or pro-resolving pathways as well as the cellular components of the inflammatory response.

11.4.3 MMPs/TIMPs and Viral Myocarditis

Similar as to what was previously observed after MI, Heymans and colleagues revealed that broad inhibition of MMP activity was protective in a mouse model of CVB3-induced viral myocarditis. More specifically, targeted deletion of uPA or adenoviral-mediated overexpression of TIMP-1 and PAI-1 reduced cardiac inflammation ($CD4^+$ and $CD8^+$ T-lymphocytes and neutrophils), necrosis, and fibrosis while preserving ventricular function without affecting viral load (Table 11.2) [43]. This potential therapeutic effect of MMP inhibition was further substantiated when Pausinger et al. [51] showed that lowering MMP activity in CVB3-infected mice through the administration of carvedilol significantly reduced cardiac inflammation and ventricular dysfunction.

Although the number of studies is still limited, mice deficient for a single MMP have yielded very contrasting findings in a mouse model of CVB3-induced viral myocarditis, as to what previously was observed after MI, thus highlighting the disease-specific context of MMP/TIMP biology (Table 11.2). Furthermore, these recent studies start to incorporate the notion of non-ECM-related targets of the MMP system during cardiac disease, inflammation, and remodeling. Whereas no differences in viral load and functional parameters were noted between MMP-8-null mice and their WT counterparts, ablation of MMP-9 increased the severity of CVB3-induced viral myocarditis in mice [32]. More specifically, lack of MMP-9 resulted in increased levels of INF-β/INF-γ and increased $CD3^+$ lymphocytic infiltration, leading to a dramatic increase in immune cell-mediated damage of the myocardium, uncontrolled viral replication, and faster transition to heart failure after CVB3 infection. The authors ascribe the protective role of MMP-9 to the inactivation of pro-inflammatory cytokines IFN-β/IFN-γ and hence the dampening of the immune response [32]. In addition, preliminary studies also suggest a protective role for MMP-2 and MMP-12 in CVB3-induced myocarditis. Upon CVB3 infection, MMP-2-null hearts displayed increased levels of pro-inflammatory mediators MCP-3, IL-1β, TNF-α, and fractalkine—all known substrates of MMP-2—leading to an exaggerated inflammatory response and immune-mediated cardiac damage [17]. MMP-12-null mice, on the other hand, produced the most dramatic response to CVB3, as high rates of mortality were observed 3 days post infection. However, more in-depth investigations are mandatory to elucidate the exact mechanistic role of MMP-12 during innate antiviral immunity [39].

Concordant with the detrimental effects observed in MMP-null mice, TIMP-1 deficiency leads to improved survival rates when challenged with CVB3 (Table 11.2) [42]. This was not attributed to changes in viral replication or T-cell response, but this rather indicated that local TIMP-1 expression is mandatory to retain and concentrate the immune response at sites of CVB3 infection, resulting in myocarditis and tissue injury [42].

11.5 Conclusions and Future Directions

It has long been presumed that the functional contribution of MMPs in the development of cardiac disease, remodeling, and failure was limited to their capacity to degrade and turnover the structural components of the ECM. Unfortunately, this functional misconception led clinical studies to evaluate MMP inhibitors, for example, in patients with MI, that failed to meet their endpoints and casted a shadow on MMPs as valid therapeutic targets [52]. In addition, these studies underscore our lack of knowledge within the MMP field and the complexity of translating basic studies into clinical therapeutics in general.

However, accumulating evidence from a wide variety of studies is changing this functional dogma. The growing number of new substrates discovered for MMPs and TIMPs increases their significance in regulating many localized signaling pathways and cellular responses. In this respect, a complex bidirectional cross talk between the MMP system and inflammation is emerging, where location, timing, stimulation, or inhibition will be the key to prevent adverse cardiac remodeling and heart failure in response to cardiac stress and disease. Particularly striking is that the analysis of single-MMP-mutant mice seems to identify MMP inhibition as a potential tool to dampen cardiac inflammation and therefore as cardioprotective after MI, whereas the same mutants exaggerate inflammation and proved to be detrimental during CVB3-induced viral myocarditis. However, further efforts to delete the remaining mouse MMPs and TIMPs and the continuing analyses of existing mutants will be essential to see how all MMPs are deployed within the wide range of cardiovascular diseases. Another major challenge will be to carefully map the complete spatiotemporal and cell-specific expression profiles of the many inflammatory mediators, MMPs, and their inhibitors in a cardiac disease-specific manner to confirm known biological mechanisms and to uncover novel functions. Finally, we will need to identify the full portfolio of MMP- and TIMP-specific substrates using approaches and systems relevant to cardiac health and disease while keeping an open mind towards both intra- and extracellular functions of MMPs as well as the emerging MMP-independent functional properties of TIMPs. This will, for example, provide a detailed view into the complex interactions of the MMP system with the pro-inflammatory and pro-resolving immune cascades as well as the cell-specific reparative pathways that are activated during cardiac repair, and vice versa.

Taken together, further in-depth studies should lay the foundations for the development of new therapeutic strategies that specifically inhibit cardiac disease-related MMPs and/or their effectors that cause deleterious effects while sparing or even enhancing MMP functions that are protective.

Acknowledgments The author apologizes to the many colleagues whose work could not be cited because of space limitations. S.H. received a Vidi grant from the Netherlands Organization for Scientific Research (NWO, 91796338) and research grants from the Netherlands Heart Foundation (2008B011); Fund of Scientific Research Flanders (FWO-Vlaanderen, G074009N); European Union, FP7-HEALTH-2010, MEDIA, large-scale integrating project; and European Union, FP7-HEALTH-2011, EU-Mascara. D.V. is supported by a fellowship of the Fund of Scientific Research Flanders (FWO-Vlaanderen, Belgium, 1208910N).

References

1. Marchant DJ, Boyd JH, Lin DC et al (2012) Inflammation in myocardial diseases. Circ Res 110:126–144
2. Frangogiannis NG (2008) The immune system and cardiac repair. Pharmacol Res 58:88–111
3. Papageorgiou AP, Heymans S (2012) Interactions between the extracellular matrix and inflammation during viral myocarditis. Immunobiology 217:503–510
4. Mann DL (2011) The emerging role of innate immunity in the heart and vascular system: for whom the cell tolls. Circ Res 108:1133–1145
5. Soehnlein O, Lindbom L (2010) Phagocyte partnership during the onset and resolution of inflammation. Nat Rev Immunol 10:427–439
6. Corsten MF, Schroen B, Heymans S (2012) Inflammation in viral myocarditis: friend or foe? Trends Mol Med 18:426–437

7. Hu Y, Zhang H, Lu Y et al (2011) Class A scavenger receptor attenuates myocardial infarction-induced cardiomyocyte necrosis through suppressing M1 macrophage subset polarization. Basic Res Cardiol 106:1311–1328
8. Li K, Xu W, Guo Q et al (2009) Differential macrophage polarization in male and female BALB/c mice infected with coxsackievirus B3 defines susceptibility to viral myocarditis. Circ Res 105:353–364
9. Vanhoutte D, Schellings M, Pinto Y, Heymans S (2006) Relevance of matrix metalloproteinases and their inhibitors after myocardial infarction: a temporal and spatial window. Cardiovasc Res 69:604–613
10. Bouchentouf M, Forner KA, Cuerquis J et al (2010) Induction of cardiac angiogenesis requires killer cell lectin-like receptor 1 and alpha4beta7 integrin expression by NK cells. J Immunol 185:7014–7025
11. Tang TT, Yuan J, Zhu ZF et al (2012) Regulatory T cells ameliorate cardiac remodeling after myocardial infarction. Basic Res Cardiol 107:232
12. Rodriguez D, Morrison CJ, Overall CM (2010) Matrix metalloproteinases: what do they not do? New substrates and biological roles identified by murine models and proteomics. Biochim Biophys Acta 1803:39–54
13. Spinale FG (2007) Myocardial matrix remodeling and the matrix metalloproteinases: influence on cardiac form and function. Physiol Rev 87:1285–1342
14. Nagase H, Visse R, Murphy G (2006) Structure and function of matrix metalloproteinases and TIMPs. Cardiovasc Res 69:562–573
15. Vanhoutte D, Heymans S (2010) TIMPs and cardiac remodeling: 'Embracing the MMP-independent-side of the family'. J Mol Cell Cardiol 48:445–453
16. Lindsey ML, Zamilpa R (2012) Temporal and spatial expression of matrix metalloproteinases and tissue inhibitors of metalloproteinases following myocardial infarction. Cardiovasc Ther 30:31–41
17. Westermann D, Savvatis K, Schultheiss HP, Tschope C (2010) Immunomodulation and matrix metalloproteinases in viral myocarditis. J Mol Cell Cardiol 48:468–473
18. Parks WC, Wilson CL, Lopez-Boado YS (2004) Matrix metalloproteinases as modulators of inflammation and innate immunity. Nat Rev Immunol 4:617–629
19. Deschamps AM, Spinale FG (2006) Pathways of matrix metalloproteinase induction in heart failure: bioactive molecules and transcriptional regulation. Cardiovasc Res 69:666–676
20. Kandasamy AD, Chow AK, Ali MA, Schulz R (2010) Matrix metalloproteinase-2 and myocardial oxidative stress injury: beyond the matrix. Cardiovasc Res 85:413–423
21. Rutschow S, Li J, Schultheiss HP, Pauschinger M (2006) Myocardial proteases and matrix remodeling in inflammatory heart disease. Cardiovasc Res 69:646–656
22. Siwik DA, Chang DL, Colucci WS (2000) Interleukin-1beta and tumor necrosis factor-alpha decrease collagen synthesis and increase matrix metalloproteinase activity in cardiac fibroblasts in vitro. Circ Res 86:1259–1265
23. Mauviel A (1993) Cytokine regulation of metalloproteinase gene expression. J Cell Biochem 53:288–295
24. Wahl SM, Allen JB, Weeks BS et al (1993) Transforming growth factor beta enhances integrin expression and type IV collagenase secretion in human monocytes. Proc Natl Acad Sci USA 90:4577–4581
25. Van Lint P, Libert C (2007) Chemokine and cytokine processing by matrix metalloproteinases and its effect on leukocyte migration and inflammation. J Leukoc Biol 82:1375–1381
26. McQuibban GA, Gong JH, Wong JP et al (2002) Matrix metalloproteinase processing of monocyte chemoattractant proteins generates CC chemokine receptor antagonists with anti-inflammatory properties in vivo. Blood 100:1160–1167
27. Dean RA, Cox JH, Bellac CL et al (2008) Macrophage-specific metalloelastase (MMP-12) truncates and inactivates ELR+ CXC chemokines and generates CCL2, -7, -8, and −13 antagonists: potential role of the macrophage in terminating polymorphonuclear leukocyte influx. Blood 112:3455–3464

28. Sheu BC, Hsu SM, Ho HN et al (2001) A novel role of metalloproteinase in cancer-mediated immunosuppression. Cancer Res 61:237–242
29. Adair-Kirk TL, Senior RM (2008) Fragments of extracellular matrix as mediators of inflammation. Int J Biochem Cell Biol 40:1101–1110
30. Dobaczewski M, Gonzalez-Quesada C, Frangogiannis NG (2010) The extracellular matrix as a modulator of the inflammatory and reparative response following myocardial infarction. J Mol Cell Cardiol 48:504–511
31. Stefanidakis M, Koivunen E (2006) Cell-surface association between matrix metalloproteinases and integrins: role of the complexes in leukocyte migration and cancer progression. Blood 108:1441–1450
32. Cheung C, Marchant D, Walker EK et al (2008) Ablation of matrix metalloproteinase-9 increases severity of viral myocarditis in mice. Circulation 117:1574–1582
33. Matsumura S, Iwanaga S, Mochizuki S et al (2005) Targeted deletion or pharmacological inhibition of MMP-2 prevents cardiac rupture after myocardial infarction in mice. J Clin Invest 115:599–609
34. Hayashidani S, Tsutsui H, Ikeuchi M et al (2003) Targeted deletion of MMP-2 attenuates early LV rupture and late remodeling after experimental myocardial infarction. Am J Physiol Heart Circ Physiol 285:H1229–H1235
35. Heymans S, Luttun A, Nuyens D et al (1999) Inhibition of plasminogen activators or matrix metalloproteinases prevents cardiac rupture but impairs therapeutic angiogenesis and causes cardiac failure. Nat Med 5:1135–1142
36. Ducharme A, Frantz S, Aikawa M et al (2000) Targeted deletion of matrix metalloproteinase-9 attenuates left ventricular enlargement and collagen accumulation after experimental myocardial infarction. J Clin Invest 106:55–62
37. Lindsey ML, Escobar GP, Mukherjee R et al (2006) Matrix metalloproteinase-7 affects connexin-43 levels, electrical conduction, and survival after myocardial infarction. Circulation 113:2919–2928
38. Koenig GC, Rowe RG, Day SM et al (2012) MT1-MMP-dependent remodeling of cardiac extracellular matrix structure and function following myocardial infarction. Am J Pathol 180:1863–1878
39. Marchant D, McManus BM (2009) Matrix metalloproteinases in the pathogenesis of viral heart disease. Trends Cardiovasc Med 19:21–26
40. Creemers EE, Davis JN, Parkhurst AM et al (2003) Deficiency of TIMP-1 exacerbates LV remodeling after myocardial infarction in mice. Am J Physiol Heart Circ Physiol 284:H364–H371
41. Ikonomidis JS, Hendrick JW, Parkhurst AM et al (2005) Accelerated LV remodeling after myocardial infarction in TIMP-1-deficient mice: effects of exogenous MMP inhibition. Am J Physiol Heart Circ Physiol 288:H149–H158
42. Crocker SJ, Frausto RF, Whitmire JK et al (2007) Amelioration of coxsackievirus B3-mediated myocarditis by inhibition of tissue inhibitors of matrix metalloproteinase-1. Am J Pathol 171:1762–1773
43. Heymans S, Pauschinger M, De Palma A et al (2006) Inhibition of urokinase-type plasminogen activator or matrix metalloproteinases prevents cardiac injury and dysfunction during viral myocarditis. Circulation 114:565–573
44. Kandalam V, Basu R, Abraham T et al (2010) TIMP2 deficiency accelerates adverse post-myocardial infarction remodeling because of enhanced MT1-MMP activity despite lack of MMP2 activation. Circ Res 106:796–808
45. Ramani R, Nilles K, Gibson G et al (2011) Tissue inhibitor of metalloproteinase-2 gene delivery ameliorates postinfarction cardiac remodeling. Clin Transl Sci 4:24–31
46. Hammoud L, Lu X, Lei M, Feng Q (2011) Deficiency in TIMP-3 increases cardiac rupture and mortality post-myocardial infarction via EGFR signaling: beneficial effects of cetuximab. Basic Res Cardiol 106:459–471
47. Kandalam V, Basu R, Abraham T et al (2010) Early activation of matrix metalloproteinases underlies the exacerbated systolic and diastolic dysfunction in mice lacking TIMP3 following myocardial infarction. Am J Physiol Heart Circ Physiol 299:H1012–H1023

48. Creemers E, Cleutjens J, Smits J et al (2000) Disruption of the plasminogen gene in mice abolishes wound healing after myocardial infarction. Am J Pathol 156:1865–1873
49. Askari AT, Brennan ML, Zhou X et al (2003) Myeloperoxidase and plasminogen activator inhibitor 1 play a central role in ventricular remodeling after myocardial infarction. J Exp Med 197:615–624
50. Kandalam V, Basu R, Moore L et al (2011) Lack of tissue inhibitor of metalloproteinases 2 leads to exacerbated left ventricular dysfunction and adverse extracellular matrix remodeling in response to biomechanical stress. Circulation 124:2094–2105
51. Pauschinger M, Rutschow S, Chandrasekharan K et al (2005) Carvedilol improves left ventricular function in murine coxsackievirus-induced acute myocarditis association with reduced myocardial interleukin-1beta and MMP-8 expression and a modulated immune response. Eur J Heart Fail 7:444–452
52. Fingleton B (2007) Matrix metalloproteinases as valid clinical targets. Curr Pharm Des 13:333–346

Chapter 12
Role of Chymase in Matrix and Myocardial Remodeling Due to Mitral Regurgitation: Implications for Therapy

Spencer J. Melby, Carlos M. Ferrario, Chih-Cheng Wei, and Louis J. Dell'Italia

Abstract The pure volume overload of mitral regurgitation (MR) has many unique features including matrix metalloproteinase (MMP) activation, increased bradykinin, extracellular matrix loss, disruption of the focal adhesion complex, and cardiomyocyte myofibrillar loss—all of which either directly or indirectly are beneficially affected by inhibition of chymase. Cardiomyocyte myofibrillar loss and cytoskeletal disruption may be related to intracellular oxidative stress and/or increased chymase production within the cardiomyocyte. Increased adrenergic drive is also an important underlying pathophysiologic feature, which, like chymase activation, is present both early and late in course of MR. There is now both dog and human data

S.J. Melby
Birmingham Veteran Affairs Medical Center, University of Alabama at Birmingham, 434 BMRII, 901 19th Street South, Birmingham, AL 35294-2180, USA

Department of Cardiovascular Surgery, University of Alabama Medical Center, Alabama, AL 35294, USA

UAB Comprehensive Cardiovascular Center, University of Alabama at Birmingham, Birmingham, AL, USA

C.M. Ferrario
Wake Forest University School of Medicine, Winston-Salem, NC 27157, USA

C.-C. Wei
Division of Cardiovascular Disease, Department of Medicine, Birmingham Veteran Affairs Medical Center, University of Alabama at Birmingham, 434 BMRII, 901 19th Street South, Birmingham, AL 35294-2180, USA

UAB Comprehensive Cardiovascular Center, University of Alabama at Birmingham, Birmingham, AL, USA

L.J. Dell'Italia, M.D. (✉)
Division of Cardiovascular Disease, Department of Medicine, Birmingham Veteran Affairs Medical Center, University of Alabama at Birmingham, 434 BMRII, 901 19th Street South, Birmingham, AL 35294-2180, USA
e-mail: loudell@uab.edu

B.I. Jugdutt and N.S. Dhalla (eds.), *Cardiac Remodeling: Molecular Mechanisms*, Advances in Biochemistry in Health and Disease 5, DOI 10.1007/978-1-4614-5930-9_12,

demonstrating the benefit of β_1-receptor blockade in isolated MR. However, neither chymase inhibition nor β_1-receptor blockade alone attenuates left ventricular (LV) dilatation. These data raise the intriguing question whether the combination of a chymase inhibitor and β_1-receptor blocker would have a synergistic effect in preventing LV remodeling, especially if started early in the course of isolated MR.

Keywords Mitral regurgitation • Volume overload • Chymase • Matrix metalloproteinase • Focal adhesion kinase

12.1 Introduction

Mitral regurgitation (MR) creates a unique hemodynamic stress by inducing a low-pressure form of volume overload caused by ejection into the low-pressure left atrium. Mechanisms of left ventricular (LV) dysfunction in isolated MR are not well understood. Vasodilator therapy in other forms of LV dysfunction reduces LV wall stress and improves LV function; however, studies in isolated MR show no beneficial effect on LV remodeling using standard medical therapy with renin–angiotensin system (RAS) blockade [1, 2]. As opposed to other forms of heart failure in which RAS blockade has been highly successful, chronic isolated/pure volume overload is associated with loss of interstitial collagen surrounding cardiomyocytes and upregulation of the cardiac kallikrein–kinin system, which further promotes collagen loss as well as inflammatory cell infiltration that is not improved by RAS blockade [3–6].

The study of LV remodeling in isolated MR was initially fueled by questions related to (1) difficulty in determining optimal timing for surgical intervention because of the common incidence of heart failure and LV dyofunction post-valve repair/replacement and (2) the failure of conventional medical therapy to attenuate LV remodeling and to improve LV function. These important clinical questions provided the impetus to better understand the myocardial mechanisms of LV remodeling of an isolated volume overload which are totally distinct from pressure overload.

12.2 Studies in Isolated Volume Overload

To better understand why standard therapy with RAS blockade was ineffective, we performed gene arrays in both the dog with isolated MR and in the rat with the pure LV volume overload produced by aortocaval fistula (ACF) [7–9]. These animals developed LV and cardiomyocyte eccentric remodeling with a loss of interstitial collagen throughout a 4-month time course [6–9]. In a dog model with isolated MR, the loss of collagen weave that connects individual cardiomyocytes was also associated with a significant downregulation of matrix genes and profibrotic growth factors, in particular transforming growth factor-beta (TGF-β), in addition to

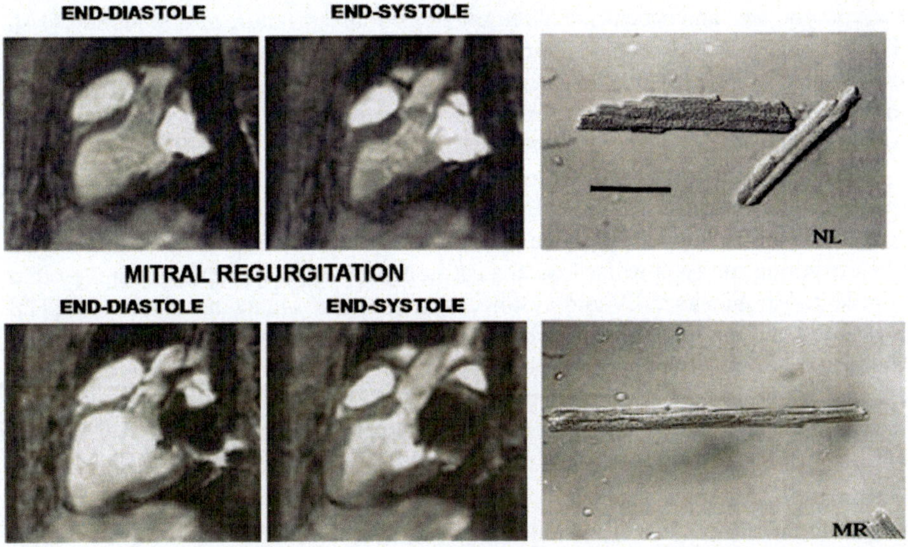

Fig. 12.1 Magnetic resonance images of a dog at baseline and 4 months after induction of MR with representative isolated cardiomyocytes. The LV demonstrates an increase in LV end-diastolic volume with wall thinning resulting in a decrease in the LV mass-to-volume ratio. There is also a corresponding increase in the length-to-width ratio of cardiomyocytes after 4 months of MR

upregulation of matrix metalloproteinase (MMP) genes and the antifibrotic kallikrein–kinin system [6–9]. The MR heart has a less robust hypertrophy response (despite a significant increase in LV angiotensin II levels) than the volume overload of aortic regurgitation where as opposed to ejection into the low-pressure left atrium in isolated MR, the excess volume overload of aortic regurgitation is ejected into the high-pressure aorta. However, the hemodynamic response cannot completely explain the LV remodeling of isolated MR because there is a decrease in protein synthesis that has been attributed to the toxic effects of excessive adrenergic drive throughout the course of MR in the dog [10]. The loss of collagen and noncollagen extracellular matrix combined with a weak hypertrophy response produces LV wall thinning and a decrease in the LVEDV mass/volume ratio that is characteristic of isolated MR in the dog model and in humans (Fig. 12.1) [11]. Taken together, the loss of collagen combined with a decrease in profibrotic growth factors is a poor therapeutic match for the antifibrotic and antihypertrophic effects of RAS blockade or vasodilators in general.

The loss of surface cellular proteins that connects collagen weave to the cell surface results in the dephosphorylation of focal adhesion kinase (FAK) in the dog with isolated MR [12]. An intact focal adhesion complex is essential for cell growth, survival, and myofibril assembly. As opposed to the volume-overloaded MR heart, increased FAK phosphorylation has been documented in animal models of pressure overload [13]. Interestingly, mice with selective inactivation of cardiomyocyte FAK demonstrate increased myofibrillar degeneration, elongation and thinning of cardiomyocytes, and eccentric LV remodeling and heart failure after 6 months [14]. The disruption of FAK and cellular integrins is also associated with disruption, disorganization, and decreased respiratory capacity of subsarcolemmal mitochondria early in the course of volume overload in the rat with ACF, which is prevented by MMP inhibition in the acute 24 h ACF [15].

Another key factor in pure volume overload is the role of increased adrenergic drive, which is central to both early- and late-stage MR [16–18]. However, prolonged excessive adrenergic stimulation has a cytotoxic effect on cardiomyocytes [19], resulting in myofibrillar loss and cardiomyocyte elongation and thinning that is reflected in increased LV end-systolic dimension or volume and wall stress [11]. In the dog and human with isolated MR, β_1-receptor blockade results in significant improvements in myofibrillar loss, calcium homeostasis, and cardiomyocyte and LV function; however, there is no improvement in extracellular matrix loss, LV dilatation, or cardiomyocyte elongation [20–22]. The discordance between LV remodeling and improved LV systolic function with β_1-receptor blockade is most likely related to the failure of β_1-receptor blockade to improve the loss of extracellular matrix components, which are essential to cardiac geometry. Thus, the hallmark features of early and persistent adrenergic drive and extracellular matrix loss of a pure volume overload elude a completely effective medical therapy that both attenuates LV dilatation and improves LV function.

The pure stretch of volume overload also demonstrates a molecular and cellular inflammatory response that contributes to MMP activation and matrix degradation. However, as with studies in human heart failure, prolonged inflammatory cytokine or inflammatory cell blockade in pure volume overload has not met with success [23, 24]. Although mast cell stabilizer therapy has reported success in the rat model, in the dog with isolated MR, mast cell stabilizer results in worsening LV and cardiomyocyte function that is attributed to the inherent calcium entry-blocking effects of this class of drugs [24]. Thus, no clear target of the extracellular matrix degradation (either direct or indirect) has been found, especially in view of the results in other models which demonstrated that prolonged MMP inhibition resulted in excessive collagen deposition (fibrosis) [25] and that blockade of the bradykinin type 2 receptor resulted in a marked increase in blood pressure [5]. It is of interest that, as opposed to the beneficial effects of kallikrein delivery in pressure overload, blockade of kallikrein results in matrix preservation and improved LV function in the chronic volume overload of ACF in the rat [6]. However, at this point in time, there are no existing drugs that target kallikrein blockade.

12.3 Chymase in Cardiovascular Disease and in Isolated Mitral Regurgitation

The interest in chymase in isolated MR emanates from the significant increase in LV mast cell infiltration and degranulation in the dog with isolated MR [26, 27], resulting in release of chymase, a proteolytic enzyme which is increased at both early and late stages of MR [26]. Chymase belongs to a family of chymotrypsin-like serine proteases stored in secretory granules of mast cells, which was first described in the human heart in 1990 [28, 29]. Based on amino acid sequence homology, chymase is divided into two groups: α and β [30]. Only α-chymase has been identified in human, dog, and hamster, while multiple isoforms of α- and β-chymase are present in rodents. It is thought that chymase is tonically released from mast cells and released into the interstitium of the heart. In the mast cell-deficient mouse, there is no chymase [31, 32]. In addition, chymase mRNA is at a very low level in the human and mouse heart, supporting the notion that mast cell release is the predominant source of chymase protein in the heart.

In addition to an Ang II-forming capacity from Ang I that is 20-fold higher than ACE [28, 29], chymase directly activates MMPs [33, 34], degrades fibronectin [35], and activates kallikrein [36]. The addition of chymase to isolated smooth muscle cells [37] or neonatal cardiomyocytes [38] results in cell death due to chymase degradation of cell surface connections to extracellular matrix and disruption of the focal adhesion complex. In addition, chymase activates kallikrein [38] and transforming growth factor (TGF)-β and interleukin (IL)-1β [39]. A newly described propeptide cleaved from angiotensinogen, Ang-(1–12), may be an alternate substrate for the formation of biologically active angiotensins [40, 41]. Ang-(1–12) is increased in cardiac myocytes of adult spontaneously hypertensive rats (SHR) [42]. In the anephric rat, cardiac Ang-(1–12), Ang I, and Ang II increase, while plasma levels decrease [43], suggesting a role of ANG-(1–12) as an intermediate substrate for cardiac Ang II formation. Although the enzymatic mechanism responsible for cardiac Ang-(1–12) formation from angiotensinogen remains an open question, a recent study suggested a chymase-mediated mechanism in the ischemia–reperfusion rat heart [44]. Recently, Ferrario and coworkers have reported that this chymase-mediated Ang II formation from Ang-(1–12) represents the predominant mechanism of Ang II formation in the left atria of humans having surgery for correction of chronic atrial fibrillation [45] and in the human LV (data submitted for publication). Thus, the chymase/Ang-(1–12) axis may represent a totally renin-independent mechanism of Ang II formation in the heart (Fig. 12.2).

Chymase has multiple mechanisms to mediate LV myocardial remodeling by its activation in the cardiac interstitium and after release along with other constituents of activated mast cells. Blockade of chymase has implications beyond the effects on Ang II formation because chymase activates MMPs and directly degrades fibronectin, which promotes tissue remodeling and apoptosis by disrupting matricellular connections. The positive effects of ACE inhibition on progression of cardiovascular disease and heart failure have obscured the fact that chymase rather than ACE may be the major Ang II-forming enzyme in humans. Studies by Dell'Italia and collaborators demonstrated that cardiac Ang II production and/or degradation in the interstitial and intravascular spaces is compartmentalized and mediated by different enzymatic mechanisms—by ACE in the intravascular space and by chymase in the interstitium [32, 46]. In addition, the presence of bradykinin $(BK)_2$ receptors on mast cells may contribute to the mechanism of "ACE escape" via an ACE inhibitor-mediated chymase release from mast cells into the interstitium (Fig. 12.3) [32].

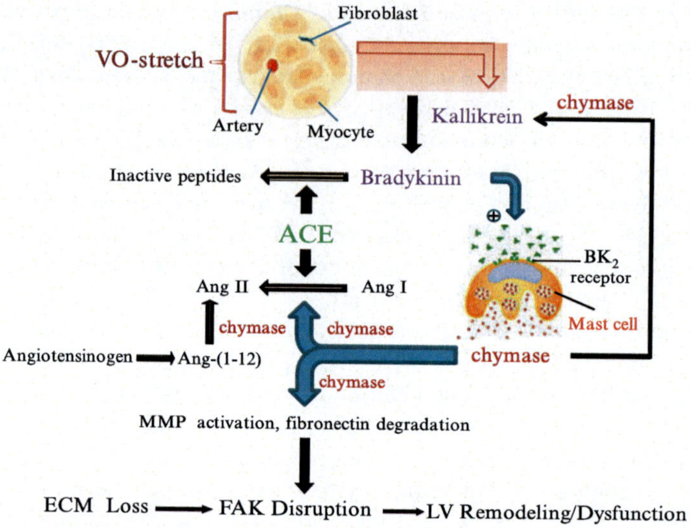

Fig. 12.2 Chymase-related mechanisms in pure volume overload. Volume overload causes an increase in cardiac kallikrein expression and mast cell degranulation and recruitment. Chymase release also has a myriad of effects on factors that promote the inflammatory response and tissue remodeling including kallikrein activation, MMP activation, and fibronectin degradation, resulting in disruption of focal adhesion kinase (FAK) complex. Chronic inhibition of ACE increases BK_2-dependent chymase release from mast cells, which counteracts the direct effect of ACE inhibition on Ang II formation through the chymase pathway of Ang II formation via Ang I and Ang-(1–12)

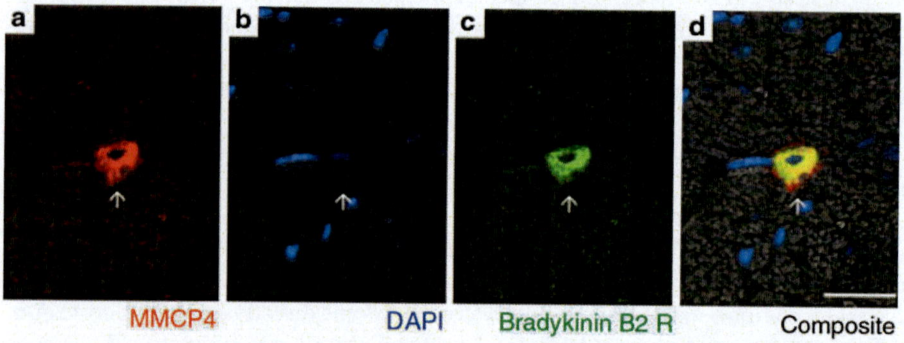

Fig. 12.3 Bradykinin BK2 receptors in LV MMCP4+ MCs. Photomicrographs of a LV section from an 8-week-old wild-type C57B6 mouse stained for (**a**) mast cell protease 4 (MCP4) to identify mast cells, (**b**) DAPI to detect nuclei, (**c**) BK_2 receptor, and (**d**) composite image. *Arrows* show the location of the MMCP4+MC. Scale bar: 20 μm (Reproduced with permission *J Clin Invest*)

The multiple actions of chymase on tissue remodeling, in particular MMP activation, have confirmed a role for cardiac chymase in the adverse LV remodeling following myocardial infarction [32,47,48], atherosclerosis [49–52], and in the pathophysiology of abdominal aortic aneurysm in the human [53].

There is now mounting evidence that chymase may provide an important intracellular mechanism for Ang II formation. Cardiac myocytes, fibroblasts, renal mesangial, and vascular smooth muscle cells have been shown to synthesize intracellular Ang II under hyperglycemic conditions [54–57], and chymase was upregulated in coronary and renal arteries in patients with diabetes [58]. Rapid field electrical stimulation of HL-1 atrial cells induces a sustained augmentation of intracellular Ca^{2+} that is associated with increases in ACE and chymase activity and angiotensinogen expression [59].

Chymase inhibition has been reported to improve hemodynamics and prevent fibrosis in animal models of heart failure due to pacing tachycardia, myocardial infarction, and hypoxia/reoxygenation [60–63]. A more recent study in the pig demonstrated that acute intravenous administration of a chymase inhibitor significantly decreased infarct size after 1 h of ischemia and 2 h of reperfusion. The protective role of chymase inhibition was related to its attenuation of MMP-9 activation and a decrease in inflammatory cytokines [64]. Exposure of mice to intermittent hypoxia increased LV chymase activity, Ang II, oxidative stress, and IL-6, TNF-α, and TGF-β activity—all of which were significantly decreased by treatment with a chymase inhibitor [65]. Treatment also resulted in a decrease in LV fibrosis and hypertrophy as well as a decrease in RV systolic pressure. In addition, the combination of ACE inhibitor and chymase inhibitor decreased infarct size and LV dilatation and improved LV function compared to ACE inhibitor alone in hamsters 1 month after coronary occlusion (Fig. 12.4) [32]. These results suggest that chymase inhibitors, which target MMPs [33, 34] and fibronectin [36], prevented cell death through attenuation of the loss of cell–ECM connections in the pathophysiology of LV remodeling [36, 37]. This suggests that chymase inhibitors might be a useful addition to ACE inhibition or AT_1-receptor blockade in the treatment of heart failure.

12.4 Effects of Chymase Inhibition in the MR Dog

Chymase activity is increased early and throughout the course of MR in the dog. In dogs with 4 months of isolated MR, chymase inhibition normalizes the significant decrease in fibronectin and FAK phosphorylation and prevents cardiomyocyte myofibrillar loss (Figs. 12.5 and 12.6) [66]. One of the major myofibrillar proteins in the heart is the large sarcomeric protein titin, which exists as two isoforms—N2BA and N2B [67–71]. Total titin and its stiffer isoform N2B are increased in the LV epicardium and parallel the changes in fibronectin and FAK phosphorylation in MR dogs treated with chymase inhibitor. Chymase inhibition also improves depressed isolated cardiomyocyte function without improving intracellular calcium transients in dogs with MR. Although the increase in cardiomyocyte function did not translate to an improved LV ejection fraction, magnetic resonance imaging with tissue tagging demonstrated an increase in LV torsion angle in chymase treated compared to untreated MR dogs [66].

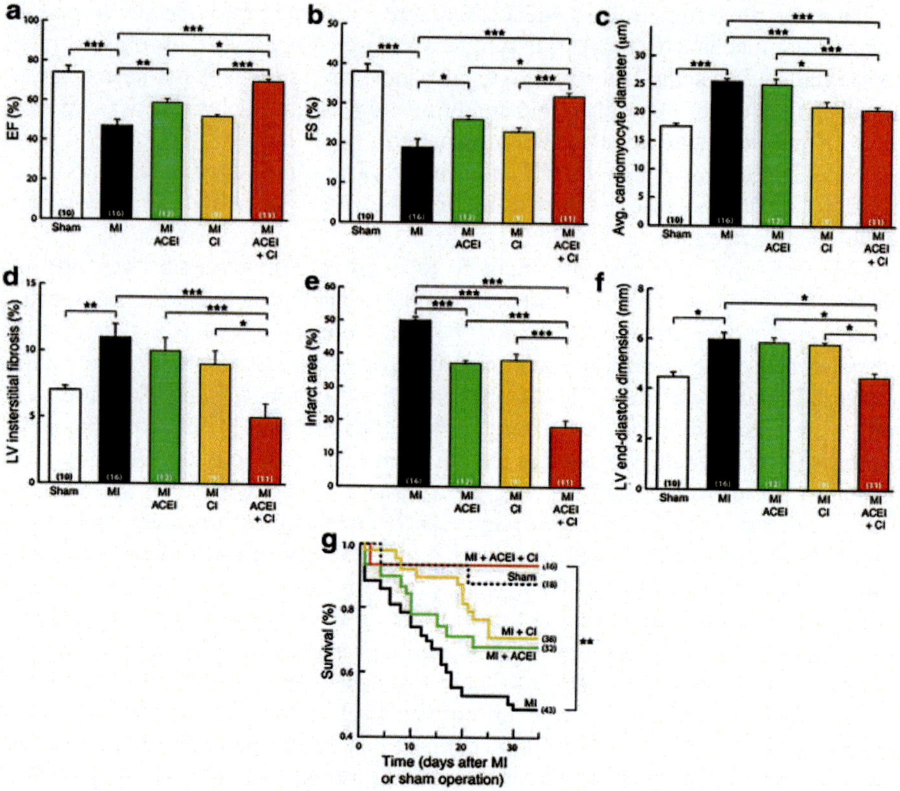

Fig. 12.4 Effect of chymase inhibition on LV remodeling and function and mortality after coronary occlusion in the hamster. The combination of ACE inhibitor and chymase inhibitor produces the greatest improvement in LV remodeling, systolic function, and mortality. Graphs demonstrating LV ejection fraction (EF) (**a**), fractional shortening (FS) (**b**), average cardiomyocyte diameter (**c**), LV interstitial fibrosis (**d**), infarct area (**e**), LV end-diastolic dimension (**f**), and survival (**g**) are all significantly improved by ACE inhibitor + chymase inhibitor in hamsters 1 month after total left anterior descending artery occlusion. Drugs were started within 24 h after total occlusion. MI myocardial infarction; *ACEI* ACE inhibitor; *CI* chymase inhibitor (Reproduced with permission *J Clin Invest*)

Torsion is the wringing motion of the LV along its long axis during systole, induced by contracting myofibers, which are aligned 180° from endocardium to epicardium. Studies in the intact animal demonstrate that acute increases in preload and contractility are associated with increased LV torsion [72, 73]. Preservation of the fibronectin–FAK complex that tracks titin N2B expression from endocardium to epicardium in the chymase inhibitor-treated MR dogs may indeed mediate the higher torsion angle. Further, FAK is located at the Z disk at the insertion of titin in the cardiomyocyte, and there is evidence that the N2B region of titin interacts with integrins and FAK [71]. The improvement in cardiomyocyte shortening and LV torsion with chymase inhibition may be related to the maintenance of a functional

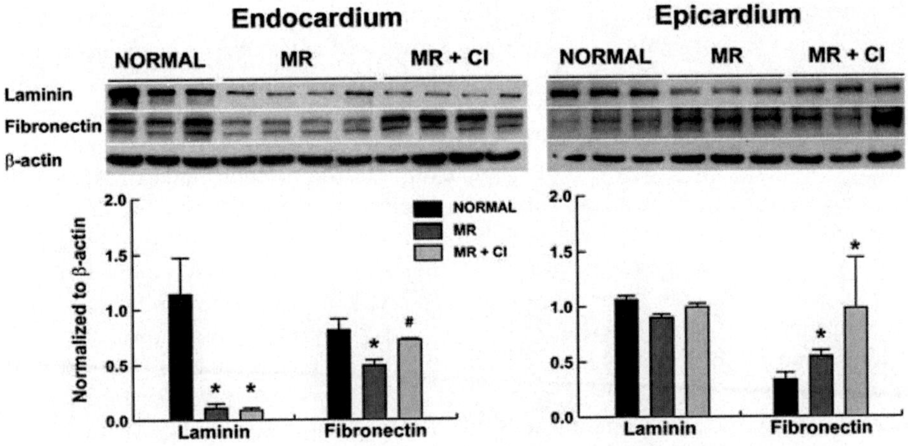

Fig. 12.5 Effect of chymase inhibition on noncollagen extracellular matrix (ECM) proteins in dog model of MR. Representative Western blots of laminin and fibronectin protein in normal ($n=7$), MR ($n=7$), and MR+CI ($n=8$) LV endocardium and epicardium. Chymase inhibition attenuated the loss of laminin and fibronectin in dogs with MR. Densitometric analysis of laminin and fibronectin protein expression is normalized to β-actin. Values mean±SEM. $^{*}P<0.05$ vs. Normal; $^{\#}P<0.05$ vs. MR. *CI* chymase inhibitor (Reproduced with permission *Circulation*)

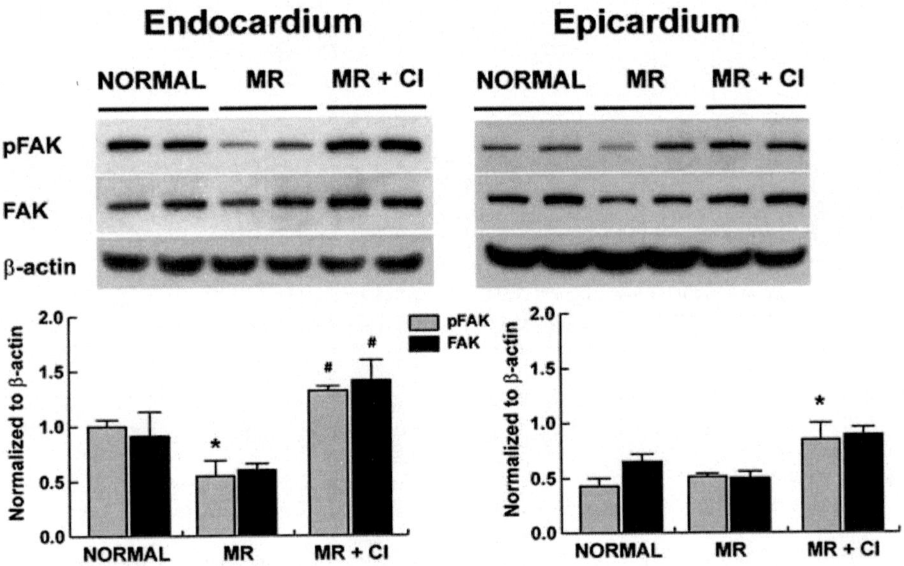

Fig. 12.6 Effect of chymase inhibition on focal adhesion kinase (FAK) phosphorylation in a dog model of MR. Representative Western blot of phospho-FAK (tyrosine 397) and total FAK (band size 125 kD) in normal ($n=7$), MR ($n=7$), and MR+CI ($n=8$) LV endocardium and epicardium. Chymase inhibition attenuated the decrease of phosphorylation of FAK. Densitometric analysis of phospho-FAK and total FAK protein expression is normalized to β-actin. Values mean±SEM. $^{*}P<0.05$ vs. Normal; $^{\#}P<0.05$ vs. MR (Reproduced with permission *Circulation*)

fibronectin–FAK interaction that promotes an increase in titin-dependent cardiomyocyte shortening in MR. Future studies will elucidate the mechanism that causes a titin isoform shift in the myocardium and whether this is dependent on preservation of an intact fibronectin–FAK interaction with chymase inhibition.

Another important feature that can contribute to the negative extracellular matrix homeostatic balance is that the isolated volume overload is associated with an increase in the cardiac kallikrein–kinin system in both the dog model of isolated MR [66] and in the rat model of aortocaval fistula [6]. Adult rat cardiomyocytes undergoing cyclical stretch increase kallikrein mRNA expression and bradykinin release which are inhibited by aprotinin, a kallikrein inhibitor, further supporting the connection of stretch and kallikrein upregulation. It is of interest that inhibition of chymase normalized a 16-fold increase of bradykinin in the MR dog [66], which is consistent with the reported activation of kallikrein by chymase in bronchial tissue [36]. Bradykinin reduced collagen I and fibronectin mRNA and protein expression by adult cardiac fibroblasts [74], while knockdown of kallikrein increased fibronectin expression in kidney mesangial cells [75]. Thus, chymase may mediate fibronectin loss by direct degradation and/or indirectly by inhibiting its synthesis through increased bradykinin formation in the MR heart.

Acknowledgments This study is supported by the Office of Research and Development, Medical Service, Department of Veteran Affairs (LJD), and Specialized Centers of Clinically Orientated Research grant in Cardiac Dysfunction P50HL077100 and in part by Teijin Pharmaceuticals Ltd, Tokyo, Japan (LJD).

References

1. Borer JS, Bonow RO (2003) Contemporary approach to aortic and mitral regurgitation. Circulation 108:2432–2438
2. Bonow RO, Carabello BA, Chatterjee K et al (2006) ACC/AHA/American Heart Association Task Force on Practice Guidelines. J Am Coll Cardiol 3:e1–e148
3. Dell'Italia LJ, Balcells E, Meng QC et al (1997) Volume overload cardiac hypertrophy is unaffected by ACE inhibitor treatment in the dog. Am J Physiol Heart Circ Physiol 273:H961–H970
4. Perry GJ, Wei CC, Hankes GH et al (2002) Angiotensin II receptor blockade does not improve left ventricular function and remodeling in subacute mitral regurgitation in the dog. JACC 39:1374–1379
5. Ryan TD, Rothstein E, Aban I et al (2007) Left ventricular eccentric remodeling is mediated by bradykinin and precedes myocyte elongation in rats with volume overload. J Am Coll Cardiol 49:811–821
6. Wei CC, Chen Y-W, Shi K et al (2012) Cardiac kallikrein kinin system is upregulated in chronic volume overload and mediates inflammatory induced collagen loss. PLoS One 7:e40110–e40113
7. Zheng J, Chen Y, Pat B et al (2009) Microarray identifies extensive downregulation of noncollagen extracellular matrix and profibrotic growth factor genes in chronic isolated mitral regurgitation in the dog. Circulation 119:2086–2095

8. Chen Y-W, Pat B, Zheng J et al (2010) Tumor necrosis factor-α produced in cardiomyocytes mediates a predominant myocardial inflammatory response to stretch in early volume overload. J Mol Cell Cardiol 49:70–78
9. Chen Y, Pat B, Gladden JD, Zheng J et al (2011) Dynamic molecular and histopathological changes in inflammation and extracellular matrix turnover in the transition to heart failure in isolated volume overload. Am J Physiol Heart Circ Physiol 300:H2251–H2260
10. Urabe Y, Mann DL, Kent RL et al (1992) Cellular and ventricular contractile dysfunction in experimental canine mitral regurgitation. Circ Res 70:131–147
11. Schiros CG, Dell'Italia LJ, Gladden JD et al (2012) Magnetic resonance imaging with three-dimensional analysis reveals important left ventricular remodeling in isolated mitral regurgitation: implications beyond dimensions. Circulation 125:2334–2342
12. Sabri A, Rafiq K, Kolpakov MA et al (2008) Impaired focal adhesion signaling in the course of volume overload due to mitral regurgitation in the dog. Effect of Beta-1 Adrenergic Receptor Blockade. Circ Res 102:1127–1136
13. Samuel JL, Barrieux A, Dufour S et al (1991) Accumulation of fetal fibronectin mRNAs during the development of rat cardiac hypertrophy induced by pressure overload. J Clin Invest 88:1737–1746
14. Peng X, Kraus MS, Wei H et al (2006) Inactivation of focal adhesion kinase in cardiomyocytes promotes eccentric cardiac hypertrophy and fibrosis in mice. J Clin Invest 116:217–227
15. Ulasova E, Gladden JD, Zheng J et al (2010) Extracellular matrix loss in acute volume overload causes structural alterations and dysfunction in cardiomyocyte subsarcolemmal mitochondria. J Mol Cell Cardiol 50:147–156
16. Nagatsu M, Zile MR, Tsutsui H et al (1994) Native β-adrenergic support for left ventricular dysfunction in experimental mitral regurgitation normalizes indexes of pump and contractile function. Circulation 89:818–826
17. Hankes GH, Ardell JL, Tallaj J et al (2006) Beta1-adrenoceptor blockade mitigates excessive norepinephrine release into cardiac interstitium in mitral regurgitation in dog. Am J Physiol Heart Circ Physiol 291:H147–H151
18. Mehta RH, Supiano MA, Oral H et al (2003) Compared with control subjects, the systemic sympathetic nervous system is activated in patients with mitral regurgitation. Am Heart J 145:1078–1085
19. Mann DL, Kent RL, Parsons B, Cooper G (1992) Adrenergic effects on the biology of the adult mammalian cardiocyte. Circulation 85:790–804
20. Pat B, Killingsworth C, Denney T et al (2008) Dissociation between cardiomyocyte function and remodeling with β-adrenergic receptor blockade in isolated canine mitral regurgitation. Am J Physiol Heart Circ Physiol 295:H2321–H2327
21. Tsutsui H, Spinale FG, Nagatsu M et al (1994) Effects of chronic beta-adrenergic blockade on the left ventricular and cardiocyte abnormalities of chronic canine mitral regurgitation. J Clin Invest 93:2639–2648
22. Ahmed MI, Aban I, Lloyd SG et al (2012) A randomized controlled trial of Beta-1 receptor blockade in isolated degenerative mitral regurgitation. J Am Coll Cardiol 60(9):833–838
23. Kolpakov MA, Seqqat R, Rafiq K et al (2009) Pleiotropic effects of neutrophils on myocyte apoptosis and left ventricular remodeling during early volume overload. J Mol Cell Cardiol 47:634–645
24. Pat B, Killingsworth C, Denney T et al (2010) Mast cell stabilizer worsens left ventricular function and cardiomyocyte function and calcium homeostasis in dogs with isolated mitral regurgitation. J Card Fail 16:769–777
25. Spinale FG, Coker ML, Krombach SR et al (1999) Matrix metalloproteinase inhibition during the development of congestive heart failure: effects on left ventricular dimensions and function. Circ Res 85:364–376
26. Stewart JA, Wei C-C, Brower GL et al (2003) Cardiac mast cell and chymase mediated matrix metalloproteinase activity and left ventricular remodeling in mitral regurgitation in the dog. J Mol Cell Cardiol 35:311–319

27. Dell'Italia LJ, Meng QC, Balcells E et al (1995) Increased ACE and chymase-like activity in cardiac tissue of dogs with chronic mitral regurgitation. Am J Physiol Heart Circ Physiol 269:H2065–H2071
28. Urata H, Kinoshita A, Misono KS et al (1990) Identification of a highly specific chymase as the major angiotensin II-forming enzyme in the human heart. J Biol Chem 265:22348–22357
29. Urata H, Healy B, Stewart RW et al (1990) Angiotensin II-forming pathways in normal and failing human hearts. Circ Res 66:883–890
30. Chandrasekharan UM, Sanker S, Glynias MJ et al (1996) Angiotensin II-forming activity in a reconstructed ancestral chymase. Science 271:502–505
31. Kakizoe E, Shiota N, Tanabe Y et al (2001) Isoform-selective upregulation of mast cell chymase in the development of skin fibrosis in scleroderma model mice. J Invest Dermatol 16:118–123
32. Li M, Liu K, Michalicek J, Angus JA et al (2004) Involvement of chymase-mediated angiotensin II generation in blood pressure regulation. J Clin Invest 114:112–120
33. Wei CC, Hase N, Inoue Y et al (2010) Mast cell chymase limits the cardiac efficacy of Ang I-converting enzyme inhibitor therapy in rodents. J Clin Invest 120:1229–1239
34. Fang KC, Raymond WW, Blount JL, Caughey GH (1997) Dog mast cell alpha-chymase activates progelatinase B by cleaving the Phe88-Gln89 and Phe91-Glu92 bonds of the catalytic domain. J Biol Chem 272:25628–25635
35. Tchougounova E, Lundequist A, Fajardo I et al (2005) A key role for mast cell chymase in the activation of pro-matrix metalloprotease-9 and pro-matrix metalloprotease-2. J Biol Chem 280:9291–9296
36. Okumura K, Takai S, Muramatsu M et al (2004) Human chymase degrades human fibronectin. Clin Chim Acta 347:223–2235
37. Forteza R, Lauredo I, Abraham WM, Conner GE (1999) Bronchial tissue kallikrein activity is regulated by hyaluronic acid binding. Am J Respir Cell Mol Biol 21:666–674
38. Leskinen MJ, Lindstedt KA, Wang Y, Kovanen PT (2003) Mast cell chymase induces smooth muscle cell apoptosis by a mechanism involving fibronectin degradation and disruption of focal adhesions. Arterioscler Thromb Vasc Biol 23:238–243
39. Hara M, Ono K, Hwang MW, Iwasaki A et al (2002) Evidence for a role of mast cells in the evolution to congestive heart failure. J Exp Med 195:375–381
40. Caughey GH (2007) Mast cell tryptases and chymases in inflammation and host defense. Immunol Rev 217:141–154
41. Nagata S, Kato J, Sasaki K et al (2006) Isolation and identification of proangiotensin-12, a possible component of the renin-angiotensin system. Biochem Biophys Res Commun 350:1026–1031
42. Jessup JA, Trask AJ, Chappell MC, Nagata S, Kato J, Kitamura K, Ferrario CM (2008) Localization of the novel angiotensin peptide, angiotensin-(1-12), in heart and kidney of hypertensive and normotensive rats. Am J Physiol Heart Circ Physiol 294:H2614–H2618
43. Trask AJ, Jessup JA, Chappell MC, Ferrario CM (2008) Angiotensin-(1-12) is an alternate substrate for angiotensin peptide production in the heart. Am J Physiol Heart Circ Physiol 294:H2242–H2247
44. Ferrario C, Varagic J, Hanini J et al (2009) Differential regulation of angiotensin-(1-12) in plasma and cardiac tissue in response to bilateral nephrectomy. Am J Physiol Heart Circ Physiol 296:H1184–H1192
45. Prosser HC, Forster ME, Richards AM, Pemberton CJ (2009) Cardiac chymase converts rat proAngiotensin-12 (PA12) to angiotensin II: effects of PA12 upon cardiac haemodynamics. Cardiovasc Res 82:40–50
46. Ahmad S, Simmons T, Varagic J et al (2011) Chymase-dependent generation of angiotensin II from angiotensin-(1-12) in human atrial tissue (2011). PLoS One 6:e28501
47. Dell'Italia LJ, Meng QC, Balcells E et al (1997) Compartmentalization of angiotensin II generation in the dog heart. Evidence for independent mechanisms in intravascular and interstitial spaces. J Clin Invest 100:253–258

48. Hoshino F, Urata H, Inoue Y et al (2003) Chymase inhibitor improves survival in hamsters with myocardial infarction. J Cardiovasc Pharmacol 41(Suppl 1):S11–S18
49. Ihara M, Urata H, Shirai K et al (2003) High cardiac angiotensin-II-forming activity in infarcted and non-infarcted human myocardium. Cardiology 94:247–253
50. Ihara M, Urata H, Kinoshita A et al (1999) Increased chymase-dependent angiotensin II formation in human atherosclerotic aorta. Hypertension 33:1399–1405
51. Arakawa K, Urata H (2000) Hypothesis regarding the pathophysiological role of alternative pathways of angiotensin II formation in atherosclerosis. Hypertension 36:638–641
52. Uehara Y, Urata H, Sasaguri M et al (2000) Increased chymase activity in internal thoracic artery of patients with hypercholesterolemia. Hypertension 35:55–60
53. Uehara Y, Urata H, Ideishi M et al (2002) Chymase inhibition suppresses high-cholesterol diet-induced lipid accumulation in the hamster aorta. Cardiovasc Res 55:870–876
54. Swedenborg J, Mäyränpää MI, Kovanen PT (2011) Mast cells: important players in the orchestrated pathogenesis of abdominal aortic aneurysms. Arterioscler Thromb Vasc Biol 31:734–740
55. Rafiq K, Sherajee SJ, Fan YY et al (2011) Blood glucose level and survival in streptozotocin-treated human chymase transgenic mice. Chin J Physiol 54:30–35
56. Singh VP, Le B, Bhat VB et al (2007) High glucose induced regulation of intracellular angiotensin II synthesis and nuclear redistribution in cardiac myocytes. Am J Physiol Heart Circ Physiol 293:H939–H948
57. Singh VP, Baker KM, Kumar R (2008) Activation of the intracellular renin angiotensin system in cardiac fibroblasts by high glucose: role in extracellular matrix production. Am J Physiol Heart Circ Physiol 294:H1675–H1684
58. Lavrentyev EN, Estes AM, Malik KU et al (2007) Mechanism of high glucose induced angiotensin II production in rat vascular smooth muscle cells. Circ Res 101:455–464
59. Koka V, Wang W, Huang XR et al (2006) Advanced glycation end products activate a chymase-dependent angiotensin II-generating pathway in diabetic complications. Circulation 113:1353–1360
60. Tsai CT, Lai LP, Hwang JJ et al (2008) Renin-angiotensin system component expression in the HL-1 atrial cell line and in a pig model of atrial fibrillation. J Hypertens 26:570–582
61. Matsumoto T, Wada A, Tsutamoto T et al (2003) Chymase inhibition prevents cardiac fibrosis and improves diastolic dysfunction in the progression of heart failure. Circulation 107:2555–2558
62. Jin D, Takai S, Sakaguchi M et al (2004) An antiarrhythmic effect of a chymase inhibitor after myocardial infarction. J Pharmacol Exp Ther 309:490–497
63. Jin D, Takai S, Yamada M et al (2003) Impact of chymase inhibitor on cardiac function and survival after myocardial infarction. Cardiovasc Res 60:413–420
64. Matsumoto C, Hayashi T, Kitada K et al (2009) Chymase plays an important role in left ventricular remodeling influenced by intermittent hypoxia in mice. Hypertension 54:164–171
65. Oyamada S, Bianchi C, Takai S et al (2011) Chymase inhibition reduces infarction and matrix metalloproteinase-9 activation and attenuates inflammation and fibrosis after acute myocardial ischemia/reperfusion. J Pharmacol Exp Ther 339:143–151
66. Pat B, Chen Y, Killingsworth C et al (2011) Chymase inhibition prevents fibronectin and myofibrillar loss and improves cardiomyocyte function and LV torsion angle in dogs with isolated mitral regurgitation. Circulation 122:1488–1495
67. Fukuda N, Granzier HL, Ishiwata S, Kurihara S (2008) Physiological functions of the giant elastic protein titin in mammalian striated muscle. J Physiol Sci 58:151–159
68. Bell SP, Nyland L, Tischler MD et al (2000) Alterations in the determinants of diastolic suction during pacing tachycardia. Circ Res 87:235–240
69. Warren CM, Joradan MC, Roos KP et al (2003) Titin isoform expression in normal and hypertensive myocardium. Cardiovasc Res 59:86–94
70. Williams L, Howell N, Pagano D, Andreka P, Vertesaljai M, Pecor T, Frenneaux M, Granzier H (2009) Titin isoform expression in aortic stenosis. Clin Sci (Lond) 117:237–242

71. Kruger M, Linkwa (2009) Titin-based mechanical signaling in normal and failing myocardium. J Mol Cell Cardiol 46:490–498
72. Russel IK, Gotte MJW, Bronzwaer JG et al (2009) Left ventricular torsion. JACC Cardiovasc Imaging 2:648–655
73. Dong S-J, Hees PS, Huang W-M et al (1999) Independent effects of preload, afterload and contractility on left ventricular torsion. Am J Physiol Heart Circ Physiol 277:H1053–H1060
74. Kim NN, Villegas S, Summerour SR, Villarreal FJ (1999) Regulation of cardiac fibroblast extracellular matrix production by bradykinin and nitric oxide. J Mol Cell Cardiol 31:457–466
75. Pawluczyk IZ, Tan EK, Lodwick D, Harris KP (2008) Kallikrein gene 'knock-down' by small interfering RNA transfection induces a profibrotic phenotype in rat mesangial cells. J Hypertens 26:93–101

Chapter 13
Cardiac Remodeling Due to Aortic Regurgitation and Mitral Regurgitation

Blase A. Carabello

Abstract Mitral regurgitation and aortic regurgitation (MR, AR) both impose a volume overload on the left ventricle (LV). Despite this similarity, the two lesions and loads they present are quite different from one another. In MR, the extra volume pumped by the LV is ejected into the left atrium, while in AR the extra volume is delivered into the aorta where it increases pulse pressure and systolic pressure. Thus MR is a "pure" volume overload, while AR is a combined pressure and volume overload, and afterload excess plays a much greater role in AR. As a consequence, the ventricle of AR demonstrates both eccentric and concentric hypertrophy, while the MR ventricle is a thin-walled dilated chamber. These different pathologies are associated with different mechanisms of hypertrophy and different mechanisms that lead to LV dysfunction.

Thus, while once MR and AR were lumped together as volume-overload lesions, their differences are considerable so that each lesion should be approached as a uniquely separate entity with differing pathophysiologies and differing clinical outcomes.

Keywords Volume overload • Mitral regurgitation • Aortic regurgitation • Remodeling

---

B.A. Carabello, M.D. (✉)
Department of Medicine, Center for Heart Valve Disease, St. Luke's Episcopal Hospital, Baylor College of Medicine, Houston, TX 77030, USA

Veterans Affairs Medical Center,Medical Service (111), 2002 Holcombe Boulevard, Houston, TX 77030, USA
e-mail: blaseanthony.carabello@va.gov

B.I. Jugdutt and N.S. Dhalla (eds.), *Cardiac Remodeling: Molecular Mechanisms*, Advances in Biochemistry in Health and Disease 5, DOI 10.1007/978-1-4614-5930-9_13,

13.1 Introduction

All valvular heart diseases place a hemodynamic overload upon the left and/or right ventricle, and for each valve lesion, this overload is unique. Uniqueness extends to mitral and aortic regurgitation (MR, AR). While MR and AR were once lumped together as volume-overloading lesions, it is now clear that mitral regurgitation and aortic regurgitation are actually quite different from one another, presenting the left ventricle (LV) with different hemodynamic overloads, in turn resulting in different geometric adaptation and remodeling [1]. Decades ago, Grossman et al. postulated that hemodynamic stress (σ) was the progenitor of cardiac hypertrophy [2]. Laplacian stress (σ) = $P \times r/2\text{th}$ where P = ventricular pressure, r = ventricular radius, and th = wall thickness. In their formulation, an increase in systolic stress, mediated by increased systolic pressure, caused the myocardium to lay down new sarcomeres in parallel, increasing myocyte thickness and thus increasing LV wall thickness. In this formulation, the increase in the pressure term in the numerator of the Laplace equation would be offset by an increase in the thickness term of the denominator, in turn normalizing systolic wall stress (afterload), an adaptation that is beneficial in maintaining normal LV ejection. Conversely in volume overload, increased diastolic stress mediated by increased LV filling pressure would signal sarcomeres to be laid down in series, causing each myocyte to elongate [3], increasing LV volume so that it could accommodate the volume and cardiac output needs of the patient. While many biologic systems modulate how stress results in remodeling and hypertrophy, the Grossman's hypothesis serves as a guide to how the ventricle responds to an overload.

It is generally held that the processes creating and maintaining LV remodeling and hypertrophy can be both adaptive and maladaptive. Since there is limited ability for myocytes to divide after birth, most cardiac growth occurs through the hypertrophic process. Clearly if the heart failed to enlarge, it could not generate the cardiac output needed to compensate for the volume demands of normal body growth let alone the ability to compensate for a volume overload. As such, hypertrophy and remodeling are not only adaptive but also necessary for life. On the other hand, increased LV wall thickness is associated with increased mortality, an obviously maladaptive consequence of concentric hypertrophy. Likewise, increased LV volume may decrease LV efficiency as the radius term in the Laplace equation increases in turn increasing afterload, limiting LV ejection, a maladaptive feature of the eccentric hypertrophic process [4].

This chapter will review the unique loads of mitral regurgitation and aortic regurgitation, contrasting the major differences that exist between the two entities. Going forward, it is important to contrast the terms hypertrophy and remodeling which are not synonymous. Ventricular hypertrophy indicates that the ventricle has gained weight, while remodeling indicates that there has been a change in ventricular geometry. While both usually occur together, they do not always do so. For instance, in a patient whose LV increases in volume but decreases in wall thickness, it is possible for there to be no increase in LV mass. In such a case, there would be eccentric remodeling without eccentric hypertrophy.

13.2 Mitral Regurgitation (A Pure Volume Overload)

13.2.1 Definitions

In discussing mitral regurgitation (MR), it is necessary to distinguish primary (organic) MR from secondary (functional) MR. In the former condition, one or more of the components of the mitral valve malfunction causing it to leak, imparting a volume overload on the LV [5]. Conversely in functional MR, the mitral valve itself is usually normal, but disease of the LV leads to ventricular dilatation, papillary muscle displacement, and annular dilatation in turn causing MR. In organic MR, valve disease leads to a hemodynamic overload, eventually leading to ventricular damage, heart failure, and death. In functional MR, there is already severe LV damage caused by myocardial infarction or cardiomyopathy in turn leading to MR as a secondary phenomenon.

13.2.2 Organic MR

13.2.2.1 Load and Remodeling

In most types of volume overload including aortic insufficiency (see below) anemia, hyperthyroidism, and heart block, the extra volume pumped by the LV to compensate the pathologic volume requirements of the disease is delivered into the aorta. Because pulse pressure is in part dependent upon stroke volume, increased stroke volume in most volume overloads widens pulse pressure, tending to increase systolic pressure. Thus most such volume overloads are in fact combined pressure and volume overloads. In this context, mitral regurgitation is nearly unique among LV volume overloads. In MR, the extra volume is pumped into the left atrium as opposed to the aorta. Thus aortic flow is normal or reduced, not increased. As a consequence, systolic pressure in patients with MR tends to be normal or even lower than normal, averaging 110 mm Hg in one study [4].

Because the development of remodeling and hypertrophy responds to hemodynamic loading, this characteristic loading in MR leads to characteristic LV geometry. The typical LV in the normotensive MR patient is thin walled and dilated [1] having the largest LV radius-to-thickness (r/h) ratio and the smallest ratio of mass-to-volume ration (m/v) of the left-sided overloading lesions. Thus, in MR, there is least amount of LV mass to pump the required increased stroke volume.

13.2.3 Consequences of Remodeling in MR

The thin-walled, enlarged LV of MR is conducive to LV filling as it allows the LV to accept the increased volume stored in the LA during systole at a normal filling pressure. In fact, MR is one of the few heart diseases in which diastolic function is actually supernormal, at least in compensated disease [6, 7].

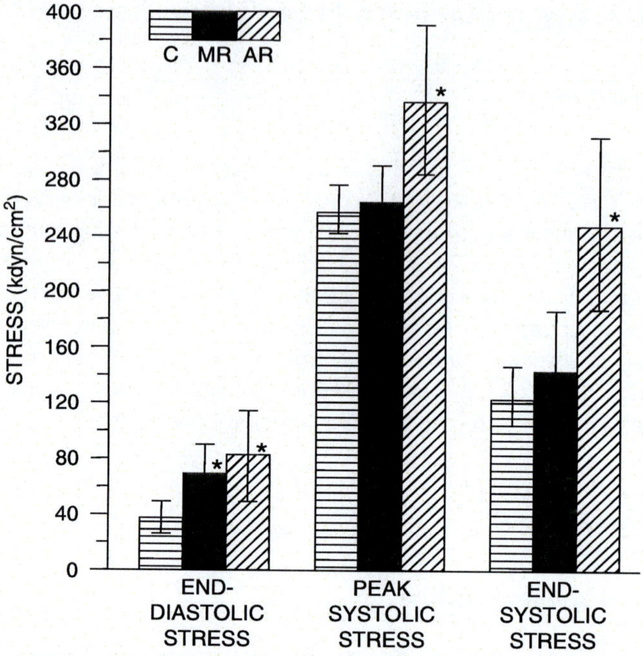

Fig. 13.1 End-diastolic and peak and end-systolic wall stress for normal subject (C) patients with mitral regurgitation (MR) and matched patients with aortic regurgitation (AR) is demonstrated. Preload represented by end-diastolic stress is elevated in both volume-overload lesions, but afterload (systolic stress) is elevated only in AR (taken from Wisenbaugh et al. [4])

While the typical MR geometry is beneficial to diastolic function, it is detrimental to systolic function. As noted above, systolic wall stress $(\sigma) = P \times r/2h$. Thus the r/h ratio forms two thirds of the Laplace equation, and r/h is increased in MR. While the leak of output into the LA during systole provides a second and low-impedance pathway for LV ejection and must in part unload the LV, the MR pattern of LV remodeling tends to increase systolic wall stress (afterload) (Fig. 13.1) [4]. Indeed, in decompensated MR, afterload is increased not decreased [8] as is often held erroneously about the pathophysiology of MR. The stages of remodeling in MR are shown in Fig. 13.2 [9]. In acute MR as might occur following rupture of a chorda tendinea, volume overload leads to increased sarcomere stretch; LV pumping ability is increased via the Frank–Starling mechanism, and

Fig. 13.2 (continued) compensated MR (CCMR). Eccentric hypertrophy and remodeling in CCMR have led to a large increase in EDV, permitting a large increase in total and forward stroke volume despite afterload returning to normal which increases ESV modestly. Increased SL still maintains a high EF. An enlarged LA now accommodates the regurgitant volume at a lower pressure. Bottom panel compares CCMR to chronic decompensated MR (CDMR). Contractile dysfunction and increased afterload impair LV emptying, increasing ESV while decreasing total and forward stroke volume. EF is also reduced but may remain within the normal range. Additional left ventricular dilatation my increase MR and RF, and LA pressure is again elevated (taken from Carabello [9])

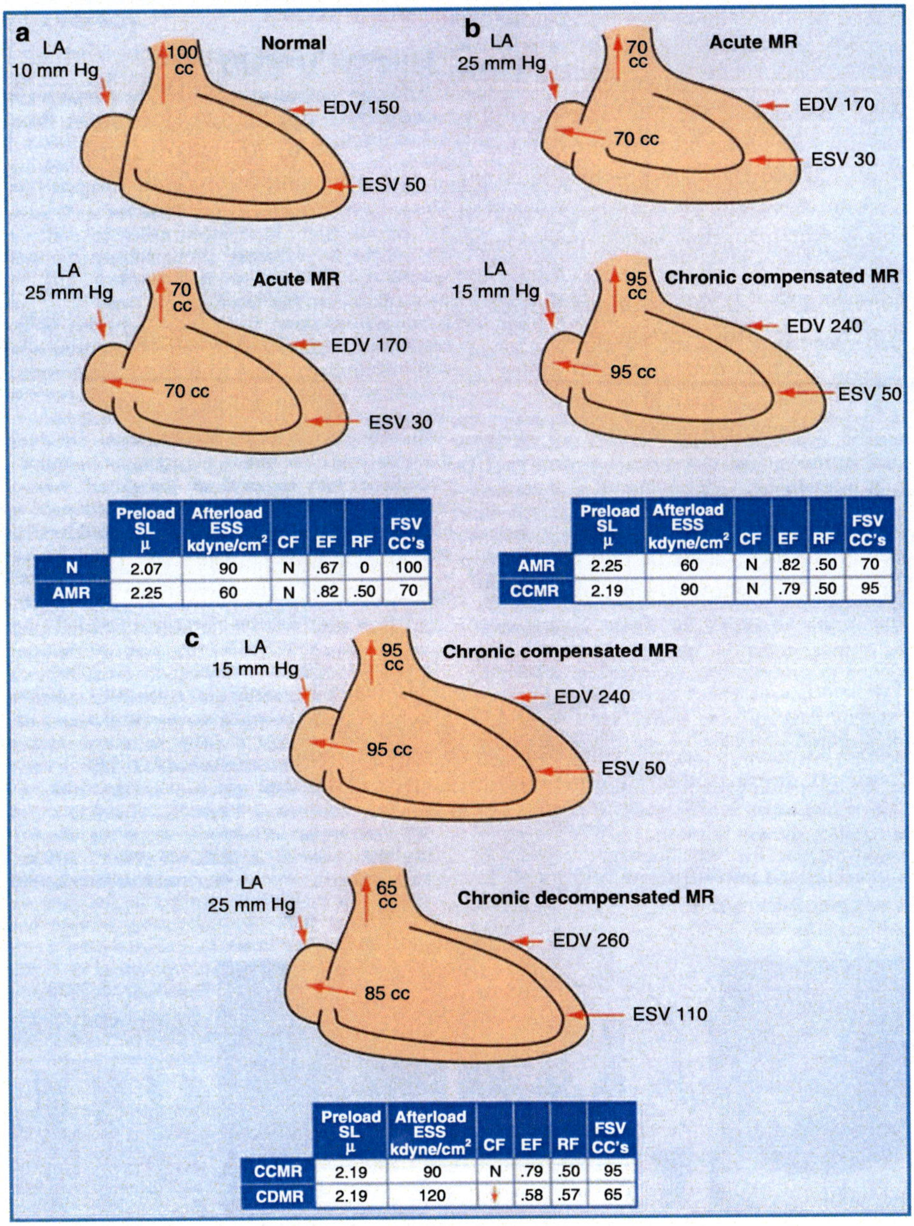

	Preload SL μ	Afterload ESS kdyne/cm²	CF	EF	RF	FSV CC's
N	2.07	90	N	.67	0	100
AMR	2.25	60	N	.82	.50	70

	Preload SL μ	Afterload ESS kdyne/cm²	CF	EF	RF	FSV CC's
AMR	2.25	60	N	.82	.50	70
CCMR	2.19	90	N	.79	.50	95

	Preload SL μ	Afterload ESS kdyne/cm²	CF	EF	RF	FSV CC's
CCMR	2.19	90	N	.79	.50	95
CDMR	2.19	120	↓	.58	.57	65

Fig. 13.2 The stages of MR are depicted. Upper left panel compares normal physiology (N) with the pathophysiology of acute MR (AMR). Preload (sarcomere length, SL) is increased, while afterload (end-systolic stress, ESS) is decreased, and contractile function (CF) is normal. These changes increase end-diastolic volume (EDV) modestly and decrease end-systolic volume (ESV) so that ejection fraction (EF) increases. Total stroke volume increases to 140 cc, but because 50 % (regurgitant fraction, RF) of the stroke volume is regurgitated into the left atrium (LA), forward stroke volume (FSV) decreases to 70 cc. In the *upper right panel*, AMR is compared to chronic

end-diastolic volume increases modestly. The extra pathway for LV ejection into the left atrium (LA) acutely reduces afterload, reducing end-systolic volume. These changes act in concert to increase total stroke volume. However, since 50 % or more of the total stroke volume is ejected into the LA, forward stroke becomes less than normal, and cardiac output is reduced. Volume overload on the small LA increases LA pressure inducing pulmonary congestion. Thus the patient suffers heart failure even though ejection fraction is supernormal. If MR develops more gradually, the patient may enter a chronic compensated phase in which eccentric hypertrophy and remodeling allow increased total stroke volume. Supernormal diastolic compliance and enlargement of the LA permit operation at near normal filling pressure. Increased LV radius returns afterload to normal, while increased diastolic filling maintains increased preload and increased ejection fraction (EF). In this phase, the patient may be completely asymptomatic despite severe MR. However, eventually chronic volume overload leads to LV decompensation in which LV contractility is impaired. End-systolic volume and afterload increase reducing ejection fraction although EF may remain in the "normal" range. Contractile impairment is in part predicated upon loss of contractile elements and is also in part due to abnormal calcium handling [3, 10]. What causes these abnormalities is uncertain, but increased catecholamines probably play a role. Support for this concept comes from evidence that catecholamines are increased in MR patients with LV dysfunction and that beta adrenergic blockade reverses the abnormalities in contractility [11, 12]. Thus a likely sequence in the pathophysiology of MR is that remodeling, while initially compensatory, eventually causes increased afterload, forcing adrenergic overdrive to maintain cardiac output. Increased catecholamines in turn cause LV damage and reduced contractility compounding the heart failure syndrome.

It is unlikely that myocyte elongation alone can explain LV enlargement in MR. Indeed there must be some rearrangement of LV connective tissues. This aspect of remodeling is in part dependent upon matrix metalloproteinases which in turn are activated by LV stretch and also by chymase and mast cell activity [13, 14].

13.2.4 The Hypertrophic Process

The myocardial contractile proteins turn over approximately every 10 days, and LV mass is maintained by an equilibrium between protein synthesis rate (k_s) and protein degradation rate (k_d). For hypertrophy (increased muscle mass) to develop, k_s must exceed k_d, either because k_s increases or because k_d decreases. Because the myocardial proteins turn over relatively rapidly, uptake rate of a labeled amino acid as it becomes incorporated into the myocardium can be used to calculate k_s. Myosin heavy chain synthesis rate increased by 35 % within 6 h of imposition of a pressure overload (Fig. 13.3) [16]. However, no increase could be detected in k_s after imposition of the severe volume overload of MR. Nor could any increase be detected at 2 weeks, 4 weeks, or 3 months after severe MR was created in dogs, yet LV mass did

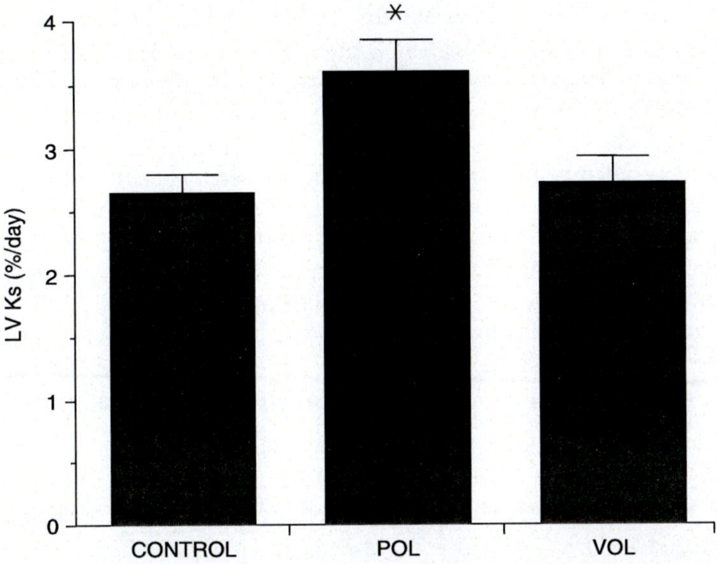

Fig. 13.3 Left ventricular (LV) myosin heavy chain synthesis rate (k_s) for control dogs, those with severe pressure overload (POL) and those with severe MR (volume overload, VOL), is demonstrated. Acute pressure overload but not acute volume overload causes a marked almost immediate (within 6 h) increase in k_s (taken from Imamura [15])

increase [17]. These data suggest that hypertrophy in MR accrues from a decrease in degradation rate [15] rather than an increase in protein synthesis rate, opposite to the mechanism of mass accrual in pressure overload hypertrophy.

13.2.5 Functional MR

In functional MR, the pathologic processes that have led to valve incompetence (myocardial infarction, ischemia, and primary cardiomyopathy) have already produced substantial eccentric LV modeling. It is highly likely that once functional MR develops, it accentuates this remodeling as shown in an ovine model of functional MR [18]. However, in any given patient, it is impossible to know how much the MR is driving the remodeling process versus how much is driven by the muscle disease independent of the MR. This fact has made it very difficult to know in whom correction of MR is likely to be effective. It does seem that once extreme eccentric remodeling has developed, mitral surgery becomes ineffective [19].

In summary, organic MR places a nearly unique hemodynamic load on the LV that of a pure volume overload. The LV responds to this burden by developing eccentric remodeling and eccentric hypertrophy, generating a thin-walled, enlarged chamber. This pattern increases the volume pumping capabilities of the LV and also enhances diastolic filling. However, this pattern of remodeling also eventually impairs systolic emptying. While MR is often viewed as a lesion that unloads the LV via the low-impedance ejection pathway into the LA, remodeling increases systolic wall stress (afterload) and thus actually impairs ejection, reducing forward output. These factors may activate the sympathetic nervous system leading to myocardial damage and dysfunction which can be reversed by beta blocker administration and/or mitral valve repair [20, 21]. Increased muscle mass in the MR LV probably accrues not from increased protein synthesis but rather from decreased protein degradation. What effect the presence of older contractile proteins might have on the natural history of MR is currently unknown.

In functional MR, the contribution of the MR to remodeling versus that of the muscle disease that caused the MR is under intense investigations that may help to determine what therapeutic strategies will be useful in treating that entity.

13.3 Aortic Regurgitation (A Mixed Pressure and Volume Overload)

13.3.1 Load and Remodeling

As noted above for many years, aortic regurgitation and mitral regurgitation were lumped together as volume-overloading lesions. Indeed, aortic regurgitation (AR) does require the LV to increase the amount of volume that it must pump in order to compensate for volume lost to regurgitation. However, unlike MR, in AR, the entire compensatory stroke volume is ejected into the aorta. While a large portion leaks back into the LV during diastole, during systole, the large stroke volume increases systolic pressure. Recalling the $\sigma = P \times \mathrm{R}/2\mathrm{th}$, the increased LV systolic pressure in conjunction with the large LV radius increases systolic wall stress (Fig. 13.1). While LV thickness is indeed greater in AR than it is in MR [1, 22] (as the LV responds to systolic pressure load), increased thickness often does not compensate the increased numerator of the Laplace equation. Indeed, wall stress in AR can be as high as is seen in the pressure overload of aortic stenosis and is consistently higher than the afterload of MR [1, 4, 23, 24]. Accordingly, LV remodeling in AR differs from that of MR. There is combined eccentric and concentric LV hypertrophy, making the typical LV mass in AR the largest of human valvular heart disease.

13.3.2 Load and LV Function

Whereas afterload excess in MR only occurs in late-stage disease and is partially offset by the low-impedance LA ejection pathway, afterload excess plays

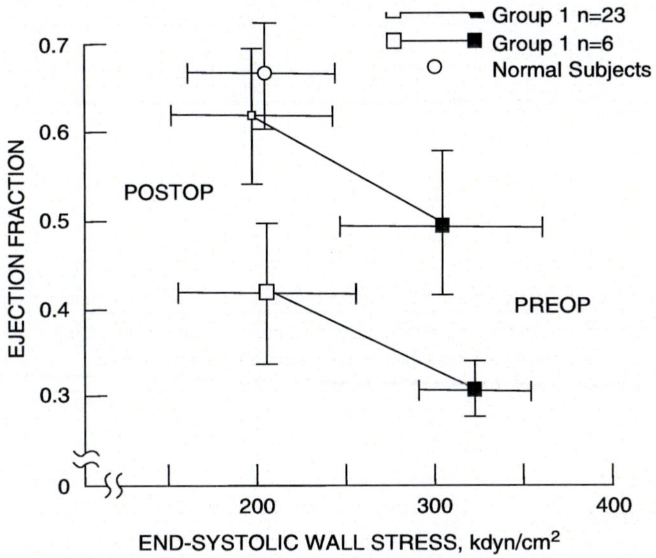

Fig. 13.4 Afterload (end-systolic wall stress) and ejection fraction are shown for normal subjects and for aortic regurgitation patients before (preop) and after (post-op) aortic valve replacement. Reduced post-op afterload both in patients with mild left ventricular dysfunction (group 1) and those with severe dysfunction (group 2) improves post-op ejection fraction (taken from Taniguchi [25])

a much more prominent role in the systolic dysfunction of AR. As can be seen in Fig. 13.4 [25], preoperative afterload excess impairs ejection fraction substantially, while aortic valve replacement leads to afterload reduction and improved postoperative EF. However, AR reduces EF not only by increasing afterload but also because persistent overload eventually impairs contractility. Borer et al. examined the contributions of afterload and contractile reserve to changes in EF during exercise [26]. If EF decreased out of proportion to exercise-induced increases in load (indicating impaired contractile reserve), prognosis was reduced. The mechanisms of LV dysfunction in AR seem different from those in MR. In both the experimental animal and in man, contractile elements of the myocardium are displaced by collagen (man) and/or fibronectin (rabbit) [27, 28], whereas in MR, collagen and other elements of the interstitium are usually normal. Also unlike MR, the hypertrophic process in AR is initiated by increased protein synthesis as is seen in pressure overload but maintained by a decrease in protein degradation as seen in the pure volume overload of MR [29]. Still other differences in AR versus MR that stem from remodeling relate to coronary blood flow reserve. Reserve is reduced in the thicker AR left ventricle [30] while normal in the thinner MR ventricle [31].

Table 13.1 Comparison of mitral (MR) and aortic (AR) regurgitation

	MR	AR
Preload	++++	++++
Afterload	+	+++
ECC hypertrophy	+	++
ECC remodeling	+++	++++
CONC hypertrophy	0	++
MECH SYS DYS	AE, Ca++	F, CBF
Diastolic PROP	++	–

AE adrenergic excess, *Ca++* abnormality in calcium handling, *CBF* coronary blood flow abnormality, *DYS* dysfunction, *ECC* eccentric, *F* fibrosis, *MECH* mechanism, *SYS* systole

13.4 Summary

Aortic regurgitation and mitral regurgitation hold in common the creation of a volume overload on the left ventricle. However, apart from this similarity, many differences exist between the character of these two overloads and the ventricular remodeling and hypertrophy that they produce (Table 13.1). MR is a "pure" volume overload leading to eccentric remodeling and a thin-walled dilated LV with modest hypertrophy and supernormal diastolic properties. "Paradoxically," this pattern of LV geometry tends to increase afterload despite the low-impedance pathway for LV ejection into the LA. On the other hand, AR causes a combined pressure and volume overload on the LV, leading to the heaviest chamber in valvular heart disease. Afterload excess plays a prominent role in the mechanism of LV dysfunction in AR, and geometry and architecture are far different from those found in patients with MR.

References

1. Carabello BA (1995) The relationship of left ventricular geometry and hypertrophy to left ventricular function in valvular heart disease. J Heart Valve Dis 4:S132–S138, discussion S138–S139
2. Grossman W, Jones D, McLaurin LP (1975) Wall stress and patterns of hypertrophy in the human left ventricle. J Clin Invest 53:332–341
3. Spinale FG, Ishihara K, Zile M et al (1993) Structural basis for changes in left ventricular function and geometry because of chronic mitral regurgitation and after correction of volume overload. J Thorac Cardiovasc Surg 106:1147–1157
4. Wisenbaugh T, Spann JF, Carabello BA (1984) Differences in myocardial performance and load between patients with similar amounts of chronic aortic versus chronic mitral regurgitation. J Am Coll Cardiol 3:916–923

5. Anyanwu AC, Adams DH (2007) Etiologic classification of degenerative mitral valve disease: Barlow's disease and fibroelastic deficiency. Semin Thorac Cardiovasc Surg 19:90–96
6. Zile MR, Tomita M, Nakano K et al (1991) Effects of left ventricular volume overload produced by mitral regurgitation on diastolic function. Am J Physiol 261:H1471–H1480
7. Corin WJ, Murakami T, Monrad ES et al (1991) Left ventricular passive diastolic properties in chronic mitral regurgitation. Circulation 83:797–807
8. Corin WJ, Monrad ES, Murakami T et al (1987) The relationship of afterload to ejection performance in chronic mitral regurgitation. Circulation 76:59–67
9. Carabello BA (1998) Mitral regurgitation. Part I. Basic pathophysiological principles. Mod Concepts Cardiovasc Dis 57:53–58
10. Mulieri LA, Leavitt BJ, Martin BJ et al (1993) Myocardial force-frequency deft in mitral regurgitation heart failure is reversed by the forskolin. Circulation 88:2700–2704
11. Tsutsui H, Spinale FG, Nagatsu M et al (1994) Effects of chronic β-adrenergic blockade on the left ventricular and cardiocyte abnormalities of chronic canine mitral regurgitation. J Clin Invest 93:2639–2648
12. Nagatsu M, Zile MR, Tsutsui H et al (1994) Native β-adrenergic support for left ventricular dysfunction in experimental mitral regurgitation normalizes indexes of pump and contractile function. Circulation 89:818–826
13. Stewart JA Jr, Wei CC, Brower GL et al (2003) Cardiac mast cell- and chymase-mediated matrix metalloproteinase activity and left ventricular remodeling in mitral regurgitation in the dog. J mol Cell Cardiol 35:311–319
14. Nagatomo Y, Carabello BA, Coker ML et al (2000) Differential effects of pressure or volume overload on myocardial MMP levels and inhibitory control. Am J Physiol Heart Circ Physiol 278:H151–H161
15. Borer JS, Carter JN, Jacobson MH et al (1997) Myofibrillar protein synthesis rates in mitral regurgitation. Circulation 96(Suppl 1):1–469
16. Imamura T, McDermott PJ, Kent RL et al (1994) Acute changes in myosin heavy chain synthesis rate in pressure versus volume overload. Circ Res 75:418–425
17. Matsuo T, Carabello BA, Nagatomo Y et al (1998) Mechanisms of cardiac hypertrophy in canine volume overload. Am J Physiol 275:H65–H74
18. Beeri R, Yosefy C, Guerrero JL et al (2008) Mitral regurgitation augments post-myocardial infarction remodeling failure of hypertrophic compensation. J Am Coll Cardiol 51:476–486
19. Braun J, Bax JJ, Versteegh MI et al (2005) Preoperative left ventricular dimensions predict reverse remodeling following restrictive mitral annuloplasty in ischemic mitral regurgitation. Eur J Cardiothorac Surg 27:847–853
20. Nakano K, Swindle MM, Spinale F et al (1991) Depressed contractile function due to canine mitral regurgitation improves after correction of the volume overload. J Clin Invest 87:2077–2086
21. Starling MR (1995) Effects of valve surgery on left ventricular function in patients with long-term mitral regurgitation. Circulation 92:811–818
22. Feiring AJ, Rumberger JA (1992) Ultrafast computed tomography analysis of regional radius-to-wall thickness ratios in normal and volume-overloaded human left ventricle. Circulation 85:1423–1432
23. Sutton M, Plappert T, Spegel A et al (1987) Early postoperative changes in left ventricular chamber size, architecture, and function in aortic stenosis and aortic regurgitation and their relation to intraoperative changes in afterload: a prospective two-dimensional echocardiographic study. Circulation 76:77–89
24. Carabello BA (1900) Aortic regurgitation. A lesion with similarities to both aortic stenosis and mitral regurgitation. Circulation 82:1051–1053
25. Taniguchi K, Nakano S, Hirose H et al (1987) Preoperative left ventricular function: minimal requirement for successful late results of valve replacement for aortic regurgitation. J Am Coll Cardiol 10:510–518

26. Borer JS, Hochreiter C, Herrold EM et al (1998) Prediction of indications for valve replacement among asymptomatic or minimally symptomatic patients with chronic aortic regurgitation and normal left ventricular performance. Circulation 97:525–534
27. Borrer JS, Truter S, Herrold EM et al (2002) Myocardial fibrosis in chronic aortic regurgitation: molecular and cellular response to volume overload. Circulation 105:1837–1842
28. Schwarz F, Flameng W, Schaper J et al (1978) Myocardial structure and function in patients with aortic valve disease and their relation to postoperative results. Am J Cardiol 41:661–669
29. Magid NM, Wallerson DC, Borer JS (1993) Myofibrillar protein turnover in cardiac hypertrophy due to aortic regurgitation. Cardiology 82:20–29
30. Gascho JA, Mueller TM, Eastham C et al (1982) Effect of volume-overload hypertrophy on the coronary circulation in awake dogs. Cardiovasc Res 16:288–292
31. Carabello BA, Nakano K, Ishihara K et al (1992) Coronary blood flow in dogs with contractile dysfunction due to experimental volume overload. Circulation 83:1063–1075

Chapter 14
Reducing Oxidative Stress and Manipulating Molecular Signaling Events Using Resveratrol as a Therapy for Pathological Cardiac Hypertrophy

Shereen M. Hamza, Miranda M. Sung, and Jason R.B. Dyck

Abstract Left ventricular hypertrophy is a significant independent risk factor for sudden cardiac death, dysrhythmia, heart failure, ventricular ischemia, and coronary heart disease. However, the pathophysiological mechanisms which lead to the development of cardiac hypertrophy and eventually heart failure are not completely understood. Although the specific steps that lead to hypertrophy are not altogether clear, it appears that oxidative stress plays a significant role. As such, the possibility of a treatment approach involving antioxidant therapies such as the naturally occurring polyphenol resveratrol is a potentially fruitful strategy that can be used to target several signaling and pathogenic events in the etiology of cardiac hypertrophy and may be useful alone or in combination with other drugs. Over the past 15 years, a great deal of evidence supporting the amelioration of hypertrophy and improvement of cardiac function by resveratrol has been documented. However, this is largely in cell-based and animal models with limited evidence from human studies. Clinical trials utilizing resveratrol as a supplementary therapy in heart failure patients are just underway, and much is left to be understood regarding the distinct cellular mechanism of action of this compound in physiological and pathological states. Thus modulation of oxidative stress pathways and cardiac function by resveratrol remains an exciting and undoubtedly rewarding area of research with important clinical implications.

Keywords Oxidative stress • Resveratrol • Hypertrophy • Cardiac metabolism • AMPK • MAPK • NOS • Nitric oxide • Pressure overload

S.M. Hamza • M.M. Sung • J.R.B. Dyck, Ph.D. (✉)
Department of Pediatrics, 458 Heritage Medical Research Centre,
University of Alberta, Edmonton, AB, Canada T6G 2S2
e-mail: jason.dyck@ualberta.ca

B.I. Jugdutt and N.S. Dhalla (eds.), *Cardiac Remodeling: Molecular Mechanisms*, Advances in Biochemistry in Health and Disease 5, DOI 10.1007/978-1-4614-5930-9_14,

14.1 Introduction

Increased hemodynamic load on the left ventricle initiates a cascade of events that leads to ventricular remodeling, cardiac hypertrophy, and eventually heart failure [1]. This ventricular remodeling is a myocardial response to various stresses and involves gradual and progressive changes in chamber architecture, cardiomyocyte phenotype, as well as non-myocytic components [2, 3]. Clinical conditions that produce the pressure-overload stimulus include essential hypertension, aortic stenosis, aortic coarctation, and drug-induced hypertension. The pattern of ventricular remodeling and hypertrophy is dependent upon the type of cardiac stress. In the case of pressure overload, the left ventricle responds with concentric hypertrophy in contrast to the eccentric hypertrophy which occurs in response to volume overload. The process of hypertrophy is initially considered to be an adaptive response because it serves to normalize wall stress and helps to mechanically compensate heart function. However, with progression of the stimulus, the initial compensatory hypertrophy can lead to maladaptive responses that can eventually lead to heart failure. Although the exact point at which hypertrophy becomes maladaptive is difficult to identify, there may be a clear delineation of compensated and decompensated phases of hypertrophy [4–6]. Alternatively, it has also been proposed that the initial hypertrophic process involves a combination of both adaptive and maladaptive cardiac processes and that initiation of any pathological hypertrophy will eventually compromise heart function [7].

Left ventricular hypertrophy is recognized as a major independent risk factor for cardiovascular morbidity and mortality from sudden cardiac death, dysrhythmia, heart failure, ventricular ischemia, and coronary heart disease [6, 8–11]. A growing body of evidence suggests that oxidative stress, which occurs when the generation of reactive oxygen species (ROS) exceeds the cell's intrinsic antioxidant defenses, is an important contributor to the development and progression of cardiac hypertrophy, in part by activating multiple intracellular signaling pathways [12]. Therefore, ameliorating oxidative stress has been proposed as a potential strategy for preventing and/or treating pathological cardiac hypertrophy and heart failure [13, 14]. Our current understanding of the pathophysiology of cardiac hypertrophy as well as potential new therapies targeting oxidative stress will be reviewed herein.

14.2 Physiological Process of Ventricular Hypertrophy

It is generally accepted that mechanical stress begins the cascade of events that leads to ventricular remodeling. However, it is important to note that in vivo, there is rarely, if ever, an isolated increase in pressure independent of any neurohormonal activation. As a result, pressure overload involves mechanical as well as neurohormonal activation that then modulates the ventricular response to stress.

Left ventricular hypertrophy begins with normal ventricle and myocyte function; however, with prolonged hypertrophy this situation eventually decompensates into a hypertrophied ventricle with corresponding myocyte dysfunction. These changes influence both structural and mechanical properties of the myocardial tissue and coronary circulation [1].

At the cellular level, concentric left ventricular hypertrophy is identified by an increase in cardiomyocyte cell size, parallel addition of new sarcomeres, and enhanced protein synthesis arising from alterations in gene transcription and translation [15]. In addition to changes to the cardiomyocyte, left ventricular hypertrophy is also characterized by proliferation of fibroblasts, increased generation of extracellular matrix proteins, and fibrosis [8, 16, 17]. This increased fibroblast growth in combination with the deposition of collagen and fibrosis leads to myocardial stiffening, which on its own is an important contributor to cardiac dysfunction [18, 19]. In the adult mammalian heart, α-myosin heavy chain (MHC) is the predominant isoform expressed and possesses a high ATPase activity and faster rate of contractility than the fetal β-MHC isoform. During hypertrophy, α-MHC expression is reduced, and β-MHC expression is increased which is characteristic of the re-induction of the fetal gene program [20]. Indeed, the transition from α- to β-MHC isoforms that occurs during pathological cardiac hypertrophy may play a role in the development of cardiac dysfunction by reducing myocardial contractility [19].

Generally there are three broad steps regarding the molecular mechanism of ventricular hypertrophy. Initially, heart remodeling is activated by biomechanical stress induced by increased pressure. The left ventricle is stretched during diastole and overloaded during systole which initiates both local myocardial and systemic neurohormonal responses. The mechanical stretch is sensed by myocyte membrane and sarcomere-coupled stretch receptors [21], whereas neurohormonal signals [i.e., endothelin-1 (ET-1), angiotensin-II (Ang-II), catecholamines] are sensed by G protein-coupled receptors that lead to activation of $G\alpha_{q/11}$ proteins [22]. In the second stage of the hypertrophic process, these mechanical and neurohormonal stimuli activate intracellular signaling pathways which eventually transmit a signal to the myocyte nucleus [7, 21]. Finally these signaling pathways activate transcription factors to induce expression of particular genes in the cardiac myocyte which ultimately causes phenotypic changes in the myocytes as well as altered gene transcription that includes re-expression of the fetal gene profile [i.e., atrial natriuretic peptide (ANP), B-type natriuretic peptide (BNP)] and new cellular growth [21, 23]. Activation of specific pathways leads to beneficial and adaptive remodeling, while activation of other pathways eventually has maladaptive and harmful results [21, 24]. Ultimately the imbalance between maladaptive and beneficial signaling results in the development of pathological hypertrophy and the progression to heart failure. Since multiple molecular signaling pathways have been shown to be involved in the development of pathological hypertrophy, these pathways are becoming potential targets for cardiovascular treatments [25].

14.3 Transition from Compensated Hypertrophy to Heart Failure

As mentioned, the initial stimulus for hypertrophy may also stimulate maladaptive events in the cardiac tissue [2, 18, 26]. Several of the initial changes occurring at the level of the cardiomyocyte may be detrimental in left ventricular hypertrophy. For instance, the aforementioned re-expression of the fetal gene profile also induces a change in energy metabolism that results in enhanced glycolysis and a reduction in overall oxidative metabolism [27]. Rates of glycolysis are accelerated in the hypertrophied heart likely as an initial adaptive mechanism to increase ATP production in the presence of impaired oxidative metabolism [28–31]. As a result of this metabolic reprogramming, there is a decrease in myocardial ATP production, leading to energy deprivation [23, 32]. In addition, when glucose oxidation is not correspondingly increased along with glycolysis, there is an accumulation of protons and ROS [28, 33, 34]. Moreover, the reduction in fatty acid utilization itself may lead to lipid accumulation within cardiomyocytes, which has been shown to result in contractile dysfunction [35]. In addition to metabolic changes, apoptosis of cardiomyocytes is augmented in hypertensive heart disease, which leads to a reduction in contractile mass and affects contractility [36]. Other events which occur during hypertrophy such as collagen deposition may promote ventricular stiffness and diastolic dysfunction and heart failure [37–39]. As a result of these and a myriad of other changes that occur [1], it is apparent that myocardial hypertrophy in the setting of pressure overload is ultimately a maladaptive process that should be prevented or reversed in order to prevent the development of impaired cardiac contractile function and heart failure.

14.4 Oxidative Stress in Cardiac Hypertrophy and Heart Failure

Although the fundamental mechanisms underlying the development and progression of cardiac hypertrophy and heart failure are not fully understood, evidence from experimental models of heart failure and clinical studies demonstrates increased oxidative stress in the failing myocardium and strongly supports the concept that it plays a major role in the development of cardiac hypertrophy and progression to heart failure [40–43]. Oxidative stress is defined as the excessive production of ROS relative to the body's natural antioxidant systems. ROS are a group of molecules including oxygen and its derivatives that are produced in all aerobic cells; this includes highly reactive free radicals such as superoxide anion ($^{\bullet}O_2^-$) and hydroxyl radical ($OH^{\bullet}$), as well as compounds such as hydrogen peroxide (H_2O_2) that can be converted to reactive radical species [13]. Nitric oxide (NO) is an important molecule that regulates normal cardiac function by activating soluble guanylyl cyclase and synthesis of intracellular cGMP to inhibit activation of hypertrophic signaling.

In the setting of oxidative stress, $^{\bullet}O_2^-$ can react with NO to form highly reactive species such as peroxynitrite, which has cytotoxic effects and further reduces NO bioavailability.

There are several enzymatic sources of ROS in mammalian cells including mitochondrial respiration, arachidonic acid pathway enzymes (lipoxygenase and cyclooxygenase), cytochrome p450s, xanthine oxidase, NADH/NADPH oxidases, nitric oxide synthase (NOS), peroxidases, and other hemoproteins [12]. The three key producers of ROS in the cardiovascular system are xanthine oxidase, NADH/NADPH oxidase, and NOS [12]. To combat ROS, the body has many specific and nonspecific antioxidant defense mechanisms that scavenge and degrade ROS to nontoxic molecules, including scavenging enzymes such as superoxide dismutase (SOD), glutathione peroxidase, and catalase as well as nonenzymatic antioxidants, such as vitamins E and C [44]. Under normal conditions, SOD converts $^{\bullet}O_2^-$ to H_2O_2, which is then further detoxified by catalase and glutathione peroxidase to H_2O [13]. Excessive production of ROS causes cell dysfunction, protein and lipid peroxidation, and DNA damage—all of which lead to cell damage and death [12]. In addition, ROS appear to participate in several steps involved in the development and progression of myocardial remodeling and failure, including modification of proteins critical for proper contractile function [45], activation of hypertrophy signaling kinases or transcription factors [46] and apoptosis [47], as well as stimulation of cardiac fibroblast proliferation and activation of matrix metalloproteinases leading to extracellular matrix remodeling [48–50].

14.4.1 Cardiac Sources of Reactive Oxygen Species

Increased oxidative stress in the failing heart may be reflective of an increase in the production of ROS from many sources and also a deficiency in antioxidant capacity. The mechanisms responsible for increased oxidative stress in cardiac hypertrophy are unknown, and little is known about the mechanism by which stimuli such as mechanical strain, neurohormones, or cytokines lead to increased ROS. However, cardiac sources of ROS include cardiac myocytes, endothelial cells, and neutrophils. Within cardiac myocytes, ROS can be produced by mitochondria due to leakage of electrons from the respiratory chain which results in the conversion of oxygen to $^{\bullet}O_2^-$ [51]. Mitochondria from the failing heart produce more superoxide than normal mitochondria, suggesting that mitochondrial electron transport could be a predominant source of $^{\bullet}O_2^-$ production [52]. In addition, these mitochondria were associated with decreased complex enzyme activity, making mitochondria an important source of ROS in failing hearts and indicating a link between mitochondrial dysfunction and oxidative stress [52].

Regarding ROS production via oxidative enzymes, NADPH oxidase activity was found to be increased by stimuli relevant to the development of heart failure (i.e., mechanical stretch, Ang II, ET-1, tumor necrosis factor-α) [53]. This increased cardiac NADPH oxidase activity has also been observed in human heart failure [54];

increased xanthine oxidase expression and activity have been similarly noted [55]. In addition to these sources of ROS, uncoupled NOS has also been shown to produce ROS in a variety of pathological conditions [56, 57]. Under normal conditions, endothelial NOS (eNOS) also known as NOS3 catalyzes the formation of NO from L-arginine and oxygen. Uncoupling of NOS due to absence of the necessary cofactor tetrahydrobiopterin (BH4), for example, produces $^{\bullet}O_2^-$ which is known to contribute to the progression of atherosclerosis and hypertension [58, 59]. eNOS is also uncoupled and figures importantly in pathological remodeling and progression to heart failure in response to chronic pressure overload in mice [60].

In addition to being a potential source of ROS during cardiac hypertrophy, mitochondria themselves can also be damaged by ROS. As ROS damage mitochondrial components, the $^{\bullet}O_2^-$ produced within the mitochondria is unable to pass through the membranes, causing still more damage to this organelle, thus compromising mitochondrial function [61]. Increased ROS generation in failing hearts was associated with mitochondrial damage and dysfunction with increased lipid peroxidation in the mitochondria and reduced oxidative capacity [62]. This establishes a vicious cycle in which mitochondrial functional decline promotes further ROS generation, more mitochondrial damage, and cellular injury.

14.5 Hypertrophic Signaling Pathways

A number of complex signaling cascades have been identified that regulate the hypertrophic process in the heart. There appear to be distinct signaling cascades that distinguish physiological versus pathological hypertrophy (see [63] for review of physiological hypertrophy). At a molecular level, pathological cardiac hypertrophy is mediated by several molecular growth signals and induction of fetal gene expression programs (for proteins involved in cardiac contractility and calcium handling) [22, 63]. Some of the better characterized signaling pathways involved in the development of pathological cardiac hypertrophy will be discussed in the section below, with a focus on those pathways where ROS have been shown to be important in activation of these signaling pathways. Indeed, ROS are known to regulate the activity of a variety of signaling kinases and transcription factors to mediate the cardiac hypertrophic response, including calcineurin–NFAT, mitogen-activated protein kinases (MAPKs), and AMP-activated protein kinases (AMPK) among others [63].

14.5.1 Calcineurin–NFAT Signaling

Calcineurin is a ubiquitously expressed calcium (Ca^{2+})–calmodulin-regulated serine/threonine phosphatase that consists of a catalytic A and regulatory B subunit. In response to sustained increases in intracellular Ca^{2+} levels, Ca^{2+}/calmodulin complexes bind to the B subunit, inducing a conformational change allowing calcineurin to dephosphorylate downstream effector proteins, most notably the members of the nuclear factor of activated T-cells (NFAT) family of transcription factors [64]. NFAT transcription factors are normally sequestered in the cytoplasm; however, upon calcineurin-mediated dephosphorylation of NFAT, these proteins rapidly translocate into the nucleus and activate gene expression [65, 66]. Calcineurin activation is observed in hypertrophied hearts following pressure overload induced by constriction of the abdominal aorta [67]. Moreover, partial inhibition of calcineurin by nonspecific immunosuppressive drugs cyclosporin A (CsA) and FK506 has been shown to prevent hypertrophy in some studies [67, 68].

Using a more targeted genetic approach, several transgenic mice have been generated and characterized in order to better define the role of calcineurin in the development and progression of cardiac hypertrophy. Mice expressing a constitutively active form of calcineurin in the heart are sufficient to induce cardiac hypertrophy (two- to threefold increase in heart size) and the expression cardiac fetal hypertrophic genes (i.e., BNP, β-MHC), a phenotype which then rapidly progresses to dilated cardiomyopathy, interstitial fibrosis, and heart failure [66]. Furthermore, targeted inhibition of calcineurin by cardiac overexpression of the inhibitory domains of Cain/Cabin-1 and A-kinase anchoring protein (AKAP) 79, which are endogenous calcineurin inhibitors, reduced cardiac hypertrophy in response to both isoproterenol infusion and pressure overload [69].

Calcineurin activity is negatively regulated by myocyte-enriched calcineurin-interacting protein (MCIP1), which is highly expressed in striated muscle and capable of binding and inhibiting the activity of calcineurin [70]. The cardiac hypertrophic response induced by aortic banding was significantly blunted in transgenic mice overexpressing the calcineurin inhibitory domain from MCIP1 [71]. Mice expressing either a cardiac-specific dominant-negative mutant of calcineurin [72] or a deletion of calcineurin Aβ [73] both displayed significantly less cardiac hypertrophy following aortic banding. Lastly, similar protection from pressure-overload-induced hypertrophy is observed in mice with targeted disruption of NFATc3, a downstream transcription factor effector of calcineurin-mediated hypertrophy [74]. Collectively these different animal models provide strong evidence that the calcineurin–NFAT pathway plays a fundamental role in regulating sustained hypertrophic growth of the heart. Moreover, it is thought that activation of the calcineurin–NFAT pathway is specific to pathological hypertrophy, such as that produced by hypertension or aortic constriction, and does not contribute to the development of physiological cardiac hypertrophy (i.e., exercise), which does not lead to deleterious consequences, thus creating a distinction between signaling pathways responsible for the two different types of cardiac hypertrophy [64].

Alterations in calcium homeostasis, as a result of changes in expression and/or activity of Ca^{2+} handling proteins, are a common phenomenon observed in the setting of cardiac hypertrophy and heart failure [75]. Since cardiac contractile function is highly dependent upon intracellular Ca^{2+} concentration, modifications in Ca^{2+} handling proteins can have a profound effect on cardiac contractility and function [76]. Oxidative stress plays a major role in the indirect regulation of intracellular Ca^{2+} homeostasis and in general is thought to lead to a rapid increase in cytosolic Ca^{2+} levels.

Both the sarcoplasmic reticulum ATPase (SERCA2), an ATP-dependent Ca^{2+} pump that transports Ca^{2+} from the cytoplasm into the endoplasmic/sarcoplasmic reticulum, and the ryanodine receptor (RyR) SR Ca^{2+} release channels are redox sensitive [45]. ROS have been shown to reduce expression and/or activity of the SERCA2A [77] and increase the activation of the RyR [78]. Indeed, SERCA protein expression and activity are found to be decreased in pressure-overload-induced cardiac hypertrophy [79]. As well, oxidants have been shown to stimulate reverse mode of the sarcolemmal sodium–calcium exchanger (NCX) to facilitate Ca^{2+} entry into the cell [45]. Therefore, together these mechanisms may lead to an overall increase in intracellular Ca^{2+} levels that then may result in activation of the calcineurin–NFAT signaling pathway to trigger hypertrophy of the myocardium [80].

Since the cardiac myocyte is constantly beating with cyclic Ca^{2+} transients mediating excitation–contraction coupling, it is likely that Ca^{2+}-dependent activation of the calcineurin–NFAT pathway is more complex than in other cell types where sustained increases in total intracellular Ca^{2+} are sufficient to trigger activation of this pathway [81]. More recently, it has been proposed the existence of specialized cellular microdomains where Ca^{2+} concentration is locally regulated and sensed by Ca^{2+}-dependent proteins localized within these microdomains [81]. Transient receptor potential (TRP) channels, which are G protein-coupled receptor (GPCR)-operated Ca^{2+} channels that are located on the sarcolemma and membrane of the SR/ER, are now emerging as potential key regulators of Ca^{2+} entry during hypertrophy [82]. Indeed, TRP channel expression and activity are upregulated in pathological hypertrophy and heart failure [83–85]. Transgenic mice overexpressing various forms of TRP channels (i.e., TRPC3 and TRPC6) display marked ventricular hypertrophy at baseline and are more sensitive to pressure-overload-induced hypertrophy, which is associated with increased calcineurin–NFAT activity [85, 86]. Moreover, targeted disruption of the calcineurin Aβ gene blocked the exaggerated hypertrophic response in TRPC3 transgenic mice when subjected to TAC [86], supporting a link between TRPC and activation of the calcineurin–NFAT pathway. ROS generated during pressure-overload hypertrophy may also contribute to activation of TRP channels in this pathological setting [87]. Collectively these data suggest that TRP channels may play an important role in mediating pathological cardiac hypertrophy, likely by increasing local Ca^{2+} concentrations in microdomains resulting in activation of the calcineurin–NFAT signaling pathway and subsequent gene expression.

14.5.2 Mitogen-Activated Protein Kinases (MAPKs)

In general, mitogen-activated protein kinases (MAPKs) are divided into three classes based on the terminal kinase in the pathway: the extracellular signal-regulated kinases (ERKs), the *c-Jun* amino-terminal kinase (JNKs), and the p38 MAPKs [88–90]. There is evidence for activation of all three families of MAPKs in cultured cardiac myocytes in response to hypertrophic stimuli (i.e., mechanical stress, GPCR agonists, Ca^{2+}) [91–93], in the experimental pressure-overloaded heart

[94, 95], and in the human failing heart [96]. Although these MAPKs have been largely considered to be pro-hypertrophic, the precise role of each of these MAPKs in the pathophysiology of cardiac hypertrophy and whether they are necessary mediators and/or modulators of the hypertrophic process remains to be clearly identified. However, early during the development of cardiac hypertrophy, it has been thought that ROS may contribute to activation of each these MAPK signaling pathways [97]. As these kinases are considered pro-hypertrophic, antioxidant therapies that target these signaling pathways may also prove to be advantageous in the treatment of pathological cardiac hypertrophy and the progression to heart failure.

14.5.3 Extracellular Signal-Regulated Kinases (ERKs)

ERK1/2 are ubiquitously expressed protein kinases that are activated in the setting of pathological cardiac hypertrophy and heart failure [63, 90, 98, 99]. In cultured cardiac myocytes, ERK1/2 are activated in response to pro-hypertrophic agonist stimulation, oxidative stress, or mechanical loading [91, 97, 100, 101]. Furthermore, activation of ERK 1/2 is required for increased protein synthesis in isolated cardiac myocytes in response to hypertrophic stimuli, such as ET-1 and α-adrenergic agonists [102]. Consistent with these in vitro findings, expression of a dominant-negative mutant of Raf-1, an upstream kinase of Erk1/2, inhibited the activation of ERK1/2 and blunted development of pressure-overload-induced cardiac hypertrophy [103]. Interestingly, mice with expression of cardiac-specific constitutively active MEK1, an immediate upstream MAPK kinase of ERK1/2, led to the development concentric physiological hypertrophy associated with enhanced cardiac systolic function without decompensation over time [104]. Furthermore, neither a global deletion of ERK1 or ERK1 −/− and ERK2 +/− mice led to a reduction in cardiac hypertrophy in mice subjected to TAC [105], suggesting that ERK1/2 may be sufficient, but not critical for mediating pathological pressure-overload-induced cardiac hypertrophy.

14.5.4 c-Jun Amino-Terminal Kinase (JNKs)

The JNK family consists of at least ten isoforms derived from three mammalian genes: JNK1, JNK2, and JNK3 [106]. JNK is rapidly activated in vitro in response to multiple cellular stresses, including mechanical stress, oxidative stress, and proinflammatory cytokines [106]. Activation of JNK is mediated via phosphorylation by MAPK kinase 4 (MEK4) and MEK7, which in turn are activated by MAPK kinase kinase (MEKK1) [107, 108]. Although several in vitro studies suggest that JNKs may contribute to the regulation of pathological hypertrophy [109, 110], data from in vivo studies have proven inconclusive. Studies in transgenic mouse models with blunted JNK activation have provided conflicting results with some showing that cardiac hypertrophy in response to pressure overload is either attenuated [111] or enhanced [112]. Moreover, TAC in mice with selective deletions of JNK1, JNK2, or JNK3 show that the cardiac hypertrophy developed to a similar extent as in wild-type mice [113], suggesting that either JNK isoforms perform similar functions and are redundant or that JNK activation may not be required for cardiac growth. At present, the exact role of JNK in the development of pathological cardiac hypertrophy remains unclear.

14.5.5 p38 MAPKs

The p38 MAPK family is an important regulator of a diverse array of biological functions including cell growth, cell proliferation, metabolism, and cell death [108, 114]. Of the four p38 isoforms, it appears that only p38α and p38β are expressed in the heart. Similar to JNK, p38 is a stress-activated protein kinase that is activated by a multitude of external stimuli including cytokines, oxidative stress, osmotic stress, and growth factors among others (see [63] for review). In vitro studies have largely supported a key role for p38 in promoting cardiac growth and hypertrophy with small molecular inhibitors of p38 and dominant-negative p38 adenovirus inhibiting hypertrophic growth [115]. Similar to the MAPKs described above, myocardial p38 activity is found to be increased by pressure overload [95] and ET-1/phenylephrine stimulation [116, 117]. Despite these findings, targeted activation of p38 in the heart by transgenic overexpression of upstream kinases MKK3 or MKK6 did not produce a significant degree of cardiac hypertrophy in the basal state. However, these mice did have increased interstitial fibrosis and ventricular wall thinning and died prematurely at 7–9 weeks as a consequence of heart failure [118], suggesting that p38 may be important in mediating late-stage ventricular remodeling as opposed to hypertrophy. Furthermore, studies in cardiac-specific p38 dominant-negative transgenic mice (DN-p38) showed that loss of p38 activity either had no effect on the development of cardiac hypertrophy [119] or enhanced the hypertrophic response [120] to pressure overload by aortic banding. In support of the latter finding, p38 MAPK has been shown to inhibit NFAT-transcriptional activity and nuclear translocation by directly phosphorylating NFAT [120, 121], suggesting that p38 may in fact be a negative regulator of hypertrophy. As several studies performed by different groups even in the same transgenic mice have given contradictory results, the precise role of p38 in the development and progression of cardiac hypertrophy and heart failure remains unclear.

14.5.6 AMP-Activated Protein Kinase (AMPK)

AMPK is a key metabolic sensing serine/threonine kinase that is activated by cellular and metabolic stresses that deplete ATP (i.e., ischemia). AMPK responds to increases in the AMP/ATP ratio by activating energy-producing metabolic pathways and inhibiting energy-consuming pathways [122]. The exact role of AMPK in cardiac hypertrophy has not yet been clearly defined. Protein translation and synthesis are known to be necessary mediators of increased myocardial cell size, and pharmacological activation of AMPK (i.e., AICAR, resveratrol) has been shown to inhibit protein synthesis associated with cardiac hypertrophy via numerous molecular pathways (see [122, 123] for review). Thus, while AMPK may act as a negative regulator of cardiac hypertrophy [124–127], it is not certain if inactivation of AMPK is a necessary step of the hypertrophic process or if it simply creates a permissive environment for cardiac growth. Supporting this idea, recent evidence has shown that impaired AMPK activity makes the heart more susceptible to pro-hypertrophic stimuli, such as hemodynamic overload [128–130]. In the SHR, increased oxidative stress [specifically the lipid peroxidation by-product, 4-hydroxy-2-nonenal (HNE)] leads to a reduction in AMPK activity via inhibition of its upstream activating kinase LKB1 [128], and this was associated with the development of cardiac hypertrophy. Consistent with studies in isolated myocytes, impaired AMPK activity was associated with activation of the mTOR/p70S6 kinase pathway that regulates protein synthesis in the heart [128]. Furthermore, cardiomyocyte-specific deletion of LKB1 results in left ventricular hypertrophy [131], suggesting that modification of LKB1 activity may contribute to the hypertrophic process. Although some studies have shown that the activation of AMPK is associated with the development of pressure-overload-induced cardiac hypertrophy [132], this is likely due to the heart becoming energetically compromised in the later stages of the disease and requiring AMPK to restore depleted ATP levels. Therefore, AMPK may play dual roles in the cardiac hypertrophic process whereby early in the disease, inactivation of AMPK may be necessary for cardiac growth, whereas activation of AMPK occurs during the later stages in order to maintain adequate ATP supply to the heart (see [122] for review).

14.5.7 JAK–STAT Pathway

Janus kinase (JAK) proteins are a family of cytosolic tyrosine kinases associated with the intracellular domain of membrane-bound receptors, which act to transduce signals from extracellular ligands such as cytokines, growth factors, and hormones to the nucleus to elicit cellular responses [133]. The JAK family of proteins rapidly transduces signals by recruitment of signal transducers and activators of transcription pathway (STAT) transcription factors. The JAK–STAT pathway plays a critical role in cardiac hypertrophy and the transition from hypertrophy to failure as well as mediates signal transduction from gp130 cytokine receptor [134] and GPCR [135] to the nucleus. Gp130 is a promiscuous receptor for several cytokines including interleukin 6/11 and transduces its signal mainly through induction of STAT3. Specifically, STAT3 is translocated to the nucleus in response to gp130 activation, which results in the induction of genes involved in hypertrophy [136]. Overexpression of STAT3 is sufficient to induce cardiomyocyte hypertrophy under both in vitro [137] and in vivo [138] settings. Transgenic mice with a deletion of gp130 in the myocardium have reduced STAT3 activity and display normal cardiac structure and function at baseline; however, these mice fail to develop compensatory hypertrophy following acute pressure overload and develop heart failure [139]. Therefore, the JAK–STAT pathway may be a novel therapeutic target for developing agents that prevent the development and/or progression of pathological cardiac hypertrophy.

The signaling pathways involved in cardiac hypertrophy are numerous and complex. The generation and characterization of transgenic rodent models have allowed investigators to better delineate the potential molecular mechanisms responsible for mediating distinct forms of cardiac growth. However, future research is still needed to clearly identify the pathways that are most critical for the development of pathological cardiac hypertrophy, and in doing so this may provide new targets for drug discovery in the management and treatment of cardiac hypertrophy and heart failure.

14.6 Antioxidants: Implications for Therapy

Since oxidative stress has been strongly implicated in the pathophysiology of cardiac hypertrophy induced by pressure overload, antioxidant therapies have received much attention as a potential therapeutic strategy to prevent and/or regress cardiac hypertrophy and/or prevent the transition to heart failure. To date, this therapeutic approach using antioxidants such as vitamin E and *n*-acetylcysteine, while having had some success in experimental animal models, has not successfully translated to the clinic for use in humans [140]. However, the natural polyphenol resveratrol that is found in a number of dietary food sources is emerging as a potential novel new therapy that may be a treatment option for those with cardiovascular disease, in particular cardiac hypertrophy.

14.6.1 Vitamin E

The antioxidant vitamin E is a naturally occurring lipid-soluble vitamin found in a variety of food sources, including nuts and sunflower seeds, and is a nonenzymatic part of the cell's intrinsic antioxidant system to counter the accumulation of ROS [141]. Indeed, symptoms in humans with a vitamin E deficiency further support that vitamin E plays an important role in protecting membranes and the nervous system from oxidative stress [142]. Of the various forms of vitamin E, α-tocopherol is the most abundant and biologically active form of vitamin E, with the acetate and synthetic forms of α-tocopherol being the primary constituents of vitamin E supplements [141, 143]. Vitamin E is a potent peroxyl radical scavenger that breaks radical-propagated chain reactions and due to its lipid solubility is present in cell membranes and plays an integral role in protecting cell membranes and plasma lipoproteins from lipid peroxidation [142, 143]. Indeed, peroxyl radicals (ROO^-) react at a rate 1,000 times faster with vitamin E than with polyunsaturated fatty acids [144].

In experimental models, treatment with vitamin E has been shown to reduce oxidative stress and prevent the development of cardiac hypertrophy and heart failure [145]. In the guinea pig model of pressure overload, chronic vitamin E treatment improved cardiac function and blunted the progression of heart failure over 20 weeks. Interestingly, however, this occurred in the absence of improvements in hypertrophy [145]. In contrast, clinical studies in humans have produced conflicting results regarding the cardiovascular benefit of vitamin E supplementation with some studies showing a significant reduction in cardiovascular risk [146–151], as well as a number of studies having failed to show a benefit to vitamin E supplementation [152–155]. Therefore, simply supplementing the antioxidant defenses of the heart using vitamin E may be insufficient to prevent cardiac hypertrophy and cardiovascular disease.

14.6.2 N-Acetylcysteine

N-acetylcysteine is a widely used thiol-containing antioxidant that is a pharmacological precursor of L-cysteine [156]. Reduced thiols are molecules with a sulfhydryl group whose biological properties include scavenging of oxygen free radicals, acting as cofactors for enzymatic reactions and potentiating the half-life and activity of NO by forming NO adducts (*S*-nitrosothiols) which are more stable than NO itself [157, 158]. Oral treatment with *N*-acetylcysteine (500 mg/kg/day) administered in the drinking water for 7 days led to a marked reduction in left ventricular weight induced by 2 weeks of abdominal aortic constriction in mice [159], which is consistent with a major role for ROS in the generation of pressure-overload-induced left ventricular hypertrophy in vivo. Interestingly, *N*-acetylcysteine administration (1.5 g/kg/day) to adult spontaneously hypertensive rats with established hypertension is both unable to reduce blood pressure and cardiac hypertrophy [160, 161]. Due to a lack of large-scale clinical trials, it is not clear whether *N*-acetylcysteine is effective in reducing cardiac hypertrophy in humans. However, *N*-acetylcysteine may be beneficial following myocardial infarction to limit infarct size [162, 163]. As well, *N*-acetylcysteine treatment has been shown to improve endothelial function and reduce systolic blood pressure in hypertensive diabetic patients via a reduction in oxidative stress and increased NO bioavailability [164]. This suggests that *N*-acetylcysteine can reduce oxidative stress in vivo and may be a useful antioxidant strategy to supplement current therapies in a variety of cardiovascular conditions.

14.7 Resveratrol as an Antioxidant

Resveratrol (3,5,4′-trihydroxystilbene) is a polyphenol that is found in a variety of berries, grapes, peanuts, and medicinal plants. Scientific interest in resveratrol has increased over the past 15 years since it has been shown to be a calorie-restriction mimetic and increase overall health in mammals [165]. Of relevance, resveratrol also appears to prevent the development of pressure-overload-induced cardiac hypertrophy [128, 166, 167] as well as show benefit in a variety of other cardiovascular diseases [168]. Indeed, resveratrol has many properties which are considered the basis of its cardiovascular effects and protection. Some of these include antioxidant, $^{\bullet}O_2^-$ scavenging, ischemic preconditioning, and angiogenic actions [169]. The exact mechanism of action of resveratrol in vivo has not been unequivocally determined; however, many cell components and processes are thought to be involved such as surface receptors, signaling pathways, metabolic pathways, nuclear receptors, and gene transcription and translation [170]. The antioxidant and $^{\bullet}O_2^-$ scavenging properties of resveratrol are of interest as they directly relate to the potential therapeutic benefits of this compound to reduce oxidative stress in the setting of cardiac hypertrophy. Some experiments suggest that resveratrol can directly scavenge $^{\bullet}O_2^-$ in potassium superoxide and xanthine oxidase-based antioxidant systems [171], with added evidence of direct inhibition of xanthine oxidase. Resveratrol has also been shown to inhibit NADPH oxidase and subsequently reduce $^{\bullet}O_2^-$ production as a mechanism for vasorelaxation in isolated rat aorta [172]. Additional evidence supports direct scavenging of ROS by resveratrol [173, 174] as well as increased expression of antioxidant enzymes such as SOD, catalase, and glutathione peroxidase [174–176]. Thus a key protective mechanism of resveratrol appears to be an augmentation of endogenous antioxidant systems in addition to direct inhibition/scavenging of ROS to counter the characteristic increase in oxidative stress during the development of cardiac hypertrophy.

In vivo experiments suggest that resveratrol is indeed a potent antioxidant via upregulation of NOS and increased NO production or by modulation of thioredoxin and heme oxygenase systems [177]. In fact, NO may be an important component in resveratrol-induced changes in intracellular redox state because NO has a greater affinity for $^{\bullet}O_2^-$ radicals compared to superoxide dismutase and thus can rapidly lower $^{\bullet}O_2^-$ concentrations [178]. In support of this concept, platelet eNOS was activated at physiologically achievable doses of resveratrol and blunted the proinflammatory pathway linked to p38 MAPK to inhibit production of ROS [179]. The proposed mechanism for increased NO production by resveratrol is thought to involve activation of the sirtuin SIRT1 protein deacetylase [180] which in turn has been shown to deacetylate eNOS at lysine residues to stimulate NO production [181, 182]. Related to this, resveratrol also activates AMPK [165, 183], which can then directly phosphorylate eNOS to stimulate NO production [184, 185]. As NO possesses vasodilatory, anti-inflammatory, anti-hypertrophic properties [186], increasing NO production may be an important pathway by which resveratrol ameliorates pathological cardiac hypertrophy.

As mentioned, mitochondria contribute importantly to the increased oxidative stress in cardiac hypertrophy. Interestingly, resveratrol has also been shown to attenuate mitochondrial oxidative damage [187], which could be another potential mechanism by which this compound is cardioprotective. Resveratrol-induced overexpression of SIRT1 attenuates mitochondrial oxidative damage in endothelial cells [188, 189], suggesting that prevention of mitochondrial oxidative damage alone could be an indirect mechanism for reduction of ROS by resveratrol. Furthermore, resveratrol also reduces mitochondrial $^{\bullet}O_2^-$ production in many cell types, including human coronary arterial endothelial cells which have been attributed to direct stimulation of mitochondrial antioxidant systems [189].

14.8 Modulation of Cell Signaling by Resveratrol

Although a currently accepted mode of action of resveratrol is as an antioxidant, this polyphenol also triggers other mechanistic pathways, which include modulation of cell signaling, apoptosis, and gene expression [190]. For example, resveratrol modulates phorbol-ester-induced signal transduction pathways, which leads to elevated COX2 expression [191] as well as other signals such as NFkB, MAP kinases, transcription factor activating protein-1 (AP-1), and ERK [192]. In addition, resveratrol interferes with many intracellular signaling pathways that regulate cell survival and apoptosis [193].

With specific relevance to cardiac pathophysiology, it is known that resveratrol reduces phenylephrine-induced hypertrophy of rat cardiac myocytes through the activation of AMPK in an AMPK kinase, LKB1-dependent manner [194]. Resveratrol acts as a negative regulator of cardiac hypertrophy via inhibition of the mTOR–p70S6 kinase protein synthesis pathway [195, 196]. Furthermore, resveratrol has been shown to inhibit calcineurin activity, NFAT translocation to the nucleus, and NFAT-dependent transcription in phenylephrine-induced hypertrophy of isolated rat cardiac myocytes [194]. As the calcineurin–NFAT signaling pathway is a major contributor to the development of pathological hypertrophy, resveratrol's ability to suppress activation of this pathway may be an important mechanism by which resveratrol mediates its anti-hypertrophic effects.

As discussed, it is widely accepted that oxidative stress due to excessive production of ROS is a prominent factor in triggering the events that result in cardiomyocyte death. Cardiac SIRT1 is the mammalian ortholog of the silent information regulator 2 (Sir 2) family and is upregulated in response to oxidative stress [197]. Resveratrol has also been shown to be an important activator of SIRT1 [198], and in neonatal rat ventricular cardiomyocytes with simulated ischemia–reperfusion injury, resveratrol-induced SIRT1 activation and overexpression protected cardiomyocytes from oxidative injury, mitochondrial dysfunction, and cell death [198]. Resveratrol-induced SIRT1 overexpression in turn affects the MAPK pathway by reducing p38 and JNK phosphorylation [198]. In a porcine coronary artery preparation, resveratrol treatment countered ET-1 enhancement of MAPK activity by directly inhibiting MAPK activity and similarly reducing JNK-1 and p38 phosphorylation, as well as phosphorylation of ERK 1/2 was also reduced [199]. In support of this finding, resveratrol pretreatment protected against oxidative stress (H_2O_2)-induced cell proliferation and ERK 1/2 activation in human coronary smooth muscle cells [200]. In addition, resveratrol strongly inhibited Ang-II-induced hypertrophy and produced dose-dependent reductions in Akt 1 protein

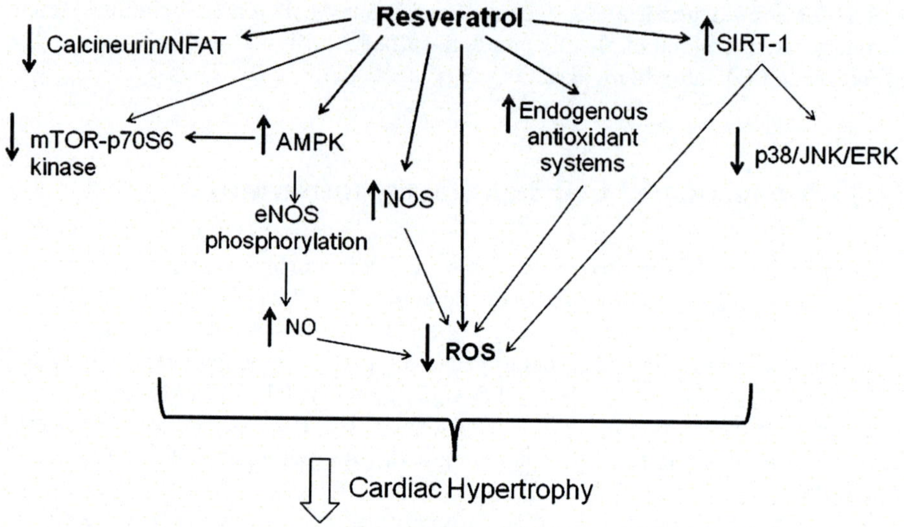

Fig. 14.1 Schematic representation of multiple cell signaling events modulated by resveratrol during the development of cardiac hypertrophy. Resveratrol affects several signaling events, many of which lead to a reduction in ROS and ultimately ameliorate cardiac hypertrophy. *NFAT* nuclear factor of activated T-cells, *mTOR* mammalian target of rapamycin, *AMPK* AMP-activated protein kinase, *eNOS* endothelial nitric oxide synthase, *NOS* nitric oxide synthase, *NO* nitric oxide, *ROS* reactive oxygen species, *SIRT1* sirtuin (silent mating type information regulation 2 homolog) 1, *p38*: p38 mitogen-activated kinase, *JNK*: c-Jun N-terminal kinases, *ERK* extracellular signal-regulated kinase

kinase and ERK 1/2 phosphorylation (two mediators thought to be essentially involved in Ang-II-mediated hypertrophy) [201].

Although activation of the p38 MAPK pathway is thought to contribute to the development of cardiac hypertrophy, recent but limited studies support a beneficial role for resveratrol-mediated activation of this pathway [202]. In H9c2 embryonic rat heart-derived cells treated with H_2O_2 to induce oxidative stress, resveratrol protected cells from oxidative damage, increased autophagy, and reduced peroxide-induced apoptosis [202]. All of these protective effects of resveratrol were prevented by prior treatment with a p38 MAPK inhibitor [202]. Overall resveratrol protected these cells from oxidative stress by upregulating autophagy via the p38 MAPK pathway [202]. Altogether these studies emphasize that the cellular actions of resveratrol are varied and dependent on the particular tissue or system under study (Fig. 14.1).

14.9 Therapeutic Applications for Resveratrol in Cardiac Hypertrophy

To complement direct antioxidant effects which could attenuate oxidative stress and the progression of hypertrophy, resveratrol also modulates other aspects of the pathophysiology of cardiac hypertrophy. As increased arterial pressure is often an initiating factor for cardiac remodeling, the ability of resveratrol to prevent increases in blood pressure in vivo is protective against left ventricular hypertrophy [203]. In several rat models of hypertension, resveratrol prevented left ventricular hypertrophy due to reduced systolic blood pressure [204–206]. Resveratrol also appears to have direct effects on cardiac myocyte growth since it can prevent cardiac hypertrophy in the absence of changes in blood pressure in spontaneously hypertensive rats or rats subjected to abdominal aortic banding as a model of pressure overload [128, 166, 167, 207]. This direct effect of resveratrol on cardiomyocytes is also supported by the modulation of signaling pathways that regulate growth and protein synthesis [194, 195]. Resveratrol also protects against other hypertrophy-induced events such as preventing the reduction in eNOS and iNOS [207] and presumably preventing a subsequent reduction in NO production. In addition, resveratrol inhibits critical steps in cardiac collagen deposition such as cardiac fibroblast proliferation which is thought to involve the cardiac NO–cGMP signaling pathway [196, 208]. Overall, evidence demonstrates that resveratrol protects cardiomyocytes from oxidative stress and death via ROS reduction, increased expression of antioxidant enzymes, and mitochondrial protection [209, 210], with ultimate reduction of fibrosis and preservation of cardiac function and survival in rodent models of heart failure [211].

14.10 Clinical Uses of Resveratrol

Although experimental models suggest that resveratrol may be an effective therapy for cardiac hypertrophy, to date there is a paucity of published human trials in this area. One study in healthy humans showed that a supplement which included 100 mg of resveratrol reduced the oxidative and inflammatory responses normally induced by a high-fat, high-carbohydrate meal [212]. A recent study focused on patients after myocardial infarction who were given a 10 mg/day resveratrol supplement for 3 months. In this study, resveratrol appeared to improve left ventricular diastolic function and endothelial function as measured by flow-mediated dilation of the brachial artery with no change in blood pressure [213]. However, despite the limited published evidence in humans, there are several clinical trials underway with this compound, and resveratrol may be a promising new therapeutic strategy for the treatment of cardiac hypertrophy and heart failure.

14.11 Conclusion

A thorough grasp of the pathophysiological mechanisms underlying the development of cardiac hypertrophy and the progression to heart failure is essential considering that left ventricular hypertrophy is a significant independent risk factor for sudden cardiac death, dysrhythmia, heart failure, ventricular ischemia, and coronary heart disease [214]. Although the specific steps that lead to hypertrophy are not completely clear, it appears that oxidative stress does play an important role in this condition. It is also promising that a treatment approach involving antioxidant therapies, including the natural polyphenol resveratrol, appears to target several signaling and pathogenic events in the etiology of this condition and may well prove to be effective in the prevention and treatment of cardiac hypertrophy alone or in combination with other drugs. Although we have a great deal of evidence supporting resveratrol treatment of hypertrophy, this is largely in cell-based and animal models with limited evidence from human studies—particularly focused on oxidative stress and cardiac function. Therefore, this is a critical and undoubtedly fruitful avenue of research in the future that may have significant clinical implications.

References

1. Frohlich ED, Susic D (2012) Pressure overload. Heart Fail Clin 8:21–32
2. Opie LH, Commerford PJ, Gersh BJ et al (2006) Controversies in ventricular remodelling. Lancet 367:356–367
3. Pfeffer JM, Pfeffer MA, Braunwald E (1985) Influence of chronic captopril therapy on the infarcted left ventricle of the rat. Circ Res 57:84–95
4. Meerson FZ (1962) Compensatory hyperfunction of the heart and cardiac insufficiency. Circ Res 10:250–258
5. Grossman W, Jones D, McLaurin LP (1975) Wall stress and patterns of hypertrophy in the human left ventricle. J Clin Invest 56:56–64
6. Frohlich ED, Apstein C, Chobanian AV et al (1992) The heart in hypertension. N Engl J Med 327:998–1008
7. Barry SP, Townsend PA (2010) What causes a broken heart–molecular insights into heart failure. Int Rev Cell Mol Biol 284:113–179
8. Diez J, Frohlich ED (2010) A translational approach to hypertensive heart disease. Hypertension 55:1–8
9. Frohlich ED (2001) Fibrosis and ischemia: the real risks in hypertensive heart disease. Am J Hypertens 14:194S–199S
10. Kannel WB, Castelli WP, McNamara PM et al (1972) Role of blood pressure in the development of congestive heart failure. The Framingham study. N Engl J Med 287:781–787
11. Dunn FG, Pringle SD (1987) Left ventricular hypertrophy and myocardial ischemia in systemic hypertension. Am J Cardiol 60:19I–22I
12. Cai H, Harrison DG (2000) Endothelial dysfunction in cardiovascular diseases: the role of oxidant stress. Circ Res 87:840–844
13. Takimoto E, Kass DA (2007) Role of oxidative stress in cardiac hypertrophy and remodeling. Hypertension 49:241–248
14. Tsutsui H, Kinugawa S, Matsushima S (2011) Oxidative stress and heart failure. Am J Physiol Heart Circ Physiol 301:H2181–H2190

15. Anversa P, Ricci R, Olivetti G (1986) Quantitative structural analysis of the myocardium during physiologic growth and induced cardiac hypertrophy: a review. J Am Coll Cardiol 7:1140–1149
16. Jalil JE, Doering CW, Janicki JS et al (1989) Fibrillar collagen and myocardial stiffness in the intact hypertrophied rat left ventricle. Circ Res 64:1041–1050
17. Camelliti P, Borg TK, Kohl P (2005) Structural and functional characterisation of cardiac fibroblasts. Cardiovasc Res 65:40–51
18. Weber KT, Janicki JS, Shroff SG et al (1988) Collagen remodeling of the pressure-overloaded, hypertrophied nonhuman primate myocardium. Circ Res 62:757–765
19. Weber KT, Brilla CG (1991) Pathological hypertrophy and cardiac interstitium. Fibrosis and renin-angiotensin-aldosterone system. Circulation 83:1849–1865
20. Morkin E (2000) Control of cardiac myosin heavy chain gene expression. Microsc Res Tech 50:522–531
21. Xu J, Carretero OA, Liao TD et al (2010) Local angiotensin II aggravates cardiac remodeling in hypertension. Am J Physiol Heart Circ Physiol 299:H1328–H1338
22. Frey N, Olson EN (2003) Cardiac hypertrophy: the good, the bad, and the ugly. Annu Rev Physiol 65:45–79
23. Fouad FM, Slominski JM, Tarazi RC (1984) Left ventricular diastolic function in hypertension: relation to left ventricular mass and systolic function. J Am Coll Cardiol 3:1500–1506
24. Hannenhalli S, Putt ME, Gilmore JM et al (2006) Transcriptional genomics associates FOX transcription factors with human heart failure. Circulation 114:1269–1276
25. Luedde M, Katus HA, Frey N (2006) Novel molecular targets in the treatment of cardiac hypertrophy. Recent Pat Cardiovasc Drug Discov 1:1–20
26. Diez J (2009) Towards a new paradigm about hypertensive heart disease. Med Clin North Am 93:637–645
27. Ingwall JS (2009) Energy metabolism in heart failure and remodelling. Cardiovasc Res 81:412–419
28. Allard MF, Schonekess BO, Henning SL et al (1994) Contribution of oxidative metabolism and glycolysis to ATP production in hypertrophied hearts. Am J Physiol 267:H742–H750
29. Takeyama D, Kagaya Y, Yamane Y et al (1995) Effects of chronic right ventricular pressure overload on myocardial glucose and free fatty acid metabolism in the conscious rat. Cardiovasc Res 29:763–767
30. Massie BM, Schaefer S, Garcia J et al (1995) Myocardial high-energy phosphate and substrate metabolism in swine with moderate left ventricular hypertrophy. Circulation 91: 1814–1823
31. Leong HS, Brownsey RW, Kulpa JE et al (2003) Glycolysis and pyruvate oxidation in cardiac hypertrophy–why so unbalanced? Comp Biochem Physiol A Mol Integr Physiol 135: 499–513
32. Beer M, Seyfarth T, Sandstede J et al (2002) Absolute concentrations of high-energy phosphate metabolites in normal, hypertrophied, and failing human myocardium measured noninvasively with (31)P-SLOOP magnetic resonance spectroscopy. J Am Coll Cardiol 40: 1267–1274
33. Sambandam N, Lopaschuk GD, Brownsey RW et al (2002) Energy metabolism in the hypertrophied heart. Heart Fail Rev 7:161–173
34. Doenst T, Pytel G, Schrepper A et al (2010) Decreased rates of substrate oxidation ex vivo predict the onset of heart failure and contractile dysfunction in rats with pressure overload. Cardiovasc Res 86:461–470
35. Wende AR, Abel ED (2010) Lipotoxicity in the heart. Biochim Biophys Acta 1801:311–319
36. Gonzalez A, Lopez B, Ravassa S et al (2002) Stimulation of cardiac apoptosis in essential hypertension: potential role of angiotensin II. Hypertension 39:75–80
37. Berk BC, Fujiwara K, Lehoux S (2007) ECM remodeling in hypertensive heart disease. J Clin Invest 117:568–575

38. Houser SR, Piacentino V 3rd, Weisser J (2000) Abnormalities of calcium cycling in the hypertrophied and failing heart. J Mol Cell Cardiol 32:1595–1607
39. Weber KT (1989) Cardiac interstitium in health and disease: the fibrillar collagen network. J Am Coll Cardiol 13:1637–1652
40. Belch JJ, Bridges AB, Scott N et al (1991) Oxygen free radicals and congestive heart failure. Br Heart J 65:245–248
41. Hill MF, Singal PK (1996) Antioxidant and oxidative stress changes during heart failure subsequent to myocardial infarction in rats. Am J Pathol 148:291–300
42. Hill MF, Singal PK (1997) Right and left myocardial antioxidant responses during heart failure subsequent to myocardial infarction. Circulation 96:2414–2420
43. Mallat Z, Philip I, Lebret M et al (1998) Elevated levels of 8-iso-prostaglandin F2alpha in pericardial fluid of patients with heart failure: a potential role for in vivo oxidant stress in ventricular dilatation and progression to heart failure. Circulation 97:1536–1539
44. Chen AF, Chen DD, Daiber A et al (2012) Free radical biology of the cardiovascular system. Clin Sci (Lond) 123:73–91
45. Zima AV, Blatter LA (2006) Redox regulation of cardiac calcium channels and transporters. Cardiovasc Res 71:310–321
46. Sabri A, Hughie HH, Lucchesi PA (2003) Regulation of hypertrophic and apoptotic signaling pathways by reactive oxygen species in cardiac myocytes. Antioxid Redox Signal 5: 731–740
47. Cesselli D, Jakoniuk I, Barlucchi L et al (2001) Oxidative stress-mediated cardiac cell death is a major determinant of ventricular dysfunction and failure in dog dilated cardiomyopathy. Circ Res 89:279–286
48. Rajagopalan S, Meng XP, Ramasamy S et al (1996) Reactive oxygen species produced by macrophage-derived foam cells regulate the activity of vascular matrix metalloproteinases in vitro. Implications for atherosclerotic plaque stability. J Clin Invest 98:2572–2579
49. Kinugawa S, Tsutsui H, Hayashidani S et al (2000) Treatment with dimethylthiourea prevents left ventricular remodeling and failure after experimental myocardial infarction in mice: role of oxidative stress. Circ Res 87:392–398
50. Spinale FG, Coker ML, Thomas CV et al (1998) Time-dependent changes in matrix metalloproteinase activity and expression during the progression of congestive heart failure: relation to ventricular and myocyte function. Circ Res 82:482–495
51. Montezano AC, Touyz RM (2012) Molecular mechanisms of hypertension-reactive oxygen species and antioxidants: a basic science update for the clinician. Can J Cardiol 28:288–295
52. Sawyer DB, Colucci WS (2000) Mitochondrial oxidative stress in heart failure: "oxygen wastage" revisited. Circ Res 86:119–120
53. Landmesser U, Dikalov S, Price SR et al (2003) Oxidation of tetrahydrobiopterin leads to uncoupling of endothelial cell nitric oxide synthase in hypertension. J Clin Invest 111:1201–1209
54. Heymes C, Bendall JK, Ratajczak P et al (2003) Increased myocardial NADPH oxidase activity in human heart failure. J Am Coll Cardiol 41:2164–2171
55. Cappola TP, Kass DA, Nelson GS et al (2001) Allopurinol improves myocardial efficiency in patients with idiopathic dilated cardiomyopathy. Circulation 104:2407–2411
56. Sullivan JC, Pollock JS (2006) Coupled and uncoupled NOS: separate but equal? Uncoupled NOS in endothelial cells is a critical pathway for intracellular signaling. Circ Res 98:717–719
57. Kwon SH, Pimentel DR, Remondino A et al (2003) H(2)O(2) regulates cardiac myocyte phenotype via concentration-dependent activation of distinct kinase pathways. J Mol Cell Cardiol 35:615–621
58. Cosentino F, Patton S, d'Uscio LV et al (1998) Tetrahydrobiopterin alters superoxide and nitric oxide release in prehypertensive rats. J Clin Invest 101:1530–1537
59. Ozaki M, Kawashima S, Yamashita T et al (2002) Overexpression of endothelial nitric oxide synthase accelerates atherosclerotic lesion formation in apoE-deficient mice. J Clin Invest 110:331–340

60. Takimoto E, Champion HC, Li M et al (2005) Oxidant stress from nitric oxide synthase-3 uncoupling stimulates cardiac pathologic remodeling from chronic pressure load. J Clin Invest 115:1221–1231
61. Ballinger SW, Patterson C, Yan CN et al (2000) Hydrogen peroxide- and peroxynitrite-induced mitochondrial DNA damage and dysfunction in vascular endothelial and smooth muscle cells. Circ Res 86:960–966
62. Ide T, Tsutsui H, Hayashidani S et al (2001) Mitochondrial DNA damage and dysfunction associated with oxidative stress in failing hearts after myocardial infarction. Circ Res 88:529–535
63. Bernardo BC, Weeks KL, Pretorius L et al (2010) Molecular distinction between physiological and pathological cardiac hypertrophy: experimental findings and therapeutic strategies. Pharmacol Ther 128:191–227
64. Wilkins BJ, Dai YS, Bueno OF et al (2004) Calcineurin/NFAT coupling participates in pathological, but not physiological, cardiac hypertrophy. Circ Res 94:110–118
65. Aramburu J, Rao A, Klee CB (2000) Calcineurin: from structure to function. Curr Top Cell Regul 36:237–295
66. Molkentin JD, Lu JR, Antos CL et al (1998) A calcineurin-dependent transcriptional pathway for cardiac hypertrophy. Cell 93:215–228
67. Shimoyama M, Hayashi D, Takimoto E et al (1999) Calcineurin plays a critical role in pressure overload-induced cardiac hypertrophy. Circulation 100:2449–2454
68. Sussman MA, Lim HW, Gude N et al (1998) Prevention of cardiac hypertrophy in mice by calcineurin inhibition. Science 281:1690–1693
69. De Windt LJ, Lim HW, Bueno OF et al (2001) Targeted inhibition of calcineurin attenuates cardiac hypertrophy in vivo. Proc Natl Acad Sci USA 98:3322–3327
70. Kingsbury TJ, Cunningham KW (2000) A conserved family of calcineurin regulators. Genes Dev 14:1595–1604
71. Hill JA, Rothermel B, Yoo KD et al (2002) Targeted inhibition of calcineurin in pressure-overload cardiac hypertrophy. Preservation of systolic function. J Biol Chem 277: 10251–10255
72. Zou Y, Hiroi Y, Uozumi H et al (2001) Calcineurin plays a critical role in the development of pressure overload-induced cardiac hypertrophy. Circulation 104:97–101
73. Bueno OF, Wilkins BJ, Tymitz KM et al (2002) Impaired cardiac hypertrophic response in Calcineurin Abeta -deficient mice. Proc Natl Acad Sci USA 99:4586–4591
74. Wilkins BJ, De Windt LJ, Bueno OF et al (2002) Targeted disruption of NFATc3, but not NFATc4, reveals an intrinsic defect in calcineurin-mediated cardiac hypertrophic growth. Mol Cell Biol 22:7603–7613
75. Bito V, Heinzel FR, Biesmans L et al (2008) Crosstalk between L-type Ca2+ channels and the sarcoplasmic reticulum: alterations during cardiac remodelling. Cardiovasc Res 77:315–324
76. Ermak G, Davies KJ (2002) Calcium and oxidative stress: from cell signaling to cell death. Mol Immunol 38:713–721
77. Castilho RF, Carvalho-Alves PC, Vercesi AE et al (1996) Oxidative damage to sarcoplasmic reticulum Ca(2+)-pump induced by Fe2+/H2O2/ascorbate is not mediated by lipid peroxidation or thiol oxidation and leads to protein fragmentation. Mol Cell Biochem 159:105–114
78. Xu L, Eu JP, Meissner G et al (1998) Activation of the cardiac calcium release channel (ryanodine receptor) by poly-S-nitrosylation. Science 279:234–237
79. de la Bastie D, Levitsky D, Rappaport L et al (1990) Function of the sarcoplasmic reticulum and expression of its Ca2(+)-ATPase gene in pressure overload-induced cardiac hypertrophy in the rat. Circ Res 66:554–564
80. Wilkins BJ, Molkentin JD (2004) Calcium-calcineurin signaling in the regulation of cardiac hypertrophy. Biochem Biophys Res Commun 322:1178–1191
81. Houser SR, Molkentin JD (2008) Does contractile Ca2+ control calcineurin-NFAT signaling and pathological hypertrophy in cardiac myocytes? Sci Signal 1:31

82. Eder P, Molkentin JD (2011) TRPC channels as effectors of cardiac hypertrophy. Circ Res 108:265–272
83. Seth M, Zhang ZS, Mao L et al (2009) TRPC1 channels are critical for hypertrophic signaling in the heart. Circ Res 105:1023–1030
84. Bush EW, Hood DB, Papst PJ et al (2006) Canonical transient receptor potential channels promote cardiomyocyte hypertrophy through activation of calcineurin signaling. J Biol Chem 281:33487–33496
85. Kuwahara K, Wang Y, McAnally J et al (2006) TRPC6 fulfills a calcineurin signaling circuit during pathologic cardiac remodeling. J Clin Invest 116:3114–3126
86. Nakayama H, Wilkin BJ, Bodi I et al (2006) Calcineurin-dependent cardiomyopathy is activated by TRPC in the adult mouse heart. FASEB J 20:1660–1670
87. Poteser M, Graziani A, Rosker C et al (2006) TRPC3 and TRPC4 associate to form a redox-sensitive cation channel. Evidence for expression of native TRPC3-TRPC4 heteromeric channels in endothelial cells. J Biol Chem 281:13588–13595
88. Clerk A, Sugden PH (1999) Activation of protein kinase cascades in the heart by hypertrophic G protein-coupled receptor agonists. Am J Cardiol 83:64H–69H
89. Pearson G, Robinson F, Beers Gibson T et al (2001) Mitogen-activated protein (MAP) kinase pathways: regulation and physiological functions. Endocr Rev 22:153–183
90. Muslin AJ (2008) MAPK signalling in cardiovascular health and disease: molecular mechanisms and therapeutic targets. Clin Sci (Lond) 115:203–218
91. Yamazaki T, Tobe K, Hoh E et al (1993) Mechanical loading activates mitogen-activated protein kinase and S6 peptide kinase in cultured rat cardiac myocytes. J Biol Chem 268:12069–12076
92. Sadoshima J, Qiu Z, Morgan JP et al (1995) Angiotensin II and other hypertrophic stimuli mediated by G protein-coupled receptors activate tyrosine kinase, mitogen-activated protein kinase, and 90-kD S6 kinase in cardiac myocytes. The critical role of Ca(2+)-dependent signaling. Circ Res 76:1–15
93. Komuro I, Kudo S, Yamazaki T et al (1996) Mechanical stretch activates the stress-activated protein kinases in cardiac myocytes. FASEB J 10:631–636
94. Esposito G, Prasad SV, Rapacciuolo A et al (2001) Cardiac overexpression of a G(q) inhibitor blocks induction of extracellular signal-regulated kinase and c-Jun NH(2)-terminal kinase activity in in vivo pressure overload. Circulation 103:1453–1458
95. Takeishi Y, Huang Q, Abe J et al (2001) Src and multiple MAP kinase activation in cardiac hypertrophy and congestive heart failure under chronic pressure-overload: comparison with acute mechanical stretch. J Mol Cell Cardiol 33:1637–1648
96. Cook SA, Sugden PH, Clerk A (1999) Activation of c-Jun N-terminal kinases and p38-mitogen-activated protein kinases in human heart failure secondary to ischaemic heart disease. J Mol Cell Cardiol 31:1429–1434
97. Tanaka K, Honda M, Takabatake T (2001) Redox regulation of MAPK pathways and cardiac hypertrophy in adult rat cardiac myocyte. J Am Coll Cardiol 37:676–685
98. Meng R, Pei Z, Zhang A et al (2011) AMPK activation enhances PPARalpha activity to inhibit cardiac hypertrophy via ERK1/2 MAPK signaling pathway. Arch Biochem Biophys 511:1–7
99. Ramos JW (2008) The regulation of extracellular signal-regulated kinase (ERK) in mammalian cells. Int J Biochem Cell Biol 40:2707–2719
100. Bogoyevitch MA, Glennon PE, Andersson MB et al (1994) Endothelin-1 and fibroblast growth factors stimulate the mitogen-activated protein kinase signaling cascade in cardiac myocytes. The potential role of the cascade in the integration of two signaling pathways leading to myocyte hypertrophy. J Biol Chem 269:1110–1119
101. Post GR, Goldstein D, Thuerauf DJ et al (1996) Dissociation of p44 and p42 mitogen-activated protein kinase activation from receptor-induced hypertrophy in neonatal rat ventricular myocytes. J Biol Chem 271:8452–8457

102. Wang L, Proud CG (2002) Ras/Erk signaling is essential for activation of protein synthesis by Gq protein-coupled receptor agonists in adult cardiomyocytes. Circ Res 91:821–829
103. Harris IS, Zhang S, Treskov I et al (2004) Raf-1 kinase is required for cardiac hypertrophy and cardiomyocyte survival in response to pressure overload. Circulation 110:718–723
104. Bueno OF, De Windt LJ, Tymitz KM et al (2000) The MEK1-ERK1/2 signaling pathway promotes compensated cardiac hypertrophy in transgenic mice. EMBO J 19:6341–6350
105. Purcell NH, Wilkins BJ, York A et al (2007) Genetic inhibition of cardiac ERK1/2 promotes stress-induced apoptosis and heart failure but has no effect on hypertrophy in vivo. Proc Natl Acad Sci USA 104:14074–14079
106. Waetzig V, Herdegen T (2005) Context-specific inhibition of JNKs: overcoming the dilemma of protection and damage. Trends Pharmacol Sci 26:455–461
107. Haeusgen W, Herdegen T, Waetzig V (2011) The bottleneck of JNK signaling: molecular and functional characteristics of MKK4 and MKK7. Eur J Cell Biol 90:536–544
108. Rose BA, Force T, Wang Y (2010) Mitogen-activated protein kinase signaling in the heart: angels versus demons in a heart-breaking tale. Physiol Rev 90:1507–1546
109. Wang Y, Su B, Sah VP et al (1998) Cardiac hypertrophy induced by mitogen-activated protein kinase kinase 7, a specific activator for c-Jun NH2-terminal kinase in ventricular muscle cells. J Biol Chem 273:5423–5426
110. Choukroun G, Hajjar R, Kyriakis JM et al (1998) Role of the stress-activated protein kinases in endothelin-induced cardiomyocyte hypertrophy. J Clin Invest 102:1311–1320
111. Choukroun G, Hajjar R, Fry S et al (1999) Regulation of cardiac hypertrophy in vivo by the stress-activated protein kinases/c-Jun NH(2)-terminal kinases. J Clin Invest 104:391–398
112. Liang Q, Bueno OF, Wilkins BJ et al (2003) c-Jun N-terminal kinases (JNK) antagonize cardiac growth through cross-talk with calcineurin-NFAT signaling. EMBO J 22:5079–5089
113. Tachibana H, Perrino C, Takaoka H et al (2006) JNK1 is required to preserve cardiac function in the early response to pressure overload. Biochem Biophys Res Commun 343:1060–1066
114. Zarubin T, Han J (2005) Activation and signaling of the p38 MAP kinase pathway. Cell Res 15:11–18
115. Liang Q, Molkentin JD (2003) Redefining the roles of p38 and JNK signaling in cardiac hypertrophy: dichotomy between cultured myocytes and animal models. J Mol Cell Cardiol 35:1385–1394
116. Ueyama T, Kawashima S, Sakoda T et al (1999) Endothelin-1 activates p38 mitogen-activated protein kinase via endothelin-A receptor in rat myocardial cells. Mol Cell Biochem 199:119–124
117. Clerk A, Michael A, Sugden PH (1998) Stimulation of the p38 mitogen-activated protein kinase pathway in neonatal rat ventricular myocytes by the G protein-coupled receptor agonists, endothelin-1 and phenylephrine: a role in cardiac myocyte hypertrophy? J Cell Biol 142:523–535
118. Liao P, Georgakopoulos D, Kovacs A et al (2001) The in vivo role of p38 MAP kinases in cardiac remodeling and restrictive cardiomyopathy. Proc Natl Acad Sci USA 98: 12283–12288
119. Zhang S, Weinheimer C, Courtois M et al (2003) The role of the Grb2-p38 MAPK signaling pathway in cardiac hypertrophy and fibrosis. J Clin Invest 111:833–841
120. Braz JC, Bueno OF, Liang Q et al (2003) Targeted inhibition of p38 MAPK promotes hypertrophic cardiomyopathy through upregulation of calcineurin-NFAT signaling. J Clin Invest 111:1475–1486
121. Molkentin JD (2004) Calcineurin-NFAT signaling regulates the cardiac hypertrophic response in coordination with the MAPKs. Cardiovasc Res 63:467–475
122. Dyck JR, Lopaschuk GD (2006) AMPK alterations in cardiac physiology and pathology: enemy or ally? J Physiol 574:95–112
123. Chan AY, Dyck JR (2005) Activation of AMP-activated protein kinase (AMPK) inhibits protein synthesis: a potential strategy to prevent the development of cardiac hypertrophy. Can J Physiol Pharmacol 83:24–28

124. Chan AY, Soltys CL, Young ME et al (2004) Activation of AMP-activated protein kinase inhibits protein synthesis associated with hypertrophy in the cardiac myocyte. J Biol Chem 279:32771–32779
125. Horman S, Beauloye C, Vertommen D et al (2003) Myocardial ischemia and increased heart work modulate the phosphorylation state of eukaryotic elongation factor-2. J Biol Chem 278:41970–41976
126. Horman S, Browne G, Krause U et al (2002) Activation of AMP-activated protein kinase leads to the phosphorylation of elongation factor 2 and an inhibition of protein synthesis. Curr Biol 12:1419–1423
127. Cheng SW, Fryer LG, Carling D et al (2004) Thr2446 is a novel mammalian target of rapamycin (mTOR) phosphorylation site regulated by nutrient status. J Biol Chem 279: 15719–15722
128. Dolinsky VW, Chan AY, Robillard Frayne I et al (2009) Resveratrol prevents the prohypertrophic effects of oxidative stress on LKB1. Circulation 119:1643–1652
129. Zarrinpashneh E, Beauloye C, Ginion A et al (2008) AMPKalpha2 counteracts the development of cardiac hypertrophy induced by isoproterenol. Biochem Biophys Res Commun 376:677–681
130. Zhang P, Hu X, Xu X et al (2008) AMP activated protein kinase-alpha2 deficiency exacerbates pressure-overload-induced left ventricular hypertrophy and dysfunction in mice. Hypertension 52:918–924
131. Ikeda Y, Sato K, Pimentel DR et al (2009) Cardiac-specific deletion of LKB1 leads to hypertrophy and dysfunction. J Biol Chem 284:35839–35849
132. Tian R, Musi N, D'Agostino J et al (2001) Increased adenosine monophosphate-activated protein kinase activity in rat hearts with pressure-overload hypertrophy. Circulation 104:1664–1669
133. O'Shea JJ, Gadina M, Schreiber RD (2002) Cytokine signaling in 2002: new surprises in the Jak/Stat pathway. Cell 109(Suppl):S121–S131
134. Kunisada K, Hirota H, Fujio Y et al (1996) Activation of JAK-STAT and MAP kinases by leukemia inhibitory factor through gp130 in cardiac myocytes. Circulation 94:2626–2632
135. Pelletier S, Duhamel F, Coulombe P et al (2003) Rho family GTPases are required for activation of Jak/STAT signaling by G protein-coupled receptors. Mol Cell Biol 23:1316–1333
136. Yamauchi-Takihara K, Kishimoto T (2000) A novel role for STAT3 in cardiac remodeling. Trends Cardiovasc Med 10:298–303
137. Kunisada K, Tone E, Fujio Y et al (1998) Activation of gp130 transduces hypertrophic signals via STAT3 in cardiac myocytes. Circulation 98:346–352
138. Kunisada K, Negoro S, Tone E et al (2000) Signal transducer and activator of transcription 3 in the heart transduces not only a hypertrophic signal but a protective signal against doxorubicin-induced cardiomyopathy. Proc Natl Acad Sci USA 97:315–319
139. Hirota H, Chen J, Betz UA et al (1999) Loss of a gp130 cardiac muscle cell survival pathway is a critical event in the onset of heart failure during biomechanical stress. Cell 97:189–198
140. Tinkel J, Hassanain H, Khouri SJ (2012) Cardiovascular antioxidant therapy: a review of supplements, pharmacotherapies, and mechanisms. Cardiol Rev 20:77–83
141. Saremi A, Arora R (2010) Vitamin E and cardiovascular disease. Am J Ther 17:e56–e65
142. Traber MG, Stevens JF (2011) Vitamins C and E: beneficial effects from a mechanistic perspective. Free Radic Biol Med 51:1000–1013
143. Dutta A, Dutta SK (2003) Vitamin E and its role in the prevention of atherosclerosis and carcinogenesis: a review. J Am Coll Nutr 22:258–268
144. Buettner GR (1993) The pecking order of free radicals and antioxidants: lipid peroxidation, alpha-tocopherol, and ascorbate. Arch Biochem Biophys 300:535–543
145. Dhalla AK, Hill MF, Singal PK (1996) Role of oxidative stress in transition of hypertrophy to heart failure. J Am Coll Cardiol 28:506–514
146. Stephens NG, Parsons A, Schofield PM et al (1996) Randomised controlled trial of vitamin E in patients with coronary disease: Cambridge Heart Antioxidant Study (CHAOS). Lancet 347:781–786

147. Boaz M, Smetana S, Weinstein T et al (2000) Secondary prevention with antioxidants of cardiovascular disease in endstage renal disease (SPACE): randomised placebo-controlled trial. Lancet 356:1213–1218
148. Salonen JT, Nyyssonen K, Salonen R et al (2000) Antioxidant Supplementation in Atherosclerosis Prevention (ASAP) study: a randomized trial of the effect of vitamins E and C on 3-year progression of carotid atherosclerosis. J Intern Med 248:377–386
149. Stampfer MJ, Hennekens CH, Manson JE et al (1993) Vitamin E consumption and the risk of coronary disease in women. N Engl J Med 328:1444–1449
150. Rimm EB, Stampfer MJ, Ascherio A et al (1993) Vitamin E consumption and the risk of coronary heart disease in men. N Engl J Med 328:1450–1456
151. Salonen RM, Nyyssonen K, Kaikkonen J et al (2003) Six-year effect of combined vitamin C and E supplementation on atherosclerotic progression: the Antioxidant Supplementation in Atherosclerosis Prevention (ASAP) Study. Circulation 107:947–953
152. Yusuf S, Dagenais G, Pogue J et al (2000) Vitamin E supplementation and cardiovascular events in high-risk patients. The Heart Outcomes Prevention Evaluation Study Investigators. N Engl J Med 342:154–160
153. Hodis HN, Mack WJ, LaBree L et al (2002) Alpha-tocopherol supplementation in healthy individuals reduces low-density lipoprotein oxidation but not atherosclerosis: the Vitamin E Atherosclerosis Prevention Study (VEAPS). Circulation 106:1453–1459
154. Heart Protection Study Collaborative Group (2002) MRC/BHF Heart Protection Study of antioxidant vitamin supplementation in 20,536 high-risk individuals: a randomised placebo-controlled trial. Lancet 360:23–33
155. Steinhubl SR (2008) Why have antioxidants failed in clinical trials? Am J Cardiol 101:14D–19D
156. Aruoma OI, Halliwell B, Hoey BM et al (1989) The antioxidant action of N-acetylcysteine: its reaction with hydrogen peroxide, hydroxyl radical, superoxide, and hypochlorous acid. Free Radic Biol Med 6:593–597
157. Andrews NP, Prasad A, Quyyumi AA (2001) N-acetylcysteine improves coronary and peripheral vascular function. J Am Coll Cardiol 37:117–123
158. Stamler JS, Simon DI, Osborne JA et al (1992) S-nitrosylation of proteins with nitric oxide: synthesis and characterization of biologically active compounds. Proc Natl Acad Sci USA 89:444–448
159. Byrne JA, Grieve DJ, Bendall JK et al (2003) Contrasting roles of NADPH oxidase isoforms in pressure-overload versus angiotensin II-induced cardiac hypertrophy. Circ Res 93: 802–805
160. Pechanova O, Zicha J, Paulis L et al (2007) The effect of N-acetylcysteine and melatonin in adult spontaneously hypertensive rats with established hypertension. Eur J Pharmacol 561:129–136
161. Pechanova O, Zicha J, Kojsova S et al (2006) Effect of chronic N-acetylcysteine treatment on the development of spontaneous hypertension. Clin Sci (Lond) 110:235–242
162. Sochman J (2002) N-acetylcysteine in acute cardiology: 10 years later: what do we know and what would we like to know?! J Am Coll Cardiol 39:1422–1428
163. Sochman J, Vrbska J, Musilova B et al (1996) Infarct Size Limitation: acute N-acetylcysteine defense (ISLAND trial): preliminary analysis and report after the first 30 patients. Clin Cardiol 19:94–100
164. Martina V, Masha A, Gigliardi VR et al (2008) Long-term N-acetylcysteine and L-arginine administration reduces endothelial activation and systolic blood pressure in hypertensive patients with type 2 diabetes. Diabetes Care 31:940–944
165. Baur JA, Pearson KJ, Price NL et al (2006) Resveratrol improves health and survival of mice on a high-calorie diet. Nature 444:337–342
166. Thandapilly SJ, Wojciechowski P, Behbahani J et al (2010) Resveratrol prevents the development of pathological cardiac hypertrophy and contractile dysfunction in the SHR without lowering blood pressure. Am J Hypertens 23:192–196

167. Wojciechowski P, Juric D, Louis XL et al (2010) Resveratrol arrests and regresses the development of pressure overload- but not volume overload-induced cardiac hypertrophy in rats. J Nutr 140:962–968
168. Dolinsky VW, Dyck JR (2011) Calorie restriction and resveratrol in cardiovascular health and disease. Biochim Biophys Acta 1812:1477–1489
169. Vidavalur R, Otani H, Singal PK et al (2006) Significance of wine and resveratrol in cardiovascular disease: French paradox revisited. Exp Clin Cardiol 11:217–225
170. Pervaiz S, Holme AL (2009) Resveratrol: its biologic targets and functional activity. Antioxid Redox Signal 11:2851–2897
171. Jia Z, Zhu H, Misra BR et al (2008) EPR studies on the superoxide-scavenging capacity of the nutraceutical resveratrol. Mol Cell Biochem 313:187–194
172. Orallo F, Alvarez E, Camina M et al (2002) The possible implication of trans-Resveratrol in the cardioprotective effects of long-term moderate wine consumption. Mol Pharmacol 61:294–302
173. Soylemez S, Sepici A, Akar F (2009) Resveratrol supplementation gender independently improves endothelial reactivity and suppresses superoxide production in healthy rats. Cardiovasc Drugs Ther 23:449–458
174. Ungvari Z, Orosz Z, Rivera A et al (2007) Resveratrol increases vascular oxidative stress resistance. Am J Physiol Heart Circ Physiol 292:H2417–H2424
175. Spanier G, Xu H, Xia N et al (2009) Resveratrol reduces endothelial oxidative stress by modulating the gene expression of superoxide dismutase 1 (SOD1), glutathione peroxidase 1 (GPx1) and NADPH oxidase subunit (Nox4). J Physiol Pharmacol 60(Suppl 4):111–116
176. Brito PM, Mariano A, Almeida LM et al (2006) Resveratrol affords protection against peroxynitrite-mediated endothelial cell death: A role for intracellular glutathione. Chem Biol Interact 164:157–166
177. Hung LM, Su MJ, Chen JK (2004) Resveratrol protects myocardial ischemia-reperfusion injury through both NO-dependent and NO-independent mechanisms. Free Radic Biol Med 36:774–781
178. Inoue M, Sato EF, Park AM et al (2000) Cross-talk between NO and oxyradicals, a supersystem that regulates energy metabolism and survival of animals. Free Radic Res 33:757–770
179. Gresele P, Pignatelli P, Guglielmini G et al (2008) Resveratrol, at concentrations attainable with moderate wine consumption, stimulates human platelet nitric oxide production. J Nutr 138:1602–1608
180. Lagouge M, Argmann C, Gerhart-Hines Z et al (2006) Resveratrol improves mitochondrial function and protects against metabolic disease by activating SIRT1 and PGC-1alpha. Cell 127:1109–1122
181. Mattagajasingh I, Kim CS, Naqvi A et al (2007) SIRT1 promotes endothelium-dependent vascular relaxation by activating endothelial nitric oxide synthase. Proc Natl Acad Sci USA 104:14855–14860
182. Arunachalam G, Yao H, Sundar IK et al (2010) SIRT1 regulates oxidant- and cigarette smoke-induced eNOS acetylation in endothelial cells: role of resveratrol. Biochem Biophys Res Commun 393:66–72
183. Um JH, Park SJ, Kang H et al (2010) AMP-activated protein kinase-deficient mice are resistant to the metabolic effects of resveratrol. Diabetes 59:554–563
184. Chen ZP, Mitchelhill KI, Michell BJ et al (1999) AMP-activated protein kinase phosphorylation of endothelial NO synthase. FEBS Lett 443:285–289
185. Robich MP, Osipov RM, Nezafat R et al (2010) Resveratrol improves myocardial perfusion in a swine model of hypercholesterolemia and chronic myocardial ischemia. Circulation 122:S142–S149
186. Chatterjee A, Black SM, Catravas JD (2008) Endothelial nitric oxide (NO) and its pathophysiologic regulation. Vascul Pharmacol 49:134–140
187. Page MM, Robb EL, Salway KD et al (2010) Mitochondrial redox metabolism: aging, longevity and dietary effects. Mech Ageing Dev 131:242–252

188. Kao CL, Chen LK, Chang YL et al (2010) Resveratrol protects human endothelium from H(2)O(2)-induced oxidative stress and senescence via SirT1 activation. J Atheroscler Thromb 17:970–979
189. Ungvari Z, Labinskyy N, Mukhopadhyay P et al (2009) Resveratrol attenuates mitochondrial oxidative stress in coronary arterial endothelial cells. Am J Physiol Heart Circ Physiol 297:H1876–H1881
190. Kovacic P, Somanathan R (2010) Multifaceted approach to resveratrol bioactivity: focus on antioxidant action, cell signaling and safety. Oxid Med Cell Longev 3:86–100
191. Lin HY, Tang HY, Davis FB et al (2011) Resveratrol and apoptosis. Ann N Y Acad Sci 1215:79–88
192. Kundu JK, Shin YK, Surh YJ (2006) Resveratrol modulates phorbol ester-induced pro-inflammatory signal transduction pathways in mouse skin in vivo: NF-kappaB and AP-1 as prime targets. Biochem Pharmacol 72:1506–1515
193. Fulda S, Debatin KM (2006) Resveratrol modulation of signal transduction in apoptosis and cell survival: a mini-review. Cancer Detect Prev 30:217–223
194. Chan AY, Dolinsky VW, Soltys CL et al (2008) Resveratrol inhibits cardiac hypertrophy via AMP-activated protein kinase and Akt. J Biol Chem 283:24194–24201
195. Hannan RD, Jenkins A, Jenkins AK et al (2003) Cardiac hypertrophy: a matter of translation. Clin Exp Pharmacol Physiol 30:517–527
196. Olson ER, Naugle JE, Zhang X et al (2005) Inhibition of cardiac fibroblast proliferation and myofibroblast differentiation by resveratrol. Am J Physiol Heart Circ Physiol 288:H1131–H1138
197. Alcendor RR, Gao S, Zhai P et al (2007) Sirt1 regulates aging and resistance to oxidative stress in the heart. Circ Res 100:1512–1521
198. Becatti M, Taddei N, Cecchi C et al (2012) SIRT1 modulates MAPK pathways in ischemic-reperfused cardiomyocytes. Cell Mol Life Sci 69:2245–2260
199. El-Mowafy AM, White RE (1999) Resveratrol inhibits MAPK activity and nuclear translocation in coronary artery smooth muscle: reversal of endothelin-1 stimulatory effects. FEBS Lett 451:63–67
200. El-Mowafy AM, Alkhalaf M, El-Kashef HA (2008) Resveratrol reverses hydrogen peroxide-induced proliferative effects in human coronary smooth muscle cells: a novel signaling mechanism. Arch Med Res 39:155–161
201. Haider UG, Sorescu D, Griendling KK et al (2002) Resveratrol suppresses angiotensin II-induced Akt/protein kinase B and p70 S6 kinase phosphorylation and subsequent hypertrophy in rat aortic smooth muscle cells. Mol Pharmacol 62:772–777
202. Lv XC, Zhou HY (2012) Resveratrol protects H9c2 embryonic rat heart derived cells from oxidative stress by inducing autophagy: role of p38 mitogen-activated protein kinase. Can J Physiol Pharmacol 90:655–662
203. Li H, Xia N, Forstermann U (2012) Cardiovascular effects and molecular targets of resveratrol. Nitric Oxide 26:102–110
204. Liu Z, Song Y, Zhang X et al (2005) Effects of trans-resveratrol on hypertension-induced cardiac hypertrophy using the partially nephrectomized rat model. Clin Exp Pharmacol Physiol 32:1049–1054
205. Toklu HZ, Sehirli O, Ersahin M et al (2010) Resveratrol improves cardiovascular function and reduces oxidative organ damage in the renal, cardiovascular and cerebral tissues of two-kidney, one-clip hypertensive rats. J Pharm Pharmacol 62:1784–1793
206. Chan V, Fenning A, Iyer A et al (2011) Resveratrol improves cardiovascular function in DOCA-salt hypertensive rats. Curr Pharm Biotechnol 12:429–436
207. Juric D, Wojciechowski P, Das DK et al (2007) Prevention of concentric hypertrophy and diastolic impairment in aortic-banded rats treated with resveratrol. Am J Physiol Heart Circ Physiol 292:H2138–H2143
208. Wang S, Wang X, Yan J et al (2007) Resveratrol inhibits proliferation of cultured rat cardiac fibroblasts: correlated with NO-cGMP signaling pathway. Eur J Pharmacol 567:26–35

209. Danz ED, Skramsted J, Henry N et al (2009) Resveratrol prevents doxorubicin cardiotoxicity through mitochondrial stabilization and the Sirt1 pathway. Free Radic Biol Med 46: 1589–1597
210. Tatlidede E, Sehirli O, Velioglu-Ogunc A et al (2009) Resveratrol treatment protects against doxorubicin-induced cardiotoxicity by alleviating oxidative damage. Free Radic Res 43: 195–205
211. Tanno M, Kuno A, Yano T et al (2010) Induction of manganese superoxide dismutase by nuclear translocation and activation of SIRT1 promotes cell survival in chronic heart failure. J Biol Chem 285:8375–8382
212. Ghanim H, Sia CL, Korzeniewski K et al (2011) A resveratrol and polyphenol preparation suppresses oxidative and inflammatory stress response to a high-fat, high-carbohydrate meal. J Clin Endocrinol Metab 96:1409–1414
213. Magyar K, Halmosi R, Palfi A et al (2012) Cardioprotection by resveratrol: a human clinical trial in patients with stable coronary artery disease. Clin Hemorheol Microcirc 50:179–187
214. Frohlich ED, Gonzalez A, Diez J (2011) Hypertensive left ventricular hypertrophy risk: beyond adaptive cardiomyocytic hypertrophy. J Hypertens 29:17–26

Chapter 15
Angiogenesis, Arteriogenesis, and Mitochondrial Dysfunction

M.S. McMurtry

Abstract Angiogenesis and arteriogenesis are key processes involved in the response to occlusive arterial disease and the pathology of cancer. Angiogenesis remodels the circulatory bed by new capillary formation, while arteriogenesis remodels existing arterioles by increasing their diameter. The main molecular mechanism of angiogenesis is initiation by activation of hypoxia-inducible factor (HIF)-1, and arteriogenesis, initiated mainly by shear stress, is modulated by HIF-1 activation. Mitochondria can modulate HIF-1 activation by release of mitochondrial reactive oxygen species and alpha-ketoglutarate, a required cofactor for prolyl hydroxylation and destruction of HIF components, and thus, mitochondria can influence HIF-1-dependent angiogenesis and arteriogenesis. Mitochondria may serve as metabolic sensors that link metabolic derangements to appropriate neovascularization in health and in chronic ischemia and inappropriate neovascularization in the context of cancer. Mitochondrial remodeling or dysfunction may impair angiogenesis and contribute to the pathology of diseases, including diabetes and myocardial dysfunction. Drugs that alter mitochondrial function may alter HIF-1-dependent neovascularization and may represent novel therapies for chronic ischemic diseases and cancer.

Keywords Angiogenesis • Arteriogenesis • Neovascularization • Mitochondria

M.S. McMurtry, M.D., Ph.D., F.R.C.P.C. (✉)
Department of Medicine, University of Alberta, 8440 112th Street,
Edmonton, AB, Canada T6G 2B7
e-mail: mcmurtry@ualberta.ca

B.I. Jugdutt and N.S. Dhalla (eds.), *Cardiac Remodeling: Molecular Mechanisms*,
Advances in Biochemistry in Health and Disease 5, DOI 10.1007/978-1-4614-5930-9_15,

15.1 Background: The Treatment Gap for Chronic Ischemic Diseases

Treatments for chronic refractory ischemia are needed for large numbers of patients that have failed conventional therapy. Cardiovascular diseases, including coronary disease (CAD), cerebrovascular, and lower extremity peripheral arterial disease (PAD), are leading causes of mortality and morbidity worldwide and in Canada [1–3]. Despite advances in medical therapy and revascularization, 10% of coronary artery disease patients have chronic refractory angina [4, 5], and 20–30% of critical limb ischemia patients need amputation [6]. Amputation rates for critical limb ischemia reach 1 per thousand of the general population [7–9], and the annual risk of amputation for those with PAD is approximately 1% [10]. New methods to revascularize ischemic tissues represent an important clinical need.

15.2 Neovascularization: A Therapeutic Target

Targeting innate mechanisms of vascular remodeling and repair, including angiogenesis and arteriogenesis, is an attractive but elusive therapeutic strategy. Three processes of new blood vessel development in humans have been described, including angiogenesis, arteriogenesis, and vasculogenesis [11–13]. Angiogenesis is the formation of new capillaries, but not large conduit vessels [12, 14]. Angiogenesis is initiated by hypoxia-induced stabilization of hypoxia-inducible factor 1 alpha (HIF-1α), causing nuclear translocation of HIF-1, a heterodimer of HIF-1α and HIF-1β, with subsequent transcription of angiogenic growth factors such as vascular endothelial growth factor (VEGF)-A (Fig. 15.1) [15]. Arteriogenesis is the maturation of preexisting arterioles into larger conduits and is the way arterial collaterals form [16]. The primary stimulus for arteriogenesis is thought to be shear stress-induced endothelial cell activation, causing proliferation and migration of endothelial and smooth muscle cells and monocytes, resulting in outward remodeling of the arteriole [17]. Vasculogenesis is the de novo blood vessel formation by circulating progenitor cells [18, 19], but the importance of this process in adults is unknown [14]. Despite promising preclinical work, trials to improve angiogenesis and arteriogenesis in human coronary and peripheral arterial disease utilizing protein or gene formulations of angiogenic growth factors such as VEFG-A, fibroblast growth factor (FGF) 2, FGF-4, and granulocyte–macrophage colony-stimulating factor have fallen short of demonstrating significant efficacy [10, 20–23]. Antiangiogenic treatments for cancer are also in their infancy [24, 25]. Despite challenges, targeting neovascularization remains attractive, and fundamental studies of mechanisms and treatments beyond single-factor replacement are required.

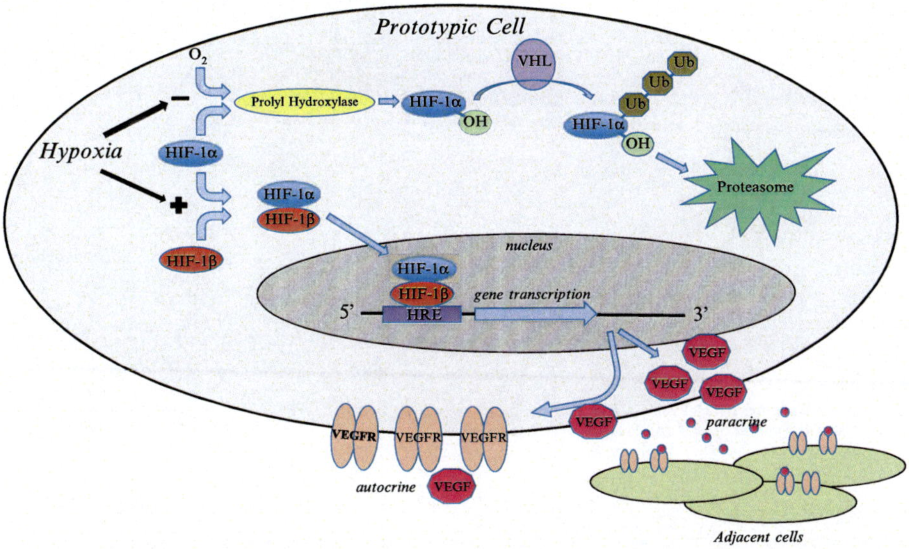

Fig. 15.1 The hypoxia-inducible factor (HIF)-1 system is activated by hypoxia, which promotes dimerization of HIF-1α and HIF-1β to form the active HIF-1. HIF-1 moves to the nucleus, activating transcription of proteins that induce angiogenesis, such as vascular endothelial growth factors (VEGFs)

15.3 Angiogenesis Is HIF-1 Dependent

Angiogenesis, or new capillary formation, is initiated by hypoxia-induced activation of HIF-1 [15]. Low local pO_2 inhibits the prolyl hydroxylation of HIF-1α, a reaction that requires both oxygen and α-ketoglutarate as substrates [26]. Decreased prolyl hydroxylation of HIF-1α prevents an interaction with the von Hippel–Lindau E3 ubiquitin ligase complex that targets HIF-1α for destruction [27–29]. Stabilized HIF-1α forms a heterodimer with HIF-1β to form HIF-1, which binds DNA and induces transcription of many angiogenic factors [30], including VEGFs [31], nitric oxide synthase [31], VEGF receptors (KDR/Flk and Flt) [32], angiopoietins [33], matrix metalloproteases [34], platelet-derived growth factor (PDGF) [35, 36], fibroblast growth factor (FGF) [36], and monocyte chemoattractant protein [37]. The VEGFs, including VEGF-A, -B, -C, and -D and placental growth factor (PlGF), bind to tyrosine kinase VEGF receptors including VEGF receptor 1 (VEGFR-1; Flt) and VEGFR-2 (KDR/Flk) on endothelial cells [38]. Bound ligands induce VEGF receptor dimerization and autophosphorylation, inducing signaling via the phosphatidylinositol 3-kinase/Akt pathway [39], the extracellular-regulated kinase pathway, and the mitogen-activated protein kinase pathway [40]. These pathways culminate in enhanced endothelial cell survival [39], endothelial cell proliferation [40], enhanced migration of endothelial cells, vascular smooth muscle cells

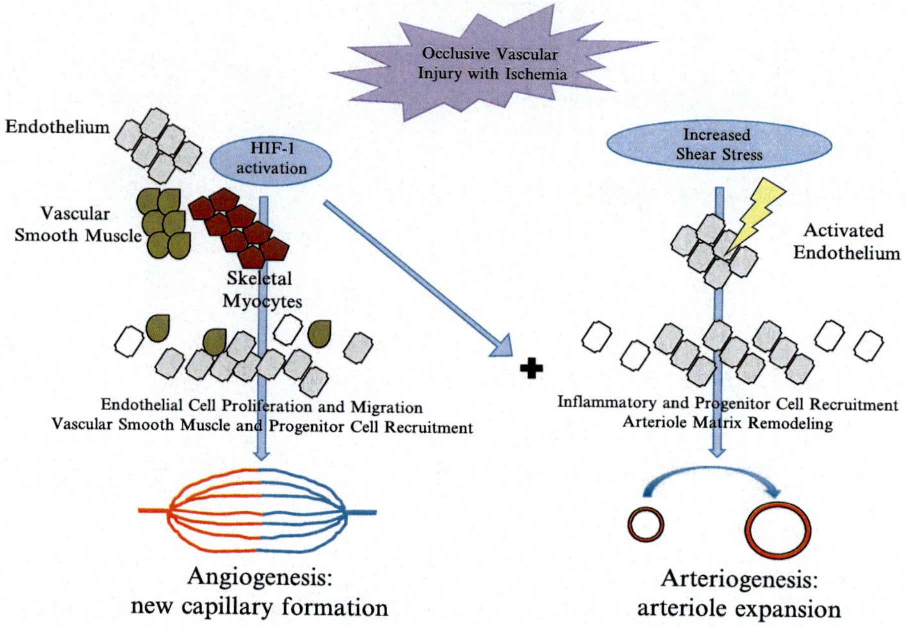

Fig. 15.2 Both angiogenesis, which is formation of new capillaries, and arteriogenesis, which is expansion of existing capillaries, are influenced by activation of HIF-1

and circulating inflammatory or progenitor cells [40], as well as enhanced angiogenesis in vivo (Fig. 15.2) [38, 41, 42]. Activation of HIF-1 can occur in endothelium [43], as well as vascular smooth muscle cells [44] and tissues, including cardiomyocytes [45] and skeletal myocytes [46, 47], fibroblasts [48], and bone marrow-derived cells [49]. HIF-1-mediated VEGF signaling and angiogenesis may be both autocrine [50, 51] and paracrine [51, 52], though the relative importance of autocrine versus paracrine signaling is incompletely understood [51]; Lee et al. suggest paracrine signaling predominates in developmental angiogenesis [52], but if this is true in pathologic models is not known. HIF-1 activation by hypoxia-induced inhibition of HIF-1α prolyl hydroxylation is the central trigger for angiogenesis.

15.4 Arteriogenesis Is Modulated by HIF-1 Activation

Arteriogenesis, or outward enlargement of preexisting arterioles, may also be partially HIF-1 dependent. Evidence suggests the primary stimulus for arteriogenesis is shear stress acting on endothelial cells [53–55], and not hypoxia [16]. Inflammatory markers and cells play a supporting role [56–62], as do growth factors such as basic fibroblast growth factor (FGF) and FGF receptors [63, 64], macrophage colony-stimulating factor [65], granulocyte–macrophage colony-stimulating factor

[66], and transforming growth factor beta [67]. At least three signaling mechanisms are activated by shear stress, including the RAS–ERK pathway, the Rho pathway, and nitric oxide [55]. Though HIF-1 is thought not to be a primary arteriogenic stimulus, variations in HIF-α genotype influence the development of human coronary collaterals in patients with coronary atherosclerosis [68]. HIF-1α expression in monocytes and lymphocytes in patients with coronary atherosclerosis also correlated with coronary collaterals in one study [69]. VEGF-A, a major effector of HIF-1, appears to enhance ischemic collateral formation in mice [70, 71] and rats [72], though the data in rabbits conflicts [73, 74]. In addition, genetically altered mice underexpressing or overexpressing VEGF-A demonstrate that native collateral density in the hind limb correlates with VEGF-A expression [75]. Blockade of the VEGF receptors decreases arteriogenesis [76], and VEGFR-2 is part of a complex that transduces endothelial cell shear stress, the key arteriogenic stimulus [77]. While experimental data supports that the major stimulus for arteriogenesis is shear stress [74], arteriogenesis also appears to be susceptible to modulation by changes in HIF-1 activation (Fig. 15.2).

15.5 Mitochondria Sense Hypoxia

Hypoxia reduces mitochondrial oxidative phosphorylation and changes Krebs cycle substrates and ROS production, allowing mitochondria to function as oxygen sensors. Mitochondria are the major ATP source in most cells [78]. Metabolites from glucose and fatty acids enter the Krebs cycle reducing NAD to NADH and FAD to $FADH_2$. As these reoxidize, they supply electrons to the electron transport chain (complexes I–IV). These electrons travel through the enzyme complexes to O_2, producing H_2O. During this process, protons translocate out of the inner mitochondrial membrane, generating a potential of −150 to −200 mV ($\Delta\Psi_m$) that reflects the level of the respiratory activity in the mitochondria [78]. ATP synthesis takes place at the ATP synthase (complex V), which contains a proton channel (F_0) and an ATP synthase (F_1). This enzyme is driven by the inward movement of protons through F_0 and phosphorylates ADP, producing ATP. It uses the stored energy of $\Delta\Psi_m$ to synthesize ATP, coupling respiration to ATP synthesis. Decreased $\Delta\Psi_m$, due to changes in substrate flux or the presence of uncoupling agents such as carbonyl cyanide-4(trifluoromethoxy)phenylhydrazone (FCCP), causes decreased ATP synthase activity [78]. Decreased electron transport chain complex activity would disrupt the normal electron flow and proton pumping, decreasing $\Delta\Psi_m$. Decreased $\Delta\Psi_m$ can also directly result from protonophores, such as FCCP or uncoupling proteins, generating heat [79]. As part of this process, ROS such as superoxide (O_2^-) are formed, which are rapidly dismutated by superoxide dismutase enzymes to hydrogen peroxide (H_2O_2). Vascular cells intricately balance energy supply and demand [80] and, with pO_2 reductions, shift from oxidative phosphorylation to glycolysis [81], resulting in changes of $\Delta\Psi_m$, ROS, and levels of Krebs cycle intermediates, such as α-ketoglutarate and succinate. In addition to other metabolic sensors like

AMP-activated protein kinase [82], mitochondria behave as sensors, with substrate flux, mitochondrial membrane potential, mitochondrial ROS, and Krebs cycle substrate levels shifting according surrounding metabolic state and pO_2 [78, 83, 84]. Endothelial and vascular smooth muscle cells are predominantly glycolytic, consistent with mitochondria being oxygen sensors [80, 81, 85]. Endothelial diversity [86, 87] is also consistent with EC mitochondria being fundamental in oxygen sensing, as different tissues respond to local conditions in different ways [81, 83]. Mitochondria are important sensors of intracellular hypoxia, the fundamental stimulus for HIF-1 activation [88].

15.6 Mitochondria Modulate HIF-1 Activation and Neovascularization

Changes in mitochondrial Krebs cycle substrates and ROS enhance HIF-1 activation [89], even in normoxia, linking mitochondria to HIF-1 and angiogenesis. Mitochondria interact with the HIF pathway in at least three ways. Mitochondria may act as a sink for oxygen [90] and in hypoxia may keep oxygen from prolyl hydroxylases. Mitochondrial ROS, frequently reported as increased by hypoxia [91–93], may also directly stabilize HIF-1α and therefore enhance activation of HIF-1 [92–97], as well as reinforce the glycolytic phenotype by enhancing PDK [98]. It should be noted, however, that controversy exists whether mitochondrial ROS increase [99] or decrease [100] in hypoxia, and discrepant findings might relate to differences between the systemic and pulmonary vasculature [83]. Furthermore, nitric oxide (NO) also modulates HIF-1, with prolonged NO exposure, decreasing HIF-1 activation by up to three different mechanisms, including direct inhibition of mitochondria by NO [101]. In addition, Krebs cycle substrates can directly interact with prolyl hydroxylation. The substrate α-ketoglutarate is required for prolyl hydroxylation of HIF-1α [26], and other Krebs cycle substrates, including succinate and fumarate, can directly inhibit prolyl hydroxylation [102], an effect that can be reversed with exogenous α-ketoglutarate [103, 104]. Mutations of succinate dehydrogenase lead to paragangliomas and pheochromocytomas, and mutations in fumarate hydratase lead to leiomyomas and renal cell carcinoma, all vascular tumors [105, 106]. It is debated in the literature whether mitochondrial ROS or Krebs cycle substrates are more important for HIF-1 activation, but both appear to play a role [89, 107–109]. These changes can occur when the metabolic balance shifts from oxidative phosphorylation to glycolysis, leading to activation of HIF-1 [110], even in normoxia (pseudohypoxia) [111]. Exogenous ROS can cause HIF-1 activation in normoxia [95, 97], and blocking prolyl hydroxylases, even in normoxia, activates HIF-1 [112]. Dysfunction of the Krebs cycle enzymes fumarate hydratase and succinate dehydrogenase causes accumulation of fumarate and succinate and activates HIF-1 in normoxia by inhibition of prolyl hydroxylases [106]. This normoxic HIF-1 activation can be reduced by replacement of α-ketoglutarate [103]. Since both mitochondrial ROS, in the form of hydrogen peroxide [81], and substrates such as α-ketoglutarate [103] are diffusible and can be modulated

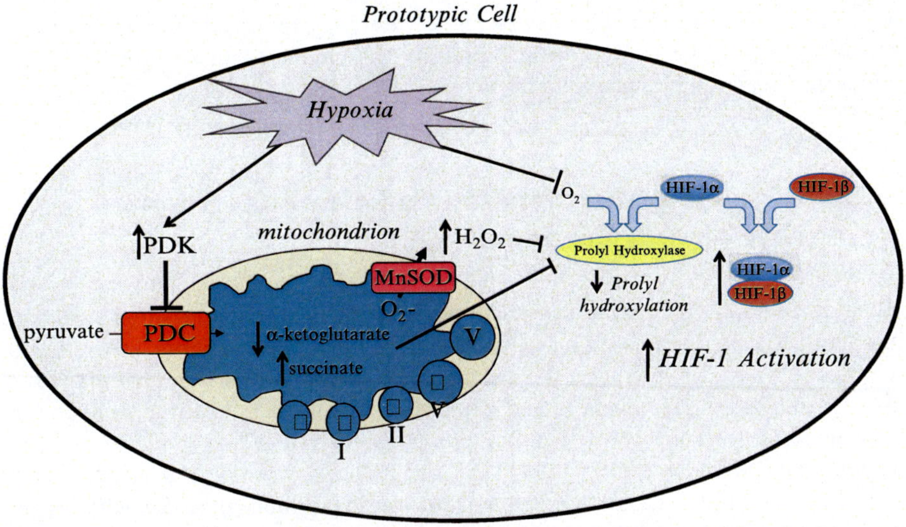

Fig. 15.3 Mitochondria respond to hypoxia by changes in release of the mitochondrial reactive oxygen species H_2O_2 and the prolyl hydroxylase cofactor α-ketoglutarate. These changes result in activation of HIF-1, supporting HIF-1-dependent neovascularization

independent of hypoxia, they represent possible mechanisms for both autocrine and paracrine activation of angiogenesis proximal to HIF-1 and VEGF, different from the usual mechanism of reduced pO_2 (Fig. 15.3).

15.7 Mitochondrial Dysfunction: An Effector of Impaired Neovascularization in Disease

The ability of mitochondria to modulate HIF-1-dependent processes, including neovascularization, has implications for the mechanisms of human diseases [89]. For example, an inhibitor of mitochondrial redox signaling, thioredoxin-interacting protein, is induced by glucose and may contribute to disordered angiogenesis in diabetes mellitus [113]. This is supported by evidence from thioredoxin-2 transgenic mice, which demonstrated enhanced arteriogenesis and angiogenesis in a murine hind limb ischemia model [114]. The glucose-derived oxoaldehyde glyoxal, present in hyperglycemia in the context of diabetes, also has been associated with mitochondrial dysfunction and impaired angiogenesis [115]. In aged rats, dysfunctional mitochondria, with lower expression of superoxide dismutase and other mitochondrial proteins, have been associated with impaired VEGF expression and angiogenesis in the rat kidney and may be a mechanism for exacerbation of renal injury [116]. Thyroid hormone (3,5,3′-Levo-triiodothyronine) replacement post-myocardial infarction has been demonstrated to improve left ventricular function in rats by improving both mitochondrial function and angiogenesis [117]. Conversely, well-functioning mitochondria can promote pathology in the context of cancer by enhancing angiogenesis [118]. Mitochondrial dysfunction as an effector of pathology is just beginning to be explored.

15.8 Modulating Mitochondrial Function with Drugsto Alter Neovascularization

Hypoxia reinforces reductions in mitochondrial oxidative phosphorylation by inducing pyruvate dehydrogenase kinase. Oxidative phosphorylation is not only reduced by direct effects of hypoxia on the mitochondrial electron transport chain [84] but also by enhanced glycolysis at the expense of decreased substrate delivery to mitochondria. In many cell types, hypoxia activates HIF-1 and induces pyruvate dehydrogenase kinase (PDK) [119–121] and reduces expression of Krebs cycle enzymes [122]. PDK is a key gatekeeping enzyme that inhibits pyruvate dehydrogenase complex (PDH) [123], the major pathway for entry of pyruvate into mitochondria [124]. PDH is a multienzyme complex [125] that is redox sensitive [126] and inhibited by isoforms of PDK by phosphorylation of the E1a subunit [123]. As PDK is induced by hypoxia, there is less flux of substrate through into the mitochondria to the Krebs cycle and therefore lower oxidative phosphorylation [127]. There are four isoforms of PDK, and the expression is variable between organs [128]. Expression of PDK in endothelial cells is understudied, but it is known that PDK expression is increased in a HIF-1-dependent manner in ischemic skeletal muscle [129]. PDK isoforms are additionally regulated by nuclear receptors such as peroxisome proliferator-activated receptor alpha (PPARα) and gamma (PPARγ), liver X receptor, PPAR gamma coactivator alpha (PGC-1α), and insulin [130, 131]. Drugs are available that inhibit PDK (dichloroacetate) [132], indirectly increase PDK expression (WY-14,564; a PPARα agonist) [133], decrease expression of PDK (GW7845 [131], GW1929 [134]), or inhibit PDH (bromopyruvate and fluoropyruvate) [135]. In addition to hypoxia, other metabolic factors, including but not limited to fatty acids, can reduce oxidative phosphorylation, an effect previously described as the Randle cycle [136]. For example, the antianginal medication trimetazidine increases oxidative phosphorylation supposedly by inhibition of long-chain 3-ketoacyl-CoA thiolase, a mitochondrial enzyme [137]. Both cancer cells (the Warburg effect) and vascular cells are predominantly glycolytic, an observation that is incompletely explained [80, 81, 85]. Since the hypoxia-induced glycolytic shift mediated by increased PDK is associated with proliferation and apoptosis resistance in cancer, this phenotype might promote endothelial proliferation in angiogenesis [138, 139]. Drugs that activate PDK or inhibit PDH may promote angiogenesis and could be used as therapy for chronic ischemic diseases. Drugs that inhibit PDK or activate PDH may be useful as antiangiogenic therapies for cancer (Fig. 15.4). The concept that mitochondrial function can be pharmacologically manipulated is supported by reports of dichloroacetate suppressing cancer-associated angiogenesis as part of its therapeutic anticancer effect [140, 141].

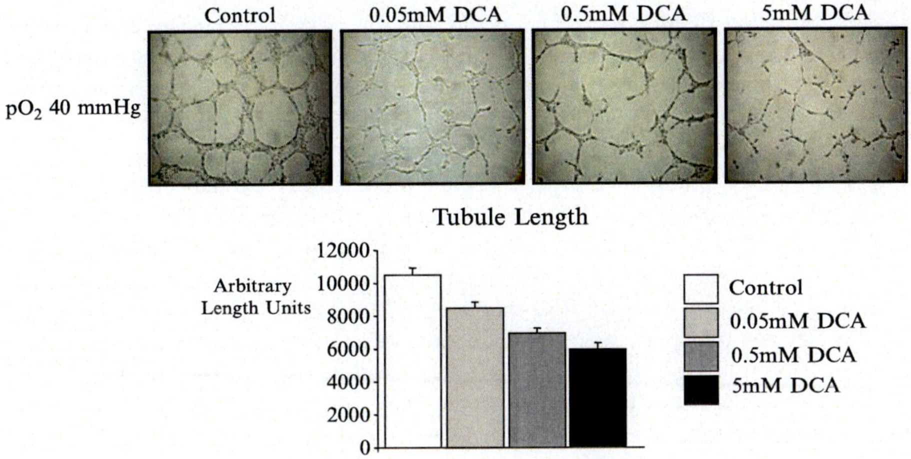

Fig. 15.4 Dichloroacetate (DCA), which inhibits pyruvate dehydrogenase kinase and promotes glucose oxidation, decreases angiogenesis in vitro in a dose-dependent manner. *Reproduced with permission from Michelakis et al.* [140]

15.9 Thiazolidinediones, Mitochondria, and Impaired Angiogenesis

Drugs that affect mitochondria can modulate HIF-1 activation and angiogenesis and may represent new avenues for therapies for diseases in which angiogenesis plays an important role. In contrast, there is a growing, yet somewhat conflicting, literature evaluating the effects of angiogenesis of agonists of peroxisome proliferator-activated receptors (PPARs) and angiogenesis [142–145]. In vitro PPARγ ligands inhibit angiogenesis on matrigel and on rat corneas [146], as well as human choroidal endothelial cells in vitro and in vivo [147]. Rosiglitazone reduces angiogenesis in human umbilical vein endothelial cells [148] and the human uterus [149], and both rosiglitazone and pioglitazone inhibit angiogenesis in the chick chorioallantoic membrane model [150]. While generally the literature suggests that PPARγ ligands, including thiazolidinediones (TZDs), reduce angiogenesis, rosiglitazone [151], troglitazone [152], and pioglitazone [153] have all been associated with increased angiogenesis in various reports. Clinical reports of TZD use also suggest that angiogenesis in vivo may be inhibited. Rosiglitazone has been associated with delayed onset of proliferative diabetic retinopathy [154], and the PROactive trial of pioglitazone in type 2 diabetics demonstrated an association between pioglitazone and increased rates of lower extremity revascularization [155]. Rosiglitazone and pioglitazone are associated with increased rates of heart failure and bone fractures, both of which could be plausibly related to impaired angiogenesis in myocardium and bone, respectively [156, 157]. There is interest in developing PPARγ agonists as antiangiogenic therapy for cancer [158]. Effects of TZDs may be mediated by both

PPARγ and non-PPARγ mechanisms. PPARγ appears to decrease PDK, while PPARα or PPARδ increases PDK [131]. PPARγ agonism decreases PDK expression in skeletal muscle [123, 131, 134] and other insulin-sensitive tissues, though one report demonstrated that rosiglitazone increased PDK-4 in fat [159]. TZD-mediated decreases in PDK may increase mitochondrial oxidative phosphorylation and via increases in α-ketoglutarate or decreases in mitochondrial ROS may decrease HIF-1 activation and angiogenesis. One experiment showed that GW1929, a selective PPARγ agonist, was associated with increased in vitro and in vivo angiogenesis, while rosiglitazone was not [160], suggesting that rosiglitazone's reduction on angiogenesis may be a non-PPARγ-mediated effect. TZDs directly inhibit complex I of the electron transport chain [161], and inhibition of complex I of the mitochondrial electron transport chain blocks HIF-1 activation [162]; therefore, TZDs may inhibit angiogenesis also by direct effects on the electron transport chain. Since the prevalence of peripheral arterial disease is as high as 1 in five diabetics [163, 164], TZDs are used in thousands of patients [165], and TZDs might increase in limb ischemia in human diabetics [155]; the association of TZDs and impaired angiogenesis strongly merits further study.

15.10 Perspective

Mitochondrial remodeling and changes in mitochondrial function, caused by hypoxia or a glycolytic shift in metabolism, enhance HIF-1 activation in via changes in α-ketoglutarate and mitochondrial ROS in endothelial cells and vascular smooth muscle cells, augmenting HIF-1-dependent autocrine and paracrine molecular pathways that result in increased angiogenesis. Though arteriogenesis is thought to be shear stress dependent, mitochondrial-dependent HIF-1 activation may improve arteriogenesis as well, both directly as well as indirectly by enhancing shear stress. Dysfunctional mitochondria that produce less α-ketoglutarate and mitochondrial ROS may be an effector for impaired neovascularization and vascular remodeling in disease. Mitochondria represent therapeutic targets to modulate neovascularization, either to enhance favorable vascular remodeling in chronic ischemic diseases or to decrease it in cancer.

References

1. Statistics-Canada (2008) http://www40.statcan.ca/l01/cst01/health30a.htm
2. Lopez AD, Mathers CD, Ezzati M, Jamison DT, Murray CJ (2006) Global and regional burden of disease and risk factors, 2001: systematic analysis of population health data. Lancet 367:1747–1757
3. Lloyd-Jones D, Adams R, Carnethon M, De Simone G, Ferguson TB, Flegal K et al (2009) Heart disease and stroke statistics–2009 update: a report from the American Heart Association Statistics Committee and Stroke Statistics Subcommittee. Circulation 119:480–486

4. Mukherjee D, Bhatt DL, Roe MT, Patel V, Ellis SG (1999) Direct myocardial revascularization and angiogenesis – how many patients might be eligible? Am J Cardiol 84:598–600, A598
5. Mannheimer C, Camici P, Chester MR, Collins A, DeJongste M, Eliasson T et al (2002) The problem of chronic refractory angina; report from the ESC Joint Study Group on the Treatment of Refractory Angina. Eur Heart J 23:355–370
6. Palmer-Kazen U, Wahlberg E (2003) Arteriogenesis in peripheral arterial disease. Endothelium 10:225–232
7. Carmona GA, Hoffmeyer P, Herrmann FR, Vaucher J, Tschopp O, Lacraz A et al (2005) Major lower limb amputations in the elderly observed over ten years: the role of diabetes and peripheral arterial disease. Diabetes Metab 31:449–454
8. Heller G, Gunster C, Schellschmidt H (2004) How frequent are diabetes-related amputations of the lower limbs in Germany? An analysis on the basis of routine data. Dtsch Med Wochenschr 129:429–433
9. Senior PA, McMurtry MS, Tsuyuki RT (2007) Diabetes and lower limb amputations in Alberta. In: Alberta diabetes atlas 2007. Institute of Health Economics, Alberta, pp 73–81
10. Rissanen TT, Vajanto I, Yla-Herttuala S (2001) Gene therapy for therapeutic angiogenesis in critically ischaemic lower limb - on the way to the clinic. Eur J Clin Invest 31:651–666
11. Risau W (1997) Mechanisms of angiogenesis. Nature 386:671–674
12. Carmeliet P (2003) Angiogenesis in health and disease. Nat Med 9:653–660
13. Schaper W, Ito WD (1996) Molecular mechanisms of coronary collateral vessel growth. Circ Res 79:911–919
14. Simons M (2005) Angiogenesis: where do we stand now? Circulation 111:1556–1566
15. Pugh CW, Ratcliffe PJ (2003) Regulation of angiogenesis by hypoxia: role of the HIF system. Nat Med 9:677–684
16. Schaper W (2009) Collateral circulation: past and present. Basic Res Cardiol 104:5–21
17. Heil M, Schaper W (2004) Influence of mechanical, cellular, and molecular factors on collateral artery growth (arteriogenesis). Circ Res 95:449–458
18. Asahara T, Masuda H, Takahashi T, Kalka C, Pastore C, Silver M et al (1999) Bone marrow origin of endothelial progenitor cells responsible for postnatal vasculogenesis in physiological and pathological neovascularization. Circ Res 85:221–228
19. Luttun A, Carmeliet P (2003) De novo vasculogenesis in the heart. Cardiovasc Res 58:378–389
20. Baklanov D, Simons M (2003) Arteriogenesis: lessons learned from clinical trials. Endothelium 10:217–223
21. Lekas M, Lekas P, Latter DA, Kutryk MB, Stewart DJ (2006) Growth factor-induced therapeutic neovascularization for ischaemic vascular disease: time for a re-evaluation? Curr Opin Cardiol 21:376–384
22. Schirmer SH, van Nooijen FC, Piek JJ, van Royen N (2009) Stimulation of collateral artery growth: travelling further down the road to clinical application. Heart 95:191–197
23. Schirmer SH, van Royen N (2004) Stimulation of collateral artery growth: a potential treatment for peripheral artery disease. Expert Rev Cardiovasc Ther 2:581–588
24. Grimm D, Bauer J, Schoenberger J (2009) Blockade of neoangiogenesis, a new and promising technique to control the growth of malignant tumors and their metastases. Curr Vasc Pharmacol 7:347–357
25. Carmeliet P, De Smet F, Loges S, Mazzone M (2009) Branching morphogenesis and antiangiogenesis candidates: tip cells lead the way. Nat Rev Clin Oncol 6:315–326
26. Semenza GL (2007) Hypoxia-inducible factor 1 (HIF-1) pathway. Sci STKE 2007:cm8
27. Ivan M, Kondo K, Yang H, Kim W, Valiando J, Ohh M et al (2001) HIF alpha targeted for VHL- mediated destruction by proline hydroxylation: implications for O2 sensing. Science 292:464–468
28. Semenza GL (2004) Hydroxylation of HIF-1: oxygen sensing at the molecular level. Physiology (Bethesda) 19:176–182
29. Wang GL, Jiang BH, Rue EA, Semenza GL (1995) Hypoxia-inducible factor 1 is a basic-helix-loop- helix-PAS heterodimer regulated by cellular O2 tension. Proc Natl Acad Sci USA 92:5510–5514

30. Yamakawa M, Liu LX, Date T, Belanger AJ, Vincent KA, Akita GY et al (2003) Hypoxia-inducible factor-1 mediates activation of cultured vascular endothelial cells by inducing multiple angiogenic factors. Circ Res 93:664–673
31. Liu Y, Cox SR, Morita T, Kourembanas S (1995) Hypoxia regulates vascular endothelial growth factor gene expression in endothelial cells. Identification of a 5' enhancer. Circ Res 77:638–643
32. Tuder RM, Flook BE, Voelkel NF (1995) Increased gene expression for VEGF and the VEGF receptors KDR/Flk and Flt in lungs exposed to acute or to chronic hypoxia. Modulation of gene expression by nitric oxide. J Clin Invest 95:1798–1807
33. Currie MJ, Gunningham SP, Turner K, Han C, Scott PA, Robinson BA et al (2002) Expression of the angiopoietins and their receptor Tie2 in human renal clear cell carcinomas; regulation by the von Hippel-Lindau gene and hypoxia. J Pathol 198:502–510
34. Ben-Yosef Y, Lahat N, Shapiro S, Bitterman H, Miller A (2002) Regulation of endothelial matrix metalloproteinase-2 by hypoxia/reoxygenation. Circ Res 90:784–791
35. Gleadle JM, Ebert BL, Firth JD, Ratcliffe PJ (1995) Regulation of angiogenic growth factor expression by hypoxia, transition metals, and chelating agents. Am J Physiol 268:C1362–C1368
36. Kuwabara K, Ogawa S, Matsumoto M, Koga S, Clauss M, Pinsky DJ et al (1995) Hypoxia-mediated induction of acidic/basic fibroblast growth factor and platelet-derived growth factor in mononuclear phagocytes stimulates growth of hypoxic endothelial cells. Proc Natl Acad Sci USA 92:4606–4610
37. Negus RP, Turner L, Burke F, Balkwill FR (1998) Hypoxia down-regulates MCP-1 expression: implications for macrophage distribution in tumors. J Leukoc Biol 63:758–765
38. Ferrara N, Gerber HP, LeCouter J (2003) The biology of VEGF and its receptors. Nat Med 9:669–676
39. Gerber HP, McMurtrey A, Kowalski J, Yan M, Keyt BA, Dixit V et al (1998) Vascular endothelial growth factor regulates endothelial cell survival through the phosphatidylinositol 3'- kinase/Akt signal transduction pathway. Requirement for Flk-1/KDR activation. J Biol Chem 273:30336–30343
40. Byrne AM, Bouchier-Hayes DJ, Harmey JH (2005) Angiogenic and cell survival functions of vascular endothelial growth factor (VEGF). J Cell Mol Med 9:777–794
41. Leung DW, Cachianes G, Kuang WJ, Goeddel DV, Ferrara N (1989) Vascular endothelial growth factor is a secreted angiogenic mitogen. Science 246:1306–1309
42. Plouet J, Schilling J, Gospodarowicz D (1989) Isolation and characterization of a newly identified endothelial cell mitogen produced by AtT-20 cells. EMBO J 8:3801–3806
43. Tang N, Wang L, Esko J, Giordano FJ, Huang Y, Gerber HP et al (2004) Loss of HIF-1 alpha in endothelial cells disrupts a hypoxia-driven VEGF autocrine loop necessary for tumorigenesis. Cancer Cell 6:485–495
44. Jung YD, Liu W, Reinmuth N, Ahmad SA, Fan F, Gallick GE et al (2001) Vascular endothelial growth factor is upregulated by interleukin-1 beta in human vascular smooth muscle cells via the P38 mitogen-activated protein kinase pathway. Angiogenesis 4:155–162
45. Giordano FJ, Gerber HP, Williams SP, VanBruggen N, Bunting S, Ruiz-Lozano P et al (2001) A cardiac myocyte vascular endothelial growth factor paracrine pathway is required to maintain cardiac function. Proc Natl Acad Sci USA 98:5780–5785
46. van Weel V, Seghers L, de Vries MR, Kuiper EJ, Schlingemann RO, Bajema IM et al (2007) Expression of vascular endothelial growth factor, stromal cell-derived factor-1, and CXCR4 in human limb muscle with acute and chronic ischemia. Arterioscler Thromb Vasc Biol 27:1426–1432
47. Tuomisto TT, Rissanen TT, Vajanto I, Korkeela A, Rutanen J, Yla-Herttuala S (2004) HIF-VEGF- VEGFR-2, TNF-alpha and IGF pathways are upregulated in critical human skeletal muscle ischemia as studied with DNA array. Atherosclerosis 174:111–120
48. Noma K, Smalley KS, Lioni M, Naomoto Y, Tanaka N, El-Deiry W et al (2008) The essential role of fibroblasts in esophageal squamous cell carcinoma-induced angiogenesis. Gastroenterology 134:1981–1993

49. O'Neill TJ, Wamhoff BR, Owens GK, Skalak TC (2005) Mobilization of bone marrow-derived cells enhances the angiogenic response to hypoxia without transdifferentiation into endothelial cells. Circ Res 97:1027–1035
50. Aitkenhead M, Christ B, Eichmann A, Feucht M, Wilson DJ, Wilting J (1998) Paracrine and autocrine regulation of vascular endothelial growth factor during tissue differentiation in the quail. Dev Dyn 212:1–13
51. Fong GH (2009) Regulation of angiogenesis by oxygen sensing mechanisms. J Mol Med 87:549–560
52. Lee S, Chen TT, Barber CL, Jordan MC, Murdock J, Desai S et al (2007) Autocrine VEGF signaling is required for vascular homeostasis. Cell 130:691–703
53. Topper JN, Gimbrone MA Jr (1999) Blood flow and vascular gene expression: fluid shear stress as a modulator of endothelial phenotype. Mol Med Today 5:40–46
54. Pipp F, Boehm S, Cai WJ, Adili F, Ziegler B, Karanovic G et al (2004) Elevated fluid shear stress enhances postocclusive collateral artery growth and gene expression in the pig hind limb. Arterioscler Thromb Vasc Biol 24:1664–1668
55. Eitenmuller I, Volger O, Kluge A, Troidl K, Barancik M, Cai WJ et al (2006) The range of adaptation by collateral vessels after femoral artery occlusion. Circ Res 99:656–662
56. Bergmann CE, Hoefer IE, Meder B, Roth H, van Royen N, Breit SM et al (2006) Arteriogenesis depends on circulating monocytes and macrophage accumulation and is severely depressed in op/op mice. J Leukoc Biol 80:59–65
57. Ito WD, Arras M, Winkler B, Scholz D, Schaper J, Schaper W (1997) Monocyte chemotactic protein-1 increases collateral and peripheral conductance after femoral artery occlusion. Circ Res 80:829–837
58. Hoefer IE, Grundmann S, van Royen N, Voskuil M, Schirmer SH, Ulusans S et al (2005) Leukocyte subpopulations and arteriogenesis: specific role of monocytes, lymphocytes and granulocytes. Atherosclerosis 181:285–293
59. Stabile E, Burnett MS, Watkins C, Kinnaird T, Bachis A, la Sala A et al (2003) Impaired arteriogenic response to acute hindlimb ischemia in CD4-knockout mice. Circulation 108:205–210
60. Scholz D, Ito W, Fleming I, Deindl E, Sauer A, Wiesnet M et al (2000) Ultrastructure and molecular histology of rabbit hind-limb collateral artery growth (arteriogenesis). Virchows Arch 436:257–270
61. Hoefer IE, van Royen N, Rectenwald JE, Bray EJ, Abouhamze Z, Moldawer LL et al (2002) Direct evidence for tumor necrosis factor-alpha signaling in arteriogenesis. Circulation 105:1639–1641
62. Grundmann S, Hoefer I, Ulusans S, van Royen N, Schirmer SH, Ozaki CK et al (2005) Anti-tumor necrosis factor-{alpha} therapies attenuate adaptive arteriogenesis in the rabbit. Am J Physiol Heart Circ Physiol 289:H1497–H1505
63. Deindl E, Hoefer IE, Fernandez B, Barancik M, Heil M, Strniskova M et al (2003) Involvement of the fibroblast growth factor system in adaptive and chemokine-induced arteriogenesis. Circ Res 92:561–568
64. Werner GS, Jandt E, Krack A, Schwarz G, Mutschke O, Kuethe F et al (2004) Growth factors in the collateral circulation of chronic total coronary occlusions: relation to duration of occlusion and collateral function. Circulation 110:1940–1945
65. Okazaki T, Ebihara S, Takahashi H, Asada M, Kanda A, Sasaki H (2005) Macrophage colony- stimulating factor induces vascular endothelial growth factor production in skeletal muscle and promotes tumor angiogenesis. J Immunol 174:7531–7538
66. Schneeloch E, Mies G, Busch HJ, Buschmann IR, Hossmann KA (2004) Granulocyte-macrophage colony-stimulating factor-induced arteriogenesis reduces energy failure in hemodynamic stroke. Proc Natl Acad Sci USA 101:12730–12735
67. Wiseman DM, Polverini PJ, Kamp DW, Leibovich SJ (1988) Transforming growth factor-beta (TGF beta) is chemotactic for human monocytes and induces their expression of angiogenic activity. Biochem Biophys Res Commun 157:793–800

68. Resar JR, Roguin A, Voner J, Nasir K, Hennebry TA, Miller JM et al (2005) Hypoxia-inducible factor 1alpha polymorphism and coronary collaterals in patients with ischemic heart disease. Chest 128:787–791
69. Chen SM, Li YG, Zhang HX, Zhang GH, Long JR, Tan CJ et al (2008) Hypoxia-inducible factor-1alpha induces the coronary collaterals for coronary artery disease. Coron Artery Dis 19:173–179
70. Couffinhal T, Silver M, Kearney M, Sullivan A, Witzenbichler B, Magner M et al (1999) Impaired collateral vessel development associated with reduced expression of vascular endothelial growth factor in ApoE–/– mice. Circulation 99:3188–3198
71. Jacobi J, Tam BY, Wu G, Hoffman J, Cooke JP, Kuo CJ (2004) Adenoviral gene transfer with soluble vascular endothelial growth factor receptors impairs angiogenesis and perfusion in a murine model of hindlimb ischemia. Circulation 110:2424–2429
72. Toyota E, Warltier DC, Brock T, Ritman E, Kolz C, O'Malley P et al (2005) Vascular endothelial growth factor is required for coronary collateral growth in the rat. Circulation 112:2108–2113
73. Deindl E, Buschmann I, Hoefer IE, Podzuweit T, Boengler K, Vogel S et al (2001) Role of ischemia and of hypoxia-inducible genes in arteriogenesis after femoral artery occlusion in the rabbit. Circ Res 89:779–786
74. Schierling W, Troidl K, Troidl C, Schmitz-Rixen T, Schaper W, Eitenmuller IK (2009) The role of angiogenic growth factors in arteriogenesis. J Vasc Res 46:365–374
75. Clayton JA, Chalothorn D, Faber JE (2008) Vascular endothelial growth factor-A specifies formation of native collaterals and regulates collateral growth in ischemia. Circ Res 103:1027–1036
76. Lloyd PG, Prior BM, Li H, Yang HT, Terjung RL (2005) VEGF receptor antagonism blocks arteriogenesis, but only partially inhibits angiogenesis, in skeletal muscle of exercise- trained rats. Am J Physiol Heart Circ Physiol 288:H759–H768
77. Tzima E, Irani-Tehrani M, Kiosses WB, Dejana E, Schultz DA, Engelhardt B et al (2005) A mechanosensory complex that mediates the endothelial cell response to fluid shear stress. Nature 437:426–431
78. Duchen MR (2004) Mitochondria in health and disease: perspectives on a new mitochondrial biology. Mol Aspects Med 25:365–451
79. Michelakis ED (2008) Mitochondrial medicine: a new era in medicine opens new windows and brings new challenges. Circulation 117:2431–2434
80. Barron JT, Parrillo JE (1995) Production of lactic acid and energy metabolism in vascular smooth muscle: effect of dichloroacetate. Am J Physiol 268:H713–H719
81. Michelakis ED, Weir EK (2008) The metabolic basis of vascular oxygen sensing: diversity, compartmentalization and lessons from cancer. Am J Physiol Heart Circ Physiol 295(3):H928–H930
82. Fisslthaler B, Fleming I (2009) Activation and signaling by the AMP-activated protein kinase in endothelial cells. Circ Res 105:114–127
83. Michelakis ED, Hampl V, Nsair A, Wu X, Harry G, Haromy A et al (2002) Diversity in mitochondrial function explains differences in vascular oxygen sensing. Circ Res 90:1307–1315
84. Goldenthal MJ, Marin-Garcia J (2004) Mitochondrial signaling pathways: a receiver/integrator organelle. Mol Cell Biochem 262:1–16
85. Davidson SM, Duchen MR (2007) Endothelial mitochondria: contributing to vascular function and disease. Circ Res 100:1128–1141
86. Aird WC (2007) Phenotypic heterogeneity of the endothelium: II. Representative vascular beds. Circ Res 100:174–190
87. Aird WC (2007) Phenotypic heterogeneity of the endothelium: I. Structure, function, and mechanisms. Circ Res 100:158–173
88. Bell EL, Emerling BM, Chandel NS (2005) Mitochondrial regulation of oxygen sensing. Mitochondrion 5:322–332
89. Chavez A, Miranda LF, Pichiule P, Chavez JC (2008) Mitochondria and hypoxia-induced gene expression mediated by hypoxia-inducible factors. Ann N Y Acad Sci 1147:312–320 [Review]

90. Hagen T, Taylor CT, Lam F, Moncada S (2003) Redistribution of intracellular oxygen in hypoxia by nitric oxide: effect on HIF1 alpha. Science 302:1975–1978
91. Guzy RD, Schumacker PT (2006) Oxygen sensing by mitochondria at complex III: the paradox of increased reactive oxygen species during hypoxia. Exp Physiol 91:807–819
92. Brunelle JK, Bell EL, Quesada NM, Vercauteren K, Tiranti V, Zeviani M et al (2005) Oxygen sensing requires mitochondrial ROS but not oxidative phosphorylation. Cell Metab 1:409–414
93. Simon MC (2006) Mitochondrial reactive oxygen species are required for hypoxic HIF alpha stabilization. Adv Exp Med Biol 588:165–170
94. Chandel NS, Maltepe E, Goldwasser E, Mathieu CE, Simon MC, Schumacker PT (1998) Mitochondrial reactive oxygen species trigger hypoxia-induced transcription. Proc Natl Acad Sci USA 95:11715–11720
95. Chandel NS, McClintock DS, Feliciano CE, Wood TM, Melendez JA, Rodriguez AM et al (2000) Reactive oxygen species generated at mitochondrial complex III stabilize hypoxia-inducible factor-1alpha during hypoxia: a mechanism of O2 sensing. J Biol Chem 275:25130–25138
96. Schroedl C, McClintock DS, Budinger GR, Chandel NS (2002) Hypoxic but not anoxic stabilization of HIF-1 alpha requires mitochondrial reactive oxygen species. Am J Physiol Lung Cell Mol Physiol 283:L922–L931
97. Mansfield KD, Guzy RD, Pan Y, Young RM, Cash TP, Schumacker PT et al (2005) Mitochondrial dysfunction resulting from loss of cytochrome c impairs cellular oxygen sensing and hypoxic HIF-alpha activation. Cell Metab 1:393–399
98. Sun W, Zhou S, Chang SS, McFate T, Verma A, Califano JA (2009) Mitochondrial mutations contribute to HIF1alpha accumulation via increased reactive oxygen species and up-regulated pyruvate dehydrogenease kinase 2 in head and neck squamous cell carcinoma. Clin Cancer Res 15:476–484
99. Guzy RD, Hoyos B, Robin E, Chen H, Liu L, Mansfield KD et al (2005) Mitochondrial complex III is required for hypoxia-induced ROS production and cellular oxygen sensing. Cell Metab 1:401–408
100. Archer SL, Gomberg-Maitland M, Maitland ML, Rich S, Garcia JG, Weir EK (2008) Mitochondrial metabolism, redox signaling, and fusion: a mitochondria-ROS-HIF-1alpha-Kv1.5 O2-sensing pathway at the intersection of pulmonary hypertension and cancer. Am J Physiol Heart Circ Physiol 294:H570–H578
101. Berchner-Pfannschmidt U, Tug S, Kirsch M, Fandrey J (2010) Oxygen-sensing under the influence of nitric oxide. Cell Signal 22:349–356
102. Koivunen P, Hirsila M, Remes AM, Hassinen IE, Kivirikko KI, Myllyharju J (2007) Inhibition of hypoxia-inducible factor (HIF) hydroxylases by citric acid cycle intermediates: possible links between cell metabolism and stabilization of HIF. J Biol Chem 282:4524–4532
103. MacKenzie ED, Selak MA, Tennant DA, Payne LJ, Crosby S, Frederiksen CM et al (2007) Cell-permeating alpha-ketoglutarate derivatives alleviate pseudohypoxia in succinate dehydrogenase-deficient cells. Mol Cell Biol 27:3282–3289
104. Mailloux RJ, Puiseux-Dao S, Appanna VD (2009) Alpha-ketoglutarate abrogates the nuclear localization of HIF-1alpha in aluminum-exposed hepatocytes. Biochimie 91:408–415
105. Hao HX, Khalimonchuk O, Schraders M, Dephoure N, Bayley JP, Kunst H et al (2009) SDH5, a gene required for flavination of succinate dehydrogenase, is mutated in paraganglioma. Science 325:1139–1142
106. Pollard PJ, Briere JJ, Alam NA, Barwell J, Barclay E, Wortham NC et al (2005) Accumulation of Krebs cycle intermediates and over-expression of HIF1alpha in tumours which result from germline FH and SDH mutations. Hum Mol Genet 14:2231–2239
107. Vaux EC, Metzen E, Yeates KM, Ratcliffe PJ (2001) Regulation of hypoxia-inducible factor is preserved in the absence of a functioning mitochondrial respiratory chain. Blood 98:296–302
108. Pan Y, Mansfield KD, Bertozzi CC, Rudenko V, Chan DA, Giaccia AJ et al (2007) Multiple factors affecting cellular redox status and energy metabolism modulate hypoxia-inducible factor prolyl hydroxylase activity in vivo and in vitro. Mol Cell Biol 27:912–925

109. Taylor CT (2008) Mitochondria and cellular oxygen sensing in the HIF pathway. Biochem J 409:19–26
110. Bellance N, Lestienne P, Rossignol R (2009) Mitochondria: from bioenergetics to the metabolic regulation of carcinogenesis. Front Biosci 14:4015–4034
111. Lu H, Forbes RA, Verma A (2002) Hypoxia-inducible factor 1 activation by aerobic glycolysis implicates the Warburg effect in carcinogenesis. J Biol Chem 277:23111–23115
112. Hagg M, Wennstrom S (2005) Activation of hypoxia-induced transcription in normoxia. Exp Cell Res 306:180–191
113. Dunn LL, Buckle AM, Cooke JP, Ng MK (2010) The emerging role of the thioredoxin system in angiogenesis. Arterioscler Thromb Vasc Biol 30:2089–2098 [Research Support, N.I.H., Extramural Research Support, Non-U.S. Gov't Review]
114. Dai S, He Y, Zhang H, Yu L, Wan T, Xu Z et al (2009) Endothelial-specific expression of mitochondrial thioredoxin promotes ischemia-mediated arteriogenesis and angiogenesis. Arterioscler Thromb Vasc Biol 29(4):495–502 [Research Support, N.I.H., Extramural Research Support, Non-U.S. Gov't]
115. Sliman SM, Eubank TD, Kotha SR, Kuppusamy ML, Sherwani SI, Butler ES et al (2010) Hyperglycemic oxoaldehyde, glyoxal, causes barrier dysfunction, cytoskeletal alterations, and inhibition of angiogenesis in vascular endothelial cells: aminoguanidine protection. Mol Cell Biochem 333(1–2):9–26 [Research Support, N.I.H., Extramural Research Support, Non-U.S. Gov't]
116. Satoh M, Fujimoto S, Horike H, Ozeki M, Nagasu H, Tomita N et al (2011) Mitochondrial damage-induced impairment of angiogenesis in the aging rat kidney. Lab Invest 91:190–202 [Research Support, Non-U.S. Gov't]
117. Forini F, Lionetti V, Ardehali H, Pucci A, Cecchetti F, Ghanefar M et al (2011) Early long-term L-T3 replacement rescues mitochondria and prevents ischemic cardiac remodelling in rats. J Cell Mol Med 15(3):514–524 [Research Support, N.I.H., Extramural Research Support, Non-U.S. Gov't]
118. Comito G, Calvani M, Giannoni E, Bianchini F, Calorini L, Torre E et al (2011) HIF-1 alpha stabilization by mitochondrial ROS promotes Met-dependent invasive growth and vasculogenic mimicry in melanoma cells. Free Radic Biol Med 51:893–904 [Research Support, Non-U.S. Gov't]
119. Papandreou I, Cairns RA, Fontana L, Lim AL, Denko NC (2006) HIF-1 mediates adaptation to hypoxia by actively downregulating mitochondrial oxygen consumption. Cell Metab 3:187–197
120. Kim JW, Tchernyshyov I, Semenza GL, Dang CV (2006) HIF-1-mediated expression of pyruvate dehydrogenase kinase: a metabolic switch required for cellular adaptation to hypoxia. Cell Metab 3:177–185
121. Richards JG, Sardella BA, Schulte PM (2008) Regulation of pyruvate dehydrogenase in the common killifish, Fundulus heteroclitus, during hypoxia exposure. Am J Physiol Regul Integr Comp Physiol 295:R979–R990
122. De Palma S, Ripamonti M, Vigano A, Moriggi M, Capitanio D, Samaja M et al (2007) Metabolic modulation induced by chronic hypoxia in rats using a comparative proteomic analysis of skeletal muscle tissue. J Proteome Res 6:1974–1984
123. Holness MJ, Sugden MC (2003) Regulation of pyruvate dehydrogenase complex activity by reversible phosphorylation. Biochem Soc Trans 31:1143–1151
124. Stacpoole PW, Owen R, Flotte TR (2003) The pyruvate dehydrogenase complex as a target for gene therapy. Curr Gene Ther 3:239–245
125. Mattevi A, Obmolova G, Schulze E, Kalk KH, Westphal AH, de Kok A et al (1992) Atomic structure of the cubic core of the pyruvate dehydrogenase multienzyme complex. Science 255:1544–1550
126. Patel MS, Korotchkina LG (2006) Regulation of the pyruvate dehydrogenase complex. Biochem Soc Trans 34:217–222
127. Semenza GL (2007) Oxygen-dependent regulation of mitochondrial respiration by hypoxia-inducible factor 1. Biochem J 405:1–9

128. Roche TE, Hiromasa Y (2007) Pyruvate dehydrogenase kinase regulatory mechanisms and inhibition in treating diabetes, heart ischemia, and cancer. Cell Mol Life Sci 64:830–849
129. Mason SD, Rundqvist H, Papandreou I, Duh R, McNulty WJ, Howlett RA et al (2007) HIF-1 alpha in endurance training: suppression of oxidative metabolism. Am J Physiol Regul Integr Comp Physiol 293:R2059–R2069
130. Sugden MC, Holness MJ (2006) Mechanisms underlying regulation of the expression and activities of the mammalian pyruvate dehydrogenase kinases. Arch Physiol Biochem 112:139–149
131. Abbot EL, McCormack JG, Reynet C, Hassall DG, Buchan KW, Yeaman SJ (2005) Diverging regulation of pyruvate dehydrogenase kinase isoform gene expression in cultured human muscle cells. FEBS J 272:3004–3014
132. Stacpoole PW, Gilbert LR, Neiberger RE, Carney PR, Valenstein E, Theriaque DW et al (2008) Evaluation of long-term treatment of children with congenital lactic acidosis with dichloroacetate. Pediatrics 121:e1223–e1228
133. Huang B, Wu P, Bowker-Kinley MM, Harris RA (2002) Regulation of pyruvate dehydrogenase kinase expression by peroxisome proliferator-activated receptor-alpha ligands, glucocorticoids, and insulin. Diabetes 51:276–283
134. Way JM, Harrington WW, Brown KK, Gottschalk WK, Sundseth SS, Mansfield TA et al (2001) Comprehensive messenger ribonucleic acid profiling reveals that peroxisome proliferator-activated receptor gamma activation has coordinate effects on gene expression in multiple insulin-sensitive tissues. Endocrinology 142:1269–1277
135. Brown A, Nemeria N, Yi J, Zhang D, Jordan WB, Machado RS et al (1997) 2-Oxo-3-alkynoic acids, universal mechanism-based inactivators of thiamin diphosphate-dependent decarboxylases: synthesis and evidence for potent inactivation of the pyruvate dehydrogenase multienzyme complex. Biochemistry 36:8071–8081
136. Hue L, Taegtmeyer H (2009) The Randle cycle revisited: a new head for an old hat. Am J Physiol Endocrinol Metab 297:E578–E591
137. Kantor PF, Lucien A, Kozak R, Lopaschuk GD (2000) The antianginal drug trimetazidine shifts cardiac energy metabolism from fatty acid oxidation to glucose oxidation by inhibiting mitochondrial long-chain 3-ketoacyl coenzyme A thiolase. Circ Res 86:580–588
138. Cosse JP, Michiels C (2008) Tumour hypoxia affects the responsiveness of cancer cells to chemotherapy and promotes cancer progression. Anticancer Agents Med Chem 8:790–797
139. McFate T, Mohyeldin A, Lu H, Thakar J, Henriques J, Halim ND et al (2008) Pyruvate dehydrogenase complex activity controls metabolic and malignant phenotype in cancer cells. J Biol Chem 283:22700–22708
140. Michelakis ED, Sutendra G, Dromparis P, Webster L, Haromy A, Niven E et al (2010) Metabolic modulation of glioblastoma with dichloroacetate. Sci Transl Med 2(31):31ra34 [In Vitro Research Support, Non-U.S. Gov't]. 12;2(31):31–34
141. Sutendra G, Dromparis P, Kinnaird A, Stenson TH, Haromy A, Parker JM et al (2012) Mitochondrial activation by inhibition of PDKII suppresses HIF1a signaling and angiogenesis in cancer. Oncogene. doi: 10.1038/onc.2012.198
142. Biscetti F, Straface G, Pitocco D, Zaccardi F, Ghirlanda G, Flex A (2009) Peroxisome proliferator-activated receptors and angiogenesis. Nutr Metab Cardiovasc Dis 19(11):751–759
143. Giaginis C, Margeli A, Theocharis S (2007) Peroxisome proliferator-activated receptor-gamma ligands as investigational modulators of angiogenesis. Expert Opin Investig Drugs 16:1561–1572
144. Panigrahy D, Huang S, Kieran MW, Kaipainen A (2005) PPAR gamma as a therapeutic target for tumor angiogenesis and metastasis. Cancer Biol Ther 4:687–693
145. Margeli A, Kouraklis G, Theocharis S (2003) Peroxisome proliferator activated receptor-gamma (PPAR-gamma) ligands and angiogenesis. Angiogenesis 6:165–169
146. Xin X, Yang S, Kowalski J, Gerritsen ME (1999) Peroxisome proliferator-activated receptor gamma ligands are potent inhibitors of angiogenesis in vitro and in vivo. J Biol Chem 274:9116–9121
147. Murata T, He S, Hangai M, Ishibashi T, Xi XP, Kim S et al (2000) Peroxisome proliferator-activated receptor-gamma ligands inhibit choroidal neovascularization. Invest Ophthalmol Vis Sci 41:2309–2317

148. Kim KY, Cheon HG (2006) Antiangiogenic effect of rosiglitazone is mediated via peroxisome proliferator-activated receptor gamma-activated maxi-K channel opening in human umbilical vein endothelial cells. J Biol Chem 281:13503–13512
149. Peeters LL, Vigne JL, Tee MK, Zhao D, Waite LL, Taylor RN (2005) PPAR gamma represses VEGF expression in human endometrial cells: implications for uterine angiogenesis. Angiogenesis 8:373–379
150. Aljada A, O'Connor L, Fu YY, Mousa SA (2008) PPAR gamma ligands, rosiglitazone and pioglitazone, inhibit bFGF- and VEGF-mediated angiogenesis. Angiogenesis 11:361–367
151. Gealekman O, Burkart A, Chouinard M, Nicoloro SM, Straubhaar J, Corvera S (2008) Enhanced angiogenesis in obesity and in response to PPARgamma activators through adipocyte VEGF and ANGPTL4 production. Am J Physiol Endocrinol Metab 295:E1056–E1064
152. Chintalgattu V, Harris GS, Akula SM, Katwa LC (2007) PPAR-gamma agonists induce the expression of VEGF and its receptors in cultured cardiac myofibroblasts. Cardiovasc Res 74:140–150
153. Huang PH, Sata M, Nishimatsu H, Sumi M, Hirata Y, Nagai R (2008) Pioglitazone ameliorates endothelial dysfunction and restores ischemia-induced angiogenesis in diabetic mice. Biomed Pharmacother 62:46–52
154. Shen LQ, Child A, Weber GM, Folkman J, Aiello LP (2008) Rosiglitazone and delayed onset of proliferative diabetic retinopathy. Arch Ophthalmol 126:793–799
155. Dormandy JA, Charbonnel B, Eckland DJ, Erdmann E, Massi-Benedetti M, Moules IK et al (2005) Secondary prevention of macrovascular events in patients with type 2 diabetes in the PROactive Study (PROspective pioglitAzone Clinical Trial In macroVascular Events): a randomised controlled trial. Lancet 366:1279–1289
156. Home PD, Pocock SJ, Beck-Nielsen H, Curtis PS, Gomis R, Hanefeld M et al (2009) Rosiglitazone evaluated for cardiovascular outcomes in oral agent combination therapy for type 2 diabetes (RECORD): a multicentre, randomised, open-label trial. Lancet 373:2125–2135
157. Erdmann E, Charbonnel B, Wilcox RG, Skene AM, Massi-Benedetti M, Yates J et al (2007) Pioglitazone use and heart failure in patients with type 2 diabetes and preexisting cardiovascular disease: data from the PROactive study (PROactive 08). Diabetes Care 30:2773–2778
158. Giaginis C, Tsantili-Kakoulidou A, Theocharis S (2008) Peroxisome proliferator-activated receptor-gamma ligands: potential pharmacological agents for targeting the angiogenesis signaling cascade in cancer. PPAR Res 2008:431763
159. Cadoudal T, Distel E, Durant S, Fouque F, Blouin JM, Collinet M et al (2008) Pyruvate dehydrogenase kinase 4: regulation by thiazolidinediones and implication in glyceroneogenesis in adipose tissue. Diabetes 57:2272–2279
160. Biscetti F, Gaetani E, Flex A, Aprahamian T, Hopkins T, Straface G et al (2008) Selective activation of peroxisome proliferator-activated receptor (PPAR)alpha and PPAR gamma induces neoangiogenesis through a vascular endothelial growth factor-dependent mechanism. Diabetes 57:1394–1404
161. Brunmair B, Staniek K, Gras F, Scharf N, Althaym A, Clara R et al (2004) Thiazolidinediones, like metformin, inhibit respiratory complex I: a common mechanism contributing to their antidiabetic actions? Diabetes 53:1052–1059
162. Agani FH, Pichiule P, Carlos Chavez J, LaManna JC (2002) Inhibitors of mitochondrial complex Iattenuate the accumulation of hypoxia-inducible factor-1 during hypoxia in Hep3B cells. Comp Biochem Physiol A Mol Integr Physiol 132:107–109
163. Rabia K, Khoo EM (2007) Prevalence of peripheral arterial disease in patients with diabetes mellitus in a primary care setting. Med J Malaysia 62:130–133
164. Guan H, Li YJ, Xu ZR, Li GW, Guo XH, Liu ZM et al (2007) Prevalence and risk factors of peripheral arterial disease in diabetic patients over 50 years old in China. Chin Med Sci J 22:83–88
165. Starner CI, Schafer JA, Heaton AH, Gleason PP (2008) Rosiglitazone and pioglitazone utilization from January 2007 through May 2008 associated with five risk-warning events. vJ Manag Care Pharm 14:523–531

Part II
Molecular Mechanisms of Remodeling After Myocardial Injury and Infarction

Chapter 16
Subcellular Remodeling and Cardiac Dysfunction Due to Ischemia–Reperfusion Injury

Naranjan S. Dhalla, Vijayan Elimban, Larry Hryshko, and Darren H. Freed

Abstract Ischemic heart disease as a consequence of the blockade of coronary flow is associated with dramatic changes in cardiac function, metabolism, and ultrastructure. A wide variety of subcellular defects have been observed in ischemic and ischemia–reperfusion (I/R) hearts. There is evidence that various subcellular organelles become remodeled during the development of I/R injury and oxidative stress may be intimately involved in producing these abnormalities. In view of the direct participation of the sarcoplasmic reticulum (SR) and myofibrils in heart function, it appears that cardiac contraction and relaxation abnormalities in ischemic heart disease are due to remodeling of the SR and myofibrils, whereas remodeling of the sarcolemma membrane may determine the extent of intracellular Ca^{2+} overload, subsequent proteolysis, and irreversible injury to the heart. Furthermore, the acute effects of I/R injury on cardiac function are thought to be due to changes in the activities of subcellular organelles as a consequence of functional group modification, whereas the chronic effects of I/R yielding delayed recovery of cardiac function may be the consequence of changes in cardiac gene expression and subcellular remodeling. Although female hearts are less susceptible to I/R injury, in comparison to males, the basis for this gender difference in cardiac ischemic injury and protection needs to be defined. As females lose their resistance to different cardiovascular diseases after menopause, it appears that gender differences in cardiac susceptibility

N.S. Dhalla • V. Elimban (✉) • L. Hryshko
Department of Physiology, Institute of Cardiovascular Sciences,
St. Boniface Hospital Research Centre, 351 Tache Avenue, Winnipeg, MB, Canada R2H 2A6
e-mail: velimban@sbrc.ca

D.H. Freed
Department of Surgery, Faculty of Medicine, University of Manitoba,
Winnipeg, MB, Canada R2H 2A6

B.I. Jugdutt and N.S. Dhalla (eds.), *Cardiac Remodeling: Molecular Mechanisms*, Advances in Biochemistry in Health and Disease 5, DOI 10.1007/978-1-4614-5930-9_16,

to I/R injury may be mediated through the participation of ovarian hormones. On the other hand, it is possible that the male sex hormone, testosterone, exacerbates I/R-induced cardiac dysfunction in adult males. Notably, in comparison to males, there is very little information in the literature on subcellular remodeling or on the mechanisms which regulate cardiac function during the development of I/R injury in female hearts.

Keywords Ischemia–reperfusion injury • Subcellular remodeling • Oxidative stress • Intracellular Ca^{2+} overload • Proteases • Gender differences

16.1 Introduction

Ischemic heart disease as a consequence of the blockade of coronary flow is associated with dramatic changes in cardiac function, metabolism, and ultrastructure [1, 2]; however, the exact cellular and molecular events leading to contractile dysfunction and derangement of cardiac structure are not fully understood. Although restitution of coronary flow to the ischemic heart is essential for the recovery of cardiac pump function, reperfusion after a certain period of ischemia has been shown to further aggravate the myocardial abnormalities [2–5]. As cardiac contractile defects due to I/R are almost invariably associated with situations such as angioplasty, thrombolytic therapy, cardiac surgery, and cardiac transplantation, studies on I/R injury are highly relevant for understanding the pathophysiology of an important clinical problem, namely, myocardial stunning. Both myocardial ischemia and I/R have been shown to generate different oxyradicals and oxidants such as H_2O_2, peroxynitrite, and HOCl, and these are suggested to be responsible for the occurrence of intracellular Ca^{2+} overload due to I/R injury [5–9]. Various active oxygen species such as superoxide radicals, hydroxyl radicals, and H_2O_2, which are formed during the development of I/R injury, produce electrical abnormalities [10–12], ultrastructural damage [13], intracellular Ca^{2+} overload [14], and cardiac dysfunction [15]. Both H_2O_2 and peroxynitrite have also been reported to activate some proteases and induce cardiac dysfunction [16–19]. Likewise, the intracellular Ca^{2+} overload may induce cardiac dysfunction and cell damage by activating different proteases and phospholipases [20, 21] and thus may modify the activities of various subcellular organelles such as sarcolemma (SL), sarcoplasmic reticulum (SR), myofibrils, and mitochondria. While both intracellular Ca^{2+} overload and oxidative stress have been shown to be involved in producing changes in cardiac gene expression as well as remodeling of subcellular organelles [20–22], oxidative stress seems to play a critical role in the genesis of intracellular Ca^{2+} overload and thus may induce cardiac dysfunction by remodeling of subcellular organelles during the development of I/R injury. A schematic representation of the events involving oxidative stress and intracellular Ca^{2+} overload due to I/R injury is given in Fig. 16.1. This view does not exclude the role of either lipid metabolites or oxidative stress/intracellular Ca^{2+} overload in apoptosis and necrosis commonly seen in ischemic and I/R hearts.

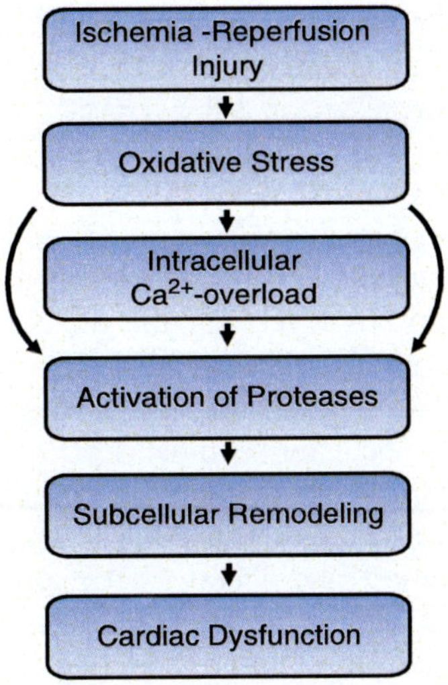

Fig. 16.1 Involvement of oxidative stress in inducing subcellular remodeling and cardiac dysfunction due to ischemia–reperfusion injury

16.2 Subcellular Remodeling and Molecular Abnormalities in I/R Hearts

Over the past 30 years, a wide variety of membrane defects have been observed in both ischemic and I/R hearts [5, 20–24]. It is now clear that the SR Ca^{2+} pump and associated regulatory mechanisms become defective due to changes in the molecular composition of the SR membrane as a consequence of I/R injury [25–35]. Several investigators have reported a reduction in the density of SR Ca^{2+}-release channels during I/R [36–39]. Various oxidants as well as hydroxyl radicals were also observed to depress the SR Ca^{2+}-pump activity [40, 41]. Although the efficiency of mitochondrial ATP production is impaired at the late stages of I/R injury, depression in both electron transport chain activity and Ca^{2+} transport in mitochondria also occurs at moderate degree of I/R injury [42, 43]. The biochemical activities of several SL membrane proteins including the Na^+–Ca^{2+} exchanger, Ca^{2+}-stimulated ATPase, Na^+–K^+ ATPase, and phosphoinositol turnover are markedly altered during myocardial I/R as well as during hypoxia–reoxygenation phases [44–55]. I/R, as well as reactive oxygen species and oxidants, have also been shown to reduce the sensitivity of myofilaments to Ca^{2+} by causing proteolysis of myofibrils [56–62]. These observations provide evidence that various subcellular organelles become remodeled or altered in I/R heart and that oxidative stress may be intimately involved in producing these abnormalities. In view of the direct participation of the SR and myofibrils in heart function, it appears that cardiac contraction and relaxation abnormalities in ischemic heart disease are due to remodeling of SR and myofibrils, whereas remodeling of the SL membrane may determine the extent of intracellular Ca^{2+} overload, subsequent proteolysis, and irreversible injury in the myocardium. It should also be noted that both I/R and oxidative stress have been shown to produce dramatic effects on cardiac gene expression. We have demonstrated that mRNA levels for the SR Ca^{2+} pump, Ca^{2+} channels, phospholamban, and calsequestrin proteins were depressed in I/R hearts [25]. As H_2O_2 and I/R were observed to produce similar changes in SR gene expression, we suggested that these effects of I/R may be due to oxidative stress [25]. We have also observed that I/R produced differential changes in gene expression for SL Na^+–K^+ ATPase isoforms, and these alterations were simulated by perfusing the hearts with an oxyradical-generating system or H_2O_2 [63]. mRNA levels for the SL Na^+–Ca^+ exchanger were also depressed by I/R [55]. Although intracellular Ca^{2+} overload was demonstrated to occur in I/R hearts [64, 65], it is not known whether alterations in cardiac gene expression are affected by both intracellular Ca^{2+} overload and oxidative stress. Nonetheless, the acute effects of I/R injury on cardiac function are considered to be due to changes in the activities of subcellular organelles and proteins as a consequence of functional group modification, whereas the chronic effects of I/R including delayed recovery of cardiac function may be the consequence of changes in cardiac gene expression and subcellular remodeling.

16.3 Oxidative Stress and Development of Intracellular Ca^{2+} Overload in I/R Hearts

Some investigators have demonstrated the generation of oxygen free radicals in I/R hearts by employing electron paramagnetic resonance spectroscopy [66–68]. Oxidative stress has been shown to result in the development of intracellular Ca^{2+} overload due to I/R injury [69–73] because the activities of both SL Na^+–Ca^{2+} exchanger and SL Ca^{2+} pump were depressed following hypoxia or I/R as well as upon exposure of heart membranes to oxyradicals [74–78]. Oxyradicals were also reported to alter other SL activities such as Na^+–K^+ ATPase, Na^+–Ca^{2+} exchanger, phospholipid methyltransferase [46, 47, 78–82], Ca^{2+}/Mg^{2+} ecto-ATPase, superficial store of Ca^{2+} [83], and ATP receptors [84], which are considered to affect Ca^{2+} movements in the cell. The SL changes in different cation currents have also been observed upon exposure to oxyradicals and oxidants [85–87]. Several other defects such as changes in membrane permeability, loss of dystrophin, and alterations in phospholipases due to I/R injury have also been found in the SL membrane [88–92]. Oxidative stress has also been shown to produce marked alterations in myofibrils, mitochondria, and SR as well as induce autophagy during the development of I/R injury [62, 93, 94]. Thus, the increased formation of oxyradicals and oxidants in I/R hearts may induce a complex set of subcellular alterations with respect to their biochemical composition and functional activities related to Ca^{2+} movements, and these on balance may result in the development of intracellular Ca^{2+} overload and subcellular remodeling.

Oxidative stress seems to alter the subcellular activities by oxidizing different functional groups of subcellular organelles/proteins, and these changes seem to explain the development of intracellular Ca^{2+} overload and cardiac dysfunction due to I/R injury [3, 5, 21]. The effects of oxidative stress favoring the development of

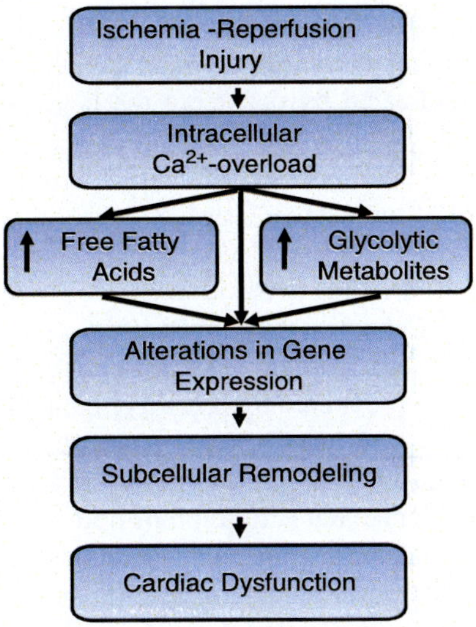

Fig. 16.2 Role of intracellular Ca^{2+} overload and changes in gene expression in inducing subcellular remodeling and cardiac dysfunction due to ischemia–reperfusion injury

intracellular Ca^{2+} overload are shown to be mediated through the activation of protein kinase C (PKC), mitogen-activated protein kinase (MAPK), and/or stress-activated protein kinase [95–97], as well as translocation of PKC in ischemic heart [98, 99]. It is also noteworthy that besides the production of oxygen-derived free radicals, changes in nitric oxide (NO) metabolism have been observed in I/R hearts [67]. Although NO is known to regulate various events, its action becomes toxic by reaction with superoxide anion forming a potent oxidant, peroxynitrite, which has also been demonstrated to impair cardiac function [100, 101]. Thus, it appears that oxidative stress generated by different sources in the I/R heart plays an important role in the genesis of subcellular remodeling and cardiac dysfunction. A general scheme involving changes in gene expression for inducing subcellular remodeling due to I/R is shown in Fig. 16.2.

16.4 Activation of Proteases and Subcellular Remodeling in I/R Hearts

Activation of different types of proteases including calpain I and calpain II as well as metalloproteases (MMP-2 and MMP-9) has been suggested to be intimately involved in the pathophysiology of several forms of cardiac diseases [6–19, 102–106]. These proteases are activated by Ca^{2+}-dependent and Ca^{2+}-independent mechanisms and have been shown to cleave subcellular proteins and depress or alter their activities. While calpain I and calpain II are activated by Ca^{2+}, both MMP-2 and MMP-9 are activated by oxidative stress as well as by proteolysis. Pretreatment of heart with calpain inhibitors, MDL-28170 and A-70523, was observed to attenuate I/R-induced cardiac stunning and infarct size [24, 106, 107]. The activation of calpain as well as defects in the SR Ca^{2+}-uptake and Ca^{2+}-release activities due to I/R injury was also attenuated by perfusing the heart with calpain inhibitors, leupeptin and E64d, or exercise training [103, 108, 109]. Preventing the activation of calpain by nitrosylation upon perfusing the heart with L-arginine was associated with improvements of SR function and cardiac performance in I/R hearts [51, 110]. Calpain-mediated depression in the activity of SL Na^{+}–K^{+} ATPase and the loss of cytoskeleton protein α-fodrin due to I/R injury were prevented by a calpain inhibitor, MDL-7943, as well as ischemic preconditioning [111–114]. Likewise, the activation of MMP-2 [115, 116], changes in myosin light chain, and cardiac dysfunction due to I/R injury were prevented by doxycycline, an inhibitor of MMP-2 [116]. The influence of I/R injury on MMP-2 was observed to be mediated through the phosphoinositide 3 kinase (PI3K)/protein kinase B (Akt kinase) pathway [117]. These results provide evidence that activation of both calpain and MMP-2 due to I/R injury may depress cardiac performance due to subcellular defects; however, it remains to be examined if the observed changes in the activation of metalloproteases due to I/R injury are mediated directly through oxidative stress and/or indirectly through intracellular Ca^{2+} overload. It should be mentioned that both MMP-2 and MMP-9 are localized within cardiomyocytes [104, 105, 118] and their endogenous inhibitors, TIMP-2 and TIMP-1, are also found in these cells [104, 118]. Further, calpastatin serves as an endogenous inhibitor of both I and II isoforms of calpain in the heart [103]. Thus, the I/R-induced activation of MMP-2 and MMP-9 could also be due to a reduction in the TIMP-2 and TIMP-1 contents, whereas that of calpain I and II may be associated with the reduction of calpastatin content in cardiomyocytes. Nonetheless, the activation of different proteases under conditions of I/R injury would disrupt myocardial structure, remodel different subcellular organelles with respect to their protein content, and produce irreversible cardiac dysfunction.

16.5 I/R-Induced Subcellular Remodeling and Gender Difference

To date, the majority of work done to describe I/R-induced development of oxidative stress, occurrence of intracellular Ca^{2+} overload, activation of proteases, and defects in subcellular function has employed male animals, and little information is available for the female hearts. As female hearts are less susceptible to I/R injury, the basis for gender difference in cardiac ischemic injury and protection remains to defined [119, 120]. Several epidemiological studies have revealed sex differences with respect to the incidence of coronary artery disease, atherosclerosis, apoptosis, hypertension, and heart failure [120–134]. Various experimental investigations have also reported gender differences in the development of cardiac hypertrophy and heart failure due to myocardial infarction, pressure overload, and volume overload [135–146]. Gender difference in the properties of cardiac Na^{+}–K^{+} ATPase due to hypertension [147], SR Ca^{2+} loading due to catecholamines [148], and Na^{+}–Ca^{2+} exchanger due to heart failure [149] has also been observed. As females lose their resistance to different cardiovascular diseases after menopause, it appears that the intrinsic cardioprotection observed in females may be mediated through the participation of ovarian hormones [150]. Dramatic changes in Akt and NO synthase signaling as well as protein kinase A-mediated changes in Ca^{2+} handling involving SL Ca^{2+} channels, Na^+–Ca^{2+} exchanger, SR Ca^{2+} uptake, and Ca^{2+}-release channels have been observed upon ovariectomy [151–154]. Furthermore, estrogen, a major ovarian hormone, has been demonstrated to affect different Ca^{2+}-handling proteins, β-adrenoceptors, and Na^+–H^+ exchanger as a consequence of its action on various kinase-mediated signal pathways [155–160]. However, a detailed study regarding the mechanisms of subcellular remodeling responsible for the resistance to ischemic insult in females, by employing ovariectomized animals with or without estrogen treatment, needs additional investigation before overarching conclusions can be made.

Several investigators have attempted to investigate the mechanisms of gender difference in cardioprotection against I/R injury in adult hearts. The postischemic recovery of cardiac function was greater and infarct size was smaller in female hearts in comparison to males, and these changes were attributed to differences in the Akt and PKC signal transduction [161, 162]. The loss of ischemic preconditioning effect on contractile function, infarct size, and enzyme leakage in I/R hearts was associated with impaired PKC phosphorylation [163]. Gender differences with respect to improved cardiac function in female I/R hearts were shown to be due to alterations in the regulation of $[Na^+]_i$ by a NO synthase-dependent mechanism [164–166]. Differences in the gender-dependent I/R-induced infarct size were associated with increased expression of SL K_{ATP} channels, and in fact, blockade of these channels was observed to abolish this difference [167, 168]. Conversely, gender differences with respect to resistance of female heart to I/R injury have also been attributed to difference in mitochondrial Ca^{2+} uptake [169] and tumor necrosis factor receptor signaling [170, 171]. By using ovariectomized animals, estrogen was found to attenuate I/R-induced changes in cardiac function and reduce infarct size as a consequence of changes in calpain and p38 MAP kinase activities [172–176]. Female mouse cardiomyocytes were protected against oxidative stress due to H_2O_2 as a consequence of Akt activation [177]. On the other hand, castration was found to decrease mRNA levels for L-type Ca^{2+} channels and Na^+–Ca^{2+} exchanger, and these alterations were reversed by testosterone [178–180]. Testosterone was also observed to modify I/R-induced changes in Akt signal transduction and apoptotic pathway [181–183]. Overall, there is a paucity of information on gender differences in subcellular remodeling as well as changes in mechanisms, which regulate cardiac function during I/R injury.

16.6 Conclusions

It is now known that the SR Ca^{2+} pump and associated regulatory mechanisms are defective due to changes in the molecular composition of SR membrane as a consequence of I/R injury. Various oxidants as well as hydroxyl radicals are observed to depress the SR Ca^{2+}-pump activity in I/R hearts. It should be noted that ischemic

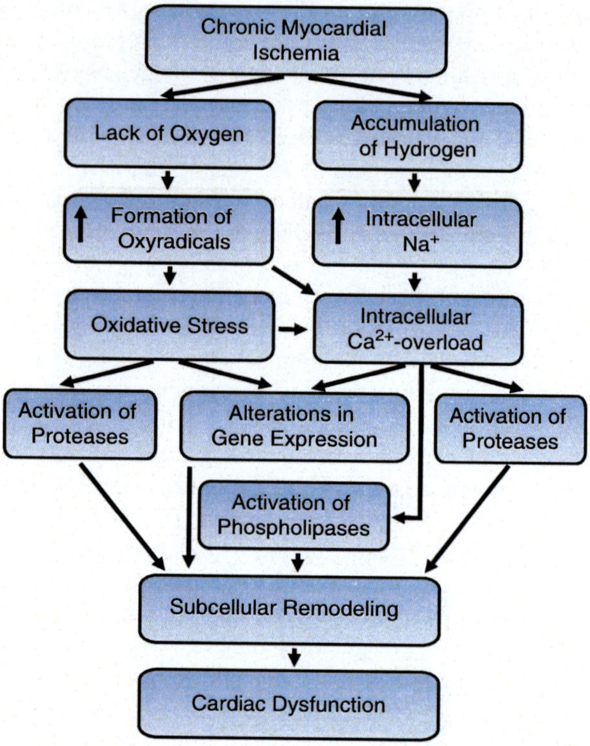

Fig. 16.3 Mechanisms of subcellular remodeling and cardiac dysfunction involving both oxidative stress and intracellular Ca^{2+} overload in ischemic heart disease

insult produces oxidative stress due to the generation of oxyradicals generating in addition to accumulating protons in cardiomyocytes. The magnitude of oxidative stress is amplified upon reperfusion of the ischemic myocardium, whereas protons are exchanged for Na^+ via the Na^+–H^+ exchanger. Furthermore, oxidative stress rapidly oxidizes the functional groups of Na^+–K^+ ATPase and augments the development of intracellular Na^+ overload. Intracellular Na^+ is exchanged with Ca^{2+} via the Na^+–Ca^{2+} exchanger and favors the occurrence of intracellular Ca^{2+} overload in the I/R heart. Thus, it is emphasized that alterations in the activities of Na^+-handling proteins (SL Na^+–K^+ ATPase and Na^+–Ca^{2+} exchanger) are critical for the net gain of Ca^{2+} within the cardiomyocytes. Oxidative stress and subsequent intracellular Ca^{2+} overload result in the activation of different proteases and induce dramatic changes in the composition of subcellular organelles/proteins in the I/R hearts. Accordingly, oxidative stress as well as changes in SL Na^+-handling proteins and protease activation plays an important role in inducing cardiac dysfunction due to I/R injury. Various events involved in subcellular remodeling during the development of cardiac dysfunction in ischemic heart disease are depicted in Fig. 16.3. It is also evident that females are more resistant to I/R-induced injury than males and the recovery of cardiac function upon reperfusing the male ischemic hearts is less than that of the female. Although I/R-induced cardiac dysfunction in male hearts has been shown to be associated with the occurrence of oxidative stress, increase in development of intracellular Ca^{2+} overload, activation of proteases, cleavage of subcellular proteins, and alterations in subcellular activities, very little information regarding these changes in female hearts is available.

Acknowledgments The research reported in this chapter was supported by Canadian Institute of Health Research. Infrastructural support for the project was provided by the St. Boniface Hospital Research Foundation.

References

1. Jennings RB, Reimer KA (2007) The cell biology of acute myocardial ischemia. Annu Rev Med 42:225–246
2. Bolli R, Marban E (1999) Molecular and cellular mechanisms of myocardial stunning. Physiol Rev 79:609–634
3. Dhalla NS, Elmoselhi AB, Hata T, Makino N (2000) Status of myocardial antioxidants in ischemia-reperfusion injury. Cardiovasc Res 47:446–456
4. Kim SJ, Depre C, Vatner SF (2003) Novel mechanisms mediating stunned myocardium. Heart Fail Rev 8:143–153
5. Dhalla NS, Saini HK, Tappia PS et al (2007) Potential role and mechanisms of subcellular remodeling in cardiac dysfunction due to ischemic heart disease. J Cardiovasc Med 8:238–250
6. Vanden Hoek TL, Li C, Shao Z et al (1997) Significant levels of oxidants are generated by isolated cardiomyocytes during ischemia prior to reperfusion. J Mol Cell Cardiol 29:2571–2583
7. Slezak J, Tribulova N, Pristacova J et al (1995) Hydrogen peroxide changes in ischemic and reperfused heart. Cytochemistry and biochemical and X-ray microanalysis. Am J Pathol 147:772–781
8. Kukreja RC, Loesser KE, Kearns AA et al (1993) Protective effects of histidine during ischemia-reperfusion in isolated perfused rat hearts. Am J Physiol Heart Circ Physiol 264:H137–1381H
9. Lee WH, Gounarides JS, Roos ES, Wolin MS (2003) Influence of peroxynitrite on energy metabolism and cardiac function in a rat ischemia-reperfusion model. Am J Physiol Heart Circ Physiol 285:H1385–H1395
10. Barrington PL, Meier CF Jr, Weglicki WB (1988) Abnormal electrical activity induced by free radical generating systems in isolated cardiocytes. J Mol Cell Cardiol 20:1163–1178
11. Tarr M, Valenzeno DP (1991) Modification of cardiac ionic currents by photosensitizer-generated reactive oxygen. J Mol Cell Cardiol 23:639–649
12. Coetzee WA, Opie LH (1992) Effects of oxygen free radicals on isolated cardiac myocytes from guinea-pig ventricle: electrophysiological studies. J Mol Cell Cardiol 24:651–663
13. Burton KP, McCord JM, Ghai G (1984) Myocardial alterations due to free-radical generation. Am J Physiol Heart Circ Physiol 246:H776–H783
14. Kaminishi K, Yanagishita T, Kako KJ (1989) Oxidant injury to isolated heart cells. Can J Cardiol 5:168–174
15. Prasad K, Kalra J, Chan WP, Chaudhary AK (1989) Effect of oxygen free radicals on cardiovascular function at organ and cellular levels. Am Heart J 117:1196–1202
16. Leon H, Bautista-Lopaz N, Sawicka J, Schulz R (2007) Hydrogen peroxide causes cardiac dysfunction independent from its effects on matrix metalloproteinase-2 activation. Can J Physiol Pharmacol 85:341–348
17. Sung M, Schulz CG, Wang W et al (2007) Matrix metalloproteinase-2 degrades the cytoskeletal protein alpha-actinin in peroxynitrite mediated myocardial injury. J Mol Cell Cardiol 43:429–436

18. Leon H, Baczko I, Sawicki G et al (2008) Inhibition of matrix metalloproteinases prevents peroxynitrite-induced contractile dysfunction in the isolated cardiac myocyte. Br J Pharmacol 153:676–683
19. Müller AL, Hryshko LV, Dhalla NS (2012) Extracellular and intracellular proteases in cardiac dysfunction due to ischemia-reperfusion injury. Int J Cardiol. doi:10.1016/j.ijcard.2012.01.103
20. Dhalla NS, Saini HK, Duhamel TA (2008) Strategies for the regulation of intracellular calcium in ischemic heart disease. Future Cardiol 4:339–345
21. Dhalla NS, Temsah RM, Netticadan T (2000) Role of oxidative stress in cardiovascular diseases. J Hypertens 18:655–673
22. Dhalla NS, Temsah RM (2001) Sarcoplasmic reticulum and cardiac oxidative stress: an emerging target for heart disease. Expert Opin Ther Targets 5:205–17
23. Saini HK, Machackova J, Dhalla NS (2004) Role of reactive oxygen species in ischemic preconditioning of subcellular organelles in the heart. Antioxid Redox Signal 6:393–404
24. Singh RB, Hryshko L, Freed D, Dhalla NS (2012) Activation of proteolytic enzymes and depression of the sarcolemmal Na+/K+–ATPase in ischemia-reperfused heart may be mediated through oxidative stress. Can J Physiol Pharmacol 90(2):249–60. doi:10.1139/y11-128
25. Temsah RM, Netticadan T, Chapman D et al (1999) Alterations in sarcoplasmic reticulum function and gene expression in ischemic-reperfused rat heart. Am J Physiol Heart Physiol 277:H584–H594
26. Netticadan T, Temsah R, Osada M, Dhalla NS (1999) Status of Ca2+/calmodulin protein kinase phosphorylation of cardiac SR proteins in ischemia-reperfusion. Am J Physiol Cell Physiol 277:C384–C391
27. Temsah RM, Dyck C, Chapman D et al (2000) Effect of beta-adrenoceptor blockers on sarcoplasmic reticular function and gene expression in the ischemic-reperfused heart. J Pharmacol Exp Ther 293:15–23
28. Osada M, Netticadan T, Tamura K, Dhalla NS (1998) Modification of ischemia-reperfusion-induced changes in cardiac sarcoplasmic reticulum by preconditioning. Am J Physiol Heart Circ Physiol 274:H2025–H2034
29. Osada M, Netticadan T, Kawabata K et al (2000) Ischemic preconditioning prevents I/R-induced alterations in SR calcium-calmodulin protein kinase II. Am J Physiol Heart Circ Physiol 278:H1791–H1791
30. Kawabata K, Osada M, Netticadan T, Dhalla NS (1998) Beneficial effect of ischemic preconditioning on Ca2+ paradox in the rat heart. Life Sci 63:685–692
31. Kawabata KI, Netticadan T, Osada M et al (2000) Mechanisms of ischemic preconditioning effects on Ca(2+) paradox-induced changes in heart. Am J Physiol Heart Circ Physiol 278:H1008–H1015
32. Schoutsen B, Blom JJ, Verdouw PD, Lamers JM (1989) Calcium transport and phospholamban in sarcoplasmic reticulum of ischemic myocardium. J Mol Cell Cardiol 21:719–727
33. Yoshida Y, Shiga T, Imai S (1990) Degradation of sarcoplasmic reticulum calcium-pumping ATPase in ischemic-reperfused myocardium: role of calcium-activated neutral protease. Basic Res Cardiol 85:495–507
34. Krause SM (1991) Effect of increased free [Mg2+]i with myocardial stunning on sarcoplasmic reticulum Ca(2+)-ATPase activity. Am J Physiol Heart Physiol 261:H229–H235
35. Fukumoto K, Takenaka H, Onitsuka T et al (1991) Effect of hypothermic ischemia and reperfusion on calcium transport by myocardial sarcolemma and sarcoplasmic reticulum. J Mol Cell Cardiol 23:525–535
36. Zucchi R, Ronca-Testoni S, Yu G et al (1994) Effect of ischemia and reperfusion on cardiac ryanodine receptors–sarcoplasmic reticulum Ca2+ channels. Circ Res 74:271–280
37. Zucchi R, Yu G, Ghelardoni S et al (2001) A3 adenosine receptor stimulation modulates sarcoplasmic reticulum Ca(2+) release in rat heart. Cardiovasc Res 50:56–64
38. Ikeda Y, Gohra H, Hamano K et al (2001) Effects of cardioplegic arrest and reperfusion on rabbit cardiac ryanodine receptors. Jpn Circ J 65:330–334
39. Zucchi R, Ronca F, Ronca-Testoni S (2001) Modulation of sarcoplasmic reticulum function: a new strategy in cardioprotection? Pharmacol Ther 89:47–65

40. Kukreja RC, Weaver AB, Hess ML (1989) Stimulated human neutrophils damage cardiac sarcoplasmic reticulum function by generation of oxidants. Biochim Biophys Acta 990:198–205
41. Xu KY, Zweier JL, Becker LC (1997) Hydroxyl radical inhibits sarcoplasmic reticulum Ca(2+)-ATPase function by direct attack on the ATP binding site. Circ Res 80:76–81
42. Sordahl LA, Stewart ML (1980) Mechanism(s) of altered mitochondrial calcium transport in acutely ischemic canine hearts. Circ Res 47:814–820
43. Park Y, Kanekal S, Kehrer JP (1991) Oxidative changes in hypoxic rat heart tissue. Am J Physiol Heart Circ Physiol 260:H1395–H1405
44. Dixon IM, Eyolfson DA, Dhalla NS (1987) Sarcolemmal Na+–Ca2+ exchange activity in hearts subjected to hypoxia reoxygenation. Am J Physiol Heart Circ Physiol 253: H1026–H1034
45. Mouton R, Huisamen B, Lochner A (1991) The effect of ischaemia and reperfusion on sarcolemmal inositol phospholipid and cytosolic inositol phosphate metabolism in the isolated perfused rat heart. Mol Cell Biochem 105:127–135
46. Dixon IM, Kaneko M, Hata T et al (1990) Alterations in cardiac membrane Ca2+ transport during oxidative stress. Mol Cell Biochem 99:125–135
47. Samouilidou EC, Levis GM, Darsinos JT et al (1991) Effect of low calcium on high-energy phosphates and sarcolemmal Na+/K(+)-ATPase in the infarcted-reperfused heart. Biochim Biophys Acta 1070:343348
48. Venter H, Genade S, Mouton R et al (1991) Myocardial membrane cholesterol: effects of ischaemia. J Mol Cell Cardiol 23:1271–1286
49. Andres J, Moczarska A, Stepkowski D, Kakol I (1991) Contractile proteins in globally "stunned" rabbit myocardium. Basic Res Cardiol 86:219–226
50. Dhalla NS, Panagia V, Singal PK et al (1988) Alterations in heart membrane calcium transport during the development of ischemia-reperfusion injury. J Mol Cell Cardiol 20(Suppl 2):3–13
51. Chohan PK, Singh RB, Dhalla NS, Netticadan T (2006) L-arginine administration recovers sarcoplasmic reticulum function in ischemic reperfused hearts by preventing calpain activation. Cardiovac Res 69:152–163
52. Hohl CM, Garleb AA, Altschuld RA (1992) Effects of simulated ischemia and reperfusion on the sarcoplasmic reticulum of digitonin-lysed cardiomyocytes. Circ Res 70:716–723
53. Davis MD, Lebolt W, Feher JJ (1992) Reversibility of the effects of normothermic global ischemia on the ryanodine-sensitive and ryanodine-insensitive calcium uptake of cardiac sarcoplasmic reticulum. Circ Res 70:163–171
54. Ostadal P, Elmoselhi AB, Zdobnicka I et al (2004) Role of oxidative stress in ischemia-reperfusion-induced changes in Na+, K(+)-ATPase isoform expression in rat heart. Antioxid Redox Signal 6:91932
55. Elmoselhi AB, Lukas A, Ostadal P, Dhalla NS (2003) Preconditioning attenuates ischemia-reperfusion-induced remodeling of Na+–K+–ATPase in hearts. Am J Physiol Heart Circ Physiol 285:H1055–1063
56. Van Eyk JE, Powers F, Law W et al (1998) Breakdown and release of myofilament proteins during ischemia and ischemia/reperfusion in rat hearts: identification of degradation products and effects on the pCa-force relation. Circ Res 82:261–271
57. Gao WD, Atar D, Liu Y et al (1997) Role of troponin I proteolysis in the pathogenesis of stunned myocardium. Circ Res 80:393–399
58. White MY, Cordwell SJ, McCarron HC et al (2003) Modifications of myosin-regulatory light chain correlate with function of stunned myocardium. J Mol Cell Cardiol 35:838–840
59. Foster DB, Noguchi T, VanBuren P et al (2003) C-terminal truncation of cardiac troponin I causes divergent effects on ATPase and force: implications for the pathophysiology of myocardial stunning. Circ Res 93:917–924
60. Canton M, Neverova I, Menabò R et al (2004) Evidence of myofibrillar protein oxidation induced by postischemic reperfusion in isolated rat hearts. Am J Physiol Heart Circ Physiol 286:H870–H877
61. Powell SR, Gurzenda EM, Wahezi SE (2001) Actin is oxidized during myocardial ischemia. Free Radic Biol Med 30:1171–1176

62. Maddika S, Elimban V, Chapman D, Dhalla NS (2009) Role of oxidative stress in ischemia-reperfusion-induced alterations in myofibrillar ATPase activities and gene expression in the heart. Can J Physiol Pharmacol 87:120–129
63. Ostadal P, Elmoselhi AB, Zdobnicka I et al (2003) Ischemia-reperfusion alters gene expression of Na+–K+ATPase isoforms in rat heart. Biochem Biophys Res Commun 306:457–462
64. Saini HK, Dhalla NS (2005) Defective calcium handling in cardiomyocytes isolated from hearts subjected to ischemia-reperfusion. Am J Physiol Heart Circ Physiol 288:H2260–H2270
65. Saini HK, Elimban V, Dhalla NS (2005) Attenuation of extracellular ATP response in cardiomyocytes isolated from hearts subjected to ischemia-reperfusion. Am J Physiol Heart Circ Physiol 289:H614–H623
66. Flaherty JT, Weisfeldt ML (1988) Reperfusion injury. Free Radic Biol Med 5:409–419
67. Arroyo CM, Kramer JH, Dickens BF, Weglicki WB (1987) Identification of free radicals in myocardial ischemia/reperfusion by spin trapping with nitrone DMPO. FEBS Lett 221:101–104
68. Zweier JL, Flaherty JT, Weisfeldt ML (1987) Direct measurement of free radical generation following reperfusion of ischemic myocardium. Proc Natl Acad Sci USA 84:1404–1407
69. Murphy JG, Smith TW, Marsh JD (1988) Mechanisms of reoxygenation-induced calcium overload in cultured chick embryo heart cells. Am J Physiol Heart Circ Physiol 254:H1133–1141
70. Nayler WG, Panagiotopoulos S, Elz JS, Daly MJ (1988) Calcium-mediated damage during post-ischemic reperfusion. J Mol Cell Cardiol 20(Suppl 2):41–54
71. Billman GE, McIlroy B, Johnson JD (1991) Elevated myocardial calcium and its role in sudden cardiac death. FASEB J 5:2586–2592
72. Lee JA, Allen DG (1991) Mechanisms of acute ischemic contractile failure of the heart. Role of intracellular calcium. J Clin Invest 88:361–367
73. Saini-Chohan HK, Dhalla NS (2009) Attenuation of ischemia-reperfusion-induced alterations in intracellular Ca2+ in cardiomyocytes from hearts treated with N-acetylcysteine and N-mercaptopropionylglycine. Can J Physiol Pharmacol 87:1110–1119
74. Meno H, Jarmakani JM, Philipson KD (1989) Effect of ischemia on sarcolemmal Na+–Ca2+ exchange in neonatal hearts. Am J Physiol Heart Circ Physiol 256:H1615–H1620
75. Kaneko M, Beamish RE, Dhalla NS (1989) Depression of heart sarcolemmal Ca2+–pump activity by oxygen free radicals. Am J Physiol Heart Circ Physiol 256:H368–H3674
76. Kaneko M, Elimban V, Dhalla NS (1989) Mechanism for depression of heart sarcolemmal Ca2+ pump by oxygen free radicals. Am J Physiol Heart Circ Physiol 257:H804–H811
77. Hata T, Kaneko M, Beamish RE, Dhalla NS (1991) Influence of oxygen free radicals on heart sarcolemmal Na^+- Ca^{2+} exchange. Coron Artery Dis 2:397–407
78. Xie ZJ, Wang YH, Askari A et al (1990) Studies on the specificity of the effects of oxygen metabolites on cardiac sodium pump. J Mol Cell Cardiol 22:91920
79. Kramer JH, Mak IT, Weglicki WB (1984) Differential sensitivity of canine cardiac sarcolemmal and microsomal enzymes to inhibition by free radical-induced lipid peroxidation. Circ Res 55:120–124
80. Kim MS, Akera T (1987) O^2 free radicals: cause of ischemia-reperfusion injury to cardiac Na+–K+–ATPase. Am J Physiol Heart Circ Physiol 252:H252–H257
81. Kukreja RC, Weaver AB, Hess ML (1990) Sarcolemmal Na(+)-K(+)-ATPase: inactivation by neutrophil-derived free radicals and oxidants. Am J Physiol Heart Circ Physiol 259: H1330–H1336
82. Kaneko M, Panagia V, Paolillo G et al (1990) Inhibition of cardiac phosphatidylethanolamine N-methylation by oxygen free radicals. Biochim Biophys Acta 1021:33–38
83. Kaneko M, Singal PK, Dhalla NS (1990) Alterations in heart sarcolemmal Ca2(+)-ATPase and Ca2(+)-binding activities due to oxygen free radicals. Basic Res Cardiol 85:45–54
84. Musat S, Dhalla NS (1996) Alteration in cardiac sarcolemmal ATP receptors by oxyradicals. Ann N Y Acad Sci 793:1–12
85. Kato K, Shao Q, Elimban V et al (1998) Mechanism of depression in cardiac sarcolemmal Na+–K+–ATPase by hypochlorous acid. Am J Physiol Cell Physiol 275:C826–C831
86. Shao Q, Matsubara T, Bhatt SK, Dhalla NS (1995) Inhibition of cardiac sarcolemma Na(+)-K+ATPase by oxyradical generating systems. Mol Cell Biochem 147:139–144

87. Shattock MJ, Matsuura H (1993) Measurement of Na(+)-K+pump current in isolated rabbit ventricular myocytes using the whole-cell voltage-clamp technique. Inhibition of the pump by oxidant stress. Circ Res 72:91–101
88. Askenasy N, Vivi A, Tassini M et al (2001) NMR spectroscopic characterization of sarcolemmal permeability during myocardial ischemia and reperfusion. J Mol Cell Cardiol 33: 1421–1433
89. Kyoi S, Otani H, Sumida T et al (2003) Loss of intracellular dystrophin: a potential mechanism for myocardial reperfusion injury. Circ J 67:725–727
90. Asemu G, Tappia PS, Dhalla NS (2003) Identification of the changes in phospholipase C isozymes in ischemic-reperfused rat heart. Arch Biochem Biophys 411:174–182
91. Asemu G, Dent M, Singal T et al (2005) Differential changes in phospholipase D and phosphatidate phosphohydrolase activities in ischemia-reperfusion of rat heart. Arch Biochem Biophys 436:136–144
92. Munakata M, Stamm C, Friehs I et al (2002) Protective effects of protein kinase C during myocardial ischemia require activation of phosphatidyl-inositol specific phospholipase C. Ann Thorac Surg 73:1236–1245
93. Hariharan N, Zhai P, Sedoshima J (2011) Oxidative stress stimulates autophagic flux during ischemia/reperfusion. Antioxid Redox Signal 14:21792190
94. Makazan Z, Saini HK, Dhalla NS (2007) Role of oxidative stress in alterations of mitochondrial function in ischemic-reperfused hearts. Am J Physiol Heart Circ Physiol 292: H1986–H1994
95. Ward CA, Moffat MP (1995) Role of protein kinase C in mediating effects of hydrogen peroxide in guinea-pig ventricular myocytes. J Mol Cell Cardiol 27:1089–1097
96. Sabri A, Byron KL, Samarel AM et al (1998) Hydrogen peroxide activates mitogen-activated protein kinases and Na+–H+exchange in neonatal rat cardiac myocytes. Circ Res 82: 1053–1062
97. Yue TL, Ma XL, Wang X et al (1998) Possible involvement of stress-activated protein kinase signaling pathway and Fas receptor expression in prevention of ischemia/reperfusion-induced cardiomyocyte apoptosis by carvedilol. Circ Res 82:166–174
98. Uecker M, Da Silva R, Grampp T et al (2003) Translocation of protein kinase C isoforms to subcellular targets in ischemic and anesthetic preconditioning. Anesthesiology 99:138–147
99. Yoshida K, Hirata T, Akita Y et al (1996) Translocation of protein kinase C-alpha, delta and epsilon isoforms in ischemic rat heart. Biochim Biophys Acta 1317:36–44
100. Schulz R, Dodge KL, Lopaschuk GD, Clanachan AS (1997) Peroxynitrite impairs cardiac contractile function by decreasing cardiac efficiency. Am J Physiol Heart Circ Physiol 272:H1212
101. Xie YW, Kaminski PM, Wolin MS (1998) Inhibition of rat cardiac muscle contraction and mitochondrial respiration by endogenous peroxynitrite formation during posthypoxic reoxygenation. Circ Res 82:891–897
102. Singh RB, Dandekar SP, Elimban V et al (2004) Role of proteases in the pathophysiology of cardiac disease. Mol Cell Biochem 263:241–256
103. Singh RB, Chohan PK, Dhalla NS, Netticadan T (2004) The sarcoplasmic reticulum proteins are targets for calpain action in the ischemic-reperfused heart. J Mol Cell Cardiol 37: 101–110
104. Schulz R (2007) Intracellular targets of matrix metalloproteinase-2 in cardiac disease: rationale and therapeutic approaches. Ann Rev Pharmacol Toxicol 47:211–242
105. Chow AK, Cena J, Schulz R (2007) Acute actions and novel targets of matrix metalloproteinases in the heart and vasculature. Br J Pharmacol 152:189–205
106. Khalil PN, Neuhof C, Huss R et al (2005) Calpain inhibition reduces infarct size and improves global hemodynamics and left ventricular contractility in a porcine myocardial ischemia/reperfusion model. Eur J Pharmacol 528:124–131
107. Urthaler F, Wolkowicz PE, Digerness SB et al (1997) MDL-28170, a membrane-permeant calpain inhibitor, attenuates stunning and PKC epsilon proteolysis in reperfused ferret hearts. Cardiovasc Res 35:60–67

108. Pedrozo Z, Sanchez G, Torrealba N et al (2010) Calpains and proteasomes mediate degradation of ryanodine receptors in a model of cardiac ischemic reperfusion. Biochim Biophys Acta 1802:356–362
109. French JP, Quindry JC, Falk DJ et al (2006) Ischemia-reperfusion-induced calpain activation and SERCA2a degradation are attenuated by exercise training and calpain inhibition. Am J Heart Circ Physiol 290:H128–H136
110. Singh RB, Elimban V, Dhalla NS (2008) Differences in ischemia-reperfusion-induced endothelial changes in hearts perfused at constant flow and constant pressure. J Appl Physiol 105:1779–1787
111. Inserte J, Dorado-Garcia D, Hernando V, Soler-Soler J (2005) Calpain-mediated impairment of Na+/K+–ATPase activity during early reperfusion contributes to cell death after myocardial ischemia. Circ Res 97:465–473
112. Inserte J, Dorado-Garcia D, Hernando V et al (2006) Ischemic preconditioning prevents calpain-mediated impairment of Na+/K+–ATPase activity during early reperfusion. Cardiovasc Res 70:364–373
113. Yoshikawa Y, Zhang G-X, Obata K et al (2010) Cardioprotective effects of a novel calpain inhibitor SNJ-1945 for reperfusion injury after cardioplegic cardiac arrest. Am J Heart Circ Physiol 298:H643–H651
114. Singh RB, Dhalla NS (2010) Ischemia-reperfusion-induced changes in sarcolemmal Na+/K+–ATPase are due to the activation of calpain in the heart. Can J Physiol Pharmacol 88:388–397
115. Lalu MM, Pasini E, Schulze CJ et al (2005) Ischaemia-reperfusion injury activates matrix metalloproteinases in the human heart. Eur Heart J 26:27–35
116. Sawicki G, Leon H, Sawicka J et al (2005) Degradation of myosin light chain in isolated rat hearts subjected to ischemia-reperfusion injury: a new intracellular target for matrix metalloproteinase-2. Circulation 112:544–552
117. Spanikova A, Ivanova M, Matejíková J et al (2010) Influence of ischemia/reperfusion and modulation of PI3K/Akt kinase pathway on matrix metalloproteinase-2 in rat hearts. Gen Physiol Biophys 29:31–40
118. Kundasamy AD, Chow AK, Ali MA, Schulz R (2010) Matrix metalloproteinase-2 and myocardial oxidative stress injury: beyond the matrix. Cardiovasc Res 85:413–423
119. Murphy E, Steenbergen C (2007) Gender-based differences in mechanisms of protection in myocardial ischemia-reperfusion injury. Cardiovasc Res 75:478–486
120. Ostadal B, Netuka I, Maly J et al (2009) Gender differences in cardiac ischemic injury and protection–experimental aspects. Exp Biol Med 234:1011–1019
121. Arain FA, Kuniyoshi FH, Abdalrhim AD, Miller VM (2009) Sex/gender medicine. The biological basis for personalized care in cardiovascular medicine. Circ J 73:1774–1782
122. Mercuro G, Deidda M, Piras A et al (2010) Gender determinants of cardiovascular risk factors and diseases. J Cardiovasc Med 11:207–220
123. Banos G, Medina-Campos ON, Maldonado PD et al (2005) Antioxidant enzymes in hypertensive and hypertriglyceridemic rats: effect of gender. Clin Exp Hypertens 1:45–57
124. Safar ME, Smulyan H (2004) Hypertension in women. Am J Hypertens 17:82–87
125. Regitz-Zagrosak V, Lehmkuhl E (2005) Heart failure and its treatment in women. Role of hypertension, diabetes, and estrogen. Herz 30:356–367
126. Belo N, Mosca L (2004) Epidemiology of coronary heart disease in women. Prog Cardiovasc Dis 4:287–295
127. Hoppe BL, Hermann DD (2003) Sex differences in the causes and natural history of heart failure. Curr Cardiol Rep 5:193–199
128. Jessup M, Pina IL (2004) Is it important to examine gender differences in the epidemiology and outcome of severe heart failure? J Thor Cardiovasc Surg 127:1247–1252
129. Stromberg A, Martensson J (2003) Gender differences in patients with heart failure. Eur J Cardiovasc Nurs 2:7–18
130. Schonfelder G (2005) The biological impact of estrogens on gender differences in congestive heart failure. Cardiovasc Res 67:573–574

131. Biondi-Zoccai GGL, Baldi A, Biasucci LM, Abbate A (2004) Female gender, myocardial remodeling and cardiac failure: are women protected from increased myocardial apoptosis? Ital Heart J 5:498
132. Solimene MC (2010) Coronary heart disease in women: a challenge for the 21st century. Clinics (Sao Paulo) 65:99–106
133. Campbell DJ, Somaratne JB, Jenkins AJ et al (2011) Differences in myocardial structure and coronary microvasculature between men and women with coronary artery disease. Hypertension 57:186–192
134. Mackay MH, Ratner PA, Johnson JL et al (2011) Gender differences in symptoms of myocardial ischaemia. Eur Heart J 32:3107–3114
135. Litwin SE, Katz SE et al (1999) Gender differences in postinfarction left ventricular remodeling. Cardiology 91:173–183
136. Douglas PS, Katz SE et al (1998) Hypertrophic remodeling: gender differences in the early response to left ventricular pressure overload. J Am Coll Cardiol 32:1118
137. Skavdahl M, Steenbergen C, Litwin CM et al (2005) Gender differences in postinfarction left ventricular remodeling. Am J Physiol Heart Circ Physiol 288:H469
138. Alyono D, Ring WS, Anderson MR, Anderson RW (1984) Left ventricular adaptation to volume overload from large aortocaval fistula. Surgery 96:360–367
139. Flaim SF, Minteer WJ, Nellis SH, Clark DP (1979) Chronic arteriovenous shunt: evaluation of a model for heart failure in rat. Am J Physiol Heart Circ Physiol 236:H698–H704
140. Gardner JD, Brower GL, Janicki JS (2002) Gender differences in cardiac remodeling secondary to chronic volume overload. J Cardiac Fail 8:101–107
141. Gardner JD, Brower GL, Voloshenyuk TG, Janicki JS (2008) Cardioprotection in female rats subjected to chronic volume overload: synergistic interaction of estrogen and phytoestrogens. Am J Physiol Heart Circ Physiol 294:H198–H204
142. Gardner JD, Brower GL, Janicki JS (2005) Effects of dietary phytoestrogens on cardiac remodeling secondary to chronic volume overload in female rats. J App Physiol 99:1378–1383
143. Dent MR, Tappia PS, Dhalla NS (2010) Gender differences in cardiac dysfunction and remodeling due to volume overload. J Cardiac Fail 16:439–449, erratum: J Cardiac Fail 17: 179, 2011
144. Dent MR, Tappia PS, Dhalla NS (2010) Gender differences in apoptotic signaling in heart failure due to volume overload. Apoptosis 15:499–510, erratum: Apoptosis 16: 757, 2011
145. Dent MR, Tappia PS, Dhalla NS (2011) Gender differences in β-adrenoceptor system in cardiac hypertrophy due to arteriovenous fistula. J Cell Physiol 226:181–186
146. Dent MR, Tappia PS, Dhalla NS (2012) Gender related alterations of β-adrenoceptor mechanisms in heart failure due to arteriovenous fistula. J Cell Physiol 227:3080–3087
147. Vlkovicova J, Javorkova V, Pechánová O, Vrbjar N (2005) Gender difference in functional properties of Na. K-ATPase in the heart of spontaneously hypertensive rats. Life Sci 76:971–982
148. Chen J, Petranka J, Yamamura K et al (2003) Gender differences in sarcoplasmic reticulum calcium loading after isoproterenol. Am J Physiol Heart Circ Physiol 285:H2657–H2662
149. Wei S-K, McCurley JM, Hanlon SU, Haigney MC (2007) Gender differences in Na/Ca exchanger current and beta-adrenergic responsiveness in heart failure in pig myocytes. Ann N Y Acad Sci 1099:183–189
150. Brower GL, Gardner JD, Janicki JS (2003) Gender mediated cardiac protection from adverse ventricular remodeling is abolished by ovariectomy. Mol Cell Biochem 251:89–95
151. Bhuiyan MS, Shioda N, Fukunaga K (2007) Ovariectomy augments pressure overload-induced hypertrophy associated with changes in Akt and nitric oxide synthase signaling pathways in female rats. Am J Physiol Endocrinol Metab 293:E1606–E1614
152. Kam KW, Kravtsov GM, Liu J, Wong TM (2005) Increased PKA activity and its influence on isoprenaline-stimulated L-type Ca2+ channels in the heart from ovariectomized rats. Brit J Pharmacol 144:972–981
153. Kravtsov GM, Kam KW, Liu J et al (2007) Altered Ca(2+) handling by ryanodine receptor and Na(+)-Ca(2+) exchange in the heart from ovariectomized rats: role of protein kinase A. Am J Physiol Heart Circ Physiol 292:C1625–C1635

154. Bupha-Intr T, Wattanapermpool J (2006) Regulatory role of ovarian sex hormones in calcium uptake activity of cardiac sarcoplasmic reticulum. Am J Physiol Heart Circ Physiol 291: H1101–H1108
155. Ma Y, Cheng WT, Wu S, Wong TM (2009) Oestrogen confers cardioprotection by suppressing Ca2+/calmodulin-dependent protein kinase II. Brit J Pharmacol 157:705–715
156. Kam KW, Qi JS, Chen M, Wong TM (2004) Estrogen reduces cardiac injury and expression of beta1-adrenoceptor upon ischemic insult in the rat heart. J Pharmacol Exp Ther 309:8–15
157. Patten RD, Pourati I, Aronovitz MJ et al (2004) 17beta-estradiol reduces cardiomyocyte apoptosis in vivo and in vitro via activation of phospho-inositide-3 kinase/Akt signaling. Circ Res 95:692–699
158. Marni F, Wang Y, Morishima M et al (2009) 17 beta-estradiol modulates expression of low-voltage-activated Ca(V)3.2 T-type calcium channel via extracellularly regulated kinase pathway in cardiomyocy. Endocrinology 150:879–888
159. Chu SH, Goldspink P, Kowalski J et al (2006) Effect of estrogen on calcium-handling proteins, beta-adrenergic receptors, and function in rat heart. Life Sci 79:1257–1267
160. Kilic A, Javadov S, Karmazyn M (2009) Estrogen exerts concentration-dependent pro-and anti-hypertrophic effects on adult cultured ventricular myocytes. Role of NHE-1 in estrogen-induced hypertrophy. J Mol Cell Cardiol 46:360–369
161. Bae S, Zhang L (2005) Gender differences in cardioprotection against ischemia/reperfusion injury in adult rat hearts: focus on Akt and protein kinase C signaling. J Pharmacol Exp Ther 315:1125–1135
162. Xue Q, Zhang L (2009) Prenatal hypoxia causes a sex-dependent increase in heart susceptibility to ischemia and reperfusion injury in adult male offspring: role of protein kinase C epsilon. J Pharamcol Exp Ther 330:624–632
163. Shinmura K, Nagai M, Tamaki K, Bolli R (2008) Loss of ischaemic preconditioning in ovariectomized rat hearts: possible involvement of impaired protein kinase C epsilon phosphorylation. Cardiovasc Res 79:387–394
164. Cross HR, Lu L, Steenbergen C et al (1998) Overexpression of the cardiac Na+/Ca2+ exchanger increases susceptibility to ischemia/reperfusion injury in male, but not female, transgenic mice. Circ Res 83:1215–1223
165. Cross HR, Murphy E, Steenbergen C (2002) Ca(2+) loading and adrenergic stimulation reveal male/female differences in susceptibility to ischemia-reperfusion injury. Am J Physiol Heart Circ Physiol 283:H481–H489
166. Imahashi K, London RE, Steenbergen C, Murphy E (2004) Male/female differences in intracellular Na+regulation during ischemia/reperfusion in mouse heart. J Mol Cell Cardiol 37:747–753
167. Brown DA, Lynch JM, Armstrong CJ et al (2005) Susceptibility of the heart to ischaemia-reperfusion injury and exercise-induced cardioprotection are sex-dependent in the rat. J Physiol 564:619–630
168. Johnson MS, Moore RL, Brown DA (2006) Sex differences in myocardial infarct size are abolished by sarcolemmal KATP channel blockade in rat. Am J Physiol Heart Circ Physiol 290:H2644–H2647
169. Arieli Y, Gursahani H, Eaton MM et al (2004) Gender modulation of Ca(2+) uptake in cardiac mitochondria. J Mol Cell Cardiol 37:507–513
170. Wang M, Tsai BM, Crisostomo PR, Meldrum DR (2006) Tumor necrosis factor receptor 1 signaling resistance in the female myocardium during ischemia. Circulation 114(suppl I):I 282–I 289
171. Zeller CN, Wang Y, Markel TA et al (2009) Role of tumor necrosis factor receptor 1 in sex differences of stem cell mediated cardioprotection. Ann Thorac Surg 87:812–819
172. Beer S, Reincke M, Kral M et al (2002) Susceptibility to cardiac ischemia/reperfusion injury is modulated by chronic estrogen status. J Cardiovasc Pharmacol 40:420–428
173. Jeanes HL, Wanikiat P, Sharif I, Gray GA (2006) Medroxyprogesterone acetate inhibits the cardioprotective effect of estrogen in experimental ischemia-reperfusion injury. Menopause 13:80–86

174. Chae S-U, Ha K-C, Piao CS et al (2007) Estrogen attenuates cardiac ischemia-reperfusion injury via inhibition of calpain-mediated bid cleavage. Arch Pharm Res 30:1225–1235
175. Kim JK, Pedram A, Razandi M, Levin ER (2006) Estrogen prevents cardiomyocyte apoptosis through inhibition of reactive oxygen species and differential regulation of p38 kinase isoforms. J Biol Chem 281:6760–6767
176. Wang M, Tsai BM, Reiger KM et al (2006) 17-beta-Estradiol decreases p38 MAPK-mediated myocardial inflammation and dysfunction following acute ischemia. J Mol Cell Cardiol 40:205–212
177. Wang F, He Q, Sun Y et al (2010) Female adult mouse cardiomyocytes are protected against oxidative stress. Hypertension 55:1172–1178
178. Golden KL, Marsh JD, Jiang Y (2002) Castration reduces mRNA levels for calcium regulatory proteins in rat heart. Endocrine 19:339–344
179. Golden KL, Marsh JD, Jiang Y et al (2003) Gonadectomy of adult male rats reduces contractility of isolated cardiac myocytes. Am J Physiol Endocrinol Metab 285:E449–E453
180. Golden KL, Marsh JD, Jiang Y (2004) Testosterone regulates mRNA levels of calcium regulatory proteins in cardiac myocytes. Horm Metab Res 36:197–202
181. Huang C, Gu H, Zhang W et al (2010) Testosterone-down-regulated Akt pathway during cardiac ischemia/reperfusion: a mechanism involving BAD, Bcl-2 and FOXO3a. J Surg Res 164:e1–11
182. Tsang S, Wu S, Liu J, Wong TM (2008) Testosterone protects rat hearts against ischaemic insults by enhancing the effects of alpha(1)-adrenoceptor stimulation. Br J Pharmacol 153:693–709
183. Wang M, Wang Y, Abarbanell A et al (2009) Both endogenous and exogenous testosterone decrease myocardial STAT3 activation and SOCS3 expression after acute ischemia and reperfusion. Surgery 146:138–144

Chapter 17
Role of MicroRNAs in Cardiac Hypertrophy and Postinfarction Remodeling

Jian Ding and Da-Zhi Wang

Abstract A family of small noncoding RNAs, termed microRNAs (miRNAs), plays critical roles in multiple physiological and pathological processes by negatively modulating gene expression at the posttranscriptional level. Cardiac hypertrophy and myocardial infarction, which are among the major causes of heart failure, have been found to be accompanied by dysregulated expression of miRNAs, suggesting that miRNA-mediated gene regulation is involved in cardiovascular pathogenesis. Loss- and gain-of-function studies by genetic or pharmacological manipulation of individual miRNAs further reveal that these small RNAs play critical regulatory roles in hypertrophy and postinfarction remodeling. Numerous miRNAs are cardiac protective, while some others are apparently detrimental during cardiac remodeling. Intriguingly, the functional consequences of some miRNAs in the heart appear to be cell type specific. Although the biological functions and molecular mechanisms of most miRNAs still remain elusive, their key regulatory roles in cardiac pathogenesis make them promising targets for therapeutic intervention. Here, we summarize the recent progress made in understanding the function and mechanisms of individual miRNAs in cardiac hypertrophy and postinfarction remodeling. The biological and clinical implications of miRNAs in the cardiovascular system are also discussed.

Keywords MicroRNA • Gene regulation • Cardiac hypertrophy • Myocardial infarction • Cardiac remodeling

J. Ding • D.-Z. Wang, Ph.D. (✉)
Department of Cardiology, Boston Children's Hospital, Harvard Medical School, 320 Longwood Avenue, Boston, MA 02115, USA
e-mail: dwang@enders.tch.harvard.edu

B.I. Jugdutt and N.S. Dhalla (eds.), *Cardiac Remodeling: Molecular Mechanisms*, Advances in Biochemistry in Health and Disease 5, DOI 10.1007/978-1-4614-5930-9_17,

17.1 Introduction

miRNAs have emerged as pivotal regulatory factors of gene expression. This class of small RNAs inhibits the expression of downstream genes through translational repression or mRNA degradation by base-pairing with mRNA transcripts. The human genome may encode more than 1,500 miRNAs (http://www.mirbase.org), and most of them are conserved across species [1]. Additional miRNAs are continually being identified. Conversely, at least 60% of mammalian genes are directly regulated by miRNAs. An individual mRNA transcript usually contains binding sites for multiple miRNAs, while a single miRNA can also target numerous mRNA transcripts. These miRNAs and mRNAs form a complex regulatory network and modulate multiple physiological/pathological processes. Unlike transcription factors, which often act via a crude "on–off" mechanism, miRNAs function as "fine-tuners" rather than "master-switches" [2, 3]. However, growing evidence suggests that such "fine-tuning" regulation of gene expression by miRNAs can significantly influence pathogenesis in human diseases.

miRNA involvement in the regulation of cardiac remodeling has been implicated by numerous observations. First, cardiovascular diseases are often accompanied by dramatically altered miRNA expression [4–7]. Second, a cardiac-specific block of miRNA biogenesis in mice resulted in severe heart defects [8, 9]. Furthermore, a growing number of functional studies using genetic or pharmacological manipulation have demonstrated that many individual miRNAs significantly influence the pathogenesis of various cardiovascular diseases [10–12].

Cardiac hypertrophy and myocardial infarction are among the major causes of heart failure with a high risk of morbidity and mortality. Many regulatory networks, including multiple signal cascades, are involved in pathogenesis. Recent studies have uncovered that miRNAs play critical roles in cardiac hypertrophy and postinfarction remodeling. Due to their relative ease of pharmacological manipulation, miRNAs are considered an intriguing and promising target for therapeutic intervention. In this review, we summarize recent studies on the functions of numerous individual miRNAs in cardiac hypertrophy and postinfarction remodeling. The molecular mechanisms as well as the biological/clinical implications of miRNAs in the cardiovascular system are also discussed.

17.2 miRNAs in Cardiac Hypertrophy

Cardiac hypertrophy, primarily defined as an increase in the size of cardiomyocytes without a change in myocyte numbers, is further grouped in two different types: physiological hypertrophy and pathological hypertrophy. Pathologic cardiac hypertrophy is one of the major health problems in developed countries. It is usually accompanied with enhanced protein synthesis, an "adult to fetal" switch in cardiac gene expression (switch of adult α- to fetal β-myosin heavy chains in rodents), as well as reorganization of the cytoskeleton. It has long been speculated that the switch from α- to β-myosin heavy chain composition in myocytes may contribute

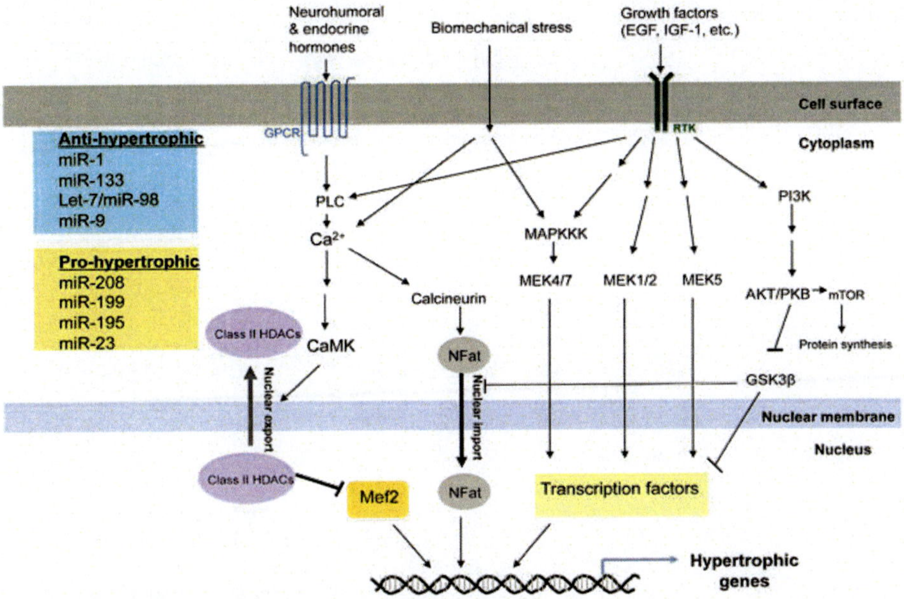

Fig. 17.1 Major cellular signal pathways that modulate cardiac hypertrophy. Signal cues, including biomechanical stress, neurohumoral, and endocrine hormones, can induce cellular hypertrophy responses by triggering a complex signal network involving MAPK, Ca^{2+}–calcineurin–NFAT, and PI3K-Akt pathways. miRNAs play critical roles in modulating cardiac hypertrophy by regulating the expression of multiple components in the signal network. Particularly, miR-1, miR-133, miR98, and miR-9 are apparently anti-hypertrophic, while miR-208, miR-199, miR-195, and miR-23 appear to be pro-hypertrophic miRNAs

to hypertrophic pathogenesis [13]. Alteration of the actin cytoskeleton not only affects the morphological integrity and mechanical resistance of the myocytes but also influences the transduction of stress signals, leading to changes in downstream gene expression and function [5]. Multiple cellular pathways and molecular events have been found to modulate cardiac hypertrophy. Signal cues, such as biomechanical stress, neurohumoral, and endocrine hormones (Endothelin-1, Insulin-like growth factor-1, etc.), could induce hypertrophic growth of cardiomyocytes by triggering numerous intracellular pathways, including the MAPK, Ca^{2+}–calcineurin–NFAT, and PI3K-Akt signal cascades [14–19]. Intriguingly, dysregulation of numerous miRNAs has been observed in cardiac hypertrophy in recent years, and direct evidence linking miRNAs to hypertrophy is emerging (Fig. 17.1).

17.2.1 Anti-hypertrophic miRNAs

Many miRNAs are shown anti-hypertrophic (Fig. 17.1) (Table 17.1). The bicistronic miRNA clusters miR-1-1/miR133a-2 and miR-1-2/miR-133a-1 are preferentially expressed in cardiac and skeletal muscle. miR-1-1 and miR-1-2 are probably the

Table 17.1 miRNAs in cardiac hypertrophy

	miRNA	Expression in cardiac hypertrophy	Targets (biological functions)
Anti-hypertrophic	miR-1	Downregulated	Calm1, Calm2, Mef2C, Gata4 (Ca^{2+} signaling)
			Twf1 (cytoskeleton regulation)
			IGF-1 (IGF-1 signaling)
	miR-133	Downregulated	RhoA, Cdc42 (cytoskeleton regulation)
			CnAβ, NFATc (Ca^{2+} signaling)
	let-7/miR-98	Downregulated	Cyclin D2 (MAPK signaling)
	miR-9	Downregulated	Myocardin (Ca^{2+} signaling)
Pro-hypertrophic	miR-208	miR-208a downregulated	Thrap1 (thyroid hormone signaling)
		miR-208b upregulated	Myostatin (inhibitor of Akt/PKB pathways)
	miR-195	Upregulated	
	miR-199	Upregulated	Dyrk1a (inhibitor of Ca^{2+} signaling)
	miR-23	Upregulated	Murf1, Foxo3A (negative regulator of cell growth)

most abundant miRNAs in the heart [20–22]. Downregulation of miR-1 levels in the heart has been consistently observed in different animal models of hypertrophy as well as in patients with hypertrophic myopathy [4, 23]. Inhibition of miR-1 activity with antagomirs appears to be sufficient to induce hypertrophic growth of rat ventricle cardiomyocytes in vitro, suggesting that downregulation of miR-1 may be causative in the pathogenesis of cardiac hypertrophy. Consistently, overexpression of miR-1 attenuates hypertrophy both in vitro and in vivo, further implying that this small noncoding RNA may be an important anti-hypertrophy regulator [5, 23]. miR-1 may exert its anti-hypertrophic function by repressing numerous pro-hypertrophic targets. A study by Ikeda et al. showed that miR-1 can repress the translation of the calmodulin-encoding mRNAs Calm1 and Calm2 by targeting their 3′ untranslated regions (3′ UTRs). miR-1 also has inhibitory effects on the expression of Mef2a and Gata4, key downstream transcription factors in calcium pathways. Together, these studies demonstrated that miR-1 could repress hypertrophic growth by negatively regulating calcium signaling [23]. miR-1 was also reported to target twinfilin-1 (Twf1), a cytoskeletal regulator. Overexpression of miR-1 represses the expression of endogenous Twf1, while attenuation of miR-1 results in an elevated level of Twf1 protein in cardiomyocytes. Thus, miR-1 could repress hypertrophy through regulation of the cardiac cytoskeleton [5]. Insulin-like growth factor-1 (IGF-1) is another pro-hypertrophic factor repressed by miR-1. It has been demonstrated that the expression levels of miR-1 and the IGF-1 protein are inversely correlated in models of cardiac hypertrophy and heart failure. Intriguingly, IGF-1 signaling reciprocally regulates miR-1 expression. Thus, miR-1 and IGF-1 form a signal transduction feedback loop and together regulate hypertrophic growth [24].

Like miR-1, miR-133 has been found to be downregulated in hypertrophic cardiomyocytes. Overexpression of miR-133 in endothelin-1 (ET1) or phenylephrine (PE)-treated rat neonatal cardiomyocytes repressed hypertrophy, while loss of function of miR-133 by antagomir treatment was sufficient to induce hypertrophic growth of myocytes both in vitro and in vivo [25]. Studies have identified multiple targets of miR-133, many of which act as hypertrophic regulators. The RhoA GTPase and Cdc42, which function in modulating cytoskeletal and myofibrillar rearrangements during hypertrophy, have been shown to be targeted by miR-133 [25]. miR-133 also exhibits inhibitory effects on the expression of NFATc and CnA, two key players in calcium signal-mediated cardiac hypertrophy pathways [26].

Manipulation of miR-133 levels via adenovirus-mediated overexpression or antagomir-mediated knockdown altered the expression of its pro-hypertrophic targets and repressed or induced hypertrophic growth of cardiomyocytes, respectively. However, transgenic overexpression of miR-133 in cardiac muscle, though it inhibited myocardial fibrosis, did not affect the extent of hypertrophy in mice with transverse aortic constriction (TAC). Mice with a genetic deletion of miR-133a-1, miR-133a-2, or both did not exhibit a cardiac hypertrophy phenotype [27]. The discrepancy of the phenotypic outcomes may be attributed to the different approaches used to manipulate the level of miR-133. For instance, genetic deletion of miR-133 results in complete loss of 5p and 3p miR-133 species, both of which could have certain functions. In contrast, antagomirs only inhibit the function of one miRNA strand (5p or 3p). Moreover, conventional knockout or transgenic expression of miR-133 alters the gene expression program during cardiogenesis and, simultaneously or subsequently, could also initiate potential compensatory responses in the embryonic/fetal heart, which could be preserved and thus blunt miR-133's effects on hypertrophy in the adult heart. Alternatively, overexpression or antagomir inhibition of miR-133 could result in an "off-target" issue; therefore, caution should be used in interpreting the results.

Let-7/miR-98 are another set of miRNAs downregulated in cardiac hypertrophy that appears to be anti-hypertrophic. Adenovirus-mediated expression of miR-98 in cardiomyocytes resulted in reduced cell size and the repression of angiotensin II (ANG II)-induced hypertrophic growth. The anti-hypertrophic function of miR-98 may be partially attributed to its inhibitory effects on the expression of cyclin D2, a member of the G1-phase cyclin family, which is reportedly involved in cardiac hypertrophy. Intriguingly, expression of let-7/miR-98 is positively regulated by thioredoxin 1 (Trx1), which inhibits pathologic cardiac hypertrophy. Knockdown of miR-98 attenuated the inhibitory effects of Trx1 in Ang II-induced hypertrophic growth [28]. Thus, miR-98 appears to act downstream of Trx1 to mediate its anti-hypertrophic function.

Expression of miR-9 is downregulated in cardiomyocytes upon stimulation of hypertrophy with isoproterenol and aldosterone. On the other hand, administration of miR-9 attenuated cardiac hypertrophy, suggesting that this miRNA functions as a key regulator in hypertrophy signaling. miR-9 appears to exert its function in hypertrophy by repressing myocardin, a key downstream mediator of NFATc3 in calcium-induced hypertrophic cascades [29]. In this manner, miR-9 appears to be anti-hypertrophic. However, the cardiac functions of this small RNA might be more complicated. As recently reported by Zhang et al. expression of miR-9 was induced by PDGF signaling in cardiomyocytes, while miR-9 reciprocally repressed the expression of PDGFR-β, a key component of the PDGF signaling cascade, which plays a pivotal role in stress-induced cardiac angiogenesis during hypertrophy [30]. Therefore, miR-9 may also function in modulating the paracrine angiogenic capacity of cardiomyocytes in response to stress. miR-9 appears to be a versatile miRNA, with its functional role(s) in hypertrophy remains in need of further characterization in vivo.

17.2.2 *Pro-hypertrophic miRNAs*

Conversely, many miRNAs are known pro-hypertrophic (Fig. 17.1) (Table 17.1). The cardiac sarcomeric myosin heavy chains (MHCs) are the primary regulators of muscle strength and contractility. In mouse, the Myh7 and myh7b genes encode the β-isoforms of MHC, with relatively low actomyosin ATPase activity, while myh6 encodes the fast-switch α-MHC. Cardiac hypertrophy is usually accompanied by an altered ratio of these two myosin isoforms [13]. Intriguingly, three miRNAs, miR-208a, miR-208b, and miR-499, termed "myomiRs," are encoded within the introns of myh6, myh7, and myh7b genes, respectively. These myomiRs function in modulating myosin gene expression and muscle contractility [31–33]. Given its unique expression pattern, it is not surprising that miR-208a is a key regulator of cardiac hypertrophy. In response to pressure overload and calcineurin signaling, miR-208a knockout mice failed to upregulate β-MHC and exhibited resistance to cardiac hypertrophy [32]. Overexpression of miR-208a results in increased expression of β-MHC and induction of hypertrophic growth in cardiomyocytes [31]. Recent studies have demonstrated that the functions of miR-208a in regulating myosin expression may be attributed to its direct repression of numerous transcription regulators, including thyroid hormone receptor associated protein 1 (Thrap1), Sox6, Hp-1β, and Sp3 [32, 33]. However, the mechanisms by which miR-208a modulates the upregulation of β-MHC in vivo and in hypertrophy need to be further investigated. For instance, it will be important to test the inhibitory effects of miR-208 on its targets: Thrap1, Sox6, and Hp-1β in hypertrophic heart vs. normal heart. miR-208 potently induced the expression of β-MHC in cardiomyocytes, but intriguingly, the induced expression of β-MHC in myocardium is heterogeneous in miR-208a transgenic mice [31]. This may be due to the "mosaicism" of transgene expression. However, the possibility that individual cardiomyocytes might differentially respond to miR-208 overexpression, or that the upregulation of α-MHC in vivo could be secondary to miR-208 induced cardiac abnormalities, cannot be excluded.

Previous studies have demonstrated that elevated expression of β-MHC alone is not sufficient to induce hypertrophy. Thus, the pro-hypertrophic function of miR-208a may not be solely attributed to its upregulation of β-MHC expression. The study by van Rooij et al. suggested that miR-208a might also regulate how the heart responds to stress during hypertrophy. Genetic deletion of miR-208a significantly altered the mRNA levels of genes encoding early response factors and heat shock proteins. However, none of these mRNAs contained predicted binding sites of miR-208a. Thus, the regulation on these cardiac stress response genes by miR-208a may be indirect [32]. Myostatin, a negative regulator of muscle growth, has been shown to be directly targeted by miR-208a. Inhibition of myostatin may contribute to the function of miR-208a in hypertrophy [31]. On the other hand, miR-208a also repressed the expression of Gata4, which is required for cardiac hypertrophy and appears to be pro-hypertrophic. Thus, miR-208a could target both anti-hypertrophic and pro-hypertrophic factors. The effects of miR-208 on these targets with opposing functions are likely dependent on cellular context. Growing evidence has suggested that miRNAs may regulate the expression of different subsets of downstream genes in different conditions. Comprehensive and systemic identification of targets of miR-208a in relevant physiological/pathological contexts is a key step to identify and understand the molecular mechanisms underlying its function in cardiac hypertrophy.

The three identified myomiRs, miR-208a, miR-208b, and miR-499, contain nearly identical seed sequences and may share certain common downstream targets and act redundantly. Consistent with this idea, transgenic expression of miR-499 in the hearts of miR-208a null mice was sufficient to restore the expression of β-MHC in the presence of propylthiouracil [33]. miR-499, like miR-208a, when overexpressed at a high level, can induce cardiac hypertrophy in vivo [34]. However, miR-499 was also shown to have anti-hypertrophic effects. CnA, a strong pro-hypertrophic factor, was found to be one of the direct targets of miR-499 in this setting. As shown by Wang et al. inhibition of miR-499 with an antagomir significantly elevated the signaling activity of the CnA–NFAT pathway and enhanced isoproterenol (ISO)-induced hypertrophy in vivo [35]. Thus, the functional roles of miR-499 in cardiac hypertrophy are still elusive.

More interestingly, miR-208a appears to act upstream of and regulate the expression of two other myomiRs in the adult mouse heart. These miRNAs form a "myomiR network" and regulate muscle myosin content and myofiber identity in the heart [33]. However, it is still a mystery how these myomiRs act together to modulate muscle strength and contractility in the process of cardiac hypertrophy. Are they friends or foes? In the myomiR network, we might predict that miR-499 cooperates with miR-208 to repress the expression of a set of common target genes, therefore synergistically or additively regulating muscle performance. On the other hand, miR-208 and miR-499 could also inhibit subsets of distinct downstream targets, which might be functionally unrelated or opposite, therefore resulting in complicated consequences. The study by van Rooij et al. demonstrated that the three myomiRs, though sharing a nearly identical seed sequence, might differentially repress their common target genes. The differences may be attributed to the significant dissimilarity in sequences outside the seed region of mature miR-208a, miR-208b, and miR-499, which may influence their targeting capacities and preferences [33]. The mechanisms of the myomiR network could be even more complicated than mentioned above, since miR-499 and miR-208a/b, having different targeting capacities on their common targets, could each function as a "dominant negative miRNA" of the other. Obviously, such complexity will confound the study of individual myomiRs.

miR-195 is a stress-responsive miRNA that has been found upregulated in mouse models of cardiac hypertrophy induced by TAB or active CnA overexpression as well as in human patients with cardiac hypertrophy [4, 6]. Overexpression of miR-195 induced hypertrophic growth of isolated rat neonatal cardiomyocytes in vitro. Consistently, transgenic expression of miR-195 in cardiomyocytes driven by the α-MHC promoter also resulted in hypertrophy and cardiac dysfunction [6, 36]. Though the mechanisms are still not clear, the pro-hypertrophic effects of miR-195 suggest that this miRNA is a key regulator of hypertrophy and could be a potential therapeutic target of anti-hypertrophic treatment in the heart.

The miR-199 family has two members: miR-199a and miR-199b. miR199b has been found to be upregulated in heart failure of both mice and humans. It has also been demonstrated that the level of miR-199a was elevated in hypertrophic hearts but decreased in the later stages of heart failure in mice [36]. Overexpression of miR-199a in neonatal rat cardiomyocytes was sufficient to induce hypertrophic growth in vitro, whereas inhibition of miR-199a attenuated PE-induced hypertrophy [36]. A pro-hypertrophic function was also confirmed for miR-199b. Martins et al. reported that silencing of endogenous miR-199b by antagomirs significantly inhibited hypertrophy and fibrosis in vivo in mice hearts. Intriguingly, the promoter region (~2.7 kb upstream) of the miR-199b gene contained a conserved and functional NFAT binding site, so transcription of this miRNA can be regulated by CnA–NFAT signaling. miR-199b could directly target the dual-specificity tyrosine (Y) phosphorylation-regulated kinase 1a (Dyrk1a), which has been reported to downregulate NFAT activity [37]. Thus, miR-199b apparently forms a signal circuit with NFAT to promote cardiac hypertrophy. Such a notion is not unique to miR-199b, given increasing evidence that many miRNAs and their targets function in a feedback or feed-forward regulatory network to modulate gene expression.

miR-23a is another miRNA that has been consistently found to be upregulated in numerous hypertrophic models. miR-23a is a direct target of NFATc and may act as a feed-forward factor to elevate the signaling output of the CnA–NFAT pathway in hypertrophy. Lin et al. reported that a miR-23a antagomir could abolish CnA–NFATc3-induced cardiomyocyte hypertrophy in vitro. Transgenic expression of miR-23a, though not sufficient to cause obvious cardiac abnormality, significantly enhanced PE- or TAC-induced hypertrophy. Conversely, downregulation of miR-23a significantly attenuated isoproterenol-induced cardiac hypertrophic growth in vivo [38]. Thus, miR-23a appears to be a key regulator for the induction of cardiac hypertrophy. It has been reported that miR-23a can target and repress Murf1 and Foxo3a, negative regulators of cell growth. This may contribute to the pro-hypertrophic function of miR-23a [39]. Clearly, genetic studies in which miR-23a is mutated will allow us to further define the function of this miRNA.

Another question is how much redundancy miR-23a has with other members of the miR-23 family (miR-23b and miR-23C). Nevertheless, miR-23a appears to be a potential therapeutic target for the development of anti-hypertrophic drugs.

17.3 miRNAs in the Postinfarction Heart

Myocardial infarction (MI) is among the major causes of both debilitating illness and death in the world. Interruption of the local blood supply in heart ventricle, usually due to coronary artery occlusion, results in death of cardiomyocytes. The loss of myocardium induces a unique pattern of postinfarction remodeling, including ventricle dilation, eccentric hypertrophy, scar formation in the infarct area, interstitial fibrosis of adjacent myocardium, and alteration of ventricle geometry. At the cellular level, the remodeling process is usually accompanied with cell death (necrosis and apoptosis), hypertrophic growth of myocytes, as well as altered proliferation/migration of fibroblasts and endothelial cells (Fig. 17.2). Upregulation or downregulation of miRNAs has been found in myocardial infarction and postinfarction remodeling. Multiple studies have shown that alteration of miRNA expression contributes to cardiac injury and/or protection in MI by regulating key signaling elements (Fig. 17.2) (Table 17.2).

Numerous studies have documented the significantly altered expression pattern of miR-1 in myocardial infarction and/or ischemia/reperfusion (I/R). miR-1 is dramatically elevated in rat hearts upon acute myocardial I/R [40]. In the hearts of patients who died of myocardial infarction, miR-1 levels appear to be reduced in the infarcted area but elevated in remote non-infarcted regions [41]. The altered temporal and spatial expression pattern suggests that miR-1 might play key role(s) in myocardial infarction. Mechanistically, miR-1 appears to function as a proapoptotic miRNA. Overexpression of miR-1 is sufficient to induce apoptosis in H9c2 cells, while myocytes treated with a miR-1 inhibitor were more resistant to oxidative stress. miR-1 might exert its proapoptotic functions cumulatively by repressing the expression of multiple antiapoptotic regulators, including HSP60, HSP70, and Bcl-2 [40, 42].

As discussed earlier, miR-133 and miR-1 may function cooperatively or synergistically to regulate hypertrophic growth in myocytes. However, during myocyte apoptosis, these two miRNAs appear to play opposite roles. Unlike miR-1, miR-133 blunted oxidative stress responses and inhibited apoptosis when overexpressed in H9c2 cells. Inhibition of miR-133 promoted apoptotic cell death in H9c2 cells as well as in isolated neonatal rat ventricular cardiomyocytes. miR-133 repressed the expression of caspase-9, a key regulator of apoptosis, likely by directly targeting the 3′ UTR of this gene. Thus, miR-133 could inhibit proapoptotic factors, including caspase-9, and protect cardiomyocyte from apoptosis [42]. Together, these studies suggest that miR-1 and miR-133 could play pivotal roles in maintaining myocardial homeostasis by balancing cardiomyocyte death and survival. However, further investigations in vivo are needed to properly define their functional roles in infarcted hearts.

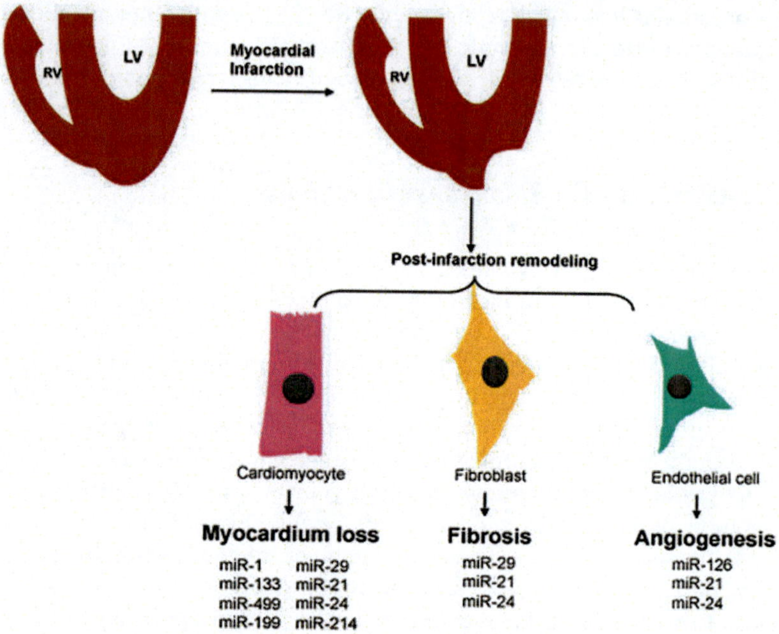

Fig. 17.2 The schematic of postinfarction remodeling. The loss of myocardium induces a unique pattern of postinfarction remodeling accompanied by cell death and altered cell growth and proliferation and migration of cardiomyocytes, fibroblasts, and endothelial cells. miR-1, miR-133, miR-499, miR-199, miR-29, miR-21, miR-24, and miR-214 are involved in cardiomyocyte death and survival. miR-29, miR-21, and miR-24 in fibroblasts have also been found to modulate fibrosis. miR-126, miR-21, and miR-24 in endothelial cells may function in regulating angiogenesis during postinfarction remodeling

Table 17.2 miRNAs in myocardial infarction

Cell type	miRNA	Biological functions	Targets
Cardiomyocyte	miR-1	Proapoptotic	HSP60, HSP70, and Bcl-2
	miR-133	Antiapoptotic	Caspase-9
	miR-499	Antiapoptotic	CnAα and CnAβ
	miR-29	Proapoptotic	Mcl-1, DNMT3, Tcl-1, YY1, P85α, CDC42
	miR-21	Antiapoptotic	FasL, Pcd4
	miR-24	Antiapoptotic	Bim
	miR-199	Regulate hypoxia-induced cell death	Hifα, Sirt1
	miR-214	Antiapoptotic, regulate Ca^{2+} influx	Ncx1
Fibroblast	miR-29	Antifibrotic	Fibrillins and elastins
	miR-21	Antifibrotic	Spry1
	miR-24	Antifibrotic	Furin
Endothelial cells	miR-126	Proangiogenic	Spred-1
	miR-21	Antiangiogenic	RhoB
	miR-24	Antiangiogenic	Gata2 and PAK4
	miR-214	Antiangiogenic	Quaking

miR-499 was found downregulated in the heart in ischemic conditions. As demonstrated by Wang et al. this muscle-enriched miRNA regulates left ventricular remodeling in hearts subjected to I/R injury. miR-499 represses the expression of CnAα and CnAβ by directly targeting their 3′ UTRs, resulting in reduced dephosphorylation of Drp-1, thereby inhibiting mitochondria fission and apoptosis of cardiomyocytes. Transgenic overexpression of miR-499 reduced deleterious cardiac remodeling and exhibited beneficial effects in infarcted heart, whereas knockdown of endogenous miR-499 enhanced I/R-induced collagen deposition and resulted in more pronounced cardiac dysfunction [35]. It is intriguing to note that miR-499, when overexpressed at a high level, can induce cardiac hypertrophy in vivo [34]. Thus, it is evident that the cellular steady-state level of a miRNA can dramatically influence cellular functions.

After myocardial infarction, the expression of the miR-29 family, miR-29a, miR-29b, and miR-29c, is downregulated in the border zone, suggesting that miR-29 family members may have regulatory functions in postinfarction cardiac remodeling [43]. Ye et al. reported that downregulation of miR-29 by antagomirs inhibited I/R-induced myocyte apoptosis/necrosis and reduced myocardial infarct size. miR-29 may repress the expression of numerous apoptotic factors. Mcl-1, a member of the antiapoptotic Bcl-2 family, appears to be directly targeted by miR-29 [44]. Consistently, inhibition of miR-29 elevated the level of Mcl-1 in mouse hearts [45]. Transcripts of several other antiapoptotic genes, including DNMT3, Tcl-1, YY1, P85α, and CDC42, contain putative miR-29 binding sites and are likely direct targets of miR-29 [46, 47]. Given its proapoptotic property in cardiomyocytes, miR-29 could be a potential postinfarct anti-remodeling therapeutic target. Interestingly, higher miR-29 expression is detected in cardiac fibroblasts. Numerous fibrotic genes, including collagens, fibrillins, and elastins, appear to be direct targets of miR-29. These observations suggest that miR-29 not only regulates myocyte apoptosis/survival, but it also modulates cardiac fibrosis in hearts subjected to myocardial injury. Indeed, inhibition of miR-29 in the heart elevated the expression of collagens and induced fibrosis in vivo in mice [43]. Thus, global inhibition of miR-29 in the heart may represent a double-edged sword during postinfarction cardiac remodeling [48]. It could be beneficial by inhibiting cardiomyocyte apoptosis, but it could also be detrimental due to the enhanced fibrosis in the infarcted heart.

miR-21 is widely expressed in various cell types in the heart, including fibroblasts, endothelial cells, and cardiomyocytes. Altered expression of miR-21 in myocardial remodeling has been observed in numerous studies; however, the functional roles of this miRNA in cardiac stress response are still controversial. In cardiomyocytes, miR-21 has been shown to target several proapoptotic genes, including FasL and Pcd4, thereby functioning as a pro-survival, cardiac-protective miRNA [49–51]. Indeed, knockdown of miR-21 resulted in elevated expression of proapoptotic genes and enhanced oxidative stress-triggered cell death. miR-21 also exhibited cardioprotective effects in vivo in that mice with cardiac-specific overexpression of miR-21 displayed more resistance to I/R-induced myocardial injury [51]. Consistently, Dong et al. reported that adenovirus-mediated overexpression of miR-21 in hearts improved left ventricle remodeling in the early phase of acute myocardial infarction [52].

In spite of its beneficial effects in cardiomyocytes, miR-21, when upregulated in endothelial and fibroblast cells, could be detrimental in cardiac remodeling, as shown in several other studies [53–55]. In endothelial cells, miR-21 might function as an antiangiogenic factor, since miR-21 represses the expression of RhoB, a positive regulator of vessel growth. Knockdown of miR-21 results in enhanced endothelial migration and tube formation, whereas overexpression of miR-21 inhibits angiogenesis both in vitro and in vivo [55]. However, the function of miR-21 in cardiac angiogenesis in postinfarction remodeling remains unknown.

Enhanced cardiac fibrosis appears to be closely correlated with an elevated expression of miR-21 in infarcted heart. miR-21 is upregulated in the fibroblast-rich region adjacent to the infarct after myocardial I/R [54]. Thum et al. showed that knockdown of miR-21 with a specific antagomir inhibited interstitial fibrosis in a mouse pressure overload-induced heart disease model [53]. However, such a view is not supported by Patrick et al. who showed that downregulation of miR-21 either by genetic deletion or LNA administration did not result in any alteration of cardiac fibrosis after pressure overload or myocardial infarction [56]. Thus, the functional roles of miR-21 in cardiac fibrosis are still controversial.

miR-126 is an endothelial cell-enriched miRNA. Genetic deletion of this miRNA resulted in leaky vessels and partial embryonic lethality [57]. Compared to their wild-type littermates, the surviving miR-126 null mice were more susceptible to cardiac injury and exhibited severe ventricular rupture and defective cardiac neovascularization following MI [57]. Consistently, a study by van Solingen et al. demonstrated that knockdown of miR-126 impaired ischemia-induced angiogenesis [58]. Thus, this endothelial cell-specific miRNA, which is required for normal vessel growth, may function as a proangiogenic factor and appears to be cardiac protective in postinfarction remodeling. Sprouty-related protein-1 (Spred-1) has been found to be one of the direct targets of miR-126 [57, 59]. Spred-1 is an inhibitor of the Ras/MAP kinase pathway and can repress vascular endothelial growth factor (VEGF)-/fibroblast growth factor (FGF)-mediated angiogenesis. Thus, Spred-1 appears to be a downstream mediator of the angiogenic action of miR-126. Intriguingly, miR-126 is also highly expressed in endothelial cell-derived apoptotic bodies, where it mediates the induction of CXCL12 expression, likely by repressing RGS16. In response to tissue injury, CXCL12, a CXC chemokine, functions as a positive regulator of regeneration by counteracting apoptosis and recruiting progenitor cells to repair the damaged tissue [60]. Although the role of miR-126 in CXCL12 induction needs to be further investigated in heart injury models, the potential cardiac-protective effects of miR-126 in MI may be due to its direct role in angiogenesis as well as in regulating apoptosis and progenitor cell mobilization. Thus, miR-126 may rise up as a promising cardiac protection reagent or drug.

Like miR-21, miR-24 can be detected in endothelial cells, cardiomyocytes, and fibroblasts of the heart. I/R dramatically upregulates the expression of miR-24 in endothelial cells of the parainfarcted region, but not in cardiomyocytes or fibroblasts [61]. Overexpression of miR-24 in endothelial cells triggers apoptosis and represses angiogenesis in vitro. Consistently, endothelial miR-24 antagonism in vivo exhibited inhibitory effects on apoptosis and resulted in increased capillary and arteriolar density in the parainfarcted region, thus significantly reducing the infarct size and improving heart function after MI [61]. Intriguingly, the function of miR-24 in cardiomyocytes appears to be different from that in endothelial cells. As shown by Qian et al. in vivo delivery of miR-24 significantly inhibited cardiomyocyte apoptosis, attenuated infarct size, and reduced cardiac dysfunction in a mouse MI model [62]. Recently, another study showed that miR-24 regulated cardiac fibrosis after myocardial infarction. In vivo delivery of miR-24 after MI attenuated fibrosis in the infarct border zone and improved heart function. Overexpression of miR-24 by synthetic miR-24 precursors also decreased the differentiation and migration of cardiac fibroblasts in vitro. The antifibrotic effects of miR-24 could be at least partially attributed to its repression of furin, a protease which controls processing which activates latent TGF-β [63]. Thus, the functional effect of miR-24 on myocardial infarction appears to be cell type dependent. The presence of miR-24 appears to be beneficial to cardiomyocytes and fibroblasts but detrimental to endothelial cells. This miRNA could be a promising therapeutic target; however, careful design will be required to ensure miR-24 is targeted in desired cell types, not others. Furthermore, the clinical consequences of miR-24 on different cardiac cell types need to be thoroughly evaluated.

miR-199 is another important regulator of MI. A study by Rane et al. demonstrated that miR-199a could repress the expression of Hif-1α in cardiomyocytes. Hypoxia acutely downregulated miR-199a expression in cardiomyocytes. This reduction of miR-199 causes rapid induction of Hif-1α and is apparently required for hypoxia-induced apoptosis. Intriguingly, knockdown of miR-199a during normoxia recapitulated HPC (hypoxia precondition) and protected cardiomyocytes against hypoxia-induced cell death [64]. Additionally, miR-199 appears to be a key oxygen sensor as well as a sophisticated regulator of apoptosis in cardiomyocytes. Rane et al. also showed that Sirt1 is another direct target of miR-199a. Sirt1 can downregulate prolyl hydroxylase2 (PHD2) and is required for the stabilization of Hif-1α. Thus, miR-199 could repress Hif-1α directly by targeting its 3′ UTR or indirectly by inhibiting Sirt1 and consequently inducing PHD2 [64]. Recently, several groups uncovered a "competitive endogenous RNA" (CeRNA) mechanism, in which mRNA targets sharing common miRNA binding sites cross talk with each other and form a large-scale regulatory network [65–68]. Although CeRNAs are yet to be discovered in the cardiovascular system, it is very interesting to speculate that Sirt1 and Hif-1α, which are functionally related and both directly targeted by miR-199a in cardiomyocytes, could act as CeRNAs to modulate each other's expression.

Upregulation of miR-214 has been observed during ischemic injury and in heart failure [6, 43]. However, the potential role of miR-214 in cardiac remodeling remained unknown until very recently. Aurora et al. showed that genetic deletion of miR-214 in mouse resulted in loss of cardiac contractility, increased apoptosis, and excessive fibrosis in response to I/R injury, indicating that miR-214 may play a role in cardiac protection. miR-214 appears to exert its cardioprotective function by regulating cardiomyocyte Ca^{2+} homeostasis and cellular survival during myocardial injury [69]. While the study by Aurora et al. provides the genetic evidence for miR-214's function in vivo, additional investigation is necessary to determine in which cell types miR-214 functions to exhibit its cardiac protection. Since miR-214 appears to be expressed not only in cardiomyocytes but also in endothelial cells and fibroblasts, it may modulate angiogenesis and fibrosis. For instance, a study by van Mil et al. has shown that miR-214 in endothelial cells could repress the expression of Quaking to inhibit the secretion of proangiogenic growth factors and negatively regulate angiogenesis [70]. Aurora et al. used ubiquitously expressed CAG-Cre to drive the global deletion of miR-214 in all cell types, which could result in tangled phenotypic effects and prevent the precise assessment of its cell type-specific functions. Thus, it would be interesting to assay the phenotypic consequences of a cardiomyocyte-specific deletion of miR-214 or test if cardiomyocyte-specific transgenic overexpression of miR-214 would result in cardioprotective effects and/or rescue the defects observed in miR-214 mutants after I/R injury.

17.4 Biological Implications and Future Prospects

Identification of miRNAs as novel modulators of gene expression adds a new dimension to our understanding of gene-regulatory networks. However, several fundamental questions regarding the basic biological principles of this class of small noncoding RNAs still need to be addressed. For instance, despite the global contribution of miRNAs to development and disease, most miRNAs appear to be "nonessential," since animals with genetic deletions of individual miRNAs appear to be phenotypically normal in unstressed conditions [12]. This apparent "phenotype gap" may be attributed to functional redundancy among related miRNAs in a family, which may share similar seed sequences and/or repress common downstream target genes. From an evolutionary standpoint, such functional dispensability and potential redundancy may imply less stringent selective constraints acting on miRNA genes. However, this does not seem to be the case, because most miRNAs are highly conserved across species [1]. The evolutionary selective advantages of this class of small RNAs appear elusive.

Although "nonessential" in normal conditions, many miRNAs play crucial roles in stress responses and disease models. In some cases, animals with genetic deletion of individual miRNAs are more susceptible to stress. For instance, genetic deletion of miR-214 in mice did not cause detectable cardiac defects at physiological baseline, but resulted in loss of cardiac contractility, increased apoptosis, and excessive fibrosis in response to I/R injury [69]. In some other cases, miRNAs appear to be required for stress-induced pathogenesis. As shown by van Rooij et al. miR-208a null mice did not display obvious abnormalities in size, shape, or structure of the heart at normal physiological conditions. Intriguingly, mutant mice exhibited strong resistance to cardiac hypertrophy upon pressure-overload stress [32]. Given the fact that most miRNAs are "nonessential" in normal conditions but play pivotal roles in stress responses, it is speculated that most, if not all, of the selective advantages of miRNA genes during evolution may be contributed by their significant functions in regulating pathogenesis.

Previous studies have suggested that regulation of the processing of a single pre-miRNA hairpin could not only affect the level of mature miRNA but also result in the modulated production of numerous miRNA isoforms with distinct biological functions [71]. Humphreys et al. reported that miRNAs in HL-1 cardiomyocytes exhibit pronounced sequence diversity [72]. It will be important to profile the cardiac miRNA population at single nucleotide resolution during heart development and/or cardiac pathogenesis in order to understand the functional relevance of such variation.

17.5 Clinical Implications and Prospects

As consistently shown in multiple studies, miRNAs are also present in mammalian serum and plasma. In many cases, human pathogenesis, including cardiac remodeling, appears to be accompanied with an altered population of miRNAs in these biological fluids, suggesting that the circulating miRNAs may potentially serve as biomarkers for disease diagnosis [73]. Although it is still debated whether the dysregulation of circulating miRNAs is adaptive or causative and whether they have potential functions in cardiac pathogenesis, growing evidence suggests that miRNAs are promising and attractive biomarkers for cardiac defects. First, an altered population of circulating miRNAs has been found to be associated with multiple heart diseases, including myocardial infarction and heart failure [73]. Second, using circulating miRNAs as biomarkers would allow for minimally invasive diagnosis of cardiac pathogenesis. Third, many miRNAs in serum or plasma, often protected in vesicles or exosomes, are very stable [74]. The high stability of these small RNAs can minimize potential technical variations and provide a firm grounding for clinical diagnosis. Moreover, circulating miRNAs in collected samples also appear quite resistant to temperature variation and repetitive freeze–thaw cycles, therefore allowing storage and comparison of patient samples between point-of-care sites [75]. Together, the unique characteristics of miRNAs make them promising biomarkers for diagnosis of human diseases, including cardiovascular disorders.

Numerous studies have suggested that miRNAs could also serve as potential therapeutic targets for the treatment of heart diseases. Given the fact that miRNAs play crucial roles in cardiac remodeling and stress responses, with little to no functional effect under normal unstressed physiological conditions, miRNA inhibition or delivery may provide an efficient and potent means to modulate cardiac pathogenesis with minimal side effects on normal tissues. Moreover, pharmaceutical manipulation of miRNAs is becoming more and more feasible. Systemic delivery of LNA-modified antimiRs, synthetic miRNA duplexes/miRNA mimics, or shRNAs by adeno-associated virus (AAV-shRNA) has been successfully developed and applied in mouse models of human diseases [76–78].

Despite the promise of miRNA-based therapeutics, one should be aware of the limitations and challenges that need to be overcome prior to full clinical application. For instance, functions of certain miRNAs (miR-29, miR-21), whether beneficial or detrimental in cardiac remodeling, are apparently cell type dependent. Thus, our current miRNA-manipulating strategies need to be optimized in order to achieve efficient and selective delivery of miRNA inhibitors or mimics into the desired tissue types or specific population of cells.

17.6 Summary

Cardiac hypertrophy and postinfarction remodeling are regulated by complex signal networks and involve multiple cellular processes and molecular events (Figs. 17.1 and 17.2). miRNAs, the novel noncoding small RNA modulators of gene expression, add a new dimension to the regulatory networks governing cardiac remodeling (Tables 17.1 and 17.2). miRNAs have clearly reshaped our understanding of the pathogenesis of heart diseases. Studies on miRNAs also provide important insight into the development of innovative clinical approaches. It is expected that in the future miRNAs will become ideal biomarkers for disease diagnosis and will have great potential as strategic targets for clinical therapeutics in human diseases, including cardiovascular disorders.

Acknowledgments We thank Ms. Xuefei Wang for careful reading of this manuscript. Work in Dr. Wang's laboratory was supported by the March of Dimes Foundation and the National Institutes of Health. D. Z. Wang is an Established Investigator of the American Heart Association.

References

1. Berezikov E (2011) Evolution of microRNA diversity and regulation in animals. Nat Rev Genet 12:846–860
2. Kim VN, Han J, Siomi MC (2009) Biogenesis of small RNAs in animals. Nat Rev Mol Cell Biol 10:126–139
3. Gregory RI, Chendrimada TP, Cooch N, Shiekhattar R (2005) Human RISC couples microRNA biogenesis and posttranscriptional gene silencing. Cell 123:631–640
4. Ikeda S, Kong SW, Lu J (2007) Altered microRNA expression in human heart disease. Physiol Genom 31:367–373
5. Sayed D, Hong C, Chen IY et al (2007) MicroRNAs play an essential role in the development of cardiac hypertrophy. Circ Res 100:416–424
6. van Rooij E, Sutherland LB, Liu N et al (2006) A signature pattern of stress-responsive microRNAs that can evoke cardiac hypertrophy and heart failure. Proc Natl Acad Sci USA 103:18255–18260
7. Tatsuguchi M, Seok H, Callis TE et al (2007) Expression of microRNAs is dynamically regulated during cardiomyocyte hypertrophy. J Mol Cell Cardiol 42:1137–1141
8. Chen JF, Murchison EP, Tang R et al (2008) Targeted deletion of dicer in the heart leads to dilated cardiomyopathy and heart failure. Proc Natl Acad Sci USA 105:2111–2116
9. da Costa Martins PA, Bourajjaj M, Gladka M et al (2008) Conditional dicer gene deletion in the postnatal myocardium provokes spontaneous cardiac remodeling. Circulation 118:1567–1576
10. Mendell JT, Olson EN (2012) MicroRNAs in stress signaling and human disease. Cell 148:1172–1187
11. Small EM, Frost RJ, Olson EN (2010) MicroRNAs add a new dimension to cardiovascular disease. Circulation 121:1022–1032

12. Small EM, Olson EN (2011) Pervasive roles of microRNAs in cardiovascular biology. Nature 469:336–342
13. Childs TJ, Adams MA, Mak AS (1990) Regression of cardiac hypertrophy in spontaneously hypertensive rats by enalapril and the expression of contractile proteins. Hypertension 16:662–668
14. Heineke J, Molkentin JD (2006) Regulation of cardiac hypertrophy by intracellular signalling pathways. Nat Rev Mol Cell Biol 7:589–600
15. Molkentin JD, Lu JR, Antos CL et al (1998) A calcineurin-dependent transcriptional pathway for cardiac hypertrophy. Cell 93:215–228
16. Sussman MA, Lim HW, Gude N et al (1998) Prevention of cardiac hypertrophy in mice by calcineurin inhibition. Science 281:1690–1693
17. Braz JC, Bueno OF, Liang Q et al (2003) Targeted inhibition of p38 MAPK promotes hypertrophic cardiomyopathy through upregulation of calcineurin-NFAT signaling. J Clin Invest 111:1475–1486
18. Sugden PH (2003) Ras, Akt, and mechanotransduction in the cardiac myocyte. Circ Res 93:1179–1192
19. Streicher JM, Ren S, Herschman H, Wang Y (2010) MAPK-activated protein kinase-2 in cardiac hypertrophy and cyclooxygenase-2 regulation in heart. Circ Res 106:1434–1443
20. Zhao Y, Samal E, Srivastava D (2005) Serum response factor regulates a muscle-specific microRNA that targets Hand2 during cardiogenesis. Nature 436:214–220
21. Wang Z, Lu Y, Yang B (2011) MicroRNAs and atrial fibrillation: new fundamentals. Cardiovasc Res 89:710–721
22. Rao PK, Toyama Y, Chiang HR et al (2009) Loss of cardiac microRNA-mediated regulation leads to dilated cardiomyopathy and heart failure. Circ Res 105:585–594
23. Ikeda S, He A, Kong SW et al (2009) MicroRNA-1 negatively regulates expression of the hypertrophy-associated calmodulin and Mef2a genes. Mol Cell Biol 29:2193–2204
24. Elia L, Contu R, Quintavalle M et al (2009) Reciprocal regulation of microRNA-1 and insulin-like growth factor-1 signal transduction cascade in cardiac and skeletal muscle in physiological and pathological conditions. Circulation 120:2377–2385
25. Care A, Catalucci D, Felicetti F et al (2007) MicroRNA-133 controls cardiac hypertrophy. Nat Med 13:613–618
26. Li Q, Lin X, Yang X, Chang J (2010) NFATc4 is negatively regulated in miR-133a-mediated cardiomyocyte hypertrophic repression. Am J Physiol Heart Circ Physiol 298:1340–1347
27. Liu N, Bezprozvannaya S, Williams AH et al (2008) MicroRNA-133a regulates cardiomyocyte proliferation and suppresses smooth muscle gene expression in the heart. Genes Dev 22:3242–3254
28. Yang Y, Ago T, Zhai P et al (2011) Thioredoxin 1 negatively regulates angiotensin II-induced cardiac hypertrophy through upregulation of miR-98/let-7. Circ Res 108:305–313
29. Wang K, Long B, Zhou J, Li PF (2010) miR-9 and NFATc3 regulate myocardin in cardiac hypertrophy. J Biol Chem 285:11903–11912
30. Zhang J, Chintalgattu V, Shih T et al (2011) MicroRNA-9 is an activation-induced regulator of PDGFR-beta expression in cardiomyocytes. J Mol Cell Cardiol 51:337–436
31. Callis TE, Pandya K, Seok HY et al (2009) MicroRNA-208a is a regulator of cardiac hypertrophy and conduction in mice. J Clin Invest 119:2772–2786
32. van Rooij E, Sutherland LB, Qi X et al (2007) Control of stress-dependent cardiac growth and gene expression by a microRNA. Science 216:575–579
33. van Rooij E, Quiat D, Johnson BA et al (2009) A family of microRNAs encoded by myosin genes governs myosin expression and muscle performance. Dev Cell 17:662–673
34. Shieh JT, Huang Y, Gilmore J, Srivastava D (2011) Elevated miR-499 levels blunt the cardiac stress response. PLoS One 6:e19481
35. Wang JX, Jiao JQ, Li Q et al (2011) miR-499 regulates mitochondrial dynamics by targeting calcineurin and dynamin-related protein-1. Nat Med 17:71–78
36. Rane S, He M, Sayed D et al (2010) An antagonism between the AKT and beta-adrenergic signaling pathways mediated through their reciprocal effects on miR-199a-5p. Cell Signal 22:1054–1062

37. da Costa Martins PA, Salic K, Gladka MM et al (2010) MicroRNA-199b targets the nuclear kinase Dyrk1a in an auto-amplification loop promoting calcineurin/NFAT signalling. Nat Cell Biol 12:1220–1227
38. Lin Z, Murtaza I, Wang K et al (2009) miR-23a functions downstream of NFATc3 to regulate cardiac hypertrophy. Proc Natl Acad Sci USA 106:12103–12108
39. Wang K, Lin Z, Long B et al (2012) Cardiac hypertrophy is positively regulated by MicroRNA miR-23a. J Biol Chem 287:589–599
40. Tang Y, Zhen J, Sun Y et al (2009) MicroRNA-1 regulates cardiomyocyte apoptosis by targeting Bcl-2. Inter Heart J 50:377–387
41. Bostjancic E, Zidar N, Glavac D (2009) MicroRNA microarray expression profiling in human myocardial infarction. Dis Markers 27:255–268
42. Xu C, Lu Y, Pan Z et al (2007) The muscle-specific microRNAs miR-1 and miR-133 produce opposing effects on apoptosis by targeting HSP60, HSP70 and caspase-9 in cardiomyocytes. J Cell Sci 120:3045–3052
43. van Rooij E, Sutherland LB, Thatcher JE et al (2008) Dysregulation of microRNAs after myocardial infarction reveals a role of miR-29 in cardiac fibrosis. Proc Natl Acad Sci USA 105:13027–13032
44. Mott JL, Kobayashi S, Bronk SF, Gores GJ (2007) mir-29 regulates Mcl-1 protein expression and apoptosis. Oncogene 26:6133–6140
45. Ye Y, Hu Z, Lin Y et al (2010) Downregulation of microRNA-29 by antisense inhibitors and a PPAR-g agonist protects against myocardial ischaemia–reperfusion injury. Cardiovasc Res 87:535–544
46. Wang H, Garzon R, Sun H et al (2008) NF-kappaB-YY1-miR-29 regulatory circuitry in skeletal myogenesis and rhabdomyosarcoma. Cancer Cell 14:369–381
47. Park SY, Lee JH, Ha M et al (2009) miR-29 miRNAs activate p53 by targeting p85a and CDC42. Nat Struct Mol Biol 16:23–29
48. Zhu H, Fan G (2012) Role of microRNAs in the reperfused myocardium towards post-infarct remodelling. Cardiovasc Res 94:284–292
49. Cheng Y, Liu X, Zhang S et al (2009) MicroRNA-21 protects against the H2O2-induced injury on cardiac myocytes via its target gene PDCD4. J Mol Cell Cardiol 47:5–14
50. Cheng Y, Zhu P, Yang J et al (2010) Ischaemic preconditioning-regulated miR-21 protects heart against ischaemia/reperfusion injury via anti-apoptosis through its target PDCD4. Cardiovasc Research 87:431–439
51. Sayed D, He M, Hong C et al (2010) MicroRNA-21 is a downstream effector of AKT that mediates its antiapoptotic effects via suppression of Fas ligand. J Biol Chem 285:20281–20290
52. Dong S, Cheng Y, Yang J et al (2009) MicroRNA expression signature and the role of microRNA-21 in the early phase of acute myocardial infarction. J Biol Chem 284:29514–29525
53. Thum T, Gross C, Fiedler J et al (2008) MicroRNA-21 contributes to myocardial disease by stimulating MAP kinase signalling in fibroblasts. Nature 456:980–984
54. Roy S, Khanna S, Hussain SR et al (2009) MicroRNA expression in response to murine myocardial infarction: miR-21 regulates fibroblast metalloprotease-2 via phosphatase and tensin homologue. Cardiovasc Res 82:21–29
55. Sabatel C, Malvaux L, Bovy N et al (2011) MicroRNA-21 exhibits antiangiogenic function by targeting RhoB expression in endothelial cells. PLoS One 6:e16979
56. Patrick DM, Montgomery R, Qi X et al (2010) Stress-dependent cardiac remodeling occurs in the absence of microRNA-21 in mice. J Clin Invest 120:3912–3916
57. Wang S, Aurora A, Johnson BA et al (2008) The endothelial-specific microRNA miR-126 governs vascular integrity and angiogenesis. Dev Cell 15:261–271
58. van Solingen C, Seghers L, Bijkerk R et al (2009) Antagomir-mediated silencing of endothelial cell specific microRNA-126 impairs ischemia-induced angiogenesis. J Cell Mol Med 13:1577–1585
59. Fish JE, Santoro MM, Morton SU et al (2008) miR-126 regulates angiogenic signaling and vascular integrity. Dev Cell 15:272–284

60. Zernecke A, Bidzhekov K, Noels H et al (2009) Delivery of microRNA-126 by apoptotic bodies induces CXCL12-dependent vascular protection. Sci Signal 2:ra81
61. Fiedler J, Jazbutyte V, Kirchmaier BC et al (2011) MicroRNA-24 Regulates Vascularity After Myocardial Infarction. Circulation 124:720–730
62. Qian L, Van Laake L, Huang Y et al (2011) miR-24 inhibits apoptosis and represses Bim in mouse cardiomyocytes. J Exp Med 208:549–560
63. Wang J, Huang W, Xu R et al (2012) MicroRNA-24 Regulates Cardiac Fibrosis after Myocardial Infarction. J Cell Mol Med 16(9):2150–2160. doi:10.1111/j.1582-4934.2012.01523.x
64. Rane S, He M, Sayed D et al (2009) Downregulation of miR-199a derepresses hypoxia-inducible factor-1alpha and Sirtuin 1 and recapitulates hypoxia preconditioning in cardiac myocytes. Circ Res 104:879–886
65. Karreth FA, Tay Y, Perna D et al (2011) In vivo identification of tumor- suppressive PTEN ceRNAs in an oncogenic BRAF-induced mouse model of melanoma. Cell 147:382–395
66. Tay Y, Kats L, Salmena L et al (2011) Coding-independent regulation of the tumor suppressor PTEN by competing endogenous mRNAs. Cell 147:344–357
67. Sumazin P, Yang X, Chiu HS et al (2011) An extensive microRNA-mediated network of RNA-RNA interactions regulates established oncogenic pathways in glioblastoma. Cell 147: 370–381
68. Salmena L, Poliseno L, Tay Y et al (2011) A ceRNA hypothesis: the Rosetta Stone of a hidden RNA language? Cell 146:353–358
69. Aurora AB, Mahmoud AI, Luo X et al (2012) MicroRNA-214 protects the mouse heart from ischemic injury by controlling Ca2+ overload and cell death. J Clin Invest 122:1222–1232
70. van Mil A, Grundmann S, Goumans MJ et al (2012) MicroRNA-214 inhibits angiogenesis by targeting Quaking and reducing angiogenic growth factor release. Cardiovasc Res 93: 655–665
71. Chiang HR, Schoenfeld L, Ruby JG et al (2010) Mammalian microRNAs: Experimental evaluation of novel and previously annotated genes. Genes Dev 24:992–1009
72. Humphreys DT, Hynes CJ, Patel HR et al (2012) Complexity of murine cardiomyocyte miRNA biogenesis, sequence variant expression and function. PLoS One 7:e30933
73. McManus DD, Ambros V (2011) Circulating MicroRNAs in cardiovascular disease. Circulation 124:1908–1910
74. Turchinovich A, Weiz L, Langheinz A, Burwinkel B (2011) Characterization of extracellular circulating microRNA. Nucleic Acids Res 39:7223–7233
75. Seto AG, van Rooij E (2012) Circulating microRNAs to identify human heart failure. Eur J Heart Fail 13:118–119
76. Montgomery RL, Hullinger T, Semus HM et al (2011) Therapeutic inhibition of miR-208a improves cardiac function and survival during heart failure. Circulation 124:1537–1547
77. Hullinger TG, Montgomery R, Seto AG et al (2011) Inhibition of miR-15 protects against cardiac ischemic injury. Circ Res 110:71–78
78. Kota J, Chivukula RR, O'Donnell KA et al (2009) Therapeutic microRNA delivery suppresses tumorigenesis in a murine liver cancer model. Cell 137:1005–1017

Chapter 18
Negative Regulators of Inflammation as Endogenous Protective Mechanisms in Postinfarction Remodeling

Amit Saxena and Nikolaos G. Frangogiannis

Abstract Myocardial infarction triggers an intense inflammatory reaction that serves to clear the wound from dead cells and matrix debris while promoting cardiac repair and formation of a scar. Dysregulated inflammation following cardiac injury has adverse consequences on the reparative response enhancing dilative remodeling and causing contractile dysfunction. Inhibitory molecular signals and suppressive pathways that prevent excessive or uncontrolled inflammation are activated in the infarcted myocardium and may protect from the development of adverse remodeling. This chapter discusses the cellular effectors and molecular signals responsible for suppression and containment of the postinfarction inflammatory response. Neutrophils, monocytes/macrophages, and lymphocyte subpopulations with suppressive properties, dendritic cells, vascular cells, fibroblasts, and extracellular matrix proteins contribute to inhibition of the inflammatory signals by producing soluble suppressive mediators (such as transforming growth factor-β, interleukin-10, and lipid-derived mediators) and through activation of intracellular inhibitory signals. We propose that dilative remodeling in patients with myocardial infarction may reflect impairment of suppressive anti-inflammatory pathways. Selective inhibition of inflammatory mediators in patient subpopulations with overactive postinfarction inflammation may protect from the development of heart failure.

Keywords Myocardial infarction • Remodeling • Transforming growth factor (TGF)-β • Interleukin-10 • Toll-like receptors • Cytokine • Regulatory T cells • Monocytes • Fibroblasts • Neutrophils

---

A. Saxena • N.G. Frangogiannis (✉)
Division of Cardiology, Department of Medicine, The Wilf Family Cardiovascular Research Institute, Albert Einstein College of Medicine, 1300 Morris Park Avenue Forchheimer G46B, Bronx, NY 10461, USA
e-mail: nikolaos.frangogiannis@einstein.yu.edu

B.I. Jugdutt and N.S. Dhalla (eds.), *Cardiac Remodeling: Molecular Mechanisms*, Advances in Biochemistry in Health and Disease 5, DOI 10.1007/978-1-4614-5930-9_18,

18.1 Introduction

18.1.1 The Postinfarction Inflammatory Response and Cardiac Remodeling

Over the last 30 years, advances in acute coronary care and development of early reperfusion strategies significantly improved survival rates in patients suffering an acute myocardial infarction leading to a decline of over 60% in all-cause mortality during the first 30 days after the acute event [1]. This impressive success in care of patients with acute infarction has resulted in a growing pool of patients who, having survived the acute infarction, are at risk of developing heart failure [2]. Thus, in patients with acute myocardial infarction, the decreasing acute mortality rates are associated with an increased incidence of postinfarction heart failure [1]. The pathogenesis of postinfarction heart failure is linked with profound alterations in cardiac geometry, function, and structure, a process termed "ventricular remodeling." Adverse dilative remodeling is a predictor of arrhythmias, heart failure, and mortality in patients surviving an acute infarction [3]. The extent of postinfarction remodeling is dependent not only on the amount of lost myocardium but also on the qualitative characteristics of the reparative process [4].

The adult mammalian myocardium has negligible endogenous regenerative capacity; as a result, loss of a large number of cardiomyocytes leads to their replacement with collagen-based scar. In acute infarction, sudden necrosis of millions of cardiomyocytes triggers an intense inflammatory reaction that not only serves to clear the wound from dead cells and matrix debris but also provides key molecular signals for cardiac repair. In healing wounds timely repression and containment of inflammatory signals are needed to ensure optimal tissue repair. Because cardiac function is dependent on preservation of architectural integrity, defective temporal and spatial regulation of the postinfarction inflammatory reaction may have catastrophic consequences on cardiac geometry and function. A dysregulated and overactive inflammatory reaction may result in impaired cardiac repair through several distinct mechanisms. First, exaggerated inflammation may augment matrix degradation causing cardiac rupture. Second, prolongation of the inflammatory reaction may perturb collagen deposition leading to formation of a scar with reduced tensile strength, thus promoting chamber dilation and adverse remodeling. Third, enhanced expression of proinflammatory mediators may activate proapoptotic signals inducing further loss of cardiomyocytes in the non-infarcted remodeling myocardium. Finally, defective containment of the inflammatory reaction may lead to extension of the inflammatory infiltrate into the non-infarcted myocardium enhancing fibrosis and worsening diastolic function [5].

A growing body of evidence has demonstrated that inhibition and resolution of inflammation are biosynthetically active processes that require recruitment of cellular effectors and activation of molecular mediators that suppress inflammation. This chapter will discuss the pathways responsible for negative regulation, resolution, and containment of inflammation in the infarcted myocardium. These STOP signals are crucial for cardiac repair following myocardial infarction. Defects in repression and resolution of the postinfarction inflammatory reaction may be responsible for adverse remodeling and development of heart failure in a large number of patients surviving a myocardial infarction.

18.2 Molecular Signals Involved in Regulation of Postinfarction Inflammation

18.2.1 Initiation and Activation of the Innate Immune Response Following Myocardial Infarction

In the infarcted myocardium, necrotic cardiomyocytes release their intracellular contents initiating an intense inflammatory response through activation of innate immune mechanisms. Generation of damage-associated molecular patterns by necrotic cells and matrix fragments activates membrane-bound Toll-like receptors (TLRs) [6]. A growing body of evidence implicates TLR signaling in initiation of the postinfarction inflammatory response. TLR4- and TLR2-mediated pathways promote inflammation in the infarcted myocardium [7, 8], thus contributing to the development of adverse remodeling [9]. High-mobility group box 1, the receptor for advanced glycation end products [10], and the complement system are also activated after necrotic myocardial injury and participate in the early stages of the inflammatory response in the infarcted heart [11]. Reactive oxygen species (ROS) are also generated and induce inflammatory signals while exerting direct inhibitory effects on myocardial function.

18.2.2 Negative Regulation of the Innate Immune Response

Several distinct pathways have been identified as negative regulators of innate immune responses [12, 13]; however, their involvement in limiting myocardial inflammation following infarction remains unknown. Disruption of some of these negative regulators results in uncontrolled inflammation suggesting their essential role in restraining TLR-mediated immune responses. The TLR-inducible zinc finger proteins A20 and Zc3h12a have also been identified as essential negative regulators of TLR signaling in vivo [14]. A20 is dynamically regulated in the infarcted and pressure-overloaded myocardium [15]. Myocardial A20 overexpression attenuates postinfarction inflammation by reducing nuclear factor (NF)-κB activity and exerts potent anti-hypertrophic effects [16]. Interleukin-1 receptor-associated kinase (IRAK)-M is the only member of the IRAK family that lacks kinase activity, thus functioning as a decoy that negatively regulates TLR- and IL-1-mediated inflammatory responses. IRAK-M null mice exhibit accentuated inflammatory responses following bacterial and viral infections [17, 18] and IRAK-M-deficient macrophages display enhanced activation of IL-1/TLR signaling [19]. Ongoing research in our laboratory has identified IRAK-M as an important negative regulator of the postinfarction inflammatory response; in addition to its established role in limiting macrophage-derived inflammation, IRAK-M expression in cardiac fibroblasts may limit their matrix-degrading potential. Several other pathways are involved in negative regulation of the innate immune response; however, their role in myocardial infarction has not been investigated. MyD88s, a lipopolysaccharide-inducible spliced form of MyD88 that lacks the IRAK-4-interacting domain, is involved in negative regulation of TLR signaling by inhibiting IRAK-4-mediated IRAK-1 phosphorylation [20]. Receptor-interacting protein 3 (RIP3 - a kinase that inhibits RIP1-induced NF-κB signaling) [21], activating transcription factor 3 (ATF3), the autophagy-related molecule Atg16L1, and the TAM receptor protein tyrosine kinases are also involved in inhibition of TLR signaling.

18.2.3 Cytokine and Chemokine Signaling Following Myocardial Infarction

Innate immune signaling pathways ultimately activate the NF-κB system driving expression of proinflammatory cytokines and chemokines [22, 23]. As prominent and early mediators of inflammation, proinflammatory cytokines critically regulate the response to cardiac injury. Interleukin (IL)-1, the prototypical proinflammatory cytokine, mediates chemokine synthesis in the infarcted myocardium and promotes infiltration of the infarct with leukocytes [24]. Generation of active IL-1β requires processing of pro-IL-1β, an inactive precursor, by the converting enzyme caspase-1. Caspase-1 activity is tightly regulated within the "inflammasomes," multiprotein molecular platforms that control maturation and processing of IL-1β [25]. Inflammasome activation in the infarcted myocardium is localized in both leukocytes and resident fibroblasts and drives IL-1-mediated inflammatory activity [26, 27].

The proinflammatory cytokine tumor necrosis factor (TNF)-α is also released in the infarcted myocardium [28] and may play an important role in regulation of the postinfarction inflammatory response. TNF-α null mice were protected from cardiac rupture and chronic dysfunction following infarction and exhibited suppressed inflammation and attenuated matrix metalloproteinase (MMP) activity in comparison with wild-type animals [29]. However, TNF-α may also induce cytoprotective signals capable of preventing or delaying the development of myocyte apoptosis following myocardial infarction [30]. It has been suggested that TNF-α may exert distinct biological effects through the TNFR1 and TNFR2 receptor. Actions mediated through TNFR1 may be deleterious, inducing cardiac dysfunction, whereas TNFR2-mediated actions may be protective by attenuating adverse remodeling [31].

Chemokine induction is also a consistent and prominent feature of the postinfarction inflammatory response [32]. Chemokines provide directional signals for recruitment of leukocyte subpopulations into the infarcted myocardium.

CC chemokines, such as monocyte chemoattractant protein (MCP)-1 and the macrophage inflammatory proteins (MIP-1α and MIP-1β) [33], function as potent mononuclear cell chemoattractants, whereas ELR motif-containing CXC chemokines (such as IL-8) play an important role in neutrophil chemoattraction [34]. Through interactions with chemokine receptors expressed by leukocyte subpopulations, the chemokine expression profile in the infarcted myocardium determines the composition of the leukocytic infiltrate.

18.2.4 Negative Regulation of Proinflammatory Cytokine and Chemokine Responses

Negative regulation of cytokine signaling may involve several distinct mechanisms:

(a) Upregulation of endogenous cytokine inhibitors may suppress specific cytokine responses. IL-1Ra, an endogenous competitive inhibitor of IL-1-driven inflammation, plays an important role in preventing autoinflammatory responses. Although the role of endogenous IL-1Ra in suppression and resolution of the postinfarction inflammatory response has not been dissected, IL-1Ra overexpression exerts protective actions on the infarcted heart [35].
(b) Decoy receptors (such as the type II IL-1 receptor) may be upregulated in the infarct serving as molecular sinks for cytokines.
(c) Endocytosis of signaling-competent receptors followed by proteasomal degradation may play an important role in preventing uncontrolled cytokine actions.
(d) Active termination of cytokine signaling through dephosphorylation of Janus-activated kinases (JAKs) [36]. Most of the cytokine-activated intracellular proteins belong to the family of signal transducers and activators of transcription (STAT) and after tyrosine phosphorylation by JAKs, then dimerize and translocate to the nucleus where they initiate transcription. Several phosphatases have been implicated in negative regulation of cytokine signaling, including SHP-1, the protein tyrosine phosphatase PTP1B and the T cell protein tyrosine phosphatase (TCPTP).
(e) Perhaps the best-studied family of cytokine inhibitors is the Suppressor of Cytokine Signaling (SOCS) family of inhibitory proteins. Although members of the SOCS family are likely to play an important role in regulation of the postischemic myocardial inflammatory response, their contribution to suppression of inflammation in the infarcted heart has not been investigated.

Termination of chemokine signaling may also play an important role in regulating the postinfarction inflammatory reaction [32]. Suppression of chemokine synthesis by inhibitory mediators, such as transforming growth factor (TGF)-β [37], IL-10, or lipid-derived substances, may be the predominant mechanism for deactivation of chemokinetic signals. Two additional pathways may operate for negative regulation of chemokine responses. First, posttranslational modifications of mature chemokine proteins through protease-mediated actions may generate molecules with reduced activity or endogenous chemokine antagonists [38]. Second, silent chemokine receptors that bind to their ligand without signaling may function as decoy or scavenging receptors [39]. Whether chemokine decoy receptors are expressed and play a crucial role in myocardial inflammation remains unknown.

18.2.5 Soluble Inhibitors of the Postinfarction Inflammatory Response: The Role of IL-10 and TGF-β

Induction of soluble inhibitory mediators may play an important role in suppression and resolution of the postinfarction inflammatory response. Extensive evidence has identified IL-10 and TGF-β as important negative regulators of immune responses [40]. Both mediators have been implicated in regulation of the postinfarction inflammatory response; however, their role is complex as they affect most cell types involved in cardiac repair in a context-dependent manner.

IL-10 mRNA and protein expression are markedly induced in reperfused canine and murine myocardial infarction [34, 41]. T cell subsets and macrophage subpopulations appear to be responsible for IL-10 synthesis in the infarcted myocardium [41, 42]. In a canine model, IL-10 upregulation in the infarcted myocardium was associated with decreased expression of proinflammatory cytokines; in vitro, IL-10 released in the cardiac interstitium was responsible for induction of tissue inhibitors of metalloproteinases (TIMP-1) mRNA by isolated canine mononuclear cells suggesting a role in matrix stabilization [41]. Two independent loss-of-function studies in mouse models of reperfused infarction have produced somewhat contradictory results. Yang and coworkers suggested that IL-10–/– mice had markedly increased mortality and exhibited an enhanced inflammatory response following reperfused myocardial infarction showing accentuated neutrophil recruitment, elevated plasma levels of TNF-α, and higher tissue expression of intercellular adhesion molecule 1 (ICAM-1) [43]. In contrast, Zymek and coworkers found much more subtle differences in the postinfarction inflammatory response between IL-10 null and wild-type animals [44]. Infarcted IL-10–/– mice and wild-type littermates exhibited comparable survival following infarction. Although IL-10 disruption was associated with higher peak TNF-α and MCP-1 mRNA levels in the infarcted heart, repression of proinflammatory cytokine and chemokine synthesis and resolution of the neutrophil infiltrate were not affected. IL-10 gene disruption did not alter fibrous tissue deposition and dilative remodeling of the infarcted heart [44]. Thus, IL-10 induction in the infarcted heart may play a limited role in controlling the acute inflammatory response, but is not involved in clearance of the inflammatory infiltrate and resolution of inflammation.

TGF-β may regulate immune and inflammatory responses by modulating leukocyte phenotype and activity. The effects of TGF-β on lymphocytes, monocytes, and macrophages can be either stimulatory or inhibitory, depending on microenvironmental cues, the state of cellular differentiation, and the tissue origin of the cells highlighting the pleiotropic nature of the cytokine. TGF-β has a deactivating effect on macrophages, suppressing proinflammatory cytokine and chemokine synthesis [45] and decreasing ROS generation. These TGF-β-mediated effects may be important in suppression and resolution of the inflammatory response. In addition to its effects on leukocytes, TGF-β is a potent modulator of fibroblast phenotype and function. TGF-β stimulation induces myofibroblast transdifferentiation [46] and enhances synthesis of extracellular matrix proteins. Moreover, TGF-β suppresses the activity of proteases that degrade extracellular matrix by inhibiting MMP expression and by inducing synthesis of protease inhibitors, such as plasminogen activator inhibitor 1 (PAI-1) and TIMPs [47, 48]. Thus, through its broad immunomodulatory and anti-inflammatory properties and its profibrotic/matrix-preserving actions, TGF-β is an excellent candidate for a role as the "master switch" mediating suppression of postinfarction inflammation while promoting fibroblast activation and scar formation [49].

TGF-β is markedly upregulated in experimental models of myocardial infarction [50] where it is predominantly localized in the infarct border zone, associated with expression and activation of its downstream intracellular effectors, Smad2, Smad3, and Smad4 [51, 52]. Systemic inhibition of TGF-β signaling by injection of an adenovirus harboring soluble TGF-β type II receptor in the hind limb muscles resulted in attenuated left ventricular remodeling by modulating cardiac fibrosis [53, 54]. However, early TGF-β inhibition increased mortality and exacerbated left ventricular dilation enhancing cytokine expression, suggesting that during the resolution phase of the inflammatory response, TGF-β signaling may play an important role in suppression of inflammatory mediator synthesis [53].

Because of the pleiotropic effects of TGF-β on the inflammatory and reparative response following infarction, identification of therapeutic targets requires dissection of the signaling pathways responsible for its actions in the infarcted myocardium. TGF-β is capable of activating the canonical Smad-mediated cascade as well as a variety of Smad-independent pathways. Smad3 null animals had markedly suppressed peak chemokine expression and decreased neutrophil recruitment in the infarcted myocardium and showed timely repression of inflammatory gene synthesis and resolution of the inflammatory infiltrate. Although myofibroblast density was higher in Smad3 null infarcts, interstitial deposition of matrix proteins in the peri-infarct zone and in the non-infarcted myocardium was markedly reduced. In vitro, Smad3 signaling not only was essential for TGF-β-mediated myofibroblast differentiation and matrix synthesis but also exerted antiproliferative effects [55] on cardiac fibroblasts. In the absence of Smad3, the infarct was filled with a large number of dysfunctional fibroblasts. Thus, TGF-β/Smad3 signaling does not appear to be involved in suppression and resolution of inflammation in healing infarcts, but mediates interstitial fibrosis in the infarct border zone and in the non-infarcted myocardium. The potential role of Smad-independent signaling in mediating TGF-β-induced suppression of postinfarction inflammation has not been investigated.

Growth differentiation factor (GDF)-15, another member of the TGF-β superfamily, has been recently identified as a crucial endogenous mediator involved in suppression of the postinfarction inflammatory response [56]. GDF-15 restrains inflammation by counteracting conformational activation of neutrophil β2 integrins, thus preventing excessive chemokine-activated leukocyte arrest on the endothelium [57].

18.2.6 The Potential Involvement of Lipid Mediators in Resolution of Postinfarction Inflammation

Lipid-derived mediators play an important role in regulation of the inflammatory response: prostaglandins and leukotrienes are released early following tissue injury and exert proinflammatory actions [58], whereas specialized proresolving mediators (SPM) are induced during the resolution phase of inflammation [59]. Four distinct families of SPMs have been identified: the lipoxins, resolvins, protectins, and maresins. These mediators have potent and direct anti-inflammatory and proresolving functions, limiting inflammatory leukocyte recruitment while promoting clearance of apoptotic neutrophils by macrophages. Administration of exogenous resolvin E1 exerted protective actions on the ischemic and reperfused myocardium [60]. However, the role of endogenous lipid mediators in negative regulation of the postinfarction inflammatory response has not been studied.

18.3 The Cellular Effectors of the Anti-inflammatory Response

Timely repression of inflammatory gene synthesis and resolution of the inflammatory infiltrate following myocardial infarction require the coordinated actions of several different cell types and involve the participation of the extracellular matrix (Fig. 18.1). Neutrophils, mononuclear cells/macrophages, vascular endothelial cells, fibroblasts, and pericytes undergo dynamic phenotypic changes contributing to suppression and resolution of the inflammatory reaction. Moreover, the extracellular matrix dynamically participates in regulation of the inflammatory cascade.

18.3.1 The Neutrophils

One of the most important cellular mechanisms responsible for resolution of the inflammatory response is clearance of apoptotic cells in the injured tissue [61]; this process is associated with active suppression of inflammation, as phagocytes

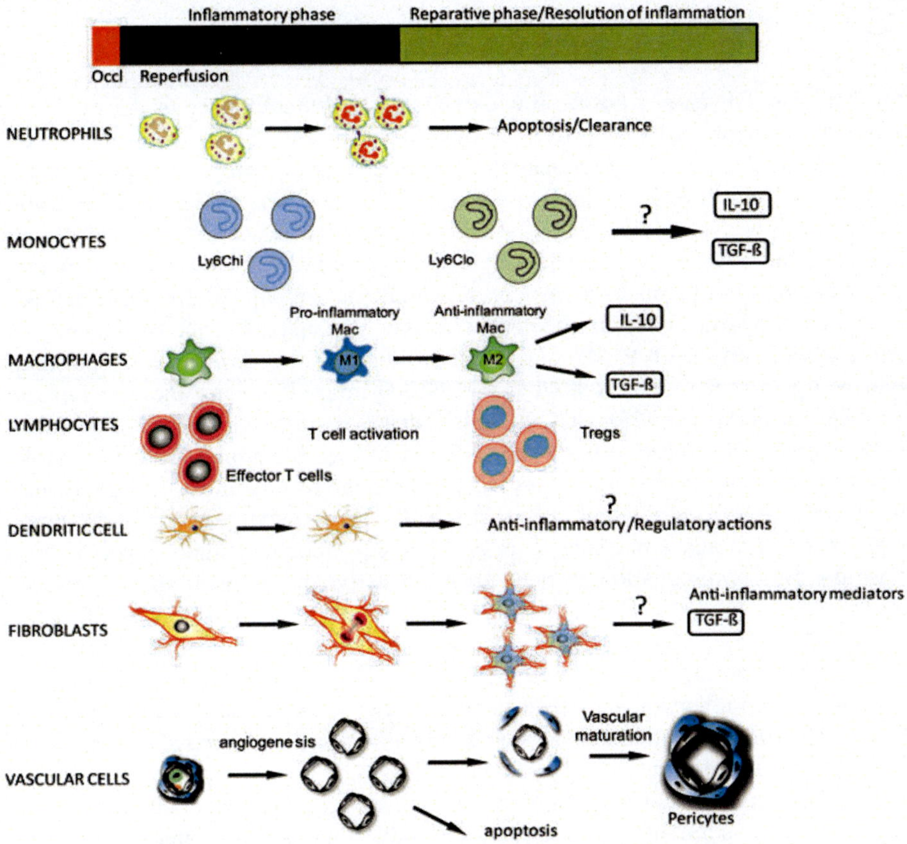

Fig. 18.1 All cells involved in cardiac repair contribute to resolution of postinfarction inflammation. During the reparative phase of infarct healing neutrophils, monocytes/macrophages, lymphocytes, dendritic cells, fibroblasts, and vascular cells undergo phenotypic changes secreting soluble inhibitory mediators and suppressing the inflammatory reaction

ingesting apoptotic cells release large amounts of inhibitory mediators [62, 63]. The infarcted myocardium is infiltrated by abundant neutrophils; these short-lived inflammatory cells represent a cellular population programmed to undergo apoptosis. Their clearance by professional phagocytes is likely to activate an anti-inflammatory program in phagocytotic macrophages, characterized by increased expression and release of IL-10, TGF-β, and proresolving lipid mediators [64]. However, this hypothesis has not been tested in experimental models of myocardial infarction.

18.3.2 Monocyte Subpopulations and Macrophages

Over the last 10 years, it has been widely recognized that monocytes are functionally heterogeneous cells [65]. In mice two major subsets of monocytes have been documented: (a) Ly6C^{hi} inflammatory monocytes that migrate into injured tissues and exhibit high levels of expression of the CC chemokine receptor CCR2 while expressing low levels of the fractalkine receptor CX3CR1 (Ly6C^{hi}CCR2^{hi}CX3CR1^{low}) and (b) Ly6C^{low}/CCR2$^{low/neg}$/CX3CR1^{hi} "resident" monocytes [66] that exhibit less pronounced proinflammatory responses. Monocyte subpopulations with distinct functional properties have also been described in humans: human CD16 monocytes express high levels of CCR2 and have proinflammatory properties resembling murine Ly6C^{hi} cells [65]. Following myocardial infarction, early induction of the CC chemokine MCP-1 [33], a CCR2 ligand, drives recruitment of proinflammatory Ly6C^{hi} monocytes; these cells scavenge debris and secrete proinflammatory cytokines and matrix-degrading proteases [67]. Repression of inflammatory signals may be associated with recruitment of Ly6C^{lo}/CX3CR1^{hi} monocytes that produce angiogenic mediators and contribute to infarct healing. Whether timely repression of postinfarction inflammation is dependent on active recruitment of suppressive monocyte subpopulations in the healing myocardium, driven by specific chemokine/chemokine receptor interactions, remains unknown. Monocyte heterogeneity has also been demonstrated in human patients with acute myocardial infarction. Circulating proinflammatory CD14+/CD16– cells show an early peak in patients with myocardial infarction and are negatively associated with recovery of function [68] after the acute event.

In the proinflammatory environment of the healing infarct, upregulation of macrophage colony-stimulating factor (M-CSF) induces monocyte to macrophage differentiation [69]. In the infarcted myocardium, the spatially and temporally regulated expression of cytokines, chemokines, and growth factors is likely to affect macrophage phenotype leading to sequential generation of many distinct subsets. During resolution of postinfarction inflammation, macrophages are capable of exerting both inhibitory actions by secreting mediators that suppress inflammation and proresolving effects through the removal of inflammatory leukocytes. Our knowledge on the role of macrophage-mediated interactions in repression of inflammatory signals and in resolution of the leukocyte infiltrate in the infarcted myocardium remains extremely limited. During the resolution phase of inflammation, macrophages that phagocytose apoptotic neutrophils may acquire a regulatory phenotype, expressing large amounts of inhibitory mediators and promoting suppression of the inflammatory response.

18.3.3 T Lymphocytes

Effector T cells are recruited in sites of injury where they release proinflammatory mediators. In addition, the immune system also produces regulatory T cells (Tregs), a population of T cells with suppressive properties that serves to limit the activation and expansion of the immune response [70]. Recent evidence suggests that Tregs may play a role in suppression of the postinfarction inflammatory response [71]. Mice with genetic disruption of the chemokine receptor CCR5 exhibited accentuated inflammatory response and enhanced MMP activity following reperfused myocardial infarction. Increased inflammation in CCR5-null animals was associated with reduced infiltration of the infarcted myocardium with CD4+/CD25+ Tregs [71]. Although a causative relation between defective recruitment of Tregs and enhanced remodeling has not been established, Tregs may modulate macrophage phenotype and suppress postinfarction inflammation by secreting soluble mediators, such as IL-10 and TGF-β, or through contact-dependent pathways [71].

18.3.4 Dendritic Cells

Recent studies have demonstrated that dendritic cell ablation enhanced the postinfarction inflammatory reaction [72]. The anti-inflammatory effects of the dendritic cells were presumed due to regulation of monocytes and macrophage phenotype in the infarcted heart.

18.3.5 Vascular Cells

As the inflammatory infiltrate is replaced by granulation tissue, extensive angiogenesis occurs resulting in formation of neovessels that provide oxygen and nutrients to the highly dynamic and metabolically active cells of the healing wound. Early neovessels in the infarcted heart lack a pericyte coat and are hyperpermeable and proinflammatory [73], perhaps due to the high levels of vascular endothelial growth factor (VEGF) in their microenvironment. As the scar matures, the infarct vasculature undergoes a maturation process that may play an important role in suppression of the inflammatory response in the wound [74]. Some of the infarct neovessels acquire a coat comprised of pericytes and smooth muscle cells, whereas uncoated endothelial cells undergo apoptosis. Acquisition of a mural cell coat reduces vascular permeability and decreases angiogenic potential of the vessels, contributing to the formation of a stable scar. The interactions between pericytes and endothelial cells that result in vascular coating involve platelet-derived growth factor (PDGF)-BB-PDGFR-β signaling [75]. Pericyte coating is an important step for suppression of granulation tissue formation following myocardial infarction and promotes resolution of inflammation and stabilization of the scar. Whether pericytes secrete soluble mediators, or directly signal to promote endothelial cell quiescence, is unknown.

18.3.6 Fibroblasts

Due to their abundance in normal and infarcted hearts and their phenotypic plasticity, fibroblasts are capable of playing an important role as regulators of the postinfarction inflammatory response [76]. Through activation of the inflammasome, resident cardiac fibroblasts play an important role in initiation of the inflammatory phase following cardiac injury [26]. Whether activated infarct myofibroblasts can participate in repression of inflammation and resolution of the leukocyte infiltrate in the healing wound remains unknown.

18.3.7 The Extracellular Matrix

In addition to its structural role, the extracellular matrix proteins also acts as a key regulator of the inflammatory response [77]. In the early stages following infarction, generation of matrix fragments enhances cytokine and chemokine synthesis, while deposition of a plasma-derived provisional matrix provides a scaffold for leukocyte infiltration and may transduce proinflammatory signals. Clearance of matrix fragments from the injured tissue appears to be an essential step for resolution of inflammation and, much like removal of apoptotic cells, may activate inhibitory pathways [78].

The temporally and spatially restricted expression of fibronectin and matricellular proteins in the infarcted heart may provide an additional matrix-related mechanism for regulation of inflammation [79]. The matricellular protein thrombospondin (TSP)-1, a crucial TGF-β activator with potent angiostatic and anti-inflammatory properties, is strikingly upregulated in the border zone of the infarct (Fig. 18.2) [80]. TSP-1−/− mice exhibited enhanced and prolonged expression of chemokines in the infarcted heart and showed expansion of the inflammatory infiltrate into the non-infarcted area. Prolonged and expanded postinfarction inflammation in TSP-1-null animals resulted in increased adverse remodeling of the ventricle [80]. These findings suggest an important role for TSP-1 in suppression and containment of the postinfarction inflammatory reaction. Localized induction of TSP-1 in the infarct border zone may form a "barrier" preventing expansion of the inflammatory infiltrate in the non-infarcted myocardium. The mechanisms involved in TSP-1-mediated inhibition of inflammation remain unknown and may involve impaired TGF-β activation or direct anti-inflammatory actions.

18.4 Anti-inflammatory Therapeutic Strategies in Myocardial Infarction

The idea of targeting the inflammatory response in patients with myocardial infarction is not new. Twenty to thirty years ago, extensive experimental evidence derived primarily from research in large animal models suggested that early infiltration of

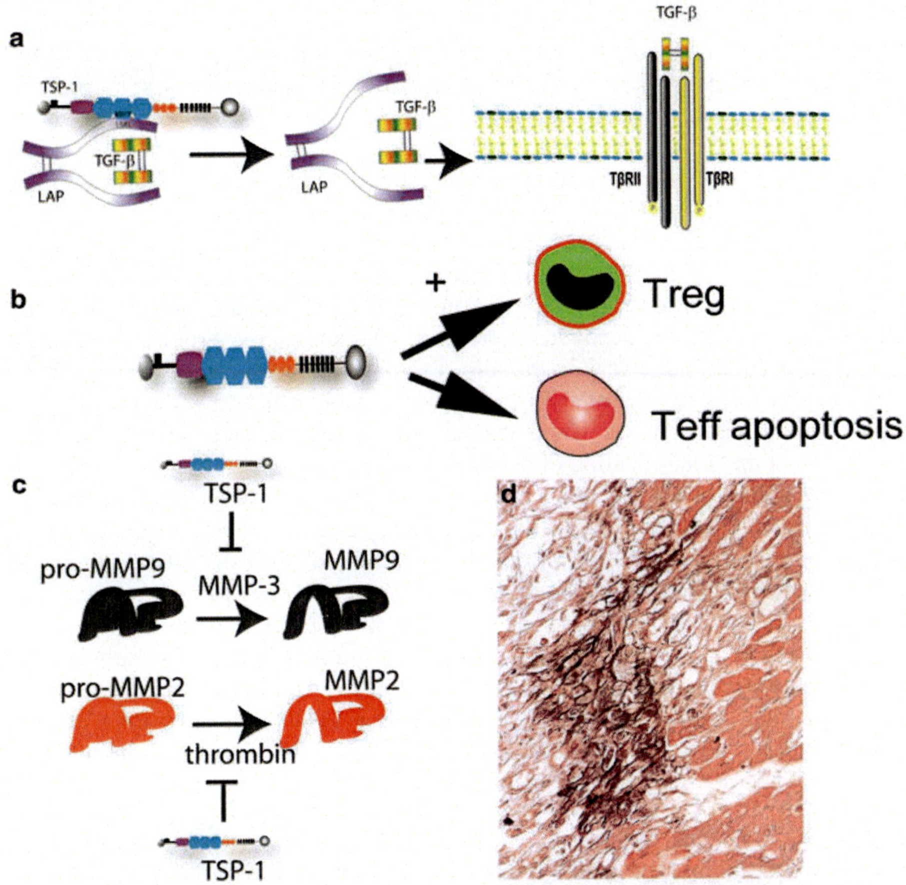

Fig. 18.2 Matricellular proteins play an important role in regulation of the postinfarction inflammatory reaction. Thrombospondin (TSP)-1, a prototypical matricellular protein, activates TGF-β (**a**), promotes regulatory T cell (Treg) formation and induces effector T cell apoptosis (**b**), and inhibits matrix metalloproteinase (MMP) activity (**c**). These actions may be important in modulating the postinfarction inflammatory response. TSP-1 shows a strikingly selective localization in the border zone of myocardial infarction (D-immunohistochemical staining for TSP-1 in canine infarcted myocardium after 1 h ischemia and 7 days of reperfusion, counterstained with eosin) where it may play a role in containment of the inflammatory reaction

the infarcted myocardium with leukocytes induces cytotoxic injury on viable cardiomyocytes extending ischemic damage [81]. These findings fueled several clinical trials aimed at protecting the ischemic myocardium through early inhibition of key inflammatory signals. Unfortunately, both anti-CD18 integrin approaches [82] and complement inhibition strategies [83] produced disappointing results raising concerns regarding the potential effectiveness of inflammatory targets in myocardial infarction. Later studies in murine models have challenged the concept of "leukocyte-mediated cytotoxic injury" demonstrating that animals with genetic disruption of genes with a critical role in the inflammatory response (such as ICAM-1/P-selectin and IL-1RI) [24, 84] had no reduction in infarct size despite marked attenuation of leukocyte infiltration.

Thus, direct inflammatory cardiomyocyte injury may not be significant; however, a growing body of evidence suggests that disruption of inflammatory pathways may protect the heart by preventing adverse remodeling through effects on reparative cells and on matrix metabolism [5, 85]. Targeting the inflammatory reaction remains promising; however, the focus of our therapeutic strategies should be shifted to a new direction: to ensure optimal temporal and spatial regulation of the inflammatory response preventing uncontrolled or prolonged inflammation. From this perspective, the early disruption of key inflammatory signals attempted in some clinical trials may in fact prolong inflammation by preventing removal of dead cells and matrix debris leading to impaired wound healing, accentuated remodeling, and adverse outcome. Identification of novel therapeutic targets requires intensification of our research efforts in two main areas. First, using carefully selected experimental models, we need to understand the mechanisms involved in repression and resolution of the postinfarction inflammatory response and study the relation between specific defects in negative regulation of inflammation and cardiac remodeling. Second, using suitable biomarkers, we need to identify patients exhibiting defective resolution of inflammation following myocardial infarction; these subpopulations may be ideal candidates for targeted anti-inflammatory approaches to inhibit excessive or prolonged inflammation. Understanding the STOP signals involved in suppression of inflammation following infarction may be the key to prevent adverse remodeling and progression to heart failure.

Acknowledgments The Frangogiannis' laboratory is supported by R01 HL76246, R01HL85440, the Wilf Family Cardiovascular Research Institute, and the Edmond J. Safra/Republic National Bank of New York Chair in Cardiovascular Medicine.

References

1. Velagaleti RS, Vasan RS (2007) Heart failure in the twenty-first century: is it a coronary artery disease or hypertension problem? Cardiol Clin 25:487–495, v
2. Lewis EF, Moye LA, Rouleau JL et al (2003) Predictors of late development of heart failure in stable survivors of myocardial infarction: the CARE study. J Am Coll Cardiol 42:1446–1453
3. White HD, Norris RM, Brown MA et al (1987) Left ventricular end-systolic volume as the major determinant of survival after recovery from myocardial infarction. Circulation 76:44–51
4. Frangogiannis NG (2008) The immune system and cardiac repair. Pharmacol Res 58:88–111
5. Frangogiannis NG (2012) Regulation of the inflammatory response in cardiac repair. Circ Res 110:159–173
6. Mann DL (2011) The emerging role of innate immunity in the heart and vascular system: for whom the cell tolls. Circ Res 108:1133–1145
7. Oyama J, Blais C Jr, Liu X et al (2004) Reduced myocardial ischemia-reperfusion injury in toll-like receptor 4-deficient mice. Circulation 109:784–789

8. Arslan F, Smeets MB, O'Neill LA et al (2010) Myocardial ischemia/reperfusion injury is mediated by leukocytic toll-like receptor-2 and reduced by systemic administration of a novel anti-toll-like receptor-2 antibody. Circulation 121:80–90
9. Riad A, Jager S, Sobirey M et al (2008) Toll-like receptor-4 modulates survival by induction of left ventricular remodeling after myocardial infarction in mice. J Immunol 180:6954–6961
10. Andrassy M, Volz HC, Igwe JC et al (2008) High-mobility group box-1 in ischemia-reperfusion injury of the heart. Circulation 117:3216–3226
11. Rossen RD, Michael LH, Kagiyama A et al (1988) Mechanism of complement activation after coronary artery occlusion: evidence that myocardial ischemia in dogs causes release of constituents of myocardial subcellular origin that complex with human C1q in vivo. Circ Res 62:572–584
12. O'Neill LA (2008) When signaling pathways collide: positive and negative regulation of toll-like receptor signal transduction. Immunity 29:12–20
13. Akira S, Takeda K (2004) Toll-like receptor signalling. Nat Rev Immunol 4:499–511
14. Kawai T, Akira S (2010) The role of pattern-recognition receptors in innate immunity: update on Toll-like receptors. Nat Immunol 11:373–384
15. Cook SA, Novikov MS, Ahn Y, Matsui T, Rosenzweig A (2003) A20 is dynamically regulated in the heart and inhibits the hypertrophic response. Circulation 108:664–667
16. Li HL, Zhuo ML, Wang D et al (2007) Targeted cardiac overexpression of A20 improves left ventricular performance and reduces compensatory hypertrophy after myocardial infarction. Circulation 115:1885–1894
17. Kobayashi K, Hernandez LD, Galan JE et al (2002) IRAK-M is a negative regulator of Toll-like receptor signaling. Cell 110:191–202
18. Seki M, Kohno S, Newstead MW et al (2010) Critical role of IL-1 receptor-associated kinase-M in regulating chemokine-dependent deleterious inflammation in murine influenza pneumonia. J Immunol 184:1410–1418
19. Flannery S, Bowie AG (2010) The interleukin-1 receptor-associated kinases: critical regulators of innate immune signalling. Biochem Pharmacol 80:1981–1991
20. Burns K, Janssens S, Brissoni B et al (2003) Inhibition of interleukin 1 receptor/Toll-like receptor signaling through the alternatively spliced, short form of MyD88 is due to its failure to recruit IRAK-4. J Exp Med 197:263–268
21. Meylan E, Burns K, Hofmann K et al (2004) RIP1 is an essential mediator of Toll-like receptor 3-induced NF-kappa B activation. Nat Immunol 5:503–507
22. Stancovski I, Baltimore D (1997) NF-kappaB activation: the I kappaB kinase revealed? Cell 91:299–302
23. Lenardo MJ, Baltimore D (1989) NF-kappa B: a pleiotropic mediator of inducible and tissue-specific gene control. Cell 58:227–229
24. Bujak M, Dobaczewski M, Chatila K et al (2008) Interleukin-1 receptor type I signaling critically regulates infarct healing and cardiac remodeling. Am J Pathol 173:57–67
25. Schroder K, Tschopp J (2010) The inflammasomes. Cell 140:821–832
26. Kawaguchi M, Takahashi M, Hata T et al (2011) Inflammasome activation of cardiac fibroblasts is essential for myocardial ischemia/reperfusion injury. Circulation 123:594–604
27. Mezzaroma E, Toldo S, Farkas D et al (2011) The inflammasome promotes adverse cardiac remodeling following acute myocardial infarction in the mouse. Proc Natl Acad Sci USA 108:19725–19730
28. Frangogiannis NG, Lindsey ML, Michael LH et al (1998) Resident cardiac mast cells degranulate and release preformed TNF-alpha, initiating the cytokine cascade in experimental canine myocardial ischemia/reperfusion. Circulation 98:699–710
29. Sun M, Dawood F, Wen WH et al (2004) Excessive tumor necrosis factor activation after infarction contributes to susceptibility of myocardial rupture and left ventricular dysfunction. Circulation 110:3221–3228
30. Kurrelmeyer KM, Michael LH, Baumgarten G et al (2000) Endogenous tumor necrosis factor protects the adult cardiac myocyte against ischemic-induced apoptosis in a murine model of acute myocardial infarction. Proc Natl Acad Sci USA 97:5456–5461

31. Monden Y, Kubota T, Inoue T et al (2007) Tumor necrosis factor-alpha is toxic via receptor 1 and protective via receptor 2 in a murine model of myocardial infarction. Am J Physiol Heart Circ Physiol 293:H743–H753
32. Frangogiannis NG (2007) Chemokines in ischemia and reperfusion. Thromb Haemost 97:738–747
33. Dewald O, Zymek P, Winkelmann K et al (2005) CCL2/Monocyte Chemoattractant Protein-1 regulates inflammatory responses critical to healing myocardial infarcts. Circ Res 96:881–889
34. Dewald O, Ren G, Duerr GD et al (2004) Of mice and dogs: species-specific differences in the inflammatory response following myocardial infarction. Am J Pathol 164:665–677
35. Suzuki K, Murtuza B, Smolenski RT et al (2001) Overexpression of interleukin-1 receptor antagonist provides cardioprotection against ischemia-reperfusion injury associated with reduction in apoptosis. Circulation 104:I308–I313
36. Naka T, Fujimoto M, Tsutsui H, Yoshimura A (2005) Negative regulation of cytokine and TLR signalings by SOCS and others. Adv Immunol 87:61–122
37. Frangogiannis NG, Mendoza LH, Lewallen M et al (2001) Induction and suppression of interferon-inducible protein 10 in reperfused myocardial infarcts may regulate angiogenesis. FASEB J 15:1428–1430
38. Mortier A, Gouwy M, Van Damme J, Proost P (2011) Effect of posttranslational processing on the in vitro and in vivo activity of chemokines. Exp Cell Res 317:642–654
39. Mantovani A, Bonecchi R, Locati M (2006) Tuning inflammation and immunity by chemokine sequestration: decoys and more. Nat Rev Immunol 6:907–918
40. Li MO, Flavell RA (2008) Contextual regulation of inflammation: a duet by transforming growth factor-beta and interleukin-10. Immunity 28:468–476
41. Frangogiannis NG, Mendoza LH, Lindsey ML et al (2000) IL-10 is induced in the reperfused myocardium and may modulate the reaction to injury. J Immunol 165:2798–2808
42. Curato C, Slavic S, Dong J et al (2010) Identification of noncytotoxic and IL-10-producing CD8+AT2R+ T cell population in response to ischemic heart injury. J Immunol 185:6286–6293
43. Yang Z, Zingarelli B, Szabo C (2000) Crucial role of endogenous interleukin-10 production in myocardial ischemia/reperfusion injury. Circulation 101:1019–1026
44. Zymek P, Nah DY, Bujak M et al (2007) Interleukin-10 is not a critical regulator of infarct healing and left ventricular remodeling. Cardiovasc Res 74:313–322
45. Kitamura M (1997) Identification of an inhibitor targeting macrophage production of monocyte chemoattractant protein-1 as TGF-beta 1. J Immunol 159:1404–1411
46. Desmouliere A, Geinoz A, Gabbiani F, Gabbiani G (1993) Transforming growth factor-beta 1 induces alpha-smooth muscle actin expression in granulation tissue myofibroblasts and in quiescent and growing cultured fibroblasts. J Cell Biol 122:103–111
47. Mauviel A (2005) Transforming growth factor-beta: a key mediator of fibrosis. Methods Mol Med 117:69–80
48. Schiller M, Javelaud D, Mauviel A (2004) TGF-beta-induced SMAD signaling and gene regulation: consequences for extracellular matrix remodeling and wound healing. J Dermatol Sci 35:83–92
49. Bujak M, Frangogiannis NG (2007) The role of TGF-beta signaling in myocardial infarction and cardiac remodeling. Cardiovasc Res 74:184–195
50. Thompson NL, Bazoberry F, Speir EH et al (1988) Transforming growth factor beta-1 in acute myocardial infarction in rats. Growth Factors 1:91–99
51. Dean RG, Balding LC, Candido R et al (2005) Connective tissue growth factor and cardiac fibrosis after myocardial infarction. J Histochem Cytochem 53:1245–1256
52. Bujak M, Ren G, Kweon HJ et al (2007) Essential role of Smad3 in infarct healing and in the pathogenesis of cardiac remodeling. Circulation 116:2127–2138
53. Ikeuchi M, Tsutsui H, Shiomi T et al (2004) Inhibition of TGF-beta signaling exacerbates early cardiac dysfunction but prevents late remodeling after infarction. Cardiovasc Res 64:526–535
54. Okada H, Takemura G, Kosai K et al (2005) Postinfarction gene therapy against transforming growth factor-beta signal modulates infarct tissue dynamics and attenuates left ventricular remodeling and heart failure. Circulation 111:2430–2437

55. Dobaczewski M, Bujak M, Li N et al (2010) Smad3 signaling critically regulates fibroblast phenotype and function in healing myocardial infarction. Circ Res 107:418–428
56. Kempf T, Eden M, Strelau J et al (2006) The transforming growth factor-beta superfamily member growth-differentiation factor-15 protects the heart from ischemia/reperfusion injury. Circ Res 98:351–360
57. Kempf T, Zarbock A, Widera C et al (2011) GDF-15 is an inhibitor of leukocyte integrin activation required for survival after myocardial infarction in mice. Nat Med 17:581–588
58. Ricciotti E, FitzGerald GA (2011) Prostaglandins and inflammation. Arterioscler Thromb Vasc Biol 31:986–1000
59. Serhan CN (2010) Novel lipid mediators and resolution mechanisms in acute inflammation: to resolve or not? Am J Pathol 177:1576–1591
60. Keyes KT, Ye Y, Lin Y et al (2010) Resolvin E1 protects the rat heart against reperfusion injury. Am J Physiol Heart Circ Physiol 299:H153–H164
61. Nathan C, Ding A (2010) Nonresolving inflammation. Cell 140:871–882
62. Huynh ML, Fadok VA, Henson PM (2002) Phosphatidylserine-dependent ingestion of apoptotic cells promotes TGF-beta1 secretion and the resolution of inflammation. J Clin Invest 109:41–50
63. Devitt A, Marshall LJ (2011) The innate immune system and the clearance of apoptotic cells. J Leukoc Biol 90(3):447–457
64. Soehnlein O, Lindbom L (2010) Phagocyte partnership during the onset and resolution of inflammation. Nat Rev Immunol 10:427–439
65. Robbins CS, Swirski FK (2010) The multiple roles of monocyte subsets in steady state and inflammation. Cell Mol Life Sci 67:2685–2693
66. Geissmann F, Jung S, Littman DR (2003) Blood monocytes consist of two principal subsets with distinct migratory properties. Immunity 19:71–82
67. Nahrendorf M, Swirski FK, Aikawa E et al (2007) The healing myocardium sequentially mobilizes two monocyte subsets with divergent and complementary functions. J Exp Med 204:3037–3047
68. Tsujioka H, Imanishi T, Ikejima H et al (2009) Impact of heterogeneity of human peripheral blood monocyte subsets on myocardial salvage in patients with primary acute myocardial infarction. J Am Coll Cardiol 54:130–138
69. Frangogiannis NG, Mendoza LH, Ren G et al (2003) MCSF expression is induced in healing myocardial infarcts and may regulate monocyte and endothelial cell phenotype. Am J Physiol Heart Circ Physiol 285:H483–H492
70. Littman DR, Rudensky AY (2010) Th17 and regulatory T cells in mediating and restraining inflammation. Cell 140:845–858
71. Dobaczewski M, Xia Y, Bujak M, Gonzalez-Quesada C, Frangogiannis NG (2010) CCR5 signaling suppresses inflammation and reduces adverse remodeling of the infarcted heart, mediating recruitment of regulatory T cells. Am J Pathol 176:2177–2187
72. Anzai A, Anzai T, Nagai S et al (2012) Regulatory role of dendritic cells in postinfarction healing and left ventricular remodeling. Circulation 125:1234–1245
73. Li J, Brown LF, Hibberd MG et al (1996) VEGF, flk-1, and flt-1 expression in a rat myocardial infarction model of angiogenesis. Am J Physiol 270:H1803–H1811
74. Ren G, Michael LH, Entman ML, Frangogiannis NG (2002) Morphological characteristics of the microvasculature in healing myocardial infarcts. J Histochem Cytochem 50:71–79
75. Zymek P, Bujak M, Chatila K et al (2006) The role of platelet-derived growth factor signaling in healing myocardial infarcts. J Am Coll Cardiol 48:2315–2323
76. Souders CA, Bowers SL, Baudino TA (2009) Cardiac fibroblast: the renaissance cell. Circ Res 105:1164–1176
77. Dobaczewski M, Gonzalez-Quesada C, Frangogiannis NG (2010) The extracellular matrix as a modulator of the inflammatory and reparative response following myocardial infarction. J Mol Cell Cardiol 48:504–511
78. Teder P, Vandivier RW, Jiang D et al (2002) Resolution of lung inflammation by CD44. Science 296:155–158
79. Frangogiannis NG (2012) Matricellular proteins in cardiac adaptation and disease. Physiol Rev 92:635–688

80. Frangogiannis NG, Ren G, Dewald O et al (2005) The critical role of endogenous Thrombospondin (TSP)-1 in preventing expansion of healing myocardial infarcts. Circulation 111:2935–2942
81. Entman ML, Michael L, Rossen RD et al (1991) Inflammation in the course of early myocardial ischemia. FASEB J 5:2529–2537
82. Faxon DP, Gibbons RJ, Chronos NA, Gurbel PA, Sheehan F (2002) The effect of blockade of the CD11/CD18 integrin receptor on infarct size in patients with acute myocardial infarction treated with direct angioplasty: the results of the HALT-MI study. J Am Coll Cardiol 40:1199–1204
83. Armstrong PW, Granger CB, Adams PX et al (2007) Pexelizumab for acute ST-elevation myocardial infarction in patients undergoing primary percutaneous coronary intervention: a randomized controlled trial. JAMA 297:43–51
84. Briaud SA, Ding ZM, Michael LH et al (2001) Leukocyte trafficking and myocardial reperfusion injury in ICAM-1/P-selectin-knockout mice. Am J Physiol Heart Circ Physiol 280:H60–H67
85. Abbate A, Kontos MC, Grizzard JD et al (2010) Interleukin-1 blockade with anakinra to prevent adverse cardiac remodeling after acute myocardial infarction (Virginia Commonwealth University Anakinra Remodeling Trial [VCU-ART] Pilot study). Am J Cardiol 105:1371–1377 e1

Chapter 19
TLR-Dependent Pathways and Akt/mTOR/P70S6K Pathways in Cardiac Remodeling After Myocardial Infarction

Lina Badimon and Gemma Vilahur

Abstract Left ventricular remodeling is a progressive process which starts immediately after acute myocardial infarction (MI) and evolves in the chronic phase of heart failure. The molecular and cellular changes associated with ventricular remodeling affect both the cardiomyocytes and the interstitial space and manifest clinically as increased ventricular size, altered shape of the ventricle, and worsened cardiac function. Specific therapy to optimize healing and prevent adverse post-MI remodeling is currently lacking. Understanding of the specific events (time and space) that occur in response to MI and are involved in infarct healing is crucial to abrogate postinfarction ventricular remodeling and/or stimulate the healing process. Inflammation and extracellular matrix remodeling are two key components associated with LV remodeling after MI. Myocardial injury upon MI activates an inflammatory response that orchestrates the cardiac repair process through a complex cascade involving cytokines, other inflammatory mediators, and growth factors. In this regard, the innate immune system, such as that mediated by the Toll-like receptors (TLRs), has appeared during the past decade as a major contributor to the pathogenesis of MI by modulating cell survival and tissue injury. Cardiac remodeling

---

L. Badimon (✉)
Cardiovascular Research Center, CSIC-ICCC, Hospital de la Santa Creu i Sant Pau, IIB-Sant Pau, c/Sant Antoni MªClaret 167, 08025 Barcelona, Spain

CIBEROBN-Pathophysiology of Obesity and Nutrition, c/Sant Antoni MªClaret 167, 08025 Barcelona, Spain

UAB, Barcelona, Spain
e-mail: lbadimon@csic-iccc.org

G. Vilahur
Cardiovascular Research Center, CSIC-ICCC, Hospital de la Santa Creu i Sant Pau, IIB-Sant Pau, c/Sant Antoni MªClaret 167, 08025 Barcelona, Spain

CIBEROBN-Pathophysiology of Obesity and Nutrition, c/Sant Antoni MªClaret 167, 08025 Barcelona, Spain

B.I. Jugdutt and N.S. Dhalla (eds.), *Cardiac Remodeling: Molecular Mechanisms*, Advances in Biochemistry in Health and Disease 5, DOI 10.1007/978-1-4614-5930-9_19,

also encompasses many changes in the myocardial interstitium leading to fibrosis and hypertrophy. The mammalian target of rapamycin (mTOR) signaling has shown to exert a broad spectrum of functions including cell cycle progression, hypertrophy, protein synthesis, autophagy, and angiogenesis, all of which largely contribute to the fibrotic reparative scar.

This chapter aims to provide an in-depth analysis of the contribution of both the innate immune system (via TLR activation) and the activation of Akt/mTOR/P70S6K in the cardiac remodeling phase post-MI.

Keywords Myocardial infarction • Cardiac remodeling • Inflammation • Innate immune system • Toll-like receptor • mTOR • Hypertrophy • Autophagy • Fibrosis • Matrix remodeling

Abbreviations

HMGB	High-mobility group box
IL	Interleukin
IRF3	Interferon-regulatory factor-3
LV	Left ventricle
MI	Myocardial infarction
mTOR	Mammalian target of rapamycin
MyD88	Myeloid differentiation factor 88
NF-κB	Nuclear factor kappa-light-chain-enhancer of activated B cells
PI3K	Phosphoinositide 3-kinase
TLR	Toll-like receptors
TNF	Tumor necrosis factor
TORC	mTOR complex
TRAF6	Tumor necrosis factor receptor-associated factor-6
TRAM	TRIF-related adaptor molecule
TRIF	TIR domain-containing adaptor protein inducing type 1 interferons
VEGF	Vascular endothelial growth factor
WT	Wild type

19.1 Toll-Like Receptors

19.1.1 Toll-Like Receptors and the Innate Immunity

Innate immunity is our first line of defense against foreign intruders, and it is mainly driven by neutrophils, dendritic cells, and macrophages, as well as cytokines, chemokines, and the activation of the complement cascade [1]. The cells of the innate immune system recognize conserved motifs on pathogens termed pathogen-associated molecular patterns (PAMPs). Recent evidence also suggests that the innate immune system recognizes endogenous danger signals, termed damage-associated molecular patterns (DAMPs). PAMP/DAMP in turn is recognized by pattern recognition receptors (PRRs). There are currently four known families of PRR: the transmembrane protein families of Toll-like receptors (TLRs), C-type lectin receptors (CLRs), cytoplasmic protein families such as nucleotide oligomerization domain (NOD)-like receptors (NLRs) and retinoic acid-inducible gene (RIG)-like receptors (RLRs) [1].

Toll, meaning "amazing" and "wonderful" in German, was first described by Christiane Nüsslein-Volhard and colleagues in 1985 when they were studying the genetic mechanisms that control early embryonic development in the fruit fly *Drosophila* [2]. TLRs are type I transmembrane proteins composed of three structural domains: an extracellular C-terminal leucine-rich repeat (LRR) as ligand recognition domain, a central transmembrane domain, and an N-terminal cytoplasmic Toll/interleukin receptor homologous signaling domain (TIR) as an effector domain that mediates the homotypic interactions facilitating downstream signaling or substrate processing (Fig. 19.1). TLRs are expressed not only on immune cells but also in nonimmune cells including myocytes, fibroblasts, epithelial cells, keratinocytes, skeletal muscle, and neurons [3]. TLRs mostly detect viral and bacterial components; however, during the last years, increasing evidence indicates that TLRs may be turned on in the absence of infectious microorganisms by host-derived molecules (Fig. 19.1; detailed below).

There are at least 11 TLRs identified in humans (TLR1 to TLR11), and two new members TLR12 and TLR13 have been discovered in mice, but not much information is available about them. TLRs can be either located in the plasma membrane (TLR1, TLR 2, TLR4, TLR5, TLR6, and TLR10) or localized in the membranes of intracellular compartments (TLR3, TLR7, TLR8, and TLR9) such as endosomes. Newly discovered TLR11, TLR12, and TLR13 are also believed to be localized in the plasma membrane (Fig. 19.1). The location of the TLRs seems to be determined by the nature of their ligands. For instance, membrane-bound TLRs recognize pathogens and the breakdown products of extracellular matrix, whereas intracellular TLRs generally recognize nucleic acid structures (Fig. 19.1) [4]. The stimulation of TLRs leads to, through their specific intracellular signaling pathways, the activation of various downstream transcription factors and the ultimate production of inflammatory cytokines in host immune cells, a process that has been linked to myocardial infarction and cardiac remodeling. Indeed, myocardial infarction is associated with an inflammatory reaction that, although a prerequisite for healing and scar formation, when exacerbated may interfere with cardiac healing and collagen matrix deposition.

19.1.2 TLR Signaling Pathways

TLRs act as either homodimers (TLR4 and TLR9) or heterodimers with other TLRs (TLR2 with TLR1 and TLR6) [4]. The different dimmers recognize distinct molecules that include lipids, lipoproteins, nucleic acids, and proteins (Fig. 19.1).

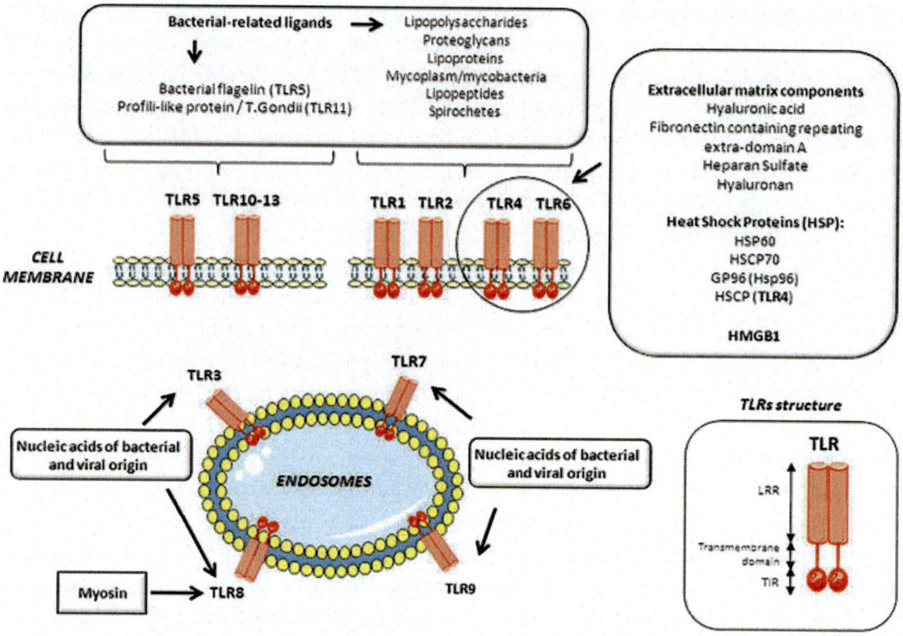

Fig. 19.1 *TLRs structure and ligands/activators of TLRs. HMGB* high-mobility group box, *HSP* heat shock protein, *HSCP* heat shock cognate protein, *LRR* extracellular C-terminal leucine-rich repeat (ligand recognition domain), *TIR* N-terminal cytoplasmic Toll/interleukin receptor homologous signaling domain

Upon dimerization of TLRs, the cytosolic TIR changes conformation to facilitate the recruitment of adaptors that initiate TLR downstream signaling. Myeloid differentiation factor 88 (MyD88) is a nearly universal TLR adapter protein (except with regard to TLR3) that also functions as an adaptor in signaling initiated by the IL-1 receptor. Thus, the signaling pathways activated by TLRs are broadly classified into MyD88-dependent or MyD88-independent signaling pathways (Fig. 19.2). TLRs, however, also involve the recruitment of other adaptors including MyD88 adaptor-like [(MAL) also known as TIR-domain-containing adaptor protein (TIRAP)], TIR-domain-containing adaptor protein inducing type 1 interferons (TRIF), TRIF-related adaptor molecule (TRAM), and sterile-α and armadillo motifs (SARM).

TLR–MyD88-dependent pathway. In some TLR signaling, such as TLR2 and TLR4, MAL is required for recruiting MyD88 to the receptor. As shown in Fig. 19.2, once recruited the MyD88/MAL complex activates the downstream interleukin-1 receptor-associated kinase 4 (IRAK4) which, in turn, induces IRAK1 phosphorylation. TRAF6 (tumor necrosis factor receptor-associated factor 6) is also recruited to the receptor complex by associating with phosphorylated IRAK1. Phosphorylated TRAF6 then dissociates from the receptor and forms a complex with TAK1 (TGF-β-activated kinase-1), TAB1 (TAK1-binding protein-1), and TAB2 (TAK1-binding

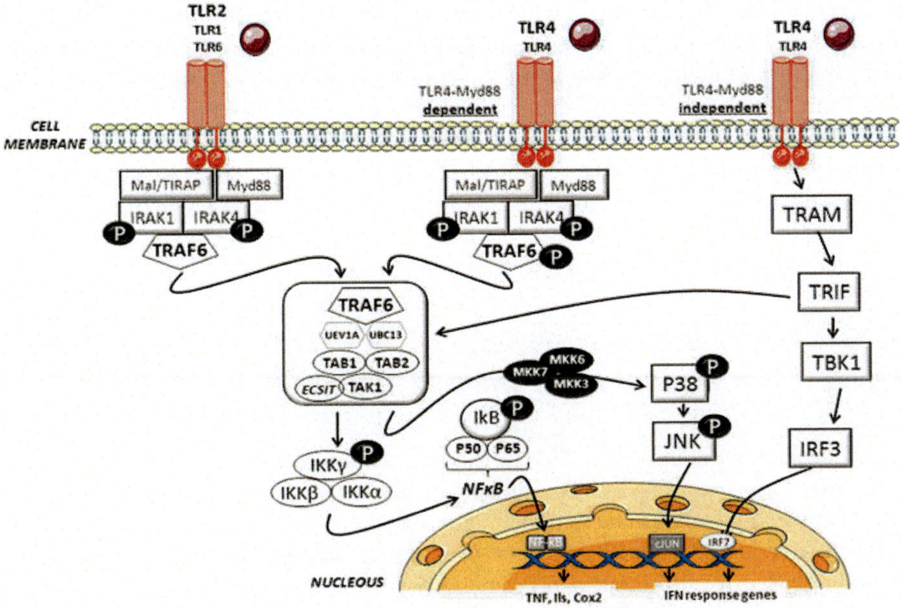

Fig. 19.2 *Both myeloid differentiation primary response protein 88 (MyD88)-dependent and independent TLR downstream pathways and eventual gene induction.* This figure depicts the signaling pathways activated by the two most important cardiac TLR receptors, TLR2 and TLR4. Only the signaling components discussed in the chapter are depicted in this figure. *IRAK1* interleukin-1 receptor-associated kinase, *TRAF6* tumor necrosis factor receptor-associated factor 6, *TAK1* TGF-β-activated kinase-1, *TAB1* TAK1-binding protein-1, *TAB2* TAK1-binding protein-2, *UBC13* ubiquitin-conjugating enzyme-13, *UEV1A* ubiquitin-conjugating enzyme E2 variant 1, *IKK* inhibitor of κ-light polypeptide gene enhancer in B-cell kinase, *NF-κB* nuclear transcription factor-κB, *MKK* mitogen-activated protein kinase, *TRIF* TIR-domain-containing adaptor protein inducing type 1 interferons, *TRAM* TRIF-related adaptor molecule, *TBK1*, TANK-binding kinase-1, *IRF* interferon regulatory factor

protein-2). TRAF6, TAK1, TAB1, and TAB2 associate with the ubiquitin ligases UBC13 (ubiquitin-conjugating enzyme-13) and UEV1A (ubiquitin-conjugating enzyme E2 variant 1). TAK1 phosphorylates the IKK complex (inhibitor of κ-light polypeptide gene enhancer in B-cell kinase), which consists of IKK-α, IKK-β, and IKK-γ. The IKK complex then phosphorylates I-κB, which leads to its ubiquitylation and subsequent degradation. This allows NF-κB to translocate to the nucleus and induce the expression of its target genes (proinflammatory cytokines) (Fig. 19.2). TAK1 also phosphorylates both p38 kinases and c-Jun N-terminal kinase (JNKs) by activating MKK3 (mitogen-activated protein kinase-3), MKK6, and MKK7. p38 and JNKs then enter the nucleus and induce the expression of their target genes (Fig. 19.2) [5].

TLR–MyD88-independent pathway. TLR4 can also signal through a MyD88-independent pathway (also known as the TRIF-dependent pathway) (Fig. 19.2). In the TLR–MyD88-independent pathway, TLR4 signals via TRIF-related adaptor molecule (TRAM) and TRIF which activates TBK1 (TANK-binding kinase-1). TBK1 then stimulates the downstream interferon regulatory factor 3 (IRF3) initiating expression of type 1 IFNs while translocation into nucleus occurs (Fig. 19.2) [5]. TRIF, in turn, may indirectly activate NF-κB through TRAF6.

TLR5, TLR7/8, TLR9, and TLR11 use only MyD88 as its signaling adaptor and MAL is not required, whereas TLR3 signals through the adaptor TIR-domain-containing adaptor protein inducing interferon-β (IFN-β)-mediated transcription factor (TRIF), which recruits and activates TNF receptor-associated factor family member-associated NF-κB activator-binding kinase 1 (TBK1).

19.1.3 TLR Ligands in the Infarcted Heart

The mechanisms involved in TLR activation remain to be fully described. Upon ischemia, TLRs can respond to various endogenous cellular components released from damaged myocytes and extracellular components whose structure is altered as a result of released cellular enzymes (Fig. 19.1). Indeed, the endogenous tissue debris released from injured and inflamed tissues including the highly conserved nuclear protein high-mobility group box (HMGB)1 and breakdown products of the cardiac extracellular matrix (hyaluronan, S100 proteins, heat shock proteins (HSP), and the spliced extra domain A of fibronectin) have been shown to activate the TLR4 signaling pathway (Fig. 19.1) [6, 7]. HMGB1 is a damage-associated molecule passively released by necrotic cells and activated macrophages capable of activating TLR2, TLR4, and TLR9 [8]. HMGB has shown to accumulate in the ischemic myocardium, and peak serum levels have shown to correlate with worse outcome after MI. Another recent study has supported that 70 kDa heat shock cognate protein (HSCP70) released from cardiac cells during ischemia is able of activating TLR4 in resident cardiac driving myocardial inflammatory recruitment [9].

19.1.4 TLRs in Myocardial Infarction

TLR-mediated signaling has been robustly associated with atherosclerotic plaque formation, vascular remodeling, and cell death. During the past years, however, evidence has emerged that clearly indicates that TLR signaling may also play a critical role in acute MI and further remodeling process by modulating the inflammatory response, myocyte survival/death, and the fibrotic scar resolution. Cardiac myocytes express six receptors involved in TLR signaling (TLR2, 3, 4, 5, 7, and 9) [10] although the most studied TLRs in the heart are TLR2 and TLR4. However, the mechanisms by which TLRs exert their cardiovascular effects are incompletely defined [11]. Therefore, defining the critical role of TLR, particularly that of cardiac origin, in the setting of MI may have important therapeutic implications for the posterior inflammatory reaction and cardiac remodeling response.

A short period (5 min) of coronary artery occlusion has not shown to trigger an inflammatory reaction [12]. Yet, a longer ischemic insult may result in a more intense inflammatory response. We have reported in a pig model of closed-chest coronary balloon occlusion that 30 min of ischemia triggers systemic cytokine release and renders platelets and circulating leukocytes more susceptible to activation (higher cycloxigenase-2, monocyte chemoattractant protein-1, and tissue factor protein expression). Such inflammatory response is enhanced upon prolongation of the ischemic period [13]. Interestingly, such systemic inflammatory reaction is associated with a marked upregulation of myocardial TLR4–MyD88-dependent and TLR4–MyD88-independent signaling pathways in the ischemic myocardium leading to NF-κB nucleus translocation and IRF3 activation [14]. These observations support a prompt activation of the cardiac innate immune response upon an ischemic insult that may rapidly translate into a proinflammatory systemic response through TLR4/cytokine networks [13]. Besides, such inflammatory response may have stimulated the detected homing of leukocytes to the site of MI which, in turn, may have contributed to the detected TLR4 increase upon reperfusion [14]. In line with these observations, hearts from TLR4 mutant mice have demonstrated lower myocardial proinflammatory cytokine (TNF-α and IL-1β) levels and NF-κB activation, reduced cardiac inflammatory infiltration, smaller infarctions, and improved postischemic functional recovery compared with their wild-type (WT) counterparts [15–17]. Moreover, mice pretreated with eritoran, a specific TLR4 antagonist, have shown reduced nuclear NF-κB translocation and proinflammatory cytokine expression, when compared with mice treated with vehicle alone, leading to significantly smaller infarcts [18].

Blockade of TLR downstream effectors has also shown to contribute to attenuate cardiac injury upon MI. For instance, rats transfected with Ad5-dn-MyD88 (inactive form) into the myocardium have shown significantly reduced infarct size after MI compared with those transfected with Ad5-GFP (control green fluorescent protein) suggesting a deleterious effect of MyD88-dependent signaling on the postischemic myocardial recovery [19]. Inhibition of NF-κB using molecular (inhibition of p65 by double-stranded oligonucleotides [20] and use of an inhibitor of NF-κB (IkB) triple mutant (S32A, S36A, Y42A) [21]) as well as pharmacological methods (IkB kinase (IKK) inhibition [22]) have also shown to reduce ischemia/reperfusion injury.

In contrast, however, other studies have supported a protective role of TLR4–MyD88 signaling in the setting of MI by preventing myocyte death [23]. Human studies have also shown conflicting results. Population-based studies designed to determine the impact of TLR4 polymorphism on the risk of MI have reported conflicting data. Some studies suggest that among symptomatic men with documented coronary artery disease, the TLR4 Asp299Gly polymorphism is associated with a lower risk of cardiovascular events compared to the control population [24, 25], whereas others suggest an increased risk of MI [26] or no association [27]. It is noteworthy that TLRs are expressed in both compartments (cardiac and vascular), and the exact contribution of TLR signaling of cardiac versus circulatory origin is not clearly defined and may help to explain these opposing findings.

19.1.5 TLR in Healing Infarcts: Cardiac Remodeling

The stimulation of cytokine secretion in response to MI has profound effects on myocardium that provokes at least four changes directly in cardiac myocytes that contribute to cardiac remodeling: progressive myocyte apoptosis, myocyte hypertrophy, defects in contractility, and inflammatory signal transduction [28]. Among the cytokines, TLR-induced TNF-α, IL-1β, and IL-6 are most commonly associated with the remodeling process post-MI [29]. Although data on myocardial remodeling and TLRs are very limited, evidence has been gathered documenting that the levels of TNF-α and IL-1β are increased in hearts of patients with heart failure as well as in the hearts of animals with experimental cardiac dysfunction [30]. Moreover, cytokine levels in the non-infarcted region correlated with the eventual LV end-diastolic diameter measured at 20-week postinfarction [29]. We further evidenced that, although both NF-κB and IRF3 activation represent a rapid and intrinsic response upon MI, there is a sustained increase, for up to 6 days post-reperfusion, in TLR4 gene and protein expression in both infarcted and non-infarcted/remote myocardium. All together, these observations suggest that timely interception of TLR4 might be critical to prevent adverse myocardial scar formation and remodeling [14]. Frantz and colleagues [31] first documented that there was an enhanced TLR4 expression in remodeling murine myocardium remote from sites of ischemic injury and in heart tissue from patients with idiopathic dilated cardiomyopathy which was not observed in control myocardium. Later studies have also shown that TLR4 is increased in the myocardium of patients with heart failure [32]. What is unclear, however, is the functional significance of TLR4 upregulation under these cardiac conditions. Studies conducted in TLR4 knockout mice have shown improved left ventricular function and attenuated adverse left ventricular remodeling as indicated by reduced levels of atrial natriuretic factor, collagen, and heart weight–body weight ratio [15, 33].

Recent data in mice has evidenced a supporting role of the TLR downstream signaling in cardiac remodeling. As such, pharmacologic MyD88 inhibition has shown to protect against pathologic LV remodeling despite no changes were detected in infarct scar formation [34]. Yet, targeted deletion of the NF-κB subunit p50 has shown conflicting results. Some reports have documented that deletion of NF-κB subunit p50 markedly increases the extent of expansive remodeling and functional deterioration following MI [35]. In fact, NF-κB subunit p50 is generally considered to be an inhibitory subunit of the NF-κB complex. However, in contrast, studies published in p50 knockout mice have shown protection against LV remodeling following MI [36]. These discrepancies may derive from the imaging technique used for cardiac remodeling assessment, using a 3D high-resolution MRI 9.4 Tesla in the former and 2D short axis echocardiography in the latter.

As to the role of TLR2 in cardiac remodeling, conflicting results have also been reported. Shishido et al. [37] reported that 4 weeks after permanent coronary ligation, myocardial fibrosis in the non-infarcted area of TLR2 knockout mice was found to be reduced as compared to their WT counterparts and was accompanied by reduced transforming growth factor 1 and collagen type 1 mRNA expression [37]. Furthermore, left ventricular dimensions at end diastole were smaller, and fractional shortening was higher in TLR2 knockout mice than in WT mice at 1 week and 4 weeks after surgery. However, infarct size and the degree of inflammatory cell infiltration in the infarcted area were similar between WT and TLR2 knockout mice. These findings suggested a critical role for TLR2 signaling in ventricular remodeling by modulating fibrous tissue deposition rather than affecting the inflammatory response [37]. In contrast to these pioneering findings, Mersmann et al. [38] found that TLR2-null animals had a marked left ventricular dilation which was associated with pronounced matrix remodeling characterized by reduced collagen and decorin density in the infarct scar as compared to their WT littermates.

In summary, during the last years emerging evidence has indicated that TLRs mediate the inflammatory response upon MI driving the subsequent remodeling process. Yet, the critical role of TLR2 and TLR4, especially that of cardiac origin, in cardioprotection and prevention of adverse left ventricle remodeling remains unclear.

19.2 PI3K/Akt/mTOR/P70S6K

As described above, following the first acute inflammatory phase, cellular and interstitial changes occur in the left ventricle, including myocyte hypertrophy and enhanced protein synthesis (interstitial collagen deposition) that govern the evolving cardiac fibrotic scar. Several studies have suggested that collagen quantity and quality in the infarcted LV plays a critical role in heart failure progression [28, 39]. An important pathway related to the activation of the protein translation machinery is the phosphoinositide 3-kinase (PI3K)/Akt/mammalian target of rapamycin (mTOR)/P70S6K signaling pathway [14]. The contribution of this pathway to cardiac remodeling is supported by several studies in mice reporting that whereas cardiac-specific overexpression of constitutively active PI3Kα results in enlarged hearts owing to increased cardiomyocyte size, overexpression of dominant-negative PI3Kα produces smaller hearts with reduced cardiomyocyte size without affecting cell number [40]. These changes in heart size are associated with corresponding alterations in Akt phosphorylation and activity and its downstream effectors mTOR/P70S6K [40].

19.2.1 PI3K/Akt/mTOR/P70S6K Signaling

As depicted in Fig. 19.3, oxidative stress-related activation of PI3K results in the activation of 3-phosphoinositide-dependent kinase-1 (PDK1) which in turn phosphorylates Akt, also called protein kinase B. Yet, full Akt activation requires Akt phosphorylation both on threonine 308 (by PDK) and on serine 473 by the mTOR

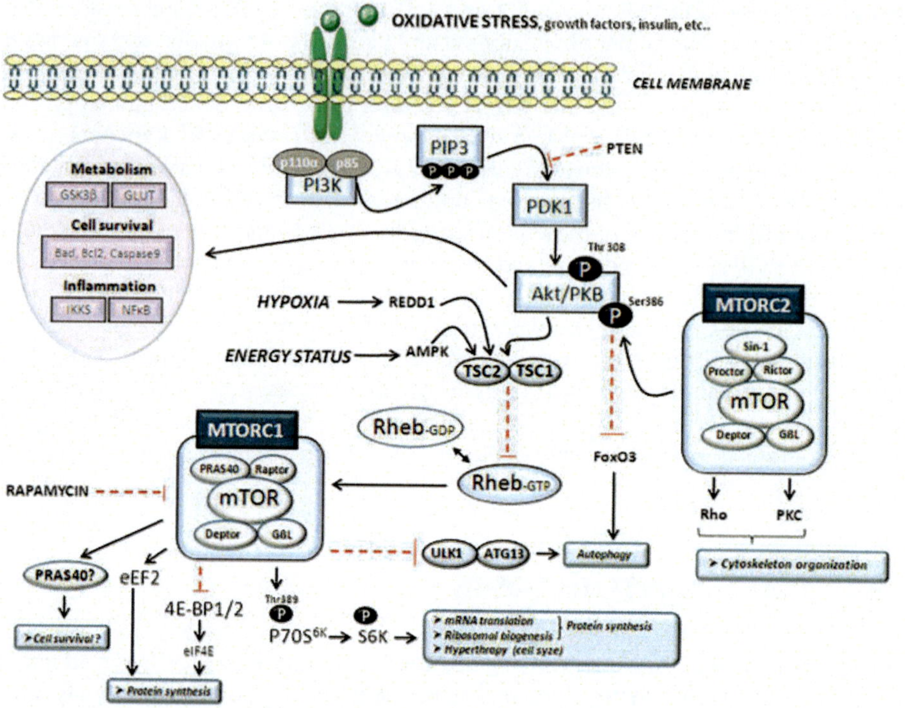

Fig. 19.3 *Diagram of key features of the PI3K/Akt/TOR signaling network.* The lipid kinase PI3K is activated downstream of oxidative stress signals (*continuous arrows*). PI3K generates the second messenger PIP3 which activates PDK1. The TOR serine–threonine kinase is present in two distinct multiprotein complexes, termed TORC1 and TORC2. Key substrates are shown, and their activities are either increased (*block arrows*) or decreased (*pointed red lines*) by TOR-mediated phosphorylation. The resultant functional outcomes are included. *PDK1* phosphoinositide-dependent kinase-1, *GSK3β* glycogen synthase kinase 3β, *GLUT* glucose transporter, *mTOR* mammalian target of rapamycin, *TSC* tuberous sclerosis complex, *mTORC* mTOR complex, *eEF2* elongation factor 2

complex 2 (see below for a more detailed description). Once activated, Akt/PKB phosphorylates several downstream targets including glycogen synthase kinase 3β (GSK3β), glucose transporter (GLUT), several antiapoptotic effectors, inflammatory mediators, and, last but not the least, the mammalian target of rapamycin (mTOR; Fig. 19.3). mTOR is a large (290 kDa) serine–threonine protein kinase belonging to the phosphatidylinositol kinase-related kinase (PIKK) family [16]. Akt/PKB regulates mTOR activation via tuberous sclerosis complex (TSC)-1/2 and Rheb. Phosphorylation of TSC2 by Akt inhibits TSC2 function, but disinhibits Rheb, resulting in mTOR activation. mTOR exists in two different complexes: TORC1 (mTOR complex 1) and TORC2 (mTOR complex 2). Both these complexes bind mLST8/GβL and deptor, whereas TORC1 contains PRAS40 and raptor, and TORC2 contains protor, rictor, and mSin1 (Fig. 19.3). mTORC2 is mainly involved in Akt activation an actin cytoskeleton organization.

mTOR also functions downstream of Akt via mTORC1 which has been implicated in mRNA translation, ribosomal biogenesis, autophagy, and hypertrophy. Akt-related activation of mTORC1 leads to P70S6 kinase (P70S6K, a short isoform of the ribosomal S6 kinase) phosphorylation at threonine 389. By phosphorylating P70S6K, this results in the phosphorylation of the ribosomal protein S6K which in turn phosphorylates the ribosomal S6 protein, a component of the 40S ribosomal subunit. S6 regulates protein translation, ribosomal biogenesis, and cell growth (hypertrophy) (Fig. 19.3). Protein synthesis is also triggered by mTOR negative regulation of the translation repressor proteins 4E-BP1/2. Blockade of 4E-BP1/2 activates the translation initiator eIF4E leading to increased protein translation and the progression of the cell from G0/G1 to S phase [41]. mTOR also enhances the cellular capacity for protein synthesis (i.e., ribosomal biogenesis) by activating elongation factor 2 (eEF2 kinase).

19.2.2 mTOR in Cardiac Remodeling

The signaling mechanism regulated through mTOR largely supports its central role in cardiac remodeling post-MI. In fact, mTOR mRNA and protein levels have found to be elevated in cardiac tissue from heart failure patients compared to healthy subjects [42]. Moreover, the data presented so far also emphasizes that control of mTOR-dependent pathway may potentially prevent post-MI adverse cardiac remodeling. In this regard, mTOR signaling pathway can be regulated by PTEN (physiological inhibition) or rapamycin (pharmacological inhibition). PTEN is a lipid phosphatase and tensin homologue deleted on chromosome 10 which dephosphorylates PIP3, preventing Akt activation (Fig. 19.3) [43]. Rapamycin is a macrocyclic antibiotic discovered in the Easter Island ("Rapa Nui") with high affinity to block mTORC1 (mTORC2 is insensitive to rapamycin) [44]. Inhibition of mTOR through systemic rapamycin treatment upon reperfusion of the ischemic heart has shown to prevent adverse LV remodeling, a protective effect not detected when rapamycin was given before the onset of ischemia or 3-day post-MI [45]. In line with these observations, we have reported in a preclinical model of MI that activation of PI3K/Akt and its downstream effectors mTOR/P70S6k occurs during early reperfusion (within the first 2.5 h) and its maintained over late phase reperfusion (up to 6 days), and not during severe/persistent ischemia [14]. Moreover, Akt/mTOR/P70S6k activation was timely associated with myocyte hypertrophy and collagen interstitial deposition in the forming reparative fibrotic scar. According to these data, rapamycin has shown to attenuate left ventricle hypertrophy and cardiac remodeling by reducing P70S6k and S6K activity in mice [46]. Moreover, rapamycin has also shown to attenuate cardiac hypertrophy in humans [47].

Finally, further data has also suggested that there may be a TLR-mediated signaling pathway that cross talks with the PI3K/Akt signaling pathway concomitantly contributing to cardiac hypertrophy. In fact, inhibition of mTOR by rapamycin in TLR4-deficient mice has shown to result in an additional decrease in the development of cardiac hypertrophy [19]. A growing body of evidence suggests that reactive oxygen species (ROS) may play a link in the pathogenesis of cardiac remodeling since ROS can activate transcription factors, such as NF-κB, and stimulate the PI3K/Akt/mTOR signaling pathway.

19.2.3 *New Horizons for mTOR*

mTOR and autophagy. Autophagy is essential in the heart for protein turnover and the homeostasis of organelles. If autophagy is reduced, abnormal proteins accumulate and lead into apoptosis, enhanced endoplasmatic reticulum stress, and cardiac dysfunction. In addition to its well-established role in translation, the mTOR pathway has been suggested to play a role in autophagy. mTOR, through both TORC complexes, has shown to negatively regulate autophagy in the context of MI and cardiac remodeling. Recent studies have revealed that this is achieved via TORC1—through the phosphorylation and subsequent repression of the autophagy-promoting factors unc-51-like kinase 1 (ULK1) and autophagy-related gene 13 (ATG13)—and/or via TORC2 by repressing the transcription factor FoxO3 (Fig. 19.3). mTOR inhibition and the subsequent upregulation of autophagy in the critical border zone (periinfarction) have shown to act as protective mechanisms in the failing heart by limiting infarct size and preventing left ventricular remodeling [48]. Nonetheless, caution in mTOR sustained repression should be taken since unrestricted augmentation of autophagy may be detrimental.

mTOR and angiogenesis. Heart failure is associated with a decrease in myocardial capillaries in part by concomitant fibrosis. Recent studies have suggested that mTOR downstream activation beneficially contributes to angiogenesis by increasing VEGF and angiopoietin (Ang)-2 expression via hypoxia-inducible factor (HIF)-1α [49, 50]. These findings indicate that disruption of the mTOR signal may promote pathological remodeling.

19.3 Conclusions and Perspectives

Due to improvement of therapeutic strategies after MI, mortality has significantly decreased in the past decades. Consecutively, there have also been an increasing number of patients whose prognosis depends on optimal treatment after MI. Robust clinical data supports that the extent of adverse myocardial remodeling contributes essentially to the prognosis after MI. Despite the widespread use of ACE inhibitors (angiotensin-converting enzyme inhibitors), β-blockers, and aldosterone antagonists, the incidence of heart failure as the end stage of left ventricular remodeling still remains high. Novel strategies to reduce cardiac remodeling are required. However, better understandings of the fundamental processes that occur upon MI/revascularization are needed to help achieve this goal. Underlying mechanisms of remodeling are manifold [28] as detailed along the book, including the activation of the innate immune system (inflammation) and protein translation machinery (fibrosis), highlighted throughout this chapter. A large body of evidence suggests that innate immunity is triggered through TLR in models of acute MI. However, the importance of TLR and its downstream effector NF-κB in the chronic context of myocardial remodeling is far from determined. Likewise, whether the activation of the innate immunity is just an attempt to repair damaged tissue and is a beneficial process counteracting ischemic damage should also be regarded. Finally, the role of innate immunity in the heart versus the periphery needs to be ascertained.

On the other hand, increasing evidence supports that the signaling mechanisms regulated through mTOR have shown to play a major role in cell growth and proliferation, protein synthesis, and autophagy, hallmarks of cardiac remodeling. However, new venues of mTOR signaling such as angiogenesis stimulation open the question as whether mTOR inhibition may mediate beneficial or negative effects in the cardiac healing response following MI.

References

1. Takeuchi O, Akira S (2010) Pattern recognition receptors and inflammation. Cell 140:805–820
2. Anderson KV, Bokla L, Nusslein-Volhard C (1985) Establishment of dorsal-ventral polarity in the Drosophila embryo: the induction of polarity by the Toll gene product. Cell 42:791–798
3. Arumugam TV, Okun E, Tang SC et al (2009) Toll-like receptors in ischemia-reperfusion injury. Shock 32:4–16
4. Wang Y, Abarbanell AM, Herrmann JL et al (2010) Toll-like receptor signaling pathways and the evidence linking toll-like receptor signaling to cardiac ischemia/reperfusion injury. Shock 34:548–557
5. Chao W (2009) Toll-like receptor signaling: a critical modulator of cell survival and ischemic injury in the heart. Am J Physiol Heart Circ Physiol 296:H1–H12
6. Scheibner KA, Lutz MA, Boodoo S et al (2006) Hyaluronan fragments act as an endogenous danger signal by engaging TLR2. J Immunol 177:1272–1281
7. Okamura Y, Watari M, Jerud ES et al (2001) The extra domain A of fibronectin activates Toll-like receptor 4. J Biol Chem 276:10229–10233
8. Scaffidi P, Misteli T, Bianchi ME (2002) Release of chromatin protein HMGB1 by necrotic cells triggers inflammation. Nature 418:191–195
9. Su X, Sykes JB, Ao L et al (2010) Extracellular heat shock cognate protein 70 induces cardiac functional tolerance to endotoxin: differential effect on TNF-alpha and ICAM-1 levels in heart tissue. Cytokine 51:60–66
10. Frantz SEG, Bauersachs J (2007) Mechanisms of disease Toll-like receptors in cardiovascular-disease. Nat Clin Pract Cardiovasc Med 4:444–454
11. Aikawa R, Nagai T, Tanaka M et al (2001) Reactive oxygen species in mechanical stress-induced cardiac hypertrophy. Biochem Biophys Res Commun 289:901–907
12. Bolli R, Marban E (1999) Molecular and cellular mechanisms of myocardial stunning. Physiol Rev 79:609–634
13. Vilahur G, Hernandez-Vera R, Molins B et al (2009) Short-term myocardial ischemia induces cardiac modified C-reactive protein expression and proinflammatory gene (cyclooxygenase-2, monocyte chemoattractant protein-1, and tissue factor) upregulation in peripheral blood mononuclear cells. J Thromb Haemost 7:485–493

14. Vilahur G, Juan-Babot O, Pena E et al (2011) Molecular and cellular mechanisms involved in cardiac remodeling after acute myocardial infarction. J Mol Cell Cardiol 50(3):522–533
15. Chong AJ, Shimamoto A, Hampton CR et al (2004) Toll-like receptor 4 mediates ischemia/reperfusion injury of the heart. J Thorac Cardiovasc Surg 128:170–179
16. Oyama J, Blais C Jr, Liu X et al (2004) Reduced myocardial ischemia-reperfusion injury in toll- like receptor 4-deficient mice. Circulation 109:784–789
17. Sakata Y, Dong JW, Vallejo JG et al (2007) Toll-like receptor 2 modulates left ventricular function following ischemia-reperfusion injury. Am J Physiol Heart Circ Physiol 292:H503–H509
18. Kim HM, Park BS, Kim JI et al (2007) Crystal structure of the TLR4-MD-2 complex with bound endotoxin antagonist Eritoran. Cell 130:906–917
19. Hua F, Ha T, Ma J et al (2005) Blocking the MyD88-dependent pathway protects the myocardium from ischemia/reperfusion injury in rat hearts. Biochem Biophys Res Commun 338:1118–1125
20. Sawa Y, Morishita R, Suzuki K et al (1997) A novel strategy for myocardial protection using in vivo transfection of cis element 'decoy' against NF-kappa B binding site: evidence for a role of NF-kappa B in ischemia-reperfusion injury. Circulation 96: II:280–284, discussion II-285
21. Brown M, McGuinness M, Wright T et al (2005) Cardiac-specific blockade of NF-kappa B in cardiac pathophysiology: differences between acute and chronic stimuli in vivo. Am J Physiol Heart Circ Physiol 289:H466–H476
22. Onai Y, Suzuki J, Kakuta T, Maejima Y et al (2004) Inhibition of I-kappa B phosphorylation in cardiomyocytes attenuates myocardial ischemia/reperfusion injury. Cardiovasc Res 63:51–59
23. Zhu X, Zhao H, Graveline AR et al (2006) MyD88 and NOS2 are essential for toll-like receptor 4-mediated survival effect in cardiomyocytes. Am J Physiol Heart Circ Physiol 291:H1900–H1909
24. Balistreri CR, Candore G, Colonna-Romano G et al (2004) Role of Toll-like receptor 4 in acute myocardial infarction and longevity. JAMA 292:2339–2340
25. Boekholdt SM, Agema WR, Peters RJ et al (2003) Variants of toll-like receptor 4 modify the efficacy of statin therapy and the risk of cardiovascular events. Circulation 107:2416–2421
26. Edfeldt K, Bennet AM, Eriksson P et al (2004) Association of hypo-responsive toll-like receptor 4 variants with risk of myocardial infarction. Eur Heart J 25:1447–1453
27. Koch W, Hoppmann P, Pfeufer A, Schomig A, Kastrati A (2006) Toll-like receptor 4 gene polymorphisms and myocardial infarction: no association in a Caucasian population. Eur Heart J 27:2524–2529
28. Jugdutt BI (2003) Ventricular remodeling after infarction and the extracellular collagen matrix: when is enough enough? Circulation 108:1395–1403
29. Nian M, Lee P, Khaper N, Liu P (2004) Inflammatory cytokines and postmyocardial infarction remodeling. Circ Res 94:1543–1553
30. Morishita S, Kinoshita T (2008) Predictors of response to sertraline in patients with major depression. Hum Psychopharmacol 23:647–651
31. Frantz S, Kobzik L, Kim YD et al (1999) Toll4 (TLR4) expression in cardiac myocytes in normal and failing myocardium. J Clin Invest 104:271–280
32. Birks EJ, Felkin LE, Banner NR et al (2004) Increased toll-like receptor 4 in the myocardium of patients requiring left ventricular assist devices. J Heart Lung Transplant 23:228–235
33. Riad A, Jager S, Sobirey M et al (2008) Toll-like receptor-4 modulates survival by induction of left ventricular remodeling after myocardial infarction in mice. J Immunol 180:6954–6961
34. Van Tassell BW, Seropian IM, Toldo S et al (2010) Pharmacologic inhibition of myeloid differentiation factor 88 (MyD88) prevents left ventricular dilation and hypertrophy after experimental acute myocardial infarction in the mouse. J Cardiovasc Pharmacol 55:385–390
35. Timmers L, van Keulen JK, Hoefer IE et al (2009) Targeted deletion of nuclear factor kappa B p50 enhances cardiac remodeling and dysfunction following myocardial infarction. Circ Res 104:699–706
36. Frantz S, Hu K, Bayer B et al (2006) Absence of NF-kappa B subunit p50 improves heart failure after myocardial infarction. FASEB J 20:1918–1920

37. Shishido T, Nozaki N, Yamaguchi S et al (2003) Toll-like receptor-2 modulates ventricular remodeling after myocardial infarction. Circulation 108:2905–2910
38. Mersmann J, Habeck K, Latsch K et al (2011) Left ventricular dilation in toll-like receptor 2 deficient mice after myocardial ischemia/reperfusion through defective scar formation. Basic Res Cardiol 106:89–98
39. Jugdutt BI (1998) Effect of nitroglycerin and ibuprofen on left ventricular topography and rupture threshold during healing after myocardial infarction in the dog. Can J Physiol Pharmacol 66:385–395
40. Crackower MA, Oudit GY, Kozieradzki I et al (2002) Regulation of myocardial contractility and cell size by distinct PI3K-PTEN signaling pathways. Cell 110:737–749
41. Guertin DA, Sabatini DM (2005) An expanding role for mTOR in cancer. Trends Mol Med 11:353–361
42. Song X, Kusakari Y, Xiao CY et al (2010) mTOR attenuates the inflammatory response in cardiomyocytes and prevents cardiac dysfunction in pathological hypertrophy. Am J Physiol Cell Physiol 299:C1256–C1266
43. Maehama T, Dixon JE (1998) The tumor suppressor, PTEN/MMAC1, dephosphorylates the lipid second messenger, phosphatidylinositol 3,4,5-trisphosphate. J Biol Chem 273: 13375–13378
44. Averous J, Proud CG (2006) When translation meets transformation: the mTOR story. Oncogene 25:6423–6435
45. Buss SJ, Muenz S, Riffel JH et al (2009) Beneficial effects of Mammalian target of rapamycin inhibition on left ventricular remodeling after myocardial infarction. J Am Coll Cardiol 54:2435–2446
46. Lajoie C, El-Helou V, Proulx C et al (2009) Infarct size is increased in female post-MI rats treated with rapamycin. Can J Physiol Pharmacol 87:460–470
47. McMullen JR, Sherwood MC, Tarnavski O et al (2004) Inhibition of mTOR signaling with rapamycin regresses established cardiac hypertrophy induced by pressure overload. Circulation 109:3050–3055
48. Buss SJ, Riffel JH, Katus HA, Hardt SE (2010) Augmentation of autophagy by mTOR-inhibition in myocardial infarction: when size matters. Autophagy 6:304–306
49. Hudson CC, Liu M, Chiang GG et al (2002) Regulation of hypoxia-inducible factor 1alpha expression and function by the mammalian target of rapamycin. Mol Cell Biol 22:7004–7014
50. Shiojima I, Sato K, Izumiya Y et al (2005) Disruption of coordinated cardiac hypertrophy and angiogenesis contributes to the transition to heart failure. J Clin Invest 115:2108–2118

Chapter 20
The STAT3 Pathway and Downstream Mechanisms in Cardiac Remodeling: Friend or Foe

Melanie Ricke-Hoch, Britta Stapel, Irina Gorst, Arash Haghikia, and Denise Hilfiker-Kleiner

Abstract Multiple in vitro and in vivo studies have shown that the signal transducer and activator of transcription 3 (STAT3) protein regulates key mechanisms in cardiac physiology (exercise, pregnancy) and pathophysiology (pressure overload, ischemia/reperfusion, myocardial infarction (MI), myocarditis, and cardiotoxic agents). STAT3 is activated in various cardiac cell types including cardiomyocytes, endothelial cells, fibroblasts, and cardiac progenitor cells by a multitude of factors including cytokines, growth factors, neurohormones, mechanical load, and ischemia. It acts as a signaling molecule, a transcription factor, and a mitochondrial protein involved in energy production, and it controls autocrine and paracrine pathways. While the majority of data imply rather beneficial roles of STAT3 in the heart, newer studies implicate that this is mainly the case when the expression and activation of STAT3 is precisely regulated. In contrast, continuous uncontrolled activation of STAT3 in cardiomyocytes seems to promote adverse cardiac remodeling processes especially after MI. Here, we provide an overview on STAT3 signaling and summarize the current understanding of the role of STAT3 for cardiac inflammation, metabolism, remodeling, and regeneration based on experimental and clinical studies. Finally, we highlight the consequences of targeting STAT3 for future therapeutic approaches in the setting of cardiac remodeling.

Keywords STAT3 • Cardiac remodeling • gp130–STAT3 signaling • Inflammation • Metabolism • Angiogenesis • Hypertrophy • Sarcomere structure

M. Ricke-Hoch (✉) • B. Stapel • I. Gorst • A. Haghikia • D. Hilfiker-Kleiner
Department of Cardiology and Angiology, Hannover Medical School,
Carl-Neuberg Str. 1, 30625 Hannover, Germany
e-mail: hoch.melanie@mh-hannover.de

B.I. Jugdutt and N.S. Dhalla (eds.), *Cardiac Remodeling: Molecular Mechanisms*, Advances in Biochemistry in Health and Disease 5, DOI 10.1007/978-1-4614-5930-9_20,

Abbreviations

ADMA	Asymmetric dimethylarginine
Bcl-3	B cell lymphoma 3
CCL2	Chemokine ligand 2
CCR2	Chemokine ligand receptor 2
CLC	Cardiotrophin-like cytokine
CNTF	Ciliary neurotrophic factor
CT-1	Cardiotrophin 1
CTGF	Connective tissue growth factor
CVF	Cobra venom factor
ECM	Extracellular matrix
eNOS	Endothelial nitric oxide synthase
EPO	Erythropoietin
ERK	Extracellular signal-regulated kinase
ETC	Electron transport chain
G-CSF	Granulocyte colony-stimulating factor
GM-CSF	Granulocyte-macrophage colony-stimulating factor
gp130	Glycoprotein-130
HK2	Hexokinase 2
IL	Interleukin
JAK	Janus kinase
LIF	Leukemia inhibitory factor
LPS	Lipopolysaccharide
LV	Left ventricle
MBL	Mannose-binding lectin
MHC	Myosin heavy chain
MI	Myocardial infarction
MMP	Matrix metalloproteinase
OSM	Oncostatin M
miRNA	microRNA
NF-κB	Nuclear factor-κB
PAI-1	Plasminogen activator-1
PFKFB3	6-phosphofructo-2-kinase/fructose-2,6-biphosphatase 3
PKB	Protein kinase B, also known as AKT
PPAR	Peroxisome proliferator-activated receptor
PPCM	Peripartum cardiomyopathy
PIAS	Protein inhibitors of activated STAT
PTPRT	Protein tyrosine phosphatase receptor T
ROS	Reactive oxygen species
SH2	Src homology 2 domain
SHP2	SH2 domain-containing cytoplasmatic protein
SOCS	Suppressor of cytokine signaling
STAT3	Signal transducer and activator of transcription 3

STAT3-KO	STAT3-knockout
TAC	Transverse aortic constriction
TIMP-1	Tissue inhibitor of metalloproteinase-1
TNC	Tenascin C
TNF-α	Tumor necrosis factor-α
TSP-1	Thrombospondin-1
Ube	Ubiquitin-conjugating enzyme
UPS	Ubiquitin–proteasome system
VEGF	Vascular endothelial growth factor

20.1 Introduction

The adult heart experiences numerous insults of physiological (aging, pregnancy, exercise) and pathophysiological [ischemia/reperfusion, myocardial infarction (MI), pressure overload, infections, and cardiotoxic agents] nature, which can induce alterations in the cardiac homeostasis and architecture. These alterations may be adaptive and protective to a certain stress situation or can be maladaptive, thereby inducing, driving, and contributing to heart failure [1].

Recent findings in experimental in vitro and in vivo models implicate important intermediate signal-transduction pathways in the coordination of remodeling processes following cardiac stress stimuli [2]. It is a key issue for the clinical arena to identify both detrimental as well as protective pathways and to transfer this knowledge into clinical practice. In this regard, the STAT3 signaling system has been shown to critically impact on changes in cardiac inflammation, vasculature, extracellular matrix (ECM) composition, energetics and metabolism, and cardiomyocyte survival and architecture [3–13].

As a transcription factor, STAT3 participates in the regulation of numerous genes involved in cell survival, hypertrophic growth, proliferation, angiogenesis and neovascularization, development and regeneration, and anti-oxidative pathways [3, 4, 6–11, 13, 14]. Furthermore, STAT3 has been reported to modulate ECM response to mechanical overload and to influence inflammatory processes [4, 6–8, 10]. While many of these effects of STAT3 are achieved by its own direct transcriptional activity, recent findings identified the novel function of STAT3 to directly interact with mitochondrial function [12], which adds to the complexity of STAT3-mediated regulatory processes.

Altogether, recent studies support the conclusion that STAT3 functions as part of an integrated signaling network in the heart. Here, we provide an overview of established STAT3-related biological mechanisms in cardiac physiology and pathophysiology, we illustrate its particular role in directing cardiac cellular processes that modulate ventricular remodeling, and finally we challenge the question whether STAT3 signaling conducts beneficial or detrimental effects with respect to cardiac function and heart failure.

20.2 Canonical gp130-STAT3 Signaling, a Central Regulatory Circuit in Cardiac Physiology and Pathophysiology

Interleukin (IL)-6-type cytokines such as IL-6, IL-11, leukemia inhibitory factor (LIF), oncostatin M (OSM), ciliary neurotrophic factor (CNTF), cardiotrophin-1 (CT-1), and cardiotrophin-like cytokine (CLC) are a family of helix bundle cytokines that play a key role in cellular and tissue homeostasis [15, 16]. They are involved in inflammatory and immunologic processes as well as in hematopoiesis, liver and neuronal regeneration, embryonic development, and cardiovascular physiology [16–18].

All IL-6-type cytokines interact with the glycoprotein-130 (gp130) receptor, and upon ligand binding, the homodimerization of gp130 (upon binding of IL-6 and IL-11) or heterodimerization of gp130 with LIF receptor (following binding of LIF, CNTF, CLC, OSM, and CT-1) or OSM receptor (for OSM), respectively, occurs [19–21] leading to the activation of three major downstream pathways: the Janus kinase (JAK)-STAT, the SH2 domain-containing cytoplasmic protein tyrosine phosphatase (SHP2)–extracellular signal-regulated kinase (ERK), and the phosphatidylinositol-3-kinase (PI3K)–protein kinase B (PKB, AKT) pathways [4, 18, 22–26]. The induction of the gp130-JAK-STAT pathway through IL-6-type cytokines leads mainly to activation of STAT3 and to a lesser extent to STAT1 activation [17]. Following dimerization of the gp130 receptor complex, JAK1, JAK2, JAK3, and tyrosine kinase 2 (Tyk2) constitutively connected to the intracytoplasmic membrane-proximal regions of the receptor subunits are catalytically activated and themselves transphosphorylate tyrosine residues in the gp130 receptor intracellular domain [27, 28]. This allows the recruitment of STAT proteins to the gp130 receptor complex via recognition of its phosphotyrosines by the STAT Src homology 2 domains (SH2) resulting in binding of monomeric STAT to the Y-X-X-Q motives and in phosphorylation of STAT proteins at a single carboxy-terminal tyrosine residue at amino acid position 705 (Y705) [29]. The phosphorylated STAT proteins then detach from their receptor and undergo homo- or heterodimerization through reciprocal interactions between the SH2 domain of one monomer and the phosphorylated tyrosine residue of the other [30]. Dimeric STAT proteins translocate in an importin-α-dependent manner to the nucleus, bind to specific DNA elements in the promoters of immediate early target genes, and activate specific gene expression programs.

Ligand-dependent STAT activation is a tightly controlled transient process, lasting for minutes up to several hours, during which nuclear STAT proteins are eventually inactivated by tyrosine dephosphorylation leading to their nuclear export [31]. Additional regulatory mechanisms are known to negatively control STAT activity, that is, gp130 receptors are rapidly internalized upon binding of their ligands and are degraded by the ubiquitin–proteasome pathway [32]. Furthermore, SHP-1/2 can interact with the intracytoplasmic portion of cytokine receptors and dephosphorylate JAK proteins, thereby lowering their activity [33, 34]. In addition, the suppressor of cytokine signaling (SOCS) family represents a specific negative regulatory feedback element of STAT signaling [35], since, for example, SOCS3 is a direct transcriptional target of STAT3. SOCS1, SOCS3, and SHP-1/2 interact with the kinase domain of various JAK proteins or the cytoplasmic phosphotyrosine residue (phospho-Y759 in humans, phospho-Y757 in mice) of the gp130 receptor resulting in the inhibition of STAT protein phosphorylation, thereby leading to termination of JAK-STAT signaling [36–38]. Additional negative regulators consisting of the group of protein inhibitors of activated STAT (PIAS-1, PIAS-3) inhibit the binding of phosphorylated STAT1 and STAT3 dimers to DNA, thereby reducing the transcriptional activation of STAT target genes. Finally, the protein tyrosine phosphatase receptor T (PTPRT) specifically dephosphorylates the Y705 residue of STAT3 and thereby regulates target gene expression but also cellular localization of STAT3 [39]. Importantly, the activation state of STAT3 seems to play a major role whether IL-6-gp130 signaling acts in a more anti-inflammatory or a more proinflammatory manner.

Not only is gp130-STAT3 signaling tightly controlled, but also activation of the other gp130-triggered signaling pathways is regulated to avoid unlimited cytokine signaling from the gp130 receptor with possible detrimental effects. For example, SHP2 limits gp130-SHP2-ERK signaling [18, 40, 41]. In addition, SOCS3 competes with SHP2 for binding to phosphorylated Y757 of the gp130 receptor, thereby limiting gp130-SHP2-ERK signaling and STAT activation [18, 40, 41]. Thus, simultaneous activation of SHP2-ERK and STAT maintains a balanced gp130 signaling, which appears crucial for beneficial responses mediated by this receptor system [40, 42]. Indeed, mice with a systemic deletion of all STAT-binding sites at the C terminus of the gp130 receptor, $gp130^{\Delta STAT}$, are characterized by impaired induction of the STAT1 and STAT3 activation and sustained activity of the SHP2-ERK pathway, which is associated with impaired intestinal wound healing and immune response and reduced life spans [18, 40, 41]. In contrast, $gp130^{Y757F}$ mice harboring a phenylalanine substitution at Y757 (Y757F) abolish binding of SHP2 and SOCS3 leading to a prolonged and continuous activation of STAT1 and STAT3 in response to IL-6 and IL-11 without activation of ERK or AKT signaling. As a consequence, $gp130^{Y757F}$ mice spontaneously develop gastric adenomas, splenomegaly, and an exaggerated immune response [18, 40, 41]. Moreover, mice with a cardiomyocyte-specific Y757F mutation ($\alpha MHC\text{-}Cre^{tg/-}$; $gp130^{fl/Y757F}$) display a high inflammatory reaction and LV rupture rate with increased mortality after MI [10].

Albeit an unrestricted high activation of gp130-STAT3 signaling seems to be detrimental in many organ systems including the heart, activation of STAT3 is absolutely required for many protective mechanisms in the heart including compensatory hypertrophy, cardiomyocyte protection, and wound healing after MI [4, 24, 43, 44].

In conclusion, it seems that a well-controlled and balanced activation of IL-6-type cytokine gp130-STAT3 signaling is beneficial for most organs including the heart, but sustained and prolonged gp130-STAT3 appears to be detrimental and is responsible for enhanced inflammation and high mortality, for example, after MI [10].

20.3 STAT3 in Cardiac Inflammation

STAT3 was initially described as a DNA-binding protein that recognizes the IL-6-responsive element in the promoter of the α_2-macroglobulin gene whose gene product belongs to the acute-phase response of the liver [45, 46]. Further evidences for STAT3 being a mediator of inflammation include the discovery of overlapping DNA-binding sites for STAT3 and the proinflammatory nuclear factor-κB (NF-κB) in the promoter of acute-phase proteins and the finding that it is activated by most proinflammatory agents, such as IL-6-type cytokines [47, 48]. Experiments with IL-6-knockout and cardiac-specific transgenic mice reveal its role in mediating cardiac inflammation after burn injury in combination with sepsis [49].

To ensure resolution of inflammation, the infiltration of inflammatory cells and the expression of proinflammatory cytokines in response to injury are balanced by the production of the anti-inflammatory cytokine IL-10 [50]. In an ischemia/reperfusion model, IL-10 is shown to be induced in the myocardium with concomitant downregulation of IL-6 mRNA expression [51, 52]. Furthermore, IL-10 suppresses the inflammatory response by inhibiting the expression of several cytokines such as IL-1, IL-6, IL-12, and TNF-α in activated macrophages and dendritic cells and mediates its effects indirectly through activation of STAT3 [52–54]. Hereafter, the induced STAT3 activates the expression of other genes, which in turn are required for a selective control of the transcription of inflammatory genes [53]. One of those intermediate genes might be *B cell lymphoma (Bcl-3)*, whose gene product blocks TNF expression after IL-10-mediated STAT3 activation in human macrophages [54].

Interestingly, infiltrating CD5-positive T cells have been identified as a predominant source of reperfusion-dependent IL-10 production, which express the cytokine upon induction by IL-6 [52, 55]. Another proinflammatory cytokine, tumor necrosis-α (TNF-α), is as well able to induce IL-10 synthesis by monocytes [55].

20.3.1 STAT3 and Inflammation Following Myocardial Ischemia

STAT3 is rapidly and transiently activated in response to myocardial ischemia per se or by reperfusion following ischemia. Mice harboring a cardiomyocyte-restricted deletion of STAT3 ($\alpha MHC\text{-}Cre^{tg/-}$; $STAT3^{fl/fl}$: STAT3-KO) display increased cardiac injury and lower survival rates after MI mainly caused by adverse remodeling suggesting that STAT3 is needed for protective mechanisms after ischemia [7]. Likewise, mice lacking the gp130 receptor and therefore having reduced STAT3 activation in the acute phase after infarction show increased mortality rates [10]. However, not only lack of STAT3 activation but also sustained uncontrolled cardiac activation of STAT3 after MI as it is present in mice carrying the mutation of $gp130^{Y757F}$ only in cardiomyocytes ($\alpha MHC\text{-}Cre^{tg/-}$; $gp130^{fl/Y757F}$ mice) leads to lower survival rates post-MI. The high mortality under unrestricted gp130-STAT3 activation derives mainly from enhanced and prolonged ventricular inflammation and high rupture rates [10].

In search for adverse downstream effectors of STAT3 signaling in infarcted *αMHC-Cre*$^{tg/-}$; *gp130*$^{fl/Y757F}$ left ventricles (LVs), the mannose-binding lectin (MBL)/lectin complement activation has been identified as a mediator of inflammation by hyperactivated STAT3 whose relevance is supported by the finding that attenuation of complement activation by the C3 antagonist cobra venom factor (CVF) reduces cardiac inflammation and improves cardiac function and survival in infarcted *αMHC-Cre*$^{tg/-}$; *gp130*$^{fl/Y757F}$ mice [10]. In addition, a murine SOCS3-knockout model, in which STAT3 is continuously activated because the negative feedback loop of the gp130-STAT3 signaling in cardiomyocytes is abolished, develops lethal cardiac arrhythmias and heart failure in response to pressure overload, supporting the hypothesis that too much of STAT3 activation seems to be harmful for the heart [56]. However, it seems that STAT3 not simply promotes inflammation, it can also act as a mediator of anti-inflammatory effects since it has been shown that in ischemia/reperfusion injury, the cardioprotective effects and the attenuation of myocardial inflammation by recombinant IL-10 given were lost if STAT3 activation was inhibited [57]. Likewise, intravenous administration of exogenous IL-11 evokes activation of STAT3 in cardiomyocytes in vivo and attenuates cardiac fibrosis and suppresses the expression of proinflammatory cytokines like IL-6 and TNF-α in a STAT3-dependent manner supporting the anti-inflammatory properties of activated STAT3 [58]. Here it is important to note that a majority of cytokine expression in the ischemic heart seems to derive from non-myocyte cells [59, 60].

Taken together, STAT3 plays an important role for protective mechanisms in the ischemic heart and after ischemia/reperfusion; however, the genetic and pharmacological dissection of gp130-mediated signaling indicates that only a well-controlled and balanced STAT3 activation is protective in the ischemic heart. Moreover, it seems specifically important in which cell type STAT3 is activated under ischemic condition, a feature that may explain the somehow contrary roles of STAT3 with regard to cardiac inflammation under ischemic conditions.

20.3.2 STAT3 and Myocarditis

It seems that STAT3 acts as a protective factor in viral myocarditis since cardiac-specific overexpression of SOCS3 increases the susceptibility to viral infection of the myocardium [61]. Further evidences for STAT3 regulating cardiac inflammation in responses to pathogens derive from STAT3-KO mice, which display a significantly higher TNF-α secretion and cardiac apoptosis rate in response to lipopolysaccharide (LPS)-induced inflammation [11].

With regard to cardiac inflammation, the question whether STAT3 is a friend or a foe cannot be answered in general. STAT3-mediated responses depend on the abundance of the particular cytokines, which activate the gp130-STAT3 signaling cascade, and on the cell type where these cytokines and the receptor system are expressed and activated in order to be either pro- or anti-inflammatory. Furthermore, the activation status of STAT3 is of major importance for the behaviors as a friend or a foe, since too much STAT3 activation after MI or cardiac deletion of STAT3 during LPS-mediated inflammation could be detrimental. Thus, the key issue seems to be a balanced and well-controlled activation of STAT3 in all cells affected by inflammatory processes in the heart.

20.4 STAT3 in Angiogenesis and Extracellular Matrix Composition

One of the earliest evidence of a potential role of STAT3 in angiogenesis was derived from observations showing that granulocyte-macrophage colony-stimulating factor (GM-CSF) via activated STAT3 contributes to vessel formation [62].

However, genetic analysis in mice with an endothelial-specific deficiency of STAT3 (*Tie2-Cre*$^{tg/-}$; *STAT3*$^{fl/fl}$), where STAT3 is depleted at E8 in angioblasts, showed that these mice survive into adulthood without major vascular defects implicating that endothelial STAT3 is optional for vessel formation during development [63].

While it seems that STAT3 is not directly required for vessel formation by endothelial cells, it has been shown that STAT3 acts as a transcription factor for angiogenic mediators such as vascular endothelial growth factor (VEGF) and as a suppressor of angiostatic genes [64]. The mechanisms of vascular formation in the postnatal heart involve both auto- and paracrine circuits where specifically secreted paracrine factors from cardiomyocytes seem to play a pivotal role for cardiac angiogenesis [65]. Indeed, it has been shown that STAT3 participates in the upregulation of VEGF and VE-cadherin expression in cardiomyocytes [64, 66], a feature that is in line with the observation that STAT3 acts as a paracrine mediator in cardiomyocytes for vascular growth as, for example, demonstrated in mice with cardiomyocyte-specific overexpression of STAT3 (*αMHC-STAT3tg*) that displays increased myocardial capillary density [64, 67].

In contrast to *αMHC-STAT3tg* mice, the STAT3-KO (*αMHC-Cre*$^{tg/-}$; *STAT3*$^{fl/fl}$) in cardiomyocytes leads to a rarification of cardiac capillaries, an enhanced fibrosis, and the development of an age-dependent heart failure suggesting a crucial role of STAT3 in controlling paracrine mechanisms involved in postnatal angiogenesis [7, 11]. Surprisingly, STAT3-KO mice show no differences in the expression of VEGF or VEGF receptors. However, they display an increase in the expression of antiangiogenic factors, which are able to inhibit endogenous VEGF activity or release such as connective tissue growth factor (CTGF) and thrombospondin-1 (TSP-1) [7, 68, 69]. Other potent antiangiogenic factors are as well upregulated like tissue inhibitor of metalloproteinase-1 (TIMP-1), tenascin C (TNC), and plasminogen activator-1 (PAI-1), which act by promoting apoptosis of endothelial cells or alter ECM composition [7, 8, 70, 71]. In this regard, supernatants of STAT3-KO cardiomyocytes are able to repress the proliferation of endothelial cells [7]. Additionally, TIMP-1 is well known to enhance fibrosis by inhibition of matrix-degrading enzymes such as matrix metalloproteinases (MMPs) [72]. This regulatory alteration may explain the age-dependent increase in interstitial fibrosis and enhanced deposition of collagens of STAT3-KO mice, indicating impaired ECM homeostasis [7, 11]. Thus, deterioration of angiogenesis in the aging myocardium of STAT3-KO mice is caused at least in part by altered structural and regulatory components of the matrix scaffold of the vascular bed and by enhanced collagen accumulation [7, 73]. Additionally, disturbance of ECM homeostasis may cause insufficient migration of angiogenic cells such as endothelial cell precursors or blood-derived primitive stem cells to the side of injured myocardium and thereby reduces the neoangiogenic capacity [74].

Moreover, STAT3 is involved in angiogenesis under pathophysiological conditions like MI. In this regard, activated STAT3 induces angiogenesis in the ischemic preconditioned infarcted myocardium [75]. In a STAT3-dependent manner, granulocyte colony-stimulating factor (G-CSF) is involved in cardiac remodeling after MI by increasing the number of endothelial cells in the border zone of the infarcted myocardium. This observation is diminished in dominant-negative STAT3 transgenic mice, suggesting that the enhanced vascularization is mediated STAT3-dependently in cardiomyocytes [76].

Furthermore, in $\alpha MHC\text{-}Cre^{tg/-}$; $gp130^{fl/Y757F}$ with a continuous activation of STAT3 in cardiomyocytes, the prolonged activation of STAT3 contributes to a higher capillary density in the nonischemic myocardium in the subacute phase after MI [10]. But the increase in capillarization is not sufficient to improve the outcome after MI. In fact, the high vessel density in the border zone of infarcted $\alpha MHC\text{-}Cre^{tg/-}$; $gp130^{fl/Y757F}$ mice may weaken the tissue or may promote further inflammation and thereby even contribute to adverse effects such as LV rupture and dilatation of the scar. During elimination of necrotic tissue and digestion of collagens in the border zone and later in the infarct scar, the adverse remodeling is mediated by MMPs released from inflammatory cells. In fact, in $\alpha MHC\text{-}Cre^{tg/-}$; $gp130^{fl/Y757F}$ mice, MMP-1 and MMP-13 expression is upregulated, but is not counterbalanced by upregulated expression of their inhibitors such as TIMP-1 [10].

Additionally, STAT3 signaling influences the fate of cardiac stem or progenitor cells by the modulation of paracrine circuits in the heart or the direct induction of differentiation of stem cells. In this context, STAT3 activation (mediated by LIF stimulation) is able to directly induce endothelial differentiation of cardiac Sca-1^{+} stem cells contributing to neovascularization during cardiac remodeling [77, 78]. Moreover, cardiomyocyte-specific deficiency of STAT3 alters the cardiac microenvironment by the release of STAT3-dependent paracrine factors such as erythropoietin (EPO). The depletion of EPO within the cardiac microenvironment attenuates the endothelial differentiation of an endogenous cardiac progenitor cell population by arresting their chemokine ligand 2 (CCL2)/chemokine ligand receptor 2 (CCR2) system, which as a consequence impairs the vasculogenic regeneration of the heart [79]. Interestingly, the endothelial differentiation seems to be VEGF independent [79]. These aspects suggest a potential positive role of STAT3 signaling for the fate of cardiac stem or progenitor cells contributing to the homeostasis of the heart, especially after cardiac injury, and harbor an endogenous source for the generation of new capillaries and vessels. The formation of new vessels could contribute to the preservation of the cardiac function.

Finally, STAT3 seems to play a key role in the regulation of cardiac angiogenesis during pregnancy-induced adaptive processes such as physiological hypertrophy of the maternal heart. In fact, cardiac-restricted deletion of STAT3 leads to peripartum cardiomyopathy (PPCM) with a massive loss of cardiac microvessels as the major phenotype [9]. This pathologic decrease in blood vessels is mainly caused by unbalanced oxidative stress in cardiomyocytes that promotes the release of activated cathepsin D, which subsequently cleaves the nursing hormone prolactin in the angiostatic factor 16 kDa prolactin. It is important to note that STAT3 expression and activation is downregulated in cardiac tissue of patients with PPCM suggesting that STAT3 activation is also crucial for cardioprotection from pregnancy-related stress in humans.

Conclusively, these data suggest that STAT3 plays an important role in protecting the vasculature as well as inducing neoangiogenesis in the heart under physiological and pathophysiological conditions by controlling paracrine circuits.

The proangiogenic features of STAT3 seem to involve paracrine mechanisms in cardiomyocytes and non-myocytes including the expression and regulation of VEGF, which indicate that activation of STAT3 controls vessel growth in vivo. In addition, STAT3 seems important to suppress the expression of antiangiogenic factors such as CTGF, TSP-1, and TIMP-1 and the generation of angiostatic peptides such as 16 kDa prolactin. Especially in cardiac adaptation processes in response to physiological stress (pregnancy) or in response to injury, STAT3 is essential for the formation and preservation of blood vessels and ultimately for the maintenance of the cardiac function. However, too much proangiogenic activity induced by a continuous activation of STAT3, for example, in the infarcted heart may be detrimental since enhanced vessel formation in the border zone of the infarcted LV may weaken the tissue and promote inflammation, ventricular rupture, and scar dilatation. Therefore, a precise regulation of STAT3 expression and activity is also necessary for beneficial cardiac angiogenesis.

20.5 STAT3 and Heart Metabolism

It is inevitable for the cardiac muscle to consistently produce high amounts of ATP to meet the demand of energy required to maintain cardiac pump function. There is ample evidence that derangements in myocardial fuel selection and energetics occur in heart failure; however, causes and consequences of these defects remain poorly understood [80]. With its important role as a mediator of cardiac vessel growth, STAT3 is important to provide the pipes to transport energy and oxygen to the cardiac muscle. Apart from its role in angiogenesis, distinct forms of STAT3 actions have been implied in the direct modulation of substrate metabolism and energy generation. However, so far data are derived from liver and from tumor biology, while effects on the role of STAT3 on cardiac energy generation are less elaborated.

STAT3 has been shown to act as a master metabolic regulator in STAT3-dependent cancer cell lines where it enhances aerobic glycolysis and downregulates mitochondrial activity [81]. For instance, in human breast cancer cells, IL-6 enhances glycolysis by STAT3-mediated upregulation of the glycolytic enzymes hexokinase 2 (HK2) and 6-phosphofructo-2-kinase/fructose-2,6-bisphosphatase 3 (PFKFB3), a result that could be confirmed in mouse embryonic fibroblast and links inflammation to alterations in cancer metabolism [82, 83]. In liver, a link between STAT3 signaling and modulation of glycolysis has been suggested since hepatocyte-specific STAT3 deletion results in hepatic insulin resistance [84]. In the heart of a transgenic mouse model with a cardiac-specific overexpression of peroxisome proliferator-activated receptor (PPAR)-α, insulin-induced STAT3 activation measured as phosphorylation at Y705 is diminished and therefore associated with an elevated oxidation of fatty acids and downregulation of genes involved in glycolysis [85].

With regard to substrate use, there is a major difference between cancer cells which strongly rely on glucose oxidation as their preferred way of energy generation and striated muscle cells which mostly favor fatty acids as the substrate for ATP production. In this context, STAT3 was attributed to mediate leptin-induced lipid oxidation in skeletal muscle, while the effect of STAT3 on cardiac fatty acid oxidation has yet to be investigated [86].

However, it is known that with the development of heart failure, the cardiac muscle switches from fatty acid utilization to glucose oxidation as the preferred substrate for energy generation [87]. Whether this switch is protective as it adapts the heart to different energy demands or contributes to the development of heart failure is a matter of debate and not really understood. A contribution of STAT3 in the regulation of genes involved in cardiac substrate utilization has not been evaluated yet. However, a direct function of STAT3 in mitochondrial ATP production in the heart and in cardiomyocytes has recently been demonstrated [12]. In fact, evidence has been presented that STAT3 is present in mitochondria of B cells and cardiomyocytes, where it binds to complex I and II of the electron transport chain (ETC) [12, 88]. This direct protein–protein interaction of STAT3 with ETC components requires phosphorylation at S727 but seems to be independent from Y705 phosphorylation or DNA binding. Deletion of STAT3 in this model leads to a drastic reduction in complex I and II activities, a lower membrane potential and ATP production, and an increase in the production of reactive oxygen species (ROS) [12]. The mitochondrial localization of STAT3 could recently be confirmed, and additionally a Tom20-mediated import of the protein was described [88]. Genetic deletion as well as pharmacologic inhibition of STAT3 phosphorylation at both sites, Y705 and S727, leads to a decreased ADP-stimulated respiration in the presence of complex I substrates [88]. No effect of STAT3 on complex II, complex IV, or uncoupled respiration was detected in cardiomyocyte mitochondria [88].

Also, an interaction between the mitochondrial protein cyclophilin D that is part of the mitochondrial permeability transition pore complex and STAT3 is reported, and STAT3 deletion or inhibition is found to increase the susceptibility of pore opening [88]. Interestingly, mice overexpressing a transgenic STAT3 protein with a mitochondrial target sequence but a mutated DNA-binding site in cardiomyocytes also display reduced complex I and II activities. However, this reduction is, unlike in the STAT3-deficient situation, not associated with augmented ROS, or decreased

ATP production and cardiac function and left ventricular dimensions are comparable to wild-type mice at 1 year of age [89]. Mitochondria from these transgenic mice are protected against ischemia-induced ROS production via complex I and show diminished cytochrome c release in response to ischemic insult [89]. It therefore seems likely that mitochondrial-targeted STAT3 on top of the endogenous protein protects against ischemic damage by partially inhibiting electron flow through complex I [89].

In the aforedescribed studies, the effect of mitochondrial-located STAT3 on cellular respiration at the level of the ETC is attributed to direct interactions of STAT3 and ETC proteins. However, to generate sufficient amounts of ATP required for maximal myocardial performance, nearly equimolar amounts of STAT3 and complex I and II proteins would be necessary. Using different proteomic approaches, it was shown that the ratio of STAT3 to complex I/II is ~10^{-5} [90]. From these experiments it can be concluded that direct regulation of cellular respiration by protein–protein interaction of STAT3 and respiratory chain proteins is not feasible and that the observed modulations in ATP production by STAT3 might rather be due to regulation of the expression of mitochondrial genes [90].

Taken together, STAT3 plays an important role in the heart and in cardiac cells for upstream processes like transport of nutrition and oxygen, while little is known about the role of STAT3 in cardiac fuel selection and substrate oxidation. However, strong evidence exists that STAT3 is essential to mediate ATP production in mitochondria via the ETC. However, the exact mechanisms by which STAT3 influences ATP generation especially at the level of complex I and II are controversially discussed. Additional studies are therefore necessary to decipher the contribution of disrupted metabolism in adverse structural remodeling of the myocardium in general and to determine the role of STAT3 in this context.

20.6 STAT3 in Hypertrophy and Sarcomere Composition

The role of the JAK-STAT pathway for the development of cardiac hypertrophy and a putative therapeutic utility still remains a matter of debate. While early studies demonstrated that under massive STAT3 overexpression ($\alpha MHC\text{-}STAT3^{tg}$), mice develop spontaneous concentric cardiac hypertrophy [67], the role of physiological expression and activation of STAT3 for cardiac hypertrophy was until recently unclear [10]. Some studies have favored a decisive contribution of STAT3 to the development of cardiac hypertrophy, particularly in response to certain pathophysiological stimuli, such as MI and ischemia/reperfusion [7], pressure overload, and hormones [91]. In this regard, upregulation of STAT3 signaling in response to MI may contribute to an adaptive component of remodeling processes in an effort to maintain function and forward flow in the acute phase. However, prolonged and unbalanced activation of JAK-STAT3 signaling appears detrimental after MI due to enhanced inflammation and left ventricular instability with increased rupture rate as mentioned above in the $\alpha MHC\text{-}Cre^{tg/-}$; $gp130^{fl/Y757F}$ mice [10].

Further evidence for the involvement of STAT3 signaling in cardiomyocyte hypertrophy is provided by in vitro experiments using the activator of gp130 receptor signaling, LIF, which induces cell growth in cardiomyocytes, a feature that is prevented by concomitant overexpression of SOCS3 [38]. However, LIF mainly induces a distinct type of cell growth that is marked by thinning of the width and elongation of the length, a phenotype that is considered as an eccentric cell growth, rather than a concentric hypertrophy. A link between STAT3 and the ubiquitin–proteasome system (UPS) was identified in a microRNA-dependent manner. In this circuit, STAT3 negatively regulates the expression of miR-199a on the transcriptional level, which in turn fine-tunes the expression of at least two ubiquitin-conjugating enzymes, the Ube2i and Ube2g, on the posttranscriptional level [5]. Under circumstances of low STAT3 expression, for example, as it occurs in the end-stage failing heart [92], the expression of miR-199a is upregulated with a subsequent impairment of the UPS. Consequently, this leads to disturbed homeostasis of certain sarcomeric proteins, such as α- and β-MHC proteins, a severe derangement of the sarcomere ultrastructure, and an eccentric type of cardiomyocyte growth—a phenotype that corresponds to the architecture of the cardiomyocyte syncytium in dilated cardiomyopathy. Moreover, the derailed UPS activity and protein turnover also result in alteration of the cardiomyocyte secretome with accumulation of the endogenous endothelial nitric oxide synthase (eNOS) inhibitor asymmetric dimethylarginine (ADMA), which in a paracrine fashion impairs endothelial function of adjacent vessels [5].

Conclusively, STAT3 regulates sarcomere composition, cardiomyocyte growth, and hypertrophy and thus determines the transition from adaptive to maladaptive remodeling depending on the extent, time point, and duration of signaling activity. While under certain conditions a transiently enhanced STAT3 activity may promote adaptive concentric hypertrophy, prolonged STAT3 activation or deficiency result in ventricular instability or eccentric ventricular dilatation, respectively. Notably the STAT3-related processes with regard to regulation of sarcomeric gene composition via microRNAs and the UPS are not restricted exclusively to the cardiomyocyte itself but may also impact via release of metabolites such as ADMA in a paracrine manner on the non-myocyte compartment especially the myocardial vasculature.

20.7 Conclusion

In conclusion, STAT3 is involved in multiple biological mechanisms in the developing, the adult, and the injured heart. A timely regulated STAT3 activation appears cardioprotective by positively modulating anti-oxidative defense, angiogenesis, eventually metabolism and energy production, hypertrophy, and survival, as reported in previous studies [7, 8, 11, 12, 67, 93]. However, upregulation and sustained activation of STAT3 promotes ventricular inflammation and rupture after ischemic insults resulting in high mortality. These antithetic effects clearly demonstrate that a precise regulation of STAT3 expression and activity is required for beneficial effects of STAT3 in the heart.

This double-faced nature of STAT3 may offer an explanation for discrepancies existing between experimental findings that demonstrated mainly beneficial effects of gp130-STAT3 signaling for the cardiovascular system and clinical studies which found that high serum levels of gp130 ligands, that is, IL-6, predict a poor outcome in patients after MI or with heart failure [94–96]. Based on the data summarized in this report, it could be speculated that high IL-6 serum levels in patients indicate a high and continuous activation state of STAT3 which has been shown to be detrimental especially after MI. Therefore, it should be explored whether there is a link between high IL-6 serum levels, activation state of cardiac STAT3, and poor prognosis in patients after MI. If this is the case, targeting STAT3 by small molecule inhibitors as currently tested in cancer therapy [97] might be a novel therapy option in these patients. In turn, in other scenarios such as heart failure due to myocarditis, peripartum cardiomyopathy, or heart failure due to cardiotoxic treatment strategies, therapies that enhance STAT3 activation may be beneficial and promote healing and regeneration. Taken together, in most cardiac physiological and pathophysiological situations, balanced STAT3 expression and activation is essential for cardioprotection and behaves as a friend, but an unbalanced STAT3 activation can turn a friend into a foe. Therefore, therapeutic success of targeting STAT3 in positive or negative ways will require the possibility to precisely control the activation status of STAT3.

Acknowledgment We thank the DFG for supporting this work.

References

1. Forrester JS, Wyatt HL, Da Luz PL et al (1976) Functional significance of regional ischemic contraction abnormalities. Circulation 54:64–70
2. Dorn GW (2009) Novel pharmacotherapies to abrogate postinfarction ventricular remodeling. Nat Rev Cardiol 6:283–291
3. Boengler K, Hilfiker-Kleiner D, Drexler H et al (2008) The myocardial JAK/STAT pathway: from protection to failure. Pharmacol Ther 120:172–185
4. Fischer P, Hilfiker-Kleiner D (2007) Survival pathways in hypertrophy and heart failure: the gp130-STAT axis. Basic Res Cardiol 102:393–411
5. Haghikia A, Missol-Kolka E, Tsikas D et al (2011) Signal transducer and activator of transcription 3-mediated regulation of miR-199a-5p links cardiomyocyte and endothelial cell function in the heart: a key role for ubiquitin-conjugating enzymes. Eur Heart J 32:1287–1297
6. Haghikia A, Stapel B, Hoch M, Hilfiker-Kleiner D (2011) STAT3 and cardiac remodeling. Heart Fail Rev 16:35–47
7. Hilfiker-Kleiner D, Hilfiker A, Fuchs M et al (2004) Signal transducer and activator of transcription 3 is required for myocardial capillary growth, control of interstitial matrix deposition, and heart protection from ischemic injury. Circ Res 95:187–195
8. Hilfiker-Kleiner D, Hilfiker A, Drexler H (2005) Many good reasons to have STAT3 in the heart. Pharmacol Ther 107:131–137
9. Hilfiker-Kleiner D, Kaminski K, Podewski E et al (2007) A Cathepsin D-Cleaved 16 kDa Form of Prolactin Mediates Postpartum Cardiomyopathy. Cell 128:589–600
10. Hilfiker-Kleiner D, Shukla P, Klein G et al (2010) Continuous glycoprotein-130-mediated signal transducer and activator of transcription-3 activation promotes inflammation, left ventricular rupture, and adverse outcome in subacute myocardial infarction. Circulation 122:145–155

11. Jacoby JJ, Kalinowski A, Liu MG et al (2003) Cardiomyocyte-restricted knockout of STAT3 results in higher sensitivity to inflammation, cardiac fibrosis, and heart failure with advanced age. Proc Natl Acad Sci USA 100:12929–12934
12. Wegrzyn J, Potla R, Chwae YJ et al (2009) Function of mitochondrial Stat3 in cellular respiration. Science 323:793–797
13. Luedde M, Spaich S, Hippe HJ et al (2011) Affixin (beta-parvin) promotes cardioprotective signaling via STAT3 activation. J Mol Cell Cardiol 50:919–923
14. Kurdi M, Booz GW (2009) JAK redux: a second look at the regulation and role of JAKs in the heart. Am J Physiol Heart Circ Physiol 297:H1545–1556
15. Betz UA, Bloch W, van den Broek M et al (1998) Postnatally induced inactivation of gp130 in mice results in neurological, cardiac, hematopoietic, immunological, hepatic, and pulmonary defects. J Exp Med 188:1955–1965
16. Keller ET, Wanagat J, Ershler WB (1996) Molecular and cellular biology of interleukin-6 and its receptor. Front Biosci 1:d340–357
17. Heinrich PC, Behrmann I, Muller-Newen G et al (1998) Interleukin-6-type cytokine signalling through the gp130/Jak/STAT pathway. Biochem J 334:297–314
18. Heinrich PC, Behrmann I, Haan S et al (2003) Principles of interleukin (IL)-6-type cytokine signalling and its regulation. Biochem J 374:1–20
19. Olivier C, Auguste P, Chabbert M et al (2000) Identification of a gp130 cytokine receptor critical site involved in oncostatin M response. J Biol Chem 275:5648–5656
20. Pennica D, Wood WI, Chien KR (1996) Cardiotrophin-1: a multifunctional cytokine that signals via LIF receptor-gp 130 dependent pathways. Cytokine Growth Factor Rev 7:81–91
21. Plun-Favreau H, Perret D, Diveu C et al (2003) Leukemia inhibitory factor (LIF), cardiotrophin-1, and oncostatin M share structural binding determinants in the immunoglobulin-like domain of LIF receptor. J Biol Chem 278:27169–27179
22. Funamoto M, Hishinuma S, Fujio Y et al (2000) Isolation and characterization of the murine cardiotrophin-1 gene: expression and norepinephrine-induced transcriptional activation. J Mol Cell Cardiol 32:1275–1284
23. Hirano T, Nakajima K, Hibi M (1997) Signaling mechanisms through gp130: a model of the cytokine system. Cytokine Growth Factor Rev 8:241–252
24. Hirota H, Chen J, Betz UA et al (1999) Loss of a gp130 cardiac muscle cell survival pathway is a critical event in the onset of heart failure during biomechanical stress. Cell 97:189–198
25. Kunisada K, Hirota H, Fujio Y et al (1996) Activation of JAK-STAT and MAP kinases by leukemia inhibitory factor through gp130 in cardiac myocytes. Circulation 94:2626–2632
26. Oh H, Fujio Y, Kunisada K et al (1998) Activation of phosphatidylinositol 3-kinase through glycoprotein 130 induces protein kinase B and p70 S6 kinase phosphorylation in cardiac myocytes. J Biol Chem 273:9703–9710
27. Kishimoto T (1994) Signal transduction through homo- or heterodimers of gp130. Stem Cells 12(Suppl 1):37–44, discussion 44–35
28. Murakami M, Hibi M, Nakagawa N et al (1993) IL-6-induced homodimerization of gp130 and associated activation of a tyrosine kinase. Science 260:1808–1810
29. Abe K, Hirai M, Mizuno K et al (2001) The YXXQ motif in gp 130 is crucial for STAT3 phosphorylation at Ser727 through an H7-sensitive kinase pathway. Oncogene 20:3464–3474
30. Chen X, Vinkemeier U, Zhao Y et al (1998) Crystal structure of a tyrosine phosphorylated STAT-1 dimer bound to DNA. Cell 93:827–839
31. Haspel RL, Salditt-Georgieff M, Darnell JE Jr (1996) The rapid inactivation of nuclear tyrosine phosphorylated Stat1 depends upon a protein tyrosine phosphatase. EMBO J 15:6262–6268
32. Heinrich PC, Bode J, Decker M et al (2001) Termination and modulation of IL-6-type cytokine signaling. Adv Exp Med Biol 495:153–160
33. David M, Chen HE, Goelz S et al (1995) Differential regulation of the alpha/beta interferon-stimulated Jak/Stat pathway by the SH2 domain-containing tyrosine phosphatase SHPTP1. Mol Cell Biol 15:7050–7058
34. Fischer P, Lehmann U, Sobota RM et al (2004) The role of the inhibitors of interleukin-6 signal transduction SHP2 and SOCS3 for desensitization of interleukin-6 signalling. Biochem J 378:449–460

35. Starr R, Hilton DJ (1999) Negative regulation of the JAK/STAT pathway. Bioessays 21:47–52
36. Lehmann U, Schmitz J, Weissenbach M et al (2003) SHP2 and SOCS3 contribute to Tyr-759-dependent attenuation of interleukin-6 signaling through gp130. J Biol Chem 278:661–671
37. Yasukawa H, Misawa H, Sakamoto H et al (1999) The JAK-binding protein JAB inhibits Janus tyrosine kinase activity through binding in the activation loop. EMBO J 18:1309–1320
38. Yasukawa H, Hoshijima M, Gu Y et al (2001) Suppressor of cytokine signaling-3 is a biomechanical stress-inducible gene that suppresses gp130-mediated cardiac myocyte hypertrophy and survival pathways. J Clin Invest 108:1459–1467
39. Zhang X, Guo A, Yu J et al (2007) Identification of STAT3 as a substrate of receptor protein tyrosine phosphatase T. Proc Natl Acad Sci USA 104:4060–4064
40. Ernst M, Jenkins BJ (2004) Acquiring signalling specificity from the cytokine receptor gp130. Trends Genet 20:23–32
41. Fischer P, Hilfiker-Kleiner D (2008) Role of gp130-mediated signalling pathways in the heart and its impact on potential therapeutic aspects. Br J Pharmacol 153(Suppl 1):414–427
42. Tebbutt NC, Giraud AS, Inglese M et al (2002) Reciprocal regulation of gastrointestinal homeostasis by SHP2 and STAT-mediated trefoil gene activation in gp130 mutant mice. Nat Med 8:1089–1097
43. Yamauchi-Takihara K, Hirota H, Kunisada K et al (1996) Roles of gp130 signaling pathways in cardiac myocytes: recent advances and implications for cardiovascular disease. J Card Fail 2(4 Suppl):S63–68
44. Yamauchi-Takihara K (2002) Gp130-mediated pathway and left ventricular remodeling. J Card Fail 8(6 Suppl):S374–378
45. Akira S, Nishio Y, Inoue M et al (1994) Molecular cloning of APRF, a novel IFN-stimulated gene factor 3 p91-related transcription factor involved in the gp130-mediated signaling pathway. Cell 77:63–71
46. Zhong Z, Wen Z, Darnell JE Jr (1994) Stat3: a STAT family member activated by tyrosine phosphorylation in response to epidermal growth factor and interleukin-6. Science 264:95–98
47. Aggarwal BB, Kunnumakkara AB, Harikumar KB et al (2009) Signal transducer and activator of transcription-3, inflammation, and cancer: how intimate is the relationship? Ann N Y Acad Sci 1171:59–76
48. Zhang Z, Fuller GM (1997) The competitive binding of STAT3 and NF-kappa B on an overlapping DNA binding site. Biochem Biophys Res Commun 237:90–94
49. Zhang H, Wang HY, Bassel-Duby R et al (2007) Role of interleukin-6 in cardiac inflammation and dysfunction after burn complicated by sepsis. Am J Physiol Heart Circ Physiol 292:H2408–2416
50. Krishnamurthy P, Rajasingh J, Lambers E et al (2009) IL-10 inhibits inflammation and attenuates left ventricular remodeling after myocardial infarction via activation of STAT3 and suppression of HuR. Circ Res 104:e9–18
51. Dewald O, Ren G, Duerr GD et al (2004) Of mice and dogs: species-specific differences in the inflammatory response following myocardial infarction. Am J Pathol 164:665–677
52. Frangogiannis NG, Mendoza LH, Lindsey ML et al (2000) IL-10 is induced in the reperfused myocardium and may modulate the reaction to injury. J Immunol 165:2798–2808
53. Murray PJ (2006) STAT3-mediated anti-inflammatory signalling. Biochem Soc Trans 34:1028–1031
54. Williams LM, Sarma U, Willets K et al (2007) Expression of constitutively active STAT3 can replicate the cytokine-suppressive activity of interleukin-10 in human primary macrophages. J Biol Chem 282:6965–6975
55. Daftarian PM, Kumar A, Kryworuchko M, Diaz-Mitoma F (1996) IL-10 production is enhanced in human T cells by IL-12 and IL-6 and in monocytes by tumor necrosis factor-alpha. J Immunol 157:12–20
56. Yajima T, Murofushi Y, Zhou H et al (2011) Absence of SOCS3 in the cardiomyocyte increases mortality in a gp130-dependent manner accompanied by contractile dysfunction and ventricular arrhythmias. Circulation 124:2690–2701
57. Manukyan MC, Alvernaz CH, Poynter JA et al (2011) Interleukin-10 protects the ischemic heart from reperfusion injury via the STAT3 pathway. Surgery 150:231–239

58. Obana M, Maeda M, Takeda K et al (2010) Therapeutic activation of signal transducer and activator of transcription 3 by interleukin-11 ameliorates cardiac fibrosis after myocardial infarction. Circulation 121:684–691
59. Deten A, Volz HC, Briest W, Zimmer HG (2003) Differential cytokine expression in myocytes and non-myocytes after myocardial infarction in rats. Mol Cell Biochem 242:47–55
60. Yue P, Massie BM, Simpson PC, Long CS (1998) Cytokine expression increases in nonmyocytes from rats with postinfarction heart failure. Am J Physiol 275:H250–258
61. Yajima T, Yasukawa H, Jeon ES et al (2006) Innate defense mechanism against virus infection within the cardiac myocyte requiring gp130-STAT3 signaling. Circulation 114:2364–2373
62. Valdembri D, Serini G, Vacca A et al (2002) In vivo activation of JAK2/STAT-3 pathway during angiogenesis induced by GM-CSF. FASEB J 16:225–227
63. Kano A, Wolfgang MJ, Gao Q et al (2003) Endothelial cells require STAT3 for protection against endotoxin-induced inflammation. J Exp Med 198:1517–1525
64. Osugi T, Oshima Y, Fujio Y et al (2002) Cardiac-specific Activation of Signal Transducer and Activator of Transcription 3 Promotes Vascular Formation in the Heart. J Biol Chem 277:6676–6681
65. Dor Y, Djonov V, Abramovitch R et al (2002) Conditional switching of VEGF provides new insights into adult neovascularization and pro-angiogenic therapy. EMBO J 21:1939–1947
66. Funamoto M, Fujio Y, Kunisada K et al (2000) Signal transducer and activator of transcription 3 is required for glycoprotein 130-mediated induction of vascular endothelial growth factor in cardiac myocytes. J Biol Chem 275:10561–10566
67. Kunisada K, Negoro S, Tone E et al (2000) Signal transducer and activator of transcription 3 in the heart transduces not only a hypertrophic signal but a protective signal against doxorubicin-induced cardiomyopathy. Proc Natl Acad Sci USA 97:315–319
68. Inoki I, Shiomi T, Hashimoto G et al (2002) Connective tissue growth factor binds vascular endothelial growth factor (VEGF) and inhibits VEGF-induced angiogenesis. FASEB J 16:219–221
69. Rodriguez-Manzaneque JC, Lane TF, Ortega MA et al (2001) Thrombospondin-1 suppresses spontaneous tumor growth and inhibits activation of matrix metalloproteinase-9 and mobilization of vascular endothelial growth factor. Proc Natl Acad Sci USA 98:12485–12490
70. Bloomston M, Shafii A, Zervos EE, Rosemurgy AS (2002) TIMP-1 overexpression in pancreatic cancer attenuates tumor growth, decreases implantation and metastasis, and inhibits angiogenesis. J Surg Res 102:39–44
71. Volpert OV, Zaichuk T, Zhou W et al (2002) Inducer-stimulated Fas targets activated endothelium for destruction by anti-angiogenic thrombospondin-1 and pigment epithelium-derived factor. Nat Med 8:349–357
72. Ahmed MS, Oie E, Vinge LE et al (2004) Connective tissue growth factor–a novel mediator of angiotensin II-stimulated cardiac fibroblast activation in heart failure in rats. J Mol Cell Cardiol 36:393–404
73. Edelberg JM, Reed MJ (2003) Aging and angiogenesis. Front Biosci 8:s1199–1209
74. Ingber DE (2002) Mechanical signaling and the cellular response to extracellular matrix in angiogenesis and cardiovascular physiology. Circ Res 91:877–887
75. Fukuda S, Kaga S, Sasaki H et al (2004) Angiogenic signal triggered by ischemic stress induces myocardial repair in rat during chronic infarction. J Mol Cell Cardiol 36:547–559
76. Harada M, Qin Y, Takano H et al (2005) G-CSF prevents cardiac remodeling after myocardial infarction by activating the Jak-Stat pathway in cardiomyocytes. Nat Med 11:305–311
77. Mohri T, Fujio Y, Maeda M et al (2006) Leukemia inhibitory factor induces endothelial differentiation in cardiac stem cells. J Biol Chem 281:6442–6447
78. Mohri T, Fujio Y, Obana M et al (2009) Signals through glycoprotein 130 regulate the endothelial differentiation of cardiac stem cells. Arterioscler Thromb Vasc Biol 29:754–760
79. Hoch M, Fischer P, Stapel B et al (2011) Erythropoietin preserves the endothelial differentiation capacity of cardiac progenitor cells and reduces heart failure during anticancer therapies. Cell Stem Cell 9:131–143

80. van Bilsen M, van Nieuwenhoven FA, van der Vusse GJ (2009) Metabolic remodelling of the failing heart: beneficial or detrimental? Cardiovasc Res 81:420–428
81. Demaria M, Giorgi C, Lebiedzinska M et al (2010) A STAT3-mediated metabolic switch is involved in tumour transformation and STAT3 addiction. Aging (Albany NY) 2:823–842
82. Ando M, Uehara I, Kogure K et al (2010) Interleukin 6 enhances glycolysis through expression of the glycolytic enzymes hexokinase 2 and 6-phosphofructo-2-kinase/fructose-2,6-bisphosphatase-3. J Nihon Med Sch 77:97–105
83. Jiang S, Zhang LF, Zhang HW et al (2012) A novel miR-155/miR-143 cascade controls glycolysis by regulating hexokinase 2 in breast cancer cells. EMBO J 31:1985–1998
84. Inoue H, Ogawa W, Ozaki M et al (2004) Role of STAT-3 in regulation of hepatic gluconeogenic genes and carbohydrate metabolism in vivo. Nat Med 10:168–174
85. Park SY, Cho YR, Finck BN et al (2005) Cardiac-specific overexpression of peroxisome proliferator-activated receptor-alpha causes insulin resistance in heart and liver. Diabetes 54:2514–2524
86. Akasaka Y, Tsunoda M, Ogata T et al (2010) Direct evidence for leptin-induced lipid oxidation independent of long-form leptin receptor. Biochim Biophys Acta 1801:1115–1122
87. Neubauer S (2007) The failing heart–an engine out of fuel. N Engl J Med 356:1140–1151
88. Boengler K, Hilfiker-Kleiner D, Heusch G, Schulz R (2010) Inhibition of permeability transition pore opening by mitochondrial STAT3 and its role in myocardial ischemia/reperfusion. Basic Res Cardiol 105:771–785
89. Szczepanek K, Chen Q, Derecka M et al (2011) Mitochondrial-targeted Signal transducer and activator of transcription 3 (STAT3) protects against ischemia-induced changes in the electron transport chain and the generation of reactive oxygen species. J Biol Chem 286:29610–29620
90. Phillips D, Reilley MJ, Aponte AM et al (2010) Stoichiometry of STAT3 and mitochondrial proteins: Implications for the regulation of oxidative phosphorylation by protein-protein interactions. J Biol Chem 285:23532–23536
91. Beckles DL, Mascareno E, Siddiqui MA (2006) Inhibition of Jak2 phosphorylation attenuates pressure overload cardiac hypertrophy. Vascul Pharmacol 45:350–357
92. Podewski EK, Hilfiker-Kleiner D, Hilfiker A et al (2003) Alterations in Janus kinase (JAK)-signal transducers and activators of transcription (STAT) signaling in patients with end-stage dilated cardiomyopathy. Circulation 107:798–802
93. Negoro S, Kunisada K, Fujio Y et al (2001) Activation of signal transducer and activator of transcription 3 protects cardiomyocytes from hypoxia/reoxygenation-induced oxidative stress through the upregulation of manganese superoxide dismutase. Circulation 104:979–981
94. Rattazzi M, Puato M, Faggin E et al (2003) C-reactive protein and interleukin-6 in vascular disease: culprits or passive bystanders? J Hypertens 21:1787–1803
95. Tsutamoto T, Hisanaga T, Wada A et al (1998) Interleukin-6 spillover in the peripheral circulation increases with the severity of heart failure, and the high plasma level of interleukin-6 is an important prognostic predictor in patients with congestive heart failure. J Am Coll Cardiol 31:391–398
96. Ohtsuka T, Hamada M, Inoue K et al (2004) Relation of circulating interleukin-6 to left ventricular remodeling in patients with reperfused anterior myocardial infarction. Clin Cardiol 27:417–420
97. Heimberger AB, Priebe W (2008) Small molecular inhibitors of p-STAT3: novel agents for treatment of primary and metastatic CNS cancers. Recent Patents CNS Drug Discov 3:179–188

Chapter 21
The Role of Growth Differentiation Factor 5 in Cardiac Repair Post-Myocardial Infarction

Eric A. Shikatani and Mansoor Husain

Abstract Progressive post-myocardial infarction (MI) remodelling of the heart is mediated by processes that include apoptosis, proliferation, fibrosis and hypertrophy. The regulation of pathways leading to these events can be mediated by diverse factors present in specific spatiotemporal expression patterns post-MI. The growth differentiation factors (GDFs) which are closely related to the bone morphogenetic protein (BMP) family of growth factors regulate diverse processes in many tissues including the heart. The GDF/BMPs regulate a cascade of intracellular signalling molecules to affect gene regulation in target cells. GDF/BMP signal transduction is mediated by specific receptor heteromer activation, extracellular inhibition and intracellular regulation of downstream signalling. Despite roles in cardiac development, the role of specific GDF/BMPs in post-MI processes is not well known. While GDF-5 does not appear to be involved in cardiac development, it is involved in a variety of processes mediating both acute and chronic remodelling post-MI. As a function of specific heteromeric receptor activation and resultant downstream signalling cascades, GDF-5 impacts specific pathways differentially in cardiomyocytes, cardiac fibroblasts and endothelial and vascular smooth muscle cells. GDF-5 spares cardiomyocyte apoptosis via a SMAD1/5/8-SMAD4-Bcl/Bcl-xL-mediated process, increases vascularity post-MI and decreases p38-mediated collagen production post-MI. Thus, a single growth factor is able to exert a multitude of cardioactive effects post-MI, which together represent a potential therapeutic target.

Keywords Growth differentiation factor • Bone morphogenetic protein • Mitogen-activated protein kinase • Transgenic mouse model of human disease • Fibrosis

E.A. Shikatani • M. Husain, M.D. (✉)
Toronto General Hospital Research Institute, University Health Network,
200 Elizabeth St, MaRS 3-909, Toronto, ON, Canada, M5G 2C4
e-mail: mansoor.husain@utoronto.ca

B.I. Jugdutt and N.S. Dhalla (eds.), *Cardiac Remodeling: Molecular Mechanisms*, Advances in Biochemistry in Health and Disease 5, DOI 10.1007/978-1-4614-5930-9_21,

21.1 Introduction

While great progress has been made in the early recognition and treatment of myocardial infarction (MI), this disease remains a leading cause of morbidity and mortality [1, 2]. The myocyte necrosis that occurs as a consequence of restricted coronary perfusion is followed by acute inflammatory and chronic fibrotic responses resulting in progressive remodelling of the heart [3–10]. Maturation of the scar, with or without infarct expansion, and hypertrophy of adjacent cardiomyocytes (CM) contribute to post-infarct remodelling, creating potential for cardiac dysfunction and heart failure due to impaired contractility and conductivity of the scar tissue. During this chronic remodelling phase, lack of nutrient and oxygen supply causes apoptosis of the hypertrophied myocytes [11, 12] which results in gradual expansion of the infarct zone, dilatation of the left ventricle and subsequent heart failure. The development of heart failure is the most important determinant of subsequent survival and quality of life [1, 2].

Bone morphogenetic proteins (BMPs) and BMP-like growth differentiation factors (GDFs) are secreted proteins which function in an autocrine/paracrine manner. GDFs have overlapping members with the BMPs, together which form a subfamily within the transforming growth factor-β (TGF-β) superfamily of cytokines. Their interactions with specific receptors on target cells initiate signal transduction resulting in the regulation of gene expression. Expression of several BMPs has been documented in the heart, some of which are essential for cardiac morphogenesis [13–20]. Some BMPs have also been shown to promote stem cell differentiation into CMs in the peri-infarct area [21, 22]. However, little is known about the effect of BMP deficiency on chronic remodelling responses post-MI. Differences in the affinity of BMPs and GDFs for their receptors, specific spatiotemporal expression patterns and the presence of extra- and intracellular inhibitors and downstream signalling molecules in target cells together result in diverse and overlapping functions in different cells and tissues (Table 21.1).

Growth differentiation factor 5 (GDF-5), also known as BMP-14 or cartilage-derived morphogenetic protein 1, is a member of the GDF/BMP family that has mainly been studied in skeletal development and chondrogenesis. While GDF-5 expression has been reported in the developing heart, its function during development as well as its expression and role in adult heart post-MI was unknown. Although GDF-5-knockout (KO) mice survive to adulthood, mutations in other BMP ligands, their receptors and downstream signalling proteins often confer embryonic lethality. While this may be taken to suggest that loss of GDF-5 does not impact cardiovascular development, it in no way addresses the role of GDF-5 in adult cardiovascular diseases. Our work to date shows that GDF-5 is expressed in the adult mouse heart and cardiac vascular smooth muscle cells (VSMCs) and endothelial cells (ECs) and that its expression is upregulated in a specific spatiotemporal pattern post-MI. Moreover, the deficiency of GDF-5 in KO mice results in impaired cardiac remodelling due to decreases in activity of downstream signalling cascades resulting in defects in cell survival, proliferation and matrix deposition [23].

Table 21.1 Role of GDF/BMP signalling molecules in cardiovascular development and disease

	Species	Cardiovascular role	References
Ligands			
BMP-2	m	KO has multiple developmental and cardiac defects	[20]
	r	Inhibits apoptosis in neonatal cardiomyocytes	[73]
		Regulates cardiomyocyte contractility	[48, 103]
BMP-4	m	Multiple developmental and cardiac defects	[15, 16, 93, 124]
	x, z	Essential in left–right asymmetry of heart	[13, 14, 18]
BMP-2/4	h	Expression found in adult calcified heart valves	[125]
		Elevated expression in atherosclerotic plaques	[126]
BMP-5	c	Involved in endocardial cushion formation	[37]
BMP-5/7	m	Cardiac defects in double mutants	[127]
BMP-6/7	m	Cardiac defects in double mutants	[38]
GDF-1/2 (BMP-9)	m	KO has left–right patterning defects, including cardiac	[27, 28]
	h	Congenital cardiac malformations	[29, 30]
BMP-10	m	Maintains cardiomyocyte proliferation in development	[39]
GDF-5	m	Reduced vascularisation and VSMC migration	[58, 63, 64]
		KO has defect in post-MI cardiac repair	[23]
GDF-8	m, b, s	Expressed in heart; upregulated after MI	[35]
	m, r	Inhibits cardiomyocyte hypertrophy	[107–109]
GDF-15	r	Increased expression post-MI	[105]
	m	Inhibits influx of inflammatory cells post-MI	[110, 111]
	h	Upregulated in cardiac pathologies	[104, 128]
Receptors			
BMPR1A	m	KO has multiple developmental and cardiac defects	[51, 92]
BMPR2	m	KO lethal due to gastrulation defects	[49]
		Outflow tract defects with hypomorphic BMPR2	[50]
ACVR2/2B	r	Increased expression post-MI	[105]
SMADs			
SMAD-3	m	KO has attenuated TGF-β-mediated fibrosis post-MI	[101, 102]
SMAD-5	m	KO lethal from defective angiogenesis	[46]
		EC- and VSMC-specific KO cause mild cardiac defects	[47]
SMAD-6	m	KO has vascular dilatation and outflow tract septation	[129]
SMAD-7	c	Misexpression causes vascular dilatation and malformations	[130]
Inhibitors			
NOG	m, c	Thicker myocardium and endocardial cushion in mutants	[40, 41]

Of the 20 identified GDF/BMPs, several of them have roles in cardiovascular development or disease. While the developmental role of many GDF/BMPs, their receptors, downstream signalling SMADs and extracellular inhibitors has been characterised, their roles post-MI remain largely unexamined in experimental systems. Species—*h* human, *m* mouse, *r* rat, *s* sheep, *b* bovine, *c* chicken, *x* xenopus, *z* zebra fish

21.2 GDF/BMPs in General and Those Expressed in the Heart

The GDF and BMP family of cytokines have various functions in embryonic development and tissue homeostasis through their roles in the regulation of cell differentiation, proliferation and survival. Like other TGF-β superfamily cytokines, GDF/BMPs are synthesised as a large precursor peptide which dimerises then has the mature cytokine cleaved from the N-terminal pro-domain [24]. Each monomer has the characteristic cystine-knot structure of TGF-β cytokines. With the exception of GDF-8 and 9, GDF/BMPs are secreted without the pro-domain, resulting in small protein dimers that are then able to exert autocrine and paracrine effects. Despite being small peptides fully capable of diffusing freely, because they are secreted as membrane-bound inactive pro-peptides, coupled with the specific localisation of peptidases that cleave their pro-domains, the actions of GDF/BMPs tend to be localised to specific tissue compartments [25].

GDF/BMPs mediate signal transduction through specific receptors that regulate target gene expression during development and adult life (Fig. 21.1). BMP signalling plays critical roles in various tissues including the cardiovascular and skeletal systems. GDF/BMPs are expressed with spatiotemporal specificity and as such only a subset of GDF/BMPs is known to be expressed in either of the developing, mature or pathological heart. GDF/BMPs are expressed in the developing heart in a distinct but partially overlapping manner, suggesting they play specific roles in cardiac morphogenesis [17, 19]. BMP-2 or BMP-4 deletion resulted in multiple developmental abnormalities including cardiac anomalies [13–16, 18, 20, 26]. Mutations in GDF-1 and 2 (also known as BMP-9) cause several cardiac defects involving left–right patterning [27, 28], which have also been characterised in human congenital cardiac malformations [29, 30]. Although their roles in cardiac development are not well characterised, under normal physiological conditions, GDF-5, GDF-8 (also known as Myostatin) and GDF-15 (also known as macrophage inhibitory protein 1) are all expressed at low levels in the developing and adult heart [31–36]. BMP-5 [37], 6 and 7 [38] have been shown to be expressed during endocardial cushion development. BMP-10 is specifically expressed in the developing heart and has a critical role in the maintenance of embryonic cardiomyocyte proliferation [39]. Hence, while several GDF/BMPs have been shown to be expressed during cardiac development, their roles in post-myocardial infarct remodelling remain largely unexplored.

In addition to the study of GDF/BMPs themselves, several studies have examined the role of inhibitors and downstream mediators of GDF/BMP function in the context of cardiac development. Misexpression of noggin, an inhibitor of BMPs, led to septal defects in the heart [40]. Additionally, noggin-KO embryos exhibited thicker myocardium and larger endocardial cushion compared to wild type [41]. Small mothers against decapentaplegic homolog 4 (SMAD-4) a downstream mediator of GDF-5 and BMP signalling in general has been implicated in endocardial cushion, outflow tract and cardiac valvular development [42–44]. Deletion of SMAD-5 caused embryonic lethality due to disorganised vasculature and decreased

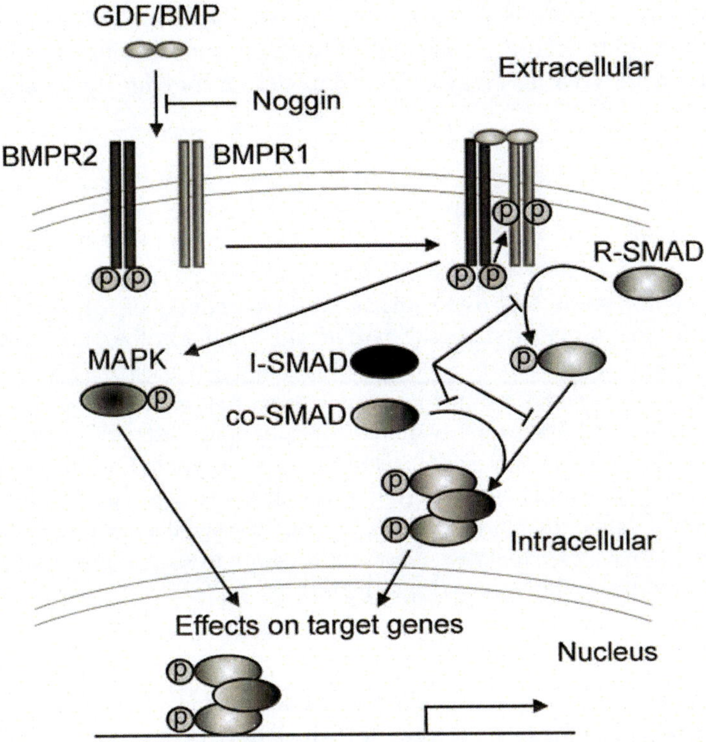

Fig. 21.1 *GDF/BMP signalling and regulation.* Homo/heterodimeric GDF/BMP binds to heteromeric complexes of type 1 BMPR (BMPR-1A and BMPR-1B) and type 2 BMPR (BMPR-2, ACVR-2A, ACVR-2B) to activate downstream signalling. Upon ligand binding, BMPR-2s transphosphorylate BMPR-1s, which then phosphorylate a regulatory SMAD (R-SMAD). The phosphorylated R-SMAD then complexes with a common-mediator SMAD (co-SMAD), and the whole complex translocates to the nucleus to target SMAD-responsive genes. BMPR activation by ligand binding can also result in the activation of MAPK pathways. Signalling by GDF/BMPs can be inhibited extracellularly by inhibitors such as noggin. SMAD signalling can be inhibited by inhibitory SMADs (I-SMADs), which can interfere with multiple signalling processes as well as enhance the degradation of signalling molecules

number of VSMCs [45, 46]. SMC- or EC-specific deletion of SMAD-5 did not affect angiogenesis during development [47]. Mice lacking SMAD-5 in both VSMC and CM exhibited increased left ventricular internal diameter and decreased fractional shortening, which manifested in adulthood. BMP-2, together with myocyte-specific enhancer factor 2A, was found to regulate cardiac ventricular contractility [48]. BMPR-1A-KO and BMPR-2-KO have been shown to cause multiple developmental defects, including cardiac malformations [8, 49–51].

These studies suggest that tight spatiotemporally controlled expression of BMPs, their receptors and their inhibitors is critical to cardiac development and possibly also normal cardiac function. In the embryonic mouse heart, GDF-5 is transiently expressed only at E13.5 [32]. While cardiac development has never been examined in GDF-5-KO mice, these animals survive without overt cardiac malformations [33]. Indeed, as one of the few BMP mutants without overt cardiac malformation, the GDF-KO may be a good model for studying dysregulated BMP signalling.

GDF-5 is expressed in a variety of tissues including the heart [32, 33, 52–57]. In 13.5-day post-coitum mouse embryos, GDF-5 expression was seen throughout the ventricles and atria [32]. We have found that GDF-5 is expressed in adult heart, CMs and arterial SMC and ECs [23]. Since identification of GDF-5 mutations in humans with skeletal disorders [58–62], GDF-5 has been mainly studied in the context of chondrogenesis and osteogenesis. Vascular effects of GDF-5 have been described in some previous studies. These include findings of reduced vascularisation post-tendon injury in GDF-5-KO mice [58], angiogenesis [63], SMC migration [64], induction of vascular endothelial growth factor expression [65–67] and activation of SMAD-1/5 in human microvascular and pulmonary artery ECs [68]. Recently, GDF-5 was shown to prevent apoptosis in mouse embryonic fibroblasts but not in vascular SMCs via stabilising XIAP [69]. Our data also suggest a role for GDF-5 in proliferation and migration of SMC and ECs (unpublished data), cardiac fibroblasts, and in CM survival [23]. Together, these studies point to several potential effects of GDF-5 on cardiovascular tissues (Fig. 21.2).

21.3 Signalling Pathways Downstream of GDF/BMPs and Their Role in Remodelling Post-MI

GDF/BMPs regulate gene expression through a complex signalling cascade mediated by the existence of various inhibitors in the extracellular microenvironment, the combination of type 1 and 2 receptor heteromers activated, intracellular regulation of downstream signalling molecules and the recruitment of transcriptional coactivators or repressor to target gene promoters. GDF/BMP receptor heteromers are made of type 1 BMPR-1A and BMPR-1B and type 2 ACVR-2A, ACVR-2B and BMPR-2 [70, 71]. These receptors are expressed differentially in various tissues including the heart and blood vessels [56, 72, 73]. Although there is considerable receptor specificity for each GDF/BMP, it remains unknown to what extent activation of different combinations of type 1 and type 2 receptor complexes results in functional signal transduction cascades. Additionally, it is known that GDF/BMPs are able to heterodimerise, increasing the complexity of possible combinations of receptor–ligand interactions and downstream effects [74].

The crystal structures of several ligand–receptor complexes have been solved, showing a close and extensive interaction of ligands to their respective receptor complexes. Type 1 receptors have been found to bind the GDF/BMP dimer at several amino acid residues in the region where the two monomers interact, whereas type 2 receptors bind to a single monomer in a region distal from the interaction site. Receptor–ligand studies have shown that the receptor–ligand geometry is similar

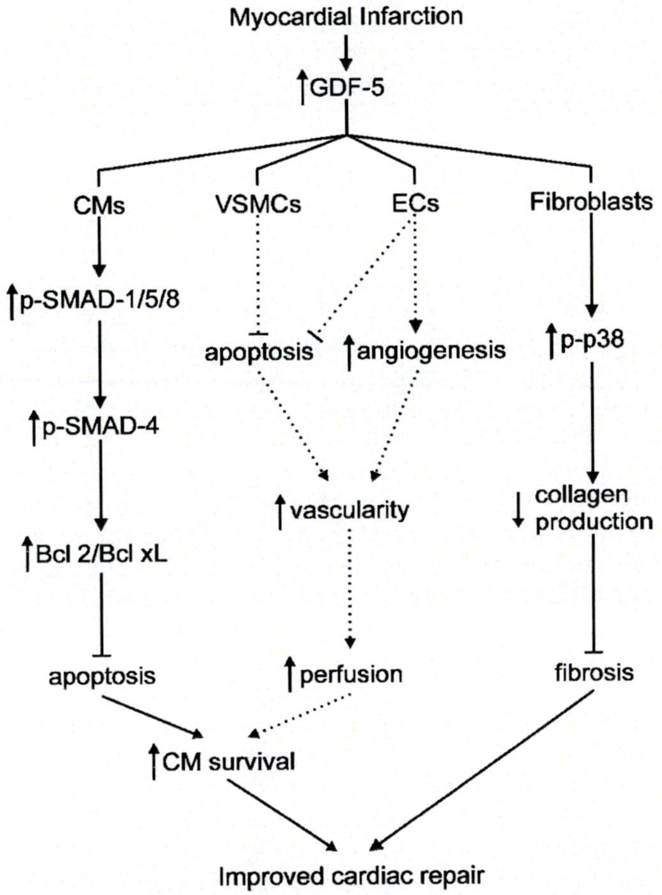

Fig. 21.2 *GDF-5 signalling affects multiple remodelling processes post-MI.* GDF-5 spares cardiomyocyte apoptosis post-MI via a SMAD-1/5/8-SMAD-4-Bcl/Bcl-xL-mediated process. In vascular cells, GDF-5 may inhibit apoptosis as well as promote angiogenesis, increasing vascularity and perfusion of infarcted myocardium post-MI. In fibroblasts, GDF-5 decreases p38-mediated collagen deposition preventing fibrosis of infarcted myocardium. Together, these effects result in improved remodelling post-MI. *CM* cardiomyocytes, *VSMC* vascular smooth muscle cells, *EC* endothelial cells. *Solid arrows* indicate experimentally validated pathways; *dashed arrows* indicate proposed mechanisms

regardless of the receptor–ligand pairing, suggesting that the specificity of receptor binding is mediated predominantly by ligand amino acid sequence [75, 76]. Since there is considerable sequence similarity of GDF/BMP family members [25], it is not surprising to find that single residue mutations result in the loss or gain of specificity of ligand–receptor specificity [75, 76]. Specific amino acids have been identified as being responsible for GDF-5 receptor specificity [77, 78], as well as interactions with specific BMP inhibitors [79]. Post-translational modifications such as glycosylation also confer further receptor specificity, as is the case with

BMP-6 where glycosylation at an asparagine residue is required for recognition by type 1 ACVR but not BMPR-1A and B [80].

Signal transduction by GDF/BMPs after binding and activation of receptor complexes is accomplished by transphosphorylation of the type 1 receptor by the type 2 receptor [81]. Upon phosphorylation, type 1 receptors phosphorylate regulatory SMAD (R-SMAD) proteins, resulting in subsequent nuclear localisation of R-SMADs. There are two antagonistic sets of R-SMADS: SMAD-1/5/8 and SMAD-2/3. Phosphorylation of R-SMADs depends on the type 1 receptor activated, with ACVR-1A, BMPR-1A and 2A phosphorylating SMAD-1/5/8, whereas ACVR-1B, ACVR-1C and ACRV-like kinase 4 phosphorylate SMAD-2/3 [82]. Upon phosphorylation, SMAD-1/5/8 binds to the common-mediator SMAD (co-SMAD) SMAD-4, where the whole complex translocates to the nucleus to interact with other transcription factors and SMAD target genes [83]. The inhibitory SMADs (I-SMADs) SMAD-6 and 7 regulate SMAD signalling by interfering with R-SMAD phosphorylation by type 1 receptors, R and co-SMAD promoter binding, as well as by regulating SMAD degradation. While members of the TGF-β superfamily (TGF-βs, GDF/BMPs and activins) share some receptors, use of different subsets of SMADs by BMPs (SMAD-1/5/8) and TGF-β, activins and GDF-15 (SMAD-2/3) segregates their pathways. Despite this complex network of potentially overlapping ligands, receptors and signal transduction pathways, our data show that specific deficiency of a single family member, namely, GDF-5, significantly hampers cardiac remodelling [23].

Mature GDF-5 preferentially signals through the heterotetrameric complex of BMPR-1A and BMPR-1B with BMPR-2, ACVR-2A and ACVR-2B with equal affinity [70, 84]. Since these receptors are expressed in the heart and type 2 ACVRs are upregulated after ischemia [85], GDF-5 could confer signalling through its preferred receptors and produce various cellular effects depending upon the presence of downstream signalling molecules as well the presence of extracellular inhibitors of GDF-5 signalling. There is also substantial evidence showing BMPs and their receptors transduce signals through SMAD-independent pathways. For example, interactions of several proteins with the cytoplasmic domain of BMPR-2 receptor suggest that this receptor, independent of SMAD, can regulate MAP kinases, PKC and PKA [86]. Indeed, GDF-5 has been shown to induce p38-MAP kinase phosphorylation during chondrogenesis [87, 88]. SMAD-independent gene regulation was also shown in pulmonary artery SMCs harbouring an inactive BMPR-2 mutant [89]. While BMP receptors are shared by several GDF/BMPs, their deletion in mice causes embryonic lethality [49, 51, 90–92], thus limiting their utility for studies of BMP signalling in cardiovascular disease.

Despite these studies on the developmental expression of various GDF/BMP family members, few studies have examined their roles in chronic post-infarction myocardial remodelling. In addition, the role of various extra- and intracellular GDF/BMP and SMAD regulatory proteins has not been examined in the context of cardiac remodelling post-MI. Understanding the role of these growth factors and associated signalling pathways will allow for the development of therapeutics to target the chronic remodelling post-MI, which could significantly improve morbidity and mortality post-MI.

21.4 Transgenic Studies of GDF/BMPs Post-MI

Although several GDF/BMPs and related regulatory proteins have been well studied in cardiac development [13, 14, 19, 40, 93, 94], their regulation in MI and their role in post-MI cardiac repair remain largely unknown. Among members of the TGF-β superfamily, the role of TGF-β in the induction of cardiac fibrosis post-MI has been studied in detail [95–98]. Although BMPR-regulated SMAD-1/5/8 expression has been observed in the infarct zone, their physiological importance has not been explored [99]. Cardiac over-expression of Smad1 in transgenic mice also conferred CM protection following ischemia–reperfusion injury [100]. Since GDF/BMPs and their receptors may regulate gene expression and confer signalling independent of SMADs [89], over-expression of SMAD-1 only partially determines the role of BMP signalling in cardiac injury. SMAD-3 deficiency has been shown to attenuate TGF-β-mediated induction of cardiac fibrosis post-MI, independent of any changes to acute inflammatory responses [101, 102]. Although fibroblast proliferation and migration were not affected, fibrosis was abrogated by inhibited collagen release by SMAD-3-KO myofibroblasts [102]. In neonatal CMs, BMP-2 was shown to suppress apoptosis [73] and regulate CM contractility [48, 103]. Among other members of the TGF-β pathway, GDF-8, activin, ACVR-2A, ACVR-2B and GDF-15 were also increased after MI [33, 35, 85, 104, 105]. It was recently shown that in MI, mesenchymal stem cell differentiation into CM-like cells required TGF-β/BMP-2 pathways [21]. Embryonic stem cells (ESC) were also shown to differentiate into functional CMs in the infarcted myocardium by signalling mediated through TGF-β/BMP-2 pathways [106]. Leukaemia inhibitor factor and BMP-2 synergistically promoted differentiation of ESC into CM and EC in a mouse model of MI [22].

GDF-8 has been shown to be expressed in murine, ovine and bovine hearts, localising to cardiomyocytes and Purkinje fibres [35], and is upregulated following MI in an ovine model of MI. Multiple studies have indicated that GDF-8 inhibits cardiomyocyte hypertrophy [107–109]; however, no studies have examined the utility of therapeutic inhibition of GDF-8 to repress post-MI cardiac hypertrophy.

After MI, GDF-15 is expressed in the myocardium and has a role in inhibiting the influx of inflammatory cells [110, 111]. In GDF-15-KO mice, lack of inhibition of the inflammatory response results in worse outcomes after MI due to increased cardiac rupture mediated by the increased recruitment of leukocytes to the infarcted myocardium [111]. Administration of recombinant GDF-15 reduced leukocyte recruitment in KO mice, but changes in post-MI survival was not reported. In this study, no long-term differences in cardiac remodelling were found in GDF-15-KO infarcted or peri-infarct areas versus wild-type (WT) mice, indicating that the role of GDF-15 is predominantly to mediate the inflammatory response in the acute post-MI phase without influencing the downstream pathways relevant in chronic post-MI remodelling events like cardiac hypertrophy and fibrosis.

21.5 The Role of GDF-5 in Chronic Remodelling Post-MI

While the expression of GDF-5 in adult heart has been previously described, the role of GDF-5 in post-MI responses had not been examined. The GDF-5-KO mouse does not develop any overt cardiac malformations, suggesting that the role of GDF-5 is not during developmental processes in the mouse heart. In order to examine the role of GDF-5 in MI, ligand as well as putative receptor expression was first determined in WT mouse hearts before and after MI. In uninfarcted WT mouse hearts, GDF-5 and the receptors ACVR-2A, ACVR-2B, BMPR and BMPR-1B were all expressed in uninfarcted tissues [23]. Following permanent ligation of the left anterior descending coronary artery causing an experimentally induced MI, GDF-5 mRNA and protein levels were most elevated 14 and 28 days post-MI, with protein levels sustained as much as 40 days post-MI, alluding to a possible involvement in chronic post-MI processes [23].

Histological analysis of infarcted hearts revealed no differences in infarct area at early time points between GDF-5-WT and KO hearts, suggesting that GDF-5 does not have a role in acute post-MI responses. This is in contrast to what is now known for GDF-15, which was shown to have a role exclusively in early, acute post-MI inflammatory processes [111]. GDF-15-KO mice did not have any differences with WT mice in terms of chronic post-MI remodelling processes. During chronic remodelling post-MI, infarcted myocardium is gradually replaced by scar tissue [112, 113]. During this process, proliferation of cardiac fibroblasts and deposition of extracellular matrix occur resulting in loss of contractility and conductivity of the infarcted area [112, 114, 115]. As a result, cardiac fibrosis contributes to the development and progression of heart failure [112, 113, 115, 116]. At 28 days, GDF-5-KO hearts had 42% greater infarct areas, 30% increase in the thinning of infarcted left ventricle (LV) and 156% greater infarct expansion compared to WT [23]. In GDF-5-KO hearts, increased fibrosis and collagen 1a1and 3a1 expression were observed post-MI compared to WT hearts [23]. Furthermore, p38-MAP kinase phosphorylation was further reduced in KO hearts post-MI. In mice, p38-MAP kinase phosphorylation suppressed COL1A1 and COL3A1 expression in the heart [117] and reduced cardiac fibrosis post-MI [118]. These data indicate that endogenous GDF-5 levels regulate collagen expression post-MI, a finding supported by recombinant GDF-5 (rGDF-5)-mediated p38-MAP kinase phosphorylation in cardiac fibroblasts [23]. GDF-5-KO infarcted hearts revealed expected losses in cardiac function by echocardiography consistent with increased scar expansion observed histologically.

Decreased scar thickness in GDF-5-KO mice and associated findings of increased infarct expansion and poor LV function [23] may be due to reduced proliferation of cardiac fibroblasts in GDF-5-KO hearts. As rGDF-5 was shown to stimulate proliferation of connective tissue fibroblasts [119] and mouse cardiac fibroblasts [23], it is necessary to examine the effect of rGDF-5 treatment on collagen gene expression and cardiac fibrosis. Despite a thinner scar in GDF-5-KO mice, no increase in mortality was seen in these mice. No differences in matrix metalloproteinase-2 and 9 levels or activity were observed in infarct tissue [23], suggesting that increased collagen may have prevented cardiac rupture in these KO mice. At this time, it is not known if the structural arrangement of the deposited collagen is more reticular than fibrillar, which may aid in preventing cardiac rupture despite thinner scar tissue.

A prominent phenotype of GDF-5-KO mice post-MI was reduction in the number of SMα-actin-stained (i.e. muscularised) precapillary arterioles in the infarct and peri-infarct areas [23]. This finding is consistent with the reported role of GDF-5 in angiogenesis [63]. In GDF-5-KO mice, revascularisation was also reduced following tendon injury [58]. Among cell types that play a role in vascularisation, GDF-5 has been shown to promote migration of bovine EC [63] and human umbilical vein SMC [64]. These data indicate that the GDF-5 deficiency may have contributed to reduced proliferation and migration of coronary arterial SMC and EC, resulting in decreased vascularisation in the infarct areas post-MI. Deletion of SMAD-5, which is a mediator of GDF-5 and other BMP signalling, resulted in defective angiogenesis and decreased number of vascular SMC [45, 46]. This is in agreement with our finding of reduced muscularised arteries in GDF-5-KO mice post-MI. GDF-5 and other BMPs were shown to induce SMAD-1/5 phosphorylation in human microvascular and pulmonary artery ECs [68]. Another major finding of the GDF-5-KO hearts post-MI was a large reduction in p38-MAP kinase phosphorylation compared to wild type [23]. Over-expression of p38-MAP kinase was shown to enhance angiogenesis in post-MI hearts [118]. Together, these observations suggest that GDF-5 may mediate vascularisation post-MI and that its deficiency results in decreased vascularisation leading to reduced supply of oxygen and nutrients and subsequent loss of CMs in the peri-infarct area. However, exhaustive characterisation of capillary density post-MI in infarct and peri-infarct areas was not determined, thus specific conclusions about the role of GDF-5 in post-MI angiogenesis may be somewhat limited.

Our finding of increased loss of CMs in GDF-5-KO mice post-MI, and our in vitro findings of rGDF-5-enhanced CM survival, via induction of the antiapoptotic Bcl2 and Bcl-xL genes mediated by SMAD4 suggests that GDF-5 also plays a role in CM survival in vivo [23]. While treatment of in vitro cultures of CMs with rGDF-5 spared them from apoptosis, the effects of rGDF-5 on apoptosis of cardiac fibroblasts, ECs and VSMCs were not examined, so it is unclear if GDF-5 also has a role in maintaining the survival of other cardiac cell types. In addition to the anti-apoptotic effect of GDF-5 on CMs, the combined loss of vascularity and resultant decrease in perfusion post-MI in GDF-5-KO mice may have increased the loss of peri-infarct tissue resulting in the expansion of the scar in GDF-5-KO mice. Examination of peri-infarct areas in post-MI hearts by cleaved caspase-3 staining showed a fivefold increase in apoptotic cells in GDF-5-KO 4 days post-MI versus WT [23], suggesting that the effects of GDF-5 also include acute post-MI sparing of cardiomyocyte apoptosis.

In contrast to GDF-15 which acts entirely as an acute post-MI anti-inflammatory stimuli [111], the aforementioned results indicate that GDF-5 acts both acutely and chronically post-MI as a pro-proliferative, antiapoptotic and antifibrotic stimuli to reduce adverse remodelling and improving healing post-MI.

21.6 Future Directions

With insights into the role of GDF-5 post-MI being involved in myocardial remodelling, future studies will be able to examine the therapeutic efficacy of post-MI treatment with exogenous GDF-5. Further studies should determine if other GDF/BMPs, BMP receptors, and intra- or extracellular inhibitors and downstream signalling molecules are also involved in chronic post-MI remodelling processes. While expression of GDF-5 does not seem to overlap with GDF-15, the spatiotemporal expression pattern and function of GDF-8 post-MI is not well known, nor is it established how the spatiotemporal specificity of the expression of these GDF and other GDF/BMP ligands impacts post-MI cardiac repair. Much attention has focused on treating acute post-MI symptoms with great success, significantly improving survivability of acute MI. Considering that chronic post-MI pathology still causes significant mortality and morbidity, therapeutics that target the remodelling resulting from fibrosis and apoptosis of surviving surrounding tissue needs to be further developed. Indeed, the use of mineralocorticoid receptor antagonists to reduce cardiac fibrosis post-MI [120, 121] has been validated as an effective means of preventing heart failure progression in clinical trials [122]. With the knowledge that GDF-5 deficiency causes defects in scar expansion and cell apoptosis, exogenous GDF-5 treatment may be a therapeutic strategy to abrogate cardiac fibrosis post-MI.

Interestingly, there was a differential effect of rGDF-5 treatment on p38 phosphorylation between CMs and fibroblasts [23], indicating that signalling cascades that are activated by GDF-5 in post-MI remodelling are cell-type specific even within the same tissues. These differences can be explored using tissue and cell-type-specific transgenic knockout systems to investigate cell-specific losses in GDF-5 to further clarify the roles of GDF-5 in post-MI repair and healing processes. Whether the specificity is derived from differential receptor complex activation or downstream signalling mediators remains to be determined. Since GDF-5 is able to bind to the heterotetrameric complex of BMPR-1A and BMPR-1B with BMPR-2, ACVR-2A and ACVR-2B with equal affinity [70, 84], activation of different receptor heteromers may be able to overcome specific defects in BMPR signalling. For example, in pulmonary hypertension mediated by haploinsufficiency of BMPR2 [123], GDF-5 treatment may be able to efficiently activate defective downstream signalling through alternate receptor complex activation. Understanding the effects of GDF-5 on various cell types of the cardiovascular system may also extend the utility of GDF-5 treatment in other vascular and cardiac pathologies.

It becomes increasingly important that pathways and genes that are involved in the late remodelling process are identified and characterised. With explicit knowledge of these factors, novel therapeutics could be developed to target the remodelling events that occur following myocardial infarction with the aim of decreasing morbidity and increasing quality of life for those that are affected by MI. Presumably, factors that are involved in the early growth and differentiation of cardiac myocytes and other cardiac tissues can be targets for intervention in the late remodelling process following an MI. We have demonstrated that the lack of GDF-5 causes adverse perturbations in chronic post-MI remodelling. Demonstrating a therapeutic application for GDF-5 would open additional avenues in treating post-MI pathology to improve morbidity and mortality due to chronic remodelling and resultant heart failure.

References

1. Ng AC, Sindone AP, Wong HS, Freedman SB (2008) Differences in management and outcome of ischemic and non-ischemic cardiomyopathy. Int J Cardiol 129:198–204
2. Kloner RA, Rezkalla SH (2004) Cardiac protection during acute myocardial infarction: where do we stand in 2004? J Am Coll Cardiol 44:276–286
3. Frangogiannis NG (2004) Chemokines in the ischemic myocardium: from inflammation to fibrosis. Inflamm Res 53:585–595
4. Koike MK, Frimm Cde C, Curi M (2007) Low coronary driving pressure early in the course of myocardial infarction is associated with subendocardial remodelling and left ventricular dysfunction. Int J Exp Pathol 88:279–290
5. LaFramboise WA, Bombach KL, Dhir RJ et al (2005) Molecular dynamics of the compensatory response to myocardial infarct. J Mol Cell Cardiol 38:103–117
6. Nian M, Lee P, Khaper N, Liu P (2004) Inflammatory cytokines and postmyocardial infarction remodelling. Circ Res 94:1543–1553
7. Ohta K, Nakajima T, Cheah AY et al (2004) Elafin-overexpressing mice have improved cardiac function after myocardial infarction. Am J Physiol Heart Circ Physiol 287:H286–H292
8. Tamura K, Nakajima H, Rakue H et al (1999) Elevated circulating levels of basic fibroblast growth factor and vascular endothelial growth factor in patients with acute myocardial infarction. Jpn Circ J 63:357–361
9. Tao ZY, Cavasin MA, Yang F et al (2004) Temporal changes in matrix metalloproteinase expression and inflammatory response associated with cardiac rupture after myocardial infarction in mice. Life Sci 74(12):1561–1572
10. Zymek P, Bujak M, Chatila K et al (2006) The role of platelet-derived growth factor signalling in healing myocardial infarcts. J Am Coll Cardiol 48:2315–2323
11. Saraste A, Pulkki K, Kallajoki M et al (1997) Apoptosis in human acute myocardial infarction. Circulation 95:320–323
12. Sugano M, Hata T, Tsuchida K et al (2004) Local delivery of soluble TNF-alpha receptor 1 gene reduces infarct size following ischemia/reperfusion injury in rats. Mol Cell Biochem 266:127–132
13. Breckenridge RA, Mohun TJ, Amaya E (2001) A role for BMP signalling in heart looping morphogenesis in Xenopus. Dev Biol 232:191–203
14. Chocron S, Verhoeven MC, Rentzsch F et al (2007) Zebrafish Bmp4 regulates left-right asymmetry at two distinct developmental time points. Dev Biol 305:577–588
15. Dunn NR, Winnier GE, Hargett LK et al (1997) Haploinsufficient phenotypes in Bmp4 heterozygous null mice and modification by mutations in Gli3 and Alx4. Dev Biol 188:235–247
16. Lawson KA, Dunn NR, Roelen BA et al (1999) Bmp4 is required for the generation of primordial germ cells in the mouse embryo. Genes Dev 13:424–436
17. Neuhaus H, Rosen V, Thies RS (1999) Heart specific expression of mouse BMP-10 a novel member of the TGF-beta superfamily. Mech Dev 80:181–184
18. Schilling TF, Concordet JP, Ingham PW (1999) Regulation of left-right asymmetries in the zebrafish by Shh and BMP4. Dev Biol 210:277–287
19. Somi S, Buffing AA, Moorman AF, Van Den Hoff MJ (2004) Dynamic patterns of expression of BMP isoforms 2, 4, 5, 6, and 7 during chicken heart development. Anat Rec A Discov Mol Cell Evol Biol 279:636–651
20. Zhang H, Bradley A (1996) Mice deficient for BMP2 are nonviable and have defects in amnion/chorion and cardiac development. Development 122:2977–2986

21. Chang SA, Lee EJ, Kang HJ et al (2008) Impact of myocardial infarct proteins and oscillating pressure on the differentiation of mesenchymal stem cells: effect of acute myocardial infarction on stem cell differentiation. Stem Cells 26:1901–1912
22. Rajasingh J, Bord E, Hamada H et al (2007) STAT3-dependent mouse embryonic stem cell differentiation into cardiomyocytes: analysis of molecular signalling and therapeutic efficacy of cardiomyocyte precommitted mES transplantation in a mouse model of myocardial infarction. Circ Res 101:910–918
23. Zaidi SH, Huang Q, Momen A et al (2010) Growth differentiation factor 5 regulates cardiac repair after myocardial infarction. J Am Coll Cardiol 55:135–143
24. Innis CA, Shi J, Blundell TL (2000) Evolutionary trace analysis of TGF-beta and related growth factors: implications for site-directed mutagenesis. Protein Eng 13:839–847
25. Rider CC, Mulloy B (2010) Bone morphogenetic protein and growth differentiation factor cytokine families and their protein antagonists. Biochem J 429:1–12
26. Sanchez NS, Barnett JV (2012) TGFbeta and BMP-2 regulate epicardial cell invasion via TGFbetaR3 activation of the Par6/Smurf1/RhoA pathway. Cell Signal 24:539–548
27. Wall NA, Craig EJ, Labosky PA, Kessler DS (2000) Mesendoderm induction and reversal of left-right pattern by mouse Gdf1, a Vg1-related gene. Dev Biol 227:495–509
28. Rankin CT, Bunton T, Lawler AM, Lee SJ (2000) Regulation of left-right patterning in mice by growth/differentiation factor-1. Nat Genet 24:262–265
29. Kaasinen E, Aittomaki K, Eronen M et al (2010) Recessively inherited right atrial isomerism caused by mutations in growth/differentiation factor 1 (GDF1). Hum Mol Genet 19:2747–2753
30. Karkera JD, Lee JS, Roessler E et al (2007) Loss-of-function mutations in growth differentiation factor-1 (GDF1) are associated with congenital heart defects in humans. Am J Hum Genet 81:987–994
31. Hsiao EC, Koniaris LG, Zimmers-Koniaris T et al (2000) Characterization of growth-differentiation factor 15, a transforming growth factor beta superfamily member induced following liver injury. Mol Cell Biol 20:3742–3751
32. Coleman CM, Loredo GA, Lo CW, Tuan RS (2003) Correlation of GDF5 and connexin 43 mRNA expression during embryonic development. Anat Rec A Discov Mol Cell Evol Biol 275:1117–1121
33. Storm EE, Huynh TV, Copeland NG et al (1994) Limb alterations in brachypodism mice due to mutations in a new member of the TGF beta-superfamily. Nature 368:639–643
34. Sullivan AM, O'Keeffe GW (2005) The role of growth/differentiation factor 5 (GDF5) in the induction and survival of midbrain dopaminergic neurones: relevance to Parkinson's disease treatment. J Anat 207:219–226
35. Sharma M, Kambadur R, Matthews KG et al (1999) Myostatin, a transforming growth factor-beta superfamily member, is expressed in heart muscle and is upregulated in cardiomyocytes after infarct. J Cell Physiol 180:1–9
36. Fairlie WD, Moore AG, Bauskin AR et al (1999) MIC-1 is a novel TGF-beta superfamily cytokine associated with macrophage activation. J Leukoc Biol 65:2–5
37. Yamagishi T, Nakajima Y, Nishimatsu S et al (2001) Expression of bone morphogenetic protein-5 gene during chick heart development: possible roles in valvuloseptal endocardial cushion formation. Anat Rec 264:313–316
38. Kim RY, Robertson EJ, Solloway MJ (2001) Bmp6 and Bmp7 are required for cushion formation and septation in the developing mouse heart. Dev Biol 235:449–466
39. Chen H, Shi S, Acosta L et al (2004) BMP10 is essential for maintaining cardiac growth during murine cardiogenesis. Development 131:2219–2231
40. Allen SP, Bogardi JP, Barlow AJ et al (2001) Misexpression of noggin leads to septal defects in the outflow tract of the chick heart. Dev Biol 235:98–109
41. Choi M, Stottmann RW, Yang YP et al (2007) The bone morphogenetic protein antagonist noggin regulates mammalian cardiac morphogenesis. Circ Res 100:220–228
42. Moskowitz IP, Wang J, Peterson MA et al (2011) Transcription factor genes Smad4 and Gata4 cooperatively regulate cardiac valve development. [corrected]. Proc Natl Acad Sci USA 108:4006–4011

43. Azhar M, Wang PY, Frugier T et al (2010) Myocardial deletion of Smad4 using a novel alpha skeletal muscle actin Cre recombinase transgenic mouse causes misalignment of the cardiac outflow tract. Int J Biol Sci 6:546–555
44. Song L, Zhao M, Wu B et al (2011) Cell autonomous requirement of endocardial Smad4 during atrioventricular cushion development in mouse embryos. Dev Dyn 240:211–220
45. Chang H, Huylebroeck D, Verschueren K et al (1999) Smad5 knockout mice die at mid-gestation due to multiple embryonic and extraembryonic defects. Development 126:1631–1642
46. Yang X, Castilla LH, Xu X et al (1999) Angiogenesis defects and mesenchymal apoptosis in mice lacking SMAD5. Development 126:1571–1580
47. Umans L, Cox L, Tjwa M et al (2007) Inactivation of Smad5 in endothelial cells and smooth muscle cells demonstrates that Smad5 is required for cardiac homeostasis. Am J Pathol 170:1460–1472
48. Wang YX, Qian LX, Liu D et al (2007) Bone morphogenetic protein-2 acts upstream of myocyte-specific enhancer factor 2a to control embryonic cardiac contractility. Cardiovasc Res 74:290–303
49. Beppu H, Kawabata M, Hamamoto T et al (2000) BMP type II receptor is required for gastrulation and early development of mouse embryos. Dev Biol 221:249–258
50. Delot EC, Bahamonde ME, Zhao M, Lyons KM (2003) BMP signalling is required for septation of the outflow tract of the mammalian heart. Development 130:209–220
51. Gaussin V, Van de Putte T, Mishina Y et al (2002) Endocardial cushion and myocardial defects after cardiac myocyte-specific conditional deletion of the bone morphogenetic protein receptor ALK3. Proc Natl Acad Sci USA 99:2878–2883
52. Buxton P, Edwards C, Archer CW, Francis-West P (2001) Growth/differentiation factor-5 (GDF-5) and skeletal development. J Bone Joint Surg Am 83-A(Suppl 1 (Pt 1)):S23–S30
53. Chang SC, Hoang B, Thomas JT et al (1994) Cartilage-derived morphogenetic proteins. New members of the transforming growth factor-beta superfamily predominantly expressed in long bones during human embryonic development. J Biol Chem 269:28227–28234
54. O'Keeffe GW, Hanke M, Pohl J, Sullivan AM (2004) Expression of growth differentiation factor-5 in the developing and adult rat brain. Brain Res Dev Brain Res 151:199–202
55. Storm EE, Kingsley DM (1996) Joint patterning defects caused by single and double mutations in members of the bone morphogenetic protein (BMP) family. Development 122:3969–3979
56. Zhang S, Fantozzi I, Tigno DD et al (2003) Bone morphogenetic proteins induce apoptosis in human pulmonary vascular smooth muscle cells. Am J Physiol Lung Cell Mol Physiol 285:L740–L754
57. Storm EE, Kingsley DM (1999) GDF5 coordinates bone and joint formation during digit development. Dev Biol 209:11–27
58. Chhabra A, Tsou D, Clark RT et al (2003) GDF-5 deficiency in mice delays Achilles tendon healing. J Orthop Res 21:826–835
59. Faiyaz-Ul-Haque M, Ahmad W, Wahab A et al (2002) Frameshift mutation in the cartilage-derived morphogenetic protein 1 (CDMP1) gene and severe acromesomelic chondrodysplasia resembling Grebe-type chondrodysplasia. Am J Med Genet 111:31–37
60. Faiyaz-Ul-Haque M, Ahmad W, Zaidi SH et al (2002) Mutation in the cartilage-derived morphogenetic protein-1 (CDMP1) gene in a kindred affected with fibular hypoplasia and complex brachydactyly (DuPan syndrome). Clin Genet 61:454–458
61. Faiyaz-Ul-Haque M, Faqeih EA, Al-Zaidan H et al (2008) Grebe-type chondrodysplasia: a novel missense mutation in a conserved cysteine of the growth differentiation factor 5. J Bone Miner Metab 26:648–652
62. Thomas JT, Kilpatrick MW, Lin K et al (1997) Disruption of human limb morphogenesis by a dominant negative mutation in CDMP1. Nat Genet 17:58–64
63. Yamashita H, Shimizu A, Kato M et al (1997) Growth/differentiation factor-5 induces angiogenesis in vivo. Exp Cell Res 235:218–226
64. Chen X, Zankl A, Niroomand F et al (2006) Upregulation of ID protein by growth and differentiation factor 5 (GDF5) through a smad-dependent and MAPK-independent pathway in HUVSMC. J Mol Cell Cardiol 41:26–33

65. Kadomatsu H, Matsuyama T, Yoshimoto T et al (2008) Injectable growth/differentiation factor-5-recombinant human collagen composite induces endochondral ossification via Sry-related HMG box 9 (Sox9) expression and angiogenesis in murine calvariae. J Periodontal Res 43:483–489
66. Sena K, Sumner DR, Virdi AS (2007) Modulation of VEGF expression in rat bone marrow stromal cells by GDF-5. Connect Tissue Res 48:324–331
67. Zeng Q, Li X, Beck G et al (2007) Growth and differentiation factor-5 (GDF-5) stimulates osteogenic differentiation and increases vascular endothelial growth factor (VEGF) levels in fat-derived stromal cells in vitro. Bone 40:374–381
68. Upton PD, Long L, Trembath RC, Morrell NW (2008) Functional characterization of bone morphogenetic protein binding sites and Smad1/5 activation in human vascular cells. Mol Pharmacol 73:539–552
69. Liu Z, Shen J, Pu K et al (2009) GDF5 and BMP2 inhibit apoptosis via activation of BMPR2 and subsequent stabilization of XIAP. Biochim Biophys Acta 1793:1819–1827
70. Nishitoh H, Ichijo H, Kimura M et al (1996) Identification of type I and type II serine/threonine kinase receptors for growth/differentiation factor-5. J Biol Chem 271:21345–21352
71. Derynck R, Zhang YE (2003) Smad-dependent and Smad-independent pathways in TGF-beta family signalling. Nature 425:577–584
72. Garg RR, Bally-Cuif L, Lee SE et al (1999) Cloning of zebrafish activin type IIB receptor (ActRIIB) cDNA and mRNA expression of ActRIIB in embryos and adult tissues. Mol Cell Endocrinol 153:169–181
73. Izumi M, Fujio Y, Kunisada K et al (2001) Bone morphogenetic protein-2 inhibits serum deprivation-induced apoptosis of neonatal cardiac myocytes through activation of the Smad1 pathway. J Biol Chem 276:31133–31141
74. Guo J, Wu G (2012) The signalling and functions of heterodimeric bone morphogenetic proteins. Cytokine Growth Factor Rev 23:61–67
75. Allendorph GP, Vale WW, Choe S (2006) Structure of the ternary signalling complex of a TGF-beta superfamily member. Proc Natl Acad Sci USA 103:7643–7648
76. Keller S, Nickel J, Zhang JL et al (2004) Molecular recognition of BMP-2 and BMP receptor IA. Nat Struct Mol Biol 11:481–488
77. Seemann P, Schwappacher R, Kjaer KW et al (2005) Activating and deactivating mutations in the receptor interaction site of GDF5 cause symphalangism or brachydactyly type A2. J Clin Invest 115:2373–2381
78. Nickel J, Kotzsch A, Sebald W, Mueller TD (2005) A single residue of GDF-5 defines binding specificity to BMP receptor IB. J Mol Biol 349:933–947
79. Seemann P, Brehm A, Konig J et al (2009) Mutations in GDF5 reveal a key residue mediating BMP inhibition by NOGGIN. PLoS Genet 5:e1000747
80. Saremba S, Nickel J, Seher A et al (2008) Type I receptor binding of bone morphogenetic protein 6 is dependent on N-glycosylation of the ligand. FEBS J 275:172–183
81. Moustakas A, Heldin CH (2002) From mono- to oligo-Smads: the heart of the matter in TGF-beta signal transduction. Genes Dev 16(15):1867–1871
82. Mazerbourg S, Hsueh AJ (2006) Genomic analyses facilitate identification of receptors and signalling pathways for growth differentiation factor 9 and related orphan bone morphogenetic protein/growth differentiation factor ligands. Hum Reprod Update 12:373–383
83. Shi Y, Hata A, Lo RS et al (1997) A structural basis for mutational inactivation of the tumour suppressor Smad4. Nature 388:87–93
84. Luyten FP (1997) Cartilage-derived morphogenetic protein-1. Int J Biochem Cell Biol 29(11):1241–1244
85. Yndestad A, Ueland T, Oie E et al (2004) Elevated levels of activin A in heart failure: potential role in myocardial remodelling. Circulation 109:1379–1385
86. Hassel S, Eichner A, Yakymovych M et al (2004) Proteins associated with type II bone morphogenetic protein receptor (BMPR-II) and identified by two-dimensional gel electrophoresis and mass spectrometry. Proteomics 4:1346–1358
87. Coleman CM, Tuan RS (2003) Functional role of growth/differentiation factor 5 in chondrogenesis of limb mesenchymal cells. Mech Dev 120:823–836

88. Nakamura K, Shirai T, Morishita S et al (1999) p38 mitogen-activated protein kinase functionally contributes to chondrogenesis induced by growth/differentiation factor-5 in ATDC5 cells. Exp Cell Res 250:351–363
89. Yang X, Lee PJ, Long L et al (2007) BMP4 induces HO-1 via a Smad-independent, p38MAPK-dependent pathway in pulmonary artery myocytes. Am J Respir Cell Mol Biol 37:598–605
90. Ahn K, Mishina Y, Hanks MC et al (2001) BMPR-IA signalling is required for the formation of the apical ectodermal ridge and dorsal-ventral patterning of the limb. Development 128:4449–4461
91. Mishina Y, Hanks MC, Miura S et al (2002) Generation of Bmpr/Alk3 conditional knockout mice. Genesis 32:69–72
92. Stottmann RW, Choi M, Mishina Y et al (2004) BMP receptor IA is required in mammalian neural crest cells for development of the cardiac outflow tract and ventricular myocardium. Development 131:2205–2218
93. Jiao K, Kulessa H, Tompkins K et al (2003) An essential role of Bmp4 in the atrioventricular septation of the mouse heart. Genes Dev 17:2362–2367
94. Schneider MD, Gaussin V, Lyons KM (2003) Tempting fate: BMP signals for cardiac morphogenesis. Cytokine Growth Factor Rev 14:1–4
95. Hao J, Ju H, Zhao S et al (1999) Elevation of expression of Smads 2, 3, and 4, decorin and TGF-beta in the chronic phase of myocardial infarct scar healing. J Mol Cell Cardiol 31:667–678
96. Hao J, Wang B, Jones SC et al (2000) Interaction between angiotensin II and Smad proteins in fibroblasts in failing heart and in vitro. Am J Physiol Heart Circ Physiol 279(6):H3020–H3030
97. Lijnen PJ, Petrov VV, Fagard RH (2000) Induction of cardiac fibrosis by transforming growth factor-beta(1). Mol Genet Metab 71:418–435
98. Wang B, Hao J, Jones SC et al (2002) Decreased Smad 7 expression contributes to cardiac fibrosis in the infarcted rat heart. Am J Physiol Heart Circ Physiol 282:H1685–H1696
99. Chuva de Sousa Lopes SM, Feijen A, Korving J et al (2004) Connective tissue growth factor expression and Smad signalling during mouse heart development and myocardial infarction. Dev Dyn 231:542–550
100. Masaki M, Izumi M, Oshima Y et al (2005) Smad1 protects cardiomyocytes from ischemia-reperfusion injury. Circulation 111:2752–2759
101. Bujak M, Ren G, Kweon HJ et al (2007) Essential role of Smad3 in infarct healing and in the pathogenesis of cardiac remodelling. Circulation 116:2127–2138
102. Dobaczewski M, Bujak M, Li N et al (2010) Smad3 signalling critically regulates fibroblast phenotype and function in healing myocardial infarction. Circ Res 107:418–428
103. Ghosh-Choudhury N, Abboud SL, Chandrasekar B (2003) Ghosh Choudhury G. BMP-2 regulates cardiomyocyte contractility in a phosphatidylinositol 3 kinase-dependent manner. FEBS Lett 544:181–184
104. Kempf T, Horn-Wichmann R, Brabant G et al (2007) Circulating concentrations of growth-differentiation factor 15 in apparently healthy elderly individuals and patients with chronic heart failure as assessed by a new immunoradiometric sandwich assay. Clin Chem 53(2):284–291
105. Xu J, Kimball TR, Lorenz JN et al (2006) GDF15/MIC-1 functions as a protective and anti-hypertrophic factor released from the myocardium in association with SMAD protein activation. Circ Res 98:342–350
106. Behfar A, Zingman LV, Hodgson DM et al (2002) Stem cell differentiation requires a paracrine pathway in the heart. FASEB J 16:1558–1566
107. Artaza JN, Reisz-Porszasz S, Dow JS et al (2007) Alterations in myostatin expression are associated with changes in cardiac left ventricular mass but not ejection fraction in the mouse. J Endocrinol 194:63–76
108. McKoy G, Bicknell KA, Patel K, Brooks G (2007) Developmental expression of myostatin in cardiomyocytes and its effect on foetal and neonatal rat cardiomyocyte proliferation. Cardiovasc Res 74:304–312

109. Rodgers BD, Interlichia JP, Garikipati DK et al (2009) Myostatin represses physiological hypertrophy of the heart and excitation-contraction coupling. J Physiol 587:4873–4886
110. Kempf T, Eden M, Strelau J et al (2006) The transforming growth factor-beta superfamily member growth-differentiation factor-15 protects the heart from ischemia/reperfusion injury. Circ Res 98:351–360
111. Kempf T, Zarbock A, Widera C et al (2011) GDF-15 is an inhibitor of leukocyte integrin activation required for survival after myocardial infarction in mice. Nat Med 17:581–588
112. Bujak M, Frangogiannis NG (2007) The role of TGF-beta signalling in myocardial infarction and cardiac remodelling. Cardiovasc Res 74:184–195
113. Okada H, Takemura G, Kosai K et al (2005) Postinfarction gene therapy against transforming growth factor-beta signal modulates infarct tissue dynamics and attenuates left ventricular remodelling and heart failure. Circulation 111:2430–2437
114. Deten A, Holzl A, Leicht M et al (2001) Changes in extracellular matrix and in transforming growth factor beta isoforms after coronary artery ligation in rats. J Mol Cell Cardiol 33:1191–1207
115. See F, Kompa A, Martin J et al (2005) Fibrosis as a therapeutic target post-myocardial infarction. Curr Pharm Des 11:477–487
116. Thune JJ, Solomon SD (2006) Left ventricular diastolic function following myocardial infarction. Curr Heart Fail Rep 3:170–174
117. Ambrosino C, Iwata T, Scafoglio C et al (2006) TEF-1 and C/EBPbeta are major p38alpha MAPK-regulated transcription factors in proliferating cardiomyocytes. Biochem J 396:163–172
118. Tenhunen O, Soini Y, Ilves M et al (2006) p38 Kinase rescues failing myocardium after myocardial infarction: evidence for angiogenic and anti-apoptotic mechanisms. FASEB J 20:1907–1909
119. Yoshimoto T, Yamamoto M, Kadomatsu H et al (2006) Recombinant human growth/differentiation factor-5 (rhGDF-5) induced bone formation in murine calvariae. J Periodontal Res 41:140–147
120. Delyani JA, Robinson EL, Rudolph AE (2001) Effect of a selective aldosterone receptor antagonist in myocardial infarction. Am J Physiol Heart Circ Physiol 281:H647–H654
121. Lal A, Veinot JP, Leenen FH (2004) Critical role of CNS effects of aldosterone in cardiac remodelling post-myocardial infarction in rats. Cardiovasc Res 64:437–447
122. Pitt B, Zannad F, Remme WJ et al (1999) The effect of spironolactone on morbidity and mortality in patients with severe heart failure. Randomized Aldactone Evaluation Study Investigators. N Engl J Med 341:709–717
123. Machado RD, Pauciulo MW, Thomson JR et al (2001) BMPR2 haploinsufficiency as the inherited molecular mechanism for primary pulmonary hypertension. Am J Hum Genet 68:92–102
124. Liu W, Selever J, Wang D et al (2004) Bmp4 signalling is required for outflow-tract septation and branchial-arch artery remodelling. Proc Natl Acad Sci USA 101(13):4489–4494
125. Mohler ER 3rd, Gannon F, Reynolds C et al (2001) Bone formation and inflammation in cardiac valves. Circulation 103:1522–1528
126. Dhore CR, Cleutjens JP, Lutgens E et al (2001) Differential expression of bone matrix regulatory proteins in human atherosclerotic plaques. Arterioscler Thromb Vasc Biol 21:1998–2003
127. Solloway MJ, Robertson EJ (1999) Early embryonic lethality in Bmp5; Bmp7 double mutant mice suggests functional redundancy within the 60A subgroup. Development 126:1753–1768
128. Wollert KC, Kempf T, Peter T et al (2007) Prognostic value of growth-differentiation factor-15 in patients with non-ST-elevation acute coronary syndrome. Circulation 115:962–971
129. Galvin KM, Donovan MJ, Lynch CA et al (2000) A role for smad6 in development and homeostasis of the cardiovascular system. Nat Genet 24:171–174
130. Vargesson N, Laufer E (2001) Smad7 misexpression during embryonic angiogenesis causes vascular dilation and malformations independently of vascular smooth muscle cell function. Dev Biol 240:499–516

Chapter 22
Extracellular Matrix Biomarkers of Adverse Remodeling After Myocardial Infarction

Kristine Y. DeLeon, Lisandra E. de Castro Brás, Yonggang Ma, Ganesh V. Halade, Jianhua Zhang, and Merry L. Lindsey

Abstract Approximately every minute, someone will die from a myocardial infarction (MI). A MI is the result of an obstruction of blood supply causing the heart to undergo complex structural and functional changes in the left ventricular wall, known as ventricular remodeling. Cardiomyocytes undergo necrosis rapidly after the onset of injury, leading to early accumulation of neutrophils, activation of metalloproteinases, and degradation of the stromal tissue, eventually leading to formation of a collagen scar. Circulating factors related to inflammatory and fibrotic responses are key predictors of extracellular matrix (ECM) changes that occur after MI. Changes in the collagen network of the ECM can alter myocardial stiffness, consequently leading to cardiac hypertrophy, fibrosis, and LV dysfunction. Proteins and peptides of the ECM are promising biomarkers for MI. This book chapter provides an overview of key ECM biomarkers involved in adverse remodeling post-MI and their practical applications.

Keywords Cardiac remodeling • Myocardial infarction • Extracellular matrix • Biomarkers

K.Y. DeLeon • L.E. de Castro Brás • Y. Ma • G.V. Halade • J. Zhang • M.L. Lindsey, Ph.D. (✉)
San Antonio Cardiovascular Proteomics Center, The University of Texas Health Science Center at San Antonio (UTHCSA), 15355 Lambda Drive, MC 7755, San Antonio, TX 78245, USA

Division of Geriatrics, Gerontology and Palliative Medicine, Department of Medicine, The University of Texas Health Science Center at San Antonio (UTHSCA), 15355 Lambda Drive, MC 7755, San Antonio, TX 78245, USA
e-mail: lindseym@uthscsa.edu

B.I. Jugdutt and N.S. Dhalla (eds.), *Cardiac Remodeling: Molecular Mechanisms*, Advances in Biochemistry in Health and Disease 5, DOI 10.1007/978-1-4614-5930-9_22,

22.1 Introduction

Cardiovascular disease (CVD) is the leading cause of morbidity and mortality in the United States accounting for approximately 40% of all deaths. CVD includes a wide range of diseases that differ considerably in time scale and relative effects of genes and environment [1]. Of the different types of CVD, myocardial infarction (MI) accounts for the majority of morbidity and mortality. According to the latest American Heart Association statistics, each year an estimated 785,000 Americans will have a new coronary attack, 470,000 will have a recurrent attack, and an additional 195,000 will have a silent first MI [2]. Approximately every 25 s, an American will have a coronary event, and approximately every minute, someone will die of one [2].

22.1.1 What Is a Myocardial Infarction?

A MI is the result of an obstruction of blood supply to a section of the heart, causing myocyte death. Post-MI, the heart undergoes complex structural and functional changes in the left ventricular wall, known as ventricular remodeling. This is reflected by wall thinning, chamber dilatation, and impaired function of the left ventricle (LV). Cardiomyocytes undergo necrosis rapidly after the onset of injury, leading to early accumulation of neutrophils, activation of metalloproteinases, and degradation of the stromal tissue, eventually leading to formation of a collagen scar [3–5]. Diagnosis of MI traditionally includes analysis of three parameters: clinical history, electrocardiographic abnormalities, and increased serum concentrations of specific cardiac enzymes such as creatine kinase, aspartate aminotransferase, and lactate dehydrogenase [6–8]. The present goal in medicine is not only to treat MI patients effectively but also to identify individuals who are at risk for adverse events. One way to accomplish this goal is to develop novel cardiovascular biomarkers to more accurately aid in screening, diagnosis, and prognosis in the post-MI setting.

22.1.2 What Are Biomarkers?

Biomarkers are biological molecules used to indicate normal or pathological biological processes or to provide an index of pharmacological response to therapy [9, 10]. A biomarker can be a measurement found in a biosample (blood, urine, or tissue), a recording (blood pressure, ECG, or Holter), or an imaging test (echocardiogram or CT scan) [10]. It does not necessarily matter if an identified plasma protein marker has a cardiac origin (which would mean it is a primary indicator of response), as long as it has a cardiac action. For example, markers from the liver increase post-MI as part of a whole-body protective response to ischemia [11]. Because of the wide variety, biomarkers can be classified into three types: type 0,

Table 22.1 Considerations for biomarkers that reflect cardiac ECM remodeling following myocardial infarction

• Changes must reflect changes in cardiac ECM scan
• Changes must parallel the variation in other parameters reflecting cardiac fibrosis and LV function
• Positive gradient of concentrations: from coronary sinus blood toward peripheral vein blood, indicating cardiac origin
• High specificity and sensitivity
• Low cost
• Off the shelf/easy access
• Pain free/noninvasive
• Accurate and reproducible

type 1, and surrogate end point or type 2 biomarker [10, 12]. A type 0 biomarker demonstrates the progression of disease and correlates with known clinical indices, whereas a type 1 biomarker describes the effect a therapeutic intervention has in accordance with its mechanism of action. A surrogate end point or type 2 biomarker is intended to anticipate the clinical benefit of treatment on the basis of epidemiological, therapeutic, pathophysiological, or other scientific evidence [10, 12]. Cardiac biomarkers have to meet several criteria to be of use for both patients and clinicians (Table 22.1). Although circulating biomarkers offer the advantage of relative ease of collection, storage, and reduced cost, they may not prove as sensitive and specific as tissue or imaging markers in the diagnosis and progression of disease. Conversely, imaging technologies can detect and assess cardiovascular diseases with a high degree of sensitivity and specificity but may be limited by availability, technical difficulties, and cost.

22.1.3 Why Do We Need Biomarkers Post-MI?

The discovery and development of biomarkers has shed light on both the pathogenesis as well as the management of MI [13, 14]. In general, biomarkers have been independently linked with various phenotypes associated with high risk of MI [15–17]. For this reason, biomarker discovery has high potential for providing a real benefit for MI patients with applications in at least five areas: screening, diagnosis, prognosis, prediction of disease recurrence, and therapeutic monitoring [9, 10]. The main requirement with regard to the potential of a candidate biomarker is the strength of the correlation between the expression patterns of the biomarker and the clinical outcome [18].

Interest in using proteins and peptides of the extracellular matrix (ECM) as promising biomarkers for MI has increased within the last decade. One obvious reason is ECM biomarkers are able to detect early changes in the structure and function of the LV to provide important prognostic information [18, 19]. Having a normal connective tissue environment is a prerequisite for having normal myocardial architecture and function. Changes in the collagen network of the ECM can alter myocardial stiffness, consequently leading to cardiac hypertrophy, fibrosis, and LV dysfunction [20, 21].

Circulating factors related to inflammatory and fibrotic responses are key predictors of ECM changes that occur after MI [15, 22]. Changes in circulating levels of proinflammatory cytokines and pro- and antifibrotic proteins have been observed in patients with heart failure [23–25]. The field of biomarker discovery holds great potential for improving clinical measurements of disease. This book chapter provides an overview of key ECM biomarkers involved in adverse remodeling post-MI and their practical applications.

22.2 Inflammatory Response

Inflammation is key in the LV response to MI. MI is associated with an inflammatory reaction which is a requirement for healing and scar formation [26–28]. While the activation of inflammatory mediators following MI is necessary for adequate wound healing, at the same time it can cause harmful effects on the LV to alter its size, shape, and function. Numerous studies have implicated several inflammatory markers, including C-reactive protein, the complement factor C5a, interleukin (IL)-8, and adhesion molecules such as intercellular adhesion molecule (ICAM)-1, in predicting the risk and progression of LV dysfunction post-MI [29–35]. These studies demonstrate the importance of anti-inflammatory strategies in post-MI treatments; however, in order to achieve the most effective treatment without interfering with healing and cardiac repair, an understanding of the molecular events associated with inflammation is required.

22.2.1 Neutrophils

Inflammation occurs in stages. In the absence of reperfusion, neutrophils are the first cells to respond to the site of injury. Neutrophils have four subsets of membrane-bound compartments: secretory vesicles, gelatinase/tertiary granules, specific/secondary granules, and azurophilic/primary granules. These granules contain antimicrobial proteins, proteases, ECM proteins, and soluble mediators of inflammation [36, 37]. The order of release is as follows: secretory vesicles, gelatinase granules, specific granules, and finally azurophilic granules [38, 39]. The strength of the stimulus will affect the release of the particular granules.

22.2.1.1 Secretory Vesicles

Contents of secretory vesicles are secreted more readily than other granules, making these molecules important in understanding the early role neutrophils have on progression of LV dysfunction post-MI. Secretory vesicles are formed by endocytosis and mostly contain serum proteins such as human serum albumin. The membrane of secretory vesicles is enriched with adhesion molecules, such as Mac-1 (CD11b/CD18), complement receptor 1, and urokinase-type plasminogen activator receptor. When incorporated into the plasma membrane in response to inflammatory mediators, secretory vesicles enable the neutrophil to interact with endothelial cells, ECM, and microorganisms [37]. It is for this reason that secretory vesicles are believed to play a major role in early neutrophil activation [40].

In addition to serum proteins and adhesion molecules, heparin-binding protein (HBP) and proteinase-3 are also shown to reside within secretory vesicles [41, 42]. HBP, also known as azurocidin, exists in both secretory vesicles and azurophilic granules [43]. An association between HBP and an increase in neutrophil influx has been observed, mostly likely caused by cytoskeleton rearrangements of endothelial cells [44]. HBP has also been shown to be elevated in plasma of patients with severe sepsis and increases before circulatory collapse [45].

Similar to HBP, proteinase-3 is expressed in both secretory vesicles and azurophilic granules [42]. Ng *et al.* indicated proteinase-3 as being a potential marker for patient prognosis post-MI as part of a multimarker strategy [46]. Although the exact role of proteinase-3 post-MI has not yet been discovered, there is evidence to suggest it can induce apoptosis through caspase-like activity [47], cleave angiotensinogen to form angiotensin I and II [48], and activate tumor necrosis factor-α [49] in addition to other pro-inflammatory mediators [50, 51]. In addition, proteinase-3 has been shown to degrade ECM components such as fibronectin, type IV collagen, and the core protein of proteoglycans [52] implicating it in LV remodeling post-MI.

22.2.1.2 Gelatinase Granules

In human neutrophils, gelatinase granules account for 25% of the total peroxidase-negative granules [53]. Exocytosis of gelatinase granules liberates matrix degrading enzymes such as matrix metalloproteinases (MMPs), signifying their importance for neutrophil extravasation and diapedesis [54]. Collectively, MMPs degrade major ECM components including collagens, fibronectin, proteoglycans, laminin, and gelatin. Neutrophils are the early source of MMP-9, also known as type IV collagenase or gelatinase B [55]. MMP-9 is secreted as a zymogen and becomes active upon cleavage of the N-terminal domain or treatment with oxygen metabolites [56]. Macrophages and endogenous LV cells including myocytes, endothelial cells, and vascular smooth muscles cells also express MMP-9 [57], making it a key player in understanding the progression and recovery of patients post-MI. For this reason, MMP-9 will be discussed in greater detail later in this chapter.

22.2.1.3 Specific Granules

The second, larger component of peroxidase-negative granules is the specific granules. These granules contain mostly antimicrobial substances and are released into the phagocytic vacuole or the exterior of the cell upon encounter with inflammatory mediators. Contents of specific granules represent an oxygen-independent antimicrobial system enhancing the oxygen-dependent activities of the NADPH oxidase [36].

Of the substances released from activated neutrophils, the majority of researchers focus on myeloperoxidase (MPO) because of its involvement in the pathogenesis of atherosclerosis. Recently, interest in lactoferrin has increased due to its unique antimicrobial, immunomodulatory, and antineoplastic properties [58–61]. Lactoferrin, an iron-binding protein, has antimicrobial activity against a broad spectrum of Gram-positive and Gram-negative bacteria [62]. *In vitro* stimulation of adherent neutrophils with phorbol myristate acetate releases a significant amount of the total cell content of lactoferrin, compared to a lower percentage of released MPO, indicating lactoferrin is a more dynamic marker of early neutrophil activation [63]. In contrast to MPO, lactoferrin inhibits the upregulation of adhesion molecules on endothelial cells and limits iron-mediated damage at sites of inflammation [59, 64, 65]. Lactoferrin also inhibits proinflammatory cytokine production in monocytes decreasing inflammation by hindering neutrophil recruitment and activation [66].

Another important peptide found in specific granules of neutrophils is neutrophil gelatinase-associated lipocalin (NGAL). NGAL plays an important role in the iron-depleting strategy of innate immunity [67] and also promotes matrix degradation and myocardial remodeling [68]. NGAL is responsible for regulation of MMP-9 activity by preventing its degradation, leading to enhanced proteolytic activity with prolonged effects on collagen degradation [69]. Recent studies have revealed a significant increase of NGAL in plasma as well as in necrotic and healthy tissue of patients diagnosed with acute MI [70, 71]. Enhanced NGAL expression may thus contribute to the increased MMP-9 activity seen post-MI, potentially promoting ECM remodeling [55, 72–74].

22.2.1.4 Azurophilic Granules

Azurophilic granules are packed with acidic hydrolases and antimicrobial proteins such as MPO and serine proteases [75]. Azurophilic granules contain many proteins in their proforms which undergo proteolytic cleavage on arrival at the granule compartment [76, 77]. Exocytosis of azurophilic granules are limited and are believed to contribute mainly in the intracellular degradation of microorganisms in the phagolysosome [78, 79].

MPO is a heme enzyme, released by activated neutrophils as well as monocytes. MPO accumulates in the infarct zone leading to the formation of reactive chlorinating species. Neutrophils have been demonstrated to release MPO into the coronary circulation, yielding elevated MPO plasma levels in patients with unstable angina and acute MI [80, 81]. Moreover, MPO release has been seen to actually precede myocardial injury, even before complete microvascular obstruction. Brennan *et al.* assessed the plasma levels of MPO as a predictor of the risk of cardiovascular events in 604 sequential patients and noted that a single initial measurement of plasma MPO independently predicts the early risk of MI as well as the risk of major adverse cardiac events in the following 30-day and 6-month periods [82].

Studies in MPO-null mice using a chronic coronary artery ligation model demonstrated a marked reduction in leukocyte infiltration and LV dilatation, associated with delayed myocardial rupture and preservation of systolic function [83]. MPO and oxidants released by activated neutrophils have been found to enhance the degradation of ECM through proteases [81] consequently promoting the development of abnormal ventricular geometry. MPO levels may thus act as a prognostic test for the likelihood of MI, revascularization, and death.

Azurophilic granules also contain three structurally related serine proteases: proteinase-3, cathepsin G, and elastase. Collectively, these serine proteases degrade a variety of ECM components, such as elastin, fibronectin, laminin, type IV collagen, and vitronectin [84]. Serine proteases are also known to stimulate coagulation during systemic infection resulting in arterial thrombosis and possibly contributing to the progression of CVDs such as MI [85]. Of note, serine elastase is a potent activator of MMP-9 and may be the primary activator during the early post-MI response [55].

In addition to azurophilic granules, proteinase-3 is also found in secretory vesicles and may induce apoptosis through caspase-like activity [47]. Elastase is a key player in protease degradation of ECM. Bidouard *et al.* determined inhibition of elastase prior to reperfusion causing reduction in the infarct size [86]. This effect could be due to the ability of elastase to activate MMP-2, MMP-3, and MMP-9 and inactivate TIMP-1 resulting in an increase in ECM degradation [87]. In contrast to proteinase-3 and elastase, cathepsin G plays a protective role post-MI by inactivating bradykinin and IL-6 both of which are well-known inflammatory mediators [88, 89]. Although cathepsin G does have this protective effect post-MI, it also is a potent platelet agonist and degranulator [90]. Thus, the ability to directly activate platelets can lead to intravascular thrombosis, irreversible ischemia, and tissue death.

22.2.2 Macrophages

Following MI, circulating monocytes are recruited to the infarct zone. Upon extravasation into the tissue, the monocytes become macrophages. Macrophages exert crucial roles in wound healing after MI [91, 92]. Macrophages phagocytose tissue and cell debris; secrete cytokines, chemokines, and growth factors; regulate fibroblast function and collagen turnover; as well as modulate the angiogenic response [93]. There are a wide range of heterogeneous macrophage phenotypes with differential gene expression profile and functional characteristics [94]. A broad but perhaps oversimplified classification is the classical M1 and alternative M2 activation groups. A study by Troidl *et al.* suggests that macrophages are differently activated during different phases of MI remodeling process. During the acute inflammatory phase, classical M1 macrophages are primarily activated, while macrophages polarize into alternative M2 subtype during the scar formation period [95].

22.2.2.1 Classical M1 Activation

Post-MI, macrophage polarization toward the classic proinflammatory M1 is induced by Th1 cytokines, including interferon (IFN)-γ and tumor necrosis factor (TNF)-α. Activated M1 macrophages increase the production of several proinflammatory cytokines and chemokines, reactive oxygen species, and nitrogen intermediates. In the early inflammatory phase post-MI, M1 macrophages remove unwanted debris, boost acute inflammatory response, and promote ECM degradation.

Persistent M1 macrophage activation exacerbates post-MI ventricular remodeling and dysfunction. Hu *et al.* demonstrated that, enhanced M1 macrophage polarization was responsible for deteriorated LV malfunction and dilatation and cardiac fibrosis in class A scavenger receptor null mice. Additionally, the authors observed increased production of M1 cytokines including IL-1β, IL-6, and TNF-α [96]. IL-1β secretion in the M1 phase induces MMP-9, which facilitates the secretion of inflammatory cytokines and the degradation of ECM. Leuschner and colleagues reported that post-MI hearts showed rapid monocyte kinetics and the average residence time of monocytes in inflamed tissue is 20 h. After 20 h, most monocytes will undergo local death, and a smaller number will regress to circulation, liver, or other organs [97]. This study indicates that monocytes' transformation to macrophages or macrophage polarization occurs rapidly in response to the surrounding microenvironment. Another important concept revealed in this chapter is that macrophages originating from the spleen migrated to the LV post-MI, making the spleen a major site for macrophage production.

22.2.2.2 Alternative M2 Activation

In contrast to M1 macrophages, polarization toward the alternatively anti-inflammatory M2 during the later phase of inflammation is mediated by Th2 cytokines, including elevated production of IL-4, IL-10, and TGF-β1 post-MI [98, 99]. To date, three distinct M2 subtypes have been recognized: IL-4 and IL-13 induce M2a phenotype, immune complexes in combination with IL-1β or LPS promote M2b subset, and IL-10, TGF-β, or glucocorticoids drive M2c macrophages [100]. M2 macrophages show increased phagocytic activity and anti-inflammatory properties [101, 102]. M2 macrophages facilitate the active resolution of inflammation, ECM deposition, fibroblast proliferation, and angiogenesis [103], all of which are crucial for tissue repair [104]. Macrophage phagocytosis of apoptotic neutrophils, a marker of the active resolution of acute inflammation, triggers TGF-β production, suggesting that phagocytosis may stimulate the conversion to the M2 status [105].

22.2.2.3 Classical Versus Alternative Activation

The timely shift and balance between M1 and M2 macrophages is crucial for stable scar formation and adequate healing response post-MI. Prolonged and excessive activation of the M1 phenotype yields uncontrolled inflammation and ECM breakdown,

Table 22.2 Markers of M1 and M2 macrophage polarization phenotypes

Classical (M1)	Alternative (M2)
Proinflammatory	*Anti-inflammatory*
CCL3 (MIP-1α)	Acyl-malonyl condensing enzyme (AMAC)-1
CCL20	Arginase (Arg) 1, 2
CCR7	CCL2
CD11c	CCL17
CD16/CD32	CCL22
COX2	CCL24
CXCL9	CD36
CXCL10	CD163
CXCL11	CD206 (mannose receptor C type 1, MRC1)
iNOS	CD209a, CD209b, CD209c
IL-1β	Fizz1
IL-6	
IL-12	F13a1
INF-β	IL-10
MCP-1	IL-1R antagonist
Proplatelet basic protein (PPBP) RANTES	Macrophage galactose *N*-acetyl-galactosamine-specific lectin (Mgl)-1, 2
TNF-α	Pdcd1Ig2
	PPARγ
	Selenoprotein-1 (SEPP1)
	SOCS3
	Stabilin-1
	Ym1 (chitinase 3-like 3, Chi313)

which in turn contributes to adverse remodeling and cardiac rupture. Persistent M2 activation stimulates excessive fibroblast proliferation and ECM deposition, which is a hallmark of cardiac fibrosis and influences structural, biochemical, mechanical, and electrical properties of the myocardium [106]. Based on the crucial role of macrophage in wound healing, the M1/M2 ratio may be a potential marker for predicting long-term outcome post-MI. However, the accurate macrophage phenotype and regulation factors post-MI remains to be investigated. With true and reliable markers of macrophage activation now defined (Table 22.2), characterization of macrophage phenotypes post-MI is not only needed but is also possible.

22.2.3 Matrix Metalloproteinase-9

MMP-9 is expressed in cardiac myocytes, fibroblasts, vascular smooth muscle cells, endothelial cells, neutrophils, lymphocytes, and macrophages [55, 72, 107, 108]. MMP-9 was first described as being able to process only collagen that was first denatured or already cleaved by collagenases such as MMP-1. Recent literature, however, has shown that MMP-9 can process full-length interstitial collagens in addition to a wide array of other substrates [109]. Furthermore, numerous predicted substrates have been identified for MMP-9 [110]. In addition, MMP-9 does not require an activation cleavage step to proteolyze substrates. Bannikov and colleagues demonstrated that pro-MMP-9, in the presence of substrate, has enzymatic activity without the loss of the 10 kDa pro-domain [111]. While this feature is likely common in other MMPs, it has only been demonstrated for MMP-9 indicating that an active MMP-9 band on a gelatin zymogram may not be as useful as we have previously considered.

Recent developments in mass spectrometry-based proteomic approaches and novel emerging technologies hold particular promise for unbiased discovery and subsequent validation of new biomarkers of MI [112]. Didangelos *et al.* applied a subfractionation proteomic method for the identification of proteins present in the vascular ECM space in addition to post-MI samples [113, 114]. MMP-9 has been detected by mass spectrometry, and the presence of MMP-9 has been linked to the degradation of fibronectin in human aortic samples [113]. Zamilpa *et al.* identified multiple proteins that are differentially expressed in the infarct region of MMP-9-null mice compared to wild-type mice. Among previously known *in vitro* MMP-9 substrates, fibronectin was validated as an *in vivo* MMP-9 substrate in the post-MI setting using ECM-targeted proteomic approaches [115].

The MMP-9 promoter has several inflammatory response elements, including activator protein-1, specificity protein-1, and NF-κB sites [109]. For this reason, MMP-9 is associated with the inflammatory response to tissue injury. A striking increase in MMP-9 activity has been shown to occur at days 1 through 4 in the infarct regions of murine LV, consistent with neutrophil and macrophage infiltration [72, 73, 116–118]. The earlier initial increase in MMP-9 protein levels seen at day 1 post-MI is due to the release of preformed MMP-9 from infiltrating neutrophils, where it is stored in gelatinase granules [119]. Studies in both MMP-9-null mice as well as MMP-9 pharmaceutical inhibitors have shown a reduction in LV dilation and preservation of cardiac function [73, 120–123]. Interestingly, MMP-9 deletion also stimulates neovascularization in the post-MI infarct region [74], suggesting that MMP-9 serves both beneficial and detrimental roles in the post-MI response.

Blankenberg and colleagues performed the first comprehensive clinical study that implicated MMP-9 as a novel prognostic biomarker for cardiovascular mortality in patients with CVD. In this study, MMP-9 was found to correlate with acute-phase reactants (IL-6, hs-CRP, and fibrinogen) indicating that MMP-9 could have its own pathophysiological significance in cardiovascular mortality independent of an association with inflammatory markers [124]. Squire and coworkers extended these studies to demonstrate that increased MMP-9 correlates with larger LV volumes and greater LV dysfunction following MI [125]. Additional studies have indicated an increase in MMP-9 levels to be associated with increased wall thickness, lower ejection fraction, as well as persistent adverse LV remodeling [126–129].

22.3 Markers of Fibrosis

Several molecules have been put forward as candidate circulating biomarkers of cardiac fibrosis, but very few survive the test of time to become a useful clinical tool. Myocardial fibrosis results from a robust increase in collagen deposition, in the setting of an unchanged or lower rate of collagen degradation. Fibrosis is regulated by a number of processes mediated by mechanical and humoral mechanisms [130]. The increase in collagen content plays an important role in the deteriorating function of the heart and in progression to heart failure. Fibrosis leads to myocardial stiffness and promotes the onset of abnormalities in diastolic function. Whereas reparative fibrosis contributes to the preservation of LV morphology and systolic function [131], reactive fibrosis may contribute to the deterioration of LV function by increased myocardial stiffness [130, 132, 133] and impaired electrical activity [134].

22.3.1 Collagen-Related Biomarkers

In the heart, collagen turnover is regulated by fibroblasts and myofibroblasts [135]. These cells synthesize and secrete preprocollagen type I and III, the two more abundant subtypes of fibrillar collagen present in the heart, as a pro-α-collagen chain. In the endoplasmic reticulum, three pro-α-chains form a triple-helix structure, known as procollagen [136]. Procollagen molecules are secreted into the interstitial space where they undergo cleavage of their end-terminal propeptide sequences to enable collagen fiber formation [137]. Procollagen molecules initially contain two terminal propeptides: the amino (N)-propeptide and the carboxy (C)-propeptide; these are released into circulation by the action of specific procollagen N- and C-proteinases [138]. The fibrotic process may be indirectly characterized by measuring collagen propeptides, since quantification of cleaved propeptides correlate with the levels of fibrillar collagen deposited. Although these propeptides reflect collagen synthesis, degradation products of collagen and the MMPs responsible for the breakdown of the ECM can also be detected in blood and are used as biomarkers of collagen turnover (Fig. 22.1).

22.3.1.1 Collagen Synthesis

Procollagen Type I Carboxy-Terminal Propeptide

The most used biomarker for the quantification of collagen type I synthesis is the 100 kDa C-terminal propeptide of collagen type I. A stoichiometric ratio of 1:1 exists between the number of collagen type I molecules produced and the procollagen type I carboxy-terminal propeptide (PICP) molecules released [139]. Elevated

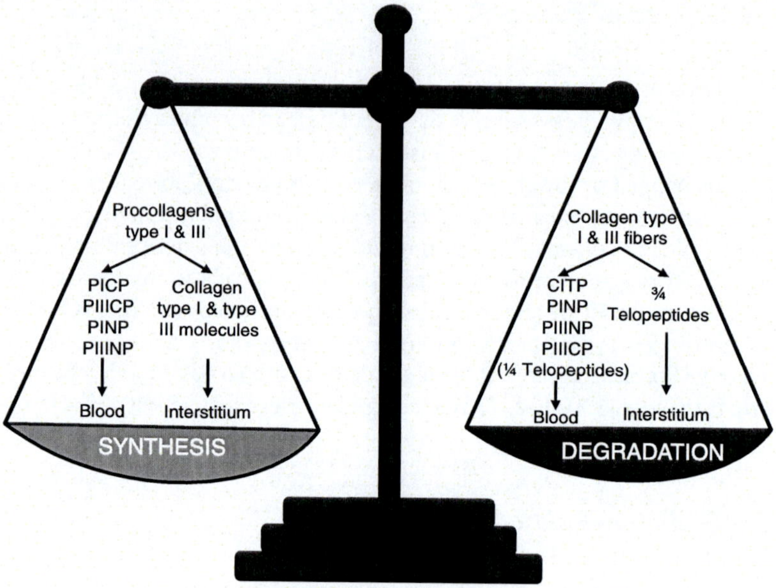

Fig. 22.1 Equilibrium between collagen synthesis and degradation dictates the quality of the LV scar after myocardial infarction. Collagen propeptides are formed during collagen synthesis and are found in circulation; the propeptide levels can be used as biomarkers for collagen synthesis. Similarly, the telopeptides originate during collagen breakdown and are present in the blood. These telopeptides can be used as biomarkers of collagen degradation

levels of PICP in serum have been associated with MI [140] and diastolic heart failure [129]. Animal models have revealed PICP is associated with the volume of myocardial tissue occupied by collagen fibers in left ventricular hypertrophy [141]. Further, changes in collagen metabolism are detected early after MI. In a study by Radovan and colleagues, increased accumulation of collagens I and III was observed within the first 48 h post-MI, and higher synthesis of collagen I persisted until 1 month post-MI [142].

Procollagen Type I Amino-Terminal Propeptide

Procollagen type I amino-terminal propeptide (PINP) is a 70 kDa propeptide that has been used as a marker for synthesis of collagen type I [143]. Contrary to PICP, PINP is not always cleaved from the procollagen molecule [144]; hence, PINP serum levels do not directly correlate to collagen synthesis. This may explain why PINP levels in the blood showed no significant difference between controls and patients with hypertrophic cardiomyopathy [145], heart failure patients [146], and hypertensive patients with or without diastolic heart failure [129].

Procollagen Type III Amino-Terminal Propeptide

Procollagen type III amino-terminal propeptide (PIIINP) is a 42 kDa N-terminal propeptide, cleaved from procollagen type III, and is widely used as a marker for collagen type III synthesis [143]. The N-terminal domain of collagen type III is sometimes removed incompletely, resulting in the incorporation of PIIINP in the collagen fibers [147]. This may lead to an underestimate of the synthesis of collagen type III. Nevertheless, PIIINP plasma levels can be used to predict cardiac events and mortality. Serum PIIINP levels were significantly correlated with relative wall thickness in hypertensive patients with LV hypertrophy [148]. Additionally, elevated serum levels of PIIINP have been positively correlated to increased mortality risk and heart failure [16, 149–151]. Patients with acute MI and with no improvement in LV function during follow-up had higher concentrations of serum PIIINP [142].

22.3.1.2 Collagen Degradation

Collagen Type I Carboxy-Terminal Telopeptide

The degradation of collagen fibrils is accomplished by interstitial collagenase as well as several other proteases. This enzyme cleaves all 3 α-chains of collagen at a single specific locus to generate a ¾ telopeptide and a ¼ telopeptide, in a 1:1 stoichiometry. Whereas the large telopeptide remains within the interstitial space for further degradation by gelatinases, the small C-terminal telopeptide (12 kDa) is found in an immunochemically intact form in the blood [152] and can be used as a biomarker of collagen degradation. In a clinical study that examined the balance between collagen degradation and synthesis 72 h post-MI, high levels of collagen type I carboxy-terminal telopeptide (CITP) in the plasma were associated with poor long-term clinical outcome [153]. A 1-year follow-up of the same clinical study revealed that poor outcome events such as death, heart failure, and recurrent MI were significantly greater for patients presenting the higher CITP levels. This trend is also observed for patients with hypertrophic cardiomyopathy [145], hypertensive heart disease [154], and heart failure with diastolic dysfunction [22, 129].

MMPs and TIMPs

Fibrillar collagens are substrates for MMPs. Of the 25 MMPs described to date, MMP-1 has the highest affinity for fibrillar collagen and preferentially degrades collagens I and III. Increased MMP-1 synthesis has been reported in the infarcted myocardium [155]. Inactivation of MMP-1 synthesis in the infarcted myocardium results in collagen accumulation, which in turn leads to a stiff and noncompliant LV and gradual heart dysfunction [156]. MMP-1 activity is dependent on the concentration of active enzyme and of a family of naturally occurring tissue inhibitors of metalloproteinases (TIMPs) [157]. TIMP-1 (29 kDa) is a glycoprotein member of the TIMP family and co-localizes with MMP-1 in normal myocardium and is expressed by cardiac fibroblasts and myocytes [158–160]. TIMP-1 levels are closely associated with established markers of heart failure severity, parameters of systemic inflammation, and markers of LV remodeling [161]. The interaction between MMP-1, which initiates the degradation of collagen fibers within the heart, and TIMP-1 is of critical relevance in the maintenance of the integrity of the cardiac collagen network post-MI. The MMP-1/TIMP-1 ratio was inversely correlated with ejection fraction in patients with either systolic or diastolic heart failure [162]; additionally, high levels of TIMP-1 have been related to diastolic dysfunction [163]. Creemers *et al.* demonstrated that TIMP-1–/null mice showed a greater degree of LV dilation post-MI [164].

22.3.2 *Galectin-3*

Galectin-3 (Gal-3) is a 29–35 kDa soluble β-galactoside-binding lectin [165] that plays an important role in the regulation of cardiac fibrosis and remodeling. Gal-3 is expressed by activated macrophages and induces proliferation of cardiac fibroblasts and increased deposition of collagen type I in the myocardium [166]. Therefore, Gal-3 has an important role post-MI. mRNA levels for Gal-3 were significantly increased in the infarct region of mice 7 days post-MI [167]. Gal-3 expression is substantially upregulated in animal models of heart failure [168] and occurs before the development of clinical heart failure [169, 170].

Presently, all cardiac biomarkers of fibrosis present limitations mainly associated with specificity and accuracy. Collagen turnover is so dynamic—estimated to occur every 80–120 days in the normal myocardium and with much increased kinetics in the MI setting [171]—that not all circulating molecules proposed as biomarkers reflect the actual collagen metabolism at the cardiac level. Furthermore, the inconsistencies reported in the acquisition of Gal-3 levels from plasma and serum cannot be dismissed. Gal-3 may be an ideal biomarker of cardiac fibrosis, but methods for quantification need to be improved. Knowledge about fibrosis in the diseased heart will provide insight into the status and prognosis of a patient. New, better biomarkers of fibrosis are needed to improve diagnosis, prognosis, and therapy. Such biomarkers need to both address fibrous tissue formation and degradation in cardiac remodeling.

22.4 Cardiac Fibroblasts

Fibroblasts synthesize several ECM proteins, including collagen, and have a pivotal role in the formation of the structural framework of most animal organs. Fibroblasts are identified by their spindle-shaped morphology and lack of cell-specific markers that reflect their multiple cell lineages [172]. While fibroblasts remain poorly defined in molecular terms, their functional importance is unmistakable. For many years, studies on cardiac function and cardiac diseases have mainly been focused on the cardiomyocyte and inflammatory cells. More recently, the cardiac fibroblast has been acknowledged as an extremely important cardiac cell with pivotal roles in cardiac physiology and pathology.

22.4.1 Configuration and Organization in the Heart

Cardiomyocytes embody approximately 75% of normal myocardial volume yet constitute only 30–40% of cell numbers [173, 174]. The remaining non-myocyte cells are predominantly fibroblasts. Cardiac fibroblasts surround myocytes and bridge the interstitial space between myocardial tissue layers; as a result, every cardiomyocyte is spatially close to a fibroblast in normal cardiac tissue [106]. Under physiological conditions, cardiac fibroblasts provide a mechanical scaffold for cardiomyocytes and coordinate pump function of the heart [175].

Cardiac fibroblasts are the primary cell type responsible for homeostatic maintenance of ECM in the normal heart. The major components of cardiac ECM are the fibrillar collagens I (80%) and III (10%) [176], with smaller amounts of collagens IV, V, VI, elastin, laminin, proteoglycans, and glycosaminoglycans [177, 178].

22.4.2 Myofibroblast

Myofibroblasts have been described as a sub-phenotype of activated fibroblasts in diseased hearts [179]. Myofibroblasts are identified by expression of contractile proteins, including α-smooth muscle actin, vimentin, and desmin [176]. Myofibroblast apoptosis has been linked to progression of the granulation tissue into a mature scar, whereas failure of myofibroblast apoptosis has been suggested to drive the progression to fibrosis [180]. At the same time, Blankesteijn and colleagues proposed that the continued presence of myofibroblasts in the infarct scar is necessary to maintain the ECM environment [181]. Cardiac myofibroblasts are particularly receptive to stimulation by proinflammatory cytokines and hormones, the levels of which are enhanced in the remodeling heart [182]. Fibroblasts respond to these stimuli by changing proliferative and migratory rates, altering secretion of ECM proteins and MMP synthesis and modulating the synthesis of cytokines (such as TNF-α, IL-1, IL-6, TGF-β) and growth factors (e.g., vascular endothelial growth factor) [182]. For the purposes of this book chapter, in the context of fibrosis post-MI, activated cardiac fibroblasts/myofibroblasts will simply be referred to as cardiac fibroblasts.

22.4.3 Function Post-MI: Remodeling and Fibrosis

Cardiac fibroblasts are activated in response to MI and have a key role in the wound healing response. The myocardium healing comprises a multitude of temporal and spatial tightly regulated steps, including hemostasis, infiltration of immune and inflammatory cells, degradation and phagocytosis of necrotic cardiomyocytes and cellular debris, proliferation of cardiac fibroblasts within the injury area, synthesis and reconstruction of granulation tissue, and subsequent ECM remodeling to produce a mature scar [176].

While collagen deposition in adult hearts is normally low, in diseased hearts collagen deposition is robustly increased [183] causing myocardial stiffening. The levels of fibrotic tissue are dependent on the proliferative potential of fibroblasts and on synthesis and deposition of ECM proteins [184]. ECM degradation, resulting from increased MMP expression, dominates the initial phase of the injury response, whereas net ECM deposition, arising from enhanced collagen synthesis, dominates the later phase of healing. The mechanisms regulating the resolution of acute injury responses versus transition to chronic activation of cardiac fibroblasts ultimately ending in heart failure are not well understood.

Fibroblasts are possibly the most underestimated cell population in the heart. They contribute to cardiac development, structure, and function. It is important to note that in addition to being the main producers of ECM proteins in the heart, fibroblasts have been named "sentinel cells" that function as local immune modulators [185, 186]. The cardiac fibroblast has also been reported to contribute to cardiac electrophysiology [187]. Cardiac fibroblasts are sources and targets of biochemical and electromechanical signaling pathways, making these cells excellent candidates for target therapies for the diseased heart.

22.5 Therapeutic Targets

Several studies demonstrate the importance of biomarkers associated with various phenotypes associated with a high risk of disease [15–17, 146]. The discovery and development of biomarkers has shed light on the pathogenesis and management of MI [188–190] and has the potential to play a major role in early diagnosis, disease prevention, drug target identification, drug response, etc. In this section, we discuss recognized and emerging drug candidates acting on a preselected therapeutic target.

22.5.1 Renin–Angiotensin–Aldosterone Inhibitors

LV remodeling post-MI is believed to be associated with prolonged angiotensin stimulation [191]. In order to prevent the harmful effects of the renin–angiotensin–aldosterone system, three distinct paths of drug development have been developed: inhibitors of the angiotensin-converting enzyme (ACE), angiotensin receptor blockers, and aldosterone antagonists.

Myocardial levels of angiotensin II (Ang II), a potent vasoconstrictor, are elevated in a number of pathologies characterized by myocardial remodeling [192]. Although Ang II can be regulated through two different receptor subtypes (AT1R and AT2R), *in vitro* cardiac fibroblasts express predominantly the AT1R. In both animal [193, 194] and human models [195, 196], Ang II was shown to induce fibrosis in the myocardium via multiple mechanisms, including enhanced ECM protein synthesis [197, 198], decreased MMP activity [199, 200], and increased TIMP activity [199, 201]. Sun and Weber have reported enhanced angiotensin production, upregulation of AT1 receptors, and increased collagen mRNA in myofibroblasts associated with healing infarct scars [202].

Clinical trials have shown that ACE inhibitors reduce morbidity and mortality, reduce the progression of established heart failure [203], lessen cardiovascular events in patients at risk but without symptomatic heart failure [204], and reduce cardiac fibrosis and improves LV diastolic function [205]. Angiotensin AT1 receptor blockers offer clinical benefits comparable to ACE inhibitors in heart failure therapy [206–208]. Aldosterone, a mineralocorticoid, has also been heavily implicated in the myocardial fibrogenic response upon stimulation with Ang II or through Ang II-independent mechanisms [209].

22.5.2 Statins

Statins are inhibitors of 3-hydroxy-3-methylglutaryl coenzyme A (HMG-CoA) reductase, the rate-limiting enzyme in the cholesterol synthesis pathway [182]. The pleiotropic anti-inflammatory properties of statins make them drugs of interest for targeting the inflammatory component in patients with advanced heart failure. Additionally, statins have unfailingly been shown to inhibit proliferation of cardiac fibroblasts, an attractive feature when targeting fibrosis. Animals treated with statins exhibit reduced LV remodeling, fibrosis, and collagen synthesis post-MI [210]. In heart failure patients undergoing statin treatment, levels of proinflammatory cytokines were reduced in plasma [211, 212], and clinical studies suggest that statins can decrease local expression of proinflammatory cytokines in the myocardium [213].

22.5.3 Cytokine Therapies

The complexity of cytokine signaling networks has posed a challenge for drug development. Initial efforts with anti-cytokine therapy in heart failure focused on TNF-α [4, 214, 215]. Etanercept is a recombinant human TNF receptor that binds to soluble TNF. Phase I safety studies showed that biweekly injections of etanercept for 3 months, in patients with advanced heart failure, resulted in a significant dose-dependent improvement in LV function and remodeling [216].

Interleukins, such as IL-6 and IL-11, have also been used as possible therapeutic targets in cardiovascular disease. Cardiotrophin-1 (CT-1) is a member of a family of cytokines including IL-6, which is capable of inducing cardiomyocyte hypertrophy in vitro [217]. CT-1 also has significant protective roles. CT-1 can protect cardiomyocytes when added either prior to simulated ischemia or at the time of reoxygenation following simulated ischemia [218]. In a murine model of MI, IL-11 treatment after coronary ligation attenuated cardiac fibrosis [219], and pretreatment with IL-11 reduced infarct area in a murine ischemia/reperfusion (I/R) model [220].

22.5.4 *MMP Inhibitors*

Whereas only a small number of studies have explored inhibitors of collagen biosynthesis to mitigate cardiac fibrosis [178, 221], numerous research has focused on MMP inhibitors' (MMPI) effects on pathologic remodeling. Rohde *et al.* showed that a nonselective MMPI decreased LV dilatation post-MI in mice [122]. In multiple animal models, MMP-9 levels increase early post-MI, and MMP-9 deletion in mice attenuates remodeling [73]. The effects of MMPI in progression to heart failure have also been examined. A porcine model of congestive heart failure showed attenuated LV remodeling and preserved cardiac function when treated with a nonselective MMPI, compared to non-treated controls [121]. Similar results were obtained in a study of chronic human heart failure [222].

22.5.5 *Nuclear Receptor Agonists*

The role of ligands of the peroxisome proliferator-activated receptors (PPAR), PPARα, PPARδ, and PPARγ, in LV remodeling post-MI is receiving widespread attention for their actions on the myocardium [223]. PPARs are nuclear receptors that regulate lipid storage and metabolism, and are expressed by multiple cell types in the cardiovascular system, including cardiomyocytes and fibroblasts [176]. Recent evidence suggests that PPARs may attenuate inflammatory responses, therefore influencing cardiac remodeling [224]. This function is mainly mediated through the ability of agonist-activated PPARs to form complexes with other transcription factors, thus cross inhibiting their transcriptional activity [225]. In animal models of acute MI, PPARα and PPARγ ligands showed cardioprotective properties [226], attenuated fibrosis, preserved diastolic function, and inhibited inflammatory activation in pressure overload cardiac hypertrophy [227].

22.6 Conclusions and Future Directions

One third of total global deaths are attributed to CVD, and the majority of deaths are due to MI [228]. The field of biomarker discovery holds great potential for improving clinical diagnosis by identifying individuals who are at risk, determining the severity of disease, and developing personalized treatment for patients diagnosed with MI. Advances in our understanding of the pathogenesis and consequences of MI have initiated the development of new biomarkers as well as expanded the role of biomarkers in classification and individualization of treatment [229]. Physicians still depend on invasive techniques for the diagnosis and treatment of MI. Recent advances in the development of gene arrays in addition to proteomic technologies have produced multiple new biomarkers providing hope for an increase in the number of new diagnostic markers. The future of biomarker use for the treatment of MI will depend on the production of cost-effective treatments, multimarker profiling to accurately stage progression of molecular disease, and the availability of simple diagnostic kits for both physicians and patients [10, 229–231].

An ideal biomarker must be accurate, fast, and more cost-effective than any current technique [230]. Because of limited healthcare budgets, only biomarkers with exceptional performance will find utility in primary care settings making a noninvasive procedure at a low cost an advantage [10, 232]. This is especially true of many current diagnostic standards, where cost often exceed $1,000 per test, and procedures are partially invasive and risk associated [232]. Developing biomarkers that are cost-effective will most likely improve health outcomes.

Development of simple tests that accurately and reliably predict the development of heart failure following MI should be a goal of biomarker research. In diseases where early diagnosis is key for survival, these tests can be vital in improving patient heath. To reach this goal, sensitivity and specificity are major issues that have to be improved. Developing diagnostic and screening tests that look at a range of biomarkers rather than relying on only one biomarker is a likely potential solution [229].

With the recent advances in technology, the use of multimarker profiling will be beneficial in the development of individualized treatment schemes for post-MI patients [10]. Although a multimarker approach has shown potential for comprehensive risk assessment, several limitations must be recognized. The ultimate test of a multimarker paradigm will be its value in therapeutic decision-making [229]. For example, the ability of a biomarker to predict relative risk and specific end points differ. The B-type natriuretic peptide (BNP) has been shown to be a potent predictor of mortality risk, but BNP exhibits a weak association with recurrent ischemic events [231].

The discovery of new potential biomarkers shows great potential for treatment post-MI. Circulating factors related to inflammatory and fibrotic responses are key predictors of early ECM changes in the structure and function of the heart providing important prognostic information for MI [15, 22]. Improvement on cost-effectiveness as well as multianalyte processing will move us in the right direction so that in the future biomarkers will not only be available to treat, diagnose, or predict recurrence but also will be able to guide physicians in decision-making. Biomarkers may one day be available in over-the-counter testing [10].

References

1. Anderson L (2005) Candidate-based proteomics in the search for biomarkers of cardiovascular disease. J Physiol 563:23–60
2. Roger VL, Go AS, Lloyd-Jones DM et al (2012) Heart disease and stroke statistics–2012 update: a report from the American Heart Association. Circulation 125:e2–e220
3. Entman ML, Youker K, Shappell SB et al (1990) Neutrophil adherence to isolated adult canine myocytes. Evidence for a CD18-dependent mechanism. J Clin Invest 85:1497–1506
4. Sivasubramanian N, Coker ML, Kurrelmeyer KM et al (2001) Left ventricular remodeling in transgenic mice with cardiac restricted overexpression of tumor necrosis factor. Circulation 104:826–831
5. Cuccurullo L, Accardo M, Agozzino L et al (2006) Ultrastructural pathology of pediatric myocardium in acute ischemia: bioptic study before and after treatment with cardioplegic solution. Ultrastruct Pathol 30:453–460
6. Elsman P, Zijlstra F, Miedema K et al (2004) The predictive value of cumulative lactate dehydrogenase release within the first 72 h of acute myocardial infarction in patients treated with primary angioplasty. Ann Clin Biochem 41:142–148
7. Puleo PR, Guadagno PA, Roberts R, Perryman MB (1989) Sensitive, rapid assay of subforms of creatine kinase MB in plasma. Clin Chem 35:1452–1455
8. Huseby NE, Grostad M, Almdahl SM, Vaage J (1990) Activities of mitochondrial aspartate aminotransferase and creatine kinase isoenzyme MB in serum following coronary bypass surgery. Clin Chim Acta 195:1–7
9. Marko-Varga G, Lindberg H, Lofdahl CG et al (2005) Discovery of biomarker candidates within disease by protein profiling: principles and concepts. J Proteome Res 4:1200–1212
10. Vasan RS (2006) Biomarkers of cardiovascular disease: molecular basis and practical considerations. Circulation 113:2335–2362
11. Liu SQ, Tefft BJ, Zhang D et al (2011) Cardioprotective mechanisms activated in response to myocardial ischemia. Mol Cell Biomech 8:319–338
12. Atkinson AJ Jr, Colburn WA, DeGruttola VG et al (2001) Biomarkers and surrogate endpoints: preferred definitions and conceptual framework. Clin Pharmacol Ther 69:89–95
13. Reichlin T, Hochholzer W, Bassetti S et al (2009) Early diagnosis of myocardial infarction with sensitive cardiac troponin assays. N Engl J Med 361:858–867
14. Ridker PM (2007) Inflammatory biomarkers and risks of myocardial infarction, stroke, diabetes, and total mortality: implications for longevity. Nutr Rev 65:S253–S259
15. Radauceanu A, Ducki C, Virion JM et al (2008) Extracellular matrix turnover and inflammatory markers independently predict functional status and outcome in chronic heart failure. J Card Fail 14:467–474
16. Klappacher G, Franzen P, Haab D et al (1995) Measuring extracellular matrix turnover in the serum of patients with idiopathic or ischemic dilated cardiomyopathy and impact on diagnosis and prognosis. Am J Cardiol 75:913–918
17. Lok DJ, Van Der Meer P, De La Porte PW et al (2010) Prognostic value of galectin-3, a novel marker of fibrosis, in patients with chronic heart failure: data from the DEAL-HF study. Clin Res Cardiol 99:323–328
18. Zannad F, Rossignol P, Iraqi W (2010) Extracellular matrix fibrotic markers in heart failure. Heart Fail Rev 15:319–329
19. Berk BC, Fujiwara K, Lehoux S (2007) ECM remodeling in hypertensive heart disease. J Clin Invest 117:568–575

20. Weber KT, Sun Y, Tyagi SC, Cleutjens JP (1994) Collagen network of the myocardium: function, structural remodeling and regulatory mechanisms. J Mol Cell Cardiol 26:279–292
21. Brilla CG (1994) The cardiac structure-function relationship and the renin-angiotensin-aldosterone system in hypertension and heart failure. Curr Opin Cardiol 9(Suppl 1):S2–S10, discussion S10–11
22. Barasch E, Gottdiener JS, Aurigemma G et al (2009) Association between elevated fibrosis markers and heart failure in the elderly: the cardiovascular health study. Circ Heart Fail 2:303–310
23. Rauchhaus M, Doehner W, Francis DP et al (2000) Plasma cytokine parameters and mortality in patients with chronic heart failure. Circulation 102:3060–3067
24. Tsutamoto T, Hisanaga T, Wada A et al (1998) Interleukin-6 spillover in the peripheral circulation increases with the severity of heart failure, and the high plasma level of interleukin-6 is an important prognostic predictor in patients with congestive heart failure. J Am Coll Cardiol 31:391–398
25. Varga J, Jimenez SA (1986) Stimulation of normal human fibroblast collagen production and processing by transforming growth factor-beta. Biochem Biophys Res Commun 138: 974–980
26. Entman ML, Smith CW (1994) Postreperfusion inflammation: a model for reaction to injury in cardiovascular disease. Cardiovasc Res 28:1301–1311
27. Frangogiannis NG, Youker KA, Rossen RD et al (1998) Cytokines and the microcirculation in ischemia and reperfusion. J Mol Cell Cardiol 30:2567–2576
28. Youker KA, Birdsall HH, Frangogiannis NG et al (1997) Phagocytes in ischemia injury. Ann N Y Acad Sci 832:243–265
29. Zhang R, Brennan ML, Fu X et al (2001) Association between myeloperoxidase levels and risk of coronary artery disease. JAMA 286:2136–2142
30. Danesh J, Collins R, Appleby P, Peto R (1998) Association of fibrinogen, C-reactive protein, albumin, or leukocyte count with coronary heart disease: meta-analyses of prospective studies. JAMA 279:1477–1482
31. Ernst E, Hammerschmidt DE, Bagge U et al (1987) Leukocytes and the risk of ischemic diseases. JAMA 257:2318–2324
32. Grigg J, Kukielka GL, Berens KL et al (1994) Induction of intercellular adhesion molecule-1 by lipopolysaccharide in canine alveolar macrophages. Am J Respir Cell Mol Biol 11:304–311
33. Kukielka GL, Youker KA, Hawkins HK et al (1994) Regulation of ICAM-1 and IL-6 in myocardial ischemia: effect of reperfusion. Ann N Y Acad Sci 723:258–270
34. Abe Y, Kawakami M, Kuroki M et al (1993) Transient rise in serum interleukin-8 concentration during acute myocardial infarction. Br Heart J 70:132–134
35. Amsterdam EA, Stahl GL, Pan HL et al (1995) Limitation of reperfusion injury by a monoclonal antibody to C5a during myocardial infarction in pigs. Am J Physiol 268:H448–H457
36. Faurschou M, Borregaard N (2003) Neutrophil granules and secretory vesicles in inflammation. Microbes Infect 5:1317–1327
37. Borregaard N, Cowland JB (1997) Granules of the human neutrophilic polymorphonuclear leukocyte. Blood 89:3503–3521
38. Sengelov H, Follin P, Kjeldsen L et al (1995) Mobilization of granules and secretory vesicles during in vivo exudation of human neutrophils. J Immunol 154:4157–4165
39. Sengelov H, Kjeldsen L, Borregaard N (1993) Control of exocytosis in early neutrophil activation. J Immunol 150:1535–1543
40. Dahlgren C, Karlsson A, Sendo F (2001) Neutrophil secretory vesicles are the intracellular reservoir for GPI-80, a protein with adhesion-regulating potential. J Leukoc Biol 69:57–62
41. Tapper H, Karlsson A, Morgelin M et al (2002) Secretion of heparin-binding protein from human neutrophils is determined by its localization in azurophilic granules and secretory vesicles. Blood 99:1785–1793
42. Witko-Sarsat V, Cramer EM, Hieblot C et al (1999) Presence of proteinase 3 in secretory vesicles: evidence of a novel, highly mobilizable intracellular pool distinct from azurophil granules. Blood 94:2487–2496

43. Pereira HA (1995) CAP37, a neutrophil-derived multifunctional inflammatory mediator. J Leukoc Biol 57:805–812
44. Gautam N, Olofsson AM, Herwald H et al (2001) Heparin-binding protein (HBP/CAP37): a missing link in neutrophil-evoked alteration of vascular permeability. Nat Med 7: 1123–1127
45. Linder A, Christensson B, Herwald H et al (2009) Heparin-binding protein: an early marker of circulatory failure in sepsis. Clin Infect Dis 49:1044–1050
46. Ng LL, Khan SQ, Narayan H et al (2011) Proteinase 3 and prognosis of patients with acute myocardial infarction. Clin Sci (Lond) 120:231–238
47. Pendergraft WF 3rd, Rudolph EH, Falk RJ et al (2004) Proteinase 3 sidesteps caspases and cleaves p21(Waf1/Cip1/Sdi1) to induce endothelial cell apoptosis. Kidney Int 65:75–84
48. Ramaha A, Patston PA (2002) Release and degradation of angiotensin I and angiotensin II from angiotensinogen by neutrophil serine proteinases. Arch Biochem Biophys 397:77–83
49. Robache-Gallea S, Morand V, Bruneau JM et al (1995) In vitro processing of human tumor necrosis factor-alpha. J Biol Chem 270:23688–23692
50. Coeshott C, Ohnemus C, Pilyavskaya A et al (1999) Converting enzyme-independent release of tumor necrosis factor alpha and IL-1beta from a stimulated human monocytic cell line in the presence of activated neutrophils or purified proteinase 3. Proc Natl Acad Sci USA 96:6261–6266
51. Sugawara S, Uehara A, Nochi T et al (2001) Neutrophil proteinase 3-mediated induction of bioactive IL-18 secretion by human oral epithelial cells. J Immunol 167:6568–6575
52. Kao RC, Wehner NG, Skubitz KM et al (1988) Proteinase 3. A distinct human polymorphonuclear leukocyte proteinase that produces emphysema in hamsters. J Clin Invest 82: 1963–1973
53. Murav'ev RA, Fomina VA, Rogovin VV (2003) Gelatinase granules of the neutrophil granulocytes. Izv Akad Nauk Ser Biol 4:389–394
54. Mollinedo F, Nakajima M, Llorens A et al (1997) Major co-localization of the extracellular-matrix degradative enzymes heparanase and gelatinase in tertiary granules of human neutrophils. Biochem J 327(Pt 3):917–923
55. Lindsey M, Wedin K, Brown MD et al (2001) Matrix-dependent mechanism of neutrophil-mediated release and activation of matrix metalloproteinase 9 in myocardial ischemia/reperfusion. Circulation 103:2181–2187
56. Peppin GJ, Weiss SJ (1986) Activation of the endogenous metalloproteinase, gelatinase, by triggered human neutrophils. Proc Natl Acad Sci USA 83:4322–4326
57. Nielsen BS, Timshel S, Kjeldsen L et al (1996) 92 kDa type IV collagenase (MMP-9) is expressed in neutrophils and macrophages but not in malignant epithelial cells in human colon cancer. Int J Cancer 65:57–62
58. Legrand D, Elass E, Carpentier M, Mazurier J (2005) Lactoferrin: a modulator of immune and inflammatory responses. Cell Mol Life Sci 62:2549–2559
59. Haversen L, Ohlsson BG, Hahn-Zoric M et al (2002) Lactoferrin down-regulates the LPS-induced cytokine production in monocytic cells via NF-kappa B. Cell Immunol 220:83–95
60. Petschow BW, Talbott RD, Batema RP (1999) Ability of lactoferrin to promote the growth of Bifidobacterium spp. in vitro is independent of receptor binding capacity and iron saturation level. J Med Microbiol 48:541–549
61. Harmsen MC, Swart PJ, De Bethune MP et al (1995) Antiviral effects of plasma and milk proteins: lactoferrin shows potent activity against both human immunodeficiency virus and human cytomegalovirus replication in vitro. J Infect Dis 172:380–388
62. Oram JD, Reiter B (1968) Inhibition of bacteria by lactoferrin and other iron-chelating agents. Biochim Biophys Acta 170:351–365
63. Xu X, Hakansson L (2002) Degranulation of primary and secondary granules in adherent human neutrophils. Scand J Immunol 55:178–188
64. Baveye S, Elass E, Fernig DG et al (2000) Human lactoferrin interacts with soluble CD14 and inhibits expression of endothelial adhesion molecules, E-selectin and ICAM-1, induced by the CD14-lipopolysaccharide complex. Infect Immun 68:6519–6525

65. Halliwell B, Gutteridge JM (1990) The antioxidants of human extracellular fluids. Arch Biochem Biophys 280:1–8
66. Crouch SP, Slater KJ, Fletcher J (1992) Regulation of cytokine release from mononuclear cells by the iron-binding protein lactoferrin. Blood 80:235–240
67. Goetz DH, Holmes MA, Borregaard N et al (2002) The neutrophil lipocalin NGAL is a bacteriostatic agent that interferes with siderophore-mediated iron acquisition. Mol Cell 10:1033–1043
68. Nymo SH, Ueland T, Askevold ET et al (2011) The association between neutrophil gelatinase-associated lipocalin and clinical outcome in chronic heart failure: results from CORONA*. J Intern Med 271(5):436–443
69. Yan L, Borregaard N, Kjeldsen L, Moses MA (2001) The high molecular weight urinary matrix metalloproteinase (MMP) activity is a complex of gelatinase B/MMP-9 and neutrophil gelatinase-associated lipocalin (NGAL). Modulation of MMP-9 activity by NGAL. J Biol Chem 276:37258–37265
70. Sahinarslan A, Kocaman SA, Bas D et al (2011) Plasma neutrophil gelatinase-associated lipocalin levels in acute myocardial infarction and stable coronary artery disease. Coron Artery Dis 22:333–338
71. Yndestad A, Landro L, Ueland T et al (2009) Increased systemic and myocardial expression of neutrophil gelatinase-associated lipocalin in clinical and experimental heart failure. Eur Heart J 30:1229–1236
72. Heymans S, Luttun A, Nuyens D et al (1999) Inhibition of plasminogen activators or matrix metalloproteinases prevents cardiac rupture but impairs therapeutic angiogenesis and causes cardiac failure. Nat Med 5:1135–1142
73. Ducharme A, Frantz S, Aikawa M et al (2000) Targeted deletion of matrix metalloproteinase-9 attenuates left ventricular enlargement and collagen accumulation after experimental myocardial infarction. J Clin Invest 106:55–62
74. Lindsey ML, Escobar GP, Dobrucki LW et al (2006) Matrix metalloproteinase-9 gene deletion facilitates angiogenesis after myocardial infarction. Am J Physiol Heart Circ Physiol 290:H232–H239
75. Arnljots K, Sorensen O, Lollike K, Borregaard N (1998) Timing, targeting and sorting of azurophil granule proteins in human myeloid cells. Leukemia 12:1789–1795
76. Garwicz D, Lindmark A, Persson AM, Gullberg U (1998) On the role of the proform-conformation for processing and intracellular sorting of human cathepsin G. Blood 92:1415–1422
77. Lindmark A, Garwicz D, Rasmussen PB et al (1999) Characterization of the biosynthesis, processing, and sorting of human HBP/CAP37/azurocidin. J Leukoc Biol 66:634–643
78. Pham CT (2006) Neutrophil serine proteases: specific regulators of inflammation. Nat Rev Immunol 6:541–550
79. Joiner KA, Ganz T, Albert J, Rotrosen D (1989) The opsonizing ligand on Salmonella typhimurium influences incorporation of specific, but not azurophil, granule constituents into neutrophil phagosomes. J Cell Biol 109:2771–2782
80. Buffon A, Biasucci LM, Liuzzo G et al (2002) Widespread coronary inflammation in unstable angina. N Engl J Med 347:5–12
81. Baldus S, Heeschen C, Meinertz T et al (2003) Myeloperoxidase serum levels predict risk in patients with acute coronary syndromes. Circulation 108:1440–1445
82. Brennan ML, Penn MS, Van Lente F et al (2003) Prognostic value of myeloperoxidase in patients with chest pain. N Engl J Med 349:1595–1604
83. Askari AT, Brennan ML, Zhou X et al (2003) Myeloperoxidase and plasminogen activator inhibitor 1 play a central role in ventricular remodeling after myocardial infarction. J Exp Med 197:615–624
84. Owen CA, Campbell EJ (1999) The cell biology of leukocyte-mediated proteolysis. J Leukoc Biol 65:137–150
85. Massberg S, Grahl L, Von Bruehl ML et al (2010) Reciprocal coupling of coagulation and innate immunity via neutrophil serine proteases. Nat Med 16:887–896
86. Bidouard JP, Duval N, Kapui Z et al (2003) SSR69071, an elastase inhibitor, reduces myocardial infarct size following ischemia-reperfusion injury. Eur J Pharmacol 461:49–52

87. Itoh Y, Nagase H (1995) Preferential inactivation of tissue inhibitor of metalloproteinases-1 that is bound to the precursor of matrix metalloproteinase 9 (progelatinase B) by human neutrophil elastase. J Biol Chem 270:16518–16521
88. Reilly CF, Schechter NB, Travis J (1985) Inactivation of bradykinin and kallidin by cathepsin G and mast cell chymase. Biochem Biophys Res Commun 127:443–449
89. Bank U, Kupper B, Reinhold D et al (1999) Evidence for a crucial role of neutrophil-derived serine proteases in the inactivation of interleukin-6 at sites of inflammation. FEBS Lett 461:235–240
90. Larosa CA, Rohrer MJ, Benoit SE et al (1994) Human neutrophil cathepsin G is a potent platelet activator. J Vasc Surg 19:306–318, discussion 318–309
91. Dos Santos L, Santos AA, Goncalves GA et al (2010) Bone marrow cell therapy prevents infarct expansion and improves border zone remodeling after coronary occlusion in rats. Int J Cardiol 145:34–39
92. Nahrendorf M, Swirski FK, Aikawa E et al (2007) The healing myocardium sequentially mobilizes two monocyte subsets with divergent and complementary functions. J Exp Med 204:3037–3047
93. Lambert JM, Lopez EF, Lindsey ML (2008) Macrophage roles following myocardial infarction. Int J Cardiol 130:147–158
94. Mosser DM, Edwards JP (2008) Exploring the full spectrum of macrophage activation. Nat Rev Immunol 8:958–969
95. Troidl C, Mollmann H, Nef H et al (2009) Classically and alternatively activated macrophages contribute to tissue remodelling after myocardial infarction. J Cell Mol Med 13:3485–3496
96. Hu Y, Zhang H, Lu Y et al (2011) Class A scavenger receptor attenuates myocardial infarction-induced cardiomyocyte necrosis through suppressing M1 macrophage subset polarization. Basic Res Cardiol 106:1311–1328
97. Leuschner F, Rauch PJ, Ueno T et al (2012) Rapid monocyte kinetics in acute myocardial infarction are sustained by extramedullary monocytopoiesis. J Exp Med 209:123–137
98. Yang Z, Zingarelli B, Szabo C (2000) Crucial role of endogenous interleukin-10 production in myocardial ischemia/reperfusion injury. Circulation 101:1019–1026
99. Timmers L, Sluijter JP, Van Keulen JK et al (2008) Toll-like receptor 4 mediates maladaptive left ventricular remodeling and impairs cardiac function after myocardial infarction. Circ Res 102:257–264
100. Chinetti-Gbaguidi G, Staels B (2011) Macrophage polarization in metabolic disorders: functions and regulation. Curr Opin Lipidol 22:365–372
101. Gordon S (2003) Alternative activation of macrophages. Nat Rev Immunol 3:23–35
102. Martinez FO, Helming L, Gordon S (2009) Alternative activation of macrophages: an immunologic functional perspective. Annu Rev Immunol 27:451–483
103. Song E, Ouyang N, Horbelt M et al (2000) Influence of alternatively and classically activated macrophages on fibrogenic activities of human fibroblasts. Cell Immunol 204:19–28
104. Gordon S, Martinez FO (2010) Alternative activation of macrophages: mechanism and functions. Immunity 32:593–604
105. Zhang X, Mosser DM (2008) Macrophage activation by endogenous danger signals. J Pathol 214:161–178
106. Camelliti P, Borg TK, Kohl P (2005) Structural and functional characterisation of cardiac fibroblasts. Cardiovasc Res 65:40–51
107. Coker ML, Jolly JR, Joffs C et al (2001) Matrix metalloproteinase expression and activity in isolated myocytes after neurohormonal stimulation. Am J Physiol Heart Circ Physiol 281:H543–H551
108. Lindsey ML, Escobar GP, Mukherjee R et al (2006) Matrix metalloproteinase-7 affects connexin-43 levels, electrical conduction, and survival after myocardial infarction. Circulation 113:2919–2928

109. Lindsey ML, Zamilpa R (2012) Temporal and spatial expression of matrix metalloproteinases and tissue inhibitors of metalloproteinases following myocardial infarction. Cardiovasc Ther 30:31–41
110. Prudova A, Auf Dem Keller U, Butler GS, Overall CM (2010) Multiplex N-terminome analysis of MMP-2 and MMP-9 substrate degradomes by iTRAQ-TAILS quantitative proteomics. Mol Cell Proteomics 9:894–911
111. Bannikov GA, Karelina TV, Collier IE et al (2002) Substrate binding of gelatinase B induces its enzymatic activity in the presence of intact propeptide. J Biol Chem 277(18):16022–16027, M110931200
112. Gerszten RE, Asnani A, Carr SA (2011) Status and prospects for discovery and verification of new biomarkers of cardiovascular disease by proteomics. Circ Res 109:463–474
113. Didangelos A, Yin X, Mandal K et al (2010) Proteomics characterization of extracellular space components in the human aorta. Mol Cell Proteomics 9:2048–2062
114. Barallobre-Barreiro J, Didangelos A, Schoendube FA et al (2012) Proteomics analysis of cardiac extracellular matrix remodeling in a porcine model of ischemia/reperfusion injury. Circulation 125:789–802
115. Zamilpa R, Lopez EF, Chiao YA et al (2010) Proteomic analysis identifies in vivo candidate matrix metalloproteinase-9 substrates in the left ventricle post-myocardial infarction. Proteomics 10:2214–2223
116. Tao ZY, Cavasin MA, Yang F et al (2004) Temporal changes in matrix metalloproteinase expression and inflammatory response associated with cardiac rupture after myocardial infarction in mice. Life Sci 74:1561–1572
117. Hudson MP, Armstrong PW, Ruzyllo W et al (2006) Effects of selective matrix metalloproteinase inhibitor (PG-116800) to prevent ventricular remodeling after myocardial infarction: results of the PREMIER (Prevention of Myocardial Infarction Early Remodeling) trial. J Am Coll Cardiol 48:15–20
118. Ramani R, Mathier M, Wang P et al (2004) Inhibition of tumor necrosis factor receptor-1-mediated pathways has beneficial effects in a murine model of postischemic remodeling. Am J Physiol Heart Circ Physiol 287:H1369–H1377
119. Mukherjee R, Colbath GP, Justus CD et al (2010) Spatiotemporal induction of matrix metalloproteinase-9 transcription after discrete myocardial injury. FASEB J 24:3819–3828
120. Chancey AL, Brower GL, Peterson JT, Janicki JS (2002) Effects of matrix metalloproteinase inhibition on ventricular remodeling due to volume overload. Circulation 105:1983–1988
121. Spinale FG, Coker ML, Krombach SR et al (1999) Matrix metalloproteinase inhibition during the development of congestive heart failure: effects on left ventricular dimensions and function. Circ Res 85:364–376
122. Rohde LE, Ducharme A, Arroyo LH et al (1999) Matrix metalloproteinase inhibition attenuates early left ventricular enlargement after experimental myocardial infarction in mice. Circulation 99:3063–3070
123. Mukherjee R, Brinsa TA, Dowdy KB et al (2003) Myocardial infarct expansion and matrix metalloproteinase inhibition. Circulation 107:618–625
124. Blankenberg S, Rupprecht HJ, Poirier O et al (2003) Plasma concentrations and genetic variation of matrix metalloproteinase 9 and prognosis of patients with cardiovascular disease. Circulation 107:1579–1585
125. Squire IB, Evans J, Ng LL et al (2004) Plasma MMP-9 and MMP-2 following acute myocardial infarction in man: correlation with echocardiographic and neurohumoral parameters of left ventricular dysfunction. J Card Fail 10:328–333
126. Sundstrom J, Evans JC, Benjamin EJ et al (2004) Relations of plasma matrix metalloproteinase-9 to clinical cardiovascular risk factors and echocardiographic left ventricular measures: the Framingham Heart Study. Circulation 109:2850–2856
127. Yan AT, Yan RT, Spinale FG et al (2006) Plasma matrix metalloproteinase-9 level is correlated with left ventricular volumes and ejection fraction in patients with heart failure. J Card Fail 12:514–519

128. Hlatky MA, Ashley E, Quertermous T et al (2007) Matrix metalloproteinase circulating levels, genetic polymorphisms, and susceptibility to acute myocardial infarction among patients with coronary artery disease. Am Heart J 154:1043–1051
129. Martos R, Baugh J, Ledwidge M et al (2007) Diastolic heart failure: evidence of increased myocardial collagen turnover linked to diastolic dysfunction. Circulation 115:888–895
130. Diez J (2007) Mechanisms of cardiac fibrosis in hypertension. J Clin Hypertens (Greenwich) 9:546–550
131. Van Den Borne SW, Diez J, Blankesteijn WM et al (2010) Myocardial remodeling after infarction: the role of myofibroblasts. Nat Rev Cardiol 7:30–37
132. Sugihara N, Genda A, Shimizu M et al (1988) Diastolic dysfunction and its relation to myocardial fibrosis in essential hypertension. J Cardiol 18:353–361
133. Mclenachan JM, Dargie HJ (1990) Ventricular arrhythmias in hypertensive left ventricular hypertrophy. Relationship to coronary artery disease, left ventricular dysfunction, and myocardial fibrosis. Am J Hypertens 3:735–740
134. Janicki JS, Brower GL (2002) The role of myocardial fibrillar collagen in ventricular remodeling and function. J Card Fail 8:S319–S325
135. Wynn TA (2008) Cellular and molecular mechanisms of fibrosis. J Pathol 214:199–210
136. Lopez B, Gonzalez A, Varo N et al (2001) Biochemical assessment of myocardial fibrosis in hypertensive heart disease. Hypertension 38:1222–1226
137. Weber KT (1997) Monitoring tissue repair and fibrosis from a distance. Circulation 96:2488–2492
138. Bradshaw AD (2009) The role of SPARC in extracellular matrix assembly. J Cell Commun Signal 3:239–246
139. Nimmi ME (1993) Extracellular Matrix. In: Zern MA, Reid LM (eds) Fibrillar collagens: their biosynthesis, molecular structure, and mode of assembly, Marcel Decker, New York, NY, pp 121–148
140. Takino T, Nakamura M, Hiramori K (1999) Circulating levels of carboxyterminal propeptide of type I procollagen and left ventricular remodeling after myocardial infarction. Cardiology 91:81–86
141. Diez J, Panizo A, Gil MJ et al (1996) Serum markers of collagen type I metabolism in spontaneously hypertensive rats. Circulation 93:1026–1032
142. Radovan J, Vaclav P, Petr W et al (2006) Changes of collagen metabolism predict the left ventricular remodeling after myocardial infarction. Mol Cell Biochem 293:71–78
143. Risteli L, Risteli J (2003) Appendix II: Extracellular Collagen Metabolites in Body Fluids, in Connective Tissue and Its Heritable Disorders: Molecular, Genetic, and Medical Aspects, Second Edition (eds P. M. Royce and B. Steinmann), John Wiley & Sons, Inc., Hoboken, NJ, USA
144. Risteli J, Risteli L (1995) Analysing connective tissue metabolites in human serum. Biochemical, physiological and methodological aspects. J Hepatol 22:77–81
145. Lombardi R, Betocchi S, Losi MA et al (2003) Myocardial collagen turnover in hypertrophic cardiomyopathy. Circulation 108:1455–1460
146. Alla F, Kearney-Schwartz A, Radauceanu A et al (2006) Early changes in serum markers of cardiac extra-cellular matrix turnover in patients with uncomplicated hypertension and type II diabetes. Eur J Heart Fail 8:147–153
147. Risteli L, Risteli J (1990) Non-invasive methods for detection of organ fibrosis. In: Rojkind M (ed) Focus on Connective Tissue in Health and Disease. CRC, Boca Raton, FL, pp 61–68
148. Nakahara T, Takata Y, Hirayama Y et al (2007) Left ventricular hypertrophy and geometry in untreated essential hypertension is associated with blood levels of aldosterone and procollagen type III amino-terminal peptide. Circ J 71:716–721
149. Sato Y, Kataoka K, Matsumori A et al (1997) Measuring serum aminoterminal type III procollagen peptide, 7S domain of type IV collagen, and cardiac troponin T in patients with idiopathic dilated cardiomyopathy and secondary cardiomyopathy. Heart 78:505–508

150. Zannad F, Alla F, Dousset B et al (2000) Limitation of excessive extracellular matrix turnover may contribute to survival benefit of spironolactone therapy in patients with congestive heart failure: insights from the randomized aldactone evaluation study (RALES). Rales Investigators. Circulation 102:2700–2706
151. Cicoira M, Rossi A, Bonapace S et al (2004) Independent and additional prognostic value of aminoterminal propeptide of type III procollagen circulating levels in patients with chronic heart failure. J Card Fail 10:403–411
152. Janicki JS (1995) Collagen degradation in the heart. In: Eghbali-Webb M (ed) Molecular biology of collagen matrix in the heart. RG Landes, Austin, TX, pp 61–76
153. Barthelemy O, Beygui F, Vicaut E et al (2009) Relation of high concentrations of plasma carboxy-terminal telopeptide of collagen type I with outcome in acute myocardial infarction. Am J Cardiol 104:904–909
154. Plaksej R, Kosmala W, Frantz S et al (2009) Relation of circulating markers of fibrosis and progression of left and right ventricular dysfunction in hypertensive patients with heart failure. J Hypertens 27:2483–2491
155. Spinale FG, Coker ML, Bond BR, Zellner JL (2000) Myocardial matrix degradation and metalloproteinase activation in the failing heart: a potential therapeutic target. Cardiovasc Res 46:225–238
156. Foronjy RF, Sun J, Lemaitre V, D'armiento JM (2008) Transgenic expression of matrix metalloproteinase-1 inhibits myocardial fibrosis and prevents the transition to heart failure in a pressure overload mouse model. Hypertens Res 31:725–735
157. Visse R, Nagase H (2003) Matrix metalloproteinases and tissue inhibitors of metalloproteinases: structure, function, and biochemistry. Circ Res 92:827–839
158. Lindsey ML, Goshorn DK, Squires CE et al (2005) Age-dependent changes in myocardial matrix metalloproteinase/tissue inhibitor of metalloproteinase profiles and fibroblast function. Cardiovasc Res 66:410–419
159. Romanic AM, Burns-Kurtis CL, Gout B et al (2001) Matrix metalloproteinase expression in cardiac myocytes following myocardial infarction in the rabbit. Life Sci 68:799–814
160. Tyagi SC, Kumar S, Glover G (1995) Induction of tissue inhibitor and matrix metalloproteinase by serum in human heart-derived fibroblast and endomyocardial endothelial cells. J Cell Biochem 58:360–371
161. Cavusoglu E, Ruwende C, Chopra V et al (2006) Tissue inhibitor of metalloproteinase-1 (TIMP-1) is an independent predictor of all-cause mortality, cardiac mortality, and myocardial infarction. Am Heart J 151:1101 e1101–1108
162. Lopez B, Gonzalez A, Querejeta R et al (2006) Alterations in the pattern of collagen deposition may contribute to the deterioration of systolic function in hypertensive patients with heart failure. J Am Coll Cardiol 48:89–96
163. Ahmed SH, Clark LL, Pennington WR et al (2006) Matrix metalloproteinases/tissue inhibitors of metalloproteinases: relationship between changes in proteolytic determinants of matrix composition and structural, functional, and clinical manifestations of hypertensive heart disease. Circulation 113:2089–2096
164. Creemers EE, Davis JN, Parkhurst AM et al (2003) Deficiency of TIMP-1 exacerbates LV remodeling after myocardial infarction in mice. Am J Physiol Heart Circ Physiol 284:H364–H371
165. Dumic J, Dabelic S, Flogel M (2006) Galectin-3: an open-ended story. Biochim Biophys Acta 1760:616–635
166. Sharma UC, Pokharel S, Van Brakel TJ et al (2004) Galectin-3 marks activated macrophages in failure-prone hypertrophied hearts and contributes to cardiac dysfunction. Circulation 110:3121–3128
167. Li X, Mikhalkova D, Gao E et al (2011) Myocardial injury after ischemia-reperfusion in mice deficient in Akt2 is associated with increased cardiac macrophage density. Am J Physiol Heart Circ Physiol 301:H1932–H1940

168. Liu YH, D'ambrosio M, Liao TD et al (2009) N-acetyl-seryl-aspartyl-lysyl-proline prevents cardiac remodeling and dysfunction induced by galectin-3, a mammalian adhesion/growth-regulatory lectin. Am J Physiol Heart Circ Physiol 296:H404–H412
169. Kamal FA, Watanabe K, Ma M et al (2011) A novel phenylpyridazinone, T-3999, reduces the progression of autoimmune myocarditis to dilated cardiomyopathy. Heart Vessels 26:81–90
170. Psarras S, Mavroidis M, Sanoudou D et al (2011) Regulation of adverse remodelling by osteopontin in a genetic heart failure model. Eur Heart J 33(15):1954–1963
171. Laurent GJ (1987) Dynamic state of collagen: pathways of collagen degradation in vivo and their possible role in regulation of collagen mass. Am J Physiol 252:C1–C9
172. Kalluri R, Zeisberg M (2006) Fibroblasts in cancer. Nat Rev Cancer 6:392–401
173. Vliegen HW, Van Der Laarse A, Cornelisse CJ, Eulderink F (1991) Myocardial changes in pressure overload-induced left ventricular hypertrophy. A study on tissue composition, polyploidization and multinucleation. Eur Heart J 12:488–494
174. Souders CA, Bowers SL, Baudino TA (2009) Cardiac fibroblast: the renaissance cell. Circ Res 105:1164–1176
175. Miragoli M, Gaudesius G, Rohr S (2006) Electrotonic modulation of cardiac impulse conduction by myofibroblasts. Circ Res 98:801–810
176. Brown RD, Ambler SK, Mitchell MD, Long CS (2005) The cardiac fibroblast: therapeutic target in myocardial remodeling and failure. Annu Rev Pharmacol Toxicol 45:657–687
177. Bosman FT, Stamenkovic I (2003) Functional structure and composition of the extracellular matrix. J Pathol 200:423–428
178. Jugdutt BI (2003) Remodeling of the myocardium and potential targets in the collagen degradation and synthesis pathways. Curr Drug Targets Cardiovasc Haematol Disord 3:1–30
179. Powell DW, Mifflin RC, Valentich JD et al (1999) Myofibroblasts. II. Intestinal subepithelial myofibroblasts. Am J Physiol 277:C183–C201
180. Zhang HY, Phan SH (1999) Inhibition of myofibroblast apoptosis by transforming growth factor beta(1). Am J Respir Cell Mol Biol 21:658–665
181. Daskalopoulos EP, Janssen BJ, Blankesteijn WM (2012) Myofibroblasts in the infarct area: concepts and challenges. Microsc Microanal 18:35–49
182. Porter KE, Turner NA (2009) Cardiac fibroblasts: at the heart of myocardial remodeling. Pharmacol Ther 123:255–278
183. Kusachi SN, Y (2003) Fibrogenesis, cellular and molecular basis. In: Razzaque MS (eds) Myocardial infarction and cardiac fibrogenesis, Kluwer Academic, New York, NY, pp 77–96
184. Mackenna D, Summerour SR, Villarreal FJ (2000) Role of mechanical factors in modulating cardiac fibroblast function and extracellular matrix synthesis. Cardiovasc Res 46:257–263
185. Smith RS, Smith TJ, Blieden TM, Phipps RP (1997) Fibroblasts as sentinel cells. Synthesis of chemokines and regulation of inflammation. Am J Pathol 151:317–322
186. Silzle T, Randolph GJ, Kreutz M, Kunz-Schughart LA (2004) The fibroblast: sentinel cell and local immune modulator in tumor tissue. Int J Cancer 108:173–180
187. Rudy Y (2004) Conductive bridges in cardiac tissue: a beneficial role or an arrhythmogenic substrate? Circ Res 94:709–711
188. Danesh J, Wheeler JG, Hirschfield GM et al (2004) C-reactive protein and other circulating markers of inflammation in the prediction of coronary heart disease. N Engl J Med 350:1387–1397
189. Ridker PM, Stampfer MJ, Rifai N (2001) Novel risk factors for systemic atherosclerosis: a comparison of C-reactive protein, fibrinogen, homocysteine, lipoprotein(a), and standard cholesterol screening as predictors of peripheral arterial disease. JAMA 285:2481–2485
190. Bernal-Mizrachi C, Gates AC, Weng S et al (2005) Vascular respiratory uncoupling increases blood pressure and atherosclerosis. Nature 435:502–506
191. Aplin M, Christensen GL, Schneider M et al (2007) Differential extracellular signal-regulated kinases 1 and 2 activation by the angiotensin type 1 receptor supports distinct phenotypes of cardiac myocytes. Basic Clin Pharmacol Toxicol 100:296–301

192. Swynghedauw B (1999) Molecular mechanisms of myocardial remodeling. Physiol Rev 79:215–262
193. Crabos M, Roth M, Hahn AW, Erne P (1994) Characterization of angiotensin II receptors in cultured adult rat cardiac fibroblasts. Coupling to signaling systems and gene expression. J Clin Invest 93:2372–2378
194. Zhou G, Kandala JC, Tyagi SC et al (1996) Effects of angiotensin II and aldosterone on collagen gene expression and protein turnover in cardiac fibroblasts. Mol Cell Biochem 154:171–178
195. Agocha A, Lee HW, Eghbali-Webb M (1997) Hypoxia regulates basal and induced DNA synthesis and collagen type I production in human cardiac fibroblasts: effects of transforming growth factor-beta1, thyroid hormone, angiotensin II and basic fibroblast growth factor. J Mol Cell Cardiol 29:2233–2244
196. Hafizi S, Wharton J, Morgan K et al (1998) Expression of functional angiotensin-converting enzyme and AT1 receptors in cultured human cardiac fibroblasts. Circulation 98:2553–2559
197. Lijnen PJ, Petrov VV, Fagard RH (2001) Angiotensin II-induced stimulation of collagen secretion and production in cardiac fibroblasts is mediated via angiotensin II subtype 1 receptors. J Renin Angiotensin Aldosterone Syst 2:117–122
198. Staufenberger S, Jacobs M, Brandstatter K et al (2001) Angiotensin II type 1 receptor regulation and differential trophic effects on rat cardiac myofibroblasts after acute myocardial infarction. J Cell Physiol 187:326–335
199. Lijnen P, Petrov V, Van Pelt J, Fagard R (2008) Inhibition of superoxide dismutase induces collagen production in cardiac fibroblasts. Am J Hypertens 21:1129–1136
200. Pan CH, Wen CH, Lin CS (2008) Interplay of angiotensin II and angiotensin(1–7) in the regulation of matrix metalloproteinases of human cardiocytes. Exp Physiol 93:599–612
201. Jiang XY, Gao GD, Du XJ et al (2007) The signalling of AT2 and the influence on the collagen metabolism of AT2 receptor in adult rat cardiac fibroblasts. Acta Cardiol 62:429–438
202. Sun Y, Weber KT (2000) Infarct scar: a dynamic tissue. Cardiovasc Res 46:250–256
203. Sleight P (2002) Angiotensin II and trials of cardiovascular outcomes. Am J Cardiol 89:11A–16A, discussion 16A–17A
204. Fox KM (2003) Efficacy of perindopril in reduction of cardiovascular events among patients with stable coronary artery disease: randomised, double-blind, placebo-controlled, multicentre trial (the EUROPA study). Lancet 362:782–788
205. Brilla CG, Funck RC, Rupp H (2000) Lisinopril-mediated regression of myocardial fibrosis in patients with hypertensive heart disease. Circulation 102:1388–1393
206. Dimopoulos K, Salukhe TV, Coats AJ et al (2004) Meta-analyses of mortality and morbidity effects of an angiotensin receptor blocker in patients with chronic heart failure already receiving an ACE inhibitor (alone or with a beta-blocker). Int J Cardiol 93:105–111
207. Pfeffer MA, Swedberg K, Granger CB et al (2003) Effects of candesartan on mortality and morbidity in patients with chronic heart failure: the CHARM-Overall programme. Lancet 362:759–766
208. Diez J, Querejeta R, Lopez B et al (2002) Losartan-dependent regression of myocardial fibrosis is associated with reduction of left ventricular chamber stiffness in hypertensive patients. Circulation 105:2512–2517
209. Stowasser M (2001) New perspectives on the role of aldosterone excess in cardiovascular disease. Clin Exp Pharmacol Physiol 28:783–791
210. Bauersachs J, Galuppo P, Fraccarollo D et al (2001) Improvement of left ventricular remodeling and function by hydroxymethylglutaryl coenzyme a reductase inhibition with cerivastatin in rats with heart failure after myocardial infarction. Circulation 104:982–985
211. Tousoulis D, Antoniades C, Vassiliadou C et al (2005) Effects of combined administration of low dose atorvastatin and vitamin E on inflammatory markers and endothelial function in patients with heart failure. Eur J Heart Fail 7:1126–1132
212. Sola S, Mir MQ, Lerakis S et al (2006) Atorvastatin improves left ventricular systolic function and serum markers of inflammation in nonischemic heart failure. J Am Coll Cardiol 47:332–337

213. Wallace CK, Stetson SJ, Kucuker SA et al (2005) Simvastatin decreases myocardial tumor necrosis factor alpha content in heart transplant recipients. J Heart Lung Transplant 24: 46–51
214. Oral H, Dorn GW 2nd, Mann DL (1997) Sphingosine mediates the immediate negative inotropic effects of tumor necrosis factor-alpha in the adult mammalian cardiac myocyte. J Biol Chem 272:4836–4842
215. Kalra D, Baumgarten G, Dibbs Z et al (2000) Nitric oxide provokes tumor necrosis factor-alpha expression in adult feline myocardium through a cGMP-dependent pathway. Circulation 102:1302–1307
216. Bozkurt B, Torre-Amione G, Warren MS et al (2001) Results of targeted anti-tumor necrosis factor therapy with etanercept (ENBREL) in patients with advanced heart failure. Circulation 103:1044–1047
217. Pennica D, King KL, Shaw KJ et al (1995) Expression cloning of cardiotrophin-1, a cytokine that induces cardiac myocyte hypertrophy. Proc Natl Acad Sci USA 92:1142–1146
218. Brar BK, Stephanou A, Liao ZH et al (2001) Cardiotrophin-1 can protect cardiac myocytes from injury when added both prior to simulated ischaemia and at reoxygenation. Cardiovasc Res 51:265–274
219. Obana M, Maeda M, Takeda K et al (2010) Therapeutic activation of signal transducer and activator of transcription 3 by interleukin-11 ameliorates cardiac fibrosis after myocardial infarction. Circulation 121:684–691
220. Kimura R, Maeda M, Arita A et al (2007) Identification of cardiac myocytes as the target of interleukin 11, a cardioprotective cytokine. Cytokine 38:107–115
221. Nwogu JI, Geenen D, Bean M et al (2001) Inhibition of collagen synthesis with prolyl 4-hydroxylase inhibitor improves left ventricular function and alters the pattern of left ventricular dilatation after myocardial infarction. Circulation 104:2216–2221
222. King MK, Coker ML, Goldberg A et al (2003) Selective matrix metalloproteinase inhibition with developing heart failure: effects on left ventricular function and structure. Circ Res 92:177–185
223. Francis GA, Annicotte JS, Auwerx J (2003) PPAR-alpha effects on the heart and other vascular tissues. Am J Physiol Heart Circ Physiol 285:H1–H9
224. Alvarez-Guardia D, Palomer X, Coll T et al (2011) PPARbeta/delta activation blocks lipid-induced inflammatory pathways in mouse heart and human cardiac cells. Biochim Biophys Acta 1811:59–67
225. Daynes RA, Jones DC (2002) Emerging roles of PPARs in inflammation and immunity. Nat Rev Immunol 2:748–759
226. Wayman NS, Hattori Y, Mcdonald MC et al (2002) Ligands of the peroxisome proliferator-activated receptors (PPAR-gamma and PPAR-alpha) reduce myocardial infarct size. FASEB J 16:1027–1040
227. Diep QN, Benkirane K, Amiri F et al (2004) PPAR alpha activator fenofibrate inhibits myocardial inflammation and fibrosis in angiotensin II-infused rats. J Mol Cell Cardiol 36:295–304
228. Mcmanus F, Freel EM, Connell JM (2007) Hypertension. Scott Med J 52:36–42, quiz 42, 54
229. Morrow DA, Braunwald E (2003) Future of biomarkers in acute coronary syndromes: moving toward a multimarker strategy. Circulation 108:250–252
230. Zhou H, Hewitt SM, Yuen PS, Star RA (2006) Acute kidney injury biomarkers—needs, present status, and future promise. Nephrol Self Assess Program 5:63–71
231. James SK, Lindahl B, Siegbahn A et al (2003) N-terminal pro-brain natriuretic peptide and other risk markers for the separate prediction of mortality and subsequent myocardial infarction in patients with unstable coronary artery disease: a Global Utilization of Strategies To Open occluded arteries (GUSTO)-IV substudy. Circulation 108:275–281
232. Good DM, Thongboonkerd V, Novak J et al (2007) Body fluid proteomics for biomarker discovery: lessons from the past hold the key to success in the future. J Proteome Res 6:4549–4555

Chapter 23
Oxidative Stress in Cardiac Repair and Remodeling: Molecular Pathways and Therapeutic Strategies

Yao Sun

Abstract Cardiac remodeling occurs in a variety of heart diseases, contributing to ventricular dysfunction. Following myocardial infarction (MI), cardiac repair and remodeling appear in both infarcted and non-infarcted myocardium. Factors regulating cardiac repair/remodeling at different stages following MI are under investigation. There is growing recognition and experimental evidence that oxidative stress mediated by reactive oxygen species (ROS) plays a role in the pathogeneses of myocardial repair/remodeling in various cardiac diseases. After acute MI, oxidative stress is developed in both infarcted and non-infarcted myocardium. Accumulating evidence has demonstrated that ROS participates in several aspects of cardiac repair/remodeling following infarction that includes cardiomyocyte apoptosis, inflammatory/fibrogenic responses, hypertrophy, and angiogenesis. The exact pathways on ROS-mediated myocardial remodeling are under investigation. The therapeutic potential of oxidative stress-directed drugs in myocardial remodeling following infarction has not been fully realized.

Keywords Oxidative stress • Myocardial infarction • Inflammation • Fibrosis • Hypertrophy • Angiogenesis

23.1 Introduction

Heart failure has emerged as a major health problem during the past two decades. It appears most commonly in patients with previous MI. Myocardial remodeling, which occurs in both infarcted and non-infarcted myocardium, contributes

---

Y. Sun, M.D., Ph.D. (✉)
Division of Cardiovascular Diseases, Department of Medicine, Health Science Center, University of Tennessee, 956 Court Ave. Rm B310, Memphis, TN 38163, USA
e-mail: yasun@uthsc.edu

B.I. Jugdutt and N.S. Dhalla (eds.), *Cardiac Remodeling: Molecular Mechanisms*, Advances in Biochemistry in Health and Disease 5, DOI 10.1007/978-1-4614-5930-9_23,

significantly to the development of heart failure [1–3]. Following MI, cardiac repair is associated with angiogenesis and inflammatory reaction, which are followed by scar formation at the site of infarction. Cardiac remodeling is represented as hypertrophy with interstitial fibrosis in non-infarcted myocardium. This leads to alterations in left ventricular size, shape, and wall thickness [4–6]. Angiogenesis and fibrous tissue that form at the site of cardiomyocyte loss preserve the structural integrity and are integral to the heart's recovery, while the structural remodeling of viable myocardium impairs tissue behavior. Multiple factors may, in fact, contribute to left ventricular remodeling at different stages post-MI. There is experimental evidence to suggest that oxidative stress mediated by ROS plays a role in the pathogeneses of myocardial repair/remodeling after MI, a hypothesis that has been earning growing recognition [7–11]. Oxidative stress results from an oxidant/antioxidant imbalance: an excess of oxidants relative to the antioxidant capacity. The heart with acute MI undergoes an increased ROS production as well as antioxidant deficit, first in the infarcted myocardium, followed by the non-infarcted myocardium. Experimental studies have also demonstrated that oxidative stress can induce most, if not all, of the changes that are thought to contribute to myocardial remodeling including angiogenesis, proinflammatory cytokine release, cardiomyocyte apoptosis [12], fibrogenesis [13], cell proliferation [14], angiogenesis [15], and hypertrophy [16]. Chronic antioxidant treatment suppresses cardiac oxidative stress; attenuates ventricular remodeling, partially preserving left ventricle function; and improves survival in rats or mice with experimental MI [17–19]. In this chapter, the potential relevance of oxidative stress on cardiac repair/remodeling in the infarcted heart will be discussed. The role of antioxidants in cardiac remodeling will be also discussed.

23.2 Occurrence of Cardiac Oxidative Stress Following Myocardial Infarction

Superoxide (O_2^-), hydroxyl (OH^-), and peroxynitrite ($ONOO^-$) are simple molecules characterized by the presence of unpaired electrons. ROS can be produced intracellularly through electron leakage from mitochondria during oxidative phosphorylation and through the activation of several cellular enzymes, including NADPH oxidase, xanthine oxidase, and nitric oxide synthase [20–22]. O_2^- can rapidly react with nitric oxide (NO) to form $ONOO^-$ or convert to H_2O_2 to form OH^- [20]. ROS in low concentrations serve as signaling molecules [23]. However, these agents elicit harmful effects when produced in excess [20]. The toxicity associated with the excessive production of these compounds is prevented by antioxidant defense systems that maintain a healthy cellular environment. Living cells have both enzymatic and nonenzymatic defense mechanisms to balance the multitude of oxidative challenges presented to them. The enzymatic subgroup includes superoxide dismutase (SOD), catalase, and glutathione peroxidase (GSHPx) [24–26]. The dismutation of O_2^- by SOD results in the generation of H_2O_2, which catalase further metabolizes into water and oxygen. The nonenzymatic group includes a variety of biologic molecules, such as vitamins E and C [27]. Oxidative stress occurs when ROS production is enhanced and/or antioxidant reserve is suppressed.

In the myocardium, as in other tissues, antioxidant enzymes protect cells by maintaining O_2^- and H_2O_2 at low levels. Following MI, oxidative stress is developed in both infarcted and non-infarcted myocardium. Singal and colleagues have shown the evidence of progressive decrease in SOD, catalase, and GSHPx activity as well as vitamin E levels in the rat infarcted heart, first in the infarcted myocardium and then followed by non-infarcted myocardium [7, 28]. Our study has shown reduced SOD gene and protein expression in the infarcted myocardium [29]. Cardiac glutathione levels are also decreased in patients with acute MI [10]. These observations raise the possibility that impaired antioxidant capacity contributes to oxidative stress in the infarcted heart. NADPH oxidase is a major source of O_2^- in the heart [30]. In the infarcted heart, NADPH oxidase expression (gp22^{phox} and gp91^{phox} subunits) is significantly increased in the infarcted myocardium [8, 29], with neutrophils and macrophages as the primary cells expressing the enzyme. These findings suggest that ROS production is also enhanced in the infarcted myocardium. Moreover, 3-nitrotyrosine, a marker of oxidative stress, is highly expressed in the inflammatory cells of the infarcted myocardium, supporting the occurrence of cardiac oxidative stress following MI [29]. Oxidative stress in non-infarcted myocardium is contributed by multiple sources. Increased mitochondrial production of ROS has been suggested in non-infarcted myocardium of mice as one of them [19]. Increased ROS levels in non-infarcted myocardium also reflect increased activity of intracellular oxidase complexes, such as NADPH oxidase, xanthine oxidase, and nitric oxide synthase [31]. In addition, reduced SOD levels were observed in the failing heart with infarction [11]. These observations indicate that the imbalance between ROS production and antioxidant defense capacity contributes to oxidative stress in non-infarcted myocardium.

23.3 Oxidative Stress and Cardiac Repair/Remodeling Following Myocardial Infarction

23.3.1 Oxidative Stress and Cardiomyocyte Apoptosis

Loss of cardiomyocytes is an important mechanism in the development of myocardial remodeling and cardiac failure [32]. Following MI, apoptotic cardiomyocyte death occurs in the infarcted myocardium as well as the surviving portions of the wall adjacent to and remote from the infarcted myocardium [33, 34]. However, the number of apoptotic cardiomyocytes is greater in the infarcted region than in the region away from infarction. The regulation of cardiomyocyte apoptosis involves multiple mechanisms. ROS have been proven to be one of the stimulators of cardiomyocyte apoptosis [35–39]. High levels of oxidative stress have been demonstrated to cause cell necrosis, while lower levels of oxidative stress can cause cell apoptosis. In vitro studies have shown that treatment with O_2^- or H_2O_2 in cardiomyocytes induces apoptosis [40]. In vivo studies have further demonstrated that oxidative stress triggers cardiomyocyte apoptosis in several cardiovascular diseases, including MI, ischemia/reperfusion injury, cardiomyopathy, atherosclerosis, and heart failure [34, 41–46]. Long-term treatment with the antioxidants, probucol or pyrrolidine dithiocarbamate, attenuates oxidative stress and cardiomyocyte apoptosis within non-infarcted myocardium in rats [18, 36]. Another antioxidant, carvedilol, is shown to attenuate apoptosis induced by ischemia–reperfusion in the rat heart [47]. Oxidant scavengers, such as SOD and vitamin E, have been demonstrated to reduce ROS and inhibit cardiomyocyte apoptosis [48].

The mechanisms responsible for oxidative stress-mediated apoptosis in the infarcted heart are not fully understood. Multiple studies have shown that ROS can induce cardiomyocyte death by one or more mechanisms. Apoptosis is tightly controlled by a number of genes, including those primarily suppressing and those promoting apoptosis. Our previous studies identified markedly increased proapoptotic Bax expression in the infarcted heart, particularly at the site of infarction. Enhanced Bax expression coexists with oxidative stress and apoptosis in the infarcted heart [49]. Overexpression of antiapoptotic Bcl-2 decreases cardiomyocyte apoptosis [50]. Cesselli et al. have shown that in dog dilated cardiomyopathy, oxidative stress-induced cardiac apoptosis is related to increased p66^{she}, cytochrome c release, and activation of caspase-9 and caspase-3 [51]. Taken together, these findings suggest that oxidative stress may trigger cardiomyocyte apoptosis via regulation of apoptotic genes.

Cytokines, such as tumor necrosis factor (TNF)-α and interleukin-6, are proven to stimulate cardiomyocyte apoptosis, which can be mediated by oxidative stress [52, 53]. TNF-α-induced apoptosis is dependent on the induction of inducible nitric oxide synthase (iNOS) [54]. NO produced by iNOS appears to cause apoptosis through the formation of $ONOO^-$, activation of the death receptor Fas through upregulation of Fas ligand expression, inhibition of mitochondrial ATP synthesis, and inactivation of serial antioxidant enzymes [55, 56]. In the infarcted mouse heart, iNOS expression is elevated in both infarcted and non-infarcted myocardium, whereas in iNOS-knockout mice, cardiomyocyte apoptosis was reduced compared to wild-type mice with MI [57]. TNF-α-induced cardiomyocyte apoptosis can be inhibited by antioxidants such as thioredoxin and *N*-acetylcysteine [58]. Following acute MI, cardiac interleukin-6 expression is also significantly increased, which initiates inflammatory response, particularly at the site of infarction. Interleukin-6 is also verified to cause cardiomyocyte apoptosis, which can be suppressed by antioxidant treatment [59]. These findings suggest that oxidative stress is involved in different pathways that induce cardiomyocyte apoptosis following infarction and can be suppressed by antioxidants.

23.3.2 *Oxidative Stress and Cardiac Inflammatory Response*

MI is associated with an inflammatory response, ultimately leading to healing and scar formation. Inflammatory response in the infarcted myocardium is related to the coordinated activation of a series of cytokine and adhesion molecule genes. A critical element in the regulation of these genes involves nuclear factor-kappa B (NF-κB), a redox-sensitive transcription factor. NF-κB maintains an inactive form bound to its inhibitory subunit Ikappa B under normal conditions. When tissue is injured, NF-κB can be activated by various local substances including ROS [60]. Upon activation, NF-κB stimulates inflammatory and immune responses and cellular growth by increasing the expression of specific cellular genes. NF-κB activation has been demonstrated in various models of myocardial ischemia and reperfusion [61, 62]. Activated NF-κB triggers gene expression of interstitial and vascular adhesion molecules, leading to leukocyte infiltration into the infarcted myocardium, as well as monocyte chemoattractant protein-1, and inducing recruitment of mononuclear cells. NF-κB also triggers gene expression of proinflammatory cytokines, such as TNF-α and interleukins, and initiates inflammatory response [63]. TNF-α is not constitutively expressed in the normal heart. In rodent models of MI, TNF-α expression is significantly upregulated in the infarcted myocardium as well as in the non-infarcted myocardium [64]. TNF-α can stimulate inflammatory protein synthesis, macrophage phagocytosis, and cell growth, differentiation, and apoptosis [65]. In the infarcted myocardium, elevated NADPH oxidase is spatially coincident with activated NF-κB and enhanced TNF-α expression in the inflammatory cells [61]. In addition to NF-κB pathway, recent studies suggest that H_2O_2 can directly induce cardiac TNF production via the p38 MARP pathway and, in turn, mediate myocardial inflammation. Free radical scavenger treatment has been demonstrated to diminish inflammatory response and cardiac remodeling [18]. The antioxidant, probucol, has been shown to attenuate cardiac inflammation and improve ventricular function [66]. Thus, these findings indicate that ROS serves as a proinflammatory mediator in the cardiac healing process following infarction and its role in cardiac inflammation involves several pathways.

In the infarcted heart, oxidative stress is mostly evident at the border zone, the area between the infarcted and non-infarcted myocardium [9]. ROS are known to cause oxidative damage to various cellular structures, including membranes, proteins, DNA sequences, and DNA repair enzymes and therefore have the potential to injure cardiomyocytes and vascular cells in the neighboring non-infarcted myocardium, causing additional cardiac damage/inflammation and infarct size extension. Studies have shown that antioxidant treatment reduces infarct size in rat MI model [67, 68]. The beneficial effect of antioxidant on infarct size may be associated with the prevention of ROS-induced damage/inflammation in the border zone of the non-infarcted myocardium.

23.3.3 Oxidative Stress and Cardiac Fibrosis

A growing bulk of evidence supports a causative role of oxidative stress in fibrogenesis in various tissues including liver, lung, arteries, nervous system, and heart [69, 70]. Extensive fibrosis is a major feature of myocardial remodeling in the infarcted heart, represented as scar at the site of infarction and interstitial fibrosis in non-infarcted myocardium [71]. Cells responsible for fibrous tissue formation at the site of infarction consist principally of phenotypically transformed fibroblast-like cells termed myofibroblasts. These cells contain α-smooth muscle actin microfilaments and are responsible for collagen synthesis and scar contraction [72]. They appear early after MI, mainly located in and around the infarcted myocardium, and persist throughout healing [73]. Interstitial fibroblasts are responsible for normal collagen turnover and are considered to be a source of myofibroblasts. Myofibroblast differentiation and proliferation as well as collagen synthesis are tightly controlled by the fibrogenic cytokine transforming growth factor-beta (TGF-β). TGF-β mRNA and concentration are significantly increased in both the infarcted and non-infarcted myocardium in rats [74]. Oxidative stress is shown to upregulate the expression of TGF and type I collagen [75]. Following MI, enhanced expressions of TGF-β and NADPH oxidase are spatially coincident at the site of infarcted myocardium. Treatment with the antioxidant taurine reduces oxidative stress, suppresses TGF-β gene expression, and attenuates hepatic fibrosis [76]. Furthermore, in vitro studies have indicated that ROS promotes fibroblast proliferation and type I collagen gene expression in cardiac fibroblasts [77]. Chronic antioxidant treatment is shown to attenuate cardiac fibrosis [17]. However, Frantz and his coworkers have shown that targeted deletion of the NADPH oxidase subunit gp91phox does not affect left ventricular remodeling including myocardial fibrosis following MI and does not decrease the production of ROS [78]. It suggests that NADPH oxidases might not be the only important pathway for the production of ROS after MI. In addition to the heart, antioxidants d-a-tocopherol and butylated hydroxytoluene are confirmed to suppress type I collagen gene expression in the repairing liver [75]. After acute MI, progressive global left ventricular dilation occurs over months [2]. MMPs favor this adverse remodeling. It was shown that inhibition of MMPs decreases the severity of remodeling in the infarcted heart [79]. In vitro studies have shown that ROS activates MMPs in cardiac fibroblasts [80]. Oxidative stress may, therefore, play a role in the pathogenesis of left ventricular dilation following infarction. However, in vivo studies on the regulatory role of ROS on MMPs and cardiac dilatation are lacking and further studies are required on this concept.

23.3.4 Oxidative Stress and Cardiac Hypertrophy

Myocardial hypertrophy is often developed in non-infarcted myocardium in weeks or months after MI. Extracellular stimuli such as mechanical strain, neurohormones, or cytokines have been well recognized to promote cardiomyocyte hypertrophy. Recent studies have further demonstrated that these extracellular stimuli may mediate cardiomyocyte hypertrophy via oxidative stress. ROS released acutely in large amounts have been implicated in the cell death associated with MI. The chronic release of ROS has been, however, linked to the development of left ventricular hypertrophy and heart failure progression [20, 81]. The chronic release of ROS appears to derive from the nonphagocytic NADPH oxidase and mitochondria [20]. The existence of nonphagocytic NADPH oxidase has been suggested in the cardiomyocytes. Experimental data suggest that abnormal activation of the nonphagocytic NADPH oxidase in response to neurohormones contributes to cardiomyocyte hypertrophy [82]. Acute MI is accompanied by systemic and local activation of numerous neurohumoral factors, including angiotensin (Ang)II, TNF-α, and norepinephrine. Increased AngII (AT1) receptors and angiotensin-converting enzyme (ACE) are co-localized with NADPH oxidase in the infarcted heart [83, 84]. Accumulating experimental evidence has demonstrated that AngII stimulates the expression of NADPH oxidase leading to the production of ROS in the repairing heart and vessels [85]. Treatment with ACE inhibitor attenuates ROS formation and prevents cardiomyocyte hypertrophy in diabetic rats to the same extent as the antioxidant *N*-acetylcysteine [86]. Moreover, in vitro studies further demonstrated that antioxidants prevent AngII-induced myocyte hypertrophy [87].

TNF-α is shown to provoke a hypertrophic cardiac phenotype [88]. In vitro studies have demonstrated that TNF-α-induced cardiomyocyte hypertrophy is mediated through NF-κB activation via the generation of ROS [89]. Norepinephrine-induced cardiac hypertrophy likewise also requires ROS. Treatment of cultured adult rat cardiomyocytes with norepinephrine increases NADPH oxidase and leads to cardiomyocyte hypertrophy.

The exact pathways of ROS-mediated hypertrophy remain to be determined. ROS can participate as second messengers in myocardial signaling events, activating stress-related nuclear transcription factors, leading to cellular hypertrophy [90]. These second messengers include MAP kinases, small GTP-binding proteins, the Src family of tyrosine kinases, and cytokine cascades. Pretreatment with a SOD mimetic inhibits both the hypertrophy and ROS production. The overexpression of GSHPx could attenuate left ventricular hypertrophy and improve cardiac function after MI [81].

23.3.5 Oxidative Stress and Cardiac Angiogenesis

Angiogenesis is the formation of new capillary blood vessels from existent microvessels. It is a tightly regulated process, which involves the degradation of the basement membrane surrounding an existing vessel, migration and proliferation of endothelial cells into the injured space, adherence of the endothelial cells to each other, and lumen formation and maturation. Angiogenesis turns on in repairing tissue. Matrix metalloproteinases (MMPs) play an important role in the degradation of the basement membranes and other extracellular matrix (ECM) components, allowing endothelial cells to detach and migrate into new tissue [91]. Angiogenic mediators such as vascular endothelial growth factor (VEGF), platelet-derived growth factor (PDGF), fibroblast growth factor (FGF), and integrins then stimulate endothelial proliferation and tube formation, while angiopoietins further promote vascular maturation and survival [92, 93].

Following acute MI, myocytes/interstitial cells and existing blood vessels in the infarcted myocardium undergo necrosis. Angiogenesis becomes evident in the infarcted myocardium particularly during the first 2 weeks post-MI. Newly formed vessels are seen first in the border zone, the region between the infarcted and non-infarcted myocardium, and then extend into the infarcted myocardium. Angiogenesis is subsequently converted into quiescence in the late stage of MI, when cardiac repair is completed. Acute MI is accompanied by increased expression of proangiogenic mediators, such as VEGF, PDGF, and FGF, in the border zone and infarcted myocardium in the very early stage of MI [94–96]. The activity of MMP-2/MMP-9 and αvβ3 integrin is also found elevated in the infarcted myocardium in the early phase of MI [97, 98]. These changes are spatially and temporally coincident with angiogenesis in the infarcted myocardium, indicating their involvement in cardiac angiogenesis following MI.

A considerable body of evidence has indicated that ROS, including O^-, H_2O_2, and NO, mediate angiogenesis in certain pathological situations [99–107]. In the infarcted heart, angiogenesis is crucial for cardiac repair. Our recent study has shown that ROS promote angiogenesis in the infarcted (necrotic) myocardium, thereby facilitating cardiac repair [15]. In addition, it has been demonstrated that ROS are required for coronary collateral development in coronary artery occlusion and reperfusion in rats [108, 109]. However, mechanisms responsible for ROS-mediated angiogenesis in the infarcted heart remain uncertain. Angiogenic response in the infarcted myocardium is extremely active and short-lived, and inflammatory cells play a major role on ROS production in the infarcted myocardium. These cells also largely release various proangiogenic growth factors and other angiogenesis-related substances such as MMPs, integrins, and cytokines. Thus, ROS may mediate angiogenesis in the infarcted myocardium through multiple pathways at the same time. ROS have both beneficial and deleterious effects in the infarcted heart. ROS induces angiogenesis in the infarcted myocardium in the early stage of MI and facilitates cardiac repair. However, ROS also contribute to cardiac remodeling, such as hypertrophy and interstitial fibrosis in non-infarcted myocardium in the late stage of MI. Thus, further studies are required to provide valuable information on the timing of antioxidant treatment, which can attenuate cardiac remodeling, but not disturb cardiac angiogenesis.

23.4 Conclusion

Experimental evidence has demonstrated that cardiac repair/remodeling is subject to oxidative stress following MI. This chapter has highlighted several mechanisms by which oxidative stress is involved in cardiac repair/remodeling in the infarcted heart. Following acute MI, oxidative stress occurs in both the infarcted and non-infarcted myocardium and ROS plays a significant role in the initiation as well as the regulation of cardiac molecular and cellular changes including angiogenesis, cardiomyocyte apoptosis, inflammatory/fibrogenic responses, and hypertrophy that contribute to myocardial repair/remodeling. These cardiac events can be significantly inhibited by antioxidants in animal models of MI. Further studies are required on the potential therapeutic interventions with antioxidants in limiting cardiac remodeling in patients with MI.

References

1. Olivetti G, Capasso JM, Meggs LG et al (1991) Cellular basis of chronic ventricular remodeling after myocardial infarction in rats. Circ Res 68:856–869
2. Jugdutt BI (2003) Ventricular remodeling after infarction and the extracellular collagen matrix: when is enough enough? Circulation 108:1395–1403
3. Ertl G, Frantz S (2005) Healing after myocardial infarction. Cardiovasc Res 66:22–32
4. Anversa P, Li P, Zhang X (1993) Ischemic myocardial injury and ventricular remodeling. Cardiovasc Res 27:145–157
5. Francis GS, Mcdonald K, Chu G, Cohn JN (1995) Pathophysiologic aspects of end-stage heart failure. Am J Cardiol 75:11A–16A
6. Weber KT (1997) Extracellular matrix remodeling in heart failure. Circulation 96: 4065–4082
7. Hill MF, Singal PK (1996) Antioxidant and oxidative stress changes during heart failure subsequent to myocardial infarction in rats. Am J Pathol 148:291–300
8. Fukui T, Yoshiyama M, Hanatani A et al (2001) Expression of p22-phox an dgp91-phox, essential components of NADPH oxidase, increases after myocardial infarction. Biochem Biophys Res Commun 281:1200–1206
9. Lu L, Quinn MT, Sun Y (2004) Oxidative stress in the infarcted heart: role of de novo angiotensin II production. Biochem Biophys Res Commun 325:943–951
10. Usal A, Acarturk E, Yuregir GT et al (1996) Decreased glutathione levels in acute myocardial infarction. Jpn Heart J 37:177–182
11. Khaper N, Kaur K, Li T et al (2003) Antioxidant enzyme gene expression in congestive heart failure following myocardial infarction. Mol Cell Biochem 251:9–15
12. Hare JM (2001) Oxidative stress and apoptosis in heart failure progression. Circ Res 89:198–201
13. Poli G, Parola M (1997) Oxidative damage and fibrogenesis. Free Radic Biol Med 22: 287–305
14. Gorlach A, Kietzmann T, Hess J (2002) Redox signaling through NADPH oxidases: involvement in vascular proliferation and coagulation. Ann N Y Acad Sci 973:505–507
15. Zhao W, Zhao T, Chen Y et al (2009) Reactive oxygen species promote angiogenesis in the infarcted rat heart. Int J Exp Pathol 90:621–629
16. Nakagami H, Takemoto M, Liao JK (2003) NADPH oxidase-derived superoxide anion mediates angiotensin II-induced cardiac hypertrophy. J Mol Cell Cardiol 35:851–859
17. Sia YT, Parker TG et al (2002) Improved post-myocardial infarction survival with probucol in rats: effects on left ventricular function, morphology, cardiac oxidative stress and cytokine expression. J Am Coll Cardiol 39:148–156
18. Sia YT, Lapointe N, Parker T et al (2002) Beneficial effects of long-term use of the antioxidant probucol in heart failure in the rat. Circulation 105:2549–2555
19. Kinugawa S, Tsutsui H, Hayashidani S et al (2000) Treatment with dimethylthiourea prevents left ventricular remodeling and failure after experimental myocardial infarction in mice: role of oxidative stress. Circ Res 87:392–398
20. Sorescu D, Griendling K (2002) Reactive oxygen species, mitochondria, and NADPH oxidases in the development and progression of heart failure. Congest Heart Fail 8:132–140

21. Droge W (2002) Free radicals in the physiological control of cell function. Physiol Rev 82:47–52
22. Halliwell B (1991) Reactive oxygen species in living systems: source, biochemistry, and role in human disease. Am J Med 91:14S–22S
23. Finkel T (1999) Signal transduction by reactive oxygen species in non-phagocytic cells. J Leukoc Biol 65:337–340
24. Dhalla AK, Singal PK (1994) Antioxidant changes in hypertrophied and failing guinea pig heart. Am J Physiol 266:H1280–H1285
25. Young IS, Woodside JV (2001) Antioxidants in health and disease. J Clin Pathol 54:176–186
26. Vaziri ND, Lin C, Farmand F, Sindhu K (2003) Superoxide dismutase, catalase, glutathione peroxidase and NADPH oxidase in lead-induced hypertension. Kidney Int 63:186–194
27. Gasparetto C, Malinverno A, Culacciati D et al (2005) Antioxidant vitamins reduce oxidative stress and ventricular remodeling in patients with acute myocardial infarction. Int J Immunopathol Pharmacol 18:487–496
28. Hill MF, Singal PK (1997) Right and left myocardial antioxidant responses during heart failure subsequent to myocardial infarction. Circulation 96:2414–2420
29. Li L, Quinn MT, Sun Y (2004) Oxidative stress in the infarcted heart: role of de novo angiotensin II production. Biochem Biophys Res Commun 325:943–951
30. Mohazzab-H KM, Kaminski PM, Wolin MS (1997) Lactate and PO2 modulate superoxide anion production in bovine cardiac myocytes: potential role of NADH oxidase. Circulation 15:614–620
31. Grieve DJ, Byrne JA, Cave AC, Shah AM (2004) Role of oxidative stress in cardiac remodeling after myocardial infarction. Heart Lung Circ 13:132–138
32. Anversa P, Cheng W, Liu Y et al (1998) Apoptosis and myocardial infarction. Basic Res Cardiol 93:8–12
33. Cheng W, Kajstura J, Nitahara JA et al (1996) Programmed myocyte cell death affects the viable myocardium after infarction in rats. Exp Cell Res 226:316–327
34. Kajstura J, Cheng W, Reiss K et al (1996) Apoptotic and necrotic myocyte cell deaths are independent contributing variables of infarct size in rats. Lab Invest 74:86–107
35. Hare JM (2001) Oxidative stress and apoptosis in heart failure progression. Cir Res 89:198–201
36. Li WG, Coppey L, Weiss RM, Oskarsson HJ (2001) Antioxidant therapy attenuates JNK activation and apoptosis in the remote noninfarcted myocardium after large myocardial infarction. Biochem Biophys Res Commun 280:353–357
37. Kumar D, Kirshenbaum LA, Li T et al (2001) Apoptosis in adriamycin cardiomyopathy and its modulation by probucol. Antioxid Redox Signal 3:135–145
38. Pagliara P, Carla EC, Caforio S et al (2003) Kupffer cells promote lead nitrate-induced hepatocyte apoptosis via oxidative stress. Comp Hepatol 2:8–16
39. von Harsdorf R, Li PF, Dietz R (1998) Signaling pathways in reactive oxygen species-induced cardiomyocyte apoptosis. Circulation 99:2934–2941
40. Li PF, Dietz R, von Harsdorf R (1999) Superoxide induces apoptosis in cardiomyocytes, but proliferation and expression of transforming growth factor-beta1 in cardiac fibroblasts. FEBS Lett 448:206–210
41. Gottlieb RA, Burleson KO, Kloner RA et al (1994) Reperfusion injury induces apoptosis in rabbit cardiomyocytes. J Clin Invest 94:1621–1628
42. Itoh G, Tamura J, Suzuki M et al (1995) DNA fragmentation of human infarcted myocardial cells demonstrated by the nick end labeling method and DNA agarose gel electrophoresis. Am J Pathol 146:1325–1331
43. Kajstura J, Zhang X, Liu Y et al (1995) The cellular basis of pacing-induced dilated cardiomyopathy: myocyte cell loss and myocyte cellular reactive hypertrophy. Circulation 92:2306–2317
44. Liu Y, Cigola E, Cheng W et al (1995) Myocyte nuclear mitotic division and programmed myocyte cell death characterize the cardiac myopathy induced by rapid ventricular pacing in dogs. Lab Invest 73:771–787

45. Sharov VG, Sabbah HN, Shimoyama H et al (1996) Evidence of cardiocyte apoptosis in myocardium of dogs with chronic heart failure. Am J Pathol 148:141–149
46. Teiger E, Than VD, Richard L et al (1996) Apoptosis in pressure overload-induced heart hypertrophy in the rat. J Clin Invest 97:2891–2897
47. Zeng H, Liu X, Zhao H (2003) Effects of carvedilol on cardiomyocyte apoptosis and gene expression in vivo after ischemia-reperfusion in rats. J Huazhong Univ Sci Technolog Med Sci 23:127–130
48. Kumar D, Jugdutt BI (2003) Apoptosis and oxidants in the heart. J Lab Clin Med 142:288–297
49. Zhao W, Lu L, Chen SS, Sun Y (2004) Temporal and spatial characteristics of apoptosis in the infarcted rat heart. Biochem Biophys Res Commun 325:605–611
50. Hockenbery DM, Oltvai ZN, Yin XM et al (1993) Bcl-2 functions in an antioxidant pathway to prevent apoptosis. Cell 75:241–251
51. Cesselli D, Jakoniuk I, Barlucchi L et al (2001) Oxidative stress-mediated cardiac cell death is a major determinant of ventricular dysfunction and failure in dog dilated cardiomyopathy. Circ Res 89:279–286
52. Kubota T, Miyagishima M, Frye CS et al (2001) Overexpression of tumor necrosis factor-alpha activates both anti- and pro-apoptotic pathways in the myocardium. J Mol Cell Cardiol 33:1331–1344
53. Haunstetter A, Izumo S (1998) Apoptosis: basic mechanisms and implications for cardiovascular disease. Circ Res 82:1111–1129
54. Song W, Lu X, Feng Q (2000) Tumor necrosis factor-alpha induces apoptosis via inducible nitric oxide synthase in neonatal mouse cardiomyocytes. Cardiovasc Res 45:595–602
55. Almeida A, Bolanos JP (2001) A transient inhibition of mitochondrial ATP synthesis by nitric oxide synthase activation triggered apoptosis in primary cortical neurons. J Neurochem 77:676–690
56. Dobashi K, Pahan K, Chahal A, Singh I (1997) Modulation of endogenous antioxidant enzymes by nitric oxide in rat C6 glial cells. J Neurochem 68:1896–1903
57. Marfella R, Di Filippo C, Esposito K et al (2004) Absence of inducible nitric oxide synthase reduces myocardial damage during ischemia reperfusion in streptozotocin-induced hyperglycemic mice. Diabetes 53:454–462
58. Buttke TM, Sandstrom PA (1994) Oxidative stress as a mediator of apoptosis. Immunol Today 15:7–10
59. Wollert KC, Drexler H (2001) The role of interleukin-6 in the failing heart. Heart Fail Rev 6:95–103
60. Kabe Y, Ando K, Hirao S et al (2005) Redox regulation of NF-kappaB activation: distinct redox regulation between the cytoplasm and the nucleus. Antioxid Redox Signal 7:395–403
61. Lu L, Chen SS, Zhang JQ et al (2004) Activation of nuclear factor-kappaB and its proinflammatory mediator cascade in the infarcted rat heart. Biochem Biophys Res Commun 321:879–885
62. Nichols TC (2004) NF-kappaB and reperfusion injury. Drug News Perspect 17:99–104
63. Schoonbroodt S, Piette J (2000) Oxidative stress interference with the nuclear factor-kappa B activation pathways. Biochem Pharmacol 15:1075–1083
64. Jacobs M, Staufenberger S, Gergs U et al (1999) Tumor necrosis factor-alpha at acute myocardial infarction in rats and effects on cardiac fibroblasts. J Mol Cell Cardiol 31: 1949–1959
65. Ceconi C, Curello S, Bachetti T et al (1998) Tumor necrosis factor in congestive heart failure: a mechanism of disease for the new millennium? Prog Cardiovasc Dis 41:25–30
66. Nakamura R, Egashira K, Machida Y et al (2002) Probucol attenuates left ventricular dysfunction and remodeling in tachycardia-induced heart failure: roles of oxidative stress and inflammation. Circulation 106:362–367
67. Sahna E, Acet A, Ozer MK, Olmez E (2002) Myocardial ischemia-reperfusion in rats: reduction of infarct size by either supplemental physiological or pharmacological doses of melatonin. J Pineal Res 33:234–238

68. Kim CH, Choi H, Chun YS et al (2002) The protective effect of resveratrols on ischaemia-reperfusion injuries of rat hearts is correlated with antioxidant efficacy. Br J Pharmacol 135:1627–1633
69. Tsukamoto H (1993) Oxidative stress, antioxidants, and alcoholic liver fibrogenesis. Alcohol 10:465–467
70. Mastruzzo C, Crimi N, Vancheri C (2002) Role of oxidative stress in pulmonary fibrosis. Monaldi Arch Chest Dis 57:173–176
71. Sun Y, Cleutjens JP, Diaz-Arias AA, Weber KT (1994) Cardiac angiotensin converting enzyme and myocardial fibrosis in the rat. Cardiovasc Res 28:1423–1432
72. Desmouliere A, Gabbiani G (1996) The role of the myofibroblast in wound healing and fibrocontractive diseases. In: Richard AF (ed) The molecular and cellular biology of wound repair. Plenum Press, New York, NY, pp 391–423
73. Sun Y, Weber KT (1996) Angiotensin converting enzyme and myofibroblasts during tissue repair in the rat. J Mol Cell Cardiol 28:851–858
74. Sun Y, Zhang JQ, Zhang J, Ramires FJ (1998) Angiotensin II, transforming growth factor-beta1 and repair in the infarcted heart. J Mol Cell Cardiol 30:1559–1569
75. Lee KS, Buck M, Houglum K, Chojkier M (1995) Activation of hepatic stellate cells by TGF alpha and collagen type I is mediated by oxidative stress through c-myb expression. J Clin Invest 96:2461–2468
76. Miyazaki T, Karube M, Matsuzaki Y et al (2005) Taurine inhibits oxidative damage and prevents fibrosis in carbon tetrachloride-induced hepatic fibrosis. J Hepatol 43:117–125
77. Murrell AC, Francis JO, Bromley L (1993) Modulation of fibroblast proliferation by oxygen free radicals. Biochem J 265:659–665
78. Frantz S, Brandes RP, Hu K et al (2005) Left ventricular remodeling after myocardial infarction in mice with targeted deletion of the NADPH oxidase subunit gp91phox. Basic Res Cardiol 100:1–6
79. Rohde LE, Ducharme A, Arroyo LH et al (1999) Matrix metalloproteinase inhibition attenuates early left ventricular enlargement after experimental myocardial infarction in mice. Circulation 99:3063–3070
80. Siwik DA, Pagano PJ, Colucci WS (2001) Oxidative stress regulates collagen synthesis and matrix metalloproteinase activity in cardiac fibroblasts. Am J Physiol Cell Physiol 280:C53–C60
81. Shiomi T, Tsutsui H, Matsusaka H et al (2004) Overexpression of glutathione peroxidase prevents left ventricular remodeling and failure after myocardial infarction in mice. Circulation 109:544–549
82. Byrne JA, Grieve DJ, Cave AC, Shah AM (2003) Oxidative stress and heart failure. Arch Mal Coeur Vaiss 96:214–221
83. Fabris B, Jackson B, Kohzuki M et al (1990) Increased cardiac angiotensin-converting enzyme in rats with chronic heart failure. Clin Exp Pharmacol Physiol 17:309–314
84. Sun Y, Weber KT (1994) Angiotensin II receptor binding following myocardial infarction in the rat. Cardiovasc Res 28:1623–1628
85. Wang HD, Xu S, Johns DG et al (2001) Role of NADPH oxidase in the vascular hypertrophic and oxidative stress response to angiotensin II in mice. Circ Res 88:947–953
86. Fiordaliso F, Cuccovillo I, Bianchi R et al (2006) Cardiovascular oxidative stress is reduced by an ACE inhibitor in a rat model of streptozotocin-induced diabetes. Life Sci 79:121–129
87. Nakamura K, Fushimi K, Kouchi H et al (1998) Inhibitory effects of antioxidants on neonatal rat cardiac myocyte hypertrophy induced by tumor necrosis factor-alpha and angiotensin II. Circulation 98:794–799
88. Mann DL (2002) Tumor necrosis factor-induced signal transduction and left ventricular remodeling. J Card Fail 8:S379–S386
89. Higuchi Y, Otsu K, Nishida K et al (2002) Involvement of reactive oxygen species-mediated NF-kappa B activation in TNF-alpha-induced cardiomyocyte hypertrophy. J Mol Cell Cardiol 34:233–240

90. Nian M, Lee P, Khaper N, Liu P (2004) Inflammatory cytokines and postmyocardial infarction remodeling. Circ Res 94:1543–1553
91. Rundhaug JE (2005) Matrix metalloproteinases and angiogenesis. J Cell Mol Med 9:267–285
92. Conway EM, Collen D, Carmeliet P (2001) Molecular mechanisms of blood vessel growth. Cardiovasc Res 49:507–521
93. Rupp PA, Little CD (2001) Integrins in vascular development. Circ Res 89:566–572
94. Zhao T, Zhao W, Chen Y et al (2010) Vascular endothelial growth factor (VEGF)-A: role on cardiac angiogenesis following myocardial infarction. Microvasc Res 80:188–194
95. Zhao W, Zhao T, Huang V et al (2011) Platelet-derived growth factor involvement in myocardial remodeling following infarction. J Mol Cell Cardiol 51:830–838
96. Zhao T, Zhao W, Chen Y et al (2011) Acidic and basic fibroblast growth factors involved in cardiac angiogenesis following infarction. Int J Cardiol 152:307–313
97. van Hinsbergh VW, Koolwijk P (2008) Endothelial sprouting and angiogenesis: matrix metalloproteinases in the lead. Cardiovasc Res 78:203–212
98. Sun M, Opavsky MA, Stewart DJ et al (2003) Temporal response and localization of integrins beta1 and beta3 in the heart after myocardial infarction: regulation by cytokines. Circulation 107:1046–1052
99. Ushio-Fukai M, Nakamura Y (2008) Reactive oxygen species and angiogenesis: NADPH oxidase as target for cancer therapy. Cancer Lett 266(1):37–52
100. Nishikawa M (2008) Reactive oxygen species in tumor metastasis. Cancer Lett 266:53–59
101. Komatsu D, Kato M, Nakayama J et al (2008) NADPH oxidase 1 plays a critical mediating role in oncogenic Ras-induced vascular endothelial growth factor expression. Oncogene 27:4724–4732
102. Xia C, Meng Q, Liu LZ et al (2007) Reactive oxygen species regulate angiogenesis and tumor growth through vascular endothelial growth factor. Cancer Res 67:10823–10830
103. Polytarchou C, Hatziapostolou M, Poimenidi E et al (2009) Nitric oxide stimulates migration of human endothelial and prostate cancer cells through up-regulation of pleiotrophin expression and its receptor protein tyrosine phosphatase beta/zeta. Int J Cancer 124:1785–1793
104. Shigyo H, Nonaka S, Katada A et al (2007) Inducible nitric oxide synthase expression in various laryngeal lesions in relation to carcinogenesis, angiogenesis, and patients' prognosis. Acta Otolaryngol 127:970–979
105. Abid MR, Kachra Z, Spokes KC, Aird WC (2000) NADPH oxidase activity is required for endothelial cell proliferation and migration. FEBS Lett 486:252–256
106. Kapila V, Sellke FW, Suuronen EJ et al (2005) Nitric oxide and the angiogenic response: can we improve the results of therapeutic angiogenesis? Expert Opin Investig Drugs 14:37–44
107. Kondo T, Kobayashi K, Murohara T (2004) Nitric oxide signaling during myocardial angiogenesis. Mol Cell Biochem 264:25–34
108. Maulik N (2002) Redox signaling of angiogenesis. Antioxid Redox Signal 4:805–815
109. Gu W, Weihrauch D, Tanaka K et al (2003) Reactive oxygen species are critical mediators of coronary collateral development in a canine model. Am J Physiol Heart Circ Physiol 285:H1582–H1589

Chapter 24
Role of SPARC in Cardiac Extracellular Matrix Remodeling After Myocardial Infarction

Davy Vanhoutte and Stephane Heymans

Abstract Research of the last decades has established that the cardiac extracellular matrix (ECM) is not a static but rather a dynamic environment in which the cellular and acellular components of the heart communicate and function. Remodeling after myocardial infarction (MI) associates with cardiac rupture or maladaptive left ventricular remodeling resulting in cardiac dilatation, dysfunction, and increased mortality. The dynamic synthesis and breakdown of the cardiac ECM-related proteins and the formation of a well-organized collagen scar play a major role in the structural and functional recovery of the infarcted heart. Secreted protein acidic and rich in cysteine (SPARC) is a prototypical collagen-binding matricellular protein and has been proven to coordinate procollagen processing and facilitate collagen fibril assembly in the pericellular environment. SPARC is vastly upregulated in response to cardiac stress, including MI. The lack of SPARC resulted in a fourfold higher incidence of mortality following MI, due to increased rates of cardiac rupture and heart failure, which was associated with a highly impaired and immature collagen scar. In contrast, adenoviral overexpression of SPARC improved collagen maturation in the infarct scar, associated with increased transforming growth factor (TGF)-β signaling, and prevented cardiac dilatation and dysfunction after MI. Therefore, this

D. Vanhoutte, Ph.D. (✉)
Department of Pediatrics, Molecular Cardiovascular Biology, Cincinnati Children's Hospital Medical Center, 240 Albert Sabin Way, MLC 7020, Cincinnati, OH, USA

Department of Cardiovascular Sciences, K.U. Leuven, Leuven, Belgium
e-mail: Davyvanhoutte@gmail.com

S. Heymans
Department of Cardiovascular Sciences, K.U. Leuven, Leuven, Belgium

Center for Heart Failure Research, Cardiovascular Research Institute Maastricht (CARIM), Maastricht University, Maastricht, The Netherlands

B.I. Jugdutt and N.S. Dhalla (eds.), *Cardiac Remodeling: Molecular Mechanisms*, Advances in Biochemistry in Health and Disease 5, DOI 10.1007/978-1-4614-5930-9_24,

chapter focuses on (1) the role of SPARC during cardiac remodeling after MI and (2) the potential mechanisms involved in SPARC-mediated extracellular post-synthetic procollagen processing in the infarcted heart.

Keywords SPARC • Matricellular proteins • Myocardial infarction • Extracellular matrix • Remodeling • Transforming growth factor-β • Procollagen processing • Collagen

24.1 Introduction

24.1.1 The Cardiac Extracellular Matrix and Myocardial Infarction

The cellular components of the heart are enmeshed in a complex network of extracellular matrix (ECM) that is primarily composed out of type I collagen and to lower extent of type III, type V collagen, fibronectin, and elastin [1, 2]. In addition to its structural elements, the cardiac ECM also harbors more specialized proteins including glycoproteins (fibronectin, periostin, fibromodulin, and vitronectin), proteoglycans (such as syndecans, lumican, and biglycan), glycosaminoglycans (hyaluronic acid and dermatan sulfate), and matricellular proteins (including SPARC and others) [3–5]. On the other hand, a large reservoir of recruitable matrix metalloproteinases (MMPs) exists within the cardiac ECM representing the main extracellular proteases involved in degradation the cardiac ECM [6].

Discoveries of the last decade have drastically changed our traditional perception of the cardiac ECM. It has become clear that the ECM not only acts as a structural support but also provides a unique environment in which the embedded cells communicate and function [1–4]. By activating signaling cascades, the ECM transduces extracellular changes into cellular and acellular responses that play central roles to maintain normal chamber geometry and ventricular function during homeostasis and in response to injury. As such, dramatic changes in the composition of the cardiac ECM occur after myocardial infarction (MI) [7, 8]. In response to MI, the left ventricle (LV) undergoes a continuum of molecular, cellular, and extracellular responses that result in LV wall thinning, dilatation, and dysfunction, collectively referred to as cardiac remodeling [8]. In summary, cardiac wound healing after MI can be divided in three distinct but overlapping phases: the inflammatory, granulation, and maturation phase [6]. The inflammatory phase is characterized by massive inflammatory infiltrates, rapid protease activation, and matrix degradation. Macrophages remove the necrotic cells and matrix debris and produce growth factors, inducing fibroblast migration, proliferation, and activation. The necrotic tissue is consequently replaced by granulation tissue, consisting of inflammatory cells, fibroblasts, myofibroblasts, and new blood vessels embedded in a fibrin/fibronectin-based provisional matrix.

Next, the provisional matrix is reabsorbed, and myofibroblasts accumulate to produce large amounts of ECM proteins. Finally, a mature collagen-based stable scar is formed, characterized by tightly aligned, well-organized, and cross-linked collagen fibers that provide mechanical support to the infarcted heart [6, 7, 9]. Importantly, excessive degradation and delayed or defective formation of the newly formed cardiac ECM could lead to cardiac rupture, a fatal complication of acute MI. In later phases, this alters the mechanical properties of the heart, resulting in exaggerated maladaptive cardiac remodeling which is associated with higher incidence of arrhythmia, increased mortality, and the development of chronic heart failure [6, 7, 9].

24.1.2 *Matricellular Protein, Essential Modulators of Cardiac Remodeling After MI*

Recently, we have witnessed increasing interest in a class of nonstructural ECM proteins, known as "matricellular proteins" (Table 24.1). Although the term describes a family of structurally unrelated extracellular matrix macromolecules with distinct functional properties, they do exhibit several general characteristics (extensively reviewed in [3] and [4]). First, unlike collagens, matricellular proteins do not directly contribute to the structural integrity of the heart but serve as a bridge between cells and the ECM. As such, they influence a broad range of critical regulatory cell–cell and cell–matrix processes through the direct binding with cell surface receptors, growth factors, proteases, and other bioactive effectors, as well as various structural ECM proteins. Secondly, expression of matricellular proteins, including SPARC (secreted protein acidic and rich in cysteine), thrombospondins, osteopontin, tenascins, periostin, and others, is generally low during normal postnatal life but is upregulated during embryogenesis and in response to tissue injury, including pathophysiological conditions of the heart (Table 24.1) [3, 4]. Concordantly, matricellular proteins are released into the infarcted heart and are degraded as the infarct scar matures. Finally, their mechanistic roles have been—at least in part—elucidated using gene-targeted mice. Surprisingly, the majority of mice that lack one of the matricellular proteins survive embryogenesis and only produce mild abnormalities under normal physiological conditions, suggesting a limited role in tissue homeostasis (Table 24.1) [3, 4]. In contrast, when subjected to pathophysiological mouse models, such as MI, they display striking phenotypes, demonstrating that the re-expression of matricellular proteins activates key molecular signaling pathways that are essential for proper cardiac healing and remodeling after MI [3, 4]. A comprehensive overview of known phenotypes in matricellular protein-deficient mice after MI is presented in Table 24.1.

This chapter presents and discusses the current knowledge on SPARC and its role proposed after MI. More specifically, it will focus on SPARC's expression, pathophysiologic relevance, and the critical role that SPARC plays in myocardial remodeling and extracellular collagen processing after MI.

Table 24.1 Matricellular proteins and myocardial infarction

Matricellular protein	Baseline ECM organization	After Myocardial Infarction			
		Increased expression	Phenotype in gene-deficient mice	Cardiac function	Proposed mechanism
TSP-1	No change	Early phase inflammatory cells	Increased inflammation, expansion of infarct, and LV dilatation	ND	Increased inflammation, MMP-9 activity, and decreased TGF-β signaling
TSP-2	Abnormal collagen fibrillogenesis	Late phase fibroblasts	Cardiac rupture > 90%	ND	Impaired maturation of infarct scar and increased MMP activity
OPN	ND	Early and late phase macrophages and fibroblasts	Increased LV dilatation, reduced collagen synthesis, and deposition in the infarcted and remote myocardium	↓	Decreased macrophage recruitment and phagocytosis, impaired fibroblast adhesion, proliferation, and myofibroblast differentiation
TN-C	ND	Early phase fibroblasts	Reduced end-diastolic pressure, LV stiffness, dilatation, and fibrosis	↑	Decreased de-adhesion, MMP production, and fibrosis
TN-X	Reduced collagen content in skin	ND	ND	ND	ND

Periostin	Aberrant collagen I fibrillogenesis in skin	Early and late phase fibroblasts	Increased incidence of cardiac rupture; reduced LV wall stiffness Decreased recruitment of myofibroblasts and impaired matrix deposition	↑	Decreased ECM integrity with smaller and less abundant collagen fibrils; and decreased adhesion-dependent FAK signaling
CCN1 (Cyr61)	ND	Cardiomyocytes and endothelial cells	Embryonic lethality associated with vascular and septal defects	ND	ND
CCN2 (CTGF)	ND	Myofibroblasts and cardiomyocytes	CCN2-null mice die within minutes after birth	ND	ND
SPARC	For an extended overview of SPARC, see Table 24.2				

The data presented in this table is more extensively reviewed by Schellings et al. [3, 46] and Frangogiannis [4]
MI myocardial infarction, *TSP* thrombospondin, *TN* tenascin, CCN (acronym for Cyr61, Connective tissue growth factor [CTGF], and Nov), *Cyr61* cysteine-rich, angiogenic inducer 61, *CTGF* connective tissue growth factor, *SPARC* secreted protein acidic and rich in cysteine, *LV* left ventricle, *TGF-β* transforming growth factor-β, *MMP* matrix metalloproteinase, *FAK* focal adhesion kinase, *ND* not determined, ↑ increased, ↓ decreased

24.2 SPARC, a Multifunctional Matricellular Protein

SPARC, secreted protein acidic and rich in cysteine (also known as osteonectin or basement membrane protein-40), is a prototypic collagen-binding, counter adhesive matricellular protein [14, 15]. It is the product of a single gene that is highly conserved in both vertebrates and invertebrates, suggesting a fundamental function for SPARC in matricellular biology [15, 16].

After the removal of a 17-amino acid signaling peptide sequence, the final mature SPARC consists out of three distinct modular domains (Fig. 24.1a) [14, 15, 17]. The 53-amino acid NH_2-terminal acidic domain contains a low-affinity, high-capacity Ca^{2+}-binding domain. The central portion includes a 24-amino acid follistatin-like domain and a 55-amino acid protease-like inhibitor region. The 151-amino acid COOH-terminal domain contains two high-affinity Ca^{2+}-binding EF hands.

Like every matricellular protein, SPARC exhibits a remarkable functional complexity that is dependent on the effectors it binds to and the cell types it associates with in the different tissues (Fig. 24.1b) [14, 16, 17]. In more than 30 years since its discovery, a number of biological activities have been associated with SPARC in vitro [16, 17]. These functions include processes that regulate (1) cell proliferation, adhesion, migration, differentiation, and survival; (2) growth factor signaling; and (3) MMP activity (Fig. 24.1b). However, phenotypic characterization of SPARC-null mice pointed towards a primary role for SPARC in collagen fibril assembly, morphology, and metabolism (reviewed in [14] and [15]). In contrast to primitive organisms where abrogation of SPARC is lethal [18], SPARC-null mice are viable and exhibit progressively milder defects [14]. The lack of SPARC resulted in significant morphological defects in fibrillar collagen and/or reduced collagen in many tissues. As such, dermis, heart, adipose tissue, and periodontal ligament of SPARC-null mice contain less fibrillar collagen compared with wild type (WT) [10, 19–21]. For example, abnormalities in the dermis were characterized by smaller collagen fibrils with a more uniform diameter and reduced tensile strength [19]. Interestingly, SPARC is able to bind fibrillar collagen I, III, and V as well as collagen type IV, the most abundant collagen in basal lamina with its COOH-terminal domain [22]. To add another level of complexity, proteolytic cleavage of SPARC into smaller fragments that produce diverse and sometimes opposing cellular responses is compared to naïve SPARC (Fig. 24.1b) [17, 23, 24]. As such, cleavage of a single peptide bond in the central module of SPARC by several MMPs (including MMP-2, MMP-3, MMP-7, and MMP-13) markedly increases its affinity for collagens [23]. MMP-3, on the other hand, generates several fragments that influence angiogenesis, whereas unprocessed SPARC plays an anti-angiogenic role (Fig. 24.1b) [24]. Nevertheless, the molecular mechanisms responsible for these functions of SPARC still remain largely unclear.

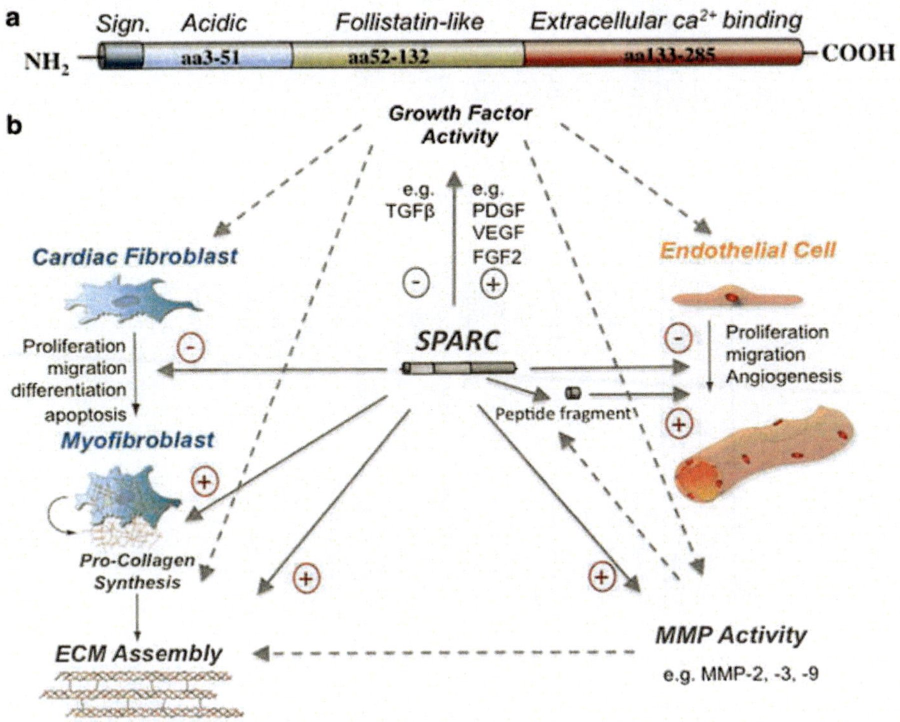

Fig. 24.1 SPARC, a pleiotropic matricellular protein in the heart. (**a**) After removal of the N-terminal signaling peptide (sign), the modular structure of SPARC contains (1) an acidic region, (2) a follistatin-like domain, and (3) a C-terminal extracellular Ca^{2+}-binding region (EC-module). (**b**) SPARC exhibits pleiotropic effects on (1) fibroblasts, (2) myofibroblasts, (3) extracellular matrix (ECM) assembly, (4) matrix metalloproteinase (MMP) activity, (5) endothelial cells, and (6) growth factor activity; all strongly involved in cardiac injury, remodeling, and repair. Moreover, proteolytic cleavage of SPARC (e.g., by MMPs) could result in peptide fragments that produce opposing biological actions. *TGF-β* transforming growth factor-β, *PDGF* platelet derived growth factor, *VEGF* vascular endothelial growth factor, *FGF2* fibroblast growth factor-2, + stimulates, – inhibits

24.3 SPARC and Myocardial Infarction

A growing body of evidence suggests a pivotal role for SPARC in a variety of cardiac pathophysiologic conditions, including aging-associated cardiomyopathy, myocardial infarction (MI), and pressure overload (Table 24.2) [10–13, 25]. Although this chapter will focus on its role during cardiac remodeling after MI, common biological features seem to be at play.

Table 24.2 SPARC in cardiac health and disease

Mouse model	SPARC expression	Phenotype in SPARC-deficient mice			References
		Pathology	Cardiac function	Proposed mechanism	
Cardiac homeostasis	High in embryonic heart; low in adult myocardium	Reduced collagen content, thinner collagen struts	Normal systolic function; decreased passive stiffness	Altered post-synthetic procollagen processing and matrix maturation	[10]
Cardiac aging	Markedly increased vs. young hearts	Reduced fibrillar collagen content and collagen cross-linking	Decreased diastolic stiffness	Impaired post-synthetic procollagen processing and decreased collagen content with aging of the heart	[10]
Myocardial infarction	Increased in early and late phase (myo)fibroblasts and leukocytes infiltrating the myocardium	Increased incidence of cardiac rupture (41% vs. 9%); increased myofibroblast infiltration, disorganized granulation tissue, and immature collagen matrix	Increased heart failure in surviving mice	Impaired collagen matrix maturation; decreased TGF-β signaling	[11]
			Attenuated early phase systolic dysfunction	Impaired fibroblast activation with differential expression of ECM and adhesion molecule genes	[12]
Pressure overload	Significantly increased 4 weeks after TAC	Decreased fibrillar collagen content and cross-linking	Decreased diastolic stiffness	Reduced post-synthetic procollagen processing and development of mature cross-linked collagen fibrils	[13]

SPARC secreted protein acidic and rich in cysteine, *TGF-β* transforming growth factor-β, *ECM* extracellular matrix, *TAC* transaortic constriction

24.3.1 Increased SPARC Expression Post-MI

Characteristic for matricellular proteins, expression of SPARC is low in the adult heart but upregulated during embryogenesis and in response to cardiac stress or injury (Table 24.2) [11, 12, 25]. Several studies using rat, canine, and mouse models of MI revealed a dramatic increase of SPARC in the infarcted LV as well as a moderate increase in the non-infarcted remote LV [11, 12, 25]. Interestingly, the spatial and temporal expression of SPARC in the infarcted heart closely paralleled that of collagens I and III deposition and ECM remodeling in those four species [11–13, 25, 26]. In mice, Schellings et al. [11] demonstrated a moderate ≈2-fold increase of SPARC protein in the infarcted and non-infarcted remote LV at 3 days after MI. At day 7 and 14 after MI, SPARC levels peaked in the infarct zone, whereas the initial increase in the non-infarcted remote LV was no longer detectable as compared to sham-operated hearts [11]. Concordantly, collagen transcription and deposition was detectable as early as day 3 and continued to rise at 7 and 14 days after MI. Further detailed analysis of infarcted mouse and canine hearts confirmed that SPARC primarily co-localized with infiltrating α-smooth muscle cell actin-positive myofibroblasts and CD45-postitive leukocytes in the infarcted LV [11, 25]. Thus, SPARC showed a localized deposition simultaneous with the formation of granulation tissue and progressive maturation of the infarct scar.

In human heart disease, a single study revealed elevated SPARC expression in cardiac tissue of patients with LV hypertrophy [27]. Based on previous animal studies, one can predict that SPARC expression will be elevated in myocardial tissue of patients with ischemic heart disease or age-related cardiomyopathy. Nevertheless, further detailed histopathological studies are mandatory and will be of great value to completely elucidate the biological and clinical relevance of SPARC expression in human heart disease.

24.3.2 Pathophysiologic Significance of Increased SPARC Expression Post-MI

The pathophysiologic relevance of increased SPARC expression in response to MI recently became clear when Schellings and colleagues (1) provided genetic evidence by subjecting SPARC-null mice to a mouse model of MI (Table 24.2) and (2) explored the effects of adenoviral-mediated overexpression of SPARC in WT mice after MI [11].

24.3.2.1 Physiologic Relevance of Increased SPARC Expression Post-MI

Surprisingly, SPARC-null mice displayed a fourfold increase in mortality post-MI, as compared to infarcted WT controls [11]. This mortality was mainly due to a higher incidence of cardiac rupture (41% vs. 9%) and failure after MI (Table 24.2). Three days after MI, increased lung weight to body weight ratio indicated increased cardiac dysfunction in absence of SPARC. In fact, hemodynamic stress induced by dobutamine increased LV contractility and relaxation in infarcted female WT hearts but to a lesser extent in female SPARC-null hearts, 14 days post-MI. In order to translate these findings into a preclinical setting, Schellings and colleagues went on and overexpressed SPARC by systemic adenoviral delivery in WT mice [11]. Enhanced SPARC protein levels in WT mice markedly protected against cardiac dilatation and dysfunction without affecting infarct size at day 14 post-MI, as compared to control-treated mice. Thus, increased SPARC expression protects the heart against cardiac rupture and failure after MI.

In contrast, a recent study revealed that the absence of SPARC is associated with attenuated systolic dysfunction during the early stages—day 3—post-MI (Table 24.2) [12]. The authors propose that lack of SPARC might switch from a positive early to a negative late effect on cardiac remodeling, given that cardiac rupture occurred later. Nevertheless, differences in genetic background and age of the mice used in the experimental design might explain the discrepancies between these studies [11, 12]. Thus, further studies comparing the effects of SPARC deletion across species and ages are warranted.

24.3.2.2 Disorganized Granulation Tissue and Deficient Maturation of the Infarct Scar in Absence of SPARC

While inflammatory cell density and the angiogenic response were similar, increased cardiac rupture and dysfunction in SPARC-null mice post-MI appeared to coincide with disorganized granulation tissue as well as abnormal collagen maturation [11]. As discussed before, a well-organized ECM plays an important role to maintain the structural and functional integrity of the heart after MI [7, 9]. The absence of SPARC did not affect collagen synthesis. No significant differences were detected between both genotypes in the transcript levels of collagen I and III nor in Sirius red stained collagen deposited in the infarcted LV [11]. However, more in-depth analysis of the newly formed infarct scar revealed a disorganized and immature collagen matrix in absence of SPARC. Sirius red polarization and transmission electron microscopy revealed a shift to smaller, disorganized, and loosely assembled collagen fibrils in SPARC-null infarcts [11]. Adenoviral-mediated overexpression of SPARC in WT mice—on the other hand—significantly enhanced collagen quantity, and maturation in the infarct zone, was associated with increased transforming growth factor-β (TGF-β)/Smad2 signaling, without influencing the amount of leukocytes, myofibroblasts, and coronary vessels compared to control-treated mice at day 14 post-MI [11]. These findings are consistent with an array of studies indicating an essential role for SPARC in post-synthetic procollagen processing and assembly, ranging from (1) abnormalities in SPARC-deficient dermal collagen fibrils [19], (2) a decrease in mature cross-linked insoluble collagen fibrils in the heart in response to pressure overload [13], to (3) alterations in fibrillar collagen content and diastolic function in the aged myocardium in absence of SPARC (Table 24.2) [10].

24.4 SPARC and MI, Proposed Mechanisms

Although we cannot exclude the importance of other molecular mechanisms, evidence supports a protective role for SPAR—at least in part—mediated through enhanced TGF-β signaling and altered fibroblast kinetics post-MI. On the other hand, SPARC has been strongly implicated as an important regulator of extracellular post-synthetic procollagen processing in the heart, and this is independent of its effect on TGF-β signaling.

24.4.1 SPARC, Modulator of TGF-β Signaling and (Myo)fibroblast Kinetics Post-MI

TGF-β is a well-established yet complex paradigm in the regulation of cardiac remodeling post-MI (reviewed by Dobaczewski et al. [28]). SPARC had previously been indicated as a positive regulator of canonical TGF-β/Smad2 signaling in epithelial and primary kidney mesangial cells in vitro [29, 30]. Interestingly, Schellings et al. [11] were able to translate these findings to isolated cardiac fibroblasts, where binding of SPARC with the TGF-β/TGF-β receptors type II complex is hypothesized to act as a cofactor that enhances its downstream signaling. In addition, the protective in vivo effect noted upon SPARC overexpression was associated with increased canonical TGF-β/Smad2 signaling in the infarcted heart [11]. In contrast, no significant differences were observed in Smad2 activation in SPARC-deficient infarcts. Nonetheless, infusion of TGF-β in SPARC-null mice decreased rupture rates (14% vs. 67%) and improved extracellular collagen remodeling in the infarct scar [11]. Together, these findings provide strong evidence that SPARC communicates—at least in part—by influencing TGF-β signaling in the infarcted heart.

Moreover, several lines of evidence suggest that SPARC might influence TGF-β-dependent and TGF-β-independent pathways that control (myo)fibroblast kinetics post-MI, and this in a time- and space-dependent fashion. In line with the observed phenotype in SPARC-null mice, a recent study showed that early TGF-β inhibition increased mortality, adverse collagen remodeling, and exacerbated LV dilatation and dysfunction post-MI [31]. In contrast, TGF-β inhibition after the inflammatory phase decreased collagen deposition but was surprisingly associated with increased numbers of myofibroblasts in the infarct scar due to impaired apoptosis [32]. Increased myofibroblast numbers were also noted in SPARC-null infarcts 7 days post-MI, indicating that increased myofibroblast differentiation alone was not sufficient to drive increased collagen deposition [11]. On the other hand, we cannot rule out TGF-β-independent changes in cardiac fibroblasts lacking SPARC. In fact, an independent study recently revealed altered expression of 22 genes encoding ECM and adhesion molecules in cardiac fibroblasts isolated from SPARC-null hearts at baseline and from remote and infarcted LV 3 days post-MI compared to WT controls [12]. However, further detailed investigations are mandatory to fully unravel the molecular constituents of these findings.

Taken together, the above indicates that SPARC probably functions through TGF-β-dependent and TGF-β-independent pathways that control (myo)fibroblast kinetics post-MI. It will be interesting to investigate how these time-related effects potentially translate into a therapeutic setting.

24.4.2 SPARC, Major Modulator of Cardiac Collagen Assembly and Remodeling

Fibrillar collagen homeostasis is regulated by at least three regulatory control mechanisms: (1) procollagen biosynthesis, (2) post-synthetic procollagen processing, and (3) collagen degradation [15]. The balance between synthesis, processing, and degradation determines the total fibrillar collagen content of the heart, and the smallest alteration could be sufficient to influence how the heart structurally and functionally recovers after MI [7]. Interestingly, SPARC has been indicated as a potential switch for all three of these processes (reviewed by Bradshaw [15]). However, in the heart, multiple studies advocate a primary role for SPARC in post-synthetic procollagen processing and assembly in the stressed heart, including MI (Table 24.2 and Fig. 24.2) [10–12, 15, 19, 33].

24.4.2.1 Procollagen Biosynthesis

Cardiac fibroblasts produce the majority or fibrillar collagen, of which collagen I and III are the most abundant after MI [15]. The hypothesis that SPARC might be directly upstream of collagen expression emerged from the fact that SPARC expression consistently parallels that of fibrillar collagens in various adult tissues undergoing wound repair or fibrotic disease [3, 14]. In addition, a diminished capacity to produce a robust fibrotic response in various pathological mouse models had been identified in SPARC-null mice [15]. However, as previously discussed, the absence of SPARC did not appear to alter the transcription of collagen I and III nor did it affect the total amount of collagen deposition in response to MI compared with WT infarcts [11]. Furthermore, no significant differences were noted in the procollagen I production by SPARC-deficient primary isolated cardiac fibroblasts upon TGF-β stimulation [33]. In contrast, McCurdy et al. [12] did reveal reduced mRNA expression levels of collagen III and IV in cardiac fibroblasts isolated from SPARC-null hearts 3 days post-MI. However, pathway analysis indicated that collagen is probably indirectly regulated by SPARC in this setting.

Taken together, these studies do collectively indicate that SPARC is not likely to be the primary contributor of elevated procollagen biosynthesis in response to MI.

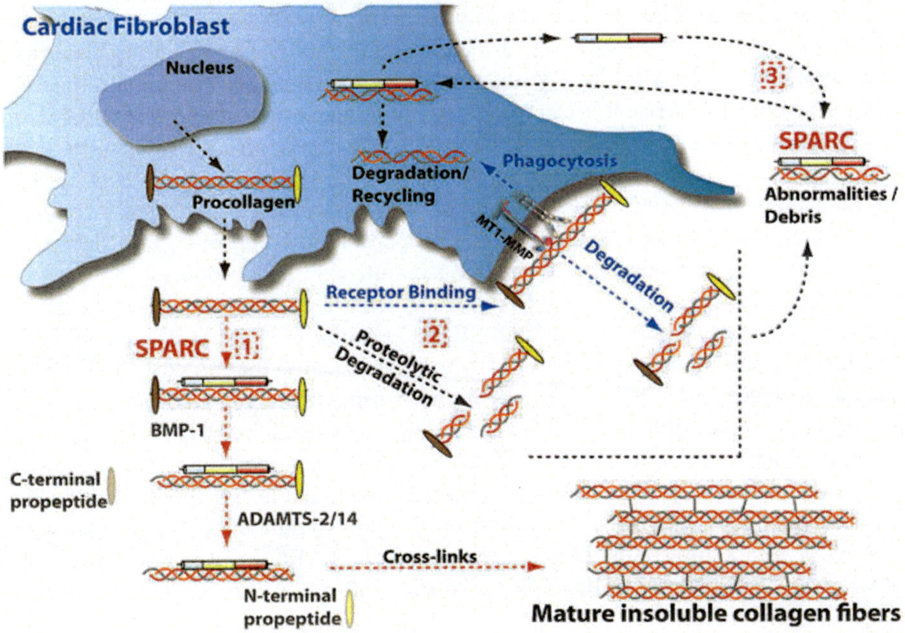

Fig. 24.2 SPARC in post-synthetic procollagen processing: "a working model" (1) SPARC binds to newly secreted procollagen and chaperones it through a series of ordered post-synthetic processing steps—including the removal of its C- and N-terminal propeptide by procollagen C- and N-terminal proteinases (e.g., BMP-1 and ADAMTS2/14, respectively)—and favors the formation of a mature cross-linked insoluble ECM. (2) SPARC thereby prevents proteolytic degradation and limits the association of procollagen with cell surface receptors that promote collagen turnover by phagocytosis or degradation by cell surface-associated collagenases (e.g., MT1-MMP). (3) In addition, SPARC might function as a scavenger chaperone that clears the extracellular space from ECM abnormalities and debris. *BMP* bone morphogenetic protein, *ADAMTS* a disintegrin-like and metalloproteinase domain with thrombospondin-type motif, *MT1-MMP* membrane type 1-matrix metalloproteinase

24.4.2.2 Post-Synthetic Procollagen Processing

After synthesis, procollagen is secreted into the extracellular space, where it must undergo a series of ordered, time- and location-sensitive post-synthetic processing steps and is predicted to have two major outcomes (Fig. 24.2) [15, 34]. Newly secreted procollagen either (1) is subjected to further processing within the extracellular space that favors the incorporation into a mature cross-linked insoluble ECM as collagen fibers or (2) associates with cell surface receptors that promote collagen turnover by phagocytosis or degradation by cell surface-associated collagenases.

The dramatic post-MI phenotype observed in absence of SPARC was in line with two recent studies that revealed impaired incorporation of cardiac fibrillar collagen into the insoluble ECM in response to pressure overload or with cardiac aging in SPARC-deficient hearts (Table 24.2) [10, 11, 13]. On the other hand, adenoviral overexpression of SPARC in WT mice improved collagen processing and maturation in response to MI [11]. Using SPARC-deficient primary cardiac fibroblasts, Harris and colleagues revealed increased cell surface-associated collagen, in addition to enhanced processing of procollagen I to collagen I [33]. These findings are similar to those previously obtained with dermal fibroblasts, which were associated with impaired mechanical and functional properties of the dermis [35]. Moreover, addition of recombinant SPARC to SPARC-null cardiac fibroblasts reduced the association of collagen I with the cell surface demonstrating a partial rescue of the phenotype by addition of exogenous SPARC protein [33]. Notably, the SPARC-binding sites within collagen Ia1 and IIIa1 are identical to that of cardiac fibroblast specific collagen cell surface receptor discoidin domain receptor 2 (DDR2) [36–38]. Therefore, it is plausible that SPARC and DDR2 compete for collagen binding after cardiac injury. Although further research is mandatory, the latter could result in increased collagen binding to DDR2 that traps collagen at the cell surface in the absence of SPARC, thereby preventing proper post-synthetic collagen processing and ECM remodeling after MI. Enhancement of SPARC levels after MI, on the other hand, further limits collagen association with the cell surface and chaperones the collagen molecule through the time- and location-sensitive processing steps within the extracellular space. SPARC thereby enhances collagen deposition and the formation of a mature cross-linked insoluble collagen scar (Fig. 24.2).

24.4.2.3 Collagen Degradation

In the adult heart, as much as 92% of the newly synthesized collagen is immediately degraded before incorporation into the insoluble ECM [39, 40]. In addition, collagen turnover—ratio of degradation to synthesis—is a tightly regulated process and significantly increases in response to MI to insure proper scar formation. The balance between collagen-degrading MMPs and their tissue inhibitors TIMPs plays a central role in this process (reviewed by Vanhoutte et al. [6] and [41]). Interestingly, in vitro studies have suggested that SPARC acts to increase the activity of certain MMPs, notably MMP-2 and MMP-9 [42, 43], both of which are widely associated with cardiac rupture and adverse ventricular remodeling after MI [6]. In contrast, a single study recently showed that in baseline conditions, the LV of SPARC-null mice contains increased levels active MMP-3, an upstream activator of several MMPs [12, 44]. However, after MI, MMP-3 activity levels were reduced comparable to those in WT infarcts [12]. Thus, although SPARC is probably not a primary contributor to MMP activity post-MI, this matter remains of significant interest for future studies.

On the other hand, SPARC might fulfill a completely different function during extracellular collagen degradation. A recent finding suggested that SPARC might function as a novel class of matricellular scavenger chaperones that shuttles from the ECM into the cell and back (Fig. 24.2) [45]. In dermal fibroblasts, Chlenski and colleagues demonstrated that at high extracellular Ca^{2+} concentrations, SPARC was able to bind multiple ECM targets (including collagen I and fibronectin), promotes the disassembly of solid networks, and directs the cellular internalization of their components [45]. At low intracellular Ca^{2+} concentrations, SPARC releases its target, and free SPARC is recycled outside the cell, serving as a shuttle that scavenges various structural and soluble ECM proteins. Whether SPARC functions as a scavenger chaperone in the stressed heart and whether this is directed through specific cell surface receptors remain intriguing questions. If so, this could imply that during the high rates of matrix turnover after MI, SPARC also functions as scavenger that clears the extracellular space from proteolytic ECM debris or abnormalities in the newly formed collagen matrix. The lack of SPARC would consequently result in premature termination of the growing extracellular collagen networks and an impaired ECM deposition [11]. Nevertheless, further understanding of this mechanism is necessary and may provide better understanding of matrix-associated abnormalities during cardiac homeostasis and disease.

24.4.2.4 SPARC, Collagen Assembly and Cardiac Remodeling: "A Working Model"

Putting previously described data together, increased expression of SPARC in response to MI—or cardiac stress in general—is very likely to bind with newly secreted procollagen, chaperones it through the processing steps within the extracellular space, and limits binding of procollagen to specific cell surface receptors that promote collagen degradation in the adult heart (Fig. 24.2). By diminishing procollagen engagement with these receptors, SPARC facilitates the optimal sequence and timing of post-synthetic processing steps and prevents procollagen from being prematurely degraded or improperly processed, thereby improving collagen deposition and maturation into well-organized and tightly cross-linked collagen fibers (Fig. 24.2) [15, 33, 35]. At the same time, SPARC might scavenge ECM debris and abnormalities within the collagen scar and clear them from the extracellular space (Fig. 24.2). Although further molecular and biochemical analysis of these mechanisms is mandatory, the regulation of SPARC and post-synthetic procollagen processing and degradation may provide important mechanism-based targets to improve matrix maturation and strengthening of the infarct scar within the first days after MI.

24.5 Conclusions

The cardiac ECM is a finely tuned and adaptable system that allows the heart to survive various pathological insults. Results obtained with SPARC-null mice have demonstrated that SPARC serves a critical role in post-synthetic collagen processing in the pericellular environment that facilitates efficient collagen incorporation into the cardiac ECM. Furthermore, it underscores the necessity of well-organized extracellular collagen remodeling to preserve proper structure and function and to prevent cardiac rupture and dilatation after MI.

However, several key questions remain to be answered to push the field forward. For example, further analysis that unravels the functional basis of SPARC in the regulation of collagen assembly and remodeling may provide important mechanism-based targets to improve matrix maturation and strengthening of the infarct scar after MI. Moreover, further dissection of SPARC's multiple biological actions and identifying the distinct structural domains and pathways responsible could allow us to design peptides that either mimic a specific protective action or act as a dominant negative in the stressed heart, while limiting undesirable side effects. In this aspect, it will also be of considerable interest to fully investigate the cell-specific biological effects of SPARC, including its effects on cardiomyocyte function and the inflammatory process. Next, it will be of equal interest to identify the complete catalogue of SPARC-binding partners. To what level does proteolytic cleavage of SPARC co-determine its binding to specific cell surface receptors and other ECM-related proteins? And to what extent does this mediate the differential context-dependent functions of SPARC? In addition, it will also be exciting to see whether members of the SPARC-related proteins, including SMOC-1, SMOC-2, and Hevin [11], display similar functional roles in the infarcted heart.

Taken together, further in-depth studies will without a doubt bring novel and exciting perspectives, not only into the biology of SPARC but also to the field of matricellular proteins, collagen, and cardiac remodeling during cardiac homeostasis and disease.

Acknowledgments S.H. received a Vidi grant from the Netherlands Organization for Scientific Research (NWO, 91796338) and research grants from the Netherlands Heart Foundation (2008B011), Fund of Scientific Research Flanders (FWO-Vlaanderen, G074009N), European Union, FP7-HEALTH-2010, MEDIA, large-scale integrating project, and European Union, FP7-HEALTH-2011, EU-Mascara. D.V. is supported by a fellowship of the Fund of Scientific Research Flanders (FWO-Vlaanderen, Belgium, 1208910N).

References

1. Jugdutt BI (2003) Remodeling of the myocardium and potential targets in the collagen degradation and synthesis pathways. Curr Drug Targets Cardiovasc Haematol Disord 3:1–30
2. Dobaczewski M, Gonzalez-Quesada C, Frangogiannis NG (2010) The extracellular matrix as a modulator of the inflammatory and reparative response following myocardial infarction. J Mol Cell Cardiol 48:504–511
3. Schellings MW, Pinto YM, Heymans S (2004) Matricellular proteins in the heart: possible role during stress and remodeling. Cardiovasc Res 64:24–31
4. Frangogiannis NG (2012) Matricellular proteins in cardiac adaptation and disease. Physiol Rev 92:635–688
5. Zamilpa R, Lindsey ML (2010) Extracellular matrix turnover and signaling during cardiac remodeling following MI: causes and consequences. J Mol Cell Cardiol 48:558–563

6. Vanhoutte D, Schellings M, Pinto Y, Heymans S (2006) Relevance of matrix metalloproteinases and their inhibitors after myocardial infarction: a temporal and spatial window. Cardiovasc Res 69:604–613
7. Jugdutt BI (2003) Ventricular remodeling after infarction and the extracellular collagen matrix: when is enough enough? Circulation 108:1395–1403
8. Gajarsa JJ, Kloner RA (2011) Left ventricular remodeling in the post-infarction heart: a review of cellular, molecular mechanisms, and therapeutic modalities. Heart Fail Rev 16:13–21
9. Lindsey ML, Mann DL, Entman ML, Spinale FG (2003) Extracellular matrix remodeling following myocardial injury. Ann Med 35:316–326
10. Bradshaw AD, Baicu CF, Rentz TJ et al (2010) Age-dependent alterations in fibrillar collagen content and myocardial diastolic function: role of SPARC in post-synthetic procollagen processing. Am J Physiol Heart Circ Physiol 298:H614–H622
11. Schellings MW, Vanhoutte D, Swinnen M et al (2009) Absence of SPARC results in increased cardiac rupture and dysfunction after acute myocardial infarction. J Exp Med 206:113–123
12. McCurdy SM, Dai Q, Zhang J et al (2011) SPARC mediates early extracellular matrix remodeling following myocardial infarction. Am J Physiol Heart Circ Physiol 301:H497–H505
13. Bradshaw AD, Baicu CF, Rentz TJ et al (2009) Pressure overload-induced alterations in fibrillar collagen content and myocardial diastolic function: role of secreted protein acidic and rich in cysteine (SPARC) in post-synthetic procollagen processing. Circulation 119:269–280
14. Bradshaw AD, Sage EH (2001) SPARC, a matricellular protein that functions in cellular differentiation and tissue response to injury. J Clin Invest 107:1049–1054
15. Bradshaw AD (2009) The role of SPARC in extracellular matrix assembly. J Cell Commun Signal 3:239–246
16. Brekken RA, Sage EH (2001) SPARC, a matricellular protein: at the crossroads of cell-matrix communication. Matrix Biol 19:816–827
17. Tai IT, Tang MJ (2008) SPARC in cancer biology: its role in cancer progression and potential for therapy. Drug Resist Updat 11:231–246
18. Fitzgerald MC, Schwarzbauer JE (1998) Importance of the basement membrane protein SPARC for viability and fertility in Caenorhabditis elegans. Curr Biol 8:1285–1288
19. Bradshaw AD, Puolakkainen P, Dasgupta J et al (2003) SPARC-null mice display abnormalities in the dermis characterized by decreased collagen fibril diameter and reduced tensile strength. J Invest Dermatol 120:949–955
20. Bradshaw AD, Graves DC, Motamed K, Sage EH (2003) SPARC-null mice exhibit increased adiposity without significant differences in overall body weight. Proc Natl Acad Sci USA 100:6045–6050
21. Trombetta JM, Bradshaw AD (2010) SPARC/osteonectin functions to maintain homeostasis of the collagenous extracellular matrix in the periodontal ligament. J Histochem Cytochem 58:871–879
22. Sasaki T, Hohenester E, Gohring W, Timpl R (1998) Crystal structure and mapping by site-directed mutagenesis of the collagen-binding epitope of an activated form of BM-40/SPARC/osteonectin. EMBO J 17:1625–1634
23. Sasaki T, Gohring W, Mann K et al (1997) Limited cleavage of extracellular matrix protein BM-40 by matrix metalloproteinases increases its affinity for collagens. J Biol Chem 272:9237–9243
24. Sage EH, Reed M, Funk SE et al (2003) Cleavage of the matricellular protein SPARC by matrix metalloproteinase 3 produces polypeptides that influence angiogenesis. J Biol Chem 278:37849–37857
25. Dobaczewski M, Bujak M, Zymek P et al (2006) Extracellular matrix remodeling in canine and mouse myocardial infarcts. Cell Tissue Res 324:475–488
26. Komatsubara I, Murakami T, Kusachi S et al (2003) Spatially and temporally different expression of osteonectin and osteopontin in the infarct zone of experimentally induced myocardial infarction in rats. Cardiovasc Pathol 12:186–194
27. Ridinger H, Rutenberg C, Lutz D et al (2009) Expression and tissue localization of beta-catenin, alpha-actinin and chondroitin sulfate proteoglycan 6 is modulated during rat and human left ventricular hypertrophy. Exp Mol Pathol 86:23–31

28. Dobaczewski M, Chen W, Frangogiannis NG (2011) Transforming growth factor (TGF)-beta signaling in cardiac remodeling. J Mol Cell Cardiol 51(4):600–606
29. Francki A, Bradshaw AD, Bassuk JA et al (1999) SPARC regulates the expression of collagen type I and transforming growth factor-beta1 in mesangial cells. J Biol Chem 274:32145–32152
30. Schiemann BJ, Neil JR, Schiemann WP (2003) SPARC inhibits epithelial cell proliferation in part through stimulation of the transforming growth factor-beta-signaling system. Mol Biol Cell 14:3977–3988
31. Frantz S, Hu K, Adamek A et al (2008) Transforming growth factor beta inhibition increases mortality and left ventricular dilatation after myocardial infarction. Basic Res Cardiol 103:485–492
32. Okada H, Takemura G, Kosai K et al (2005) Postinfarction gene therapy against transforming growth factor-beta signal modulates infarct tissue dynamics and attenuates left ventricular remodeling and heart failure. Circulation 111:2430–2437
33. Harris BS, Zhang Y, Card L et al (2011) SPARC regulates collagen interaction with cardiac fibroblast cell surfaces. Am J Physiol Heart Circ Physiol 301:H841–H847
34. Prockop DJ, Kivirikko KI (1995) Collagens: molecular biology, diseases, and potentials for therapy. Annu Rev Biochem 64:403–434
35. Rentz TJ, Poobalarahi F, Bornstein P et al (2007) SPARC regulates processing of procollagen I and collagen fibrillogenesis in dermal fibroblasts. J Biol Chem 282:22062–22071
36. Giudici C, Raynal N, Wiedemann H et al (2008) Mapping of SPARC/BM-40/osteonectin-binding sites on fibrillar collagens. J Biol Chem 283:19551–19560
37. Hohenester E, Sasaki T, Giudici C et al (2008) Structural basis of sequence-specific collagen recognition by SPARC. Proc Natl Acad Sci USA 105:18273–18277
38. Goldsmith EC, Hoffman A, Morales MO et al (2004) Organization of fibroblasts in the heart. Dev Dyn 230:787–794
39. Bishop JE, Rhodes S, Laurent GJ et al (1994) Increased collagen synthesis and decreased collagen degradation in right ventricular hypertrophy induced by pressure overload. Cardiovasc Res 28:1581–1585
40. Cleutjens JP, Kandala JC, Guarda E et al (1995) Regulation of collagen degradation in the rat myocardium after infarction. J Mol Cell Cardiol 27:1281–1292
41. Vanhoutte D, Heymans S (2010) TIMPs and cardiac remodeling: 'Embracing the MMP-independent-side of the family'. J Mol Cell Cardiol 48:445–453
42. Arnold S, Mira E, Muneer S et al (2008) Forced expression of MMP9 rescues the loss of angiogenesis and abrogates metastasis of pancreatic tumors triggered by the absence of host SPARC. Exp Biol Med (Maywood) 233:860–873
43. McClung HM, Thomas SL, Osenkowski P et al (2007) SPARC upregulates MT1-MMP expression, MMP-2 activation, and the secretion and cleavage of galectin-3 in U87MG glioma cells. Neurosci Lett 419:172–177
44. Ogata Y, Enghild JJ, Nagase H (1992) Matrix metalloproteinase 3 (stromelysin) activates the precursor for the human matrix metalloproteinase 9. J Biol Chem 267:3581–3584
45. Chlenski A, Guerrero LJ, Salwen HR et al (2011) Secreted protein acidic and rich in cysteine is a matrix scavenger chaperone. PLoS One 6:e23880
46. Schellings MW, van Almen GC, Sage EH et al (2009) Thrombospondins in the heart: potential functions in cardiac remodeling. J Cell Commun Signal 3:201–213

Chapter 25
Tissue Inhibitor of Matrix Metalloproteinases in the Pathogenesis of Heart Failure Syndromes

Dong Fan, Abhijit Takawale, and Zamaneh Kassiri

Abstract One of the characteristics of heart failure, regardless of its initial cause, is remodeling of the myocardium and the extracellular matrix (ECM). Disruption of the ECM results in structural instability as well as activation of a number of signaling pathways that could lead to fibrosis, hypertrophy, and apoptosis. The integrity of the ECM is maintained by a balance in the function of matrix metalloproteinases (MMPs) and their inhibitors, the tissue inhibitor of metalloproteinases (TIMPs). An imbalance between the activity of MMPs and TIMPs in heart disease results in adverse outcomes. In addition to their MMP-dependent functions, TIMPs possess a number of MMP-independent functions. In this chapter, we will discuss the structure, functions, and regulation of TIMPs and their role in heart failure syndromes. We will review the knowledge that we have gained from clinical studies and animal models on the contribution of TIMPs in the development and progression of heart disease.

Keywords Heart disease • Animal models • Cardiac extracellular matrix • Matrix biomarkers • Matrix metalloproteinase • Tissue inhibitor of metalloproteinases

Abbreviations

ADAM	A disintegrin and matrix metalloproteinase
CITP	Carboxy-terminal telopeptide of collagen type I
DCM	Dilated cardiomyopathy
ECM	Extracellular matrix

D. Fan • A. Takawale • Z. Kassiri, M.Sc., Ph.D. (✉)
Department of Physiology, Cardiovascular Research Centre,
Mazankowski Alberta Heart Institute, University of Alberta,
Rm 474, Heritage Medical Research Centre, Edmonton, AB, Canada T6G 2S2
e-mail: z.kassiri@ualberta.ca

B.I. Jugdutt and N.S. Dhalla (eds.), *Cardiac Remodeling: Molecular Mechanisms*, Advances in Biochemistry in Health and Disease 5, DOI 10.1007/978-1-4614-5930-9_25,

FGF	Fibroblast growth factor
HB-EGF	Heparin-bound epidermal growth factor
IL	Interleukin
LV	Left ventricle
MI	Myocardial infarction
MMP	Matrix metalloproteinase
MT-MMP	Membrane-type MMP
PDGF	Platelet-derived growth factor
PICP	Procollagen type I carboxy-terminal propeptide
PIIINP	Procollagen type III amino-terminal propeptide
PINP	Procollagen type I amino-terminal propeptides
TGFβ	Transforming growth factor-beta
TIMP	Tissue inhibitor of metalloproteinases
TNFα	Tumor necrosis factor-alpha
VEGF	Vascular endothelial growth factor
WT	Wild type

25.1 Introduction

Cardiovascular disease remains a major cause of morbidity and mortality [1]. Following an injury, the heart undergoes structural remodeling which eventually leads to maladaptive remodeling and cardiac dysfunction. The relationship among the cardiomyocytes, fibroblasts, and blood vessels is coordinated by the extracellular matrix (ECM). Cardiac ECM is connected to cardiomyocytes through a collagen–integrin–cytoskeleton–myofibril relationship (Fig. 25.1). As heart disease progresses, cardiomyocytes can undergo eccentric or concentric hypertrophy, and fibroblasts can migrate and propagate to trigger interstitial fibrosis, while the ECM can be disrupted resulting in destabilization of the vasculature and myofibrillar architecture, leading to ventricular dilation.

Matrix metalloproteinases (MMPs) are the predominant proteases responsible for degradation of the ECM proteins, the first step in ECM turnover. MMPs are Ca^{2+}- and Zn^{2+}-dependent proteases that are synthesized as inactive zymogens (pro-MMPs) and can be activated by the removal of an amino-terminal propeptide domain either by autoproteolysis or via processing by another MMP or serine protease. Activated MMPs can be inhibited by their physiological MMP inhibitors, including the tissue inhibitor of metalloproteinases (TIMPs), α2-macroglobulin, and RECK (reversion-inducing cysteine-rich protein with kazal motif) [2]. TIMPs are specific MMP inhibitors in the tissue compartment and can reversibly inhibit activated MMPs, whereas α2-macroglobulin irreversibly clears MMPs by forming a complex endocytosis only in plasma and in liquid compartments with minimal inhibitory function in the tissue [3]. Finally, RECK is a cell surface protein with MMP inhibitory capacity, which proposed a role in endothelial migration and angiogenesis since mice lacking RECK do not survive beyond day 10 of the gestation (E10.5) [4].

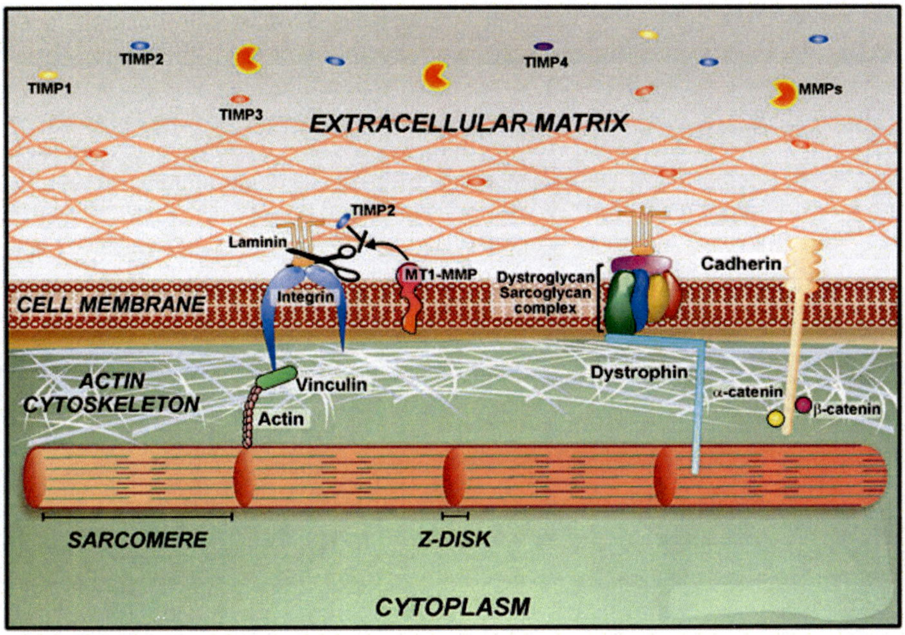

Fig. 25.1 Schematic illustration of the organization of the extracellular matrix and cardiomyocytes and localization of the TIMPs and MMPs

However, baseline expression of RECK in the heart is minimal [5], and its role in cardiac function has not yet been identified. Among these MMP inhibitors, TIMPs have been shown to be the most potent physiological inhibitors of MMPs. The role of the MMPs in heart disease has been extensively studied and reviewed [6–8]. In this chapter, we will focus on the contribution of TIMPs in development and progression of heart failure syndromes.

25.2 Cardiac Extracellular Matrix

Cardiac ECM is the supporting network that provides architectural integrity to adjoining myocytes and the neighboring vessels and capillaries and translates single myocyte contractility into a ventricular syncytium. It is comprised of highly organized protein structures, predominantly collagen type I and type III, and basement membrane components including laminin, fibronectin, collagen type IV, and fibrillin. This structural arrangement prevents myocyte slippage and overstretching and contributes to cardiac recoil during diastole.

In the myocardium, the ECM is connected to the cell membrane and to the myocytes via a number of structures. The ECM maintains alignment of myofibrils within the myocyte through a collagen–integrin–cytoskeleton–myofibril relationship (Fig. 25.1). Integrins link the ECM to the cellular cytoskeleton [9, 10]. An integrin-mediated insertion into the basal lamina localized to the Z-band of the sarcomere forms direct connections between adjacent cardiomyocytes [11]. The dystroglycan complex also works as a transmembrane linkage between the ECM and the cytoskeleton, thereby providing structural stability for the myocardium. In the myocardium, the ECM is connected to the cell membrane and to the myocytes via a number of structures. Integrins are transmembrane heterodimers comprised of alpha (α) and beta (β) subunits that link the ECM to the cellular cytoskeleton [9]. Cadherins are a large family of single-pass transmembrane proteins principally involved in Ca^{2+}-dependent homotypic cell adhesion and bind to the cytoskeleton via α-catenin and β-catenin. Cadherin is essential for optimal cardiac function and remodeling in disease [12], while increased membrane translocation of β-catenin has been shown in patients with hypertrophic cardiomyopathy [13]. Figure 25.1 shows a schematic of the arrangement and assembly of the ECM, cell membrane, and the sarcomeres.

25.3 Biology of TIMPs

TIMPs are 21–30 kDa proteins that share substantial sequence homology and structural identity. TIMPs possess 12 conserved cysteine residues forming six disulfide bonds that fold the protein into two domains each containing three loops, an MMP-inhibiting N-terminal domain (loops 1–3) and a C-terminal domain (loops 4–6) [14]. Disulfide bonds at 12 strictly conserved Cys residues maintain the six loop structures. Four TIMPs have been identified and cloned in vertebrates, TIMP1, TIMP2, TIMP3, and TIMP4. TIMPs can be present in glycosylated and unglycosylated forms [15]. Glycosylation results in an increase in the molecular weight of TIMPs but has not been shown to alter their MMP inhibitory function [16, 17]. TIMP1, but not TIMP2, has two N-linked glycosylation sites which are important in the correct folding and enhanced stability of TIMP1 [17]. TIMP3 has also been shown to be *N*-glycosylated [15]; however, this glycosylation is not required for MMP inhibition or ECM binding of this TIMP [16]. Glycosylation of TIMP4 occurs by O-linked glycosylation [18], but no report has been made on the effect of glycosylation on the inhibitory capacity of TIMP4 [19].

Classically, TIMP1, TIMP2, and TIMP4 are known to be localized extracellularly in soluble form while TIMP3 is tightly bound to ECM via the interaction of its COOH-terminal with sulfated glycosaminoglycans [16, 20]. However, a number of recent studies have provided new insights into this dogma. An in vitro study has shown that TIMP3 could also be present in the extracellular space in soluble form [21], while TIMP4 has been reported to be localized intracellularly along the thin myofilaments of sarcomeres in rat cardiomyocytes [22]. Hence, the localization of TIMPs may not be as restricted as originally believed.

In the healthy adult murine heart, TIMP2 and TIMP3 mRNA are most abundant TIMPs followed by TIMP4 and TIMP1. Although TIMP4 expression was originally believed to be restricted to the heart and the brain [5], it has become increasingly evident that TIMP4 is in fact expressed in a number of organs and tissues including kidney, lung, pancreas, colon, and testes [23, 24]. The robust expression of TIMP4 in healthy cardiomyocytes could have given rise to the original misconception.

TIMPs are inducible proteins. Expression of TIMP1 can be induced by growth factors (b-FGF, PDGF, EGF), phorbol esters, serum, cytokines (IL-6, IL-1, and IL-1β) [25], and erythropoietin [26]. Although TIMP2 was initially believed to be constitutively expressed [25], induction of TIMP2 mRNA and protein has been reported in myocardial tissue from heart failure patients [27–29], as well as in animal models of heart disease [30] and kidney disease [31, 32]. TIMP3 can be transcriptionally induced by TGFβ [33], IL-1β [34], TNFα [35], and reactive oxygen species [33]. TIMP1, TIMP2 [36, 37], and TIMP3 [38] have been shown to be transcriptionally upregulated by cAMP. Transcriptional regulation of TIMP4 has been less explored.

25.4 MMP-Dependent Functions of TIMPs

TIMPs reversibly inhibit the proteolytic activity of activated MMPs by forming non-covalent 1:1 stoichiometric complexes that are resistant to heat denaturation and proteolytic degradation [39]. The inhibitory function of TIMPs occurs through an interaction between the Zn^{2+} of the MMP active site and the N-terminal cysteine residues of the TIMP, particularly the cys1-cys70 disulfide bonds [17]. Although there is a degree of overlap among TIMPs in their inhibitory functions, each TIMP possesses unique properties and inhibits MMPs with different specificity and affinity. TIMP1 is a prototypic inhibitor for most MMPs with a selective preference to bind to and inhibit MMP9 [40] but a poor inhibitor of the membrane-type MMPs (MT-MMPs) and MMP19 [25]. TIMP2 is unique in that in addition to inhibiting MMPs, it mediates activation of MMP2 by selectively interacting with membrane type 1 (MT1)-MMP to facilitate the cell surface activation of pro-MMP2 [41]. One MT1-MMP acts as an anchor and binds to TIMP2, which then recruits pro-MMP2 forming a trimolecular complex: MT1-MMP/TIMP2/pro-MMP2. A second MT1-MMP molecule acts as an activator and cleaves pro-MMP2 first to its intermediate form (64 kDa) and then to its fully active form (62 kDa) [42]. While low levels of TIMP2 promote this cell surface activation of pro-MMP2, high levels of this TIMP prevent this activation process by inhibiting the second MT1-MMP. Pro-MMP2 also forms a complex with TIMP3 or TIMP4 in a similar manner, but not TIMP1. The pro-MMP2–TIMP4 complex interacts with MT1-MMP but does not result in MMP2 activation [43], while TIMP3 inhibits the pro-MMP2 activation in this process [21]. TIMP2 and TIMP4 have high potency against MMP2 and MT1-MMP [30, 44, 45], while TIMP3 can inhibit a broad spectrum of MMPs, MT-MMPs, and ADAMs [46, 47].

Among the ADAMs inhibited by TIMP3, ADAM12 [48] and ADAM17/TACE (TNF-alpha-converting enzyme) [49, 50] are of particular relevance to heart failure.

ADAM12 plays an integral role in the hypertrophic response upon stimulation of G-protein-coupled receptors (GPCRs) through cleavage of membrane-bound HB-EGF and subsequent transactivation of the epidermal growth factor receptor (EGFR) [51]. TACE is a potent sheddase, and its presence is required for shedding of TNFα [52] and TGFα [53]. TIMP3 deficiency led to enhanced TACE activity and elevated TNFα processing in pressure-overloaded cardiomyopathy [54], as well as in other diseases [32, 55]. TIMPs are stoichiometrically outnumbered by MMPs, but by targeting a broad spectrum of MMPs, they compensate for this numerical mismatch. In addition, MMP activities are also regulated at the transcription and activation steps [35]; thereby, a tight regulation of the proteolysis events is maintained in health.

25.5 MMP-Independent Functions of TIMPs

TIMPs possess a signaling capacity distinct from their MMP inhibitory activity. TIMP1 can interact with CD63, a member of the tetraspanin family of proteins [56]. TIMP2 can bind to integrin α3β1 [57] whereby it mediates anti-angiogenic effects by dephosphorylating the activated VEGF receptor and transforming differentiating endothelial cells into a quiescent state (Fig. 25.2) [58]. TIMP2 can also regulate neuromuscular junction development in fast-twitch muscles through transcriptional regulation of integrin β1 [59]. In vitro studies have shown that TIMP3 can exert anti-angiogenic effects by interrupting the interaction between VEGF and VEGF receptor 2 in endothelial cells (Fig. 25.3) [60]. TIMP3$^{-/-}$ mice subjected to myocardial

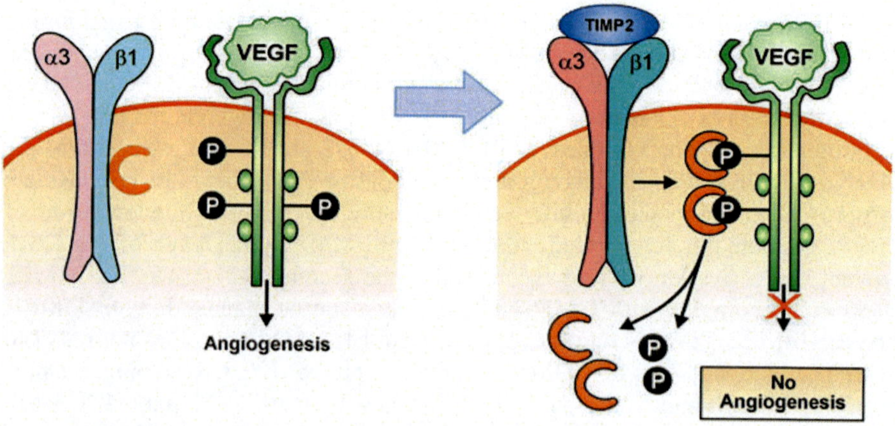

Fig. 25.2 TIMP2 and angiogenesis. TIMP2 binds to integrin α3β1 and promotes dephosphorylation of the activated VEGF receptor, thereby interrupting the VEGF-induced signaling and angiogenesis

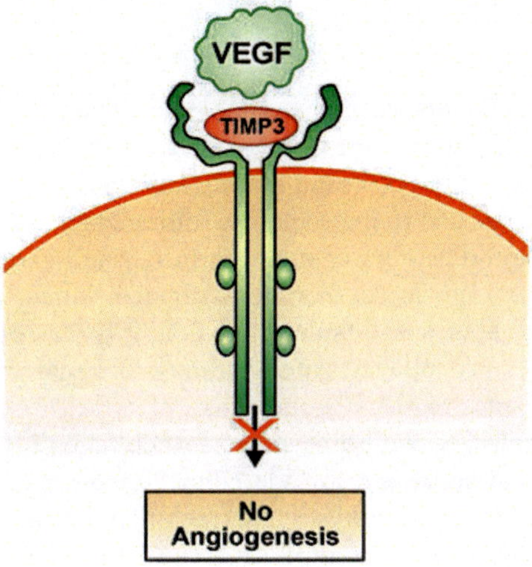

Fig. 25.3 TIMP3 and angiogenesis. TIMP3 disrupts the interaction between VEGF and VEGF receptor 2 in endothelial cells thereby inhibiting angiogenesis

infarction were reported to have enhanced angiogenesis [61], which one would predict to be beneficial to the ischemic cardiomyocytes and to limit the expansion of the infarct scar and improve cardiac function. However, TIMP3$^{-/-}$-MI mice exhibited markedly higher rates of LV rupture, greater infarct expansion, and more severe LV dilation and dysfunction [62]. In addition, TIMP3$^{-/-}$ mice generally do poorly when subjected to different diseases [32, 54, 55, 63, 64]. Hence, the role of TIMP3 in angiogenesis and the subsequent beneficial outcomes may not be simply explained by its presence or absence and certainly needs to be addressed more systematically in vivo.

A number of in vitro studies indicate that TIMPs also possess cell growth-promoting properties. Adenoviral overexpression of all TIMPs induced fibroblast proliferation, while Ad-TIMP2 additionally increased collagen synthesis, and TIMP3 triggered apoptosis [65]. TIMP1 promoted proliferation of different cell types including fibroblasts and endothelial cells [66], whereas TIMP2 promoted fibroblast growth [67] but hindered endothelial cell proliferation [68]. Growth-promoting function of TIMP1 and TIMP2 appears to be independent of their MMP-inhibiting function since reductively alkylated TIMPs, which lack MMP inhibitory activity, maintain their cell growth-promoting activity [69]. TIMP3 stimulated growth of chick embryo [70] and cardiac fibroblast [65] in addition to apoptosis in the latter group [65], whereas TIMP3 deficiency promoted proliferation of cultured neonatal mouse cardiomyocytes [71]. TIMP4 inhibited growth of tumor cells [72] and induced apoptosis in transformed fibroblasts [73]. Overall, these studies indicate that the growth-related function of TIMPs is specific to tissue and cell type and proposes potential key roles for TIMPs in cardiac repair and wound healing process which need to be further explored.

25.6 Remodeling of Myocardial ECM in Heart Failure

Maladaptive myocardial remodeling is characterized by an overall imbalance in ECM turnover, which can result in excess accumulation or disruption of the ECM structural proteins, mainly collagens. This will lead to altered spatial orientation of cells and the vasculature, diminished systolic performance, decreased compliance, and diastolic dysfunction in failing hearts. The turnover rate for collagen is about tenfold slower than non-collagen proteins [74]; thus, the myocardium is vulnerable to aberrant remodeling in conditions associated with increased ECM degradation.

Procollagen type I carboxy-terminal propeptide (PICP) and amino-terminal propeptides (PINP) and those of type III collagen peptide (PIIICP and PIIINP) are released with biosynthesis of collagen type I or type III in a stoichiometric manner and hence are considered as biomarkers of collagen synthesis. However, CITP and CIIITP, carboxy-terminal telopeptides of collagen type I and type III, are produced when collagen type I or type III are degraded and as such are considered as biomarkers of collagen degradation [75].

In a Caucasian population of hypertensive patients, the plasma levels of markers of ECM collagen turnover (CITP, PICP, PIIINP) and MMP2 were predictors of diastolic dysfunction and heart failure with preserved ejection fraction [76]. The serum PICP/CITP levels were higher in hypertensive patients [77] and in patients with dilated cardiomyopathy [78] who also exhibited increased collagen accumulation in the myocardium. Although early degradation of collagen type III has been associated with adverse LV dysfunction in later stages of the disease following MI [79], a recent study reported that in patients with acute MI, CITP was the strongest marker of outcomes among other collagen turnover markers including BNP, as a predictor of heart failure development and all cause death in these patients [80]. This is not unpredictable since MI and the subsequent "wound healing" of the myocardial tissue involve a great extent of ECM turnover. Immediately after MI, the existing ECM structure is degraded while normal collagen struts are lost, and cross-linking within the ischemic myocardium is reduced. This collagen turnover is accompanied by increased protease activity and increased probability of left ventricular rupture [81]. The degraded ECM structure allows for infiltration of inflammatory cells consistent with early elevations in MMP8 and MMP9 levels, the MMPs produced by neutrophils, in patients with MI [82]. This further leads to proliferation and maturation of fibroblasts, which synthesize new collagens and result in matrix deposition and scar formation. During the later phase of post-MI remodeling, an inefficient support of the newly formed ECM within the infarct and peri-infarct region will lead to LV wall thinning and slippage of the surviving myocytes resulting in further infarct expansion, LV dilation and dysfunction, and eventually heart failure. In addition, insufficient or impaired ECM remodeling can lead to LV rupture which has recently been shown to account for a high proportion of sudden death early post-MI [83, 84]. This is in agreement with the findings in experimental models of MI [30, 62, 85]. Hence, early prevention of adverse ECM remodeling in the infarct and the surrounding myocardium is crucial in preventing rupture and preserving cardiac function post-MI.

25.7 Alterations in TIMP Levels Have Been Linked to Different Phases of Heart Failure

A number of studies have examined the TIMP levels in plasma or the myocardium from patients with different types of heart disease (Table 25.1). Following acute myocardial infarction (AMI), plasma TIMP1 and MMP9 levels correlated directly with LV dysfunction and remodeling [86], while analysis of other TIMPs in AMI patients revealed elevated TIMP1, TIMP2, and TIMP4 levels by day 4 post-AMI which correlated positively with occurrence of major adverse cardiac events in these patients [88]. Interestingly, temporal changes in plasma MMPs and TIMP levels following myocardial infarction showed a rapid and sustained increase in MMP9 and MMP8, with a delayed increase in TIMP2 and TIMP4 levels [82]. This initial rise in proteinase activities was identified as the underlying cause of the adverse remodeling at later stages of disease progression since it was not blocked by a matching rise in TIMPs. These studies highlight the time-sensitive alterations in TIMP and MMP levels which should be considered when developing TIMP-dependent therapies for post-MI patients. In patients with ischemic or idiopathic dilated cardiomyopathy (DCM), myocardial protein levels of TIMP1, TIMP3, and TIMP4 were reduced while TIMP2 levels remained unaltered [73, 89], although another study reported a marked increase in TIMP2 levels in end-stage DCM patients [90]. Implantation of an LV assist device (LVAD) in DCM patients reduced MMP1 and MMP9 and increased TIMP1 and TIMP3 levels [91]. In patients with aortic stenosis, TIMP1 and TIMP2 levels were significantly elevated in the LV myocardium which directly correlated with the degree of fibrosis [27], while another study further reported increased TIMP3 and unaltered TIMP4 levels in similar patients [28].

Circulating levels of TIMP2 have also been correlated with systolic function in patients with hypertrophic cardiomyopathy as it was elevated in patients with systolic dysfunction but not in those with preserved systolic function, whereas TIMP1 levels were elevated in both groups [92]. Increased plasma TIMP1 levels have been correlated with myocardial hypertrophy [93], fibrosis, and diastolic dysfunction in hypertensive patients [94]. In addition to inhibiting MMPs, TIMP1 has been proposed to promote fibrosis by inducing collagen synthesis [93, 94]. Consistently, in patients with atrial fibrillation, collagen content in the atria correlated positively with elevated levels of TIMP1 [95]. TIMP1 levels are also increased in patients with cardiac amyloidosis, characterized by amyloid infiltration and ECM disruption, which correlated with diastolic dysfunction, but not with LV wall thickness [96]. TIMP1 levels were also significantly elevated in the myocardium of patients with progressive heart failure [97] and idiopathic dilated cardiomyopathy [90], although an earlier study had reported a reduction in TIMP1 levels in hearts with dilated cardiomyopathy [98]. In the Framingham heart study [99], a thorough analysis of plasma TIMP1 levels in relation to multiple cardiovascular risk factors revealed that in patients without heart failure or MI, plasma TIMP1 level was directly related to smoking, diabetes, and use of hypertensive treatment. In addition, TIMP1 was

Table 25.1 Alterations in TIMPs, MMPs, and ECM turnover markers in patients with different heart failure syndromes

Disease	Type of sample	Observations	References
Hypertensive with heart failure with preserved ejection fraction	Plasma	↑CITP, ↑PICP, ↑PIIINP ↔ TIMP1, ↔MMP1, ↑MMP2	Martos et al. [76]
Hypertensive with heart failure with preserved ejection fraction and ↑ filling pressure	Plasma	↑ TIMP1	Gonzalez et al. [100]
Hypertensive with increased myocardial collagen accumulation	Plasma	↑ PICP/CITP	Diez et al. [77]
Hypertensive with myocardial fibrosis and diastolic dysfunction	Plasma	↑ TIMP1	Lindsay et al. [94]
Hypertensive with LV hypertrophy	Plasma	↑ TIMP1, ↔ PIIINP	Timms et al. [93]
Myocardial infarction	Plasma	↑ CITP	Manhenke et al. [80]
		↑ TIMP1, ↑MMP9	Kelly et al. [86]
		↑ MMP3	Kelly et al. [87]
		↑ TIMP1, ↑ TIMP2, ↑ TIMP4	Kelly et al. [88]
Myocardial infarction	Plasma	↑TIMP1, ↑MMP8, ↑MMP2 (early) ↑TIMP1, ↑TIMP2, ↑TIMP4 (delayed)	Webb et al. [82]
Ischemic cardiomyopathy	Myocardium (protein)	↓ TIMP1, ↓TIMP3, ↓TIMP4 ↔ TIMP2	Li et al. [89]
Ischemic/idiopathic DCM	Myocardium (protein)	↓ TIMP4	Tummalapalli et al. [73]
End-stage DCM	Myocardium (protein)	↑↑ TIMP1, ↑↑TIMP2 ↓MMP1, ↔MMP2, ↑MMP3, ↑MMP9	Thomas et al. [90]
Aortic stenosis (pressure overload)	Myocardium (LV)	↑TIMP1, ↑TIMP2	Heymans et al. [27]
Aortic stenosis	Myocardium (mRNA and protein)	↓TIMP1, ↑TIMP2, ↑TIMP3, TIMP4(↓mRNA, ↑protein), ↓MMP1, ↓MMP9, ↑MMP2, ↑MMP3	Fielitz et al. [28]

Hypertrophic cardiomyopathy: 1. Systolic dysfunction 2. Preserved systolic function	Plasma	1. ↑TIMP1, ↑TIMP2, ↑MMP2,↔MMP3, ↔MMP9 2. ↔TIMP1, ↔ TIMP2, ↔MMP2	Noji et al. [92]
Cardiac amyloidosis (diastolic dysfunction, no hypertrophy)	Serum	↑ TIMP1, ↑MMP2	Biolo et al. [96]
Atrial fibrillation	Atrial tissue	↑ TIMP1, ↑ TIMP3	Mukherjee et al. [95]
Progressive heart failure	Myocardium	↑TIMP1	Barton et al. [97]
Chronic pressure overload 1. Predictor of hypertrophy 2. Predictor of diastolic HF	Plasma	1. ↑ TIMP1, ↑ MMP7, ↑ MMP9 2. ↑MMP2, ↑TIMP4, ↓ MMP8	Zile et al. [101]

positively associated with echocardiographic parameters of LV structural remodeling (LV mass, wall thickness, end-systolic and left atrial diameters) and negatively correlated with fractional shortening, a measure of LV function [99]. In hypertensive patients with heart failure with normal ejection fraction and elevated filling pressures, elevated TIMP1 levels were proposed to prevent collagen degradation, facilitating myocardial fibrosis and thereby contributing to the elevated filling pressure in these patients [100]. The rise in TIMP1 levels, irrespective of the type of heart disease, suggests that TIMP1 is likely a disease marker, and as such, targeting TIMP1 in treating heart disease may not provide significant therapeutic benefits. A gender dependence in TIMP levels has also been reported. Among patients with no heart failure or myocardial infarction, plasma TIMP1 levels were higher in men compared to women (other TIMPs were not examined) [99], whereas in AMI patients, plasma levels of TIMP1, TIMP2, and TIMP4 were elevated more in women versus men [88]. However, this difference in TIMP levels was not linked to the gender-dependent severity of heart disease.

A recent study on patients with chronic pressure overload demonstrated that increased plasma levels of TIMP1, MMP7, and MMP9 correlated with adverse ECM remodeling, while increased MMP2 and TIMP4, with decreased MMP8, best correlated with diastolic heart failure [101]. This study proposed that a combination of plasma markers of changes in ECM fibrillar collagen homeostasis as well as MMPs and TIMPs are indeed most suitable in identifying cardiac remodeling such as LV hypertrophy and diastolic heart failure [101]. While developing a multi-biomarker approach could be a more useful tool in identifying the specific aspects of cardiac remodeling and disease, identifying the alterations in TIMP levels, preferably in early stages of heart disease, could offer the opportunity to restore the imbalance in the TIMP–MMP interaction by delivering the identified TIMP(s) systemically or to the site of injury.

25.8 Animal Models Have Indicated Involvement of TIMPs in Heart Disease Syndromes

In addition to alterations in ECM regulatory proteins in patients with heart disease (discussed above), numerous studies have reported altered levels and activities of TIMPs and MMPs in different experimental models of heart disease leading to heart failure [6, 7, 74, 102]. The causal role of TIMPs in development and progression of heart disease is best demonstrated by using genetically modified mouse models with altered TIMP levels [103].

Deletion of TIMP1 gene resulted in a modest LV dilation in mice at 4 months of age, while LV systolic pressure and ejection fraction were preserved, and no myocyte hypertrophy was detected [104]. Myocardial infarction in TIMP1$^{-/-}$ mice led to greater LV dilation and increased LV end-diastolic volume compared to parallel wild-type (WT) mice [105], which were prevented with MMP inhibition [106]. This indicates that the MMP inhibitory function of TIMP1 is required for cardiac recovery from MI.

TIMP2 deficiency abolished activation of pro-MMP2 to its cleaved form in the fibroblast [107] and myocardium [30] in mice. While TIMP2$^{-/-}$ mice do not exhibit any baseline alteration in cardiac structure or function, myocardial infarction in TIMP2$^{-/-}$ mice led to greater infarct expansion but a similar rate of LV rupture incident compared to WT mice. TIMP2$^{-/-}$-MI mice exhibited exacerbated LV dilation with systolic and diastolic dysfunction, associated with elevated MT1-MMP activity [30]. Recently, we reported that biomechanical stress in mice lacking TIMP2 resulted in nonuniform ECM remodeling, integrin degradation, and impaired myocyte–ECM interaction leading to more severe LV dilation and dysfunction [108].

A recent study by Ramani et al. has reported that overexpression of TIMP2 in the peri-infarct myocardium reduced the infarct expansion and improved LV dilation and dysfunction [109]. This study is the first to examine the impact of TIMP overexpression in myocardial infarction, and it is intriguing to investigate if overexpression of individual TIMPs will impact different aspects of cardiac remodeling post-MI.

Mice lacking TIMP3 have been most extensively studied among the TIMP-deficient mice. Lack of TIMP3 led to dilated cardiomyopathy in aged mice (21-month-old) primarily due to elevated MMP9 levels [110], while young TIMP3$^{-/-}$ mice were found to be more susceptible to heart disease. Cardiac pressure overload in TIMP3$^{-/-}$ mice resulted in exacerbated LV dilation and dysfunction, excess hypertrophy, and fibrosis, with early onset of heart failure and increased mortality [54, 111]. This severe phenotype was due to an early rise in expression of tumor necrosis factor-α (TNFα) followed by elevated proteolytic activities (mostly MMP2 and MT1-MMP). Interruption of both these pathways was required for complete prevention of the disease in TIMP3$^{-/-}$ mice [54]. Pressure overload in the face of TIMP3 deficiency also disrupted an innate interaction between two key cytokines in heart disease, TNFα and TGFβ, which lead to excess interstitial and focal fibrotic lesions in the myocardium [111]. TIMP3$^{-/-}$ mice are also more susceptible to adverse complications of myocardial infarction [61, 62]. Post-MI survival was markedly compromised in TIMP3$^{-/-}$ mice primarily due to increased rate of LV rupture, between 3 days and 7 days post-MI, concomitant with an early and transient elevation in proteolytic activities [62]. Early inhibition of MMPs, during the first 2 days post-MI, was sufficient to significantly improve the post-MI outcomes in these mice at later stages [62].

TIMP4 deficiency did not impact the cardiac remodeling or dysfunction following pressure overload, while myocardial infarction leads to enhanced rate of LV rupture with unaltered LV dilation and dysfunction compared to WT-MI mice [112]. This is particularly interesting since TIMP4 was initially referred to as the "heart TIMP," and yet the impact of its absence on cardiac response to disease is quite subtle compared to other TIMPs. It is noteworthy that while TIMP4 may show a tissue-specific expression pattern in heart and brain, total levels of TIMP4 in the heart are in fact lower than TIMP2 and TIMP3 as reported in mice [5]. This could explain the lack of severe outcomes of TIMP4 deficiency in cardiac structure and function and its response to disease.

25.9 TIMP Deficiency and Cardiac Inflammation

Inflammation is one of the key characteristics of cardiac ischemic injury or infarction. High neutrophil count correlates with a high incidence of adverse cardiac events in patients with acute MI [113, 114]. In mice lacking TIMP2 [30], TIMP3 [62], or TIMP4 [112], myocardial infarction was accompanied by severe inflammation in the infarct and peri-infarct myocardium of mice. Migration of inflammatory cells into the damaged myocardium is triggered by a number of factors including the ECM degradation products [115–118] generated by the action of MMPs, which would be enhanced in the absence of the inhibitory function of TIMPs. ECM degradation products can act as chemoattractants for inflammatory cells. Fragments of fibronectin [115, 119] and elastin [116, 120] trigger migration of monocytes and neutrophils to the site of injury. Post-MI formation of laminin and fibronectin degradation products correlated with macrophage infiltrations in the infarcted myocardium [117], while mice lacking MMP2 showed less ECM degradation and less inflammation post-MI [117]. Neutrophils generate MMP8, MMP9 [121, 122], and MT6-MMP(MMP25) [123], while macrophages produce a number of MMPs, including MMP1, MMP3, MMP7, MMP8, MMP9, and MMP12 [124]. Hence, recruitment of inflammatory cells to the damaged myocardium can further exacerbate the injury. TIMP3 has also been reported to constrain neutrophil influx in the lungs by inhibiting MMPs [63] and to impact inflammation through regulating the TNFα-mediated signaling [55]. MMP-independent function of TIMPs in regulating inflammation has also been suggested, but no mechanism has been proposed [125].

Overall, mice deficient in specific TIMPs have helped us identify distinct roles for each TIMP in different types of heart disease. While TIMP3 seems to be critical in LV rupture and dysfunction following MI, TIMP2 deficiency only impacted the cardiac function while lack of TIMP4 only increased the rate of LV rupture. Further, TIMP1 and TIMP4 do not appear to play a critical role in cardiac response to pressure overload, whereas TIMP3 is required for optimal structure and function following biomechanical stress, while role of TIMP2 in this process remains undetermined.

25.10 Conclusions

TIMPs contribute to multiple aspects of cardiac remodeling and homeostasis and cardiac function. Alterations in TIMP levels in heart disease patients can be variable, depending on the type and stage of cardiomyopathy, which could pose the question if a reduction or an increase in their levels is beneficial or detrimental. TIMP-deficient mice have provided an invaluable tool in evaluating the causal role of each TIMP in different models of heart disease, clearly demonstrating that TIMPs are required for an optimal cardiac response to disease, as they each influence a different aspect of disease outcome. In addition, the role of individual TIMPs in different cell types, such as cardiomyocytes, fibroblasts, endothelial cells, and smooth muscle cells which collectively determine the cardiovascular remodeling and function, needs to be determined. While overexpressing TIMPs in an organ or cell line of interest is a valid approach to explore their function in different settings, it is important to note that excess overexpression of any gene can exert stress on the cells (endoplasmic reticulum stress or ER stress) resulting in adverse effects and cell death. Hence, overexpression of TIMPs while maintaining a balance in the TIMP–MMP interaction is critical in developing therapies for diseases where TIMP levels are reduced. It is intriguing to explore if "TIMP therapy" to replenish the TIMP levels, individually or in combination, could offer a protective outcome in heart disease.

Acknowledgments ZK is a new investigator of Heart and Stroke Foundation of Canada (HSFC) and an Alberta Innovates-Health Solutions (AIHS) Scholar.

References

1. Hunt SA, Abraham WT, Chin MH et al (2005) ACC/AHA 2005 Guideline Update for the Diagnosis and Management of Chronic Heart Failure in the Adult: a report of the American College of Cardiology/American Heart Association Task Force on Practice Guidelines. Circulation 112:e154–e235
2. Baker AH, Edwards DR, Murphy G (2002) Metalloproteinase inhibitors: biological actions and therapeutic opportunities. J Cell Sci 115:3719–3727
3. Sottrup-Jensen L, Birkedal-Hansen H (1989) Human fibroblast collagenase-alpha-macroglobulin interactions. Localization of cleavage sites in the bait regions of five mammalian alpha-macroglobulins. J Biol Chem 264:393–401
4. Oh J, Takahashi R, Kondo S et al (2001) The membrane-anchored MMP inhibitor RECK is a key regulator of extracellular matrix integrity and angiogenesis. Cell 107:789–800
5. Nuttall RK, Sampieri CL, Pennington CJ et al (2004) Expression analysis of the entire MMP and TIMP gene families during mouse tissue development. FEBS Lett 563:129–134
6. Li YY, McTiernan CF, Feldman AM (2000) Interplay of matrix metalloproteinases, tissue inhibitors of metalloproteinases and their regulators in cardiac matrix remodeling. Cardiovasc Res 46:214–224
7. Spinale FG (2007) Myocardial matrix remodeling and the matrix metalloproteinases: influence on cardiac form and function. Physiol Rev 87:1285–1342
8. Spinale FG, Wilbur NM (2009) Matrix metalloproteinase therapy in heart failure. Curr Treat Options Cardiovasc Med 11:339–346
9. Ross RS, Borg TK (2001) Integrins and the myocardium. Circ Res 88:1112–1119
10. Manso AM, Elsherif L, Kang SM, Ross RS (2006) Integrins, membrane-type matrix metalloproteinases and ADAMs: potential implications for cardiac remodeling. Cardiovasc Res 69: 574–584
11. Carver W, Molano I, Reaves TA et al (1995) Role of the alpha 1 beta 1 integrin complex in collagen gel contraction in vitro by fibroblasts. J Cell Physiol 165:425–437
12. Sheikh F, Chen Y, Liang X et al (2006) alpha-E-catenin inactivation disrupts the cardiomyocyte adherens junction, resulting in cardiomyopathy and susceptibility to wall rupture. Circulation 114:1046–1055
13. Masuelli L, Bei R, Sacchetti P et al (2003) Beta-catenin accumulates in intercalated disks of hypertrophic cardiomyopathic hearts. Cardiovasc Res 60:376–387

14. Tuuttila A, Morgunova E, Bergmann U et al (1998) Three-dimensional structure of human tissue inhibitor of metalloproteinases-2 at 2.1 A resolution. J Mol Biol 284:1133–1140
15. Apte SS, Olsen BR, Murphy G (1995) The gene structure of tissue inhibitor of metalloproteinases (TIMP)-3 and its inhibitory activities define the distinct TIMP gene family. J Biol Chem 270:14313–14318
16. Langton KP, Barker MD, McKie N (1998) Localization of the functional domains of human tissue inhibitor of metalloproteinases-3 and the effects of a Sorsby's fundus dystrophy mutation. J Biol Chem 273:16778–16781
17. Caterina NC, Windsor LJ, Bodden MK et al (1998) Glycosylation and NH2-terminal domain mutants of the tissue inhibitor of metalloproteinases-1 (TIMP-1). Biochim Biophys Acta 1388:21–34
18. Dennis JW, Granovsky M, Warren CE (1999) Protein glycosylation in development and disease. Bioessays 21:412–421
19. Stroud RE, Deschamps AM, Lowry AS et al (2005) Plasma monitoring of the myocardial specific tissue inhibitor of metalloproteinase-4 after alcohol septal ablation in hypertrophic obstructive cardiomyopathy. J Card Fail 11:124–130
20. Yu WH, Yu S, Meng Q et al (2000) TIMP-3 binds to sulfated glycosaminoglycans of the extracellular matrix. J Biol Chem 275:31226–31232
21. English JL, Kassiri Z, Koskivirta I et al (2006) Individual Timp deficiencies differentially impact pro-MMP-2 activation. J Biol Chem 281:10337–10346
22. Schulze CJ, Wang W, Suarez-Pinzon WL et al (2003) Imbalance between tissue inhibitor of metalloproteinase-4 and matrix metalloproteinases during acute myocardial ischemia-reperfusion injury. Circulation 107:2487–2492
23. Melendez-Zajgla J, Del Pozo L, Ceballos G, Maldonado V (2008) Tissue inhibitor of metalloproteinases-4. The road less traveled. Mol Cancer 7:85
24. Santos-Martinez MJ, Medina C, Jurasz P, Radomski MW (2008) Role of metalloproteinases in platelet function. Thromb Res 121:535–542
25. Lambert E, Dasse E, Haye B, Petitfrere E (2004) TIMPs as multifacial proteins. Crit Rev Oncol Hematol 49:187–198
26. Kadri Z, Petitfrere E, Boudot C et al (2000) Erythropoietin induction of tissue inhibitors of metalloproteinase-1 expression and secretion is mediated by mitogen-activated protein kinase and phosphatidylinositol 3-kinase pathways. Cell Growth Differ 11:573–580
27. Heymans S, Schroen B, Vermeersch P et al (2005) Increased cardiac expression of tissue inhibitor of metalloproteinase-1 and tissue inhibitor of metalloproteinase-2 is related to cardiac fibrosis and dysfunction in the chronic pressure-overloaded human heart. Circulation 112:1136–1144
28. Fielitz J, Leuschner M, Zurbrugg HR et al (2004) Regulation of matrix metalloproteinases and their inhibitors in the left ventricular myocardium of patients with aortic stenosis. J Mol Med 82:809–820
29. Polyakova V, Miyagawa S, Szalay Z et al (2008) Atrial extracellular matrix remodelling in patients with atrial fibrillation. J Cell Mol Med 12:189–208
30. Kandalam V, Basu R, Abraham T et al (2010) TIMP2 deficiency accelerates adverse post-myocardial infarction remodeling because of enhanced MT1-MMP activity despite lack of MMP2 activation. Circ Res 106:796–808
31. Kim H, Oda T, Lopez-Guisa J et al (2001) TIMP-1 deficiency does not attenuate interstitial fibrosis in obstructive nephropathy. J Am Soc Nephrol 12:736–748
32. Kassiri Z, Oudit GY, Kandalam V et al (2009) Loss of TIMP3 enhances interstitial nephritis and fibrosis. J Am Soc Nephrol 20:1223–1235
33. Li WQ, Qureshi HY, Liacini A et al (2004) Transforming growth factor Beta1 induction of tissue inhibitor of metalloproteinases 3 in articular chondrocytes is mediated by reactive oxygen species. Free Radic Biol Med 37:196–207

34. Hoshino Y, Mio T, Nagai S et al (2001) Fibrogenic and inflammatory cytokines modulate mRNA expressions of matrix metalloproteinase-3 and tissue inhibitor of metalloproteinase-3 in type II pneumocytes. Respiration 68:509–516
35. Clark IM, Swingler TE, Sampieri CL, Edwards DR (2008) The regulation of matrix metalloproteinases and their inhibitors. Int J Biochem Cell Biol 40:1362–1378
36. Zhong ZD, Hammani K, Bae WS, DeClerck YA (2000) NF-Y and Sp1 cooperate for the transcriptional activation and cAMP response of human tissue inhibitor of metalloproteinases-2. J Biol Chem 275:18602–18610
37. Tanaka K, Iwamoto Y, Ito Y et al (1995) Cyclic AMP-regulated synthesis of the tissue inhibitors of metalloproteinases suppresses the invasive potential of the human fibrosarcoma cell line HT1080. Cancer Res 55:2927–2935
38. Buzzio OL, Lu Z, Miller CD et al (2006) FOXO1A differentially regulates genes of decidualization. Endocrinology 147:3870–3876
39. Gomez DE, Alonso DF, Yoshiji H, Thorgeirsson UP (1997) Tissue inhibitors of metalloproteinases: structure, regulation and biological functions. Eur J Cell Biol 74:111–122
40. Birkedal-Hansen H, Moore WG, Bodden MK et al (1993) Matrix metalloproteinases: a review. Crit Rev Oral Biol Med 4:197–250
41. Strongin AY, Collier I, Bannikov G et al (1995) Mechanism of cell surface activation of 72-kDa type IV collagenase. Isolation of the activated form of the membrane metalloprotease. J Biol Chem 270:5331–5338
42. Atkinson SJ, Crabbe T, Cowell S et al (1995) Intermolecular autolytic cleavage can contribute to the activation of progelatinase A by cell membranes. J Biol Chem 270:30479–30485
43. Bigg HF, Morrison CJ, Butler GS et al (2001) Tissue inhibitor of metalloproteinases-4 inhibits but does not support the activation of gelatinase A via efficient inhibition of membrane type 1-matrix metalloproteinase. Cancer Res 61:3610–3618
44. Liu YE, Wang M, Greene J et al (1997) Preparation and characterization of recombinant tissue inhibitor of metalloproteinase 4 (TIMP-4). J Biol Chem 272:20479–20483
45. Hernandez-Barrantes S, Shimura Y, Soloway PD et al (2001) Differential roles of TIMP-4 and TIMP-2 in pro-MMP-2 activation by MT1-MMP. Biochem Biophys Res Commun 281: 126–130
46. Kashiwagi M, Tortorella M, Nagase H, Brew K (2001) TIMP-3 is a potent inhibitor of aggrecanase 1 (ADAM-TS4) and aggrecanase 2 (ADAM-TS5). J Biol Chem 276:12501–12504
47. Brew K, Nagase H (2010) The tissue inhibitors of metalloproteinases (TIMPs): an ancient family with structural and functional diversity. Biochim Biophys Acta 1803:55–71
48. Loechel F, Fox JW, Murphy G et al (2000) ADAM 12-S cleaves IGFBP-3 and IGFBP-5 and is inhibited by TIMP-3. Biochem Biophys Res Commun 278:511–515
49. Wisniewska M, Goettig P, Maskos K et al (2008) Structural determinants of the ADAM inhibition by TIMP-3: crystal structure of the TACE-N-TIMP-3 complex. J Mol Biol 381: 1307–1319
50. Amour A, Slocombe PM, Webster A et al (1998) TNF-alpha converting enzyme (TACE) is inhibited by TIMP-3. FEBS Lett 435:39–44
51. Asakura M, Kitakaze M, Takashima S et al (2002) Cardiac hypertrophy is inhibited by antagonism of ADAM12 processing of HB-EGF: metalloproteinase inhibitors as a new therapy. Nat Med 8:35–40
52. Black RA, Rauch CT, Kozlosky CJ et al (1997) A metalloproteinase disintegrin that releases tumour-necrosis factor-alpha from cells. Nature 385:729–733
53. Lee DC, Sunnarborg SW, Hinkle CL et al (2003) TACE/ADAM17 processing of EGFR ligands indicates a role as a physiological convertase. Ann N Y Acad Sci 995:22–38
54. Kassiri Z, Oudit GY, Sanchez O et al (2005) Combination of tumor necrosis factor-alpha ablation and matrix metalloproteinase inhibition prevents heart failure after pressure overload in tissue inhibitor of metalloproteinase-3 knock-out mice. Circ Res 97:380–390
55. Smookler DS, Mohammed FF, Kassiri Z et al (2006) Tissue inhibitor of metalloproteinase 3 regulates TNF-dependent systemic inflammation. J Immunol 176:721–725

56. Chirco R, Liu XW, Jung KK, Kim HR (2006) Novel functions of TIMPs in cell signaling. Cancer Metastasis Rev 25:99–113
57. Seo DW, Li H, Guedez L et al (2003) TIMP-2 mediated inhibition of angiogenesis: an MMP-independent mechanism. Cell 114:171–180
58. Stetler-Stevenson WG, Seo DW (2005) TIMP-2: an endogenous inhibitor of angiogenesis. Trends Mol Med 11:97–103
59. Lluri G, Langlois GD, McClellan B et al (2006) Tissue inhibitor of metalloproteinase-2 (TIMP-2) regulates neuromuscular junction development via a beta1 integrin-mediated mechanism. J Neurobiol 66:1365–1377
60. Qi JH, Ebrahem Q, Moore N et al (2003) A novel function for tissue inhibitor of metalloproteinases-3 (TIMP3): inhibition of angiogenesis by blockage of VEGF binding to VEGF receptor-2. Nat Med 9:407–415
61. Tian H, Cimini M, Fedak PW et al (2007) TIMP-3 deficiency accelerates cardiac remodeling after myocardial infarction. J Mol Cell Cardiol 43:733–743
62. Kandalam V, Basu R, Abraham T et al (2010) Early activation of matrix metalloproteinases underlies the exacerbated systolic and diastolic dysfunction in mice lacking TIMP3 following myocardial infarction. Am J Physiol Heart Circ Physiol 299:H1012–H1023
63. Gill SE, Huizar I, Bench EM et al (2010) Tissue inhibitor of metalloproteinases 3 regulates resolution of inflammation following acute lung injury. Am J Pathol 176:64–73
64. Mohammed FF, Smookler DS, Taylor SE et al (2004) Abnormal TNF activity in Timp3–/– mice leads to chronic hepatic inflammation and failure of liver regeneration. Nat Genet 36:969–977
65. Lovelock JD, Baker AH, Gao F et al (2005) Heterogeneous effects of tissue inhibitors of matrix metalloproteinases on cardiac fibroblasts. Am J Physiol Heart Circ Physiol 288:H461–H468
66. Hayakawa T, Yamashita K, Tanzawa K et al (1992) Growth-promoting activity of tissue inhibitor of metalloproteinases-1 (TIMP-1) for a wide range of cells. A possible new growth factor in serum. FEBS Lett 298:29–32
67. Corcoran ML, Stetler-Stevenson WG (1995) Tissue inhibitor of metalloproteinase-2 stimulates fibroblast proliferation via a cAMP-dependent mechanism. J Biol Chem 270:13453–13459
68. Murphy AN, Unsworth EJ, Stetler-Stevenson WG (1993) Tissue inhibitor of metalloproteinases-2 inhibits bFGF-induced human microvascular endothelial cell proliferation. J Cell Physiol 157:351–358
69. Wingfield PT, Sax JK, Stahl SJ et al (1999) Biophysical and functional characterization of full-length, recombinant human tissue inhibitor of metalloproteinases-2 (TIMP-2) produced in *Escherichia coli*. Comparison of wild type and amino-terminal alanine appended variant with implications for the mechanism of TIMP functions. J Biol Chem 274:21362–21368
70. Yang TT, Hawkes SP (1992) Role of the 21-kDa protein TIMP-3 in oncogenic transformation of cultured chicken embryo fibroblasts. Proc Natl Acad Sci USA 89:10676–106780
71. Hammoud L, Xiang F, Lu X et al (2007) Endothelial nitric oxide synthase promotes neonatal cardiomyocyte proliferation by inhibiting tissue inhibitor of metalloproteinase-3 expression. Cardiovasc Res 75:359–368
72. Celiker MY, Wang M, Atsidaftos E et al (2001) Inhibition of Wilms' tumor growth by intramuscular administration of tissue inhibitor of metalloproteinases-4 plasmid DNA. Oncogene 20:4337–4343
73. Tummalapalli CM, Heath BJ, Tyagi SC (2001) Tissue inhibitor of metalloproteinase-4 instigates apoptosis in transformed cardiac fibroblasts. J Cell Biochem 80:512–521
74. Kassiri Z, Khokha R (2005) Myocardial extra-cellular matrix and its regulation by metalloproteinases and their inhibitors. Thromb Haemost 93:212–219
75. Zannad F, Rossignol P, Iraqi W (2010) Extracellular matrix fibrotic markers in heart failure. Heart Fail Rev 15:319–329
76. Martos R, Baugh J, Ledwidge M et al (2009) Diagnosis of heart failure with preserved ejection fraction: improved accuracy with the use of markers of collagen turnover. Eur J Heart Fail 11:191–197

77. Diez J, Querejeta R, Lopez B et al (2002) Losartan-dependent regression of myocardial fibrosis is associated with reduction of left ventricular chamber stiffness in hypertensive patients. Circulation 105:2512–2517
78. Izawa H, Murohara T, Nagata K et al (2005) Mineralocorticoid receptor antagonism ameliorates left ventricular diastolic dysfunction and myocardial fibrosis in mildly symptomatic patients with idiopathic dilated cardiomyopathy: a pilot study. Circulation 112: 2940–2945
79. Poulsen SH, Host NB, Jensen SE, Egstrup K (2000) Relationship between serum aminoterminal propeptide of type III procollagen and changes of left ventricular function after acute myocardial infarction. Circulation 101:1527–1532
80. Manhenke C, Orn S, Squire I et al (2010) The prognostic value of circulating markers of collagen turnover after acute myocardial infarction. Int J Cardiol 150:277–282
81. Jugdutt BI, Joljart MJ, Khan MI (1996) Rate of collagen deposition during healing and ventricular remodeling after myocardial infarction in rat and dog models. Circulation 94:94–101
82. Webb CS, Bonnema DD, Ahmed SH et al (2006) Specific temporal profile of matrix metalloproteinase release occurs in patients after myocardial infarction: relation to left ventricular remodeling. Circulation 114:1020–1027
83. Pouleur AC, Barkoudah E, Uno H et al (2010) Pathogenesis of sudden unexpected death in a clinical trial of patients with myocardial infarction and left ventricular dysfunction, heart failure, or both. Circulation 122:597–602
84. Jugdutt BI (2010) Preventing adverse remodeling and rupture during healing after myocardial infarction in mice and humans. Circulation 122:103–135
85. Yang Y, Ma Y, Han W et al (2008) Age-related differences in postinfarct left ventricular rupture and remodeling. Am J Physiol Heart Circ Physiol 294:H1815–H1822
86. Kelly D, Khan S, Cockerill G et al (2008) Circulating stromelysin-1 (MMP-3): a novel predictor of LV dysfunction, remodelling and all-cause mortality after acute myocardial infarction. Eur J Heart Fail 10:133–139
87. Kelly D, Khan SQ, Thompson M et al (2008) Plasma tissue inhibitor of metalloproteinase-1 and matrix metalloproteinase-9: novel indicators of left ventricular remodelling and prognosis after acute myocardial infarction. Eur Heart J 29:2116–2124
88. Kelly D, Squire IB, Khan SQ et al (2010) Usefulness of plasma tissue inhibitors of metalloproteinases as markers of prognosis after acute myocardial infarction. Am J Cardiol 106: 477–482
89. Li YY, Feldman AM, Sun Y, McTiernan CF (1998) Differential expression of tissue inhibitors of metalloproteinases in the failing human heart. Circulation 98:1728–1734
90. Thomas CV, Coker ML, Zellner JL et al (1998) Increased matrix metalloproteinase activity and selective upregulation in LV myocardium from patients with end-stage dilated cardiomyopathy. Circulation 97:1708–1715
91. Li YY, Feng Y, McTiernan CF et al (2001) Downregulation of matrix metalloproteinases and reduction in collagen damage in the failing human heart after support with left ventricular assist devices. Circulation 104:1147–1152
92. Noji Y, Shimizu M, Ino H et al (2004) Increased circulating matrix metalloproteinase-2 in patients with hypertrophic cardiomyopathy with systolic dysfunction. Circ J 68:355–360
93. Timms PM, Wright A, Maxwell P et al (2002) Plasma tissue inhibitor of metalloproteinase-1 levels are elevated in essential hypertension and related to left ventricular hypertrophy. Am J Hypertens 15:269–272
94. Lindsay MM, Maxwell P, Dunn FG (2002) TIMP-1: a marker of left ventricular diastolic dysfunction and fibrosis in hypertension. Hypertension 40:136–141
95. Mukherjee R, Herron AR, Lowry AS et al (2006) Selective induction of matrix metalloproteinases and tissue inhibitor of metalloproteinases in atrial and ventricular myocardium in patients with atrial fibrillation. Am J Cardiol 97:532–537
96. Biolo A, Ramamurthy S, Connors LH et al (2008) Matrix metalloproteinases and their tissue inhibitors in cardiac amyloidosis: relationship to structural, functional myocardial changes and to light chain amyloid deposition. Circ Heart Fail 1:249–257

97. Barton PJ, Birks EJ, Felkin LE et al (2003) Increased expression of extracellular matrix regulators TIMP1 and MMP1 in deteriorating heart failure. J Heart Lung Transplant 22:738–744
98. Tyagi SC, Kumar S, Voelker DJ et al (1996) Differential gene expression of extracellular matrix components in dilated cardiomyopathy. J Cell Biochem 63:185–198
99. Sundstrom J, Evans JC, Benjamin EJ et al (2004) Relations of plasma total TIMP-1 levels to cardiovascular risk factors and echocardiographic measures: the Framingham heart study. Eur Heart J 25:1509–1516
100. Gonzalez A, Lopez B, Querejeta R et al (2010) Filling pressures and collagen metabolism in hypertensive patients with heart failure and normal ejection fraction. Hypertension 55: 1418–1424
101. Zile MR, Desantis SM, Baicu CF et al (2011) Plasma Biomarkers That Reflect Determinants of Matrix Composition Identify the Presence of Left Ventricular Hypertrophy and Diastolic Heart Failure. Circ Heart Fail 4:246–256
102. Iwanaga Y, Aoyama T, Kihara Y et al (2002) Excessive activation of matrix metalloproteinases coincides with left ventricular remodeling during transition from hypertrophy to heart failure in hypertensive rats. J Am Coll Cardiol 39:1384–1391
103. Gill SE, Kassim SY, Birkland TP, Parks WC (2010) Mouse models of MMP and TIMP function. Methods Mol Biol 622:31–52
104. Roten L, Nemoto S, Simsic J et al (2000) Effects of gene deletion of the tissue inhibitor of the matrix metalloproteinase-type 1 (TIMP-1) on left ventricular geometry and function in mice. J Mol Cell Cardiol 32:109–120
105. Creemers EE, Davis JN, Parkhurst AM et al (2003) Deficiency of TIMP-1 exacerbates LV remodeling after myocardial infarction in mice. Am J Physiol Heart Circ Physiol 284:H364–H371
106. Ikonomidis JS, Hendrick JW, Parkhurst AM et al (2005) Accelerated LV remodeling after myocardial infarction in TIMP-1-deficient mice: effects of exogenous MMP inhibition. Am J Physiol Heart Circ Physiol 288:H149–H158
107. Wang Z, Juttermann R, Soloway PD (2000) TIMP-2 is required for efficient activation of proMMP-2 in vivo. J Biol Chem 275:26411–26415
108. Kandalam V, Basu R, Moore L et al (2011) Lack of Tissue Inhibitor of Metalloproteinases 2 Leads to Exacerbated Left Ventricular Dysfunction and Adverse Extracellular Matrix Remodeling in Response to Biomechanical Stress. Circulation 124:2094–2105
109. Ramani R, Nilles K, Gibson G et al (2011) Tissue inhibitor of metalloproteinase-2 gene delivery ameliorates postinfarction cardiac remodeling. Clin Transl Sci 4:24–31
110. Fedak PW, Smookler DS, Kassiri Z et al (2004) TIMP-3 deficiency leads to dilated cardiomyopathy. Circulation 110:2401–2409
111. Kassiri Z, Defamie V, Hariri M et al (2009) Simultaneous transforming growth factor beta-tumor necrosis factor activation and cross-talk cause aberrant remodeling response and myocardial fibrosis in Timp3-deficient heart. J Biol Chem 284:29893–29904
112. Koskivirta I, Kassiri Z, Rahkonen O et al (2010) Mice with tissue inhibitor of metalloproteinases 4 (Timp4) deletion succumb to induced myocardial infarction but not to cardiac pressure overload. J Biol Chem 285:24487–24493
113. Takahashi T, Hiasa Y, Ohara Y et al (2008) Relationship of admission neutrophil count to microvascular injury, left ventricular dilation, and long-term outcome in patients treated with primary angioplasty for acute myocardial infarction. Circ J 72:867–872
114. Papa A, Emdin M, Passino C et al (2008) Predictive value of elevated neutrophil-lymphocyte ratio on cardiac mortality in patients with stable coronary artery disease. Clin Chim Acta 395:27–31
115. Doherty DE, Henson PM, Clark RA (1990) Fibronectin fragments containing the RGDS cell-binding domain mediate monocyte migration into the rabbit lung. A potential mechanism for C5 fragment-induced monocyte lung accumulation. J Clin Invest 86:1065–1075
116. Senior RM, Griffin GL, Mecham RP (1980) Chemotactic activity of elastin-derived peptides. J Clin Invest 66:859–862

117. Matsumura S, Iwanaga S, Mochizuki S et al (2005) Targeted deletion or pharmacological inhibition of MMP-2 prevents cardiac rupture after myocardial infarction in mice. J Clin Invest 115:599–609
118. Adair-Kirk TL, Senior RM (2008) Fragments of extracellular matrix as mediators of inflammation. Int J Biochem Cell Biol 40:1101–1110
119. Clark RA, Wikner NE, Doherty DE, Norris DA (1988) Cryptic chemotactic activity of fibronectin for human monocytes resides in the 120-kDa fibroblastic cell-binding fragment. J Biol Chem 263:12115–12123
120. Hance KA, Tataria M, Ziporin SJ et al (2002) Monocyte chemotactic activity in human abdominal aortic aneurysms: role of elastin degradation peptides and the 67-kD cell surface elastin receptor. J Vasc Surg 35:254–261
121. Van Lint P, Libert C (2006) Matrix metalloproteinase-8: cleavage can be decisive. Cytokine Growth Factor Rev 17:217–223
122. Cuadrado E, Ortega L, Hernandez-Guillamon M et al (2008) Tissue plasminogen activator (t-PA) promotes neutrophil degranulation and MMP-9 release. J Leukoc Biol 84:207–214
123. Matsuda A, Itoh Y, Koshikawa N et al (2003) Clusterin, an abundant serum factor, is a possible negative regulator of MT6-MMP/MMP-25 produced by neutrophils. J Biol Chem 278:36350–36357
124. Lambert JM, Lopez EF, Lindsey ML (2008) Macrophage roles following myocardial infarction. Int J Cardiol 130:147–158
125. Vanhoutte D, Heymans S (2010) TIMPs and cardiac remodeling: 'Embracing the MMP-independent-side of the family'. J Mol Cell Cardiol 48:445–453

Chapter 26
Intracellular Matrix Remodeling and Cardiac Function in Ischemia–Reperfusion Injury

Xiaohu Fan, Mohammad A.M. Ali, Bryan G. Hughes, Anna Laura B. Jacob-Ferreira, and Richard Schulz

Abstract The effects of ischemia on animal and human myocardium have been extensively studied during the last four decades. Myocardial ischemia followed by subsequent reperfusion can cause profound damage to cardiac myocytes through enhanced oxidative stress and intracellular Ca^{2+} overload. Ischemia–reperfusion (I/R) injury leads to structural and functional remodeling of multiple intracellular matrix components in the cardiac myocyte. The intracellular matrix of cardiac myocytes includes major cytosolic components that include the cytoskeleton, contractile myofibrils, and subcellular organelles such as mitochondria, sarcoplasmic reticulum, and the nucleus. There are several proteolytic pathways inside the cell which may participate in cell injury and/or cell repair upon reperfusion of ischemic heart muscle. These include matrix metalloproteinases, calpains, lysosomal proteases, and the proteasome system, which are a major focus of research in I/R injury. The discovery of intramyocyte matrix metalloproteinase-2 (MMP-2) and biologically relevant protein substrates of it in the intracellular matrix has shaped a new paradigm of the pathophysiological role of MMP-2 during myocardial I/R injury. Emerging evidence indicates that oxidative stress can efficiently activate intracellular MMP-2 which rapidly mediates intracellular matrix remodeling of injured myocytes. This chapter will focus on the structural and functional remodeling of the intracellular matrix including the sarcomere, cytoskeleton, mitochondria, and nucleus, by proteolytic and other processes, in the context of I/R injury, with a particular emphasis on the rapidly expanding knowledge of the biology of intracellular MMP-2.

X. Fan • M.A.M. Ali • B.G. Hughes • A.L.B. Jacob-Ferreira • R. Schulz (✉)
Departments of Pediatrics and Pharmacology, Cardiovascular Research Centre, 4–62 HMRC, Mazankowski Alberta Heart Institute, University of Alberta, Edmonton, AB, Canada, T6G 2S2
e-mail: richard.schulz@ualberta.ca

B.I. Jugdutt and N.S. Dhalla (eds.), *Cardiac Remodeling: Molecular Mechanisms*, Advances in Biochemistry in Health and Disease 5, DOI 10.1007/978-1-4614-5930-9_26,

Keywords Intracellular matrix • Proteases • Matrix metalloproteinase-2 • Calpain • Ischemia–reperfusion • Sarcomere • Cytoskeleton • MMP inhibitor • Doxycycline

26.1 Sarcomeric Protein Degradation and Contractile Dysfunction

The cardiac sarcomere (Fig. 26.1) is the basic contractile unit of the heart which contains the contractile components of the thin (actin), thick (myosin), and third (titin) myofilaments essential for cardiac function. The acute contractile defect of the heart after I/R injury [1] has been shown to involve proteolysis and/or disorganization of myofilament proteins [2, 3] including troponin I, myosin, and titin [4, 5].

26.1.1 Calpain-Mediated Remodeling in Ischemic Heart Disease

Calpains are a family of non-lysosomal cysteine proteases consisting of several isoforms. The best characterized isoforms are calpain-1 (μ-calpain), calpain-2 (m-calpain), and calpain-3 (p94 calpain). The terms μ-calpain and m-calpain indicate the

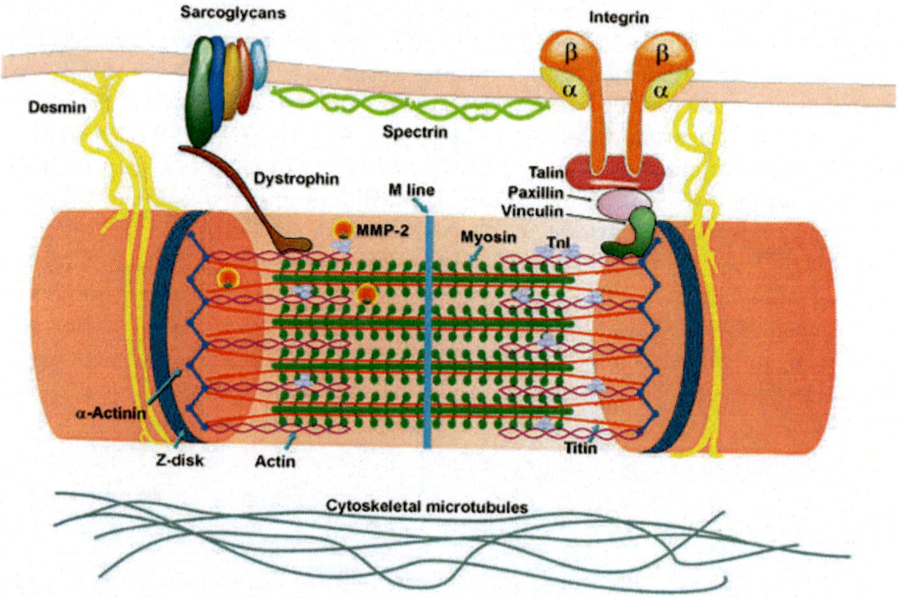

Fig. 26.1 Intracellular matrix and intracellular matrix metalloproteinase-2 (MMP-2). Schematic diagram shows the organization of sarcomeric and cytoskeletal proteins in relation to membrane anchor proteins in the cardiomyocyte. MMP-2 is localized to specific sarcomeric and cytoskeletal proteins and, upon myocardial oxidative stress injury, is able to proteolyze these substrates

required calcium concentration for in vitro activity (micromolar range for μ-calpain and millimolar range for m-calpain). Calpain-1 and -2 are considered to be ubiquitous, since they are expressed in nearly all tissues, whereas calpain-3 is expressed mainly in skeletal muscle [6]. Calpains participate in various cellular processes including remodeling of the sarcomere and cytoskeleton, signal transduction, and cell death [7]. Calpain-1, in particular, has been implicated in the pathogenesis of myocardial stunning injury, a reversible, sublethal injury of cardiac muscle which occurs upon reperfusion of the ischemic heart [8].

It has been suggested that activation of calpain-1 may lead to proteolytic degradation of sarcomeric proteins such as troponin I [9], titin, and α-actinin [10] which contributes to impaired contractile function (Fig. 26.2). In addition, calpain-1 activated during I/R injury has been shown to target various proteins involved in excitation–contraction coupling, thereby attenuating cardiac contractility. These targets include the SERCA2 pump [11], ryanodine receptor [12], and α-fodrin [13]. However, the conclusions of most of these studies rested on the use of pharmacological inhibitors of calpain, many of which were recently found to also inhibit MMP-2, another key protease in I/R injury [14]. In view of this, the relative contribution of calpain in I/R injury of the heart and other organs may need to be systemically revisited in comparison with MMP-2.

26.1.2 Intracellular Substrates of MMP-2 in Ischemic Heart Disease

Since their first description in amphibian metamorphosis [15], it has been recognized that MMPs play an active role in remodeling extracellular matrix proteins accompanying both physiological and pathological processes. Being originally described as secreted proteases, most researchers have focused on the long-term effects of MMPs on extracellular matrix remodeling following myocardial infarction (irreversible cellular injury resulting in myocyte death), hypertensive cardiac diseases, and other cardiomyopathies. In fact, MMPs have been recognized as proteases playing a pivotal role in matrix remodeling in such heart diseases (for review see [16]).

However, it has been recognized more recently that MMPs may also act on non-extracellular matrix substrates both outside [17] and inside the cell [4, 18]. This may occur within seconds to minutes rather than hours to days as occurs with so many of the extracellular matrix actions of MMPs. For example, MMP-2 was found to contribute to acute cardiac mechanical dysfunction in stunning injury before changes in extracellular matrix proteins [19]. Emerging evidence has shown that MMPs, and in particular MMP-2, are closely associated with subcellular compartments within cardiac myocytes, including the sarcomere [5, 20], cytoskeleton [21, 22], nuclei

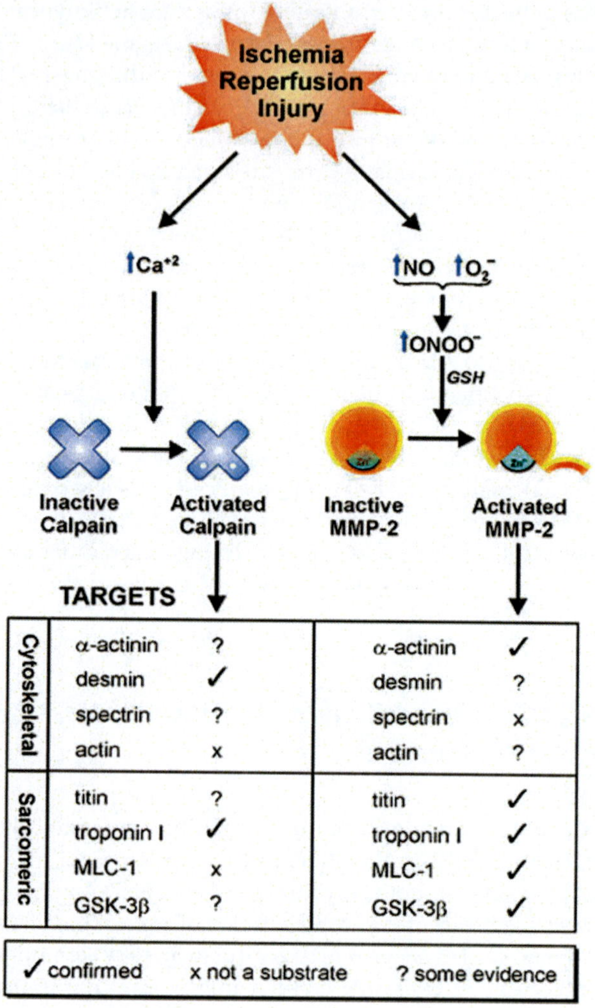

Fig. 26.2 Activation of intracellular proteases and their substrates. Mechanisms of enhanced intracellular proteolytic stress in acute myocardial ischemia/reperfusion injury include increased intracellular calcium, which leads to activation of calpains, and enhanced oxidative stress, particularly in the form of peroxynitrite biosynthesized from nitric oxide (NO) and superoxide, causing the direct activation of MMP-2. The table below summarizes potential targets for both proteases in ischemia/reperfusion injury

[23, 24], and mitochondria [20, 25]. Recently, distinct intracellular moieties of MMP-2 have been identified [25, 26], which physically confirm the concept that MMP-2 is indeed an intracellular protease [4, 18, 20].

Different mechanisms can activate and regulate MMP-2 activity, either extra- or intracellularly. MMP-2 is synthesized as a 72 kD zymogen protein which can be activated by proteolytic removal of the propeptide domain in pericellular and extracellular compartments, resulting in 64 kD MMP-2 [27]. However, 72 kD MMP-2 can also be activated without proteolysis as a direct result of oxidative stress. For example, MMP-2 is activated upon exposure to peroxynitrite, a prooxidant molecule implicated in various cardiac pathologies including I/R injury [28], via S-glutathiolation of a critical cysteine sulfhydryl moiety in the propeptide domain

[29] (Fig. 26.2). MMP-2 is also a phosphoprotein whose activity is increased by dephosphorylation [30]. MMP-2 activity can be inhibited by certain pharmacological agents, especially those which chelate the catalytic zinc found in its active site which is essential for MMP activity. Tetracyclines, especially doxycycline and minocycline, can inhibit MMPs independent of their antibacterial actions, most likely via chelation of zinc [31].

We will discuss below some cardiac sarcomeric and cytoskeletal proteins that have been shown to be targeted by MMP-2 in I/R injury (Fig. 26.2).

26.1.2.1 Troponin I

In 1999 Spinale's group showed a sarcomeric staining pattern of MMP-2 in pig heart muscle, yet did not provide an explanation of this unexpected result [21]. Our group then showed that MMP-2 is localized inside cardiac myocytes within the sarcomere and is responsible for the rapid degradation of troponin I in acute myocardial I/R injury [20]. Troponin I regulates actin–myosin interaction and is found in the thin myofilaments. Immunogold electron microscopy amongst other evidence showed that MMP-2 is an integral sarcomeric protein. Troponin I was highly susceptible to the proteolytic action of MMP-2 in vitro, and subjecting isolated rat hearts to acute I/R injury diminished myocardial troponin I content, an effect that was blocked by MMP inhibitors. In this study, we provided evidence that myocardial stunning injury is caused in part by MMP-2-mediated proteolysis of troponin I. This study was the first to recognize an intracellular biological role of a MMP as well as to identify the first intracellular target of MMP-2 in cardiac myocytes. Four years later, Lovett and Karliner's group reported that transgenic mice overexpressing a mutant, constitutively active MMP-2 in cardiomyocytes had marked derangements in the cardiac sarcomere, including troponin I degradation and reduced contractile function at the level of the myofilaments [32, 33].

26.1.2.2 Myosin Light Chain-1

Myosin light chain-1 was reported to undergo proteolytic degradation in hearts subjected to I/R injury [34]. MMP-2 activity was also found in preparations of thick myofilaments (which contain myosin light chain-1) from rat hearts, and MMP-2 was localized to the sarcomere in a pattern consistent with the known distribution of myosin light chain-1. Purified myosin light chain-1 was susceptible to proteolysis by MMP-2 in vitro. Two-dimensional gel electrophoresis followed by mass spectrometric analysis of myosin light chain-1 proteolysis products from I/R hearts identified a MMP-2 cleavage site within myosin light chain-1 at an accessible portion of the C terminus between tyrosine 189 and glutamate 190 [35].

26.1.2.3 Titin

Titin is the largest known mammalian protein (3,000–4,000 kD) and is found in cardiac and skeletal striated muscles. It spans nearly half the length of the sarcomere, from the Z-disk to the M-line region. It contains elastic segments formed by immunoglobulin-like repeats in the I band region, which allow it to act as a molecular spring, maintain the structural and functional stability of the myocyte and contribute to both active and passive stiffness of the myocyte [36].

Cardiac titin is expressed in two main isoforms: the shorter and stiffer N2B and the longer, more compliant N2BA isoforms. Hearts from adult small mammals (rats, mice, and rabbits) express predominately N2B titin, whereas large mammals including humans co-express N2BA and N2B titins at an approximate 1:1 ratio. The N2BA:N2B isoform ratio is increased in end-stage failing human hearts from chronically ischemic hearts of patients with coronary artery disease [37] and nonischemic dilated cardiomyopathy [38]. This titin remodeling decreases passive myocyte stiffness, most likely as a compensatory mechanism to counteract increased passive stiffness related to extracellular fibrosis [39]. Other forms of titin remodeling occur in heart disease, including reduced levels of titin phosphorylation, observed in dilated cardiomyopathy [40], and the formation of intramolecular disulfide bonds under oxidative stress [41], both of which could stiffen the titin spring function and contribute to impaired diastolic function following oxidative stress. The regulation of titin stiffness could affect various mechanical functions of the heart including diastolic filling, the Frank–Starling mechanism, and contractile performance in systole, the latter of which is also determined by titin [42].

In addition, earlier studies of ischemic and failing human hearts showed that titin is hydrolyzed [43] and appears highly disorganized in cardiac myocytes when analyzed by immunofluorescence microscopy [44]. We showed in rat and human myocardium that MMP-2 colocalizes with titin mainly near the Z-disk region of the cardiac sarcomere. Cleavage of titin in perfused rat hearts subjected to I/R injury, or in skinned cardiac myocytes incubated with MMP-2, was prevented by the MMP inhibitors *o*-phenanthroline or ONO-4817. Titin proteolysis in hearts was abolished in MMP-2 knockout mice subjected to I/R in vivo [5]. Thus MMP-2 appears to play an important role in titin homeostasis, which directly affects the contractile function of the heart at the sarcomeric level. Taken together, these studies reveal that MMP-2 may be a crucial protease which targets specific sarcomeric proteins as a result of oxidative stress injury to the heart.

26.1.3 MMP-2 and Calpains: A Case of Misattributed Function?

In light of the above studies, there appears to be overlap in the substrates and/or biological actions of MMP-2 and calpains in various cellular pathways (Fig. 26.2). It is now becoming evident that MMP-2 either targets a similar subset of proteins as calpain or calpain has been incorrectly identified as the protease responsible for some intracellular proteolytic activities [1, 45]. Indeed, much of the evidence for calpain degradation of substrates in cardiac cells rests on the use of pharmacological calpain inhibitors, including ALLN and PD-150606, which we recently found to efficiently inhibit MMP-2 activity at commonly used micromolar concentrations [14]. Furthermore, the exact role of calpain in acute myocardial I/R injury (stunning) is controversial as many of the earlier studies dating back several decades did not provide evidence for the subcellular co-localization of calpain with its putative substrates [1]. In a more recent study, myocardial-specific overexpression of calpain-1 in transgenic mice showed no evidence of troponin I degradation in the heart [46], whereas as mentioned above, troponin I levels were reduced in hearts from transgenic mice with myocardial-specific overexpression of constitutively active MMP-2 [33]. It is possible that MMP-2 and calpains share similar substrates. However, given the points raised above, it would be prudent to reevaluate suggested calpain substrates in the myocardium for their susceptibility to cleavage by MMP-2 and to critically evaluate the co-localization of calpains with their putative substrates. Moreover, caution is necessary in interpreting tissue calpain activity solely on the use of peptide-based enzyme assays which may not only measure calpain but also MMP activity.

26.1.4 Proteasomal and Lysosomal Degradation of the Intracellular Matrix

Proteasomes and lysosomes play important roles in the proteolysis of cardiac proteins. Lysosomes degrade the majority of endocytosed proteins when digesting expired organelles or cell debris. However, the proteasome system removes and recycles most unneeded or damaged intracellular proteins [47]. Oxidative modification of proteins affects their secondary and tertiary structures, resulting in protein unfolding, which may lead to a loss of function and enhanced proteolysis of these proteins [48]. Damaged, oxidized, and/or misfolded proteins are removed by the ubiquitin–proteasome system (UPS). The UPS is the main non-lysosomal protease complex involved in the proteolysis of intracellular proteins and therefore plays a key role in protein quality control [49].

As ATP is required for activation of the proteasome and for proper function of the UPS, ATP depletion during ischemia could be partially responsible for decreased proteasome activity in the ischemic heart. Also, accumulation of misfolded or mutated proteins as a result of oxidative stress can inhibit the cardiac UPS and may result in cardiomyopathy [48]. On the other hand, it was also reported in animal models of myocardial I/R injury that inhibition of the proteasome system significantly reduced infarct size by more than 50% and preserved ventricular contractility suggesting a role of proteasomes in the process of I/R injury [47]. Although myosin, actin, troponin C, and tropomyosin purified from skeletal muscle can be hydrolyzed by the proteasome pathway in vitro, these proteins are much less susceptible to proteasomal degradation when present in intact myofibrils or as soluble actomyosin complexes [47]. Thus, the rate-limiting step in their degradation seems to be their dissociation from the contractile filaments, where intracellular MMP-2 or calpains may play an important role in the scenario of I/R injury [47].

26.2 Cytoskeletal Protein Remodeling in Cardiac Ischemia and Reperfusion

The cardiac cytoskeleton (Fig. 26.1) consists of microfilaments, intermediate filaments (such as desmin), and the microtubular network. The cytoskeleton preserves cellular shape and enables cell migration and intracellular transport. It also maintains the proper localization of subcellular organelles such as the mitochondria, Golgi apparatus, nucleus, and sarcomere. The cardiac myocyte cytoskeleton is also specialized to transmit mechanical and electronic stimuli between cells. Investigation of early changes to cytoskeletal proteins in ischemic human myocardium suggest that they undergo degenerative alterations earlier than subcellular organelles [2]. Furthermore, disruption of the localization and integrity of the filamentous network of the cytoskeleton during prolonged ischemia accompanies I/R injury [50] (Fig. 26.3).

26.2.1 α-Actinin

α-Actinin connects actin filaments of adjacent sarcomeres and plays a substantial role in transmitting force generated by actin–myosin interaction. MMP-2 was found to colocalize with α-actinin in cardiac myocytes [21, 22]. We found that α-actinin is susceptible to degradation by MMP-2 in vitro and infusion of peroxynitrite into isolated, perfused rat hearts caused activation of MMP-2 with concomitant loss of myocardial α-actinin content. This was prevented by a selective MMP inhibitor, PD-166793 [22].

26.2.2 Desmin

Desmin is a cardiac-specific cytoskeletal protein with an average molecular weight of 53 kD. Desmin monomers assemble to form intermediate filaments 10 nm in diameter by polymerization which form in turn a transverse network that links the Z-bands of adjacent myofibrils. This maintains the integrity of cardiac myocytes and allows force transmission and mechanochemical signaling between cells [51]. In addition, mitochondria and T-tubules appear to be attached to the intermediate filament network [52].

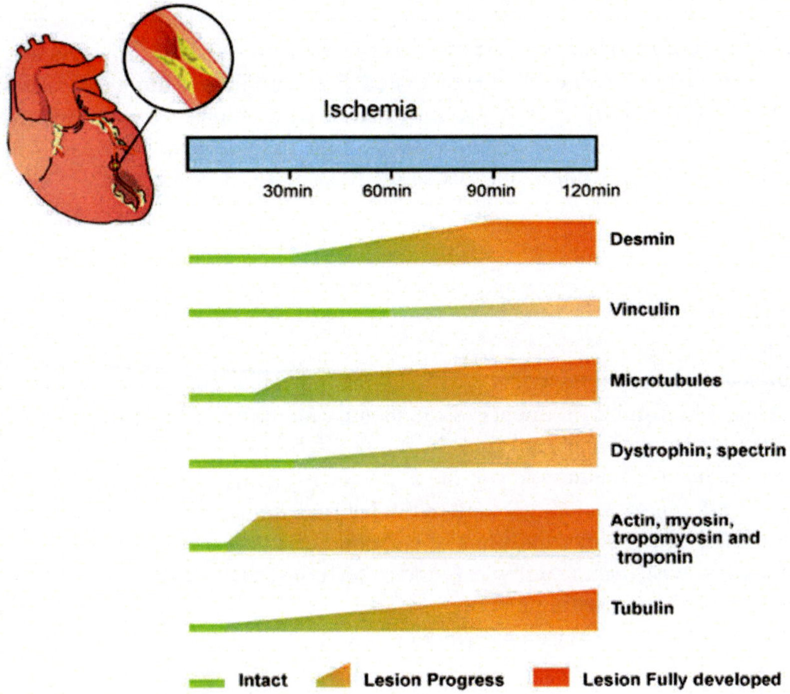

Fig. 26.3 Sarcomeric and cytoskeletal proteins have different sensitivities towards ischemic damage ("lesion"). For example, tropomyosin and troponins are among the most sensitive targets to proteolysis during myocardial ischemia, whereas vinculin resists this proteolytic damage. Reperfusion following ischemia accelerates damage to these proteins

Interestingly, although desmin is not considered to be directly involved in the generation of contractile force or the maintenance of tension, vascular smooth muscle cells from desmin knockout mice generate only 40% of the contractile force than that of wild-type controls [53]. Severe cardiac ischemia followed by reperfusion leads to intracellular Ca^{2+} overload and subsequent activation of calpain, which has the ability to proteolyze desmin [6]. Desmin content in the whole heart was shown to be decreased in myocardial I/R injury, and desmin hydrolysis by activated calpain reduces maximal force production and Ca^{2+} sensitivity in isolated cardiac myofilaments [54]. Hein et al. studied the effects of global ischemia on various cytoskeletal and contractile proteins in human left ventricles obtained from transplant recipients. Desmin was found to be affected by ischemia at later time points than the contractile filaments (Fig. 26.3). Disappearance of desmin from the cross striation pattern began after at 30 min of ischemia whereas, and the fully developed pattern of changes occurred after 90 min of ischemic injury [2]. Disruption of desmin damages the link between myofibrils and the sarcolemma and was found to contribute to increased fragility of the myofibrils [55]. Since calpain was also found to be colocalized with desmin in the Z-band of skeletal myoblast cells [56], it was postulated that calpain-mediated degradation of desmin may contribute to the increase of cell fragility which may increase the chance for rupture of the cell during reperfusion. Note that desmin is susceptible to hydrolysis by MMP-2 in vitro [22] but whether this accompanies ischemic heart injury is unknown.

26.2.3 Vinculin

Vinculin is a membrane-associated cytoskeletal protein required for the attachment of the actin-based microfilaments to the plasma membrane (Fig. 26.1). In the cardiac myocyte, vinculin participates in the formation of the attachment complex between the plasma membrane and the Z-line of myofibrils [57]. Vinculin was also localized in the intercalated disk and in the lateral sarcolemma [2].

The effects of different durations of ischemia on cardiac vinculin were assessed by immunofluorescence, and it was found to be more resistant to the effects of ischemia than desmin, with effects not being detectable until 60 min post-ischemia [2]. Similarly, in a canine heart ischemia model, Steenbergen et al. reported an unchanged pattern of vinculin staining in longitudinal sections of myocardium subjected to 60 min of total ischemia and 60 min of reperfusion. When subject to more severe myocardial ischemia (120 min or longer), there was a progressive loss of vinculin staining and increase in inulin permeability, which likely contributes to the detachment of actin from the sarcolemma, leading to the formation of blebs and rupture of the sarcolemmal membrane [58].

26.2.4 Microtubules

Microtubules are hollow tubes 25 nm in diameter that exist as a ubiquitous filamentous structure of the cytoskeleton. Polymerized α- and β-tubulin form the microtubes which surround the nucleus and spread throughout the entire cell [59]. In myocytes, microtubules are distributed along their longitudinal axis as an irregular network.

In an in situ canine heart model [50], 15 min of cardiac ischemia caused no detectable changes in the filamentous staining pattern. After 20 min of ischemia, however, small patchy lesions appeared in some myocytes in which the immunoreactivities of microtubules began to decrease in intensity. Progressive loss of microtubular staining continued to be observed over 120 min of cardiac ischemia.

Microtubules contribute significantly to the stability of cell morphology by supporting cellular architecture, plasma membranes, myofibrils, and other cellular organelles. Protection of microtubule integrity during I/R injury could therefore be a means of protecting against I/R-induced damage. For example, paclitaxel, a microtubule stabilizer, reduced myocardial I/R injury, myocardial infarct size, and the incidence of ischemic ventricular arrhythmias in perfused rat hearts [60].

26.2.5 *Other Cytoskeleton Proteins: Talin, Dystrophin, and Spectrin*

Prolonged cardiac I/R injury damages a wide array of cytoskeletal protein structures. Talin, dystrophin, and spectrin are membrane-associated cytoskeletal proteins that link the structural components of the intracellular milieu with those of the extracellular matrix via the integrins (Fig. 26.1). Dystrophin connects intracellular actin and extracellular laminin and acts as a stabilizing force and mechanotransducer for the sarcolemmal membrane [61]. Spectrin forms the backbone of the membrane skeleton providing an elastic support to the sarcolemmal membrane. Ischemia-induced loss of membrane dystrophin and spectrin was found following 30–45 min of coronary artery ligation in rabbit hearts. Loss of sarcolemmal dystrophin and spectrin seems to contribute to subsarcolemmal bleb formation and membrane fragility during the transition from reversible to irreversible ischemic myocardial injury [62]. Spectrin was not susceptible to proteolysis by MMP-2 in vitro [22].

26.2.6 *Mechanisms of Cytoskeletal Damage in I/R Injury*

Beside the direct degenerative effects of ischemia on the cytoskeleton, the process of restoring circulation to the ischemic heart, usually by acute intervention therapy, can also induce additional injury to the myocardium. The initial mechanism underlying reperfusion injury had been attributed to the generation of reactive oxygen/nitrogen species including peroxynitrite [22] and elevated intracellular Ca^{2+}. Beyond this, a variety of other biochemical abnormalities have been proposed to explain the myocardial contractile dysfunction that occurs after reperfusion, including excitation–contraction uncoupling due to dysfunction of the sarcoplasmic reticulum, altered fuel metabolism in mitochondria, inefficient energy use by myofibrils, altered ion channel activities, and decreased sensitivity of myofilaments to calcium. Importantly, these potential mechanisms are not mutually exclusive [1].

In I/R injury, the activity of cytosolic proteases such as calpain and MMPs is increased in response to increased Ca^{2+} and reactive oxygen/nitrogen species, respectively. The proteolysis of their cytoskeletal and sarcomeric substrates was found to be a major mechanism of intracellular pathology in I/R injury [4, 45]. These alterations to contractile and cytoskeletal proteins (Fig. 26.2), in addition to damaged subcellular organelles, play a key role in impairing cardiac function during I/R injury.

26.2.7 *Cytoskeletal Injury as an Indicator of Irreversible Cell Injury*

Myocardial injury is a dynamic process which, if mild, results in reversible cell injury, but if severe enough causes irreversible damage. Action to protect (or salvage) the ischemic myocardium should therefore preferably be performed before irreversible damage of myocardium occurs. Many biochemical and metabolic changes have been observed early after the onset of ischemia, but the precise cause of the transition to irreversibility is not known.

A number of hypotheses have been proposed to account for this transition to irreversible myocardial damage, including mitochondrial dysfunction, depletion of antioxidant reserves, leakage of lysosomal enzymes [63], the toxicity of metabolic end products, and lipid peroxidation caused by reactive oxygen/nitrogen species [64]. None of these postulated theories completely account for all features of the irreversible damage to cardiac myocytes. However, cytoskeletal damage can possibly explain many important biological phenomena associated with irreversible damage.

When ATP is severely depleted during ischemia, the connection between the cytoskeletal components is weakened resulting in an increased potential to dissociate under mechanical force. During the early reperfusion phase, the influx of calcium leads to calcium overload and hypercontraction [65]. Hypercontraction of the ischemia-injured myocyte will lead to cytoskeletal deformation to an extent beyond that seen under normal contraction. This will result in "cytoskeletal fracture". Since the cytoskeleton is required to maintain integrity of the cell membrane and cytosolic organelles, cytoskeletal injury is accompanied by osmotic swelling which eventually leads to rupture of the cell membrane. The membrane rupture hypothesis of irreversibility is supported by the observation that "irreversibly injured" cells can indeed recover if membrane rupture and necrosis during reperfusion are prevented [66].

26.3 Cardiac Mitochondria Remodeling in Ischemic Heart Disease

Current models of I/R injury feature mitochondria as important arbiters of cardiac myocyte survival or death [67]. The pivotal event is the opening of the mitochondrial permeability transition pore (MPTP), which results in the permeabilization of the inner mitochondrial membrane. This collapses mitochondrial membrane potential, rendering mitochondria unable to produce ATP. Responding to an osmotic gradient between the mitochondrial matrix and cytosol, water rushes into the matrix resulting in the characteristic "swollen" appearance of mitochondria in electron micrographs of cardiac tissue subject to I/R injury. This may lead to the rupture of the outer mitochondrial membrane, releasing cytochrome c and other pro-apoptotic proteins into the cytosol.

26.3.1 A Potential Role for Intracellular Matrix Remodeling in Post-I/R Mitochondrial Dysfunction

The existence of a physical relationship between mitochondria and the cytoskeleton is well established. Mitochondria have been shown to interact with microtubules, intermediate filaments, and microfilaments, with cytoskeletal elements believed to play a role in their intracellular localization and movement and possibly morphology [52]. The possibility that intracellular matrix remodeling may play a role in I/R-induced mitochondrial dysfunction remains mostly unexplored, although there are tantalizing hints that this is the case. For instance, pharmaceutical manipulation of the cytoskeleton has been shown to affect mitochondrial functions relevant to I/R. Pharmaceutical agents that either depolymerized or stabilized microtubules prevented closure of the MPTP [68], and pharmaceutical disruption of the actin cytoskeleton impaired the effectiveness of certain drugs that decrease the vulnerability of cardiac tissue to I/R injury by manipulating mitochondria [69]. It has been suggested that the cytoskeleton plays a role in the translocation of signaling molecules to the mitochondria during ischemic preconditioning, perhaps via the endosomal system [70]. Furthermore, cardiac mitochondrial function in desmin-null mice is impaired, with mitochondria exhibiting ultrastructural changes—such as swelling—strikingly similar to those observed in I/R injury [71].

More generally, it is hypothesized that the cytoskeleton modulates mitochondrial function, although the mechanisms remain unclear [52]. One possible mechanism could be via the cytoskeletal control of the subcellular localization of mitochondria. This may be particularly relevant in adult cardiac myocytes, in which most mitochondria are arranged in highly structured linear arrays between myofibrils [72]. After I/R injury, mitochondria have been observed as appearing to be detached from myofibrils [73], suggestive of a mechanism, possibly cytoskeletal in nature, that maintains this association. Indeed, disrupting cytoskeletal structure by knocking out desmin or enzymatic digestion in vitro results in a loss of the normal mitochondrial position and neat arrangement between myofibrils [71, 74]. Mitochondria appear to be localized near sites of Ca^{2+} release from the sarcoplasmic reticulum, which may facilitate Ca^{2+} movements between the two cellular subcompartments [75, 76]; it has been proposed that the cytoskeleton may play a role in maintaining this physical orientation [77]. Likewise, the close association of mitochondria with sarcomeres and the sarcoplasmic reticulum may facilitate the channeling of ADP to mitochondria; this is supported by the finding that proteolytic digestion of cytoskeletal elements in permeabilized cardiac myocytes in vitro altered the apparent binding affinity of ADP [74].

26.3.2 Mitochondrial Localization of Modifiers of the Intracellular Matrix

As described above, calpains and MMP-2 have both been shown to actively remodel the intracellular matrix in response to I/R. MMP-2 and calpain-1 have both been found in cardiac mitochondria [20, 25, 78, 79]. Mitochondrial calpain-1 is activated by I/R injury and cleaves apoptosis-inducing factor, allowing it to be released into the cytosol [78]. Overexpression of constitutively active MMP-2 in the mouse heart does not affect baseline mitochondrial function; however, the response to I/R is more severe, with mitochondrial respiration and structure being affected to a greater extent than in hearts from wild-type mice [73]. Mitochondria appear to accumulate a constitutively active, N-truncated isoform of MMP-2, overexpression of which triggers the nuclear activation of several pro-inflammatory transcriptional pathways [25]. The fact that these proteases of the sarcomere and cytoskeleton also appear to be active in mitochondria suggests that the remodeling of the intracellular matrix may actually be part of a wider program of response to I/R injury.

26.4 Nuclear Matrix Remodeling in Cardiac Disease

The nuclear matrix is the network of fibers found throughout the inside of the cell nucleus. It is analogous to the cellular cytoskeleton and provides structural and organizational support for various nuclear processes. Proteolytic cleavage of the nuclear matrix occurs in processes such as apoptosis [80], regulation of the cell cycle [81], and nuclear matrix degradation [82].

Irreversible I/R injury may lead to cellular apoptosis. During apoptosis, morphological changes such as chromatin condensation, nuclear shrinkage, and the formation of apoptotic bodies occur in the nucleus [83]. These changes are associated with numerous molecular alterations, such as DNA and RNA cleavage, posttranslational modifications of nuclear proteins, and proteolysis of several polypeptides of the nuclear matrix including topoisomerase IIa, NuMA, SAF-A, lamin B1, lamins A and C, and SATB1 [83].

Nuclear MMP-2 and/or MMP-9 activities may also contribute to the I/R injury-induced apoptotic process by processing poly-ADP-ribose polymerase [18] and X-ray cross complementary factor 1, hence, interfering with the DNA repair system [84]. Indeed, MMP-2 [23] and MMP-3 [85] carry a putative nuclear localization sequence. An active, truncated fragment of MMP-3 was localized to the nucleus of several human cancer cell lines, and it is associated with the onset of apoptosis [85]. MMP-2 and MMP-9 were found in the nucleus of human cardiac myocytes, and although MMP-2 was able to proteolyze the nuclear DNA repair enzyme, poly (ADP-ribosyl) polymerase in vitro, its precise role in the nucleus remains to be discovered [23].

26.5 Future Prospects: The Intracellular Matrix as Therapeutic Target

Ischemic heart disease is the most common cause of death in developed and, ever more so, in developing countries. Studies of the pathogenesis of myocardial I/R injury reveal structural and functional remodeling of intracellular matrix components of the cardiac myocyte. Protection of the intracellular matrix may represent a novel strategy to prevent or reduce the impact of ischemic heart disease. Many cytoskeletal and sarcomeric proteins were found to be susceptible to proteolysis by intracellularly localized MMP-2; indeed, there is substantial evidence that the cleavage of these intracellular targets by this crucial protease mediates several important pathogenic processes in myocardial I/R injury. Oxidative stress generated in I/R injury can efficiently activate intracellular MMP-2 which then mediates intracellular matrix remodeling of injured myocytes. Given that MMP-2 is readily activated by prooxidant stress, such as peroxynitrite generated in I/R injury [29], MMP-2 may represent one of the earliest mediators of the detrimental actions of oxidative stress to the heart.

Doxycycline, one of the tetracycline antibiotics, has been shown to act as an MMP inhibitor at a plasma concentration below that required for its antimicrobial action [86]. Indeed, a retrospective epidemiological study found a significant reduction in the risk of first-time acute myocardial infarction for patients who had taken tetracycline class antibiotics for prior infection. This effect was not observed in patients who had received any other classes of antibiotics [87]. Furthermore, doxycycline was also found to protect against streptozotocin-induced diabetic cardiomyopathy [88] and cardiac mechanical dysfunction triggered by endotoxic shock in rats [89]. Increased activity of cardiac matrix MMP-2 and MMP-9 was found during the acute phase of Chagasic cardiomyopathy in mice, an inflammatory heart disease triggered by infection with *Trypanosoma cruzi*. This increased MMP activity was associated with mortality following infection, and doxycycline treatment significantly improved survival [90]. Based on our current understanding of MMP-2 as an intracellular protease, the cardiovascular benefits associated with tetracycline use may be reasonably attributed to the inhibition of pathological MMP activity. These findings suggest that doxycycline may be useful as a possible therapeutic regimen for ischemic heart disease. Of note, doxycycline has already been approved by the Food and Drug Administration and Health Canada in the therapeutic treatment of periodontitis. Doxycycline may therefore emerge as a promising drug in the near future for the treatment or prevention of cardiac disease.

Increasing evidence suggests that blocking MMP-2 activity can alleviate I/R-mediated cardiac injury in animal models. It is important to consider the fact that MMP-2 plays very diverse roles in various pathological and physiological circumstances other than cell injury, such as cell cycle control, cell death, inflammation, cancer, development, and tissue remodeling. Universal MMP inhibitors could therefore have detrimental or undesirable side effects due to a lack of selectivity and/or specificity [91]. Development of pathway-specific or subcellular location "selective"

MMP-2 inhibitors, based on the advancing knowledge in this field, may help to alleviate acute myocardial I/R injury or detrimental chronic cardiac remodeling, while spare physiological MMP-2 activities inside and outside the cell.

Acknowledgments We thank Dawne Colwell for her professional assistance in preparing the illustrations. Research in the Schulz laboratory is supported by the Canadian Institutes of Health Research (MOP-77526 and MOP-66953). MAM received a graduate trainee award from Alberta Innovates—Health Solutions. ALBJF is supported by a fellowship award from Alberta Innovates—Health Solutions and an incentive award from the Mazankowski Alberta Heart Institute.

References

1. Bolli R, Marban E (1999) Molecular and cellular mechanisms of myocardial stunning. Physiol Rev 79:609–634
2. Hein S, Scheffold T, Schaper J (1995) Ischemia induces early changes to cytoskeletal and contractile proteins in diseased human myocardium. J Thorac Cardiovasc Surg 110:89–98
3. Matsumura Y, Saeki E, Inoue M et al (1996) Inhomogeneous disappearance of myofilament-related cytoskeletal proteins in stunned myocardium of guinea pig. Circ Res 79:447–454
4. Schulz R (2007) Intracellular targets of matrix metalloproteinase-2 in cardiac disease: rationale and therapeutic approaches. Annu Rev Pharmacol Toxicol 47:211–242
5. Ali MA, Cho WJ, Hudson B et al (2010) Titin is a target of matrix metalloproteinase-2: implications in myocardial ischemia/reperfusion injury. Circulation 122:2039–2047
6. Suzuki K, Hata S, Kawabata Y, Sorimachi H (2004) Structure, activation, and biology of calpain. Diabetes 53(Suppl 1):S12–S18
7. Huang Y, Wang KK (2001) The calpain family and human disease. Trends Mol Med 7:355–362
8. Gao WD, Liu Y, Mellgren R, Marban E (1996) Intrinsic myofilament alterations underlying the decreased contractility of stunned myocardium. A consequence of Ca^{2+}–dependent proteolysis? Circ Res 78:455–465
9. McDonough JL, Arrell DK, Van Eyk JE (1999) Troponin I degradation and covalent complex formation accompanies myocardial ischemia/reperfusion injury. Circ Res 84:9–20
10. Barta J, Toth A, Edes I et al (2005) Calpain-1-sensitive myofibrillar proteins of the human myocardium. Mol Cell Biochem 278:1–8
11. Gilchrist JS, Cook T, Abrenica B et al (2010) Extensive autolytic fragmentation of membranous versus cytosolic calpain following myocardial ischemia-reperfusion. Can J Physiol Pharmacol 88:584–594
12. Pedrozo Z, Sanchez G, Torrealba N et al (2010) Calpains and proteasomes mediate degradation of ryanodine receptors in a model of cardiac ischemic reperfusion. Biochim Biophys Acta 1802:356–362
13. Inserte J, Garcia-Dorado D, Hernando V, Soler-Soler J (2005) Calpain-mediated impairment of Na^+/K^+–ATPase activity during early reperfusion contributes to cell death after myocardial ischemia. Circ Res 97:465–473
14. Ali MA, Stepanko A, Fan X et al (2012) Calpain inhibitors exhibit matrix metalloproteinase-2 inhibitory activity. Biochem Biophys Res Commun 423:1–5
15. Gross J, Lapiere CM (1962) Collagenolytic activity in amphibian tissues. A tissue culture assay. Proc Natl Acad Sci USA 48:1014–1022
16. Spinale FG (2007) Myocardial matrix remodeling and the matrix metalloproteinases: influence on cardiac form and function. Physiol Rev 87:1285–1342
17. McCawley LJ, Matrisian LM (2001) Matrix metalloproteinases: they're not just for matrix anymore! Curr Opin Cell Biol 13:534–540
18. Cauwe B, Opdenakker G (2010) Intracellular substrate cleavage: a novel dimension in the biochemistry, biology and pathology of matrix metalloproteinases. Crit Rev Biochem Mol Biol 45:351–423

19. Cheung PY, Sawicki G, Wozniak M et al (2000) Matrix metalloproteinase-2 contributes to ischemia-reperfusion injury in the heart. Circulation 101:1833–1839
20. Wang W, Schulze C, Suarez-Pinzon W et al (2002) Intracellular action of matrix metalloproteinase-2 accounts for acute myocardial ischemia and reperfusion injury. Circulation 106:1543–1549
21. Coker ML, Doscher MA, Thomas CV et al (1999) Matrix metalloproteinase synthesis and expression in isolated LV myocyte preparations. Am J Physiol 277:H777–H787
22. Sung MM, Schulz CG, Wang W et al (2007) Matrix metalloproteinase-2 degrades the cytoskeletal protein alpha-actinin in peroxynitrite mediated myocardial injury. J Mol Cell Cardiol 43:429–436
23. Kwan JA, Schulze CJ, Wang W et al (2004) Matrix metalloproteinase-2 (MMP-2) is present in the nucleus of cardiac myocytes and is capable of cleaving poly (ADP-ribose) polymerase (PARP) in vitro. FASEB J 18:690–692
24. Yang Y, Candelario-Jalil E, Thompson JF et al (2010) Increased intranuclear matrix metalloproteinase activity in neurons interferes with oxidative DNA repair in focal cerebral ischemia. J Neurochem 112:134–149
25. Lovett DH, Mahimkar R, Raffai RL et al (2012) A novel intracellular isoform of matrix metalloproteinase-2 induced by oxidative stress activates innate immunity. PLoS One 7:e34177
26. Ali MA, Chow AK, Kandasamy AD et al (2012) Mechanisms of cytosolic targeting of matrix metalloproteinase-2. J Cell Physiol 227:3397–3404
27. Cao J, Sato H, Takino T, Seiki M (1995) The C-terminal region of membrane type matrix metalloproteinase is a functional transmembrane domain required for pro-gelatinase A activation. J Biol Chem 270:801–805
28. Yasmin W, Strynadka KD, Schulz R (1997) Generation of peroxynitrite contributes to ischemia-reperfusion injury in isolated rat hearts. Cardiovasc Res 33:422–432
29. Viappiani S, Nicolescu AC, Holt A et al (2009) Activation and modulation of 72 kDa matrix metalloproteinase-2 by peroxynitrite and glutathione. Biochem Pharmacol 77:826–834
30. Sariahmetoglu M, Crawford BD, Leon H et al (2007) Regulation of matrix metalloproteinase-2 (MMP-2) activity by phosphorylation. FASEB J 21:2486–2495
31. Golub LM, Ramamurthy N, McNamara TF et al (1984) Tetracyclines inhibit tissue collagenase activity. A new mechanism in the treatment of periodontal disease. J Periodontal Res 19:651–655
32. Wang GY, Bergman MR, Nguyen AP et al (2006) Cardiac transgenic matrix metalloproteinase-2 expression directly induces impaired contractility. Cardiovasc Res 69:688–696
33. Bergman MR, Teerlink JR, Mahimkar R et al (2007) Cardiac matrix metalloproteinase-2 expression independently induces marked ventricular remodeling and systolic dysfunction. Am J Physiol 292:H1847–H1860
34. Van Eyk JE, Powers F, Law W et al (1998) Breakdown and release of myofilament proteins during ischemia and ischemia/reperfusion in rat hearts: identification of degradation products and effects on the pCa-force relation. Circ Res 82:261–271
35. Sawicki G, Leon H, Sawicka J et al (2005) Degradation of myosin light chain in isolated rat hearts subjected to ischemia-reperfusion injury: a new intracellular target for matrix metalloproteinase-2. Circulation 112:544–552
36. Granzier HL, Labeit S (2004) The giant protein titin: a major player in myocardial mechanics, signaling, and disease. Circ Res 94:284–295
37. Neagoe C, Kulke M, del Monte F et al (2002) Titin isoform switch in ischemic human heart disease. Circulation 106:1333–1341
38. Makarenko I, Opitz CA, Leake MC et al (2004) Passive stiffness changes caused by upregulation of compliant titin isoforms in human dilated cardiomyopathy hearts. Circ Res 95:708–716
39. Linke WA (2010) Molecular giant vulnerable to oxidative damage: titin joins the club of proteins degraded by matrix metalloproteinase-2. Circulation 122:2002–2004
40. Borbely A, Falcao-Pires I, van Heerebeek L et al (2009) Hypophosphorylation of the stiff N2B titin isoform raises cardiomyocyte resting tension in failing human myocardium. Circ Res 104:780–786
41. Grutzner A, Garcia-Manyes S, Kotter S et al (2009) Modulation of titin-based stiffness by disulfide bonding in the cardiac titin N2-B unique sequence. Biophys J 97:825–834
42. LeWinter MM, Granzier H (2010) Cardiac titin: a multifunctional giant. Circulation 121:2137–2145

43. Morano I, Hadicke K, Grom S et al (1994) Titin, myosin light chains and C-protein in the developing and failing human heart. J Mol Cell Cardiol 26:361–368
44. Hein S, Scholz D, Fujitani N et al (1994) Altered expression of titin and contractile proteins in failing human myocardium. J Mol Cell Cardiol 26:1291–1306
45. Kandasamy AD, Chow AK, Ali MA, Schulz R (2010) Matrix metalloproteinase-2 and myocardial oxidative stress injury: beyond the matrix. Cardiovasc Res 85:413–423
46. Galvez AS, Diwan A, Odley AM et al (2007) Cardiomyocyte degeneration with calpain deficiency reveals a critical role in protein homeostasis. Circ Res 100:1071–1078
47. Zolk O, Schenke C, Sarikas A (2006) The ubiquitin-proteasome system: focus on the heart. Cardiovasc Res 70:410–421
48. Wang X, Li J, Zheng H, Powell SR (2011) Proteasome functional insufficiency in cardiac pathogenesis. Am J Physiol 301:H2207–H2219
49. Powell SR, Divald A (2010) The ubiquitin-proteasome system in myocardial ischaemia and preconditioning. Cardiovasc Res 85:303–311
50. Iwai K, Hori M, Kitabatake A et al (1990) Disruption of microtubules as an early sign of irreversible ischemic injury. Immunohistochemical study of in situ canine hearts. Circ Res 67:694–706
51. Paulin D, Li Z (2004) Desmin. A major intermediate filament protein essential for the structural integrity and function of muscle. Exp Cell Res 301:1–7
52. Anesti V, Scorrano L (2006) The relationship between mitochondrial shape and function and the cytoskeleton. Biochim Biophys Acta 1757:692–699
53. Sjuve R, Arner A, Li Z et al (1998) Mechanical alterations in smooth muscle from mice lacking desmin. J Muscle Res Cell Motil 19:415–429
54. Papp Z, van der Velden J, Stienen GJ (2000) Calpain-I induced alterations in the cytoskeletal structure and impaired mechanical properties of single myocytes of rat heart. Cardiovasc Res 45:981–993
55. Dalakas MC, Park KY, Semino-Mora C et al (2000) Desmin myopathy, a skeletal myopathy with cardiomyopathy caused by mutations in the desmin gene. N Engl J Med 342:770–780
56. Taveau M, Bourg N, Sillon G et al (2003) Calpain 3 is activated through autolysis within the active site and lyses sarcomeric and sarcolemmal components. Mol Cell Biol 23:9127–9135
57. Pardo JV, Siliciano JD, Craig SW (1983) Vinculin is a component of an extensive network of myofibril-sarcolemma attachment regions in cardiac muscle fibers. J Cell Biol 97:1081–1088
58. Steenbergen C, Hill ML, Jennings RB (1987) Cytoskeletal damage during myocardial ischemia: changes in vinculin immunofluorescence staining during total in vitro ischemia in canine heart. Circ Res 60:478–486
59. Li H, DeRosier DJ, Nicholson WV et al (2002) Microtubule structure at 8 A resolution. Structure 10:1317–1328
60. Cao HM, Wang Q, You HY et al (2011) Stabilizing microtubules decreases myocardial ischaemia-reperfusion injury. J Int Med Res 39:1713–1719
61. Klietsch R, Ervasti JM, Arnold W et al (1993) Dystrophin-glycoprotein complex and laminin colocalize to the sarcolemma and transverse tubules of cardiac muscle. Circ Res 72:349–360
62. Armstrong SC, Latham CA, Shivell CL, Ganote CE (2001) Ischemic loss of sarcolemmal dystrophin and spectrin: correlation with myocardial injury. J Mol Cell Cardiol 33:1165–1179
63. Ogawa T, Sugiyama S, Hieda N et al (1989) Biochemical and morphological changes in myocardium during coronary occlusion and reperfusion in canine hearts: effects of propranolol on myocardial damage. Cardiovasc Res 23:417–423
64. Yellon DM, Hausenloy DJ (2007) Myocardial reperfusion injury. N Engl J Med 357:1121–1135
65. Yoshikawa T, Akaishi M, Ikeda F et al (1992) Postischaemic hypercontraction is enhanced in ischaemically injured canine myocardium. Cardiovasc Res 26:337–341
66. Schluter KD, Schwartz P, Siegmund B, Piper HM (1991) Prevention of the oxygen paradox in hypoxic-reoxygenated hearts. Am J Physiol 261:H416–H423
67. Ong S, Gustafsson AB (2012) New roles for mitochondria in cell death in the reperfused myocardium. Cardiovasc Res 94:190–196
68. Evtodienko YV, Teplova VV, Sidash SS et al (1996) Microtubule-active drugs suppress the closure of the permeability transition pore in tumour mitochondria. FEBS Lett 393:86–88

69. Baines CP, Liu GS, Birincioglu M et al (1999) Ischemic preconditioning depends on interaction between mitochondrial KATP channels and actin cytoskeleton. Am J Physiol 276:H1361–H1368
70. Murphy E, Steenbergen C (2008) Mechanisms underlying acute protection from cardiac ischemia-reperfusion injury. Physiol Rev 88:581–609
71. Milner DJ, Mavroidis M, Weisleder N, Capetanaki Y (2000) Desmin cytoskeleton linked to muscle mitochondrial distribution and respiratory function. J Cell Biol 150:1283–1298
72. Hom J, Sheu S (2009) Morphological dynamics of mitochondria–a special emphasis on cardiac muscle cells. J Mol Cell Cardiol 46:811–820
73. Zhou HZ, Ma X, Gray MO et al (2007) Transgenic MMP-2 expression induces latent cardiac mitochondrial dysfunction. Biochem Biophys Res Commun 358:189–195
74. Appaix F, Kuznetsov A, Usson Y et al (2003) Possible role of cytoskeleton in intracellular arrangement and regulation of mitochondria. Exp Physiol 88:175–190
75. Wang H, Guay G, Pogan L et al (2000) Calcium regulates the association between mitochondria and a smooth subdomain of the endoplasmic reticulum. J Cell Biol 150:1489–1498
76. Filippin L, Magalhães PJ, Di Benedetto G et al (2003) Stable interactions between mitochondria and endoplasmic reticulum allow rapid accumulation of calcium in a subpopulation of mitochondria. J Biol Chem 278:39224–39234
77. Capetanaki Y, Bloch RJ, Kouloumenta A et al (2007) Muscle intermediate filaments and their links to membranes and membranous organelles. Exp Cell Res 313:2063–2076
78. Chen Q, Paillard M, Gomez L et al (2011) Activation of mitochondrial μ-calpain increases AIF cleavage in cardiac mitochondria during ischemia–reperfusion. Biochem Biophys Res Commun 415:533–538
79. Ozaki T, Tomita H, Tamai M, Ishiguro S (2007) Characteristics of mitochondrial calpains. J Biochem 142:365–376
80. Martelli AM, Bareggi R, Bortul R et al (1997) The nuclear matrix and apoptosis. Histochem Cell Biol 108:1–10
81. Ge W, Guo R, Ren J (2011) AMP-dependent kinase and autophagic flux are involved in aldehyde dehydrogenase-2-induced protection against cardiac toxicity of ethanol. Free Radic Biol Med 51:1736–1748
82. Owen CA, Campbell EJ (1995) Neutrophil proteinases and matrix degradation. The cell biology of pericellular proteolysis. Semin Cell Biol 6:367–376
83. Martelli AM, Zweyer M, Ochs RL et al (2001) Nuclear apoptotic changes: an overview. J Cell Biochem 82:634–646
84. Hadler-Olsen E, Fadnes B, Sylte I et al (2011) Regulation of matrix metalloproteinase activity in health and disease. FEBS J 278:28–45
85. Si-Tayeb K, Monvoisin A, Mazzocco C et al (2006) Matrix metalloproteinase 3 is present in the cell nucleus and is involved in apoptosis. Am J Pathol 169:1390–1401
86. Golub LM, Lee HM, Ryan ME et al (1998) Tetracyclines inhibit connective tissue breakdown by multiple non-antimicrobial mechanisms. Adv Dent Res 12:12–26
87. Meier CR, Derby LE, Jick SS et al (1999) Antibiotics and risk of subsequent first-time acute myocardial infarction. JAMA 281:427–431
88. Yaras N, Sariahmetoglu M, Bilginoglu A et al (2008) Protective action of doxycycline against diabetic cardiomyopathy in rats. Br J Pharmacol 155:1174–1184
89. Lalu MM, Gao CQ, Schulz R (2003) Matrix metalloproteinase inhibitors attenuate endotoxemia induced cardiac dysfunction: a potential role for MMP-9. Mol Cell Biochem 251:61–66
90. Gutierrez FR, Lalu MM, Mariano FS et al (2008) Increased activities of cardiac matrix metalloproteinases matrix metalloproteinase (MMP)-2 and MMP-9 are associated with mortality during the acute phase of experimental trypanosoma cruzi infection. J Infect Dis 197:1468–1476
91. Hu J, Van den Steen PE, Sang QX, Opdenakker G (2007) Matrix metalloproteinase inhibitors as therapy for inflammatory and vascular diseases. Nat Rev Drug Discov 6:480–498

Chapter 27
Aging and Markers of Adverse Remodeling After Myocardial Infarction

Bodh I. Jugdutt and Anwar Jelani

Abstract Acute myocardial infarction (MI) is a catastrophic event in patients aged 18 years and above. While improved coronary reperfusion strategies and medical therapies over the last 2 decades have increased survival, hearts of survivors undergo progressive remodeling with changes in structure, size, shape, and systolic function that lead to heart failure. Morbidity from heart failure in post-MI survivors remains high, especially after ST-segment-elevation MI (STEMI) and in older compared to younger patients. Cumulative knowledge of the biology of cardiovascular aging suggests that the aging process is progressive and results in physiological, cellular, and molecular changes that can negatively impact post-MI remodeling. Emerging evidence suggests that aging may adversely influence key factors in post-MI remodeling, including the extent of damage to the cardiac muscle and extracellular matrix (ECM) on the one hand and adequacy of the healing/repair process (with inflammation, ECM remodeling, fibrosis, hypertrophy, and angiogenesis) on the other. These effects may result in differential outcomes of therapy between younger and older survivors of STEMI. Furthermore, reperfusion therapy that is delayed beyond several minutes of acute STEMI results in significant reperfusion damage and adverse remodeling that may be amplified with aging. Biomarkers, including emerging healing-specific proteins, can be used not only to noninvasively monitor the remodeling process and responses to therapy but also to predict adverse post-STEMI remodeling and clinical outcome. Novel biomarkers of adverse remodeling hold great promise for developing novel therapeutic strategies to limit adverse remodeling and improve outcome after STEMI in patients of different ages.

B.I. Jugdutt, M.D., M.Sc., D.M. (✉)
Division of Cardiology, Department of Medicine, University of Alberta, 2C2
Walter MacKenzie Health Sciences Centre, Edmonton, AB, Canada, T6G 2R7
e-mail: bjugdutt@ualberta.ca

A. Jelani
Department of Medicine, King Abdulaziz Hospital,
Alhasa, Saudi Arabia

B.I. Jugdutt and N.S. Dhalla (eds.), *Cardiac Remodeling: Molecular Mechanisms*, Advances in Biochemistry in Health and Disease 5, DOI 10.1007/978-1-4614-5930-9_27,

Keywords Aging • Myocardial infarction • Remodeling • Biomarkers • Inflammation • Extracellular matrix • Metalloproteinases • Matricellular proteins • Perfusion • Function

27.1 Introduction: Statement of the Problem

Myocardial infarction (MI) is a major killer worldwide, and adverse remodeling is a major cause of morbidity and mortality in survivors of ST-segment-elevation MI (STEMI) [1–4]. An acute STEMI is one of the most catastrophic events in the patient with coronary artery disease. While STEMI is seen in adults aged 18 years and above, it is most prevalent in elderly (age ≥65 years) patients [5–8]. What follows after an acute STEMI, during the healing/repair phase in survivors, remains a major problem that has not been fully addressed despite many advances in molecular/subcellular mechanisms that lead to adverse remodeling during that phase ([8] for review).

27.2 Definition of STEMI

For purposes of translational research and discovery, it is important to appreciate that the commonest culprit lesion of an MI is a coronary atherosclerotic plaque with thrombosis, resulting in coronary artery occlusion [9–13]. This puts myocardium in the distal occluded bed at risk of ischemic damage [14], and persistent occlusion usually results in pathologically transmural (≥50% of the transmural extent) and electrocardiographic Q wave MI or STEMI [15–18]. Angographically, coronary obstruction is considered critical if the stenosis is ≥50% of the luminal diameter, but ischemic symptoms may occur with stenosis <50% of the luminal diameter or area depending on pathophysiological conditions. Occlusions <50% of the lumen and depending on several factors (such as collateral blood supply, hemodynamic load and oxygen demand, metabolism) usually result in subendocardial, non-Q wave or non-STEMI (NSTEMI). Acute coronary artery syndromes can result in either STEMI or NSTEMI that have different clinical outcomes needing different management strategies [15–18]. All evidences indicate that STEMI has grave prognosis and outcome. This chapter will focus on STEMI.

27.3 Morbidity and Mortality After STEMI

Effective medical therapies that reduce myocardial oxygen demand and/or increase myocardial oxygen supply result in myocardial salvage [19]. Successful early coronary reperfusion, by limiting the transmural march to necrosis [20], results in non-transmural, subendocardial, or NSTEMI that are less prone to adverse remodeling [1–4, 15, 16, 21–24]. These early concepts have withstood the test of time. However, while improved coronary reperfusion strategies and medical therapies since the 1990s have increased the number of survivors of acute STEMI, the hearts of the survivors undergo progressive remodeling with negative changes in structure, size, shape, and systolic function that lead to heart failure [21–25]. Increased survival after acute STEMI has expanded the pool of patients at risk of complications of adverse left ventricular (LV) remodeling and heart failure. Simply put, decreased early mortality has resulted in increasing numbers of post-STEMI survivors who experience prolonged morbidity. In fact, morbidity from heart failure in post-STEMI survivors remains high, especially in old compared to young patients [6]. The reasons for these discrepancies need urgent investigation [8].

27.4 Aging and Reperfusion Damage

One explanation for the discrepancies may relate to aging and the healing/repair process. Increasing knowledge of the biology of cardiovascular aging suggests that the aging process is progressive [26] and results in physiological, cellular, and molecular changes that can negatively impact post-STEMI remodeling. Emerging evidence suggests that aging may adversely influence key factors in post-STEMI remodeling, including the extent of damage to the cardiac muscle and extracellular matrix (ECM) on the one hand and adequacy of the healing/repair process (with inflammation, ECM remodeling, fibrosis, hypertrophy, and angiogenesis) on the other [8, 27]. Elderly post-STEMI survivors do poorly, with increased morbidity and greater risk of adverse LV remodeling and heart failure, compared to younger patients [5, 8]. These effects may explain differential outcomes of long-term therapy between younger and older survivors of STEMI [8].

A second explanation relates to reperfusion injury [28–36]. Cumulative evidence shows that reperfusion therapy that is delayed beyond several minutes of an acute STEMI results in significant reperfusion damage of cardiac muscle, matrix and blood vessels [28–36], and adverse LV remodeling that may be amplified with aging [8, 27]. Work in our laboratory and others indicates that the healing/repair process after STEMI is highly complex, involving multiple factors, and is associated with significant adverse LV remodeling [8]. While very early reperfusion with primary coronary intervention (PCI) and thrombolytic therapy can be very effective [5, 37], delayed reperfusion of STEMI by only a few hours is still associated with adverse outcome, persistent LV remodeling, and heart failure despite medical therapy during the healing/repair phase [38], and older patients do poorly despite prolonged optimal medical therapy recommended in guidelines [8, 39–41].

27.5 Aging and Post-Infarct Remodeling

The elderly population is increasing worldwide [5, 8, 26, 42 for review]. Concurrently, morbidity and mortality after STEMI in the elderly and related healthcare costs have been increasing [5, 8, 26]. Not only is STEMI most prevalent in the elderly [6, 7], but clinical studies show that age is a strong predictor of adverse events after STEMI [7]. Despite improved coronary reperfusion and medical therapies for acute and subsequent subacute and chronic phases after STEMI, post-STEMI mortality remains high in old patients [6]. Nearly 85% of cardiovascular deaths occur in that group [6]. As mentioned, older adults and elderly post-STEMI survivors do poorly, with increased morbidity and greater risk of adverse LV remodeling and heart failure compared to younger patients [5, 8]. While reperfusion is widely used in acute STEMI, data on healing and remodeling post-reperfused STEMI in the elderly are lacking [8].

While reperfusion with restoration of antegrade flow has undisputed benefits especially when achieved very early [8], late reperfusion is associated with significant reperfusion damage [29] with flow/function mismatch. This is mediated by microvascular damage and persistent no-reflow [29, 35, 36], excess free radicals [29], apoptosis [32, 33], necrosis [27, 34], enhanced ECM degradation [23, 30, 31, 43–48], increased MMP activity [47, 48], decreased collagen cross-links in scars [49], decreased tensile strength of scars [49, 50], enhanced inflammation [23, 51], altered healing [8], and persistent LV remodeling and dysfunction [27, 29, 38]. Post-STEMI survivors who develop heart failure on therapy have a tenfold greater risk of dying [25], and the elderly are at even greater risk [5, 39, 40]. The survivors with persistent post-ischemic LV dysfunction after reperfused STEMI remain at risk for progressive remodeling and its consequences including heart failure [22–24, 52]. In the old, impaired healing/repair mechanisms and adverse remodeling may further aggravate outcome [8, 41]. It is therefore important to study healing/repair in both non-reperfused and reperfused STEMI groups and in both young and old animals for easier translation of basic research findings to the bedside.

27.6 Aging and Healing/Repair Post-STEMI

Evidence suggests that aging-related impaired or defective healing/repair may be a major culprit resulting in adverse remodeling after STEMI [8, 41]. Physiological, cellular, and molecular changes that occur with cardiovascular aging can negatively impact the response to injury including reperfused STEMI [8]. Before 2008, most research on MI was done in young animals [8, 53]. Previous studies showed that MI triggers the healing process, which through a timed sequence of biochemical, molecular, and cellular reactions over weeks results in a firm fibrotic scar that replaces the dead tissue [8, 22, 53], at least in young animals (Fig. 27.1a, b). During healing, timed release of several factors modulate healing and repair [8, 23, 24, 53].

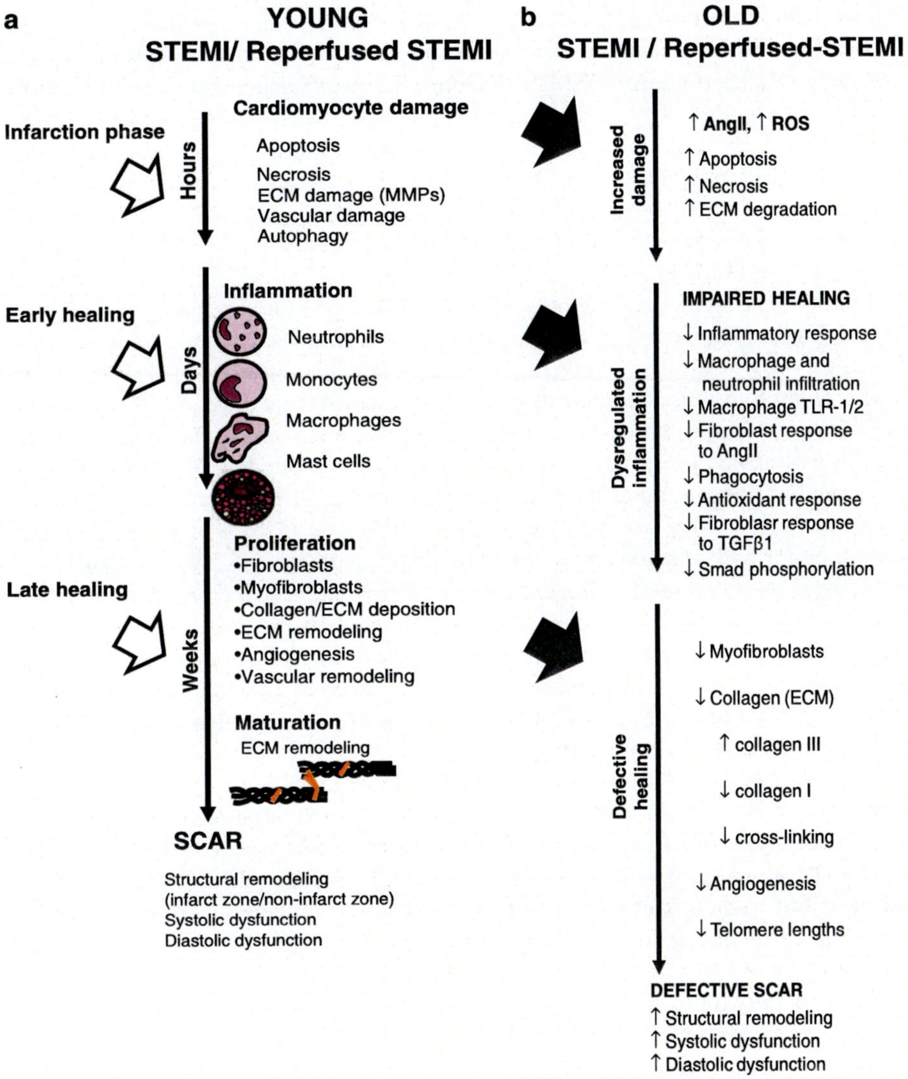

Fig. 27.1 Cellular and molecular changes in healing/repair after STEMI (**a**). The traditional concept is that an acute STEMI triggers healing; this process involves an orchestrated sequence of two highly dynamic phases: an early phase is dominated by inflammation, while a late phase is dominated by repair. The end result is formation of a fibrotic scar and fibrosis in remote myocardium, with variable dilative remodeling and dysfunction. This was based on data mostly in the young (**b**). In the twentieth century, it is recognized that in old hearts, there is amplification of early damage in the infarction phase, dysregulation of inflammation in the early healing phase, and defective healing/repair in the late phase. The end result is a defective scar and augmented adverse dilative remodeling and dysfunction. *AngII* angiotensin II, *ECM* extracellular matrix, *MMP* matrix metalloproteinase, *ROS* reactive oxygen species, *STEMI* ST-segment-elevation myocardial infarction, *TLR* toll-free receptor

These include chemokines, cytokines, growth factors, matrix metalloproteinases (MMPs), and other matrix proteins including the so-called healing-specific matrix and matricellular proteins (HSMPs) including secretory leucocyte protease inhibitor (SLPI), secreted protein acidic and rich in cysteine (SPARC), and osteopontin (OPN) that orchestrate inflammation, remodeling of the ECM, and fibrosis, under the modulatory influence of angiotensin II and other factors such as oxidative stress (Fig. 27.2). Increased MMPs relative to tissue inhibitors of MMPs (TIMPs) is one proposed major pathway leading to ECM and adverse LV remodeling during early healing post-STEMI [22–24].

In 2005, Ertl et al. proposed that SLPI, SPARC, and OPN might improve post-MI healing and remodeling on the basis of data in various transgenic/knockout young mouse models and named them healing-specific proteins [54]. This triggered more research in those proteins as potential targets and markers. SLPI is a potent inhibitor of serine proteases that upregulates production of anti-inflammatory cytokines in monocytes [55], macrophages [56], and skin wounds [57]. Ashcroft et al. reported that skin wound healing is impaired in young C57/BL6 mice lacking SLPI [57], but did not study healing after STEMI. OPN is an ECM protein that interacts with integrins and collagen and promotes collagen synthesis [58, 59], and Trueblood et al. reported decreased collagen, infarct expansion, and increased LV dilation post-MI in young B6.129S6 mice lacking OPN [60]. SPARC is an ECM protein that modulates cell-matrix interactions and regulates fibroblast migration [61–63]. Schellings et al. reported increased immature collagen, LV dilation, and cardiac rupture post-MI in young C57/BL6 mice lacking SPARC [64]. Recently, McCurdy et al. confirmed these findings [65].

In 2008, Bujak et al. first reported defective infarct healing (based on findings of suppressed inflammation and reduced collagen) and increased adverse remodeling after reperfused STEMI in old C57/BL6 mice aged >2 years versus young mice aged 8–12 weeks [66]. They also found that OPN is reduced in 1- and 3-day-old infarcts [66]. Ashcroft et al. had reported decreased collagen in skin wounds of aging mice [67] and delayed healing in skin wounds of aging humans [68], but they did not study MI or reperfused MI in aging animal models.

Since therapy for optimizing healing is lacking for both young and old age groups [8, 69], we considered the hypothesis that in aging hearts, increased myocardial angiotensin II, through its pro-inflammatory, prooxidant, and pro-remodeling effects, may amplify increases in pro-inflammatory cytokines, MMPs, and oxidative markers and contribute to impaired healing/repair and adverse LV remodeling [8] (Fig. 27.3). We postulated that aging may result in a dysregulation of the response to the novel HSMPs, leading to dysregulated inflammation and fibrosis pathways, impaired healing, and adverse LV remodeling post-reperfused STEMI (Fig. 27.3). Thus, aging-related impaired or defective healing after STEMI may be the major culprit leading to defective infarct fibrosis that in turn might result in amplified adverse maladaptive LV remodeling, increased progressive LV enlargement, and increased disability and/or death in older patients. Recent evidence in our laboratory supports the idea that aging-related adverse remodeling may be due in part to impaired healing and repair mechanisms after reperfused STEMI [8, 27, 70], although further research is needed.

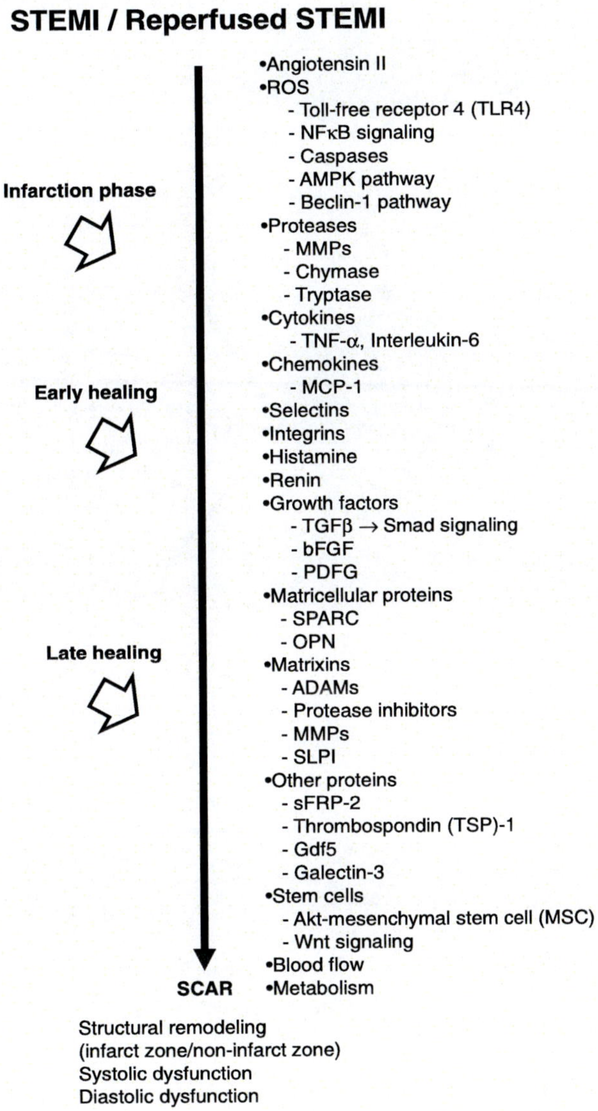

Fig. 27.2 Modulators of healing/repair after STEMI. Cellular and molecular changes during stages of post-STEMI healing and known modulators. *ADAM* a-disintegrin metalloproteinase, *AMPK* AMP-activated protein kinase, *AngII* angiotensin II, *bFGF* basic fibroblast growth factor, *ECM* extracellular matrix, *Gdf5* growth differentiation factor-5, *MCP-1* monocyte chemoattractant protein-1, *MMPs* matrix metalloproteinases, *OPN* osteopontin, *PDGF* platelet-derived growth factor, *ROS* reactive oxygen species, *sFRP-2* secreted frizzled-related protein-2, *SLPI* secretory leucocyte protease inhibitor, *SPARC* secreted protein acidic and rich in cysteine, *STEMI* ST-segment-elevation myocardial infarction, *TGF-β* transforming growth factor-β, *TNF-α* tumor necrosis factor-α

27.7 Changes in HSMPs After Reperfused STEMI

Our preliminary data in dogs, rats, and mice suggest that the 3 HSMPs (SLPI, SPARC, and OPN) are concurrently upregulated post-reperfused STEMI; may interact with concurrently upregulated ECM-proteolytic, inflammation, and fibrosis pathways; and contribute to remodeling in the young animals [27, 70]. Importantly, this upregulation of the HSMPs and proteolytic, inflammation, and fibrosis pathways was amplified in the old animals that developed more severe LV remodeling and dysfunction [27, 70, 71].

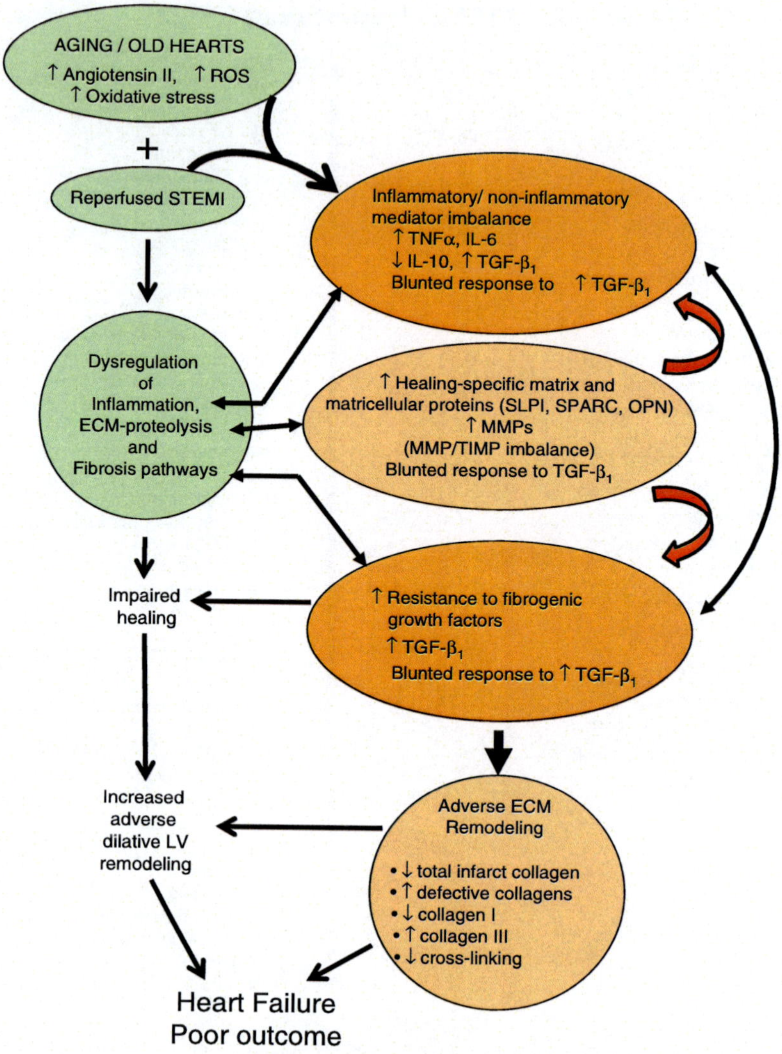

Fig. 27.3 Proposed mechanism of augmented adverse remodeling during healing/repair after reperfused STEMI in old hearts. Increased AngII and ROS augment cardiac muscle, matrix, and vascular damage. Dysregulation of the inflammatory response results from pro-inflammatory, and anti-inflammatory mediator imbalance results in blunting of inflammatory cell recruitment. Macrophages secrete MMPs, reactive oxygen species (ROS) or oxygen free radicals, and cytokines during healing. Concurrent increase in healing-specific matrix and matricellular protein (HSMP) expression (SLPI, SPARC, and OPN), resistance to fibrogenic growth factors, and an imbalance between MMPs and their endogenous tissue inhibitors (TIMPs) contribute to impaired healing with defective collagen synthesis in myofibroblasts and result in adverse ECM remodeling. The end result is augmented adverse dilative LV remodeling. ↑ increase, ↓ decrease, + increase/augment, – inhibit. Other abbreviations as in Fig. 27.2

27.8 Early Changes in the HSMPs

On day 1 in the dog model, reperfused STEMI induced early regional increases in infarct zones of young dogs in several markers of ECM remodeling including SLPI, SPARC, and OPN, MMP-9 and MMP-2, and a-disintegrin and metalloproteinase (ADAM)-10 and -17. Concurrently, markers of inflammation also increased, including pro-inflammatory cytokines tumor necrosis factor (TNF)-α and interleukin (IL)-6, the anti-inflammatory and fibrogenic cytokine transforming growth factor (TGF)-β_1 and its signaling molecule Smad-2, and inducible nitric oxide (NO) synthase (iNOS) [27]. Importantly, aging augmented these changes as well as markers of early LV structural remodeling and dysfunction [27]. More importantly, early angiotensin II type 1 receptor (AT_1R) blockade with candesartan attenuated these changes across the old groups albeit with a trend towards lesser benefit in the oldest [27], implicating angiotensin II and dysregulation of inflammatory and ECM-proteolytic pathways in the augmented remodeling during early hours after reperfused STEMI in aging hearts. Of note, old sham dogs showed the typical aging phenotype (increased LV mass, wall thickness, and fibrosis). Interestingly, levels of TGF-β_1 (fibrogenic and anti-inflammatory) were higher (suggesting enhanced upregulation), and levels of IL-10 (anti-inflammatory cytokine) were lower (suggesting impaired upregulation) in infarct zones of the senescent hearts—even during the early phase or reperfused STEMI and before substantial healing occurs [27]. In addition, there was an age-related increase in infarct size and apoptosis and evidence of progressive necrosis at day 3. These findings underscore the importance of assessing effects of aging on these markers and intersecting pathways and in vivo remodeling during both the infarction and subsequent healing phase after reperfused STEMI [8], especially since infarct size may impact healing [20, 41].

Taken together, our data suggest that the 3 HSMPs SLPI, SPARC, and OPN that are increased post-reperfused STEMI may interact with inflammation and fibrosis pathways and improve healing and LV remodeling in the young, but the pathways become dysregulated in the old [27]. Thus, aging amplifies responses in the critical cellular signaling and ECM-proteolytic pathways (MMPs, HSMPs, bioactive molecules) and interaction with the angiotensin II/AT_1R pathway after acute reperfused STEMI in the clinically relevant dog model [70].

27.9 Changes in HSMPs During Healing

Our preliminary unpublished data during healing after reperfused STEMI in rats [70, 71] also suggest that aging amplifies the increased expression of HSMPs, MMPs, inflammatory, and fibrogenic cytokines in infarct zones during healing after reperfused STEMI. We showed increased regional expression of SLPI, SPARC, OPN protein, and mRNA in infarct zones at 3 weeks that were augmented with aging. In that study [70, 71], SLPI, SPARC, and OPN were colocalized in macrophages and monocytes. We also confirmed increased HSMPs at day 2. Importantly, AT_1R blockade with candesartan over day 2 to day 23 suppressed these changes in the HSMPs and remodeling in young and old rats, implying regulation by angiotensin II with aging. Candesartan also attenuated increases in MMPs, inflammatory and fibrogenic cytokines, and iNOS in the young and old rats, implying regulation by angiotensin II during healing. The data also showed age-related increase in tissue myeloperoxidase (MPO, oxidant activity marker) and MPO-positive granulocytes and CD68 and MAC387-positive macrophages at day 25, implying persistent inflammation and granulation tissue with impaired healing in the late phase after reperfused STEMI. Of note, old sham rats showed the typical aging phenotype. Importantly, attenuation of the molecular changes and LV remodeling in young was less in the old than young rats, implying impaired inhibition of effects of angiotensin II with aging.

Interestingly, Bradshaw et al. reported increased SPARC expression and collagens in old (18–24 months) versus young (3 months) wild-type (C57.BL6/SV129) mice without MI and attenuation of increased collagen in old SPARC null mice without MI [72]. In our other studies of chronic reperfused STEMI HSMPs in the young rat model, we showed that reperfused STEMI consistently induces LV dysfunction and adverse LV remodeling that persist over 3 weeks and correlate temporally with marked increase in MMP-9 activity and MMP-9/TIMP-3 protein ratio, and less increase in MMP-2. Interestingly, the persistent increase in MMP-9 paralleled that in iNOS. Importantly, candesartan improved LV function and attenuated LV remodeling during healing, and these benefits correlated temporally with improvement in MMP-9, TIMP-3, and iNOS levels and the MMP-9/TIMP-3 ratio [72]. We looked for neutrophils, a known rich source of MMP-9 [73], in infarct or border zones between 7 and 21 days and found very few. MPO levels correlated with MMP-9, and both MPO and MMP-9 colocalized in border zone myocardium. Interestingly, immunohistochemistry showed weak staining for SLPI and strong staining for SPARC and OPN at 3 weeks in the same hearts. Candesartan normalized increases in TNF-α and TGF-β_1 as well as increases in MMP-9 and MMP-2 mRNAs [71]. Compared to reperfused STEMI, healing after non-reperfused STEMI was not associated with persistent increase in MMP-9 or LV remodeling and dysfunction at 3 weeks.

Together, the data suggest that aging upregulates two critical pathways: (1) increased ROS $\rightarrow$ iNOS-NO $\rightarrow$ peroxynitrite $\rightarrow$ MMP and HSMP activation $\rightarrow$ adverse remodeling and dysfunction and (2) increased inflammatory cytokines $\rightarrow$ MMP and HSMP activation $\rightarrow$ adverse remodeling and dysfunction (Fig. 27.3). We propose that the differential responses in the old versus young may be due to an aging-induced dysregulation of the expected response to SLPI and the other HSMPs (found in young mice), leading to amplified inflammatory, ECM-proteolytic, and pro-fibrotic pathways, blunting of expected responses to the increased HSMPs, and/or resistance in the TGF-β_1 pathway. Aging-related increase in angiotensin II results in enhanced interactions among ECM-proteolytic, inflammation, and fibrosis pathways and resistance in the TGF-β_1 signaling pathway, leading to defective healing and a cascade to amplified maladaptive LV remodeling and poor outcome in older patients.

Bujak already documented blunted fibroblast response to TGF-β_1 in old mice [66]. Simply put, depending on the stage of the healing/repair process after reperfused STEMI, a little increase in the HSMPs may be good, but too much may be harmful.

27.10 Interactions Among Different Pathways

Evidence suggests that interactions among HSMPs, inflammatory, ECM-proteolytic, and pro-fibrotic pathways occur in early and late phases of the healing process. Evidence in young animals suggests that SLPI suppresses MMP-9 and inflammation [55] and ischemia/reperfusion injury [74]. SPARC increases TGF-β_1 and MMP activity in acute MI [8, 63]. OPN interacts with integrins, activates inflammation, and increases iNOS, MMP-2, and MMP-9, thereby increasing ECM proteolysis in acute reperfused STEMI [8, 27]. After MI, OPN may decrease fibroblast response, impair healing, and decrease ECM [75].

Other evidence suggests that interactions among HSMPs, inflammatory cytokines, and NOS can aggravate the imbalance between MMPs and TIMPs, thereby leading to enhanced ECM damage and remodeling in acute stages [8, 56, 74, 75]. Angiotensin II increases inflammation and modulates damage via inflammatory cytokines that in turn regulate HSMPs and NOSs [8, 76]. By contrast in old animals, enhanced upregulation of HSMPs may contribute to changes in the ECM-proteolytic pathway either directly via MMPs/TIMPs or indirectly via interactions with pro-inflammatory cytokines and iNOS [27]. Enhanced upregulation of inflammatory cytokines leads to increased MMPs, reactive oxygen species (ROS), iNOS-derived NO, and peroxynitrite which in turn increases ECM damage and LV remodeling [27, 77].

We postulate that aging-related increases in LV angiotensin II [78], a known regulator of ECM expression and remodeling [23], and increased ROS and therefore oxidative stress [79], via dysregulation of inflammatory cell, ECM-proteolytic, and fibrotic pathways, may play major roles in aging-induced impaired healing (Fig. 27.3). More research is needed to identify potential novel therapeutic strategies, proteins, and/or pathways that may be targeted to optimize healing and limit maladaptive fibrosis, adverse LV remodeling, and dysfunction in patients post-reperfused STEMI [8].

27.11 The TGF-β_1/Smad Signaling Pathway and HSMPs

Evidence suggests that interactions between HSMPs and inflammation, ECM-proteolytic, and pro-fibrotic pathways may converge on the TGF-β_1 pathway and Smad signaling. As mentioned, we hypothesize that aging enhances these interactions and leads to dysregulation with increased resistance in this critical TGF-β_1

pathway, resulting in defective healing and ECM/LV remodeling (Fig. 27.3). Increased resistance to TGF-β_1 in old hearts may represent a "protective brake" against high TGF-β_1 levels or exaggeration of a brake gone "haywire." Data on HSMPs and healing after reperfused STEMI in old versus young wild-type mice or mice with absence of the HSMPs are lacking.

27.12 Aging and Events During Healing

We and others [23, 24, 53] have shown that post-MI healing in young animals involves 3 sets of timed molecular and cellular events over 2 highly dynamic phases: (1) an early phase dominated by inflammation—with ECM degradation and inhibition of tissue proliferation, (2) a late phase dominated by tissue repair—with fibroblast proliferation, ECM deposition by fibroblasts and myofibroblasts (containing α-actin), vascular and ECM remodeling with maturation, scar formation, and fibrosis, and (3) structural and functional remodeling spanning early and late phases, through dilation, hypertrophy, and angiogenesis. Emerging evidence suggests that aging can profoundly alter these cellular responses [27, 70, 71].

27.13 Aging and Rate of Healing

It is important to remember that healing takes time; how long depends on age, infarct size, the species and other molecular, cellular, and subcellular factors [21, 22]. Infarct size modulates the rate of healing, reparative and remodeling responses, and functional outcome irrespective of the age or species [21, 23]. Large infarcts heal slower, with more severe LV remodeling [21, 23]. Reperfusion may alter infarct healing in the young [80] and impair healing in the old [55, 66]. However, aging-related adverse remodeling occurs despite small infarct size [41, 66]. It is therefore important to assess both infarct size and LV topography during healing in research studies and in clinics.

27.14 Aging and Regional Differences in Collagen and Fibrosis

The regional distribution of fibrosis and the quality and quantity of collagen are important aspects of remodeling during healing after reperfused STEMI. Normally, the quantity of collagen and fibrosis by the end of the healing phase is several fold greater in infarct than non-infarct zones [23, 24]. Collagen type III deposited during healing is immature, weak, and distensible, and the healing area is more susceptible to dilation. As it matures to stronger and less distensible type I towards the end of the healing phase, the scar becomes less prone to dilation from increased cardiac load. Evidence suggests that aging may amplify these changes and prime the heart for adverse remodeling [27]. It is therefore important to assess regional collagens, quantity and quality, types I and III, and fibrosis in infarct and non-infarct regions in research studies.

27.15 Aging and the Window of Vulnerability

Decrease in ECM during infarct healing favors adverse LV remodeling, dysfunction, and rupture [23, 24]. Rapid ECM degradation post-STEMI or reperfused STEMI followed by slow synthesis, deposition of type III collagen, and slow maturation to type I results in a window of vulnerability for adverse remodeling. This window provides opportunity for therapeutic interventions [23, 24]. Our recent data suggests that aging may widen this window. It is important to assess remodeling at critical time points before and after collagen plateaus during healing [23, 24].

27.16 Aging and Risk of Adverse Post-MI Remodeling and Rupture

Although age is a strong predictor of adverse events post-STEMI and elderly cohorts are at high risk for heart failure and adverse remodeling [39–41], current therapies do not target these issues [8]. Studies show that the age-related increase in post-reperfused STEMI mortality involves cardiac rupture [40]. Post-MI rupture is common in young mice [81], and ECM damage by MMPs mediates rupture [82]. In humans, thrombolytics have been implicated in post-reperfused STEMI RMI rupture [83], and involves MMP activation [24, 84]. Infarct size is a key factor in rupture and LV remodeling [85, 86]. Rupture induced by high afterload in healed STEMI is linked to decreased and defective collagen [87, 88]. LV remodeling and dysfunction after reperfused STEMI [38] can be aggravated by aging-related enhanced ECM damage and defective collagen and healing. Old mice show defective healing post-reperfused STEMI (impaired phagocytosis of damaged cardiomyocytes, fibroblast resistance to TGF-β_1, decreased myofibroblasts, and infarct collagen) and develop adverse remodeling and rupture [66]. Old rats develop more remodeling and infarct-scar expansion [89], but rupture is rare. Elderly subgroups in clinical trials not only show higher post-STEMI morbidity and mortality from dilative remodeling and heart failure [39, 40] but may develop catastrophic rupture post-reperfused STEMI [83].

27.17 Aging and Impaired Post-MI Healing

As discussed, multiple defects associated with aging can impair post-STEMI healing [8, 69]. Aging affects most factors that modulate post-STEMI healing, including infarct size, neurohormones such as angiotensin II, oxidative stress via oxygen free radicals, NADPH oxidase and nuclear factor kappa B (NFκB) signaling, inflammation, ECM, HSMPs, and other matrix proteins [8]. Besides dysregulation of repair mechanisms, increased matrix degradation and decreased ECM deposition [90], increased angiotensin II and ROS, increased oxidative damage, dysregulated inflammation, and decreased endothelial NO availability may be involved (Fig. 27.3). It is important to consider effects on HSMPs as well as the inflammation, fibrosis, and ECM-proteolysis pathways when developing therapeutic strategies.

27.18 Aging and Dysregulation of Inflammation and Fibrosis Pathways

Analysis of data from us and others in mice, rats, and dogs [20, 27, 66, 70, 71, 82, 86, 87, 91–103] suggests that impaired post-STEMI healing in older hearts is associated with a healing triad that should be addressed in research studies. The triad consists of the following: (1) impaired, decreased, and/or dysregulated inflammatory cell response [66, 71], with altered expression of mediators of inflammation (increased pro-inflammatory cytokines TNF-α and IL-6), decreased anti-inflammatory cytokine IL-10, increased anti-inflammatory cytokine TGF-β_1, and blunting of inflammatory cell recruitment with decreased inflammatory cells [71]; (2) dysregulation in the fibrosis pathway (with enhanced increase in expression of HSMPs and other matrix proteins and imbalance between MMPs and TIMPs), increased ECM remodeling, delayed granulation tissue formation, decreased myofibroblast density, increased resistance of myofibroblasts (key cellular mediator of fibrosis) to TGF-β_1, and defective collagen in the infarct scar [66, 71]; and (3) adverse LV remodeling and deaths [8, 66, 71].

27.19 Aging and the ECM Post-STEMI

It is important to consider healing and remodeling after STEMI as processes that progress in parallel [22–24]. The current concept is that remodeling is a major mechanism of LV enlargement [23]. ECM disruption promotes dilation and remodeling [23, 104]. Angiotensin II is a major driver of ECM and LV remodeling [23]. ECM degradation after MI is mainly due to MMPs, and TIMPs provide post-translational control of MMP activity [24]. MMPs such as MMP-2 and MMP-9 play a major role in remodeling [8, 24]. A sharp rise in MMPs, with a high MMP/TIMP ratio, induces rapid ECM proteolysis, decreased collagen, adverse ECM and LV remodeling, and dysfunction [23, 24, 48]. Chronically high MMP/TIMP ratios may promote continued ECM degradation and contribute to progressive LV dilation during healing, whereas a low MMP/TIMP ratio may contribute to increased ECM and fibrosis and diastolic dysfunction. Increased MMPs with aging may result in increased LV remodeling and dysfunction during healing after reperfused STEMI [27, 70, 71]. It is important to assess MMP/TIMP balance.

27.20 Aging, Inflammatory Mediators, and TGF-β Signaling

Fibrogenic growth factors such as TGF-β_1 and pro-inflammatory cytokines such as angiotensin II and TNFα modulate MMPs and TIMPs, ECM degradation, fibrosis, and remodeling during healing post-MI [23, 24, 102]. Inflammatory cytokines may have favorable or unfavorable effects on remodeling [105, 106]. TGF-β signaling plays a critical role in healing and remodeling [107], under the regulatory influence of angiotensin II [108]. TGF-β null mice show excessive inflammation and early death [109]. In young animals, TGF-β exerts anti-inflammatory effects during early healing (such as repression of inflammation, macrophage deactivation, resolution of inflammatory infiltrate) and activates pro-fibrotic pathways during late healing (such as fibroblast activation, ECM deposition, myocyte hypertrophy) [107]. Early TGF-β activation protects against early myocardial damage, whereas sustained activation leads to adverse LV remodeling and failure in later stages post-MI in young mice [110]. Attention to timing and differential temporal responses is critical. In young animals, increased TGF-β_1 post-MI can promote ECM preservation during early healing through increased protease inhibitors (e.g., TIMPs, SLPI) and suppression of inflammatory cytokines (e.g., TNF-α and IL-6) and ECM degradation, and promote ECM deposition during later healing by suppressing proteases (e.g., TIMPs, SLPI) [107], and thereby lead to maladaptive remodeling [110]. Our recent data [27, 70, 71] suggest that with aging, increased inflammatory mediator expression together with increased ECM-proteolytic markers and reduced cellular responsiveness to growth factors such as TGF-β_1 may contribute to impaired healing and adverse remodeling after reperfused STEMI. These data suggest that too little or too much inflammation can be bad depending on the stage of healing (early vs. late). This important concept is becoming recognized [111, 112], underscoring the need to assess inflammatory mediators and TGF-β_1 in the infarct and non-infarct zones during both early and late phases of healing.

27.21 Cardiac Drugs and Fibrosis During Healing Post-STEMI

Most therapeutic cardiac drugs used for the management of patients after STEMI have pleiotropic effects that are pertinent during healing [81, 86, 113] and to which senescent STEMI show increased vulnerability [8, 69]. While ACE inhibitors, AT_1R blockers, statins, and beta-blockers are recommended post-STEMI therapies [5], they all decrease collagen synthesis, cytokines, MMP activation, and inflammation in infarct regions of young hearts in experimental studies [8, 54]. However, data with aging is lacking. Angiotensin II is well known to elicit several deleterious effects. ACE inhibitors reduce angiotensin II formation and AT_1R stimulation and limit remodeling post-MI [114, 115], but they do not block angiotensin II from alternative pathways [116]. AT_1R blockers inhibit

AT_1R-mediated effects of angiotensin II [117] and are beneficial for post-MI remodeling and HF [23, 113]. In old patients, both ACE inhibitors and AT_1R blockers are considered beneficial [5]. We have previously reported their beneficial effects in young dogs [88, 93, 95, 103, 118, 119]. Both attenuate TGF-β_1 [88, 105]. It is important to consider that in the real world, ACE inhibitors and AT_1R blockers as well as statins and beta-blockers are recommended therapies in post-STEMI patients, so that new therapies will need to be tested on top of those background therapies.

27.22 Biomarkers of Cardiovascular Risk

Kannel was one of the early pioneers to suggest that biomarkers may be useful for cardiovascular (CV) risk stratification and assessing benefits of therapy, and CV risk assessment may be improved by use of biomarkers, genetic markers, and vascular imaging [120]. Many studies have suggested that biomarkers [121–129] and vascular imaging [130–133] can stratify subgroups at risk of CV disease [129–133] and heart failure [124–128] and guide heart failure management [126]. Several biomarkers [121–129] including hs-CRP [121], N-terminal pro-B-type natriuretic peptide (NT-proBNP) [124–126], cardiac troponin I (cnI) [127], and cardiac troponin T (cnT) [124] can be used to predict heart failure [124–128], adverse remodeling [126, 134], and cardiovascular death in older adults and the elderly [124, 127, 128]. Elderly patients with chronic systolic heart failure (mean ages 66 and 67 years), high serum levels of cortisol, and aldosterone were shown to be independent predictors of increased mortality risk [135]. A strategy to suppress NT-proBNP levels in older adults (mean age 63 years) with chronic systolic heart failure reduced adverse events compared to the standard approach [126]. Another study showed that elderly patients (mean age 70 years) with diastolic heart failure had similar although less severe pathophysiological characteristics than those with systolic heart failure, including BNP levels [136] suggesting that BNP levels may not distinguish patients with systolic and diastolic heart failure. In predominantly elderly patients (mean age 70 years; range 18–104) presenting to the Emergency Department with congestive heart failure, BNP levels were lower in those with non-systolic heart failure than those with systolic heart failure but provided only modest discrimination of the subgroups compared to traditional parameters and felt to be best for distinguishing patients with or without congestive heart failure [137]. In another report of older adults presenting to the Emergency Department with congestive heart failure (mean age 64 years) in REDHOT, BNP levels predicted 90-day outcomes and aided stratification and triage [138]. Thus, biomarkers can be used to guide efforts to reduce CV risk on an individual basis during aging, and biomarkers such as NT-proBNP can guide measures to reduce hospitalization in elderly patients with heart failure. Whereas several markers of CV risk have been studied [139–144], the recent trend is moving towards multimarker testing for prediction of CV disease, including CRP, LpPLa2, MPO, etc., for inflammation and immune system activation; markers of adverse remodeling have not been studied in detail. Markers of collagen turnover are being evaluated for adverse remodeling in heart failure [145] and post-reperfused STEMI [146]. Other studies suggest that circulating plasma levels of the 3 HSMPs could be used as potential markers in the clinic [147–152], but more studies are needed. The requirements for biomarkers have been reviewed [153].

27.23 Future Directions

In the context of markers of adverse remodeling after STEMI, it is important to remember five pertinent points. First, although LV remodeling during healing post-reperfused STEMI may involve many mechanisms [8, 23, 66], based on recent data [27, 66, 70, 71], it is important to focus on key markers of the healing/repair process and hone in on dysregulation of the inflammatory, ECM-proteolytic, and fibrosis pathways in the infarct zone of aging myocardium and its impact on adverse LV remodeling. Second, since organized collagen deposition and scar formation in the infarct zone, and hypertrophy in the non-infarct zone are important for preserving structural and functional integrity after reperfused STEMI, it is important to focus on key markers of both global LV and differential regional infarct and non-infarct zone remodeling in aging cohorts at the end of the healing/repair process. Third, since remodeling spans phases of necrosis, early and late healing, as shown in our longitudinal studies using two-dimensional echocardiographic (2D-Echo) imaging and three-dimensional (3D) reconstruction to measure detailed in vivo geometric and functional changes [20, 51], and infarcting myocardium undergoes early expansion [4, 15, 16, 92, 94, 95], it is important to make repeated measurements of key markers of remodeling and function throughout (Fig. 27.4), from the onset of reperfusion, during the infarction process, and the healing/repair process and beyond in different aging cohorts. As remodeling extends beyond the healing/repair phase [154], it is important to extend measurements to remote STEMI. Fourth, since healing to scar formation after MI takes ~1 week in young mice [21, 22, 66] and longer in larger animals and humans [155], and may be delayed after reperfused STEMI with aging, it is important to measure markers of in vivo structural/functional remodeling through the entire time interval using noninvasive imaging and correlate these with changes in markers of collagen turnover and other key pathways in the healing/repair process. This is crucial because dynamic changes in remodeling occur during early healing before the collagen plateau and later healing after the collagen plateau [91, 97, 154], and concurrent dynamic changes also occur in inflammation, ECM-proteolytic and fibrotic pathways, and in the intersecting pathways during early versus late phases of healing/repair. Fifth, since our preliminary recent data [27, 70, 71] suggested lesser benefit from angiotensin II inhibition and downregulation of AT_2 and ACE-2 pathways in the oldest cohorts, implying differential regulation with aging by angiotensin II, it will be important to measure markers in aging cohorts with the view of formulating therapeutic strategies for the clinic.

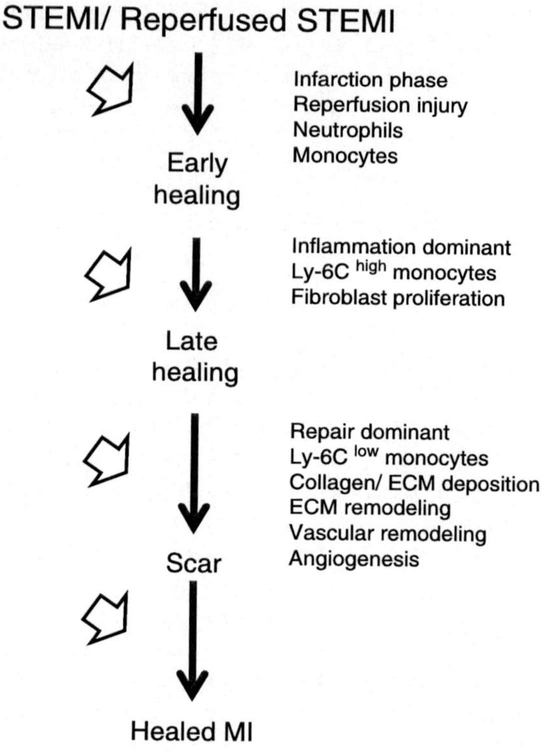

Fig. 27.4 Proposed paradigm for anti-remodeling therapy during healing/repair after STEMI The algorithm in the last 3 decades has been to use early and prolonged therapy with few pharmaceutical agents proven to improve outcome in evidence-based randomized clinical trials (RCTs). Emerging evidence suggests that there are multiple potential targets for different stages of the healing process (early, late, and remote). Pulses of therapy tailored for the different stages and novel targets may more effectively improve outcome in patients with STEMI. Multiple markers, different for the different stages, may prove more effective for monitoring pathophysiologic changes, progression and response to therapy

27.24 Conclusion

Based on emerging concepts and the recent evidence presented in this chapter, it is clear that what happens during a STEMI and the subsequent healing/repair process impacts outcome and what happens beyond. There is therefore an urgent need for more focused translational research to develop new therapeutic strategies for optimizing the healing/repair process and thereby limit adverse LV remodeling after STEMI and reperfused STEMI. In that context, emerging biomarkers may provide tools to noninvasively monitor the remodeling process, predict adverse remodeling and clinical outcome, and monitor responses to therapy in patients of different ages and at different stages after the STEMI.

Acknowledgments This work was supported in part by grant # IAP99003 (2010–2012) from the Canadian Institutes of Health Research, Ottawa, Ontario. We thank Catherine Jugdutt for assistance.

References

1. Michorowski B, Senaratne PJM, Jugdutt BI (1987) Myocardial infarct expansion. Cardiovasc Rev Rep 8:42–47
2. Michorowski B, Senaratne PJM, Jugdutt BI (1987) Deterring myocardial infarct expansion. Cardiovasc Rev Rep 8:55–62
3. Pfeffer MA, Braunwald E (1990) Ventricular remodelling after myocardial infarction. Circulation 81:1161–1172
4. Jugdutt BI (1990) Identification of patients prone to infarct expansion by the degree of regional shape distortion on an early two-dimensional echocardiogram after myocardial infarction. Clin Cardiol 13:28–40
5. Alexander KP, Newby LK, Armstrong PW et al (2007) American Heart Association Council on Clinical Cardiology; Society of Geriatric Cardiology. Acute coronary care in the elderly, Part II. ST-segment-elevation myocardial infarction. A scientific statement for healthcare professionals from the American Heart Association Council for Clinical Cardiology. Circulation 115:2570–2589
6. American Heart Association (2006) Heart disease and stroke statistics-2006 update. A report from the American Heart Association Statistics Committee and Stroke Statistics Subcommittee. Circulation 113:e85–e151
7. Cheitlin MD, Zipes DP (2001) Cardiovascular disease in the elderly. In: Braunwald E, Zipes DP, Libby P (eds) Heart Disease. WB Saunders, Philadelphia, PA, pp 2019–2037
8. Jugdutt BI (2008) Aging and remodeling during healing of the wounded heart: current therapies and novel drug targets. Curr Drug Targets 9:325–344
9. Libby P (2001) Current concepts of the pathogenesis of the acute coronary syndromes. Circulation 104:365–372
10. DeWood MA, Stifter WF, Simpson CS et al (1986) Coronary arteriographic findings soon after non-Q-wave myocardial infarction. N Engl J Med 315:417–423
11. DeWood MA, Spores J, Notske R et al (1980) Prevalence of total coronary occlusion during the early hours of transmural myocardial infarction. N Engl J Med 303:897–902
12. de Feyter PJ, van den Brand M, Serruys PW, Wijns W (1985) Early angiography after myocardial infarction: what have we learned? Am Heart J 109:194–199
13. Man J, Tymchak W, Jugdutt BI (2010) Adjunctive pharmacologic treatment for acute myocardial infarction. In: Brown DL, Jeremias A (eds) Textbook of cardiac intensive care, 2nd edn. Elsevier, Philadelphia, pp 1–72
14. Jugdutt BI, Hutchins GM, Bulkley BH, Becker LC (1979) Myocardial infarction in the conscious dog: three dimensional mapping of infarct, collateral flow and region at risk. Circulation 60:1141–1150
15. Jugdutt BI, Khan MI (1992) Impact of increased infarct transmurality on remodeling and function during healing after anterior myocardial infarction in the dog. Can J Physiol Pharmacol 70:949–958
16. Jugdutt BI, Tang SB, Khan MI et al (1992) Functional impact on remodeling during healing after non-Q-wave versus Q-wave anterior myocardial infarction in the dog. J Am Coll Cardiol 20:722–731
17. Alpert JS, Thygesen K, Antman E, Bassand JP (2000) Myocardial infarction redefined–a consensus document of The Joint European Society of Cardiology/American College of Cardiology Committee for the redefinition of myocardial infarction. J Am Coll Cardiol 36:959–969
18. Thygesen K, Alpert JS, White HD (2007) Joint ESC/ACCF/AHA/WHF task force for the redefinition of myocardial infarction. Universal definition of myocardial infarction. Eur Heart J 28:2525–2538
19. Braunwald E (1976) Protection of ischemic myocardium. Introductory remarks. Circulation 53:I-1–I-2

20. Reimer KA, Jennings RB (1979) The "wavefront phenomenon" of myocardial ischemic cell death. II. Transmural progression of necrosis within the framework of ischemic bed size (myocardium at risk) and collateral flow. Lab Invest 40:633–644
21. Jugdutt BI (1993) Prevention of ventricular remodelling post myocardial infarction: timing and duration of therapy. Can J Cardiol 9:103–114
22. Jugdutt BI (1996) Prevention of ventricular remodeling after myocardial infarction and in congestive heart failure. Heart Fail Rev 1:115–129
23. Jugdutt BI (2003) Ventricular remodeling post-infarction and the extracellular collagen matrix. When is enough enough? Circulation 108:1395–1403
24. Jugdutt BI (2003) Remodeling of the myocardium and potential targets in the collagen degradation and synthesis pathways. Curr Drug Targets Cardiovasc Haematol Disord 3:1–30
25. Lewis EF, Moye LA, Rouleau JL et al (2003) Predictors of late development of heart failure in stable survivors of myocardial infarction: the CARE study. J Am Coll Cardiol 42:1446–1453
26. Jugdutt BI (2012) Prevention of heart failure in the elderly: when, where and how to begin? Heart Fail Rev 17:531–544
27. Jugdutt BI, Jelani A, Palaniyappan A et al (2010) Aging-related changes in markers of ventricular and matrix remodelling after reperfused ST-segment elevation myocardial infarction in the canine model. Effect of early therapy with an angiotensin II type 1 receptor blocker. Circulation 122:341–351
28. Braunwald E, Kloner RA (1982) The stunned myocardium: prolonged, postischemic ventricular dysfunction. Circulation 66:1146–1149
29. Kloner RA, Ellis SG, Lange R (1983) Studies of experimental coronary artery reperfusion: effects on infarct size, myocardial function, biochemistry, ultrastructure and microvascular damage. Circulation 68(Suppl I):I-8–I-15
30. Zhao MJ, Zhang H, Robinson TF, Factor SM, Sonnenblick EH, Eng C (1987) Profound structural alterations of the extracellular collagen matrix in postischemic dysfunctional ("stunned") but viable myocardium. J Am Coll Cardiol 10:1322–1334
31. Charney RH, Takahashi S, Zhao M et al (1992) Collagen loss in the stunned myocardium. Circulation 85:1483–1490
32. Fliss H, Gattinger D (1996) Apoptosis in ischemic and reperfused rat myocardium. Circ Res 76:949–956
33. Ohno M, Takemura G, Ohno A (1998) "Apoptotic" myocytes in infarct area in rabbit hearts may be oncotic myocytes with DNA fragmentation: analysis by immunogold electron microscopy combined with In situ nick end-labeling. Circulation 98:1422–1430
34. Matsumura K, Jeremy RW, Schaper J et al (1998) Progression of myocardial necrosis during reperfusion of ischemic myocardium. Circulation 97:795–804
35. Reffelmann T, Hale SL, Dow JS, Kloner RA (2003) No-reflow phenomenon persists long-term after ischemia/reperfusion in the rat and predicts infarct expansion. Circulation 108:2911–2917
36. Orn S, Manhenke C, Greve OJ, Larsen AI, Bonarjee VVS, Edvardsen T, Dickstein K (2009) Microvascular obstruction is a major determinant of infarct healing and subsequent left ventricular remodelling following primary percutaneous coronary intervention. Eur Heart J 30:1978–1985
37. Hamdan A, Kornowski R, Solodky A et al (2006) Predictors of left ventricular dysfunction in patients with first acute anterior myocardial infarction undergoing primary angioplasty. IMAJ 8:532–535
38. Bolognese L, Neskovic AN, Parodi G et al (2002) Left ventricular remodeling after primary coronary angioplasty: patterns of left ventricular dilation and long-term prognostic implications. Circulation 106:2351–2357
39. St John Sutton M, Pfeffer MA, Moye L (1997) Cardiovascular death and left ventricular remodeling two years after myocardial infarction: baseline predictors and impact of long-term use of captopril: information from the Survival and Ventricular Enlargement (SAVE) trial. Circulation 96:3294–3299

40. Maggioni AP, Maseri A, Fresco C et al (1993) Age-related increase in mortality among patients with first myocardial infarctions treated with thrombolysis. The investigators of the gruppo Italiano per lo Studio della supravvivenza nell'Infarcto Miocardico (GISSI-2). N Engl J Med 329:1442–1448
41. Ferrari R, The PREAMI Investigators (2006) Effects of angiotensin-converting enzyme inhibition with peridopril on left ventricular remodeling and clinical outcome. Results of the randomized perindopril and remodeling elderly with acute myocardial infarction (PREAMI) Study. Arch Intern Med 166:659–666
42. Jelani A, Jugdutt BI (2010) STEMI and heart failure in the elderly: role of adverse remodeling. Heart Fail Rev 15:513–521
43. Fujiwara H, Ashraf M, Sato S et al (1982) Transmural cellular damage and blood flow distribution in early ischemia in pig heart. Circ Res 51:683–693
44. Sato S, Ashraf M, Millard RW et al (1983) Connective tissue changes in early ischemia of porcine myocardium: an ultrastructural study. J Mol Cell Cardiol 15:261–275
45. Przyklenk K, Kloner RA (1986) Superoxide dismutase plus catalase improve contractile function in the canine model of the stunned myocardium. Circ Res 58:148–156
46. Takahashi S, Barry AC, Factor SM (1990) Collagen degradation in ischaemic rat hearts. Biochem J 265:233–241
47. Danielsen CC, Wiggers H, Anderson HR (1998) Increased amounts of collagenase and gelatinase in porcine myocardium following ischemia and reperfusion. J Mol Cell Cardiol 30:1431–1442
48. Sawicki G, Menon V, Jugdutt BI (2004) Improved balance between TIMP-3 and MMP-9 after regional myocardial ischemia-reperfusion during AT_1 receptor blockade. J Card Fail 10:442–449
49. Connelly CM, Vogel WM, Wiegner AW et al (1985) Effects of reperfusion after coronary artery occlusion on post-infarction scar tissue. Circ Res 57:562–577
50. Murakami T, Kusachi S, Murakami M et al (1998) Time-dependent changes of serum carboxy-terminal peptide of type I procollagen and carboxy-terminal telopeptide of type I collagen concentrations in patients with acute myocardial infarction after successful reperfusion: correlation with left ventricular volume indices. Clin Chem 44:2453–2461
51. Vinten-Johansen J (2004) Involvement of neutrophils in the pathogenesis of lethal myocardial reperfusion injury. Cardiovasc Res 61:481–497
52. Kim CB, Braunwald E (1993) Potential benefits of late reperfusion of infarcted myocardium. The open artery hypothesis. Circulation 88:2426–2436
53. Frangogiannis NG, Smith CW, Entman ML (2002) The inflammatory response in myocardial infarction. Cardiovasc Res 53:31–47
54. Ertl G, Franz S (2005) Healing after myocardial infarction. Cardiovasc Res 66:22–32
55. Zhang Y, DeWitt DL, McNeely TB et al (1997) Secretory leukocyte protease inhibitor suppresses the production of monocyte prostaglandin H synthase-2, prostaglandin E_2, and matrix metalloproteinases. J Clin Invest 99:894–900
56. Sano C, Shimizu T, Sato K et al (2000) Effects of secretory leukocyte protease inhibitor on the production of the anti-inflammatory cytokines, IL-10 and transforming growth factor-beta (TGF-β), by lipopolysaccharide-stimulated macrophages. Clin Exp Immunol 121:77–85
57. Ashcroft GS, Lei K, Jin W et al (2000) Secretory leukocyte protease inhibitor mediates non-redundant functions necessary for normal wound healing. Nat Med 6:1147–1153
58. Denhardt DT, Guo X (1993) Osteopontin: a protein with diverse functions. FASEB J 7:1475–1482
59. Singh K, Sirokman G, Communal C et al (1999) Myocardial osteopontin expression coincides with the development of heart failure. Hypertension 33:663–670
60. Trueblood NA, Xie Z, Communal C et al (2001) Exaggerated left ventricular dilation and reduced collagen deposition after myocardial infarction in mice lacking osteopontin. Circ Res 88:1080–1087
61. Lane TF, Sage EH (1994) The biology of SPARC, a protein that modulates cell-matrix interactions. FASEB J 8:163–173

62. Masson S, Arosio B, Luvara G et al (1998) Remodelling of cardiac extracellular matrix during β-adrenergic stimulation: upregulation of SPARC in the myocardium of adult rats. J Mol Cell Cardiol 30:1501–1514
63. Wu RX, Laser M, Han H et al (2006) Fibroblast migration after myocardial infarction is regulated by transient SPARC expression. J Mol Med 84:241–252
64. Schellings MW, Vanhoutte D, Swinnen M et al (2009) Absence of SPARC results in increased cardiac rupture and dysfunction after acute myocardial infarction. J Exp Med 206:113–123
65. McCurdy SM, Dai Q, Zhang J et al (2011) SPARC mediates early extracellular matrix remodeling following myocardial infarction. Am J Physiol Heart Circ Physiol 301:H497–H505
66. Bujak M, Kweon HJ, Chatila K et al (2008) Aging-related defects are associated with adverse cardiac remodeling in a mouse model of reperfused myocardial infarction. J Am Coll Cardiol 51:1384–1392
67. Ashcroft GS, Horan MA, Ferguson MW (1997) Aging is associated with reduced deposition of specific extracellular matrix components, an upregulation of angiogenesis, and an altered inflammatory response in a murine incisional wound healing model. J Invest Dermatol 108:403–407
68. Ashcroft GS, Horan MA, Ferguson MWJ (1995) The effect of ageing on cutaneous wound healing in mammals. J Anat 187:1–26
69. Jugdutt BI, Jelani A (2008) Aging and defective healing, adverse remodeling and blunted postconditioning in the reperfused wounded heart. J Am Coll Cardiol 51:1399–1403
70. Jugdutt BI, Palaniyappan A, Uwiera RRE, Idikio H (2009) Role of healing-specific-matricellular proteins and matrix metalloproteinases in age-related enhanced early remodeling after reperfused STEMI in dogs. Mol Cell Biochem 322:25–36
71. Palaniyappan A, Idikio H, Jugdutt BI (2009) Secretory leucocyte protease inhibitor and matricellular protein modulation of post reperfused myocardial infarction healing, fibrosis and remodeling in rat model. Effect of candesartan and omapatrilat. Circulation 120(Suppl 2):S837
72. Bradshaw AD, Baicu CF, Rentz TJ et al (2010) Age-dependent alterations in fibrillar collagen content and myocardial diastolic function: role of SPARC in post-synthetic procollagen processing. Am J Physiol 298:H614–H622
73. Lindsey M, Wedin K, Brown MD, Keller C, Evans AJ, Smolen J, Burns AR, Rossen RD, Michael L, Entman M (2001) Matrix-dependent mechanism of neutrophil-mediated release and activation of matrix metalloproteinase 9 in myocardial ischemia/reperfusion. Circulation 103:2181–2187
74. Schneeberger S, Hautz T, Wahl SM et al (2008) The effect of secretory leukocyte protease inhibitor (SLPI) on ischemia/reperfusion injury in cardiac transplantation. Am J Transplant 8:773–782
75. Singh M, Foster CR, Dalal S, Singh K (2010) Role of osteopontin in heart failure associated with aging. Heart Fail Rev 5:487–494
76. Ridnour LA, Windhausen AN, Isenberg JS et al (2007) Nitric oxide regulates matrix metalloproteinases-9 activity by guanylyl-cyclase-dependent and -independent pathways. Proc Natl Acad Sci USA 104:16898–16903
77. Spinale FG (2010) Amplified bioactive signaling and proteolytic enzymes following ischemia reperfusion and aging. Remodeling pathways that are not like a fine wine. Circulation 122:322–324
78. Heymes C, Swynghedauw B, Chevalier B (1994) Activation of angiotensinogen and angiotensin-converting enzyme gene expression in the left ventricle of senescent rats. Circulation 90:1328–1333
79. Boengler K, Schulz R, Heusch G (2009) Loss of cardioprotection with ageing. Cardiovasc Res 83:247–261
80. Boyle MP, Weisman HF (1993) Limitation of infarct expansion and ventricular remodeling by late reperfusion. Study of time course and mechanism in a rat model. Circulation 88:2872–2883

81. Jugdutt BI (2010) Preventing adverse remodeling and rupture during healing after myocardial infarction in mice and humans. Circulation 122:103–105
82. Heymans S, Luttun A, Nuyens D et al (1999) Inhibition of plasminogen activators or matrix metalloproteinases prevents cardiac rupture but impairs therapeutic angiogenesis and causes cardiac failure. Nat Med 5:1135–1142
83. Honan MB, Harrell FE Jr, Reimer KA et al (1990) Cardiac rupture, mortality and the timing of thrombolytic therapy: a meta-analysis. J Am Coll Cardiol 16:359–367
84. Rohde LE, Ducharme A, Arroyo LH et al (1999) Matrix metalloproteinase inhibition attenuates early left ventricular enlargement after experimental myocardial infarction in mice. Circulation 99:3063–3070
85. Jugdutt BI, Michorowski B (1987) Role of infarction expansion in rupture of the ventricular septum after acute myocardial infarction. A two-dimensional echocardiography study. Clin Cardiol 10:641–652
86. Jugdutt BI, Basualdo CA (1989) Myocardial infarct expansion during indomethacin and ibuprofen therapy for symptomatic post-infarction pericarditis: effect of other pharmacologic agents during early remodelling. Can J Cardiol 5:211–221
87. Jugdutt BI (1987) Left ventricular rupture threshold during the healing phase after myocardial infarction in the dog. Can J Physiol Pharmacol 65:307–316
88. Jugdutt BI, Idikio H, Uwiera R (2007) Therapeutic drugs during healing after myocardial infarction modify infarct collagens and ventricular distensibility at elevated pressures. Mol Cell Biochem 304:79–91
89. Gould KE, Taffet GE, Michael LH et al (2002) Heart failure and greater infarct expansion in middle-aged mice: a relevant model of postinfarction heart failure. Am J Physiol Heart Circ Physiol 282:H625–21
90. Ballard VLT, Edelberg JM (2007) Stem cells and the regeneration of the aging cardiovascular system. Circ Res 100:1116–1127
91. Jugdutt BI, Amy RW (1986) Healing after myocardial infarction in the dog: changes in infarct hydroxyproline and topography. J Am Coll Cardiol 7:91–102
92. Jugdutt BI (1988) Effect of nitroglycerin and ibuprofen on left ventricular topography and rupture threshold during healing after myocardial infarction in the dog. Can J Physiol Pharmacol 66:385–395
93. Jugdutt BI, Humen DP, Khan MI, Schwarz-Michorowski BL (1992) Effect of left ventricular unloading with captopril on remodeling and function during healing of anterior transmural myocardial infarction in the dog. Can J Cardiol 8:151–163
94. Jugdutt BI, Khan MI, Jugdutt SJ, Blinston GE (1995) Effect of enalapril on ventricular remodeling and function during healing after anterior myocardial infarction in the dog. Circulation 91:802–812
95. Jugdutt BI (1995) Effect of captopril and enalapril on left ventricular geometry, function and collagen during healing after anterior and inferior myocardial infarction in the dog. J Am Coll Cardiol 25:1718–1725
96. Jugdutt BI, Khan MI, Jugdutt SJ, Blinston GE (1996) Effect of prolonged inotropic stimulation on ventricular remodeling during healing after myocardial infarction in the dog: mechanistic insights. J Am Coll Cardiol 27:1787–1795
97. Jugdutt BI, Joljart MJ, Khan MI (1996) Rate of collagen deposition during healing after myocardial infarction in the rat and dog models: mechanistic insights into ventricular remodeling. Circulation 94:94–101
98. Jugdutt BI (1997) Effect of reperfusion on ventricular mass, topography and function during healing of anterior infarction. Am J Physiol 272:H1205–H1211
99. Jugdutt BI, Lucas A, Khan MI (1997) Effect of ACE inhibition on infarct collagen deposition and remodeling during healing after transmural canine myocardial infarction. Can J Cardiol 13:657–668
100. Jugdutt BI (1997) Effect of amlodipine versus enalapril on left ventricular remodeling after anterior reperfused myocardial infarction. Can J Cardiol 13:945–954

101. Jugdutt BI, Musat-Marcu S (2000) Opposite effects of amlodipine and enalapril on infarct collagen during healing after reperfused myocardial infarction. Can J Cardiol 16:617–625
102. Jugdutt BI, Menon V, Kumar D, Idikio H (2002) Vascular remodeling during healing after myocardial infarction in the dog model. Effects of reperfusion, amlodipine and enalapril. J Am Coll Cardiol 39:1538–1545
103. Jugdutt BI, Idikio H, Uwiera R (2007) Angiotensin receptor blockade and ACE inhibition limit adverse collagen remodeling in the infarct zone and global diastolic dysfunction during healing after reperfused ST-elevation myocardial infarction. Mol Cell Biochem 303:27–38
104. Cleutjens JP, Kandala JC, Guarda E et al (1995) Regulation of collagen degradation in the rat myocardium after infarction. J Mol Cell Cardiol 27:1281–1292
105. Nian M, Lee P, Khaper N, Liu P (2004) Inflammatory cytokines and postmyocardial infarction remodeling. Circ Res 94:1543–1553
106. Anzai T, Yoshikawa T, Shiraki H et al (1997) C-reactive protein as a predictor of infarct expansion and cardiac rupture after a first Q-wave acute myocardial infarction. Circulation 96:778–784
107. Bujak M, Frangogiannis NG (2007) The role of TGF-β signaling in myocardial infarction and cardiac remodeling. Cardiovasc Res 74:184–195
108. Weber KT (1997) Extracelular matrix remodeling in heart failure. A role for de novo angiotensin II generation. Circulation 96:4065–4082
109. Kulkarni AB, Huh CG, Becker D et al (1993) Transforming growth factor β1 null mutation in mice causes excessive inflammatory response and early death. PNAS 90:770–774
110. Ikeuchi M, Tsutsui H, Shiomi T et al (2004) Inhibition of TGF-beta signaling exacerbates early cardiac dysfunction but prevents late remodeling after infarction. Cardiovasc Res 64:526–535
111. Nahrendorf M, Pittet MJ, Swirski FK (2010) Monocytes: protagonists of infarct inflammation and repair after myocardial infarction. Circulation 121:2437–2445
112. Vanhoutte D, Schellings M, Pinto Y, Heymans S (2006) Relevance of matrix metalloproteinases and their inhibitors after myocardial infarction: a temporal and spatial window. Cardiovasc Res 69:604–613
113. Jugdutt BI (2008) Pleiotropic effects of cardiac drugs on healing post MI. The good, bad and ugly. Heart Fail Rev 13:439–452
114. Pfeffer JM, Pfeffer MA, Braunwald E (1985) Influence of chronic captopril therapy on the infarcted left ventricle of the rat. Circ Res 57:84–95
115. Pfeffer MA, Greaves SC, Arnold MJ, The Healing and Afterload Reducing Therapy (HEART) Trial Investigators et al (1997) Early versus delayed angiotensin-converting enzyme inhibition therapy in acute myocardial infarction. The Healing and Afterload Reducing Therapy Trial. Circulation 95:2643–2651
116. Urata HB, Healy B, Stewart RW et al (1990) Angiotensin II-forming pathways in normal and failing hearts. Circ Res 66:883–890
117. Schieffer B, Wirger A, Meybrunn M et al (1994) Comparative effects of chronic angiotensin-converting enzyme inhibition and angiotensin II type 1 receptor blockade on cardiac remodeling after myocardial infarction in the rat. Circulation 89:2273–2282
118. Sawicki G, Jugdutt BI (2004) Detection of changes in protein levels in the in vivo canine model of acute heart failure following ischemia-reperfusion injury—functional proteomics studies. Proteomics 4:2195–2202
119. Sawicki G, Jugdutt BI (2007) Valsartan reverses post-translational modifications of the δ-subunit of ATP synthase during in vivo canine reperfused myocardial infarction. Proteomics 42:2100–2110
120. Kannel WB (2011) Sixty years of preventive cardiology: a framingham perspective. Clin Cardiol 34:342–343
121. Ridker PM, Rifai N, Cook NR, Bradwin G, Buring JE (2005) Non-HDL cholesterol, apolipoproteins A-1 and B_{100}, standard lipid measures, lipid ratios, and CRP as risk factors for cardiovascular disease in women. JAMA 294:326–333

122. Zethelius B, Berglund L, Sundström J et al (2008) Use of multiple biomarkers to improve the prediction of death from cardiovascular causes. N Engl J Med 358:2107–1216
123. Melander O, Newton-Cheh C, Almgren P et al (2009) Novel and conventional biomarkers for prediction of incident cardiovascular events in the community. JAMA 302:49–57
124. Daniels LB, Laughlin GA, Clopton P et al (2008) Minimally elevated cardiac troponin T and elevated N-terminal pro-B-type natriuretic peptide predict mortality in older adults: results from the Rancho Bernardo Study. J Am Coll Cardiol 52:450–459
125. DeFilippi CR, Christenson RH, Gottdiener JS (2010) Dynamic cardiovascular risk assessment in elderly people. The role of repeated N-terminal pro- B-type natriuretic peptide testing. J Am Coll Cardiol 55:441–450
126. Januzzi JL Jr, Rehman S, Mohammed AA et al (2011) Use of amino-terminal pro-B natriuretic peptide to guide outpatient therapy of patients with chronic left ventricular systolic dysfunction. J Am Coll Cardiol 58:1881–1889
127. Sundström J, Ingelsson E, Berglund L et al (2009) Cardiac troponin-I and risk of heart failure: a community-based cohort study. Eur Heart J 30:773–781
128. deFilippi CR, deLemos JA, Christenson RH (2011) Association of serial measures of cardiac troponin T using a sensitive assay with incident heart failure and cardiovascular mortality in older adults. JAMA 304:2494–2502
129. Shlipak MG, Fried LF, Cushman M et al (2005) Cardiovascular mortality risk in chronic kidney disease: comparison of traditional and novel risk factors. JAMA 293:1737–1745
130. Nasir K, Michos ED, Rumberger JA et al (2004) Coronary artery calcification and family history of premature coronary heart disease. Sibling history is more strongly associated than parental history. Circulation 110:2150–2156
131. Petretta M, Daniele S, Acampa W et al (2012) Prognostic value of coronary artery calcium score and coronary CT angiography in patients with intermediate risk of coronary artery disease. Int J Cardiovasc Imaging 28(6):1547–1556. doi:10.1007/s10554-011-9948-5
132. van Velzen JE, de Graaf FR, Jukema JW et al (2011) Comparison of the relation between the calcium score and plaque characteristics in patients with acute coronary syndrome versus patients with stable coronary artery disease, assessed by computed tomography angiography and virtual histology intravascular ultrasound. Am J Cardiol 108:658–664
133. Kerut EK (2011) Coronary risk assessment and arterial age calculation using coronary artery calcium scoring and the Framingham risk score. Echocardiography 28:686–693
134. Kim H, Januzzi JL (2010) Biomarkers in the management of heart failure. Curr Treat Options Cardiovasc Med 12:519–531
135. Guder G, Bauersachs J, Frantz S et al (2007) Complimentary and incremental mortality risk prediction by cortisol and aldosterone in chronic heart failure. Circulation 115:1754–1761
136. Kitzman DW, Little WC, Brubaker PH, Anderson RT et al (2002) Pathophysiological characterization of isolated diastolic heart failure in comparison to systolic heart failure. JAMA 288:2144–2150
137. Maisel AS, McCord J, Nowak RM et al (2003) Bedside B-type natriuretic peptide in the emergency diagnosis of heart failure with reduced or preserved ejection fraction. Results from the breathing not properly multinational study. J Am Coll Cardiol 41:2010–2017
138. Maisel AS, Hollander JE, Guss D et al (2004) Primary results of the rapid emergency department heart failure outpatient trial (REDHOT). A multicenter study of B-type natriuretic peptide levels, emergency department decision making, and outcomes in patients presenting with shortness of breath. J Am Coll Cardiol 44:1328–1333
139. Blackenberg S, Zeller T, Saarela O et al (2010) Contribution of 30 biomarkers to 10-year cardiovascular risk estimation in 2 population cohorts: the MONICA, risk, genetics, archiving, and monograph (MORGAM) biomarker project. Circulation 121:2388–2397
140. Davidson MH, Ballantyne CM, Jacobson TA et al (2011) Clinical utility of inflammatory markers and advanced lipoprotein testing: advice from an expert panel of lipid specialists. J Clin Lipidol 5:338–367

141. Greenland P, Alpert JS, Beller GA et al (2010) 2010 ACCF/AHA guideline for assessment of cardiovascular risk in asymptomatic adults. J Am Coll Cardiol 56:e50–e103
142. Karakas M, Koenig W, Zierer A et al (2012) Myeloperoxidase is associated with incident coronary heart disease independently of traditional risk factors: results from the MONICA/KORA Augsburg study. J Int Med 271:43–50
143. Hassan AKM, Bergheanu SC, Hasan-Ali H et al (2009) Usefulness of peak troponin-T to predict infarct size and long-term outcome in patients with first acute anterior myocardial infarction after primary percutaneous coronary intervention. Am J Cardiol 103:779–784
144. Lund GK, Stork A, Muellerleile K et al (2007) Prediction of left ventricular remodeling and analysis of infarct resorption in patients with reperfused myocardial infarcts by using contrast-enhanced MR imaging. Radiology 245:95–102
145. Zannad F, Rossignol P et al (2010) Extracellular matrix fibrotic markers in heart failure. Heart Fail Rev 15:319–329
146. Manhenke C, Orn S, Squire I et al (2011) The prognostic value of circulating markers of collagen turnover after acute myocardial infarction. Int J Cardiol 150:277–282
147. Rosenberg M, Zugck C, Nelles M et al (2008) Osteopontin, a new prognostic biomarker in patients with chronic heart disease. Circ Heart Fail 1:43–49
148. Ohmori R, Momiyama Y, Taniguchi H et al (2003) Plasma osteopontin levels are associated with the presence and extent of coronary artery disease. Atherosclerosis 170:333–337
149. Soejima H, Irie A, Fukunaga T et al (2007) Osteopontin expression of circulating T cells and plasma osteopontin levels are increased in relation to severity of heart failure. Circ J 71:1879–1884
150. Lentsch AB, Yoshidome H, Warner RL, Ward PA, Edwards MJ (1999) Secretory leukocyte protease inhibitor in mice regulates local and remote organ inflammatory injury induced by hepatic ischemia/reperfusion. Gastroenterology 117:953–961
151. Henriksen PA, Hitt M, Xing Z et al (2004) Adenoviral gene delivery of elafin and secretory leukocyte protease inhibitor attenuates NF-kappa B-dependent inflammatory responses of human endothelial cells and macrophages to atherogenic stimuli. J Immunol 172:4535–4544
152. Socha MJ, Manhiani M, Said N, Imig JD, Motamed K (2007) Secreted protein acidic and rich in cysteine deficiency ameliorates renal inflammation and fibrosis in angiotensin hypertension. Am J Pathol 171:1104–1112
153. Vasan RS (2006) Biomarkers of cardiovascular disease: molecular basis and practical considerations. Circulation 113:2335–2362
154. Gaudron P, Eilles C, Kugler I et al (1993) Progressive left ventricular dysfunction and remodeling after myocardial infarction. Potential mechanisms and early predictors. Circulation 87:755–763
155. Jugdutt BI (2004) Extracellular matrix and cardiac remodeling. In: Villarreal FJ (ed) Interstitial fibrosis in heart failure. New York, Springer, pp 23–55

Chapter 28
Optimizing Stem Cell Therapy for Cardiac Repair Following a Myocardial Infarction

Kaustabh Singh, Keith R. Brunt, Richard D. Weisel, and Ren-Ke Li

Abstract An extensive myocardial infarction (MI) frequently results in post-infarct remodelling, which can lead to heart failure. Emergency percutaneous coronary intervention (PCI) has reduced the mortality of ST-elevation MI, and the use of beta-blockers, angiotensin-converting enzyme (ACE) inhibitors, and anticoagulants has improved the morbidity. However, progressive ventricular dysfunction is increasingly common. New treatments are urgently needed for MI survivors to prevent pathological remodelling and functional loss. Replacing the cells lost during the infarction might limit detrimental remodelling, but none of the proposed treatments have been shown to completely restore beating cardiomyocytes. Stem cell transplantation was originally proposed to replace these lost cells, but extensive animal and human studies have suggested that the benefits of cell implantation were the result of paracrine effects. Several stem cell populations that improved ventricular function in preclinical studies have also been beneficial in clinical trials to

K. Singh
Division of Cardiovascular Surgery, Department of Surgery,
University Health Network, University of Toronto, Toronto, ON, Canada

Department of Laboratory Medicine and Pathobiology,
University of Toronto, Toronto, ON, Canada

K.R. Brunt • R.D. Weisel
Division of Cardiovascular Surgery, Department of Surgery,
University Health Network, University of Toronto, Toronto, ON, Canada

R.-K. Li (✉)
Division of Cardiovascular Surgery, Department of Surgery,
University Health Network, University of Toronto, Toronto, ON, Canada

Department of Laboratory Medicine and Pathobiology, University of Toronto,
Toronto, ON, Canada

MaRS Centre, Toronto Medical Discovery Tower, 3rd floor, Rm 3-702,
101 College Street, Toronto, ON, Canada, M5G 1L7
e-mail: renkeli@uhnres.utoronto.ca

B.I. Jugdutt and N.S. Dhalla (eds.), *Cardiac Remodeling: Molecular Mechanisms*,
Advances in Biochemistry in Health and Disease 5, DOI 10.1007/978-1-4614-5930-9_28,

treat post-MI remodelling. However, most of these trials had mixed results, highlighting the need for further research into the mechanisms responsible for improved cardiac function and the need to develop new treatment strategies to augment the beneficial effects of stem cell transplantation.

Keywords Myocardial infarction • Cell transplantation • Heart failure • Stem cells

28.1 Introduction

28.1.1 Post–Myocardial Infarction, Cardiac Remodelling, and Heart Failure

A myocardial infarction (MI) results from coronary artery occlusion and prolonged cessation of blood flow to the myocardium. The extended period of ischemia leads to myocardial damage as a result of extensive cardiac cell death, which initiates ventricular remodelling and eventually heart failure.

Heart failure is the inability of the heart to provide sufficient cardiac output to meet the metabolic demands of the body. This is a disabling and deadly condition that arises from various forms of heart disease, including acute coronary occlusion. An extensive MI frequently remodels the myocardium as a result of cardiomyocyte death and the resulting extracellular matrix (ECM) turnover and inflammation. The extent of damage determines the severity of the remodelling, including scar thinning and eventual ventricular dilation and dysfunction.

Inflammation early after coronary occlusion initiates a cascade of events that result in further cell death and more widespread tissue remodelling [1]. Inflammation in the infarct area results in not only the removal of the necrotic cells but also apoptosis in surrounding tissues. After the acute inflammatory response has resolved, additional pathophysiological remodelling begins. Following the removal of the necrotic material, fibroblasts produce ECM components, and remodelling begins with fibrosis and the loss of elasticity of the noncontractile infarcted region, which then becomes noncompliant and stiff. The loss of beating cardiomyocytes increases the physiological load on the remaining cells and promotes pathophysiological hypertrophy, including altered energy metabolism, contractile proteins, and excitation–contraction coupling. In response to high intraventricular pressures and insufficient contractility, this stiff scar thins and begins to dilate. Over time, this pathological remodelling may lead to heart failure and eventual death [2–5].

28.1.2 Current Treatment Options

Coronary artery disease (CAD) is the narrowing (stenosis) of the coronary arteries by plaques [6], which restricts blood flow to the myocardium and increases the susceptibility of rupture, thrombosis, and vessel occlusion [7]. To prevent coronary occlusion, patients can undergo percutaneous coronary intervention (PCI), which opens the narrowing by introducing a catheter into the coronary artery and inflating a balloon that crushes the plaque into the walls of the artery. During PCI, it is common to implant either a bare metal or drug-eluting stent (a wire mesh that maintains the lumen) to prevent the artery from becoming narrowed again (restenosis). An alternative to PCI is coronary artery bypass graft (CABG) surgery, in which arteries or veins from the patient's own body are employed to redirect blood around the narrowed artery. These treatments prevent the progression of CAD and reduce the risk of MI in the future.

Current therapies following an MI attempt to prevent heart failure. Beta-blockers are a class of beta-adrenoreceptor antagonists that diminish the effects of sympathetic stimulation. ACE inhibitors act on the renin–angiotensin system and also reduce the risk of heart failure and maintain the function of the blood vessels. Finally, aldosterone blockers prevent the action of aldosterone in the kidneys, which might potentiate heart failure. Both PCI and CABG surgery are very effective in relieving angina and allowing patients to return to full activity without chest pain.

There is no cure for heart failure other than heart transplantation, which is limited by donor availability. Ventricular assist devices (VADs) [8] were developed to augment the left, right, or both sides of the heart and can reverse remodelling and restore normal ventricular function and metabolism [9, 10]. Unfortunately, attempts to remove the VAD after the heart has returned to normal size and function have been associated with recurrent remodelling, resulting in the need for reinsertion of the VAD or transplantation. Given that current therapies can only prevent the progression of heart failure and do little to permanently reverse the damage caused by MI, development of new treatments is required to regenerate the heart following an extensive infarction.

28.1.3 Cell Therapy for Cardiac Repair

Although medication and interventional and surgical procedures reduce mortality and morbidity post-MI, many patients are susceptible to progressive ventricular dysfunction and become severely limited by heart failure. New treatments to prevent pathological remodelling and functional loss after surviving an MI are needed. Cell transplantation approaches that replace the lost myocardial tissue with muscle cells may reduce scar thinning and dilatation and prevent heart failure.

28.1.3.1 Fetal Cardiomyocytes

Initial studies evaluated several somatic cell types transplanted into the damaged myocardium. Adult cardiomyocytes have a limited proliferative capacity and do not reproduce in vivo to replace lost myocardium after injury. However, fetal cardiomyocytes are capable of replication. Fetal cardiomyocytes were implanted into the injured heart of rodents [11, 12], and these cells engrafted in the heart, proliferated, and integrated with the recipient myocardium. More importantly, these cells formed muscle tissue, prevented ventricular dilation, and improved cardiac function.

Although these encouraging results demonstrated the feasibility of using muscle cells to repair the heart and restore cardiac function, ethical concerns, evidence of uncontrolled propagation, and the allogeneic nature of fetal cardiomyocytes limited the enthusiasm of this cell type for clinical applications.

28.1.3.2 Skeletal Myoblasts

Skeletal myoblasts are a source of highly proliferative autologous cells that have the potential to form beating muscle in the heart. To evaluate their potential, Chiu et al. [13] implanted skeletal muscle satellite cells into the injured myocardium of dogs and provided histologic evidence of new striated muscles with intercalated discs and centrally located nuclei. Their data were consistent with the hypothesis of milieu-influenced differentiation of satellite cells into cardiac-like muscle cells. Menasche et al. [14] were the first to successfully implant skeletal myoblasts into the post-infarct scar of patients undergoing CABG. Five months after the combination of CABG and cell implantation, ventricular function improved. Subsequent studies demonstrated that transplanted myoblasts engrafted in the infarct scar, induced angiogenesis, prevented adverse matrix remodelling, and improved cardiac function [15, 16]. However, the phase II/III clinical trial (MAGIC trial) failed to demonstrate better global or regional ventricular function in the myoblast-implanted patients compared to a placebo-injected control group [17]. Although the cell transplant group had smaller ventricular volumes and improved symptomatic recovery, the trial was considered negative, and attempts to commercialize this therapy were abandoned. An additional concern related to myoblasts was the increased incidence of ventricular arrhythmias, possibly due to electrical coupling between the transplanted cells and the recipient tissue. These early clinical trials demonstrated the feasibility of implanting somatic cells into the damaged hearts of patients to restore cardiac function.

28.2 Stem Cell Therapy for Cardiac Regeneration

Stem cells hold the promise to repair the damaged myocardium and regenerate the heart. This capability has been attributed to their multipotent nature. They can form terminally differentiated cells based on their microenvironment and contribute to tissue repair. A number of stem cells have been suggested to induce cardiac repair and regeneration in both basic and clinical research studies [18].

28.2.1 Bone Marrow–Derived Stem Cells

Bone marrow (BM) transplantation has been performed following irradiation to replace hematopoietic cells. The BM houses stem cells and progenitors that are capable of regenerating cells in different tissues. The major goal of injecting BM cells into the heart was to prevent progressive pathological remodelling and heart failure by stimulating repair of the damaged tissue [19]. Initially, investigators hoped that the transplanted cells would differentiate into cardiomyocytes to replace the cells lost during the infarction. In 2004, studies were published which demonstrated that these stem cells did not differentiate into functional cardiomyocytes. Instead, the benefits of the transplanted stem cells were attributed to the release of growth factors in the heart, which promoted angiogenesis, prevented adverse matrix remodelling, and stimulated resident cardiac stem cells to initiate repair programs [20].

As the limitations of somatic cell implantation for cardiac repair became apparent (such as age and pre-existing risk factors), other types of multipotent stem cells were evaluated for regenerating the damaged myocardium. These cells are not only capable of differentiating into new cardiomyoctes but can also form endothelial cells and smooth muscle cells, which are required to form new blood vessels to regenerate the vasculature of the infarcted area. These multipotent stem cells are required to proliferate and differentiate into the resident cell types needed to constitute viable myocardial tissue. Several candidate stem cell populations have been examined for their therapeutic potential and are examined below. Recent studies have also shed light on the importance of not only the implanted cells but also the cellular microenvironment that the cells are injected into. Stem cell function is critically influenced by extrinsic signals derived from the niche or the microenvironment that the stem cells reside in. This niche plays a key role in the therapeutic modulation of stem cell behavior [21]. Therefore, in addition to providing the necessary cells for regeneration, the optimal extracellular milieu for those cells has emerged as a critical factor in stem cell therapy for cardiac regeneration.

BM has been suggested to regenerate the heart [22], as evidenced from studies showing that transplanted cells are capable of participating in post-MI repair. However, it was not established which cell source was contributing to the cardiac repair; thus, several groups have attempted to characterize the BM-derived cells that participate in post-MI repair. Although the initial reports suggested that BM cells regenerated cardiomyocytes, endothelial cells, and smooth muscle cells in the infarcted ventricle [22], subsequent studies determined that BM cells did not transdifferentiate to form new cardiomyocytes [17, 23]. Despite the controversy over cellular differentiation, the evidence is convincing that BM cells improve cardiac function and prevent remodelling after an extensive MI.

The encouraging results from preclinical animal studies prompted several clinical trials of bone marrow mononuclear cells (BMMNCs). The initial clinical trials demonstrated improved cardiac function after an extensive MI [24, 25]. BM is a rich source of stem cells, including hematopoietic stem cells (HSCs), endothelial progenitor cells (EPCs), and mesenchymal stem cells (MSCs). The following stem and progenitor cells isolated from BM have been studied for their capacity to repair the heart post-MI.

28.2.1.1 Hematopoietic Stem Cells

HSCs exist as long-term and short-term regenerating cells that cycle in the BM stem cell niche. Coculturing cardiomyocytes with HSCs has been shown to protect the cardiomyocytes from cell death via paracrine factors [26]. The differentiation of BM-derived cells to a myogenic phenotype is dependent on HSCs with intact c-kit function [27]. Furthermore, the endogenous repair of the myocardium is dependent on the function of the c-kit receptor, which is expressed on HSCs [28].

28.2.1.2 Mesenchymal Stem Cells

MSCs are multipotent stromal cells that are able to differentiate into cell types of several lineages [29]. They can be isolated from the BM or adipose tissue and can be stored long term for later implantation. In addition, MSCs suppress allo-responses and may be immuno-tolerant [30]. Therefore, they might be an ideal "off-the-shelf" product to treat MI [31, 32]. The original evaluations of allogeneic MSC transplantation demonstrated beneficial restoration of ventricular function comparable to the effects of syngeneic or autologous MSC implantation. However, the long-term effect of allogeneic MSCs was recently demonstrated to be limited by rejection [32]. Huang and colleagues demonstrated that between 3 weeks and 3 months after implantation, MSCs that differentiated became immunogenic and were rejected, resulting in deterioration of ventricular restoration and function [33].

28.2.1.3 Endothelial Progenitor Cells

EPCs are circulating stem cells involved in repair and regeneration of blood vessels. EPCs have been used for repairing the heart after an infarct in animal experiments as well as clinical trials [34]. The underlying mechanisms of cardiac regeneration are mediated by augmenting neovascularization post-MI. Enhancing revascularization reduces the extent of apoptosis in and around the infarcted region. The increase in survival of cardiomyocytes precludes the expansion of fibrosis, thus negating scar dilation.

28.2.2 Cardiac Progenitor Cells

Aside from the benefits derived from implanting noncardiac stem and progenitor cells to repair the heart, recent studies have pointed to cardiac-resident stem cell populations as also participating in cardiac repair. The benefits of cell implantation result in part from the stimulation of cardiac-resident stem cells [35]. Recently, these cardiac stem cells (CSCs) have been isolated and employed for cell transplantation. Many different types of CSCs have been described, and two have been employed in recent clinical trials. The first cell type includes the c-kit$^+$

cell populations that are believed to reside in niches within the myocardium. They were expanded from atrial appendages harvested during CABG surgery and injected into the infarct-related artery in the SCIPIO trial [36]. The other progenitor population is cardiosphere-derived cells (CDCs) produced from human endomyocardial biopsies that were expanded into spheroids in suspension cultures [37]. These cells are believed to improve heart function after an MI by paracrine mechanisms [38]. The recent CADUCEUS trial demonstrated that the intracoronary infusion of autologous CDCs after MI was safe and may be efficacious [39]. Continuing work in basic and clinical science will determine whether these cells can produce sufficient recovery in selected patient groups.

28.3 Optimizing Stem Cell Therapy to Regenerate the Infarcted Myocardium

28.3.1 Challenges of Translating Preclinical Research into Clinical Trials

Preclinical studies demonstrated profound beneficial effects of a variety of stem cells injected into the heart. However, the clinical trials have shown less robust results [40, 41]. There are several factors that contribute to the discrepancies observed between animal and human studies. Preclinical trials often do not take into consideration the comorbidities of the patients receiving this therapy. Diabetes, hypertension, chronic obstructive pulmonary disorder (COPD), dyslipidemia, and kidney disease are all strongly associated with aging, a factor that is most likely a critical determinant in the success of cell therapy for cardiac repair.

The age of the cells being injected as well as the age of the patient receiving the cells are both important determinants of the benefit achieved. The key difference between the preclinical and clinical studies is the age of the individuals treated. Preclinical studies were initially performed only in young healthy animals, but the clinical trials were performed in older patients with diffuse CAD. To improve the clinical benefits of cell therapy to the extent achieved in the preclinical studies, additional interventions will be required. These interventions include using young (allogeneic) or rejuvenated autologous stem cells, restoring the response of the recipient to stem cell injection, enhancing engraftment of the cells, and augmenting the paracrine benefits of the implanted cells.

Recent studies have demonstrated that stem cells from aged individuals have a limited capability for proliferation and differentiation, and aged recipients have a diminished capability to generate a reparative response to cell implantation [35, 42–44]. Therefore, restoration of ventricular function in aged patients by stem cell therapy will require rejuvenation of both the implanted stem cells and the response to cell implantation [35]. The methods to rejuvenate stem cells from aged individuals include cytokine therapies and BM transplantation [45]. The other approach is to use allogeneic cells from young, healthy individuals.

28.3.2 *Allogeneic Stem Cells*

To avoid the problems associated with autologous cells from aged individuals, allogeneic stem cells from young donors can be transplanted into the infarcted myocardium. Allogeneic cells have been demonstrated to improve cardiac recovery after MI, and they are not initially rejected early after implantation [46, 47]. Allogeneic MSCs are currently undergoing clinical trials [48]. However, allogeneic cells likely will lose their immunoprivileged state when they differentiate after implantation. A number of strategies have been proposed to prevent the rejection of allogeneic MSCs. Establishing a chimeric BM with donor BM cells may prevent rejection [49]. Induction of tolerance has also been proposed by transfection with interleukin 10 (IL-10) [50]. New approaches to immune modulation may produce dramatic and long-lasting improvements in ventricular function with allogeneic MSCs.

28.3.3 *Enhancing Engraftment and Paracrine Effects of Stem Cell Therapy*

Previously, the strategy for stem cell therapy for cardiac repair was one of replacing the damaged myocardium with new cells. However, with advancing research, it has become apparent that the tissue needs to be repaired with a combination of cell therapy and factors to enhance the engraftment and repair by these cells. Strategies to boost the repair capacity of implanted cells include the use of cytokines to support cell survival as well as biomaterials designed to aid in cell engraftment. The recent availability of new biodegradable biomaterials offers the opportunity to augment the survival and engraftment of cells implanted into the infarcted heart. Temperature-sensitive hydrogels that are liquid at room temperature and a gel at body temperature can enhance cell engraftment and provide prolonged cytokine delivery [51]. Biodegradable biomaterial meshes can also be employed to provide sustained cytokine release to augment stem cell survival and engraftment, and the improvement of ventricular function [52]. These biomaterials can provide sustained cytokine release to enhance the paracrine benefits of stem cell therapy to aged individuals and restore ventricular function.

28.3.4 *Strategies for Improving Stem Cells for Transplantation*

Additional strategies to enhance cell survival and engraftment include optimizing the time of delivery and preconditioning the cells prior to their implantation. The optimal time for cardiomyocyte transplantation after myocardial injury was found to be after the initial inflammatory reaction but before the onset of scar expansion [53]. However, the optimal time to implant non-cardiomyocytes, such as stem and progenitor cells, has not been conclusively determined but may be similar. Ex vivo pretreatment with growth factors or hypoxic environments may confer cytoprotection to increase the survival of implanted cells in the harsh environment of the infarct region. A variety of factors have been evaluated, including hepatocyte growth factor (HGF), vascular endothelial growth factor (VEGF) [54], insulin-like growth factor-1 (IGF-1), and transforming growth factor-β (TGF-β), to augment the efficacy of cells for myocardial repair. Alternatively, preconditioning the stem cells by genetically modifying them to express pro-survival factors may similarly increase their survival and functionality.

28.4 Perspectives

Despite the limited benefit of stem cell therapy in the initial clinical trials, the improvements in ventricular function achieved after an extensive MI were as great as those reported with PCI, beta-blockers, or ACE inhibitors [55, 56]. However, standardization of procedures for patients and further improvements in techniques are needed to reach the benefits demonstrated in preclinical animal studies. The optimal stem cell will need to be established in comparative studies. The ideal stem cell will rapidly engraft and stimulate the cardiac stem cell niche to recruit the endogenous regenerative capacity of the heart.

Further comparative preclinical studies will permit the comparison of alternate stem cell candidates, but comparative clinical trials are also necessary to design a therapy for old, debilitated patients who are most in need of cardiac regeneration. Finally, long-term studies are needed to adequately determine the efficacy of stem cell therapy. Now, more than ever, regenerative therapies to prevent heart failure will require multidisciplinary efforts from clinicians and scientists to apply the latest advances in stem cell and regenerative biology to the patients most in need of this therapy.

Acknowledgment The authors wish to acknowledge the generous support of Eileen Mercier and ING Canada Inc. This work was supported in part by the Canadian Institutes of Health Research (MOP 102535 to RKL).

References

1. Frantz S, Bauersachs J, Ertl G (2009) Post-infarct remodelling: contribution of wound healing and inflammation. Cardiovasc Res 81:474–481
2. Bolli R, Marban E (1999) Molecular and cellular mechanisms of myocardial stunning. Physiol Rev 79:609–634
3. Frangogiannis NG (2012) Matricellular proteins in cardiac adaptation and disease. Physiol Rev 92:635–688
4. Leri A, Kajstura J, Anversa P (2005) Cardiac stem cells and mechanisms of myocardial regeneration. Physiol Rev 85:1373–1416

5. Swynghedauw B (1999) Molecular mechanisms of myocardial remodeling. Physiol Rev 79:215–262
6. Libby P, Ridker PM, Hansson GK (2009) Inflammation in atherosclerosis: from pathophysiology to practice. J Am Coll Cardiol 54:2129–2138
7. Yla-Herttuala S, Bentzon JF, Daemen M et al (2011) Stabilisation of atherosclerotic plaques. Position paper of the European Society of Cardiology (ESC) Working Group on atherosclerosis and vascular biology. Thromb Haemost 106:1–19
8. Terracciano CM, Miller LW, Yacoub MH (2010) Contemporary use of ventricular assist devices. Annu Rev Med 61:255–270
9. Birks EJ, Tansley PD, Hardy J et al (2006) Left ventricular assist device and drug therapy for the reversal of heart failure. N Engl J Med 355:1873–1884
10. Rose EA, Gelijns AC, Moskowitz AJ et al (2001) Long-term use of a left ventricular assist device for end-stage heart failure. N Engl J Med 345:1435–1443
11. Soonpaa MH, Koh GY, Klug MG, Field LJ (1994) Formation of nascent intercalated disks between grafted fetal cardiomyocytes and host myocardium. Science 264:98–101
12. Li RK, Jia ZQ, Weisel RD et al (1996) Cardiomyocyte transplantation improves heart function. Ann Thorac Surg 62:654–660
13. Chiu RC, Zibaitis A, Kao RL (1995) Cellular cardiomyoplasty: myocardial regeneration with satellite cell implantation. Ann Thorac Surg 60:12–18
14. Menasche P, Hagege AA, Scorsin M et al (2001) Myoblast transplantation for heart failure. Lancet 357:279–280
15. Farahmand P, Lai TY, Weisel RD et al (2008) Skeletal myoblasts preserve remote matrix architecture and global function when implanted early or late after coronary ligation into infarcted or remote myocardium. Circulation 118:S130–S137
16. Menasche P, Alfieri O, Janssens S et al (2008) The Myoblast Autologous Grafting in Ischemic Cardiomyopathy (MAGIC) trial: first randomized placebo-controlled study of myoblast transplantation. Circulation 117:1189–1200
17. Murry CE, Soonpaa MH, Reinecke H et al (2004) Haematopoietic stem cells do not transdifferentiate into cardiac myocytes in myocardial infarcts. Nature 428:664–668
18. Ptaszek LM, Mansour M, Ruskin JN, Chien KR (2012) Towards regenerative therapy for cardiac disease. Lancet 379:933–942
19. Laflamme MA, Murry CE (2005) Regenerating the heart. Nat Biotechnol 23:845–856
20. Maltais S, Tremblay JP, Perrault LP, Ly HQ (2010) The paracrine effect: pivotal mechanism in cell-based cardiac repair. J Cardiovasc Transl Res 3:652–662
21. Wagers AJ (2012) The stem cell niche in regenerative medicine. Cell Stem Cell 10:362–369
22. Orlic D, Kajstura J, Chimenti S et al (2001) Bone marrow cells regenerate infarcted myocardium. Nature 410:701–705
23. Balsam LB, Wagers AJ, Christensen JL et al (2004) Haematopoietic stem cells adopt mature haematopoietic fates in ischaemic myocardium. Nature 428:668–673
24. Assmus B, Schachinger V, Teupe C et al (2002) Transplantation of progenitor cells and regeneration enhancement in acute myocardial infarction (TOPCARE-AMI). Circulation 106:3009–3017
25. Wollert KC, Meyer GP, Lotz J et al (2004) Intracoronary autologous bone-marrow cell transfer after myocardial infarction: the BOOST randomised controlled clinical trial. Lancet 364:141–148
26. Rosenberg M, Lutz M, Kuhl C et al (2012) Coculture with hematopoietic stem cells protects cardiomyocytes against apoptosis via paracrine activation of AKT. J Transl Med 10:115
27. Xaymardan M, Cimini M, Fazel S et al (2009) c-Kit function is necessary for in vitro myogenic differentiation of bone marrow hematopoietic cells. Stem Cells 27:1911–1920
28. Fazel S, Cimini M, Chen L et al (2006) Cardioprotective c-kit+ cells are from the bone marrow and regulate the myocardial balance of angiogenic cytokines. J Clin Invest 116:1865–1877
29. Keating A (2006) Mesenchymal stromal cells. Curr Opin Hematol 13:419–425
30. Williams AR, Hare JM (2011) Mesenchymal stem cells: biology, pathophysiology, translational findings, and therapeutic implications for cardiac disease. Circ Res 109:923–940
31. Aggarwal S, Pittenger MF (2005) Human mesenchymal stem cells modulate allogeneic immune cell responses. Blood 105:1815–1822

32. Atoui R, Shum-Tim D, Chiu RC (2008) Myocardial regenerative therapy: immunologic basis for the potential "universal donor cells". Ann Thorac Surg 86:327–334
33. Huang XP, Sun Z, Miyagi Y et al (2010) Differentiation of allogeneic mesenchymal stem cells induces immunogenicity and limits their long-term benefits for myocardial repair. Circulation 122:2419–2429
34. Richardson MR, Yoder MC (2011) Endothelial progenitor cells: quo vadis? J Mol Cell Cardiol 50:266–272
35. Li SH, Sun Z, Brunt KR et al (2012) Reconstitution of aged bone marrow with young cells repopulates cardiac-resident bone marrow-derived progenitor cells and prevents cardiac dysfunction after a myocardial infarction. Eur Heart J. Epub ahead of print. doi:10.1093/eurheartj/ehs072
36. Bolli R, Chugh AR, D'Amario D et al (2011) Cardiac stem cells in patients with ischaemic cardiomyopathy (SCIPIO): initial results of a randomised phase 1 trial. Lancet 378:1847–1857
37. Smith RR, Barile L, Cho HC et al (2007) Regenerative potential of cardiosphere-derived cells expanded from percutaneous endomyocardial biopsy specimens. Circulation 115:896–908
38. Chimenti I, Smith RR, Li TS et al (2010) Relative roles of direct regeneration versus paracrine effects of human cardiosphere-derived cells transplanted into infarcted mice. Circ Res 106:971–980
39. Makkar RR, Smith RR, Cheng K et al (2012) Intracoronary cardiosphere-derived cells for heart regeneration after myocardial infarction (CADUCEUS): a prospective, randomised phase 1 trial. Lancet 379:895–904
40. Zimmet H, Porapakkham P, Porapakkham P et al (2012) Short- and long-term outcomes of intracoronary and endogenously mobilized bone marrow stem cells in the treatment of ST-segment elevation myocardial infarction: a meta-analysis of randomized control trials. Eur J Heart Fail 14:91–105
41. Sun L, Zhang T, Lan X, Du G (2010) Effects of stem cell therapy on left ventricular remodeling after acute myocardial infarction: a meta-analysis. Clin Cardiol 33:296–302
42. Kan CD, Li SH, Weisel RD et al (2007) Recipient age determines the cardiac functional improvement achieved by skeletal myoblast transplantation. J Am Coll Cardiol 50:1086–1092
43. Zhang H, Fazel S, Tian H et al (2005) Increasing donor age adversely impacts beneficial effects of bone marrow but not smooth muscle myocardial cell therapy. Am J Physiol Heart Circ Physiol 289:H2089–H2096
44. Zhuo Y, Li SH, Chen MS et al (2010) Aging impairs the angiogenic response to ischemic injury and the activity of implanted cells: combined consequences for cell therapy in older recipients. J Thorac Cardiovasc Surg 139:1286–1294, 1294
45. Brunt KR, Wu J, Chen Z et al (2012) Ex vivo Akt/HO-1 gene therapy to human endothelial progenitor cells enhances myocardial infarction recovery. Cell Transplant 21:1443–1461
46. Amado LC, Saliaris AP, Schuleri KH et al (2005) Cardiac repair with intramyocardial injection of allogeneic mesenchymal stem cells after myocardial infarction. Proc Natl Acad Sci USA 102:11474–11479
47. Quevedo HC, Hatzistergos KE, Oskouei BN et al (2009) Allogeneic mesenchymal stem cells restore cardiac function in chronic ischemic cardiomyopathy via trilineage differentiating capacity. Proc Natl Acad Sci USA 106:14022–14027
48. Hare JM, Traverse JH, Henry TD et al (2009) A randomized, double-blind, placebo-controlled, dose-escalation study of intravenous adult human mesenchymal stem cells (prochymal) after acute myocardial infarction. J Am Coll Cardiol 54:2277–2286
49. Sykes M, Nikolic B (2005) Treatment of severe autoimmune disease by stem-cell transplantation. Nature 435:620–627
50. Nemeth K, Leelahavanichkul A, Yuen PS et al (2009) Bone marrow stromal cells attenuate sepsis via prostaglandin E(2)-dependent reprogramming of host macrophages to increase their interleukin-10 production. Nat Med 15:42–49
51. Wu J, Zeng F, Huang XP et al (2011) Infarct stabilization and cardiac repair with a VEGF-conjugated, injectable hydrogel. Biomaterials 32:579–586

52. Miyagi Y, Zeng F, Huang XP et al (2010) Surgical ventricular restoration with a cell- and cytokine-seeded biodegradable scaffold. Biomaterials 31:7684–7694
53. Li RK, Mickle DA, Weisel RD et al (2001) Optimal time for cardiomyocyte transplantation to maximize myocardial function after left ventricular injury. Ann Thorac Surg 72:1957–1963
54. Yau TM, Fung K, Weisel RD et al (2001) Enhanced myocardial angiogenesis by gene transfer with transplanted cells. Circulation 104:I218–I222
55. Reffelmann T, Konemann S, Kloner RA (2009) Promise of blood- and bone marrow-derived stem cell transplantation for functional cardiac repair: putting it in perspective with existing therapy. J Am Coll Cardiol 53:305–308
56. Weissman IL (2000) Translating stem and progenitor cell biology to the clinic: barriers and opportunities. Science 287:1442–1446

Chapter 29
Regulation of Fibrosis After Myocardial Infarction: Implications for Ventricular Remodeling

Bodh I. Jugdutt

Abstract Fibrosis plays a critical role in ventricular remodeling after myocardial infarction (MI). The normal physiologic response to tissue injury in mammals involves activation of the healing process that leads to repair of the injured area with a fibrotic scar over time. In the heart, healing of the damaged area of the left ventricular (LV) chamber after MI with a firm fibrous scar is essential for the chamber to continue to pump blood effectively into the tissues. Optimal healing and repair of the wounded heart with appropriate fibrosis is critical for restoration of chamber size, shape, integrity, and systolic function and survival with a favorable outcome. Dysregulated healing and repair mechanisms may lead to adverse remodeling and poor outcome. While reparative fibrosis in the infarct zone is desirable, reactive interstitial fibrosis in the non-infarct zone contributes to chamber stiffness and diastolic dysfunction. Evidence suggests that the regulation and temporal progression of fibrosis may differ in the two zones. In addition, fibrosis in the infarct and non-infarct zones may become dysregulated and contribute to mixed dilative and hypertrophic remodeling and mixed LV systolic and diastolic dysfunction. Besides reparative and reactive fibrosis in the infarcted LV, fibrosis that develops at remote sites such as the atrial and right ventricular chambers and the kidneys can influence outcome after MI. Understanding the differential mechanisms of fibrosis in the infarct, non-infarct, and remote zones may allow the development of new and improved therapeutic strategies for controlling fibrosis and improve outcome in survivors of MI.

Keywords Fibrosis • Healing • Inflammation • Matrix remodeling • Monocyte populations • Myocardial infarction • Reperfusion injury • Repair • Ventricular remodeling

B.I. Jugdutt, M.D., M.Sc., D.M. (✉)
Division of Cardiology, Department of Medicine, University of Alberta,
2C2 Walter MacKenzie Health Sciences Centre, Edmonton, AB, Canada, T6G 2R7
e-mail: bjugdutt@ualberta.ca

B.I. Jugdutt and N.S. Dhalla (eds.), *Cardiac Remodeling: Molecular Mechanisms*, Advances in Biochemistry in Health and Disease 5, DOI 10.1007/978-1-4614-5930-9_29,

Injury → Healing and Repair → Fibrosis → Scar

Fig. 29.1 The march from injury to fibrosis in noncardiac tissue

29.1 Introduction

Fibrosis plays a critical role in ventricular remodeling after myocardial infarction (MI). An important aspect of the normal physiologic response to tissue injury in mammals involves activation of the healing process that leads to reparative fibrosis and scar formation in the injured area over time (Fig. 29.1). While there are common features in the healing/repair process after skin and heart wounds, there are also important differences as the heart is a highly dynamic organ that contracts and relaxes many times per minute throughout life. For example, in the wounded heart, healing of the damaged area of the left ventricular (LV) chamber after an MI with a firm fibrous scar is essential for the chamber to continue to pump blood effectively into the tissues (Fig. 29.2). Optimal healing and repair of the wounded heart with appropriate fibrosis is therefore critical for restoration of chamber size, shape, integrity, and systolic function and survival with a favorable outcome [1–10]. Emerging evidence suggests that dysregulation of healing and repair mechanisms may lead to adverse ventricular remodeling and poor outcome [11–13]. Adverse ventricular remodeling during the healing/repair phase after MI continues to be an important problem in adult cardiology despite improved coronary reperfusion and medical therapies [14–18]. Despite the search for specific molecular targets that may lead to discovery of therapeutic strategies to optimize post-MI healing and repair, optimize quality and quantity of fibrosis, prevent adverse remodeling, and improve outcome since the 1980s, specific therapy to realize this goal is still lacking in the twenty-first century [11–13, 19]. This chapter focuses on the regulation of fibrosis after MI and its implications for ventricular remodeling. Understanding the differential mechanisms of fibrosis in the infarct, non-infarct, and remote zones may allow the development of new and improved therapeutic strategies for controlling fibrosis and improve outcome in survivors of MI.

29.2 Definitions and Caveats

Five important points must be considered in translating findings from the bench to the bedside. First, a coronary artery occlusion puts myocardium in the occluded coronary bed at risk of ischemic damage [20], and persistent occlusion usually results in two types of MI depending on net collateral blood supply [21–25]. Transmural MI correlates with pathological necrosis ≥50% of the transmural extent, or evidence of Q-wave MI or ST-segment elevation MI (STEMI) on electrocardiograms, whereas subendocardial MI correlates with pathological necrosis <50% of the transmural extent, or evidence of non-Q-wave or non-STEMI (NSTEMI) on

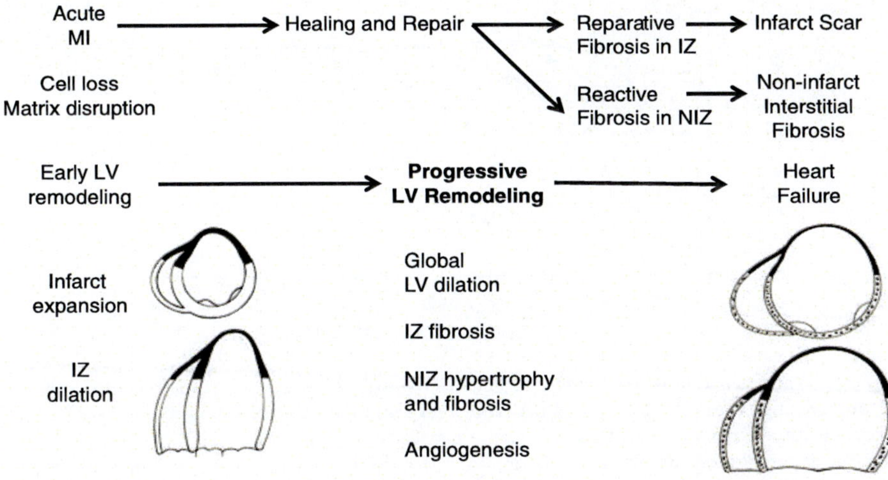

Fig. 29.2 The march from injury to fibrosis in the heart. *Upper panel*: This cartoon depicts the concept of reparative fibrosis in the infarct zone and reactive fibrosis in the non-infarct zone over time. *Lower panel*: This cartoon depicts early regional remodeling in the infarction phase and further global dilative remodeling in the healing/repair phase and beyond scar formation. Data based on anterior infarcts in the canine model. *IZ* ischemic zone, *LV* left ventricular, *MI* myocardial infarction, *NIZ* non-infarct zone

electrocardiograms [21, 22]. Clinically, an acute coronary syndrome can lead to either STEMI or NSTEMI that has different clinical outcomes and needs different management strategies [22–25]. All evidences indicate that STEMI has grave prognosis and outcome in the survivors [3, 7, 16, 18, 26]. This chapter will focus on fibrosis and remodeling after STEMI.

Second, most human MIs are the result of coronary artery occlusion caused by inflammation of an atherosclerotic plaque and thrombosis [27–30], whereas most animal models of MI lack this inflamed atherosclerotic component. In that context, Panizzi et al. recently studied MI in the hypercholesterolemic ApoE–/– mice with atherosclerosis and systemic monocytosis and showed close recapitulation of the inflammatory and cellular responses found during healing after MI in humans [31].

Third, as there are species-specific differences in the inflammatory and cellular responses during healing and repair after MI in mice and dogs [32], caution is advisable when extrapolating findings from animal studies to humans. Fourth, whereas MIs and heart failure are dominant in elderly humans [11, 33], the majority of studies of MI and fibrosis at the bench have been conducted in very young animals; caution is therefore needed in extrapolating findings from animal studies to older humans [33–35]. Bujak et al. recently showed that aging results in impaired healing/repair and adverse remodeling after reperfused MI in mice [34]. In our laboratory, studies in the dog [36, 37] and rat [38] models of reperfused STEMI showed that aging enhances adverse remodeling and the inflammatory responses [36–38].

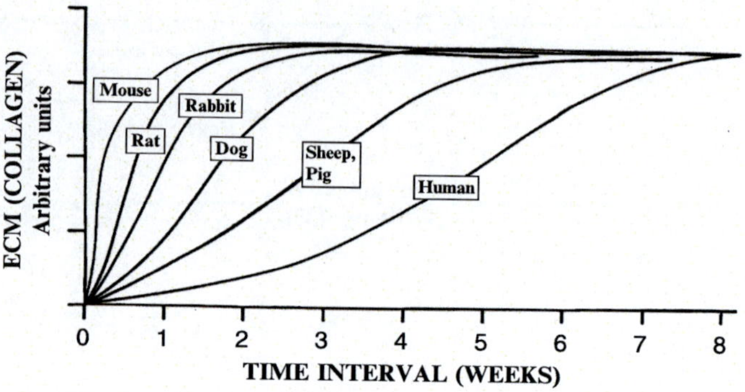

Fig. 29.3 Species and time to collagen plateau during post-MI healing and repair. Cartoon showing that healing takes time that varies between weeks and months depending on the species and infarct characteristics (size, reperfusion); large infarcts in humans may take 3–6 months to heal after MI depending on infarct size. The duration of healing is measured by the time interval to reach the collagen plateau for moderate myocardial infarct size. The duration of healing is shorter in smaller animals. *ECM* extracellular matrix, *MI* myocardial infarction

Fifth, the rate of healing and repair is influenced by several factors in any one species. These factors include infarct size, whether reperfused or not, other molecular, cellular, and subcellular factors [3–8]. Old age not only is associated with impaired healing as noted above [11, 17, 34, 35, 38] but also results in increased infarct size [36, 37]. More important, the rate of healing and repair differs among species. Thus, the march to the collagen plateau, assessed by changes in hydroxyproline concentration in the infarct zone as a marker of fibrosis, is more prolonged in larger animals (Fig. 29.3). In Fig. 29.3, the interval to the collagen plateau is between 1–2 weeks for mice, 2–3 weeks for rats, 4–6 weeks for dogs, and 8–12 weeks for humans. Accordingly, it is important to remember the time factor when interpreting data from different animal species, or extrapolating data on temporal changes from small animals to humans, or planning therapeutic interventions.

29.3 Pathophysiology of Healing, Repair, Fibrosis and Remodeling After MI

Early post-mortem and clinicopathological studies of healing/repair after myocardial infarction in the first half of the twentieth century (reviewed in [2, 6–9]) extended the simple algorithm from *injury to fibrosis of damaged tissue*, as found in the skin and other noncardiac tissues (Fig. 29.1), to the more complex one of *reparative fibrosis in the infarct zone and reactive fibrosis in the non-infarct zone* accompanied by dilation and differential structural remodeling in both zones (Fig. 29.2). Cumulative evidence from histopathological, morphological, and biochemical studies in humans and experimental animals between the late 1970s and mid-1990s expanded the algorithm further.

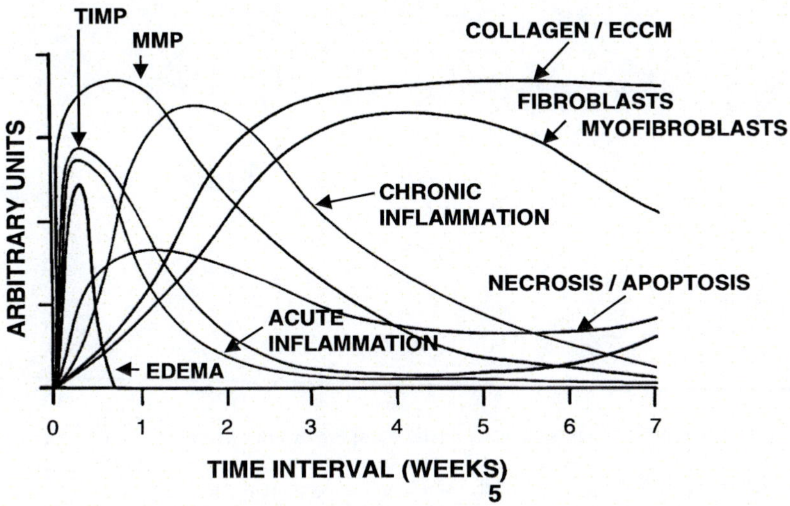

Fig. 29.4 Histopathological and extracellular matrix changes in the infarct tissue during healing and repair after MI. Cartoon depicting temporal changes in necrosis and apoptosis, edema, acute and chronic inflammation, fibroblast proliferation, myofibroblast formation, collagen, and extracellular collagen matrix (ECCM) in the infarct zone of the dog model of anterior myocardial infarction (MI). Collagen content is based on the hydroxyproline assay. The data is based on changes in the infarct zone from 194 canine hearts from previous studies. *MMP* matrix metalloproteinase, *TIMP* tissue inhibitor of MMP

In the new construct, acute MI results in early myocardial necrosis with cell loss and damage to the extracellular matrix (ECM); ECM damage drives the early dilative remodeling and progressive dilative remodeling; necrosis triggers the healing/repair process that leads to fibrosis and scar in the infarct area and maladaptive remodeling with fibrosis and hypertrophy in the non-infarct area [3–9, 11, 39–41]. Many studies attributed the ECM degradation to the release and activation of matrix metalloproteinases (MMP), resulting in an imbalance between levels of MMPs and endogenous tissue inhibitors of MMPs (TIMPs) in the infarct zone (Fig. 29.4). This elevated MMP/TIMP ratio persists during the healing/repair phase (Fig. 29.4) and is a major factor in progressive dilative remodeling [7, 8].

A host of studies between the 1990s and 2010 have addressed the molecular and cellular changes and extended the algorithm further, and identified inflammation and ECM remodeling as two key components of healing/repair, fibrosis, and remodeling after MI. The collective evidence from those studies suggests that the healing/repair process is finely orchestrated and involves tandem reactions over two highly dynamic phases: an early phase that is dominated by inflammation (with release of inflammatory cytokines, ECM proteases, and bioactive molecules, leading to ECM degradation and damping of proliferation) and a late phase dominated by repair (with release of fibrogenic cytokines, fibroblast proliferation, ECM deposition and maturation, ECM and vascular remodeling, fibrous scar formation, and remote fibrosis) [3–9, 11, 39–41].

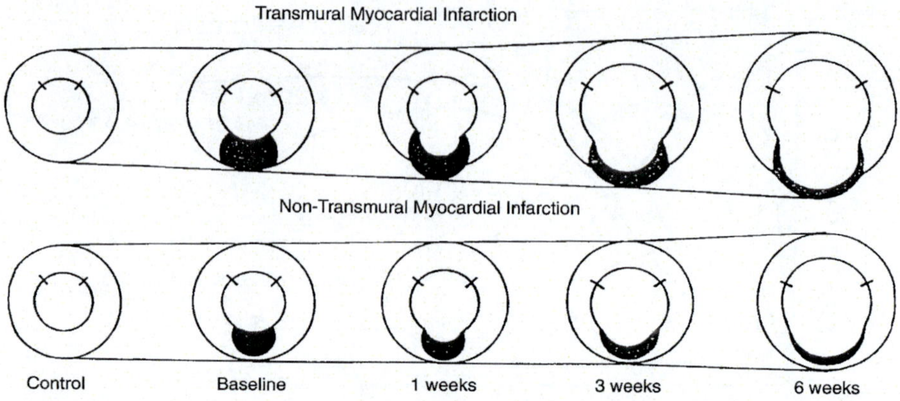

Fig. 29.5 Topographical changes during healing and repair after myocardial infarction. Cartoon depicting regional dilative remodeling and global dilative remodeling over time after anterior myocardial infarction (MI) in the dog model. Data based on systematic mapping of changes in geometry of mid-papillary transverse sections of 194 canine hearts with transmural and non-transmural infarcts at five selected time intervals using computerized morphometry. Adverse remodeling of infarct and non-infarct zones is less with subendocardial than transmural infarcts. Infarct area in black

Between the 1980s and 2010, a series of studies using repeated cardiac imaging using detailed two-dimensional echocardiography with three-dimensional reconstruction after acute MI has documented significant dilative LV remodeling after MI that spans the infarction, healing/repair phases, and beyond scar formation in the dog model and humans [42–53]. Clinical and pathological correlations were made in most of these studies [54]. Morphometric data from hearts removed at different intervals during the healing and repair phase in the dog model allowed characterization of the progressive regional infarct and global LV structural remodeling that is more severe with transmural than subendocardial MI (Fig. 29.5) and correlates with simultaneous echocardiographic imaging data [3–9].

In a longitudinal study of 70 patients with chronic heart failure after MI who were assessed prospectively, Gaudron et al. showed that progression of LV dilation and dysfunction may continue for up to 3 years [55]. This suggests that factors operating during the healing/repair phase and beyond can negatively impact LV remodeling and outcome. Importantly, they found that the high-risk subset could be predicted by initial estimates of infarct size, ejection fraction, coronary perfusion grade, as well as infarct location at 4 days [54]. These findings are in agreement with our previously reported findings in a prospective study of 244 patients with a first Q-wave MI, indicating that the degree of initial regional infarct dilation or infarct expansion (that correlated with infarct size, anterior infarct location, and low ejection fraction) predicted a high-risk subset for poor outcome [48].

Clearly large transmural infarcts would be expected to be associated with slower healing and repair, more fibrosis in both infarct and non-infarct zones, and more progressive adverse remodeling [7, 8]. Pertinent to the theme of progressive adverse remodeling at 3 years in Gaudron's study [55], infarct scars contain specialized myofibroblasts formed by phenotypic conversion of fibroblasts and containing sarcomeric α-smooth muscle actin, rendering the scar tissue capable of contraction ([9] for review; [56–64]). This emphasizes that scars should be considered live and dynamic rather than dead tissue [60], and myofibroblasts were shown to persist for up to 17 years in the infarct scars [61]. In contrast to infarct scars, myofibroblasts are transient in skin wounds [9]. This finding has important implications for therapeutic interventions to attenuate fibrosis and use of therapeutic cardiac drugs with pleiotropic effects that impact fibrosis pathways ([65, 66] for review).

29.4 Advances in Monocyte and Macrophage Biology During Healing/Repair Post-MI

It follows from the above discussions that, besides infarct size and altered composition and integrity of the supporting ECM, the adequacy or quality of healing/repair after MI is another major determinant of infarct fibrosis and ventricular remodeling. Emerging evidence over the last decade has significantly enhanced our understanding of the cellular aspects of the march of healing and repair to fibrosis after MI and further expanded the previous algorithms (Fig. 29.6). The inflammatory reaction after MI has two main functions: (1) to clear the infarct area of dead cells and matrix debris and (2) to activate the pathways that are necessary for optimal scar formation. These functions are largely modulated by recently identified different monocyte and macrophage phenotypes ([41, 67, 68] for review) and are highly relevant clinically [69–71].

In that construct, dysregulation of the inflammatory response, with prolongation or expansion of the response, results in impaired collagen deposition and weak scars, thereby leading to an augmentation of LV remodeling and dysfunction [7–9, 41]. As noted before, inflammation and ECM remodeling are closely linked. Excessive early inflammation can augment ECM degradation and promote severe early infarct expansion and precipitate cardiac rupture [72]. Increased expression of pro-inflammatory mediators, such as interleukin-6 (IL-6) and tumor necrosis factor-α (TNF-α), can activate pro-apoptotic pathways and thereby contribute to more cardiomyocyte loss.

In a recent study, early reperfused STEMI in the dog model was associated with increased pro-inflammatory markers such as inducible nitric oxide synthase (iNOS), cytokines IL-6 and TNF-α, anti-inflammatory markers such as transforming growth factor-β_1 (TGF-β_1) and IL-10, and evidence of cardiomyocyte damage (ischemic injury, infarct size, apoptosis, blood flow impairment, and no-reflow), adverse LV remodeling (LV dilation and dysfunction), and ECM remodeling with increased expression of *s*ecretory *l*eukocyte *p*rotease *i*nhibitor (SLPI), *s*ecreted *p*rotein *a*cidic and *r*ich in *c*ysteine (SPARC), osteopontin (OPN), and *a* *d*isintegrin *a*nd *m*etalloproteinase (ADAM)-10 and (ADAM)-17, as well as MMP-9 and MMP-2 [37]. Recent studies in the rat model also documented increase in all these markers in the later healing phase after reperfused STEMI [38].

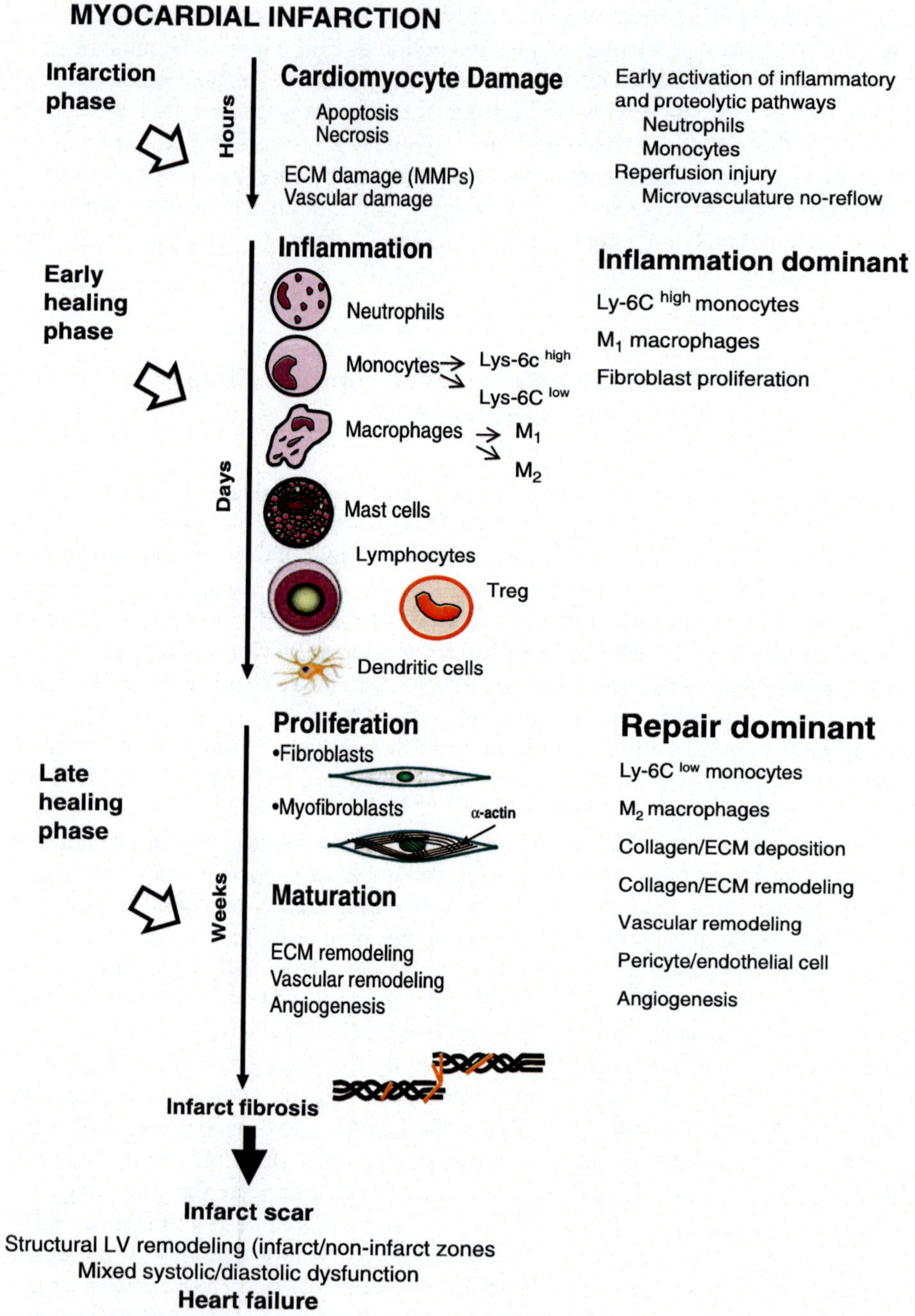

Fig. 29.6 Cellular and molecular changes during the infarction and healing/repair phases. Cartoon depicting expansion of the traditional algorithm. Acute myocardial infarction with cell loss triggers the inflammatory cascade and the healing and repair process; this process involves an orches

During the later phase of healing, resolution of the inflammatory response is critical [41] and involves three components: (1) activation followed by inhibition (negative regulation), (2) containment to the infarct zone, and (3) resolution [41]. These steps appear to involve different cellular effectors and molecular mediators that need further study [41]. The defective containment of the inflammatory reaction can lead to extension of the infiltrate into the bordering non-infarcted myocardium and thereby enhance fibrosis and worsen diastolic function. This underscores the importance of considering both the spatial and temporal aspects of regulation of the inflammatory response.

The molecular reactions after MI involve a large number of modulators and mediators [13] whose tandem release and activation are finely orchestrated [41] and converge on fibrosis and infarct formation (Fig. 29.7). Initially in the infarction phase after MI (Fig. 29.6), release of intracellular contents by necrotic cardiomyocytes and release of ECM degradation products by damaged ECM trigger activation of membrane-bound toll-like receptors (TLRs) that modulate post-ischemic inflammation ([73] for review). Concurrently, increased generation of reactive oxygen species (ROS) and reduced antioxidant reserve result in activation of the nuclear factor (NF)-κB that stimulates production and release of pro-inflammatory cytokines (such as IL-1β) and chemokines ([74] for review). Depending on the CC and CCX chemokine profile, neutrophils are attracted early and pro-inflammatory monocytes and lymphocytes later. Multiple inhibitory pathways together with coordinated actions of different cell types (such as neutrophils, mononuclears, endothelial cells, and pericytes) and interaction with the ECM lead to suppression, resolution, and containment of the inflammation [41]. At least five mechanisms have been implicated: (1) clearance of apoptotic neutrophils, (2) recruitment of inhibitory monocyte subsets, (3) macrophage differentiation, (4) recruitment of regulatory T cells, and (5) pericyte/endothelial cell interaction [41].

First, neutrophils accumulate very early after MI and peak within 1 day in the infarct zone in response to chemokines (such as IL-8), adhesion molecules (such as L- and P-selectins), and intercellular adhesion molecule 1 (ICAM-1). While they are abundant in the acute phase of MI, they undergo apoptosis (at ~3 days in mice and ~7 days in dogs), thereby releasing inhibitors of inflammation (such as annexin 1 and lactoferrin) which reduce recruitment of neutrophils and attract macrophages (through "find me" and "eat me" signals) that phagocytose the apoptotic neutrophils. This in turn activates anti-inflammatory mediators (such as IL-10 and TGF-β) and pro-resolution mediators (such as lipoxins and resolvins) ([75] for review).

Fig. 29.6 (continued) trated sequence of two highly dynamic phases: an early phase that is dominated by inflammation and a late phase that is dominated by repair. Significant remodeling of the extracellular matrix occurs in all phases. The end-result is the formation of a fibrotic infarct scar that undergoes further ventricular remodeling. Later, reactive fibrosis develops in remote myocardium. Ultimate outcome depends on the degree of dilative global remodeling and mixed systolic and diastolic dysfunction. This concept was based on data mostly in the young. *ECM* extracellular matrix, *LV* left ventricular, *MMP* matrix metalloproteinase

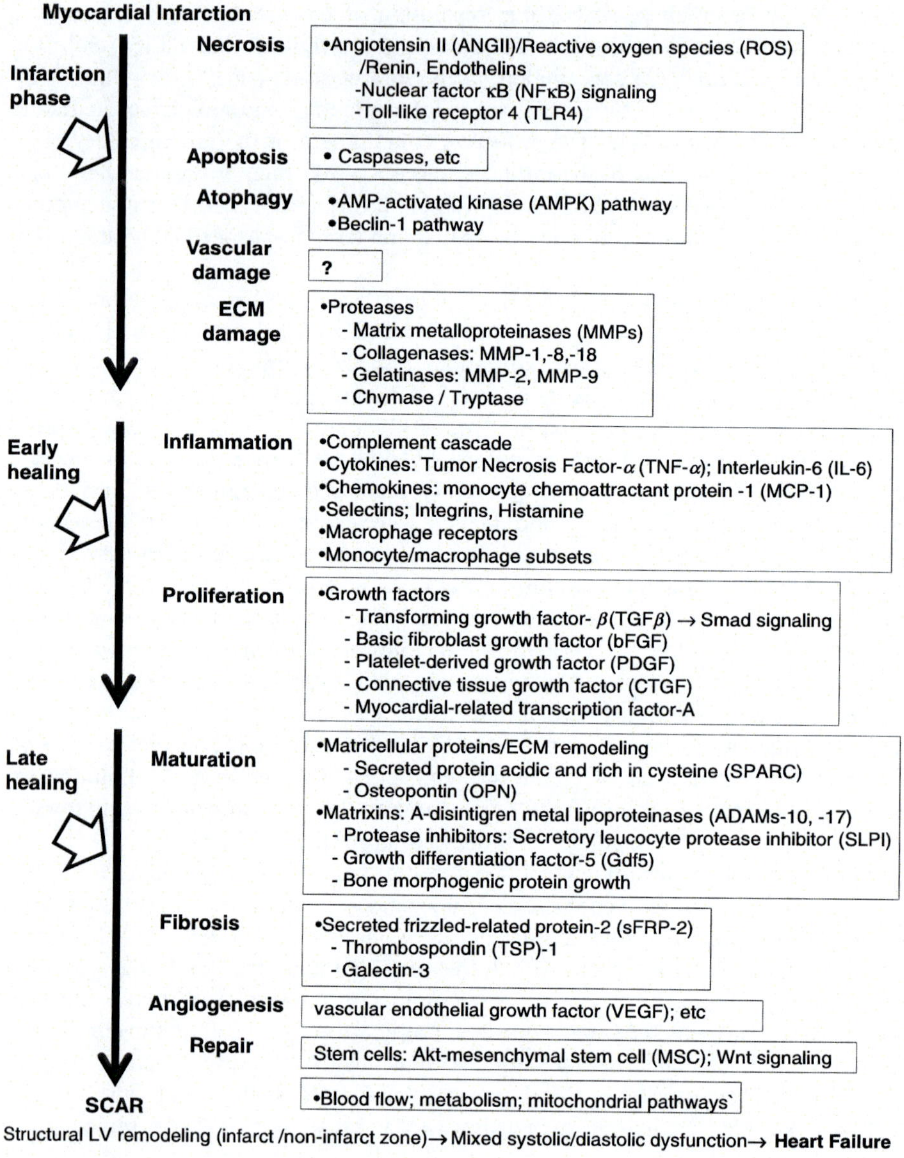

Fig. 29.7 Schematic of modulators and mediators in the healing, repair, and fibrosis processes after myocardial infarction

Second, monocytes and macrophages are present in increased numbers in the early and late phases of healing/repair and act as key regulators of the process (Fig. 29.6). Both neutrophils and monocytes belong to the myeloid lineage of cells of the innate immune system. Depending on cues from the tissue environment, monocytes differentiate into macrophages or dendritic cells. Monocytes/macrophages dominate the repair/healing phases and have at least six functions: (1) release of inflammatory mediators (such as iNOS, ROS, interferon-γ, IL-1, IL-6, TNF-α, TGF-β_1, and macrophage inflammatory protein [MIP]-1α); (2) phagocytosis of apoptotic and necrotic myocytes and neutrophils and other debris; (3) release of proteases such as MMPs, urokinase-type plasminogen activator (uPA), and cathepsins that digest collagen matrix and facilitate cell transit; (4) secretion of vascular endothelial growth factor (VEGF) and fibroblast growth factor (FGF) to promote angiogenesis; (5) transport of reparative enzymes and pro-survival factors such as transglutaminases; and (6) and release of TGF-β and FGF to stimulate myofibroblasts to synthesize and deposit collagen [68].

Evidence suggests that these functions are executed in a temporally and spatially organized manner through a biphasic response of two subsets of monocytes with distinct effects on the inflammatory response during the early and late phases ([68, 76] for review). In humans, a dominant subset in the monocyte pool expresses high levels of CD14 and low or no CD16, and a minor subset expresses low CD14 and high CD16. The CD14$^+$ monocytes also express high levels of the CC chemokine receptor (CCR2) for the inflammatory chemokine monocyte chemoattractant protein 1 (MCP-1) and can release myeloperoxidase (MPO) while CD16$^+$ monocytes produce TNF-α. In mice, the Ly-6C^{high} monocytes with high levels of CCR2 and low levels of the fractalkine receptor CX3CR1 resemble the human CD16$^-$ monocytes, whereas the Ly-6C^{low} (CCR2$^+$ CX3CR1^{high}) monocytes resemble human CD16$^+$ monocytes. The pro-inflammatory Ly-6C^{high} monocytes dominate the early phase of healing, whereas the pro-resolution and pro-repair Ly-6C^{low} monocytes dominate the late phase of healing (Fig. 29.6). Ly-6C^{high} monocytes remove necrotic debris and secrete inflammatory cytokines, OFRs, and matrix-degrading proteases. Ly-6C^{low} monocytes trigger angiogenesis, collagen/ECM synthesis by myofibroblasts, and healthy infarct scar formation. In summary, monocytes/macrophages are necessary in both phases, but their actions have to be balanced so as not to delay resolution of inflammation, impair healing, ECM remodeling, and weaken the scar. Too little or too much inflammation can therefore be harmful. Several studies have suggested that insufficient or excessive monocyte numbers can impair healing [31, 65, 68–70, 76–78].

Third, macrophages also play key roles in healing/repair and polarize to different phenotypes depending on spatially and temporally regulated upregulation of cytokines, chemokines, and growth factors in early and late phases. During the early phase of healing after MI, upregulation of macrophage colony-stimulating factor (MCSF) induces monocyte differentiation [41], with polarization to classical pro-inflammatory M1 macrophages that secrete T1 helper cell Th1 cytokines (interferon-γ, TNF-α); activated M1 macrophages in turn secrete large amounts of inflammatory mediators and release ROS and nitrogen intermediates; boost inflammatory response; promote ECM degradation; and remove debris [79]. Of note, suppression of M1 macrophages by the Class A scavenger receptor may attenuate cardiomyocyte necrosis [80]. During the late phase of healing, macrophages polarize to alternatively activated M2 macrophages (akin to Th2 cells) that show enhanced phagocytic activity and high anti-inflammatory IL-10 activity (suppressor of inflammation), the decoy type 2 IL-1 receptor and IL-1 receptor antagonist

(IL-1Ra), exerting pro-resolution effects with removal of inflammatory leukocytes [81]. Pertinent here, persistent M2 activation stimulates excessive fibroblast proliferation, ECM deposition, and fibrosis [82].

Fourth, injury sites recruit pro-inflammatory effector T cells as well as regulatory T cells called Tregs, a $CD4^+/CD25^+$ subset of T lymphocytes with suppressor properties [83]. CCR5 null mice with reperfused MI have decreased $CD4^+/CD25^+$ Tregs, enhanced inflammation and MMP activity [84] that may lead to adverse remodeling, and impaired healing/repair.

Fifth, vascular maturation through acquisition of a mural cell coat may contribute to inhibition of inflammation and suppression of granulation tissue formation [41]. During the early inflammatory phase of infarct healing, vascular endothelial cells synthesize chemokines and express surface adhesion molecules, leading to adhesive leukocyte/endothelial cell interactions that recruit inflammatory cells. Subsequent replacement of the inflammatory infiltrate by granulation tissue is associated with angiogenesis (Figs. 29.6 and 29.7). Early upregulation of VEGF after MI leads to the formation of neovessels that lack a pericyte coat and are hyperpermeable and pro-inflammatory [84]. Vascular maturation that coincides with scar maturation involves coating of endothelial cells of neovessels with pericytes and smooth muscle cells under the modulatory effect of platelet-derived growth factor (PDGF)-BB/PDGFR-β signaling; this results in decreased vascular permeability and the inflammatory and angiogenic activity of endothelial cells, thereby stabilizing the scar. Inhibition of PDGFR-β resulted in defective vascular maturation and decreased infarct collagen [85].

Besides the recognized roles of MMP in ECM degradation and adverse dilative remodeling after MI and increased TIMPs in diffuse interstitial fibrosis in chronic heart failure [8, 9], dynamic changes in the ECM during healing/repair can modulate the function and gene expression in leukocytes, endothelial cells, and fibroblasts [47, 86]. Matrix and matricellular proteins such as fibronectin and thrombospondin (TSP)-1, respectively, can regulate the inflammatory responses and contain the infiltrate to the infarct zone. TSP-1 participates in suppression and containment of the inflammatory infiltrate [41]. The roles of other molecular pathways in repression and resolution of the inflammatory response post-MI, such as toll-like receptor (TLR) signaling, lipid mediators, leukocyte adhesion cascades, have been reviewed [41].

29.5 Fibrosis Post-myocardial Infarction

It is clear from the foregoing discussion that a timed sequence of multiple molecular and cellular events during the healing/repair process modulate the march to infarct zone fibrosis and ideal formation of a mature and strong fibrotic scar (Figs. 29.6 and 29.7). The temporal and spatial organization and containment of the process in the infarct zone have been emphasized in many studies and reviews [7–9, 68, 78, 85]. The regulators and suppressors of the process have also been reviewed [41], although more research is needed to identify the "braking" and "stop" signals in the pathways. Fibrosis in the non-infarct zone has received less attention, and the mechanisms of fibrosis in the non-infarct zone remain to be fully defined. The traditional concept is that stimulation of fibroblasts by TGFβ, after the infarction phase and subsequent early healing, leads to differential deposition of ECM, collagens, and fibrosis, with greater amounts in infarct than non-infarct zones [1–22, 40–46, 49, 50, 87, 88], and increasing ratio of collagen fibril type I to type III during the later maturation process [52, 53, 89]. Concurrently during healing, differential remodeling of infarct and non-infarct zones occurs, in a bid to preserve systolic function—but this ideal scenario does not always happen [7, 8]. During early healing, net ECM degradation dominates, while later, net ECM deposition, collagen synthesis, and maturation dominate, resulting in differential ECM and LV remodeling during the two stages [7, 8].

29.6 Aging and Fibrosis

Importantly, aging can lead to dysregulation of healing and repair mechanisms and result in defective collagen deposition and excessive fibrosis [33–35]. Excessive collagen deposition, fibrosis, and diastolic dysfunction are a hallmark of aging, and post-MI remodeling is more severe in old patients and animals [11]. Defective fibrosis may be due to reduced collagen deposition or a relative paucity of mature collagen type I (which resists distension) and an abundance of immature collagen type III (which is more distensible) [11, 33–35, 90]. A key step in collagen deposition is the processing of procollagens by procollagen proteinase (pCP), under stimulation by a pCP enhancer, to form mature collagen type 1 fibrils [91]. Excessive fibrosis can lead to more diastolic and systolic dysfunction at any age. Antifibrotic therapy post-MI may decrease early collagen deposition and impair later maturation during infarct healing and thus may come at a cost [7–9, 11, 90]. New and safe therapeutic targets and strategies, above existing therapies, and aimed at differentially modulating fibrosis during healing are therefore needed, especially in the elderly.

29.7 Secreted Frizzled-Related Protein 2

One candidate is secreted frizzled-related protein 2 (sFRP-2) [90]. Wnt is linked to fibrosis with aging [92], and Wnt inhibition by sFRP-2 was suggested to limit post-MI fibrosis [93]. Recently, sFRP-2 was claimed to enhance pCP activity of bone morphogenetic protein 1 (BMP1)/tolloid (TLD)-like metalloproteinase independent of Wnt, and absence of sFRP-2 to reduce post-MI fibrosis and improve systolic function in young null mice compared to wild-type (WT) mice [94]. BMP1 can activate lysyl oxidase (the enzyme essential for collagen cross-linking leading to formation of the mature collagen type I) by removing the N-terminal peptide from the pre-enzyme of lysyl oxidase.

Preliminary data from our laboratory showed that sFRP-2 expression increases during early healing after reperfused MI in young rats and its decrease by inhibition of angiotensin II is beneficial [95]. In contrast, others claimed that enhancing rather than inhibiting sFRP-2 may be beneficial in MI [96–99], and injection into MI early at 2 days limits remodeling, systolic dysfunction, and fibrosis in young rats [100]. The controversy between these studies targeting sFRP-2 in young mice [96] and rats [100] is unexplained and needs to be resolved before clinical testing. Neither group [96, 100] considered possible differential effects of sFRP-2 during early and late healing in the mouse or rat models, nor did they study sFRP-2 after reperfused MI (which is the common clinical scenario) or sFRP-2 post-reperfused MI in aging animals (although HF and MI are more common and post-MI healing is impaired in the elderly).

Since overexpression of FrzA/sFP1 [100], a Wnt/fz inhibitor, exerted differential effects on the cellular infiltrate, MMP-2 and -9, apoptosis, infarct size, myofibroblast density, angiogenesis, total collagen, and types I and III during post-MI healing in mice, sFRP-2 may exert similar differential effects. In human skin scar, sFRP-2 increased type III collagen, which could be harmful infarct scars. Since the amounts of collagen and fibrosis are severalfold greater in the infarct than non-infarct zones, and the biology of fibrosis differs in the two zones during healing, differential regional remodeling occurs. It follows that antifibrotic agents may have differential effects on the two zones.

29.8 Other Factors in Fibrosis Post-MI

Other local factors can affect collagen and fibrosis [8, 11]. Our recent data suggest that several intersecting pathways may converge during post-MI healing [37, 38]. Besides sFRP-2, and BMP1/TLD-like metalloproteinases [94–100], pertinent factors include angiotensin II and the TGF-β/Smad pathway [7–9, 11, 41, 102, 103], the MMP/TIMP pathway [104, 105], novel healing-specific proteins [10, 11, 36–38], and BMP14 or Gdf5 [106]. Of note, Gdf5 was also linked to repair and infarct scar expansion during the later phase of healing [106]. The TLD-like proteinase activates the TGF-β superfamily members including mycostatin, Gdf8, Gdf11, and TGF-β [100]. Angiotensin II, the primary mediator of the renin–angiotensin–aldosterone system (RAAS), drives both ECM and LV remodeling [7, 8] and regulates fibrosis throughout healing post-MI and beyond [102, 103].

The so-called healing-specific proteins [10, 11, 36–38] that modulate ECM and remodeling during post-MI healing include the following: SLPI which upregulates anti-inflammatory cytokines and can improve skin wound [107–109] and post-MI [37, 38] healing; OPN, a matricellular matrix protein that promotes collagen synthesis and can improve post-MI healing [37, 38, 110–112]; and SPARC, another matricellular protein that modulates cell–matrix interactions and regulates fibroblast migration, may also improve post-MI healing [37, 38, 113–115]. Other matrix proteins such as ADAM (a disintegrin metalloproteinase) may also participate [37, 38]. In the chronic rat model of reperfused STEMI, we found that SLPI, SPARC, and OPN were upregulated in the infarct scars at 3 weeks, and the increases were attenuated by an angiotensin type 1 receptor (AT_1R) blocker [38]. In the dog model, these proteins were also upregulated after acute reperfused STEMI, and an AT_1R blocker attenuated the increases [36, 37].

We know that $TGF\beta_1$ modulates fibroblast phenotype and its conversion to myofibroblasts, thereby promoting ECM deposition [89, 116]. TGFβ is secreted as a large latent complex comprising of a mature TGFβ, TGFβ prodomain, and latent TGFβ-binding protein [89]. The large latent complex is bound to ECM via latent TGFβ-binding protein, preventing access to MMPs required for activating TGFβ through cleavage of TGFβ prodomains. By cleaving the latent TGFβ-binding protein, BMP1 releases the large latent complex from the ECM, thereby rendering prodomains susceptible to cleavage by MMPs [117]. As mentioned, $TGF\beta_1$ can upregulate BMP1 expression, thereby accelerating the LV remodeling. It follows that inhibition of BMP1 by sFRP-2 may interrupt this positive feedback loop and block the remodeling process. The possibility that inhibition of BMP-1 by sFRP-2 (via inhibition of canonical Wnt pathway) can potentially inhibit TGFβ activation and collagen cross-linking needs study.

Recent studies have shown that galectin-3 plays a major role in inflammation and fibrosis [118]. The mechanisms involve macrophage activation, chemotaxis, and activation of TGF-β–Smad-3 pathways [41, 118]. Galectin-3 is emerging as a promising marker of fibrosis in post-MI heart failure [119].

It should be noted that all four major currently recommended post-MI therapies, namely, angiotensin-converting enzyme (ACE) inhibitors, AT_1R blockers, beta blockers, and statins, can decrease collagen synthesis, cytokines, MMP activation, and inflammation and thereby modify post-infarct healing, fibrosis, and remodeling [10, 11, 68]. Aging may also result in remodeling of the RAAS [120]. Background therapy and comorbidities associated with older patients need to be considered in translational research.

29.9 Summary

While reparative fibrosis in the infarct zone is desirable, reactive interstitial fibrosis in the non-infarct zone contributes to chamber stiffness and diastolic dysfunction. In addition, fibrosis in the infarct and non-infarct zones may become dysregulated and contribute to mixed dilative and hypertrophic remodeling and mixed LV systolic and diastolic dysfunction. Besides reparative and reactive fibrosis in the infarcted LV, fibrosis that develops at remote sites such as the atrial and right ventricular chambers and the kidneys can influence outcome after MI. Evidence suggests that the regulation and temporal progression of fibrosis may differ in the two zones. Understanding the differential mechanisms of fibrosis in the infarct, non-infarct, and remote zones may allow the development of new and improved therapeutic strategies for controlling fibrosis and improve outcome in survivors of MI.

Acknowledgments This work was supported in part by grant # IAP99003 from the Canadian Institutes of Health Research, Ottawa, Ontario. I thank Catherine Jugdutt and Bernadine Jugdutt for assistance with the manuscript preparation and the figures.

References

1. Jugdutt BI (1985) Delayed effects of early infarct-limiting therapies on healing after myocardial infarction. Circulation 72:907–914
2. Jugdutt BI, Amy RW (1986) Healing after myocardial infarction in the dog: changes in infarct hydroxyproline and topography. J Am Coll Cardiol 7:91–102
3. Jugdutt BI (1993) Prevention of ventricular remodelling post myocardial infarction: timing and duration of therapy. Can J Cardiol 9:103–114
4. Jugdutt BI (1995) Modification of left ventricular remodeling after myocardial infarction. In: Dhalla NS, Beamish RE, Takeda N, Nagano M (eds) The failing heart. Lippincot-Raven, Philadelphia, PA, pp 231–245
5. Jugdutt BI (1996) Pharmacological intervention in post-infarction wound healing. In: Karmazyn M (ed) Myocardial ischemia: mechanisms, reperfusion, protection. Birkhauser Verlag, Basel, pp 501–512
6. Jugdutt BI (1996) Prevention of ventricular remodeling after myocardial infarction and in congestive heart failure. Heart Fail Rev 1:115–129
7. Jugdutt BI (2003) Ventricular remodeling post-infarction and the extracellular collagen matrix. When is enough enough? Circulation 108:1395–1403
8. Jugdutt BI (2003) Remodeling of the myocardium and potential targets in the collagen degradation and synthesis pathways. Curr Drug Targets Cardiovasc Haematol Disord 3:1–30
9. Jugdutt BI (2004) Extracellular matrix and cardiac remodeling. In: Villarreal FJ (ed) Interstitial fibrosis in heart failure. Springer, New York, NY, pp 23–55
10. Ertl G, Franz S (2005) Healing after myocardial infarction. Cardiovasc Res 66:22–32
11. Jugdutt BI (2008) Aging and remodeling during healing of the wounded heart: current therapies and novel drug targets. Curr Drug Targets 9:325–344
12. Jugdutt BI (2009) STEMI in the elderly: New insights. In: Proceedings 8th international congress on coronary artery disease: new approaches in coronary artery disease, Prague, Czech Republic, Oct 2009. Monduzzi Editore, Bologna, Italy. pp 257–263
13. Jugdutt BI (2011) Modulators of remodeling after myocardial infarction. In: Dhalla NS, Nagano M, Ostadal B (eds) Molecular defects in cardiovascular disease. Springer Science+Business media, Inc., New York, NY, pp 231–242
14. St John Sutton M, Pfeffer MA, Moye L et al (1997) Cardiovascular death and left ventricular remodeling two years after myocardial infarction: baseline predictors and impact of long-term use of captopril: information from the Survival and Ventricular Enlargement (SAVE) trial. Circulation 96:3294–3299
15. Maggioni AP, Maseri A, Fresco C et al (1993) Age-related increase in mortality among patients with first myocardial infarctions treated with thrombolysis. The investigators of the gruppo Italiano per lo Studio della supravvivenza nell'Infarcto Miocardico (GISSI-2). New Engl J Med 329:1442–1448
16. Bolognese L, Neskovic AN, Parodi G et al (2002) Left ventricular remodeling after primary coronary angioplasty: patterns of left ventricular dilation and long-term prognostic implications. Circulation 106:2351–2357

17. Ferrari R, PREAMI Investigators (2006) Effects of angiotensin-converting enzyme inhibition with peridopril on left ventricular remodeling and clinical outcome. Results of the Randomized Perindopril and Remodeling Elderly with Acute Myocardial Infarction (PREAMI) Study. Arch Intern Med 166:659–666
18. Orn S, Manhenke C, Greve OJ, Larsen AI, Bonarjee VVS, Edvardsen T, Dickstein K (2009) Microvascular obstruction is a major determinant of infarct healing and subsequent left ventricular remodelling following primary percutaneous coronary intervention. Eur Heart J 30:1978–1985
19. Jugdutt BI (2009) Limiting fibrosis after myocardial infarction. New Engl J Med 360:1567–1569
20. Jugdutt BI, Hutchins GM, Bulkley BH, Becker LC (1979) Myocardial infarction in the conscious dog: three dimensional mapping of infarct, collateral flow and region at risk. Circulation 60:1141–1150
21. Jugdutt BI, Khan MI (1992) Impact of increased infarct transmurality on remodeling and function during healing after anterior myocardial infarction in the dog. Can J Physiol Pharmacol 70:949–958
22. Jugdutt BI, Tang SB, Khan MI et al (1992) Functional impact on remodeling during healing after non-Q-wave versus Q-wave anterior myocardial infarction in the dog. J Am Coll Cardiol 20:722–731
23. Alpert JS, Thygesen K, Antman E, Bassand JP (2000) Myocardial infarction redefined–a consensus document of The Joint European Society of Cardiology/American College of Cardiology Committee for the redefinition of myocardial infarction. J Am Coll Cardiol 36:959–969
24. Thygesen K, Alpert JS, White HD (2007) Joint ESC/ACCF/AHA/WHF task force for the redefinition of myocardial infarction. Universal definition of myocardial infarction. Eur Heart J 28:2525–2538
25. Man J, Tymchak W, Jugdutt BI (2010) Adjunctive pharmacologic treatment for acute myocardial infarction. In: Brown DL, Jeremias A (eds) Textbook of cardiac intensive care, 2nd edn. Elsevier, Philadelphia, PA, pp 1–72
26. Alexander KP, Newby LK, Armstrong PW et al (2007) Acute coronary care in the elderly, Part II. ST-segment-elevation myocardial infarction. A scientific statement for healthcare professionals from the American Heart Association Council for Clinical Cardiology. Circulation 115:2570–2589
27. Libby P (2001) Current concepts of the pathogenesis of the acute coronary syndromes. Circulation 104:365–372
28. DeWood MA, Spores J, Notske R et al (1980) Prevalence of total coronary occlusion during the early hours of transmural myocardial infarction. N Engl J Med 303:897–902
29. DeWood MA, Stifter WF, Simpson CS et al (1986) Coronary arteriographic findings soon after non-Q-wave myocardial infarction. N Engl J Med 315:417–423
30. de Feyter PJ, van den Brand M, Serruys PW, Wijns W (1985) Early angiography after myocardial infarction: what have we learned? Am Heart J 109:194–199
31. Panizzi P, Swirski FK, Figueiredo JL et al (2010) Impaired infarct healing in atherosclerotic mice with Ly-6C(hi) monocytosis. J Am Coll Cardiol 55:1629–1638
32. Dewald O, Ren G, Duerr GD et al (2004) Of mice and dogs: species-specific differences in the inflammatory response following myocardial infarction. Am J Pathol 164:665–677
33. Jelani A, Jugdutt BI (2010) STEMI and heart failure in the elderly: role of adverse remodeling. Heart Fail Rev 15:513–521
34. Bujak M, Kweon HJ, Chatila K et al (2008) Aging-related defects are associated with adverse cardiac remodeling in a mouse model of reperfused myocardial infarction. J Am Coll Cardiol 51:1384–1392
35. Jugdutt BI (2008) Aging and defective healing, adverse remodeling and blunted postconditioning in the wounded heart with aging. J Am Coll Cardiol 51:1399–1403

36. Jugdutt BI, Palaniyappan A, Uwiera RRE, Idikio H (2009) Role of healing-specific-matricellular proteins and matrix metalloproteinases in age-related enhanced early remodeling after reperfused STEMI in dogs. Mol Cell Biochem 322:25–36
37. Jugdutt BI, Jelani A, Palaniyappan A et al (2010) Aging-related changes in markers of ventricular and matrix remodeling after reperfused ST-segment elevation myocardial infarction in the canine model. Effect of early therapy with an angiotensin II type 1 receptor blocker. Circulation 122:341–351
38. Palaniyappan A, Idikio H, Jugdutt BI (2009) Secretory leukocyte protease inhibitor and matricellular protein modulation of post reperfused myocardial infarction healing, fibrosis and remodeling in rat model. Effect of candesartan and omapatrilat. (Abstract). Circulation 120(Suppl 2):S837
39. Frangogiannis NG, Smith CW, Entman ML (2002) The inflammatory response in myocardial infarction. Cardiovasc Res 53:31–47
40. Frangogiannis NG (2006) The mechanistic basis of infarct healing. Antioxid Redox Signal 8:1907–1939
41. Frangogiannis NG (2012) Regulation of the inflammatory response in cardiac repair. Circ Res 110:159–173
42. Jugdutt BI, Michorowski B (1987) Role of infarction expansion in rupture of the ventricular septum after acute myocardial infarction. A two-dimensional echocardiography study. Clin Cardiol 10:641–652
43. Michorowski B, Senaratne PJM, Jugdutt BI (1987) Myocardial infarct expansion. Cardiovasc Rev Rep 8:42–47
44. Jugdutt BI (1987) Left ventricular rupture threshold during the healing phase after myocardial infarction in the dog. Can J Physiol Pharmacol 65:307–316
45. Jugdutt BI, Warnica JW (1988) Intravenous nitroglycerin therapy to limit myocardial infarct size, expansion and complications: effect of timing, dosage and infarct location. Circulation 78:906–919
46. Jugdutt BI (1988) Effect of nitroglycerin and ibuprofen on left ventricular topography and rupture threshold during healing after myocardial infarction in the dog. Can J Physiol Pharmacol 66:385–395
47. Jugdutt BI, Basualdo CA (1989) Myocardial infarct expansion during indomethacin and ibuprofen therapy for symptomatic post-infarction pericarditis: effect of other pharmacologic agents during early remodelling. Can J Cardiol 5:211–221
48. Jugdutt BI (1990) Identification of patients prone to infarct expansion by the degree of regional shape distortion on an early two-dimensional echocardiogram after myocardial infarction. Clin Cardiol 13:28–40
49. Jugdutt BI, Khan MI, Jugdutt SJ et al (1995) Impact of left ventricular unloading after late reperfusion of canine anterior myocardial infarction on remodeling and function using isosorbide-5-mononitrate. Circulation 92:926–934
50. Jugdutt BI (1997) Effect of reperfusion on ventricular mass, topography and function during healing of anterior infarction. Am J Physiol 272:H1205–H1211
51. Jugdutt BI (2006) Modification of left ventricular geometry and function during healing after acute myocardial infarction. DM Thesis, University of Glasgow. vol 1:1–223; vol 2:1–447
52. Jugdutt BI, Idikio H, Uwiera R (2007) Angiotensin receptor blockade and ACE inhibition limit adverse collagen remodeling in the infarct zone and global diastolic dysfunction during healing after reperfused ST-elevation myocardial infarction. Mol Cell Biochem 303:27–38
53. Jugdutt BI, Idikio H, Uwiera R (2007) Therapeutic drugs during healing after myocardial infarction modify infarct collagens and ventricular distensibility at elevated pressures. Mol Cell Biochem 304:79–91
54. Pfeffer MA, Braunwald E (1990) Ventricular remodelling after myocardial infarction. Circulation 81:1161–1172
55. Gaudron P, Eilles C, Kugler I et al (1993) Progressive left ventricular dysfunction and remodeling after myocardial infarction. Potential mechanisms and early predictors. Circulation 87:755–763

56. Eghbali M, Tomek R, Woods C, Bhambi B (1991) Cardiac fibroblasts are predisposed to convert into myocyte phenotype: specific effect of transforming growth factor beta. Proc Natl Acad Sci USA 88:795–799
57. Vracko R, Thorning D (1991) Contractile cells in rat myocardial scar tissue. Lab Invest 65:214–227
58. Guarda E, Katwa LC, Myers PR, Tyagi SC, Weber KT (1993) Effects of endothelins on collagen turnover in cardiac fibroblasts. Cardiovasc Res 27:2130–2134
59. Villarreal FJ, Kim NN, Ungab GD, Printz MP, Dillmann WH (1993) Identification of functional angiotensin II receptors on rat cardiac fibroblasts. Circulation 88:2849–2861
60. Sun Y, Weber KT (2000) Infarct scar: a dynamic tissue. Cradiovasc Res 46:250–256
61. Willems IE, Havenith MG, De Mey JG, Daemen MJ (1994) The alpha-smooth muscle actin-positive cells in healing human myocardial scars. Am J Pathol 145:868–875
62. Sun Y, Cleutjens JP, Diaz-Arias AA, Weber KT (1994) Cardiac angiotensin converting enzyme and myocardial fibrosis in the rat. Cardiovasc Res 28:1423–1432
63. Campbell SE, Janicki JS, Weber KT (1995) Temporal differences in fibroblast proliferation and phenotype expression in response to chronic administration of angiotensin II or aldosterone. J Mol Cell Cardiol 27:1545–1560
64. Sun Y, Weber KT (1996) Angiotensin converting enzyme and myofibroblasts during tissue repair in the rat heart. J Mol Cell Cardiol 28:851–858
65. Jugdutt BI (2007) Cyclooxygenase inhibition and ventricular remodeling after acute myocardial infarction. Circulation 115:288–291
66. Jugdutt BI (2008) Pleiotropic effects of cardiac drugs on healing post MI. The good, bad and ugly. Heart Fail Rev 13:439–452
67. Gordon S, Taylor PR (2005) Monocyte and macrophage heterogeneity. Nat Rev Immunol 5:953–964
68. Nahrendorf M, Pittet MJ, Swirski FK (2010) Monocytes: protagonists of infarct inflammation and repair after myocardial infarction. Circulation 121:2437–2445
69. Maekawa Y, Anzai T, Yoshikawa T, Asakura Y, Takahashi T, Ishikawa S, Mitamura H, Ogawa S (2002) Prognostic significance of peripheral monocytosis after reperfused acute myocardial infarction:a possible role for left ventricular remodeling. J Am Coll Cardiol 39:241–246
70. Jugdutt BI (2002) Monocytosis after acute myocardial infarction and left ventricular remodeling. J Am Coll Cardiol 39:247–250
71. Mariani M, Fetiveau R, Rossetti E et al (2006) Significance of total and differential leucocyte count in patients with acute myocardial infarction treated with primary coronary angioplasty. Eur Heart J 27:2511–2515
72. Jugdutt BI (2010) Preventing adverse remodeling and rupture during healing after myocardial infarction in mice and humans. Circulation 122:103–105
73. Mann DL (2011) The emerging role of innate immunity in the heart and vascular system: for whom the cell tolls. Circ Res 108:1133–1145
74. Gordon JW, Shaw JA, Kirshenbaum LA (2011) Multiple facets of NF-κB in the heart: to be or not to NF-κB. Circ Res 108:1122–1132
75. Soehnlein O, Lindbom L (2010) Phagocyte partnership during the onset and resolution of inflammation. Nat Rev Immunol 10:427–439
76. Nahrendorf M, Swirski FK, Aikawa E et al (2007) The healing myocardium sequentially mobilizes two monocyte subsets with divergent and complementary functions. J Exp Med 204:3037–3047
77. van Amerongen MJ, Harmsen MC, van Rooijen N, Petersen AH, van Luyn MJ (2007) Macrophage depletion impairs wound healing and increases left ventricular remodeling after myocardial injury in mice. Am J Pathol 170:818–829
78. Tsujioka H, Imanishi T, Ikejima H et al (2009) Impact of heterogeneity of human peripheral blood monocyte subsets on myocardial salvage in patients with primary acute myocardial infarction. J Am Coll Cardiol 54:130–138
79. Troidl C, Möllmann H, Nef H et al (2009) Classically and alternatively activated macrophages contribute to tissue remodelling after myocardial infarction. J Cell Mol Med 13:3485–3496

80. Hu Y, Zhang H, Lu Y et al (2011) Class A scavenger receptor attenuates myocardial infarction-induced cardiomyocyte necrosis through suppressing M1 macrophage subset polarization. Basic Res Cardiol 106:1311–1328
81. Mantovani A, Sozzani S, Locati M, Allavena P, Sica A (2002) Macrophage polarization: tumor-associated macrophages as a paradigm for polarized M2 mononuclear phagocytes. Trends Immunol 23:549–555
82. Camelliti P, Borg TK et al (2005) Structural and functional characterisation of cardiac fibroblasts. Cardiovasc Res 65:40–51
83. Littman DR, Rudensky AY (2010) Th17 and regulatory T cells in mediating and restraining inflammation. Cell 140:845–858
84. Li J, Brown LF, Hibberd MG et al (1996) VEGF, flk-1, and flt-1 expression in a rat myocardial infarction model of angiogenesis. Am J Physiol 270(5 Pt 2):H1803–H1811
85. Zymek P, Bujak M, Chatila K et al (2006) The role of platelet-derived growth factor signaling in healing myocardial infarcts. J Am Coll Cardiol 48:2315–2323
86. Vanhoutte D, Schellings M, Pinto Y, Heymans S (2006) Relevance of matrix metalloproteinases and their inhibitors after myocardial infarction: a temporal and spatial window. Cardiovasc Res 69:604–613
87. Jugdutt BI, Joljart MJ, Khan MI (1996) Rate of collagen deposition during healing after myocardial infarction in the rat and dog models: mechanistic insights into ventricular remodeling. Circulation 94:94–101
88. Frangogiannis NG (2006) Targeting the inflammatory response in healing myocardial infarcts. Curr Med Chem 13:1877–1893
89. Bujak M, Frangogiannis NG (2007) The role of TGF-β signaling in myocardial infarction and cardiac remodeling. Cardiovasc Res 74:184–195
90. Jugdutt BI (2009) Limiting fibrosis after myocardial infarction. New Engl J Med 360:1567–1569
91. Shalitin N, Schlesinger H, Levy MJ et al (2009) Expression of procollagen C-proteinase enhancer in cultured rat heart fibroblasts: evidence for co-regulation with type I collagen. J Cell Biochem 90:397–407
92. Naito AT, Shiojima I, Komuro I (2010) Wnt signaling and aging-related heart disorders. Circ Res 107:1295–1303
93. Bergmann MW (2010) WNT signaling in adult cardiac hypertrophy and remodeling: lessons learned from cardiac development. Circ Res 107:1198–1208
94. Kobayashi K, Luo M, Zhang Y et al (2009) Secreted Frizzled-related protein 2 is a procollagen C proteinase enhancer with a role in fibrosis associated with myocardial infarction. Nat Cell Biol 11(1):46–55. doi:101038/ncb1811, 2008
95. Jugdutt BI, Palaniyappan A, Idikio H (2010) Candesartan suppresses increase in secreted frizzled related protein 2 and fibrosis during healing after reperfused myocardial infarction in the rat model. (Abstract). J Am Coll Cardiol 55:A123
96. Mirotsou M, Zhang Z, Deb A et al (2007) Secreted frizzled related protein 2 (Sfrp2) is the key Akt-mesenchymal stem cell-released paracrine factor mediating myocardial survival and repair. Proc Natl Acad Sci 104:1643–1648
97. Alfaro MP, Pagni M, Vincent A et al (2008) The Wnt modulator Sfrp2 enhances mesenchymal stem cell engraftment, granulation tissue formation and myocardial repair. Proc Natl Acad Sci USA 105:18366–18371
98. Blankesteijn WM, Essers-Janssen YP, Verluyten MJ et al (1997) A homologue of Drosophila tissue polarity gene frizzled is expressed in migrating myofibroblasts in the infarcted rat heart. Nat Med 3:541–544
99. Zhang Z, Deb A, Zhang Z et al (2009) Secreted frizzled related protein 2 protects cells from apoptosis by blocking the effect of canonical Wnt3a. J Mol Cell Cardiol 46:370–377
100. He W, Zhang L, Zhang Z et al (2010) Exogenously administered secreted frizzled related protein 2 (Sfrp2) reduces fibrosis and improves cardiac function in a rat model of myocardial infarction. Proc Natl Acad Sci 107:21110–21115

101. Barandon L, Couffinhal T, Ezan J et al (2003) Reduction of infarct size and prevention of cardiac rupture in transgenic mice overexpressing FrzA. Circulation 108:2282–2289
102. Weber KT (1997) Extracellular matrix remodeling in heart failure. A role for de novo angiotensin II generation. Circulation 96:4065–4082
103. Sun Y (2009) Myocardial repair/remodelling following infarction: roles of local factors. Cardiovasc Res 81:482–490
104. Spinale FG (2002) Matrix metalloproteinases: regulation and dysregulation in the failing heart. Circ Res 90:520–530
105. Janicki JS, Brower GL, Gardner JD et al (2004) The dynamic interaction between matrix metalloproteinase activity and adverse myocardial remodeling. Heart Fail Rev 9:33–42
106. Zaidi SH, Huang Q, Momen A et al (2010) Growth differentiation factor 5 regulates cardiac repair after myocardial infarction. J Am Coll Cardiol 55:135–143
107. Zhang Y, DeWitt DL, McNeely TB et al (1997) Secretory leukocyte protease inhibitor suppresses the production of monocyte prostaglandin H synthase-2, prostaglandin E_2, and matrix metalloproteinases. J Clin Invest 99:894–900
108. Ashcroft GS, Lei K, Jin W et al (2000) Secretory leukocyte protease inhibitor mediates nonredundant functions necessary for normal wound healing. Nat Med 6:1147–1153
109. Sano C, Shimizu T, Sato K et al (2000) Effects of secretory leukocyte protease inhibitor on the production of the anti-inflammatory cytokines, IL-10 and transforming growth factor-beta (TGF-β), by lipopolysaccharide-stimulated macrophages. Clin Exp Immunol 121:77–85
110. Denhardt DT, Guo X (1993) Osteopontin: a protein with diverse functions. FASEB J 7:1475–1482
111. Singh K, Sirokman G, Communal C et al (1999) Myocardial osteopontin expression coincides with the development of heart failure. Hypertension 33:663–670
112. Trueblood NA, Xie Z, Communal C et al (2001) Exaggerated left ventricular dilation and reduced collagen deposition after myocardial infarction in mice lacking osteopontin. Circ Res 88:1080–1087
113. Lane TF, Sage EH (1994) The biology of SPARC, a protein that modulates cell-matrix interactions. FASEB J 8:163–173
114. Masson S, Arosio B, Luvara G et al (1998) Remodelling of cardiac extracellular matrix during β-adrenergic stimulation: upregulation of SPARC in the myocardium of adult rats. J Mol Cell Cardiol 30:1501–1514
115. Wu RX, Laser M, Han H et al (2006) Fibroblast migration after myocardial infarction is regulated by transient SPARC expression. J Mol Med 84:241–252
116. Wynn TA (2008) Cellular and molecular mechanisms of fibrosis. J Pathol 214:199–210
117. Ge G, Greenspan DS (2006) BMP1 controls TGFβ1 activation via cleavage of latent TGFβ-binding protein. J Cell Biol 175:111–120
118. de Boer RA, Voors AA, Muntendam P et al (2009) Galectin-3: a novel mediator of heart failure development and progression. Eur J Heart Fail 11:811–817
119. de Boer RA, Yu L, van Veldhuisen DJ (2010) Galectin-3 in cardiac remodeling and heart failure. Curr Heart Fail Rep 7:1–8
120. Jugdutt BI (2010) Aging and remodeling of the renin-angiotensin system post infarction. In: Proceedings 15th World Congress on heart disease, Vancouver, 2010. Monduzzi Editore, Bologna, Italy, pp 87–91

Chapter 30
The ACE2/Ang-(1–7) Pathway in Cardiac Fibroblasts as a Potential Target for Cardiac Remodeling

Randy T. Cowling and Barry H. Greenberg

Abstract Despite advances in its treatment, heart failure prevalence continues to increase and this condition remains a major contributor to mortality and morbidity in both industrialized and developing nations throughout the world. Cardiac remodeling that is activated by injury to the myocardium and/or increased wall stress adversely affects cardiac function over time and is known to play a critical role in the progression of heart failure. Excessive accumulation and stiffening of collagen (i.e., fibrosis) is a well-recognized component of the remodeling process, and it contributes to progression of disease and worsening heart failure by perturbing cardiac contractility, relaxation, and electrical conduction. Cardiac fibroblasts are interstitial cells that are responsible for synthesis and turnover of collagen in the myocardium. Current experimental and clinical evidence suggests that cardiac fibroblasts are an important target when inhibitors of the renin–angiotensin system (RAS) are used to treat heart failure. An alternative pathway of the RAS, involving angiotensin-converting enzyme 2 (ACE2), angiotensin-(1–7) (Ang-(1–7)), and the Mas receptor, has been found recently to be cardioprotective. Evidence suggests that Ang-(1–7), acting predominantly through the Mas receptor, can regulate maladaptive growth-promoting effects of angiotensin ll and inhibit both cardiomyocyte hypertrophy and cardiac fibrosis. This chapter highlights the potential regulatory role of the ACE2/Ang-(1–7)/Mas pathway and how it relates to cardiac fibroblasts during the remodeling process.

Keywords Angiotensin-(1–7) • Fibroblast • Angiotensin-converting enzyme 2 • Fibrosis • Heart • Collagen

R.T. Cowling • B.H. Greenberg (✉)
Division of Cardiology, Department of Medicine, University of California, San Diego, 9300 Campus Point Drive #7411, La Jolla, CA, USA
e-mail: bgreenberg@ucsd.edu

B.I. Jugdutt and N.S. Dhalla (eds.), *Cardiac Remodeling: Molecular Mechanisms*, Advances in Biochemistry in Health and Disease 5, DOI 10.1007/978-1-4614-5930-9_30,

30.1 Introduction

30.1.1 The Cardiac Fibroblast and Cardiac Remodeling

Almost 6 million people suffer from heart failure in the United States today [1]. Similar high rates of heart failure prevalence are seen in other industrialized countries, and there is evidence that heart failure is growing rapidly as a public health problem in developing nations [2–5]. Current medical and device therapies have reduced morbidity and mortality in patients with heart failure. However, the effects of available treatment strategies have plateaued, and novel targets are needed if improvements in outcome are to continue in the future. In patients with chronic heart failure, the condition develops as a consequence of progressive changes in the structure of the heart that impair its normal function [6]. Hypertrophy of cardiomyocytes is a hallmark of the remodeling process, and there is evidence that the structural changes in these cells are intimately associated with abnormalities in their function. Another change in the structure of the heart that occurs as part of the remodeling process and which plays a critical role in the progression to heart failure is the deposition of collagen within the heart (i.e., cardiac fibrosis). Cardiac fibrosis stiffens the heart wall and alters electrical contractility, leading to diminished diastolic and/or systolic function and increasing the probability of debilitating or even fatal arrhythmias. Activation of the mechanisms that cause fibrosis as well as the extent of fibrous tissue that has been deposited within the failing heart has been associated with worse outcomes in heart failure patients. Thus, minimizing the development of fibrosis is expected to improve cardiac function, slow the development of heart failure, and help reduce morbidity and mortality in the heart failure population.

30.1.2 ACE2/Ang-(1–7)/Mas Pathway

There are several current treatment strategies for heart failure that target the renin–angiotensin system (RAS). These involve inhibition of the generation of angiotensin II (Ang II) with angiotensin-converting enzyme (ACE) inhibitors or blocking activation of Ang II receptors with angiotensin receptor blockers (ARBs). The importance of the RAS in the progression of heart failure is apparent by the beneficial effects of both ACE inhibitors and ARBs in large-scale, well-designed clinical trials [7–11]. These agents have been shown to inhibit the remodeling process as well as reduce morbidity and mortality across a broad spectrum of patients ranging from those with asymptomatic left ventricular (LV) dysfunction, to MI survivors, to patients with clinically manifested heart failure including those with advanced disease.

Recently, an alternative axis of the RAS, the ACE2/Ang-(1–7)/Mas pathway, has been receiving much attention as it is believed to function as a means of modulating the effects of Ang II and other growth-promoting factors during remodeling. Angiotensin-converting enzyme 2 (ACE2) was identified in 2000 by two independent groups [12, 13]. ACE2 functions as a carboxymonopeptidase and has approximately 40% sequence similarity when compared to ACE. ACE2 can hydrolyze the carboxy-terminal phenylalanine from Ang II to form the heptapeptide angiotensin-(1–7) (Ang-(1–7)), or it can generate angiotensin-(1–9) from angiotensin I, at which point other peptidases can convert angiotensin-(1–9) to Ang-(1–7) [14]. Other substrates for ACE2 include des-Arg^9-bradykinin, apelin-13, dynorphin A (1–13), and β-casomorphin [14]. When considering only the angiotensin peptides, ACE2 has two, potentially beneficial functions. (1) It can degrade Ang II and (2) it can generate Ang-(1–7). As opposed to Ang II, which is considered to activate cardiomyocyte hypertrophy and fibrosis and, thus, to be deleterious by promoting remodeling, Ang-(1–7) is considered to have potentially beneficial antigrowth and antifibrotic effects that would act to inhibit the remodeling process. In human myocardium, Ang II has been shown to be the preferred substrate for ACE2. Moreover, ACE2 effects appear to be impervious to the ACE inhibitors that are used in clinical practice.

Some investigators have reported that Ang-(1–7) can act through well-characterized angiotensin receptors such as AT_1 or AT_2 [15–17]. However, Ang-(1–7) more frequently acts via its own dedicated receptor, which is widely considered to be the Mas receptor [18]. Mas is a G protein-coupled receptor that was originally described as a protooncogene [19, 20]. Mas becomes detectable after birth in the rat, increases its expression, and then plateaus at 4–6 months of age [21]. Mas exhibits a broad tissue expression profile, including the heart, kidney, testis, brain, retinal pigment epithelium, skeletal muscle, liver, and adipose tissue [22–24]. Within the heart, Mas is localized to the vasculature [25], cardiomyocytes [26, 27], and fibroblasts (detected by qRT-PCR, data not shown). Despite the passage of ten years since the initial characterization of Mas as the Ang-(1–7) receptor [18], the signaling pathways involved have not yet been fully delineated.

30.2 Responses of the Cardiac Fibroblast to Ang-(1–7)

30.2.1 Direct Responses to the Peptide

Direct activation of signaling pathways by Ang-(1–7) treatment of cardiac fibroblasts isolated from normal hearts has been reported. We have observed weak, transient phosphorylation of ERK1/ERK2 upon stimulation of adult rat cardiac fibroblasts with Ang-(1–7) (unpublished observations). We have also observed reduced endothelin-1 (ET-1) and leukemia inhibitory factor (LIF) mRNA levels between 1 and 2 h after stimulation of adult rat cardiac fibroblasts with Ang-(1–7), presumably via activation of an undetermined signaling pathway [28]. We have observed modest inhibition of Ang II-stimulated increases in phosphorylation of MAPKs (e.g., ERK, JNK, and p38) in fibroblasts which had been pretreated with Ang-(1–7). McCollum et al. also observed that treatment of cardiac fibroblasts with Ang-(1–7) reduced the Ang II- or ET-1-stimulated increase in phospho-ERK1 and phospho-ERK2 [29]. They observed that Ang-(1–7) increased dual-specificity phosphatase DUSP1 immunoreactivity and mRNA, suggesting that the heptapeptide hormone increases DUSP1 to reduce MAP kinase phosphorylation and activity. The mechanism through which Ang-(1–7) increased DUSP1, however, was not elucidated. Increased generation of second messengers by Ang-(1–7) stimulation has been reported, including nitric oxide in cardiomyocytes [26] and aortic endothelial cells [30], cAMP in glomerular mesangial cells [31], and arachidonic acid from Mas-transfected CHO cells [18]. However, we have failed to observe activation of these second messengers in response to Ang-(1–7) in cultured cardiac fibroblasts (data not shown).

30.2.2 Indirect Responses to the Peptide

As opposed to the direct stimulatory effects of Ang-(1–7) on cardiac fibroblasts, which tend to be mild, indirect effects of the peptide can produce a much greater response. These indirect responses include alteration of other signaling pathways and the influence of second messengers that have been generated by cardiac fibroblasts, their precursors, or other nearby cell types.

30.2.2.1 Attenuation/Augmentation of Other Receptor Signaling Pathways

Stimulation of cells with Ang-(1–7) can affect the subsequent activation of other signaling pathways. Stimulation of nitric oxide production by bradykinin in endothelial cells can be augmented by prior exposure of the cells to Ang-(1–7) [32]. There is also evidence that Ang-(1–7) stimulation can prove inhibitory to signaling pathways, such as those activated by Ang II [29, 33, 34], glucose [35], and ET-1 [29]. As noted in the previous section, preincubation with Ang-(1–7) has been reported to reduce subsequent activation of MAP kinases by Ang II and other ligands [29, 33–36]. The mechanism of these effects is not known, although activation of tyrosine phosphatases in proximal tubular cells [35] and upregulation of DUSP1 in neonatal rat cardiac fibroblasts [29] have been implicated. Although DUSP1 upregulation has been implicated in the inhibitory effect of Ang-(1–7) in neonatal rat cardiac fibroblasts [29], we were unable to detect this effect in the adult cells. In adult rat cardiac fibroblasts, we found that Ang-(1–7) preincubation inhibited subsequent Ang II-stimulated upregulation of ET-1 and LIF mRNA. We also found that Ang II-induced secretion by adult rat cardiac fibroblasts of unknown paracrine factors that induced hypertrophy in cultured cardiomyocytes could be significantly inhibited by preincubation of the fibroblasts with Ang-(1–7) (Fig. 30.1 and [28]).

30.2.2.2 Autocrine/Paracrine Second Messengers

Fibroblasts can originate from other sources in addition to the cardiac interstitium, especially when the heart is actively remodeling. Sources of nonresident fibroblasts

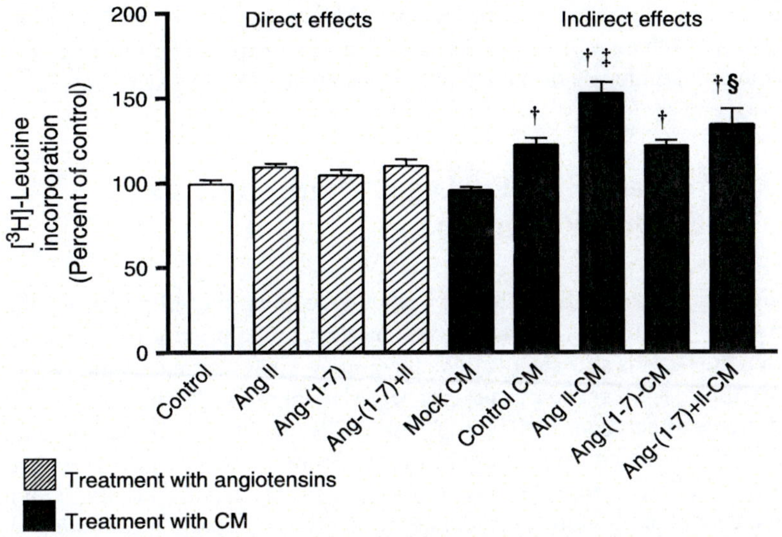

Fig. 30.1 Effects of angiotensins and conditioned media (CM) from adult rat cardiac fibroblasts (ARCFs) treated with angiotensins on [^{3}H]leucine incorporation in cardiomyocytes. Cardiomyocytes were stimulated for 24 h with Ang II, Ang-(1–7), both peptides after 1 h pretreatment with Ang-(1–7) or CM from ARCFs to assess cardiomyocyte hypertrophy measured by [^{3}H] leucine incorporation. Ang II-CM, CM from Ang II-treated ARCFs; Ang-(1–7)-CM, CM from Ang-(1–7)-treated ARCFs; Ang-(1–7) + II-CM, CM from ARCFs treated with both peptides after 1 h pretreatment with Ang-(1–7). † $P<0.001$ vs. mock CM; ‡ $P<0.001$ vs. control CM; § $P<0.05$ vs. Ang II-CM (Reproduced with permission [28])

in the remodeling heart include circulating fibrocytes and mesenchymal cells generated from epithelial-to-mesenchymal transition (EMT) and endothelial-to-mesenchymal transition (EndMT). It is still not clear whether these additional cells serve some unique purpose or simply elevate the pool of fibroblasts during increased need. However, the possibilities exist that (1) these cells may be affected by Ang-(1–7) differently than the resident fibroblasts and (2) the recruitment and/or differentiation of these cells could be altered by Ang-(1–7). Indeed, Ang-(1–7) has been shown to reduce the number, proliferative capacity, and collagen secretion of human circulating fibrocytes, possibly by increasing nitric oxide production [37]. Also, cGMP-dependent protein kinase (PKG) has been shown previously to disrupt transforming growth factor-β1 (TGF-β1)-induced nuclear translocation of pSmad3 [38]. Considering (1) the involvement of TGF-β1 in EMT (or EndMT) and myofibroblast differentiation and (2) the activation of guanylate cyclase and subsequent cGMP generation after nitric oxide exposure, there is a strong possibility that Ang-(1–7)-induced nitric oxide synthesis could inhibit fibroblast functions in a remodeling tissue. Ang-(1–7) has been shown to directly induce the release of nitric oxide in platelets, cardiomyocytes, and aortic endothelial cells [26, 30, 39], a process which involves activation of Akt and eNos in cardiomyocytes. As noted above, Ang-(1–7) has also been shown to augment the release of nitric oxide from endothelial cells that have been stimulated with bradykinin [32]. Many of these indirect effects can theoretically influence cardiac fibroblasts during cardiac remodeling, but their effects in animal models or human patients have not been studied.

30.3 Effects of the ACE2/Ang-(1–7)/Mas Pathway on Cardiac Remodeling

Germline ablation of ACE2 results in a phenotype of cardiac dilatation and dysfunction in some [40], but not all [41], mouse models, and it predisposes to cardiac failure when the heart is stressed [42]. We have shown that pharmacologic inhibition of ACE2 in the post-MI rat heart using the selective inhibitor C16 increases infarct size and adversely affects contractile function [43]. Treatment with C16 had no significant effect on the increased level of apoptosis in the infarct and border zones nor did it significantly affect capillary density surrounding the MI. It did, however, significantly reduce the number of c-kit(+) cells in the border region, a finding consistent with the possibility that Ang-(1–7) or some other product of ACE2 helped reduce infarct size by enhancing the viability of cardiomyocytes in the border zone either directly or via paracrine effects of mediators released from the c-kit(+) cells. The observation that C16 inhibited increases in wall thickness and fibrosis in non-infarcted LV, however, raises the possibility that ACE2 activity has diverse (and perhaps even some adverse) effects on post-MI remodeling. Others have reported that loss of ACE2 accelerates maladaptive post-MI LV remodeling [44], while increasing ACE2 levels by gene transfer therapy inhibits post-MI remodeling [45].

In addition to degrading Ang ll, many favorable effects of ACE2 appear to be due to increased Ang-(1–7) levels. Elevated Ang-(1–7) levels are predicted to be inhibitory to both cardiomyocyte hypertrophy and fibrosis. In rats, post-MI remodeling can be inhibited by infusing Ang-(1–7) [46] or ingestion of either AVE 0991 (a nonpeptide Ang-(1–7) analogue) [47], an Ang-(1–7) hydroxypropyl β-cyclodextrin preparation [48], or a stabilized Ang-(1–7) analogue [49]. Angiotensin II-stimulated cardiac hypertrophy in mice is inhibited by cardiac-specific Ang-(1–7) overexpression [50, 51]. Whether Ang-(1–7) cardioprotection is due to effects of the peptide within the heart or effects of the peptide on extracardiac cells/tissues, however, is uncertain as some evidence suggests that circulating rather than cardiac Ang-(1–7) produces the beneficial effects of the peptide on post-MI remodeling [52]. We have shown that ACE2 inhibition increases post-MI infarct expansion [43] so that there is also uncertainty whether Ang-(1–7) inhibits remodeling due to an early reduction in infarct size or due to later effects on the remodeling process. Among strategies for using Ang-(1–7) to treat post-MI remodeling, gene transfer therapy is one of the most appealing [53–55]. Evidence from experiments in which a lentiviral vector encoding an engineered Ang-(1–7) fusion protein was injected directly into the LV wall several weeks prior to coronary ligation supports this possibility [56].

Germline deletion of Mas in mice has been reported to cause decreased cardiac performance, which is believed to be the result of increased levels of extracellular matrix proteins in the right ventricles and AV valves [57]. Pharmacologic stimulation of the Mas receptor has shown promise in animal models of cardiac dysfunction. The nonpeptide Mas agonist, AVE 0991, improves cardiac function in rats with diabetes [58, 59], with isoproterenol treatment [60] and following experimentally induced MI [61]. AVE 0991 has also been shown to ameliorate progression of atherosclerosis in apoE-null mice [62]. Another Mas agonist peptide, CGEN-856S, which has been reported to more stable than Ang-(1–7), demonstrated anti-arrhythmogenic effects in isolated rat hearts [63], but this peptide has not been extensively studied. To our knowledge, no human data exist regarding these Mas agonists, but their clinical potential as cardioprotective agents is recognized. Based on these considerations, further exploration of the effects of stable Ang-(1–7) analogues or Mas receptor agonists in cardiovascular disease is clearly warranted.

30.4 Future Considerations

There are many questions that still need to be answered about the relevance of this relatively new axis of the RAS and its potential role in regulating cardiac remodeling. Ang-(1–7)-induced signaling pathways (whether direct or indirect) remain unclear and require further study. Although Mas is widely regarded as the Ang-(1–7) receptor, there are still indications that other receptors are utilized in certain contexts [15–17]. Regarding cardiac fibroblasts, very little study has been devoted to understanding how ACE2/Ang-(1–7)/Mas can influence the recruitment and/or production of other cell types that contribute to the interstitial cell pool in the remodeling heart. Positive outcomes have been obtained when perturbing the ACE2/Ang-(1–7)/Mas system in animal models of disease. However, to our knowledge, no clinical use of such strategies has yet been reported. Since ACE2/Ang-(1–7)/Mas is considered cardioprotective, stimulation of this pathway, rather than inhibition, is expected to be beneficial. Ang-(1–7) infusion would be clinically difficult due to its short half-life. However, synthetic genes have been engineered to enhance Ang-(1–7) secretion from transfected or transduced cells [64]. Theoretically, increased expression of Ang-(1–7) would act cooperatively with existing ACE inhibitor or ARB pharmacologic treatment regimes, but this has yet to be tested. Enhanced ACE2 activity alone could demonstrate considerable improvements over existing therapies since it would simultaneously degrade Ang II and generate the cardioprotective Ang-(1–7). Stimulation of ACE2 enzymatic activity may prove challenging using pharmacology, but overexpression of ACE2 is always possible by transfection or viral transduction. We must remain cautious, however, since ACE2 overexpression by such methods has proven to be detrimental in at least one published work [65]. Also, discrepancies as to whether cardioprotective effects are due to activation of this pathway in the myocardium or in extracardiac tissues demand further studies [52, 56].

30.5 Conclusions

Although the past several decades have witnessed substantial improvements in heart failure outcomes, morbidity and mortality still remain unacceptably high. This, along with the relentless increase in new cases of heart failure throughout the world, makes it clear that novel treatment strategies are still needed. The RAS has been a pharmacologic target for heart failure treatment for many years. A more recently discovered branch of the RAS, namely, the ACE2/Ang-(1–7)/Mas system, has demonstrated cardioprotection when activated during remodeling in animal models of heart failure. The ACE2/Ang-(1–7)/Mas system can affect many cell types, but the cardiac fibroblast has been shown to be an important target of its action. Effects of Ang-(1–7) on cardiac fibroblasts can involve direct activation of signaling pathways or indirect effects on their function that involve either the paracrine release of second messengers, alterations in the differentiation/recruitment of fibroblast "precursors," or crosstalk with other signaling pathways. Ultimately, activation of the ACE2/Ang-(1–7)/Mas system is predicted to produce an "antifibrotic" phenotype and should improve cardiac remodeling. Despite favorable results in animal models, however, adaptation of the promising strategies for enhancing the effects of this alternate pathway and proof of efficacy in well-designed human trials are still needed.

Acknowledgment Funded in part by National Institutes of Health grant 1RO1HL091191 to Dr. Greenberg.

References

1. Roger VL, Go AS, Lloyd-Jones DM et al (2012) Executive summary: heart disease and stroke statistics–2012 update: a report from the American Heart Association. Circulation 125: 188–197
2. Schocken DD, Benjamin EJ, Fonarow GC et al (2008) Prevention of heart failure: a scientific statement from the American Heart Association Councils on Epidemiology and Prevention, Clinical Cardiology, Cardiovascular Nursing, and High Blood Pressure Research; Quality of Care and Outcomes Research Interdisciplinary Working Group; and Functional Genomics and Translational Biology Interdisciplinary Working Group. Circulation 117:2544–2565
3. Albert MA (2008) Heart failure in the urban African enclave of Soweto: a case study of contemporary epidemiological transition in the developing world. Circulation 118:2323–2325
4. Shantsila E, Lip GY, Gill PS (2011) Systolic heart failure in South Asians. Int J Clin Pract 65:1274–1282
5. Khanam MA, Streatfield PK, Kabir ZN et al (2011) Prevalence and patterns of multimorbidity among elderly people in rural Bangladesh: a cross-sectional study. J Health Popul Nutr 29:406–414
6. Pfeffer MA, Braunwald E (1990) Ventricular remodeling after myocardial infarction. Experimental observations and clinical implications. Circulation 81:1161–1172
7. Effect of enalapril on survival in patients with reduced left ventricular ejection fractions and congestive heart failure. The SOLVD Investigators. N Engl J Med 325: 293–302, 1991

8. Effect of enalapril on mortality and the development of heart failure in asymptomatic patients with reduced left ventricular ejection fractions. The SOLVD Investigators. N Engl J Med 327:685–691, 1992
9. Greenberg B, Quinones MA, Koilpillai C et al (1995) Effects of long-term enalapril therapy on cardiac structure and function in patients with left ventricular dysfunction. Results of the SOLVD echocardiography substudy. Circulation 91:2573–2581
10. Konstam MA, Kronenberg MW, Rousseau MF et al (1993) Effects of the angiotensin converting enzyme inhibitor enalapril on the long-term progression of left ventricular dilatation in patients with asymptomatic systolic dysfunction. SOLVD (Studies of Left Ventricular Dysfunction) Investigators. Circulation 88:2277–2283
11. Granger CB, McMurray JJ, Yusuf S et al (2003) Effects of candesartan in patients with chronic heart failure and reduced left-ventricular systolic function intolerant to angiotensin-converting-enzyme inhibitors: the CHARM-Alternative trial. Lancet 362:772–776
12. Tipnis SR, Hooper NM, Hyde R et al (2000) A human homolog of angiotensin-converting enzyme - cloning and functional expression as a captopril-insensitive carboxypeptidase. J Biol Chem 275:33238–33243
13. Donoghue M, Hsieh F, Baronas E et al (2000) A novel angiotensin-converting enzyme-related carboxypeptidase (ACE2) converts angiotensin I to angiotensin 1–9. Circ Res 87:E1–E9
14. Vickers C, Hales P, Kaushik V et al (2002) Hydrolysis of biological peptides by human angiotensin-converting enzyme-related carboxypeptidase. J Biol Chem 277:14838–14843
15. Lara LS, Cavalcante F, Axelband F et al (2006) Involvement of the Gi/o/cGMP/PKG pathway in the AT2-mediated inhibition of outer cortex proximal tubule Na^+–ATPase by Ang-(1–7). Biochem J 395:183–190
16. Lara LS, Correa JS, Lavelle AB et al (2008) The angiotensin receptor type 1-Gq protein-phosphatidyl inositol phospholipase Cbeta-protein kinase C pathway is involved in activation of proximal tubule Na^+–ATPase activity by angiotensin(1–7) in pig kidneys. Exp Physiol 93:639–647
17. Lopez Verrilli MA, Pirola CJ, Pascual MM et al (2009) Angiotensin-(1–7) through AT receptors mediates tyrosine hydroxylase degradation via the ubiquitin-proteasome pathway. J Neurochem 109:326–335
18. Santos RAS, Silva ACSE, Maric C et al (2003) Angiotensin-(1–7) is an endogenous ligand for the G protein-coupled receptor Mas. Proc Natl Acad Sci USA 100:8258–8263
19. Kostenis E, Milligan G, Christopoulos A et al (2005) G-protein-coupled receptor Mas is a physiological antagonist of the angiotensin II type 1 receptor. Circulation 111:1806–1813
20. Young D, Waitches G, Birchmeier C et al (1986) Isolation and characterization of a new cellular oncogene encoding a protein with multiple potential transmembrane domains. Cell 45:711–719
21. Martin KA, Grant SG, Hockfield S (1992) The mas proto-oncogene is developmentally regulated in the rat central nervous system. Brain Res Dev Brain Res 68:75–82
22. Kitaoka T, Sharif M, Hanley MR, Hjelmeland LM (1994) Expression of the MAS proto-oncogene in the retinal pigment epithelium of the rhesus macaque. Curr Eye Res 13:345–351
23. Metzger R, Bader M, Ludwig T et al (1995) Expression of the mouse and rat mas proto-oncogene in the brain and peripheral tissues. FEBS Lett 357:27–32
24. Munoz MC, Giani JF, Dominici FP (2010) Angiotensin-(1–7) stimulates the phosphorylation of Akt in rat extracardiac tissues in vivo via receptor Mas. Regul Pept 161:1–7
25. Alenina N, Xu P, Rentzsch B et al (2008) Genetically altered animal models for Mas and angiotensin-(1–7). Exp Physiol 93:528–537
26. Dias-Peixoto MF, Santos RA, Gomes ER et al (2008) Molecular mechanisms involved in the angiotensin-(1–7)/Mas signaling pathway in cardiomyocytes. Hypertension 52:542–548
27. Tallant EA, Ferrario CM, Gallagher PE (2005) Angiotensin-(1–7) inhibits growth of cardiac myocytes through activation of the mas receptor. Am J Physiol Heart Circ Physiol 289:H1560–H1566
28. Iwata M, Cowling RT, Gurantz D et al (2005) Angiotensin-(1–7) binds to specific receptors on cardiac fibroblasts to initiate antifibrotic and antitrophic effects. Am J Physiol Heart Circ Physiol 289:H2356–H2363

29. McCollum LT, Gallagher PE, Tallant EA (2012) Angiotensin-(1–7) abrogates mitogen-stimulated proliferation of cardiac fibroblasts. Peptides 34:380–388
30. Sampaio WO, Santos RA Sd, Faria-Silva R et al (2007) Angiotensin-(1–7) through receptor Mas mediates endothelial nitric oxide synthase activation via Akt-dependent pathways. Hypertension 49:185–192
31. Liu GC, Oudit GY, Fang F et al (2012) Angiotensin-(1–7)-induced activation of ERK1/2 is cAMP/protein kinase A-dependent in glomerular mesangial cells. Am J Physiol Renal Physiol 302:F784–F790
32. Peiro C, Vallejo S, Gembardt F et al (2007) Endothelial dysfunction through genetic deletion or inhibition of the G protein-coupled receptor Mas: a new target to improve endothelial function. J Hypertens 25:2421–2425
33. Sampaio WO, de Henrique CC, Santos RA et al (2007) Angiotensin-(1–7) counterregulates angiotensin II signaling in human endothelial cells. Hypertension 50:1093–1098
34. Su Z, Zimpelmann J, Burns KD (2006) Angiotensin-(1–7) inhibits angiotensin II-stimulated phosphorylation of MAP kinases in proximal tubular cells. Kidney Int 69:2212–2218
35. Gava E, Samad-Zadeh A, Zimpelmann J et al (2009) Angiotensin-(1–7) activates a tyrosine phosphatase and inhibits glucose-induced signalling in proximal tubular cells. Nephrol Dial Transplant 24:1766–1773
36. Giani JF, Gironacci MM, Munoz MC et al (2008) Angiotensin-(1–7) has a dual role on growth-promoting signalling pathways in rat heart in vivo by stimulating STAT3 and STAT5a/b phosphorylation and inhibiting angiotensin II-stimulated ERK1/2 and Rho kinase activity. Exp Physiol 93:570–578
37. Wang K, Hu X, Du C et al (2012) Angiotensin-(1–7) suppresses the number and function of the circulating fibrocytes by upregulating endothelial nitric oxide synthase expression. Mol Cell Biochem 365:19–27
38. Li P, Wang D, Lucas J et al (2008) Atrial natriuretic peptide inhibits transforming growth factor beta-induced Smad signaling and myofibroblast transformation in mouse cardiac fibroblasts. Circ Res 102:185–192
39. Fraga-Silva RA, Pinheiro SV, Goncalves AC et al (2008) The antithrombotic effect of angiotensin-(1–7) involves mas-mediated NO release from platelets. Mol Med 14:28–35
40. Crackower MA, Sarao R, Oudit GY et al (2002) Angiotensin-converting enzyme 2 is an essential regulator of heart function. Nature 417:822–828
41. Gurley SB, Allred A, Le TH et al (2006) Altered blood pressure responses and normal cardiac phenotype in ACE2-null mice. J Clin Invest 116:2218–2225
42. Yamamoto K, Ohishi M, Katsuya T et al (2006) Deletion of angiotensin-converting enzyme 2 accelerates pressure overload-induced cardiac dysfunction by increasing local angiotensin II. Hypertension 47:718–726
43. Kim MA, Yang D, Kida K et al (2010) Effects of ACE2 inhibition in the post-myocardial infarction heart. J Card Fail 16:777–785
44. Kassiri Z, Zhong J, Guo D et al (2009) Loss of angiotensin-converting enzyme 2 accelerates maladaptive left ventricular remodeling in response to myocardial infarction. Circ Heart Fail 2:446–455
45. Zhao YX, Yin HQ, Yu QT et al (2010) ACE2 overexpression ameliorates left ventricular remodeling and dysfunction in a rat model of myocardial infarction. Hum Gene Ther 21:1545–1554
46. Loot AE, Roks AJ, Henning RH et al (2002) Angiotensin-(1–7) attenuates the development of heart failure after myocardial infarction in rats. Circulation 105:1548–1550
47. Zeng WT, Chen WY, Leng XY et al (2012) Impairment of cardiac function and remodeling induced by myocardial infarction in rats are attenuated by the nonpeptide angiotensin-(1–7) analog AVE 0991. Cardiovasc Ther 30:152–161
48. Marques FD, Ferreira AJ, Sinisterra RD et al (2011) An oral formulation of angiotensin-(1–7) produces cardioprotective effects in infarcted and isoproterenol-treated rats. Hypertension 57:477–483

49. Durik M, Van VR, Kuipers A et al (2012) The effect of the thioether-bridged, stabilized angiotensin-(1–7) analogue cyclic ang-(1–7) on cardiac remodeling and endothelial function in rats with myocardial infarction. Int J Hypertens 2012:536426
50. Reudelhuber TL, Mercure C, Jain DB et al (2005) Targeted over-production of angiotensin 1–7 in the heart reverses hypertension-related cardiac hypertrophy. Circulation 112(Supplement II):II-250
51. Mercure C, Yogi A, Callera GE et al (2008) Angiotensin(1–7) blunts hypertensive cardiac remodeling by a direct effect on the heart. Circ Res 103:1319–1326
52. Wang Y, Qian C, Roks AJ et al (2010) Circulating rather than cardiac angiotensin-(1–7) stimulates cardioprotection after myocardial infarction. Circ Heart Fail 3:286–293
53. Kawase Y, Ladage D, Hajjar RJ (2011) Rescuing the failing heart by targeted gene transfer. J Am Coll Cardiol 57:1169–1180
54. Tilemann L, Ishikawa K, Weber T, Hajjar RJ (2012) Gene therapy for heart failure. Circ Res 110:777–793
55. Tang T, Gao MH, Hammond HK (2012) Prospects for gene transfer for clinical heart failure. Gene Ther 19:606–612
56. Qi Y, Shenoy V, Wong F et al (2011) Lentivirus-mediated overexpression of angiotensin-(1–7) attenuated ischaemia-induced cardiac pathophysiology. Exp Physiol 96:863–874
57. Gava E, de Castro CH, Ferreira AJ et al (2012) Angiotensin-(1–7) receptor Mas is an essential modulator of extracellular matrix protein expression in the heart. Regul Pept 175:30–42
58. Ebermann L, Spillmann F, Sidiropoulos M et al (2008) The angiotensin-(1–7) receptor agonist AVE0991 is cardioprotective in diabetic rats. Eur J Pharmacol 590:276–280
59. Benter IF, Yousif MH, Cojocel C et al (2007) Angiotensin-(1–7) prevents diabetes-induced cardiovascular dysfunction. Am J Physiol Heart Circ Physiol 292:H666–H672
60. Ferreira AJ, Oliveira TL, Castro MC et al (2007) Isoproterenol-induced impairment of heart function and remodeling are attenuated by the nonpeptide angiotensin-(1–7) analogue AVE 0991. Life Sci 81:916–923
61. Ferreira AJ, Jacoby BA, Araujo CA et al (2007) The nonpeptide angiotensin-(1–7) receptor Mas agonist AVE-0991 attenuates heart failure induced by myocardial infarction. Am J Physiol Heart Circ Physiol 292:H1113–H1119
62. Toton-Zuranska J, Gajda M, Pyka-Fosciak G et al (2010) AVE 0991-angiotensin-(1–7) receptor agonist, inhibits atherogenesis in apoE-knockout mice. J Physiol Pharmacol 61:181–183
63. Savergnini SQ, Beiman M, Lautner RQ et al (2010) Vascular relaxation, antihypertensive effect, and cardioprotection of a novel peptide agonist of the MAS receptor. Hypertension 56:112–120
64. Santos RA, Ferreira AJ, Nadu AP et al (2004) Expression of an angiotensin-(1–7)-producing fusion protein produces cardioprotective effects in rats. Physiol Genomics 17:292–299
65. Masson R, Nicklin SA, Craig MA et al (2009) Onset of experimental severe cardiac fibrosis is mediated by overexpression of Angiotensin-converting enzyme 2. Hypertension 53:694–700

Index

B.I. Jugdutt and N.S. Dhalla (eds.), *Cardiac Remodeling: Molecular Mechanisms*, Advances in Biochemistry in Health and Disease 5, DOI 10.1007/978-1-4614-5930-9,